# Signal Processing Handbook

## ELECTRICAL ENGINEERING AND ELECTRONICS

*A Series of Reference Books and Textbooks*

1. Rational Fault Analysis, *edited by Richard Saeks and S. R. Liberty*
2. Nonparametric Methods in Communications, *edited by P. Papantoni-Kazakos and Dimitri Kazakos*
3. Interactive Pattern Recognition, *Yi-tzuu Chien*
4. Solid-State Electronics, *Lawrence E. Murr*
5. Electronic, Magnetic, and Thermal Properties of Solid Materials, *Klaus Schröder*
6. Magnetic-Bubble Memory Technology, *Hsu Chang*
7. Transformer and Inductor Design Handbook, *Colonel Wm. T. McLyman*
8. Electromagnetics: Classical and Modern Theory and Applications, *Samuel Seely and Alexander D. Poularikas*
9. One-Dimensional Digital Signal Processing, *Chi-Tsong Chen*
10. Interconnected Dynamical Systems, *Raymond A. DeCarlo and Richard Saeks*
11. Modern Digital Control Systems, *Raymond G. Jacquot*
12. Hybrid Circuit Design and Manufacture, *Roydn D. Jones*
13. Magnetic Core Selection for Transformers and Inductors: A User's Guide to Practice and Specification, *Colonel Wm. T. McLyman*
14. Static and Rotating Electromagnetic Devices, *Richard H. Engelmann*
15. Energy-Efficient Electric Motors: Selection and Application, *John C. Andreas*
16. Electromagnetic Compossibility, *Heinz M. Schlicke*
17. Electronics: Models, Analysis, and Systems, *James G. Gottling*
18. Digital Filter Design Handbook, *Fred J. Taylor*
19. Multivariable Control: An Introduction, *P. K. Sinha*
20. Flexible Circuits: Design and Applications, *Steve Gurley, with contributions by Carl A. Edstrom, Jr., Ray D. Greenway, and William P. Kelly*
21. Circuit Interruption: Theory and Techniques, *Thomas E. Browne, Jr.*
22. Switch Mode Power Conversion: Basic Theory and Design, *K. Kit Sum*
23. Pattern Recognition: Applications to Large Data-Set Problems, *Sing-Tze Bow*
24. Custom-Specific Integrated Circuits: Design and Fabrication, *Stanley L. Hurst*
25. Digital Circuits: Logic and Design, *Ronald C. Emery*
26. Large-Scale Control Systems: Theories and Techniques, *Magdi S. Mahmoud, Mohamed F. Hassan, and Mohamed G. Darwish*
27. Microprocessor Software Project Management, *Eli T. Fathi and Cedric V. W. Armstrong (Sponsored by Ontario Centre for Microelectronics)*
28. Low Frequency Electromagnetic Design, *Michael P. Perry*
29. Multidimensional Systems: Techniques and Applications, *edited by Spyros G. Tzafestas*
30. AC Motors for High-Performance Applications: Analysis and Control, *Sakae Yamamura*

31. Ceramic Materials for Electronics: Processing, Properties, and Applications, *edited by Relva C. Buchanan*
32. Microcomputer Bus Structures and Bus Interface Design, *Arthur L. Dexter*
33. End User's Guide to Innovative Flexible Circuit Packaging, *Jay J. Miniet*
34. Reliability Engineering for Electronic Design, *Norman B. Fuqua*
35. Design Fundamentals for Low-Voltage Distribution and Control, *Frank W. Kussy and Jack L. Warren*
36. Encapsulation of Electronic Devices and Components, *Edward R. Salmon*
37. Protective Relaying: Principles and Applications, *J. Lewis Blackburn*
38. Testing Active and Passive Electronic Components, *Richard F. Powell*
39. Adaptive Control Systems: Techniques and Applications, *V. V. Chalam*
40. Computer-Aided Analysis of Power Electronic Systems, *Venkatachari Rajagopalan*
41. Integrated Circuit Quality and Reliability, *Eugene R. Hnatek*
42. Systolic Signal Processing Systems, *edited by Earl E. Swartzlander, Jr.*
43. Adaptive Digital Filters and Signal Analysis, *Maurice G. Bellanger*
44. Electronic Ceramics: Properties, Configuration, and Applications, *edited by Lionel M. Levinson*
45. Computer Systems Engineering Management, *Robert S. Alford*
46. Systems Modeling and Computer Simulation, *edited by Naim A. Kheir*
47. Rigid-Flex Printed Wiring Design for Production Readiness, *Walter S. Rigling*
48. Analog Methods for Computer-Aided Circuit Analysis and Diagnosis, *edited by Takao Ozawa*
49. Transformer and Inductor Design Handbook, Second Edition, Revised and Expanded, *Colonel Wm. T. McLyman*
50. Power System Grounding and Transients: An Introduction, *A. P. Sakis Meliopoulos*
51. Signal Processing Handbook, *edited by C. H. Chen*

*Additional Volumes in Preparation*

Dynamic Models and Discrete Event Simulation, *William Delaney and Erminia Vaccari*

Electronic Product Design for Automated Manufacturing, *H. Richard Stillwell*

**Electrical Engineering-Electronics Software**

1. Transformer and Inductor Design Software for the IBM PC, *Colonel Wm. T. McLyman*
2. Transformer and Inductor Design Software for the Macintosh, *Colonel Wm. T. McLyman*
3. Digital Filter Design Software for the IBM PC, *Fred J. Taylor and Thanos Stouraitis*

# Signal Processing Handbook

edited by

**C. H. CHEN**

Electrical and Computer Engineering Department
Southeastern Massachusetts University
North Dartmouth, Massachusetts

**MARCEL DEKKER, INC.** **NEW YORK AND BASEL**

**Library of Congress Cataloging-in-Publication Data**

Signal processing handbook / edited by C. H. Chen.
p. cm.—(Electrical engineering and electronics ; 51)
Includes bibliographies and index.
ISBN 0-8247-7956-8
1. Signal processing—Handbooks, manuals, etc. I. Chen, C. H. (Chi-hau) II. Series.
TK5102.2.S54 1988
621.38′043—dc19 88-14676
CIP



MARCEL DEKKER, INC.
270 Madison Avenue, New York, New York 10016

Current printing (last digit):
10 9 8 7 6 5 4 3 2 1

PRINTED IN THE UNITED STATES OF AMERICA

# Preface

During the last twenty-five years, there has been enormous progress in the theory, development and applications of signal processing. This progress is due largely to rapidly increased computational power at greatly reduced costs. Because of this, most signal processing functions have been shifted from analog to digital forms. Digital signal processing thus constitutes a major part of this book.

The motivation for preparing this book is to provide readers, especially practicing engineers, with state-of-the-art knowledge in various aspects of signal processing. Although a large number of signal processing books have been published, along with journals and conference proceedings in signal processing, a handbook of this nature is much needed to provide introductory and systematic treatments of many interrelated subjects in a single volume. Each chapter begins with an introduction of a topic and then presents the latest progress. In fact, many new results are included in almost every chapter. Most topics of signal processing are examined in this volume; only certain peripheral topics such as knowledge-based signal processing and neural net signal processors are not covered. Further research progress in both topics is needed before a complete coverage can be made.

The first four chapters deal with the fundamentals of signal processing, including an overview, analog-to-digital conversion, digital filters, and the discrete Fourier transform. Chapter 5 presents some existing and new high resolution 1-D and 2-D spectral estimation algorithms. The next three chapters deal with the new signal processing architectures using systolic arrays and VLSI. The syntactic signal processing of Chapter 9 is a unique approach to signal processing. Image processing is presented in the next two chapters with emphasis on inspection and robotics applications. A brief digression from digital processing is the important subject of optical signal processing, presented in Chapter 12.

The second half of the book focuses on digital signal processing applications. The needs in practical applications have been the driving force behind the progress in signal processing. The theory and practice of adaptive signal processing are presented in Chapter 13. The two primary subjects of sonar signal processing, namely time delay beamforming and coherence/time delay estimation, are presented in the next two chapters. They are followed by radar, speech, and vibration signal analysis in Chapters 16–18. Historically, many signal processing algorithms were first developed for processing seismic signals. Chapters 19 and 20 deal with signal processing of seismic data based on land and ocean respectively. An emerging new application of signal processing is in nondestructive evaluation of materials as described in Chapter 21. In recent years, there has been a greatly increased emphasis on communications making extensive use of digital signal processing; as presented in Chapter 22, the impact of digital signal processing on telecommunications is quite evident. Although the book does not get into details of processing with biomedical waveforms such as EEG, ECG, EMG, and so forth, the last chapter, on signal processing in medical tomography, represents a major advance of digital signal processing in 3-D object reconstructions. The Appendices offer a brief survey of signal processing software packages available in the public domain and program listings for several algorithms. There is no doubt that each topic presented can be expanded into a book by itself, but a concise exposition of these topics in a single volume presents a useful source book to users of signal processing.

The timely publication of this book has resulted from the cooperation of all authors and the coordination provided by the Marcel Dekker publication staff; to all of them I would like to express my deep gratitude. In sustaining my efforts, I am grateful for the support of the Naval Underwater Systems Center, Army Materials Technology Laboratory, Analog Devices, and the Bay State Skills Corporation.

C. H. Chen

# Contents

# Contributors

HASSAN M. AHMED Department of Electrical, Computer and Systems Engineering, Boston University, Boston, Massachusetts

WOLF-EKKEHARD BLANZ Computer Science Department, IBM Almaden Research Center, San Jose, California

G. CLIFFORD CARTER Signal and Post Processing Branch, Naval Underwater Systems Center, New London, Connecticut

C. H. CHEN Electrical and Computer Engineering Department, Southeastern Massachusetts University, North Dartmouth, Massachusetts

DONALD F. DINN Department of Fisheries and Oceans, Bedford Institute of Oceanography, Dartmouth, Nova Scotia, Canada

FERIAL EL-HAWARY Faculty of Engineering, Technical University of Nova Scotia, Halifax, Nova Scotia, Canada

YESHAIAHU FAINMAN Department of Electrical and Computer Engineering, University of California at San Diego, La Jolla, California

THOMAS S. FONG Radar Systems Group, Hughes Aircraft Company, El Segundo, California

GIOVANNI GARIBOTTO Central Research Department, ELSAG S.p.A., Genoa-Sestri Ponente, Italy

HOSSEIN HAKIMMASHHADI Electrical Engineering Department, Worcester Polytechnic Institute, Worcester, Massachusetts

S. N. JEAN Department of Electrical Engineering, University of Southern California, Los Angeles, California

S. Y. KUNG* Department of Electrical Engineering, University of Southern California, Los Angeles, California

SING H. LEE Department of Electrical and Computer Engineering, University of California at San Diego, La Jolla, California

P. S. LEWIS Electronics Division, Los Alamos National Laboratory, Los Alamos, New Mexico

HSI-HO LIU† Department of Electrical and Computer Engineering, University of Miami, Coral Gables, Florida

S. C. LO‡ Department of Electrical Engineering, University of Southern California, Los Angeles, California

CHARLES EDWARD MUEHE Lincoln Laboratory, Massachusetts Institute of Technology, Lexington, Massachusetts

MOHAMMAD N. NOORI Mechanical Engineering Department, Worcester Polytechnic Institute, Worcester, Massachusetts

PEYMA OSKOUI-FARD Electrical, Computer, and Systems Engineering Department, Rensselaer Polytechnic Institute, Troy, New York

KAVEH PAHLAVAN Department of Electrical Engineering, Worcester Polytechnic Institute, Worcester, Massachusetts

HUI PENG Electrical, Computer, and Systems Engineering Department, Rensselaer Polytechnic Institute, Troy, New York

DRAGUTIN PETKOVIĆ Computer Science Department, IBM Almaden Research Center, San Jose, California

IOANNIS PITAS Department of Electrical Engineering, Aristotelean University of Thessalonica, Thessalonica, Greece

ROGER PRIDHAM Submarine Signal Division, Raytheon Company, Portsmouth, Rhode Island

JORGE L. C. SANZ Computer Science Department, IBM Almaden Research Center, San Jose, California

---

Present affiliations:

*Department of Electrical Engineering, Princeton University, Princeton, New Jersey

†Department of Electrical and Biomedical Engineering, Vanderbilt University, Nashville, Tennessee

‡Microelectronic Center, Hughes Aircraft Company, Newport Beach, California

PETER SCARLETT* Department of Electrical Engineering, University of Toronto, Toronto, Ontario, Canada

M. IBRAHIM SEZAN† Photographic Products Group, Eastman Kodak Company, Rochester, New York

HENRY STARK Electrical, Computer, and Systems Engineering Department, Rensselaer Polytechnic Institute, Troy, New York

EARL E. SWARTZLANDER, JR.‡ Digital Processing Laboratory, TRW Electronic Systems Group, Redondo Beach, California

DALE E. VEENEMAN Telecommunications Research Laboratory, GTE Laboratories, Inc., Waltham, Massachusetts

ANASTASIOS N. VENETSANOPOULOS Department of Electrical Engineering, University of Toronto, Toronto, Ontario, Canada

EITAN YUDILEVICH§ Electrical, Computer, and Systems Engineering Department, Rensselaer Polytechnic Institute, Troy, New York

---

Present affiliations:

*Advanced Systems Development, Raytheon Canada Limited, Waterloo, Ontario, Canada

†Photographic Research Laboratories, Eastman Kodak Company, Rochester, New York

‡TRW Defense Systems Group, Redondo Beach, California

§RAFAEL, Ministry of Defense, Haifa, Israel

# 1

# Introduction

C. H. CHEN Electrical and Computer Engineering Department, Southeastern Massachusetts University, North Dartmouth, Massachusetts

## 1 DESCRIPTION OF SIGNALS

The measured data in many applications are a set of waveforms, a time series, or images from which desired information is extracted. They are called signals as they provide useful information, as opposed to interfering noises, which do not carry relevant information. In practice, the measured data contain far more information than can be fully extracted by human users. Also the large volume of data will make it very difficult if not impossible for human users to obtain the desired information in a reasonable amount of time. In many cases signal processing is part of system operation. Computer-aided, interactive, and fully automatic techniques have been developed for processing the signals so that the desired information can be more readily available. Thus signal processing encompasses many different ways of manipulating the data, with the final objective of improving the human user's capability in interpreting or understanding the data.

Although a waveform from a sensor is the most popular form of signal, signals can be a collection of waveforms taken by an array of sensors spatially arranged. Thus signals can be both time- and space-dependent. Signals may also come from imaging sensors which provide pictures of generally three-dimensional (3-D) object scenes. In other situations, signals can be a sequence of recorded data such as daily stock prices or monthly sunspot numbers. Signals can be a sequence of pictures of a dynamic scene or, in stereovision, two pictures of the same scene taken from different angles. The fact that signals can take so many different forms makes signal processing indeed a very challenging and versatile field.

Analog signal processing has been used for a long time and has the major advantage of being fast in operation. Advances in computer hardware and signal processing-dedicated hardware have made digital signal processing much more effective in performance and cost. The important advantages of digital signal processing are its flexibility and accuracy. For example, digital filters meeting various specifications can be constructed easily and the precision is limited only by the computer word length. Digital signal processing has been applied to almost every area, from toys to sophisticated missile systems. The trend is continuing. Thus this book is mainly concerned with digital signal processing and applications.

## 2 OVERVIEW OF DIGITAL SIGNAL PROCESSING

Although the processing requirements are different for different applications, most signal processing functions are general enough to be applicable to many kinds of signals. A typical digital signal processor starts with the digitization (sampling) of a signal at the rate that is at least twice the highest significant frequency of the signal. The remaining part of the analog-to-digital conversion is quantization of the discrete data, which, unlike sampling, is an irreversible process. The resulting data are processed digitally by recursive or nonrecursive filtering, transformation, and the combination of various techniques so that the desired information can be extracted or enhanced. The output of the digital processor is in digital form and in many applications is converted to analog form. Both the processed digital and analog data can be displayed as needed.

Following the development of the fast Fourier transform and of digital filters in the 1960s, major areas of progress in signal processing have included high-resolution spectral analysis, notably Burg's maximum entropy spectral analysis; array processors; single-chip signal processors; highly parallel signal processing; the use of artificial intelligence knowledge-based systems in signal processing; and efficient image processing algorithms for machine vision. Theoretically, important progress has been made in the use of phase and zero-crossing information for signal reconstruction and detection. The emphasis has been on the "forward" problems in which transformation or filtering is performed. Much less progress has been made on the "inverse" problems in which the input or part of it is determined from the output.

The digital signal processing area has now matured enough that well over a hundred books have been published, a small number of which are listed in References [1–13] to give readers a general idea of book publication in the area. There are numerous journal and conference publications. A good source of new information on the progress each year is the proceedings of the IEEE Annual Conference on Acoustics, Speech and Signal Processing.

For illustrative purposes, Figure 1.1 shows some signal processing operations on two ultrasonic pulse echoes (A and B) from an aluminum plate with an artificial defect. The digital waveforms have 512 points each. The Burg power spectrum is based on 200 points with a prediction filter order of 90. The Fourier amplitude spectrum is, however, based on 512 points. The Burg spectrum is much smoother

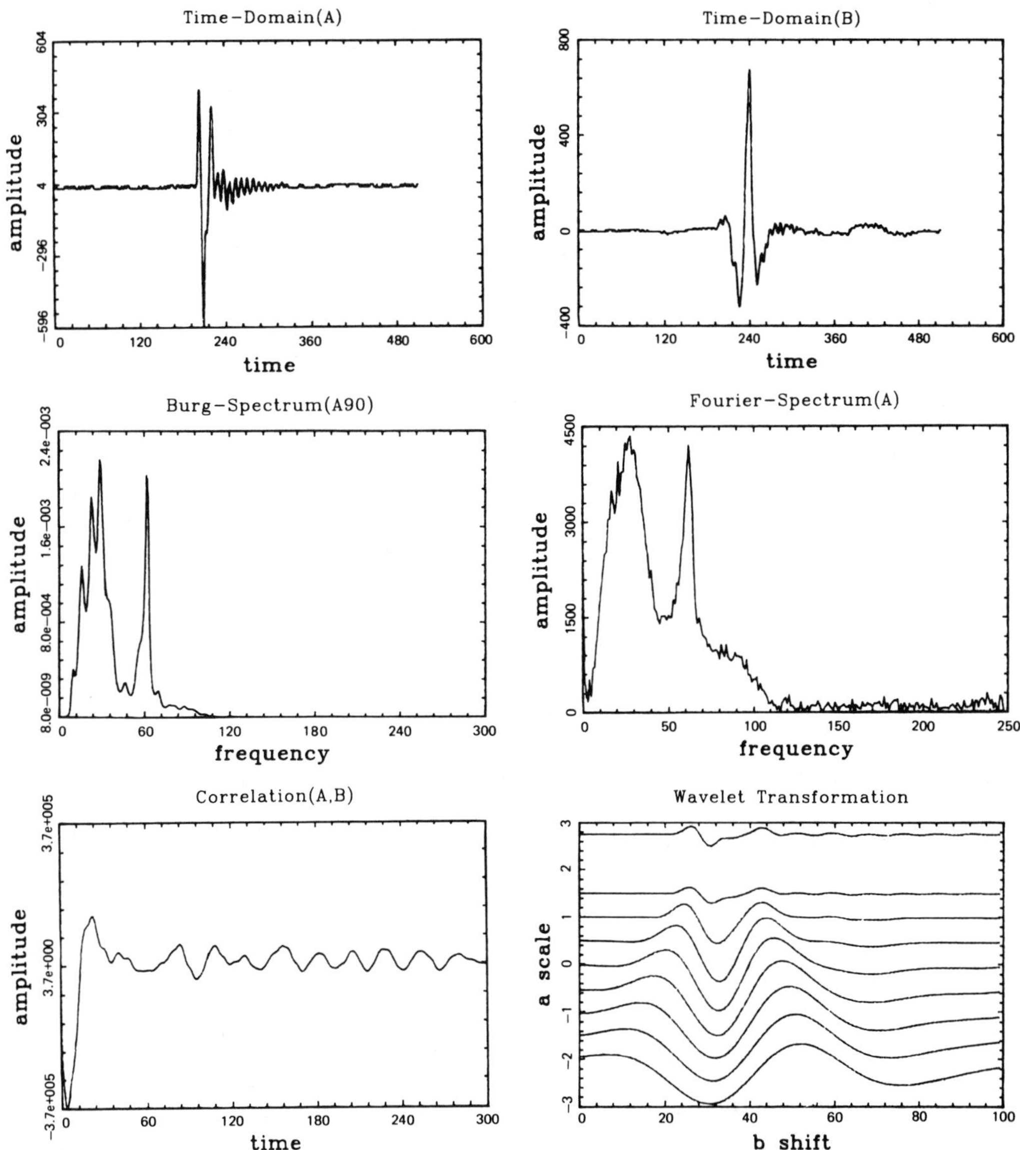


**Figure 1.1** Some signal processing operations on ultrasonic pulse echoes A and B.

with all the essential spectral details. The cross-correlation between A and B shows a negative correlation due to the fact that the peak values of A and B differ in sign. At the bottom right in the figure is the wavelet transform [14, 15] of signal A, which provides a voice representation of the ultrasonic pulse. The signal level versus time ($b$ shift) is plotted for various frequency bands ($a$ scale). The example shows that in a typical signal processing task, different kinds of information must be extracted. Application-specific signal processing operations are often needed. The next step is the proper interpretation of the processed results. The interpretation may require signal parameter estimation, signal detection, or pattern recognition. The real effectiveness of signal processing should be measured in the context of an overall system or problem at hand. The signal processor performance should also be measured in terms of its effectiveness in accomplishing its overall task.

## 3 ORGANIZATION OF THE BOOK

As a handbook, this book provides a wide range of coverage of many different topics in signal processing. The book consists basically of two parts. The first part, Chapters 2–13, is concerned with the fundamentals of signal processing, and the second part, Chapters 14–23, deals with different applications. Chapters 2–4 are on the basic topics of analog-to-digital conversion, digital filters, and Fourier transform. Chapter 5 presents some major high-resolution spectral estimation algorithms. Chapters 6–8 deal with the recent progress in signal processors. Chapter 9 presents a new syntactic approach to signal processing. The important advances in image processing are covered in Chapters 10 and 11. Chapter 12 is on optical signal processing. Chapter 13 is on adaptive signal processing principles and algorithms. Thus the first part of the book covers all essential aspects of signal processing fundamentals. The applications presented in the second part of the book include sonar, radar, speech, geophysics, nondestructive material evaluation, telecommunications, and medical tomography. Appendix A provides a brief survey of signal processing software; Appendix B supplements Chapter 5 with a FORTRAN listing for the 2-D hybrid spectral analysis method; and Appendices C and D supplement Chapter 21 with a FORTRAN listing for time domain deconvolution and an outline of the matching algorithm.

## REFERENCES

1. B. Gold and C. M. Rader. *Digital Processing of Signals*, McGraw-Hill, New York (1969).
2. S. D. Stearns, *Digital Signal Analysis*, Hayden Books, Rochelle Park, New Jersey (1975).
3. A. V. Oppenheim and R. W. Schafer, *Digital Signal Processing*, Prentice-Hall, Englewood Cliffs, New Jersey (1975).
4. L. R. Rabiner and B. Gold, *Theory and Applications of Digital Signal Processing*, Prentice-Hall, Englewood Cliffs, New Jersey (1975).
5. A. Peled and B. Liu, *Digital Signal Processing, Theory, Design and Implementation*, Wiley, New York (1976).
6. A. V. Oppenheim, ed., *Applications of Digital Signal Processing*, Prentice-Hall, Englewood Cliffs, New Jersey (1978).
7. C. H. Chen, *Digital Waveform Processing and Recognition*, CRC Press, Boca Raton, Florida (1982).
8. C. H. Chen, *Nonlinear Maximum Entropy Spectral Analysis Methods for Signal Recognition*, Research Studies Press, Letchworth, Herts, England (1982).

9. S. Haykin, ed., *Array Signal Processing*, Prentice-Hall, Englewood Cliffs, New Jersey (1984).
10. S. Haykin, ed., *Nonlinear Methods of Spectral Analysis*, Springer-Verlag, New York (1979).
11. B. Widrow and S. D. Stearns, *Adaptive Signal Processing*, Prentice-Hall, Englewood Cliffs, New Jersey (1985).
12. S. J. Orfanidis, *Optimal Signal Processing, an Introduction*, Macmillan, New York (1985).
13. S. L. Marples, Jr., *Digital Spectral Analysis with Applications*, Prentice-Hall, Englewood Cliffs, New Jersey (1987).
14. Goupillared, A. Grossman, and J. Morlet, Cycle-octave and related transforms in seismic signal analysis, *Seismic Signal Analysis and Discrimination* (C. H. Chen, ed.), Elsevier Science Publishers, Amsterdam (1984).
15. J. Molet, Sampling theory and wave propagation, *Issues in Acoustic Signal/Image Processing and Recognition*, (C. H. Chen, ed.), Springer-Verlag, New York (1983).

# 2

# Analog Signal Acquisition, Conditioning, and Conversion to Digital Format

DONALD F. DINN Department of Fisheries and Oceans, Bedford Institute of Oceanography, Dartmouth, Nova Scotia, Canada

## 1 INTRODUCTION

Analog signal conditioning is the first step in the acquisition of digital data from real-world processes. The signal conditioner is the interface between the sensor or transducer output, which represents an "analog" of the physical parameter being measured [1] and the analog-to-digital converter (ADC) input. Several important tasks must be accomplished by the signal conditioner:

1. Scaling the transducer outputs to match the ADC input range
2. Keeping induced and coupled noise to an appropriately low level while contributing little or no noise itself
3. Tailoring the frequency spectrum of the sensor outputs so that valid digital samples are taken at the selected sample rate, i.e., so that the sampling rule is followed

Approaches and methods for accomplishing these three and other tasks will be addressed in the following sections. Before proceeding, however, it is appropriate to consider some typical data acquisition systems. These systems consist of a number of generic functional circuits interconnected in arrangements which are application-dependent. A common arrangement used when many different sensors are to be interfaced and time skew between samples is not important (Figure 2.1) consists of a separate signal conditioner for each sensor followed by an analog multiplexer (AMUX) and ADC. A sample/hold (S/H) may or may not be

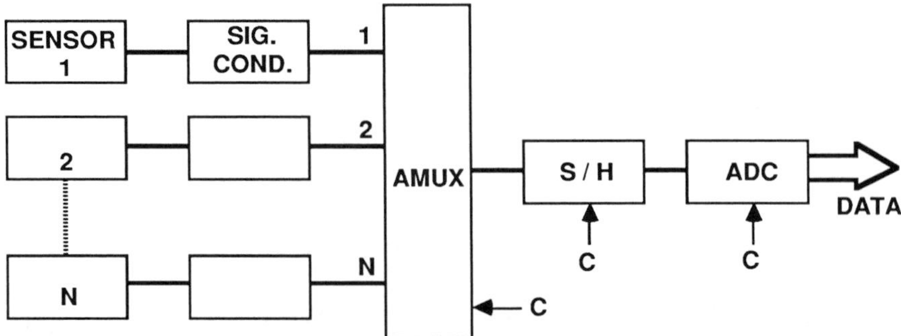


**Figure 2.1** High-quality moderate-speed data acquisition system with one sample/hold (S/H). Samples are time skewed.

required here, depending on the type of ADC used. Control signals C sequence the AMUX through its various channels (not necessarily in numerical order), cause the S/H to take an input sample, and instruct the ADC to perform a conversion.

For applications requiring no time skew between samples, there should be one S/H for each channel (Figure 2.2) or a separate ADC for each channel (Figure 2.3). The choice of one over the other may be driven by cost or by speed requirements. Clearly, the system in Figure 2.3, having dedicated components in each channel, is capable of operating at the highest speed.

The system shown in Figure 2.4 is typical of those used when a large number of sensors with similar output characteristics are involved (e.g., thermocouples, thermistors). In such cases, one signal conditioner serves all channels. This scheme is appropriate when the data acquisition rate is low because of the need to allow for the settling time of the signal conditioner each time a new input is selected by the AMUX. With a low sampling rate, this arrangement often allows the use of an integrating ADC, which can have typical conversion times of tens to hundreds of milliseconds combined with superior rejection of fixed-frequency interference and wideband noise.

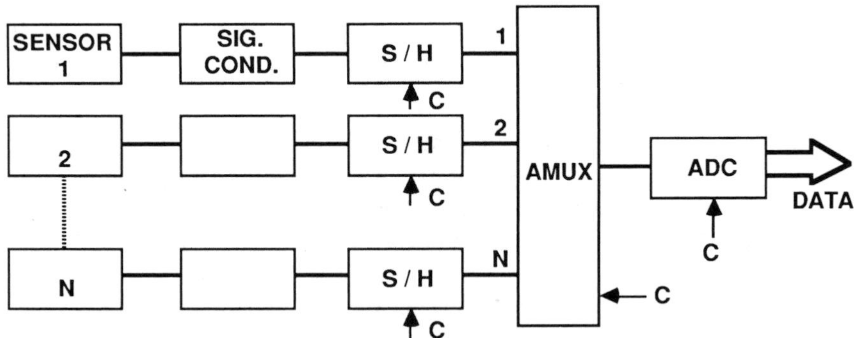


**Figure 2.2** High-quality moderate-speed data acquisition system. Using one S/H per channel eliminates time skew.

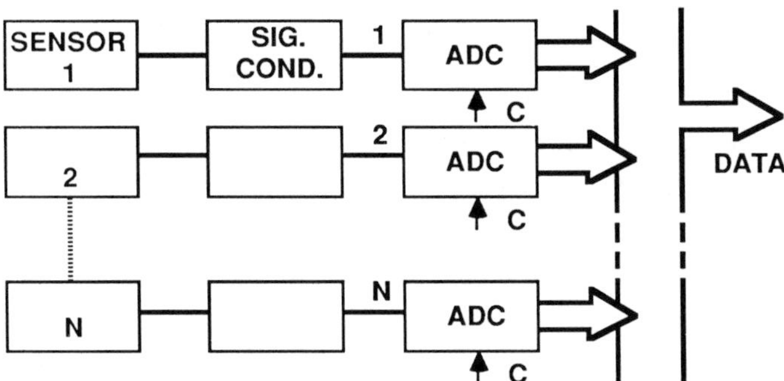


**Figure 2.3** High-speed data acquisition system using simultaneous conversion type ADCs. Samples are multiplexed digitally.

## 2 SIGNAL CONDITIONING AMPLIFIERS

A wide variety of integrated circuit and hybrid amplifiers is available to the designer of signal conditioning circuits. However, two fundamental circuit types, the operational amplifier (OA) and the instrumentation amplifier (IA), are capable of covering many of the requirements. Variations of the IA devices are available using optical or transformer coupling between input and output stages and are referred to as isolation amplifiers. They differ from the instrumentation amplifier primarily in regard to the degree of voltage isolation attainable between the input and output circuits.

In most practical data acquisition systems, particularly those with remote sensors referenced to a common or ground potential different from that of the instrumentation and data acquisition hardware, unbalanced (nondifferential) amplifiers are unable to prevent contamination of the sensor output by common mode interference which arises from differences in ground potential (Figure 2.5a). Common mode voltages are ones which appear on both the low and high side of a transducer output or amplifier input. They are normally referenced to the ground or common connection at the instrumentation side of a measuring system, as opposed to the

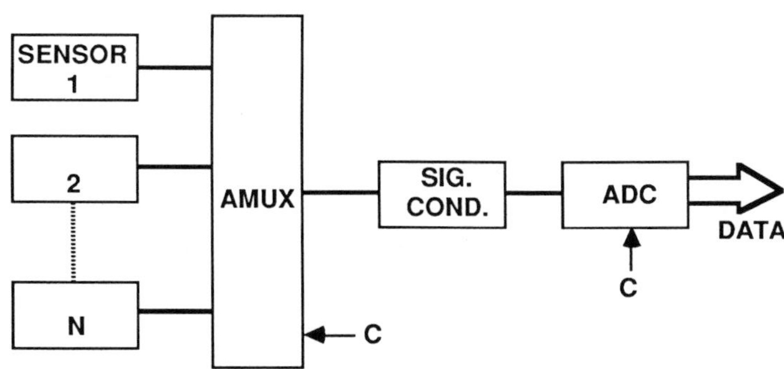


**Figure 2.4** Low-speed data acquisition system with identical sensors (e.g., thermocouples) and integrating ADC.

transducer side. Common mode voltages are readily distinguished from the wanted signal (called a normal mode voltage) because the signal is referenced to the low (or negative) input of the signal conditioning amplifier. Most often, common mode potential differences are at the power line frequency, e.g., 60 Hz. They also arise from excitation voltages (ac and dc) used in bridge circuits (Figure 2.5b). Inductive coupling (magnetic field effects) and capacitive coupling (electrical field effects) are two additional mechanisms by which interference from nearby circuits and power conductors can reduce the quality of signals presented to a data acquisition system (Figure 2.5c) [2, 3].

## 2.1 Differential Amplifier (Subtractor)

In Figure 2.5a, if the voltage $V_{ab}$ is not zero, then the voltage $V_{\text{in}}$ seen by the amplifier is

$$V_{\text{in}} = V_s + V_{ab} \tag{2.1}$$

where $V_{ab}$ is a 60-Hz interference signal and $V_s$ is the desired signal. If $V_s$ is a millivolt signal and $V_{ab}$ is hundreds of millivolts, the signal out of the amplifier will be unusable because of its poor signal-to-noise ratio (SNR). A differential amplifier can be used to minimize the effect of the interference. In Figure 2.6 the circuit gain $A_d$ for a true differential input $V_s$ is simply $n$ (neglecting $R_s$), while for a common mode input $V_{cm}$ it is theoretically zero. The common mode gain $A_{cm}$ will be zero only if the resistor ratios are precisely $n$ and if the operational amplifier is ideal. A useful parameter describing the extent to which the ideal is achieved in practice is the common mode rejection ratio (CMRR):

$$\text{CMRR} = \frac{A_d}{A_{cm}} \tag{2.2}$$

CMRR is normally expressed in decibel (dB) form

$$\text{CMRR(dB)} = A_d(\text{dB}) - A_{cm}(\text{dB}) = 20 \log |A_d| - 20 \log |A_{cm}| \tag{2.3}$$

If the circuit of Figure 2.6 is analyzed, the expressions for $A_d$ and $A_{cm}$ can be derived (assuming $R_s$ is negligible):

$$A_d = -\frac{V_0}{V_s} = -\frac{R_3 + R_4}{R_1 + R_2} \tag{2.4}$$

$$A_{cm} = \frac{V_0}{V_{cm}} = \frac{R_4}{R_2 + R_4}\,\frac{R_1 + R_3}{R_1} - \frac{R_3}{R_1} \tag{2.5}$$

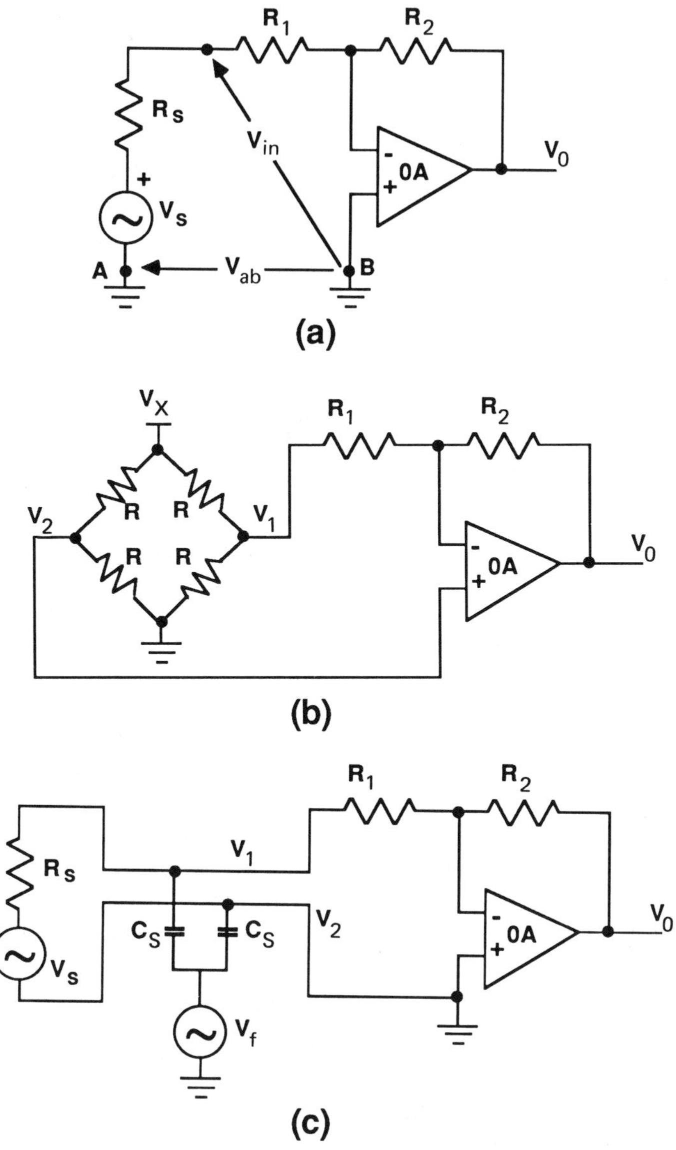


**Figure 2.5** Three mechanisms by which interference and common mode signals contaminate wanted signals: (a) differences in ground potential; (b) bridge excitation voltage $V_x$; $V_1$ and $V_2$ are common mode voltages equal to $V_x/2$; (c) stray capacitance coupling from adjacent wiring ($V_f$) onto an unbalanced source—high impedance at $V_1$, low impedance at $V_2$.

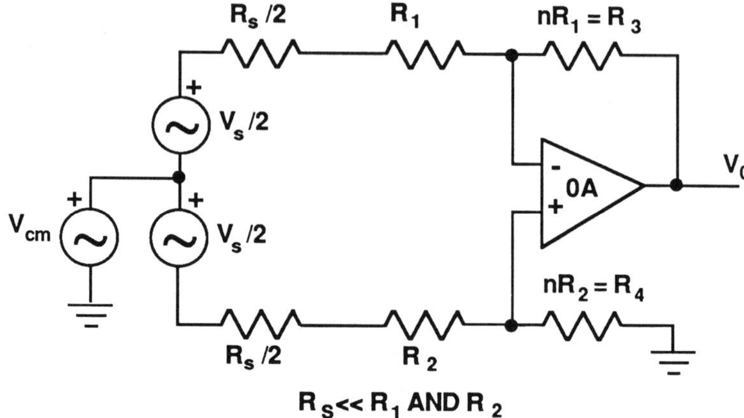


**Figure 2.6** Differential amplifier (subtractor) for common mode rejection. $V_{cm}$ is assumed to have equal effects on the low and high sides of $V_s$ and is shown at the center of $V_s$, a point not always available in an actual circuit.

The amplifier output in the presence of both $V_s$ and $V_{cm}$ is

$$V_0 = -V_s A_d + V_{cm} A_{cm} = -V_s A_d + V_{cm} \left( \frac{A_d}{\text{CMRR}} \right) \tag{2.6}$$

The effect of resistor mismatching can be examined by replacing $R$ in Eqs. (2.4) and (2.5) with $(1 \pm t)R$, where $t$ is the fractional tolerance of the resistors, i.e., for 1% resistors $t = 0.01$. Using the assumption that all the resistors are off value by the same percentage, the worst-case CMRR can be shown to be approximately

$$\text{CMRR} \simeq \frac{A_d + 1}{4t} \tag{2.7}$$

due to the resistors alone. The sign of CMRR can be positive or negative depending on which resistors are above or below their nominal value. Table 2.1 shows some typical values of CMRR achievable for the subtractor circuit.

Even with perfect matching of resistors, the CMRR cannot predictably exceed the intrinsic CMRR of the operational amplifier alone. The limitations imposed by bias/leakage current matching and gain matching for the inverting and noninverting inputs of the OA result in typical intrinsic CMRR values of about 60–100 dB. When the CMRR contribution from resistor matching equals the intrinsic CMRR, the overall CMRR is reduced to half the intrinsic value (a 6-dB reduction). This equivalent overall CMRR is analogous to the equivalent value of two resistors in parallel [5].

The intrinsic CMRR of many operational amplifiers is high (90 dB). From Eq. (2.7) a resistor tolerance of ±0.002% would be required in a unity-gain subtractor to prevent significant degradation of this CMRR. Moreover, the resistances

**Table 2.1** CMRR as a Function of Differential Gain $A_d$ and Resistor Error

| $A_d$ | ±10% Error | ±1% Error | ±0.1% Error |
|---|---|---|---|
| 1 (0 dB) | 5.0 (14 dB) | 50 (34 dB) | 500 (54 dB) |
| 10 (20 dB) | 27.5 (29 dB) | 275 (49 dB) | 2750 (69 dB) |
| 100 (40 dB) | 253 (48 dB) | 2530 (68 dB) | 25,300 (88 dB) |
| 1000 (60 dB) | 2500 (68 dB) | 25,000 (88 dB) | 250,000 (108 dB) |

would have to be temperature-insensitive or have matched temperature coefficients to maintain the CMRR. Such precise matching can best be achieved using integrated circuits with laser-trimmed resistors. Nevertheless, it is possible to use variable resistors to adjust the resistor ratios so as to obtain the highest possible CMRR. The adjustment can be carried out by applying $V_{cm}$ to both inputs and varying one of the ratios to obtain zero output from the amplifier. The adjustment can be tedious, however, and is normally avoided unless special circumstances demand it.

A drawback of the subtractor circuit is its limited input resistance. Practical maximum values of $R_1$ through $R_4$ are 1–5 megohms. For a gain of 10, for example, $R_1$ and $R_2$ could each be 500 kilohms, giving a differential input resistance of 1 megohm. Because the source resistance $R_s$ is in series with $R_1$ and $R_2$, gain is reduced depending on the value of $R_s$. Moreover, any imbalance in the source resistance seen by the amplifier inputs will limit the CMRR achievable. With $R_1 = R_2 = 500$ kilohms, a 1 kilohm imbalance in $R_s$ limits the CMRR to approximately 500 or 74 dB at dc [from Eq. (2.7) with $4t = 1/500$].

## 2.2 Instrumentation Amplifiers

When source impedance imbalance and common mode voltage are excessive, improved performance can be achieved using an instrumentation amplifier, which can provide very high intrinsic CMRR (90–120 dB) and high input resistance ($> 10^9$ ohms). Conventional IAs require that the common mode voltage remain within the bounds set by the power supplies, typically ±10 V. Isolation amplifiers must be used if larger common mode voltages are present. Isolation amplifiers have no ohmic connection between the input circuit and the output circuit. Their differential input stage normally has a CMRR rating, while the input-to-output separation is defined by an isolation mode rejection ratio (IMRR). The latter can typically be as high as 125 dB at 60 Hz and 160 dB at dc. Voltages up to ±3500 V peak can be tolerated between input and output for typical circuits [4].

A typical IA (Figure 2.7) consists of three OAs connected as a differential follower-with-gain and a subtractor. Analysis of the circuit shows that $V_1'$ follows $V_1$ and $V_2'$ follows $V_2$, thus making the differential voltage across $R_c$ equal to the input voltage $V_s$. The value of $V_1'$ is $V_s/2 + V_{cm}$ and of $V_2'$ is $-V_s/2 + V_{cm}$, that is, $I = V_s/R_c$. By writing equations for $V_a$ and $V_b$ it can easily be shown that the

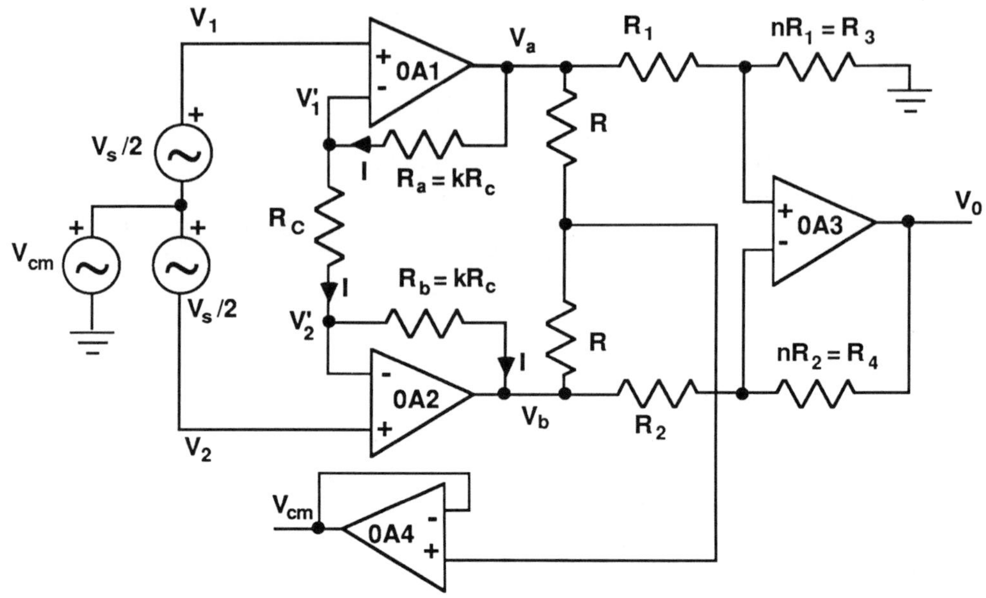


**Figure 2.7** Instrumentation amplifier (IA) made up from four operational amplifiers (OAs). Differential gain is normally controlled by varying $R_c$.

gain of the input stage is $2k+1$. Referring to Figure 2.7,

$$V_a = V_2' + I(R_c + R_a) = -\frac{V_s}{2} + V_{cm} + \left(\frac{V_s}{R_c}\right)(R_c + R_a) \tag{2.8}$$

$$V_a = V_{cm} + V_s(k + 0.5) \tag{2.9}$$

$$V_b = V_{cm} - V_s(k + 0.5) \tag{2.10}$$

therefore

$$V_a + V_b = 2V_{cm} \quad \text{or} \quad V_{cm} = \frac{V_a + V_b}{2} \tag{2.11}$$

The recovered common mode voltage $V_{cm}$ in Eq. (2.11) is available at the output of the unity-gain follower, OA4 (Figure 2.7), and is used for driving the guard or the shield in high-quality signal conditioners. This output is not available in all instrumentation amplifiers [3, 4]. Taking the difference of Eqs. (2.9) and (2.10) gives

$$V_a - V_b = V_s(2k + 1) \tag{2.12}$$

The differential gain of the followers is then

$$A_{df} = \frac{V_a - V_b}{V_s} = 2k + 1 \tag{2.13}$$

Normally $k$ is determined by the designer by choosing an appropriate value of $R_c$. Other resistors are internal in integrated circuit IAs.

Note that $V_{cm}$ is handled with unity gain by OA1 and OA2 regardless of the value of $k$. Thus $A_{cm} = 1$, and the CMRR of the followers is $\text{CMRR}_f = A_{df}/A_{cm} = A_{df}$. Because $A_{df}$ is usually much smaller than the intrinsic CMRR of the OAs, the equivalent CMRR is effectively $\text{CMRR}_f$. Clearly, for high CMRR, it is advantageous to have some gain in the followers. The subtractor amplifier OA3 provides an additional $\text{CMRR}_s$ *multiplier* determined largely by the matching of the ratio $n$ in the scaling resistors (see Table 2.1). Often $n$ is set equal to 1 to permit an overall gain of unity to be obtained when $k = 0$, that is when $R_c = \infty$. The internal resistors in integrated circuit IAs are normally laser-trimmed at the factory and have matched temperature coefficients.

The intrinsic CMRR of an instrumentation amplifier is a function of frequency because it depends on $A_d$ and on internal gain and impedance balance within the amplifier. These are functions of frequency usually with a −6 dB per octave rolloff. Thus, beyond a corner frequency which varies with gain, but which is typically between a few hertz and several hundred hertz, the CMRR decreases from its dc value at a rate of 6 dB/octave [4].

## 2.3 Input Balancing

Achieving the high available CMRR from the instrumentation amplifier requires attention to grounding, shielding, and balancing. In Figure 2.8, for example, $V_1$

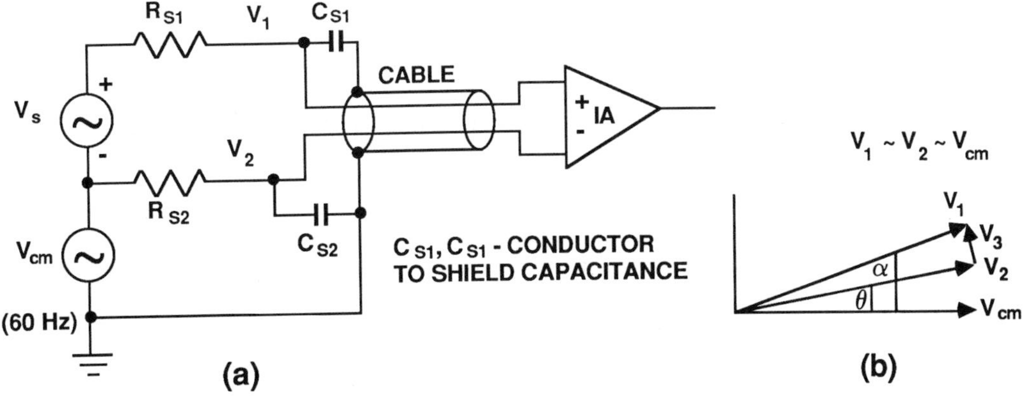


**Figure 2.8** (a) Source resistances $R_{s1}$ and $R_{s2}$ combine with cable capacitances $C_{s1}$ and $C_{s2}$ to reduce available CMRR when time constants are not equal; (b) $V_3$, a quadrature component of $V_{cm}$, is seen by the IA input.

and $V_2$ suffer phase shifts relative to $V_{cm}$ because of the cable capacitances and source resistances. If these phase shifts are not equal, the CMRR is reduced.

Assuming the phase shifts are small, the approximations $\sin\theta \simeq \theta \simeq \tan\theta$ can be employed; thus, from Figure 2.8:

$$\alpha = 2\pi f R_{s1} C_{s1} \tag{2.14}$$

$$\theta = 2\pi f R_{s2} C_{s2} \tag{2.15}$$

$$|V_1 - V_2| = |V_3| = |V_{cm}(\alpha - \theta)| \tag{2.16}$$

$$|V_3| = |V_{cm}[2\pi f(R_{s1}C_{s1} - R_{s2}C_{s2})]| \tag{2.17}$$

Equation (2.17) gives the portion of the common mode voltage which appears as a quadrature differential input. The CMRR degradation due to the input cable capacitance and source resistance alone is given by

$$\text{CMRR} = \frac{V_{cm}}{V_3} = \frac{1}{2\pi f(R_{s1}C_{s1} - R_{s2}C_{s2})} \tag{2.18}$$

For example, if $f = 60$ Hz, $R_{s1} = 800$ ohms, $R_{s2} = 1000$ ohms, $C_{s1} = 900$ pF, and $C_{s2} = 1000$ pF, the CMRR is $9.5 \times 10^3$ or 79.5 dB, which is quite low in comparison to the 100–120 dB available from typical IAs. Thus, an unbalanced cable capacitance to ground combined with an unbalanced source resistance can cause a portion of the common mode interference $V_{cm}$ to be converted into quadrature normal mode interference $V_3$, reducing the achievable CMRR. The IA cannot then distinguish $V_3$ from the signal. Several solutions are applicable:

1. Balance the $RC$ time constants by inserting appropriate values of resistance ($R_{s2} = R_{s1}$) in the signal connections or by adding capacitance in parallel with $C_{s1}$ or $C_{s2}$ at the source.
2. If possible, connect the shield to the low side of $V_s$ instead of ground; $V_{cm}$ does not then come into play with the cable capacitance or with the $RC$ time constants.
3. Connect the shield to the common mode signal output of the IA (Figure 2.7) to achieve the same effect as in item 2.
4. Locate the IA close to the sensor to minimize cable length and capacitance (and thus minimize the difference in time constants).

An important benefit of the IA is the very high input impedance provided by the input voltage followers. Note that an IA with an input resistance of only $10^8$ ohms at each of its inputs can tolerate a source resistance imbalance of $10^3$ ohms and still achieve a CMRR of $10^5$ or 100 dB at dc, assuming the IA has an intrinsic CMRR rating much greater than 100 dB [3].

## 2.4 Signal-to-Noise Improvement Using an Instrumentation Amplifier

To demonstrate the signal-to-noise improvement of an instrumentation amplifier (Figure 2.7) over a nondifferential amplifier (Figure 2.5a), consider the following application in which the signal $V_s = 10$ mV rms at 50 Hz, the interference $V_{cm} = 1$ V rms at 60 Hz (power line interference), and the instrumentation amplifier's achieved CMRR $= 10^4$ (80 dB).

Using the circuit of Figure 2.5a, the input signal-to-noise ratio is

$$\mathrm{SNR} = \frac{P_s}{P_{cm}} \left(\frac{V_s}{V_{cm}}\right)^2 = 10^{-4} \tag{2.19}$$

The output SNR is the same because $V_{cm}$ is not rejected by the amplifier. Substantial filtering would be required to improve the SNR because of the closeness of the interference frequency to the signal frequency. Using the circuit of Figure 2.7, however, the output SNR is

$$\mathrm{SNR} = \left(\frac{V_s A_d}{V_{cm} A_{cm}}\right)^2 = \left(\frac{V_s}{V_{cm}}\right)^2 (\mathrm{CMRR})^2 = 10^4 \tag{2.20}$$

Thus the SNR is improved by the factor $(\mathrm{CMRR})^2$ without resorting to filtering.

### 2.5 Amplifier Selection—Error Budget

The selection of amplifiers for signal conditioning involves assessing all sources of noise, error, or uncertainty in the amplifier transfer characteristics and determining the SNR by comparing the resulting root-sum-square (RSS) error to the rms signal referred to the input or to the output. The RSS error is determined by taking the square root of the sum of squares of all the individual uncorrelated error components. Contributions to the error include output stage voltage drift versus temperature and power supply voltage; input offset voltage drift versus temperature; offset current drift times $R_s$ versus temperature; amplifier input current noise times $R_s$ in the system bandwidth; amplifier input voltage noise in the system bandwidth; common mode effects; gain change versus temperature, supply voltage, and signal level; and so forth. This analysis technique is well documented [4, 5].

Normally, errors and noise resulting from other components in the data acquisition system, such as sample/holds, multiplexers, the ADC, and the transducer itself, are included in the RSS error calculation. Frequently, however, because input signals from many transducers are low-level, the error contributed by the transducer and the signal conditioning amplifier becomes the limiting factor controlling the overall SNR. The resulting output SNR is compared to the required SNR to determine the suitability of the amplifier and other components for the application. Very often, when high precision is required, sources of very low frequency error (due to temperature-induced effects, for example) can be rendered inconsequential by including a zero and full-scale calibration in each measurement cycle.

## 3 SIGNAL-TO-NOISE RATIO REQUIREMENTS

The quality of the data collected by a data acquisition system depends on the signal-to-noise ratio and on the resolution of the ADC used. The contributions to noise can include amplifier noise (above), aliased source noise (see Section 4), broadband source noise (which passes through the noise bandwidth of the signal conditioner), and residual common mode noise (now normal mode noise) at the output of the instrumentation amplifier. If the aliased noise or the residual common mode noise is at a single frequency it can be removed using a notch filter [6, 7], or, if appropriate (unipolar signal), an integrator can be used, normally as part of the ADC. By making the integration period $T = k(1/f)$, where $k$ is an integer and $f$ is the frequency of the interference, the integrator will function as a block averager and a low-pass filter with nulls (notches) at frequencies that are integer multiples of $f/k$, thus removing the interference. (See Section 6.)

To achieve an SNR compatible with an ADC of $n$ bits ($1/2^n$ resolution) requires the noise to be less than the minimum discernible signal, i.e., less than $0.5/2^n$ (full scale = 1). If, for example, the noise voltage $V_n$ is treated as an equivalent sine wave, that is, $V_n$ (rms) $= 0.5/(2^n 2^{1/2})$, while the signal is a full-scale dc input, the SNR can be written as

$$\text{SNR} = \frac{P_s}{P_n} = \left(\frac{V_s}{V_n}\right)^2 = \frac{1}{(0.5/2^n 2^{1/2})^2} = 2^{2n+3} \tag{2.21}$$

$$\text{SNR(dB)} = 10 \log 2^{2n+3} = 6.02n + 9 \text{ dB} \tag{2.22}$$

For 10-bit quality, the required input SNR is approximately $8 \times 10^6$ (69 dB).

If the signal is a sine wave with full-scale positive and negative excursions each equal to one-half the input range of the ADC, the SNR requirement is

$$\text{SNR} = \left(\frac{V_s}{V_n}\right)^2 = \left(\frac{0.5/2^{1/2}}{0.5/2^n 2^{1/2}}\right)^2 = 2^{2n} \tag{2.23}$$

$$\text{SNR(dB)} = 10 \log 2^{2n} = 6.02n \text{ dB} \tag{2.24}$$

For 10-bit quality this corresponds to approximately 60 dB.

Table 2.2 shows the minimum input SNR requirements of the ADC in order to achieve $n$-bit quality in the input signal. The SNR values in Table 2.2 are adequate if statistical use will eventually be made of the digital samples. When the value of each sample is individually important, however, the SNR of each sample must be considered. Noise is normally a random signal with high peak-to-rms ratio. The use of an ADC with short conversion time will cause the *instantaneous* value of the noise at sampling time to register on individual samples. The values in Table 2.2 are based on rms noise; for Gaussian noise, the probability or confidence level is 68% ($1\sigma$) that the instantaneous full-scale SNR will be equal to or greater than the value stated [8]. Achieving a better confidence level requires a reduced rms noise level, i.e., a higher SNR. The probability is 99.9% that the instantaneous Gaussian noise voltage will be less than 3.3 times the rms value. Thus, to achieve a confidence level of 99.9% ($3.3\sigma$) in individual samples requires an SNR that is $20 \log 3.3/2^{1/2} = 7.4$ dB higher than the values given in Table 2.2.

Note that if the minimum input SNR cannot be achieved, postprocessing of the digital data could be used to improve the SNR when the input signal is a repetitive one and the noise is random. This assumes that the data processor has adequate

**Table 2.2** Minimum Input SNR Requirements for ADCs for $n$-Bit Quality with Full-Scale Input

| $n$, bits[a] | Unipolar input SNR, dB | Bipolar Sinusoidal input SNR, dB |
|---|---|---|
| 4 | 33.1 | 24.1 |
| 6 | 45.1 | 36.1 |
| 8 | 57.2 | 48.2 |
| 10 | 69.2 | 60.2 |
| 12 | 81.2 | 72.2 |
| 14 | 93.3 | 84.3 |
| 16 | 105.3 | 96.3 |

[a]Includes sign bit for bipolar input.

word length for the SNR desired. The SNR improvement can only be obtained, however, by averaging over a number of cycles of the data; this is the basic time-bandwidth trade-off. Because the (repetitive) signal cycles add coherently, the processed signal voltage can be considered to be proportional to $m$, the number of cycles averaged. Random noise adds noncoherently, resulting in an rms voltage addition (signal voltage adds directly, noise power adds directly). Thus, at the input to the averager

$$\mathrm{SNR_{in}} = \left(\frac{V_s}{V_n}\right)^2 \tag{2.25}$$

and at the output

$$\mathrm{SNR_{averaged}} = \left(\frac{V_s m}{V_n m^{1/2}}\right)^2 \tag{2.26}$$

Therefore

$$\mathrm{SNR_{averaged}} = m(\mathrm{SNR_{in}}) \tag{2.27}$$

Clearly the longer one is prepared to wait ($m$ cycles), the higher the SNR that is achievable. In practice, differences between individual cycles of real data and residual coherent noise limit the achievable SNR improvement; however, improvements of 20–60 dB and more in SNR are readily achievable by signal averaging techniques. This averaging process is in a sense the digital equivalent (for repetitive signals) of the analog integration (for unipolar signals) noted above.

## 3.1 Intrinsic SNR of an Analog-to-Digital Converter

When the input signal to an ADC exceeds the minimum SNR requirements in Table 2.2, the ADC output SNR is limited by the quantizing error of the ADC, that is, by the resolution of the converter. Analog-to-digital conversion involves sampling, quantizing, and coding the analog signal. Without affecting the result, quantizing could be done first, followed by sampling and coding. Looked at in this way, the quantizing error can be considered as a noiselike error signal added to the input signal. Figure 2.9 shows the quantizing error $e(q)$ for a linear ramp input. It can be shown that, as long as the step size of the ADC is small compared to the range of the input signal, the triangular waveform representing $e(q)$ can be used without loss of generality to examine the signal-to-error ratio (SER), that is, the intrinsic output SNR of an ADC. Note that $e(q)$ has zero mean. The power $P_e$ or mean square value of $e(q)$ can be evaluated directly by integration over one period:

$$\langle e^2(q)\rangle_{\text{avg}} = \frac{1}{q}\int_{-q/2}^{+q/2} (-q)^2\, dq = \frac{q^2}{12} \tag{2.28}$$

where $q = 1/2^n$. The intrinsic output SNR or SER is

$$\text{SER} = \frac{P_s}{P_e} \tag{2.29}$$

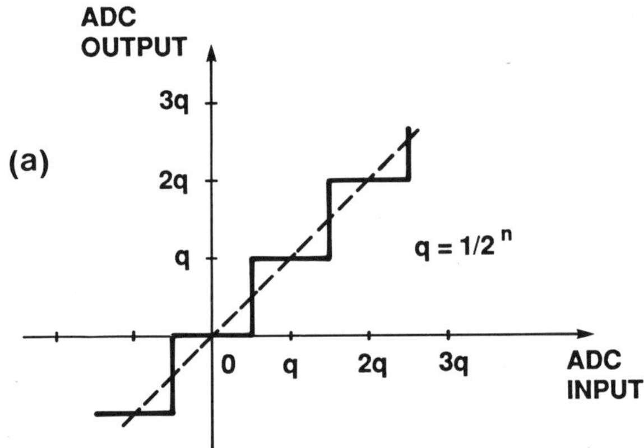


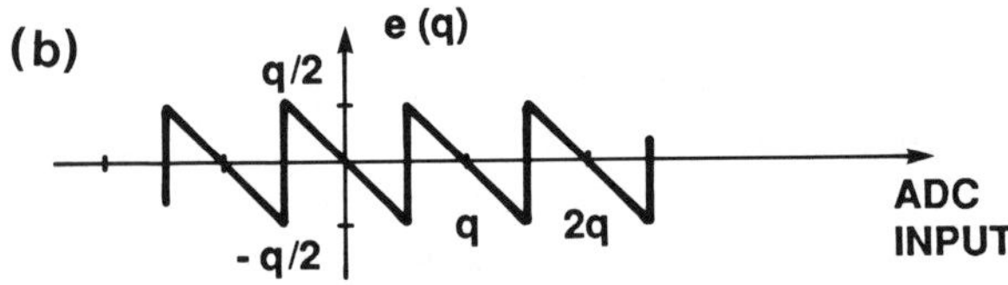


**Figure 2.9** (a) Input-output transfer function of an ADC; $q = 1/2^n$ and represents the smallest step size of the ADC; (b) quantizing error $e(q)$ corresponding to (a); $e(q)$ = output value − input value.

**Table 2.3** SER (Intrinsic Output SNR) for ADCs for Full-Scale Sine Wave Input

| ADC resolution, bits[a] | SER, dB |
|---|---|
| 4 | 25.9 |
| 6 | 37.9 |
| 8 | 50.0 |
| 10 | 62.0 |
| 12 | 74.0 |
| 14 | 86.1 |
| 16 | 98.1 |

[a]Includes sign bit.

For a sine wave input signal whose peak-to-peak amplitude fills the input range of the ADC

$$\text{SER} = \frac{(V_s)^2}{\langle e^2(q)\rangle \,\text{avg}} = \frac{(0.5/2^{1/2})^2}{q^2/12} = 1.5(2^{2n}) \tag{2.30}$$

$$\text{SER(dB)} = 10\log 1.5(2^{2n}) = 6.02n + 1.78 \text{ dB} \tag{2.31}$$

Thus, for a 10-bit ADC the SER $\cong$ 62 dB for a full-scale sine wave input. Clearly, if the signal power is reduced or if the signal has a high peak-to-rms ratio, as in speech, music, or a random process, the SER is reduced. For Gaussian signals the probability that the peak-to-rms ratio exceeds 3.3 is $\sim 10^{-3}$. To quantize these positive and negative peaks the rms input voltage must be limited to 1/3.3 times the positive or negative input range of the ADC. Note that $20\log(2^{1/2}/3.3) = -7.4$ dB, indicating that the signal-to-error ratio for this Gaussian signal case using a 10-bit ADC would be reduced to 54.6 dB compared to 62 dB for a full-scale sine wave input.

Table 2.3 gives the SER of various ADCs for full-scale sine wave input. These values represent the highest output SNRs that can be achieved regardless of how high the input SNR is.

## 4 SAMPLING AND ALIASING

According to the sampling theorem, an analog signal can be uniquely described by a set of uniformly spaced discrete samples taken at a frequency $f_s$, as long as no energy exists in the signal at a frequency equal to or greater than $f_s/2$ (Figure 2.10). If there is a base-band signal energy at a frequency $f_s/2 + \Delta f$ it will be indistinguishable (Figure 2.11) in the sampled spectrum from energy that came from a frequency $f_s/2 - \Delta f$. Energy above $f_s/2$ contributes to aliasing errors when the samples are used to reconstruct the analog signal by low-pass filtering or

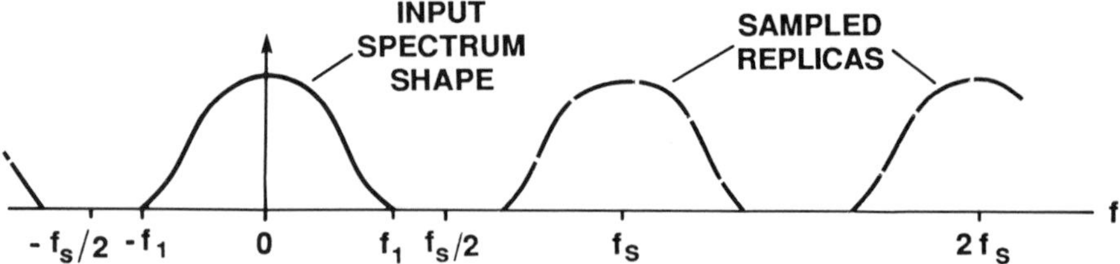


**Figure 2.10** Output spectrum of an ideal sampling circuit for an adequately sampled input signal. No energy exists in the input signal above $f_1$. By passing the sampled signal through a low-pass filter with sharp cutoff beyond $f_1$, the original input signal can be completely recovered.

when spectral analysis is performed on the samples. The term folding frequency is applied to $f_s/2$ because of the foldover effect of energy above it [9].

A knowledge of the frequency spectrum of the input signal and a method of ensuring that the signal contains no energy above $f_2/2$ are needed for proper sampling. "No energy" is a severe requirement, and in practice, the aliased spectrum levels are usually kept below some threshold, e.g., below the minimum discernible input signal. This can be accomplished in the signal conditioner by using a suitable low-pass filter with an attenuation characteristic chosen to yield an appropriately small output at $f_s/2$ and above. Because errors due to aliasing cannot be removed by signal processing, the aliased signal level should be reduced to a value in keeping with the final SNR required after any digital processing.

Many natural phenomena have inherent cutoff frequencies because of the physics of the processes, e.g., vibration in machinery, temperature variation in process liquids, and physical oceanographic parameters (Figure 2.12). Prior to analog-to-digital conversion, signals like those in Figure 2.12 are sometimes prewhitened, i.e., passed through an analog circuit such as a differentiator, in an attempt to increase the comparative signal energy at the high end of the spectrum. This eases the ADC resolution requirements but can worsen the aliasing problem.

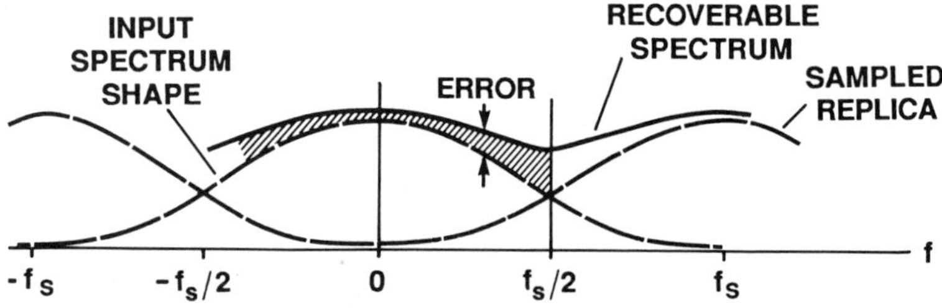


**Figure 2.11** Output spectrum of an ideal sampling circuit for an inadequately sampled input signal. A significant amount of energy is present in the input signal above the folding frequency $f_s/2$. Attempts to recover the original signal (or to use the samples to compute the power spectrum) will result in an error as shown.

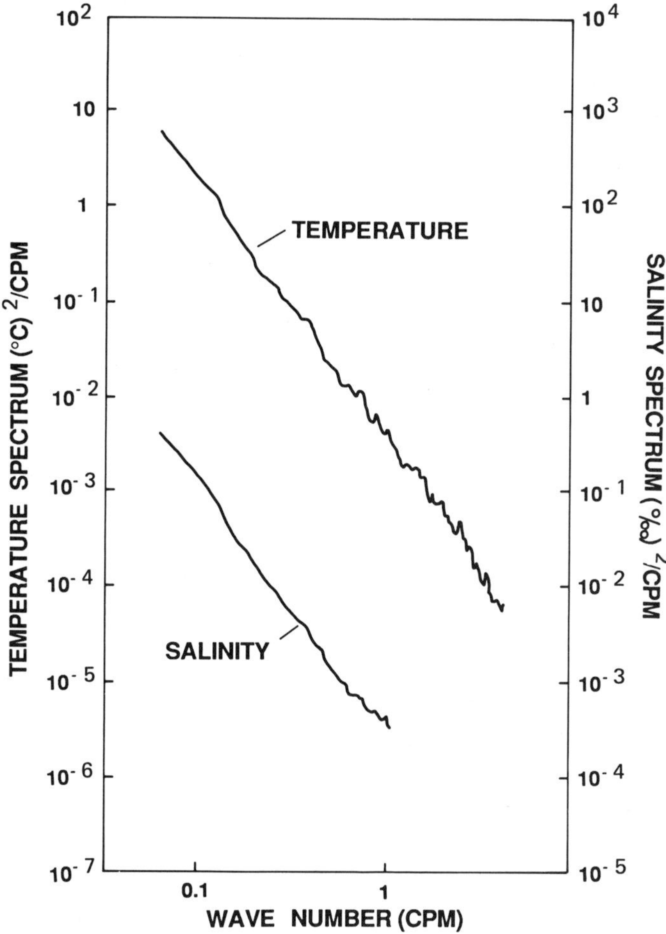


**Figure 2.12** Averaged spatial spectra of temperature and salinity fluctuations in the ocean. If the measuring probe is dropped at a speed of 1 m/s, 1 cycle per meter (cpm) will correspond to 1 Hz. (Adapted from Ref. 14.)

To cite an example of signal processing for antialiasing, consider a signal occupying the band from dc to a nominal upper −3 dB cutoff frequency of 100 Hz and having a rolloff of about 12 dB/octave. If this signal is to be digitized using a 10-bit ADC (1/1024 resolution) so that its spectrum can be computed out to 200 Hz, one can determine the amount of antialiasing protection which must be provided by a presampling filter to ensure that the input will be of 10-bit quality.

First, however, the sampling frequency $f_s$ must be selected. Clearly, $f_s$ must be greater than 400 Hz; the closer $f_s$ is to 400 Hz, the more difficult the filtering task becomes; the higher $f_s$ is made, the more samples (data) that have to be dealt with. Practical designs are a trade-off and usually have $f_s$ at 3 to 10 or more times the highest frequency of interest. Thus, if $f_s$ is selected as 800 Hz (four times the highest frequency of interest), the folding frequency will be 400 Hz (Figure 2.13). A low-pass

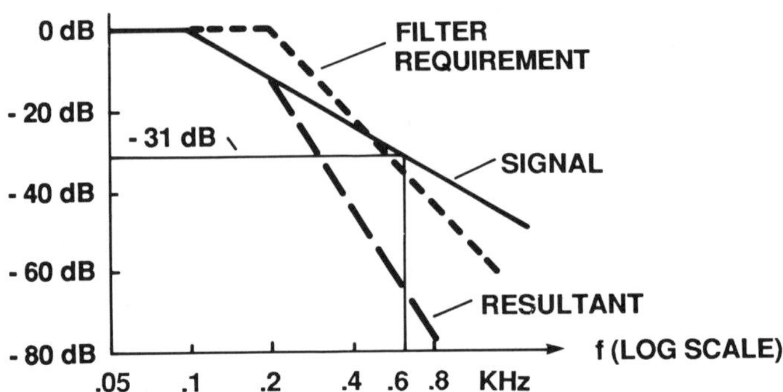


**Figure 2.13** Plot of the signal spectrum, the required filter response, and the resulting ADC input spectrum prior to sampling for the example in the text.

filter must be chosen with a cutoff frequency of approximately 200 Hz or higher. To achieve minimal aliasing error, the filter attenuation must be large enough at 600 Hz (the image frequency which folds onto 200 Hz) to make the available signal less than the amplitude corresponding to half the least significant bit (LSB) of the ADC. From Table 2.2, this represents an attenuation factor of approximately 60 dB with respect to full scale for a 10-bit converter with a sinusoidal input. However, approximately 31 dB of attenuation is available from the signal spectrum itself at 600 Hz, leaving 29 dB to be obtained from the low-pass filter. Thus, the filter response should be 3 dB down at a frequency of approximately 200 Hz and 29 dB down at a frequency of approximately 600 Hz. If a Butterworth filter were chosen, a four-pole unit would be required as $20\log(600/200)^4$ is approximately 38 dB. A three-pole filter is not quite sufficient. The required number of poles in the filter increases as $f_s$ is lowered and vice versa.

In a more general sense, if the combined rolloff of the signal spectrum and the filter response can be taken as having a normalized $-3$ dB cutoff frequency $f_c$ of 1 (Hz), the aliased spectrum level at $f_s/2$ can then be examined for various orders of rolloff and normalized sampling frequency [10]. Figure 2.14 demonstrates this process. Note from Figure 2.11 that at $f_s/2$, in the sampled spectrum, the contributions of energy from the input spectrum and from the folded spectrum are equal and can result in a sum that is 3 dB higher. Thus the plotted curves in Figure 2.14 follow $-10\log[1+(f_s/2)^{2n}]+3$ dB, in which the first term is equivalent to a Butterworth filter attenuation characteristic. The curves indicate that for high-quality analog-to-digital conversion $f_s$ must be much greater than $2f_c$ and/or a high-order spectrum rolloff must be provided.

An alternative approach to antialiasing relies on digital filtering after sampling and analog-to-digital conversion. Moderate (inexpensive) presampling filtering is carried out using analog techniques and the sampling rate chosen high enough to achieve an adequately small aliasing error. Then the sample sequence is digitally filtered to the minimum permissible bandwidth—usually a multiple of 1/2, 1/4 or 1/8 etc. of the presampling bandwidth. Finally, the output samples from the digital filter are decimated in time; i.e., every 2nd, 4th, or 8th etc. sample is taken and the rest ignored. The telephone industry often relies on this scheme because mass-

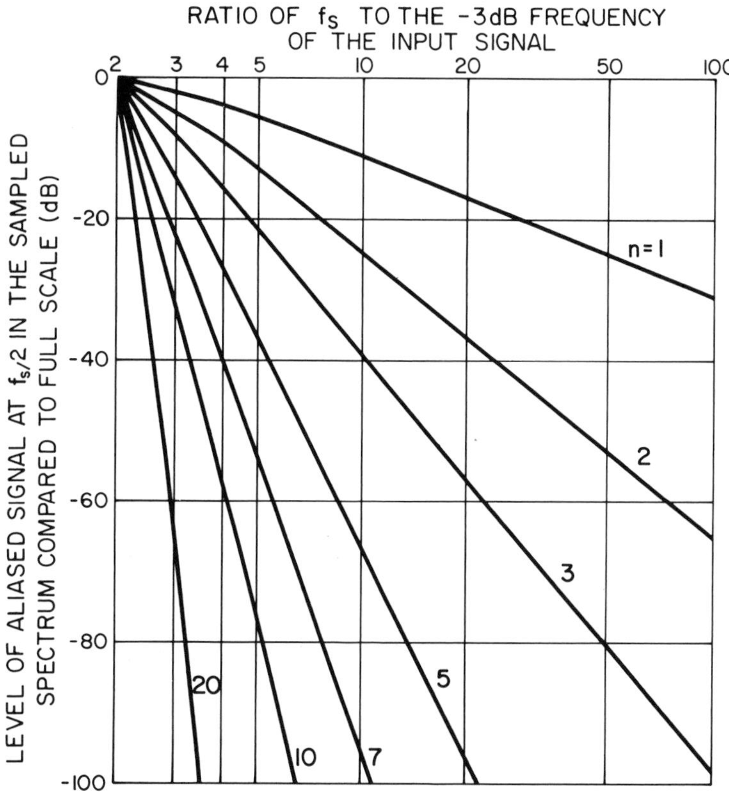


**Figure 2.14** Relationship between the level of aliased signals at $f_s/2$ and the normalized sampling frequency as a function of the order $n$ of the signal spectral rolloff.

produced digital filters, which are far more cost-effective than analog filters, can be used.

## 4.1 Antialiasing Filters

The choice of antialiasing filter is based primarily on the nature of the signal and the attenuation required. Normally, antialiasing filters for moderate bandwidth systems are implemented using active filters [5, 6, 7] placed after the instrumentation amplifier. In many applications, however, it is desirable to place some filtering (often a simple *RC* filter) in front of the IA to ensure protection from large out-of-band signals which might overload the IA. The filter frequency response in the passband must be taken into account for accurate spectral calculation. A brief description of the more common filters follows.

*RC filters* are sometimes adequate for antialiasing filters but, because of their poor attenuation characteristics near the cutoff frequency, they are usually limited to applications involving very low frequency signals (1-Hz bandwidth) and low digital sample rates. In such applications, they are useful for limiting the noise

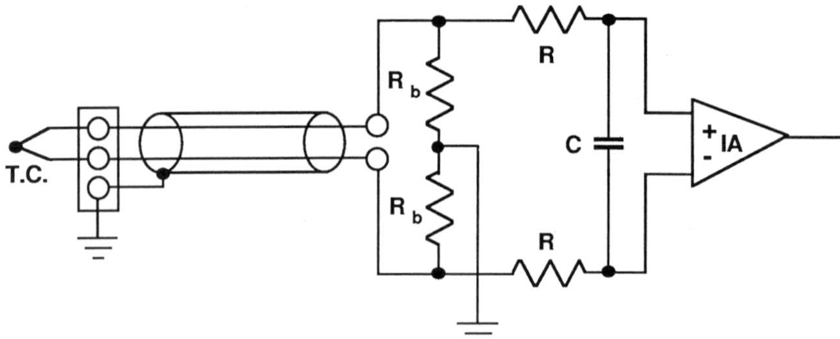


**Figure 2.15** Balanced *RC* input filter; a high value for $R_b$ provides balanced biasing for the IA inputs; $R_b$ is preferably grounded at the transducer end but can be grounded near the amplifier.

bandwidth of the system and for attenuating coupled or induced 60-Hz interference. For example, when remote thermocouples are used to monitor temperatures, the inclusion of a balanced *RC* filter (Figure 2.15) in the signal conditioner with $RC =$ 0.08 s can reduce any normal mode 60-Hz noise by a factor of approximately 35 dB while providing a bandwidth of $1/4\pi RC$ or 1 Hz. A balanced filter is important to preserve the CMRR of the system.

*Butterworth* low-pass filters are characterized by a uniform (no amplitude ripple) passband response, usually referred to as a maximally flat amplitude response. The stopband attenuation beyond 3 dB quickly approaches 6 dB per octave per pole, making it moderately good at rejecting out-of-band interference. The impulse response of the filter is not symmetrical, indicating that the Butterworth filter does not have a linear phase (constant group delay) characteristic. As a result it exhibits moderate overshoot (10% for a four-pole filter) when driven by a voltage step. It is applicable for antialiasing situations where the input signal is smoothly varying in time, e.g., a sine wave without complex modulation sidebands (which would suffer distortion because of the poor group delay characteristics of the Butterworth filter).

*Chebyshev* low-pass filters have amplitude ripples in the passband response. By accepting increasingly larger ripple values (e.g., 0.01 dB, 0.1 dB, 1 dB) the resulting stopband attenuation characteristic becomes increasingly better than that obtainable from a Butterworth filter with the same number of poles. However, the Chebyshev filter has somewhat poorer phase shift, group delay, and step response characteristics than the Butterworth filter. Thus it is not recommended as an antialiasing filter when the input signal contains important harmonics and/or sidebands, if spectral fidelity is paramount.

*Elliptic function* low-pass filters have amplitude ripples in the passband response (like Chebyshev filters) and in the stopband response. The elliptic filter has both poles and zeros in its transfer characteristic; the zeros (or notches) give rise to the stopband ripples. As the frequency of the first zero approaches the cutoff frequency (a design parameter), the stopband ripples increase in size, thus limiting the maximum attenuation available; at the same time, however, the transfer characteristic becomes steeper in the transition region between the passband and the stopband, making the elliptic filter better than the Chebyshev at rejecting interference close to the cutoff frequency. A variant of the two-pole elliptic filter, referred to as a low-pass notch filter, readily enables the notch to be freely placed within the stopband so that a single interference component can be rejected [4]. Elliptic and low-pass notch filters exhibit increasingly poorer group delay and step response characteristics as the first notch is moved closer to the cutoff frequency. Their use is recommended when sharp cutoff is required and the input signal is primarily dc or sinusoidal, i.e., containing few important harmonics or sidebands.

*Bessel* low-pass filters are characterized by maximally flat delay and consequently have a linear phase shift versus frequency characteristic. This fact makes them a good choice for handling signals containing many harmonics or modulation sidebands. The nominally constant group delay preserves the phase relationship of these harmonics and sidebands. Because of the linear phase characteristics, the step response of a Bessel filter exhibits little or no overshoot and ringing. Bessel filters have a monotonically drooping amplitude response in the passband to −3 dB and an attenuation in the stopband that is significantly less than that of a Butterworth filter of equal complexity. Attenuation asymptotically approaches 6 dB/octave per pole.

*Delay-equalized* low-pass filters can be employed when the antialiasing filter must have a good attenuation characteristic and a linear phase response. Such a filter is usually a combination of a Butterworth filter and an all-pass delay network with a delay characteristic that complements that of the Butterworth filter. The resulting composite filter has an approximately constant group delay and a linear phase response [5, 6, 16].

## 4.2 Sampling—Aperture and Resolution

The sampling process is, in theory, the multiplication of an input signal by a sequence of unit sampling pulses or infinitesimally wide ($dt$) impulses with unit area. The output of the sampler is a sequence of impulses whose area represents the amplitude of the input signal at the instant sampled. The unit impulse function is a special mathematical tool in that its frequency spectrum is constant and infinite. In practice, the sampling process is approximated by the use of narrow, constant-width ($T$) pulses whose amplitude (and hence whose area) represents the average value of the input signal. The effect of averaging over the aperture time $T$ instead of $dt$ is to make the resulting spectrum of the pulses narrower.

The Fourier transform of a train of constant-amplitude pulses of width $T$ and frequency $f_s$ indicates that the spectrum $H(j2\pi f)$ has components at multiples of the sampling frequency $f_s$ and the envelope of these components follows a $(\sin x)/x$ form, where $x = \pi f T$. The general form of the spectrum is

$$H(j2\pi f) = T f_s \exp(-j\pi f T) \left( \frac{\sin \pi f T}{\pi f T} \right) \tag{2.32}$$

where the exponential term represents a linear phase delay and $T f_s$ represents a gain factor. Note that the $(\sin x)/x$ function and the spectrum have zero value at frequencies that are integer multiples of $1/T$. The −3 dB point of the spectrum envelope is at a frequency of approximately $1/2T$. The signal spectrum at the output of the sampler is, by virtue of the averaging, multiplied by the $(\sin x)/x$

function. Thus each of the sidebands around 0, $f_s$, $2f_s$, etc., in Figure 2.10 is attenuated accordingly. Clearly, the sampling pulse width should be small enough so that the bandwidth of the signal being sampled is much smaller than $1/2T$ to avoid attenuation of the signal in the sampled spectrum.

$T$ represents the length of time the input signal must be examined to ascertain its value. If the device used to examine the signal is an integrating ADC, then $T$ corresponds to the signal integration time (not the total conversion time).

The effect of aperture time can easily be examined in the time domain. Consider an input signal $v(t) = V \sin 2\pi ft$. The maximum rate of change of the signal is $dv/dt = 2\pi fV$ at $t = 0$. If this signal is to be sampled, its value during the aperture time $T$ must not go through a fractional change of more than the resolution $(1/2^n)$ of the ADC in order to obtain a valid sample with $n$-bit resolution. Assuming the worst case, that the peak-to-peak input signal fills the ADC input range $V_{fs}$ and the frequency is at the upper band edge $f_{bw}$, it is clear the for $n$-bit resolution

$$\frac{dv}{dt} = 0.5V_{fs}(2\pi f_{bw}) \leq \frac{\Delta v}{T} \tag{2.33}$$

where

$$\Delta v = V_{fs} 2^{-n} \tag{2.34}$$

Thus

$$T \leq \frac{1}{2^n \pi f_{bw}} \tag{2.35}$$

The relationship between full-scale maximum frequency input and aperture time for a resolution of 4 to 16 bits is shown in Figure 2.16. Note that even a moderate conversion requirement, such as 8 bits resolution $(1/2^8)$ at 1 kHz input frequency, demands a short (1 $\mu$s) aperture time.

Input frequencies beyond the upper band edge $f_{bw}$ will be present in most systems. Such signals are attenuated by the antialiasing filter and/or by natural rolloff. Note that as long as there is at least first-order rolloff (20 dB/decade) in the input spectrum, the values of $T$ derived from Eq. (2.35) and Figure 2.16 will be adequate even for inputs beyond $f_{bw}$; for first-order spectral rolloff, $dv/dt$ remains nominally constant, while for second and higher orders of rolloff, $dv/dt$ decreases beyond $f_{bw}$.

Achieving a resolution equal to that of the ADC is necessary if individual samples of the input are important. However, samples taken with longer values of $T$ than indicated by Figure 2.16 are still valid but represent an averaged value of the signal over the longer time interval. The limiting value of $T$ corresponds to half the period of the maximum input frequency, i.e., the full period of the minimum sampling frequency, at which point the response to a sine wave input is down approximately 4 dB.

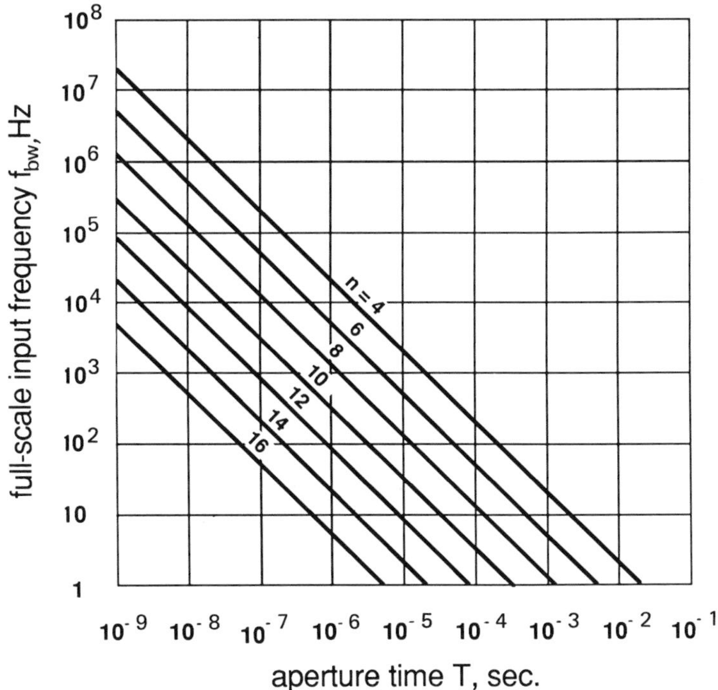


**Figure 2.16** Relationship between aperture time $T$ and full-scale input frequency $f_{bw}$ for an ADC in order to achieve a resolution of $1/2^n$, where $n$ is the number of bits including the sign bit.

### 4.3 Sample/Hold Devices

A typical sample/hold circuit (Figure 2.17) consists of an electronic switch S and a capacitor C arranged so that any detrimental effects on speed and accuracy have been reduced to an appropriate low level. While the switch S is closed, the

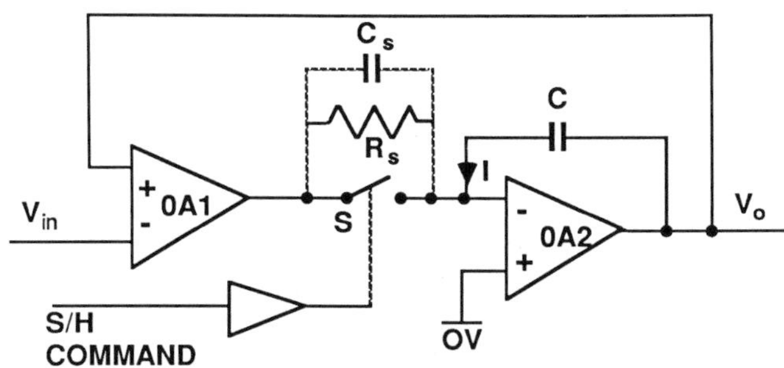


**Figure 2.17** Sample/hold circuit with the hold capacitor $C$ connected in the feedback loop of an operational amplifier.

two amplifiers function as a unity-gain follower tracking the input with some small residual gain error. The time taken to achieve a stated error after S is closed is the acquisition time $T_{aq}$. When the mode control signal changes to "hold" at $t_0$ to open S, a sample of the input signal (at $t_0$ plus the aperture delay time $T_d$ minus the signal delay time) is available from OA2 to a stated accuracy after a transient settling period $T_t$. After the settling period, the output voltage will change (drift) slowly because the input leakage current $I$, of OA2 and of the switch, must come from $C$. The rate of change of the output voltage is given by $I/C$ volts per second. The drift or droop rate must be low enough to ensure that the S/H output in the hold mode will change much less than $1/2^n$ compared to full scale during the aperture time of the ADC.

Other sample/hold output errors can be caused by parasitic capacitances within the S/H circuit allowing transients from the control signal to transfer charge onto the hold capacitor. Another source of error is the leakage of signal through the shunt resistance $R_s$ and capacitance $C_s$ of the sampling switch in the hold mode (Figure 2.17).

The conversion cycle time of a system with a sample/hold cannot be made less than $T_{aq} + T_t + T_{adc}$ where $T_{aq}$ is the S/H acquisition time, $T_t$ is the transient settling time, and $T_{adc}$ is the conversion time of the ADC [11].

Specifications for sampling devices usually refer to the aperture time, which includes an aperture delay time $T_d$ and an aperture uncertainty time $T_u$. Normally $T_d$ is constant or can be compensated by signal delays, leaving $T_u$ as the principal factor affecting resolution of the sample in repetitively sampled signals. $T_u$ can be considered as the time over which the sample switch opens, obtaining a not-quite-instantaneous sample of the input. Thus, $T_u$ represents the uncertainty about the exact time that the sample was taken [12]. Opening a semiconductor switch changes its resistance from a very low value to one of many megohms, which can be thought of as forming an $RC$ circuit (integrator), thus averaging the input value over the time $T_u$. The aperture time should not be confused with the acquisition time $T_{aq}$ of an S/H; $T_{aq}$ is the time required after closing the sampling switch for the signal at the output of the switch to become equal to the input within some tolerance, usually $\ll 1/2^n$. Analog-to-digital conversion is normally started immediately after the switch is opened and any transients have settled.

Sample/holds are available with values of $T_u$ of 10 ps to 1 ns for high-frequency applications. Figure 2.16 may be used for sample holds by using $T_u$ in place of $T$. A sample/hold should be used whenever the aperture time of the ADC is too long to yield the required amplitude resolution.

## 5 ANALOG MULTIPLEXERS

An analog multiplexer (AMUX) is a device composed of a number of switches controlled by a digital input word which can select one out of a number of input signals (Figures 2.1, 2.2, 2.4). The switches can be implemented using CMOS semiconductor devices or relays. Single-pole and double-pole multiplexers are common, the latter being employed for balanced (differential) signals [13]. For signal frequencies up to several hundred kilohertz and signal voltages above a few tens of millivolts, integrated circuit CMOS multiplexers with 8 and 16 inputs are normally employed.

At higher frequencies, output interference (crosstalk) from the input signals not selected can become a problem, depending on impedance and signal levels. Special designs or adaptations are required for high frequencies.

For very low-level signals, reed relays are recommended. They effectively provide an open circuit when off, have a very low on-resistance ($R_{on} < 1$ ohm), and have no significant common-mode or voltage offset problems. Relay operating speeds are limited to approximately 1 ms. For this reason relays are common in industrial applications in which low-frequency balanced signals are multiplexed before conditioning. CMOS multiplexers are limited to applications in which the signal plus the common mode voltage is less than the power supply voltage ($\pm 15$ V). In most applications the analog multiplexer is placed after the signal conditioning amplifier, thus eliminating multiplexer common-mode problems. In this location the AMUX is one of the more benign circuit components. It is usual, however, to keep signal source impedances less than 1–2 kilohms in order to minimize settling time and offset voltages generated by leakage current flowing through $R_{on}$ and the source resistance $R_s$. Load impedances that are greater than 10–20 megohms are recommended to minimize signal attenuation caused by $R_s + R_{on}$. A low bias current operational amplifier connected for unity gain can serve as an effective output buffer. Settling time of typical CMOS multiplexers is 1–5 $\mu$s for an output which settles within 0.01% of the input; typical on-resistance is 0.5–2 kilohms.

Several multiplexers can have their outputs connected to increase the available number of inputs; however, because of stray capacitance to ground, on-resistance, and source resistance, the settling time and leakage current effects increase more or less in proportion to the number of units connected together. Using a tiered approach, however, the output of $n$ multiplexers with $n$ inputs each can be connected to the inputs of a single $n$-input multiplexer to provide $n^2$ input channels. The settling time and offset effects for this two-tier arrangement are, very roughly speaking, improved by a factor of $n/2$ compared to the single tier arrangement with $n$ units of $n$ inputs each.

When multiplexers are used ahead of a common signal conditioning circuit (Figure 2.4), the settling time of the conditioning amplifier and any included filter must be taken into account. A single $RC$ section, for example, requires approximately $6.9RC$ s to settle to 0.1% and approximately $9.2RC$ s to settle to 0.01% after a step input. For a $-3$ dB cutoff frequency of $1/2\pi RC = 1$ Hz, these times equate to 1.1 and 1.5 s. Additional $RC$ sections or multipole filters will lengthen the settling time, particularly if the filter exhibits overshoot when driven by a step input.

## 6 ANALOG-TO-DIGITAL CONVERSION

The analog-to-digital converter performs the final function in the acquisition of digital data from real-world processes. Two primary categories of ADC are available: 1) those which directly compare the input signal to a set of quantization voltages and produce a coded output; the comparisons can be sequential, as in the successive-approximation and continuous tracking converters, or simultaneous, as in the "flash" converter, and 2) those which use integration techniques and first convert the input voltage into a proportional time duration or frequency, then quantize the time or frequency by counting, as in the integrating ADC and the voltage-to-frequency converter ADC [15].

When speed is important, ADCs from category 1 are far superior to those from 2; for noise rejection when reduced conversion speed is satisfactory, ADCs from category 2 are preferred. Digital voltmeters frequently use the integrating technique for precise measurements.

*Integrating ADCs* frequently include an automatic zero-point calibration at the start of each conversion cycle. This feature ensures that the signal-to-error ratio for small inputs is not impaired by an offset voltage. The signal is integrated for a time $T$, which thus becomes the aperture time for this conversion technique. The same integrator is used to integrate the signal and then to integrate an opposite-polarity reference voltage for a time $T_1$ (to a zero end point). Thus, component change with time and temperature has only a small effect on accuracy. Differential linearity, an important characteristic of ADCs, is excellent because the final quantization (i.e., measuring $T_1$) is carried out by a counter and clock. The absolute linearity is controlled by the quality of the analog integrator [15].

The aperture time $T$ for the integrating converter is generally 30–50% of the overall conversion time. The remainder of the time is spent performing the zero-point calibration and the reference integration. $T$ can normally be set so that it corresponds to an integer multiple $k$ of the fundamental period of any ac interference, for instance, 50, 60, or 400 Hz ac power frequencies. Because the integral of a periodic ac function over a full number of cycles is zero, the interference immunity of the integrating converter can be significant. Equation (2.32) can be used to express the frequency response of the integrator because it also represents the Fourier transform of the impulse response of an integrator in which $T$ is the integration time. If the magnitude of $H(j2\pi f)$ in Eq. (2.32) is normalized by removing $Tf_s$, the frequency response of the integrator expressed in dB form is

$$20 \log |H(j2\pi f)| = 20 \log |\sin \pi f T| - 20 \log \pi f T \tag{2.36}$$

Clearly, if $T = k/f$ the attenuation at frequency $f$ is theoretically infinite. However, the achieved attenuation depends on $T$ being precisely equal to $k/f$. If $T$ is in error by $\Delta T$, the attenuation possible at frequency $f$ is limited to $20 \log k + 20 \log(T/\Delta T)$ dB. Deep nulls (40 to 60 dB) are readily achievable in practice.

The term $-20 \log \pi f T$ in Eq. (2.36) causes an overall first-order rolloff. The nominal $-3$ dB corner frequency for the response is $1/2T$ (and is controlled by $20 \log |\sin \pi f T|$); $1/2T$ is also the nominal noise bandwidth of the integrator. If the input signal has a noise bandwidth $B$ well in excess of $1/2T$ an improvement in SNR can be achieved using an integrating converter. Assuming the noise is white, i.e., noise energy uniformly distributed over $B$, an SNR improvement of approximately $2BT$ is obtainable, corresponding to $(10 \log BT) + 3$ dB. The integrating converter requires a continuous "view" of the input signal during the aperture time in order to provide its interference and noise rejection performance. If a sample/hold is used ahead of it, the advantage is lost.

*Voltage-to-frequency (V-F) conversion ADCs* contain two levels of integration. The first is part of the V-F process in which the integration time is inversely proportional to the amplitude of the input signal. This integration does not offer a consistent noise-reducing advantage. The second integration is a function of the length of time the output frequency is counted to provide the quantized value of the input. This digital integration technique provides all of the noise reduction advantages of analog integration. The V-F approach offers some benefits in applications involving remote transducers. A dedicated V-F ADC can be used for each transducer; the output frequency, a high-level signal, can be sent over an appropriate cable (coaxial or twisted pair) to the instrumentation site where the integrating counter is located. Using this technique, good-quality (10–12 bit) digital values can be economically obtained from transducers many hundreds of meters away.

The *continuous tracking (counting)* converter offers moderate speed and good economy; as a result, it is often used in a dedicated one-per-channel application. The tracking ADC incorporates a counter whose output drives a digital-to-analog converter (DAC) to provide feedback to a comparator that determines whether the current quantized value is below (or above) the input voltage. The counter is then incremented (or decremented) one step at a time, until the quantized value exceeds (or is less than) the input by 1 LSB. The comparator output then changes state, thus completing that conversion cycle. Because the conversion goes from incomplete to complete in one count, the period of the clock pulses driving the counter becomes the aperture time $T$. The rate at which the tracking converter can follow the input is limited to a 1-LSB change per clock pulse, corresponding to an input rate of change of $dv/dt = 2^{-n}V_{fs}/T = 2^{-n}V_{fs}f_c$, where $f_c$ is the internal clock frequency. For $f_c = 1$ MHz and $n = 10$ bits, $dv/dt \simeq 10^3 V_{fs}$ V/s. This value of $dv/dt$ corresponds to a sine wave with peak amplitude of $V_{fs}/2$ and frequency of approximately 300 Hz, which is the highest full-scale input frequency that can be tracked in this example. Samples can be obtained at a rate up to $f_c$ samples per second while the input is being tracked.

If the tracking converter cannot always follow the input signal, a variable conversion time between $T$ and $(2^n - 1)T$ will result, making the time spacing of the digitized values uneven. To ensure uniformly spaced samples within $\pm T/2$, $f_c$ must be chosen high enough to allow continuous tracking within 1 LSB in a free-running mode, and the value in the counter must be read at a uniform rate. A sample/hold does not improve the conversion time but it can reduce the aperture time of this type of ADC.

*Successive-approximation ADCs* are inherently much faster than tracking types but work on a somewhat similar comparison principle. For an $n$-bit conversion, however, they require only $n$ steps versus $2^n - 1$ steps (worst case) for tracking ADCs. Conversion time is constant regardless of the input signal amplitude. The aperture time is equal to the conversion time. As indicated in Section 4.2, an appropriate sample/hold can be used with a successive-approximation ADC to ensure that the input will be resolved to $n$ bits no matter what the ADC conversion speed. Integrated circuit and hybrid successive-approximation ADCs are available with conversion times of 0.5 to 100 $\mu$s and word lengths up to 16 bits.

*Simultaneous converters* perform the analog-to-digital conversion in a parallel mode, i.e., all bits at once. The simultaneous or flash converter utilizes one voltage comparator for each quantization level except zero. A 6-bit converter requires $2^n - 1 = 63$ comparators and conversion speed is limited only by the response time of the comparators. Conversion times of 20 ns ($\sim$ 50-MHz sampling rate) are achievable at 6–8 bits resolution. The output code of the comparators is an $M$

out of $N$ representation of the signal, where $N = 2^n - 1$ and $M$ is the number of comparators whose threshold level has been exceeded by the input. This code is a unit-distance code, signifying that only one of the $N$ outputs changes at a time as the signal varies. Comparator outputs are normally stored in a buffer before being encoded in an $n$-bit gray code (another unit-distance code). In this way, if bits are changing while being stored, an error of only 1 LSB (maximum) is likely. The $n$-bit gray code is usually converted to $n$-bit binary, the final output. The aperture uncertainty $T_u$ of the simultaneous converter is normally a small fraction of the conversion time. Values of $T_u$ of approximately 25 ps are achievable in commercial ADCs of this type.

## 6.1 ADC Signal Levels: Avoiding Corruption

It is instructive to examine the change in the ADC input signal corresponding to a one-step (LSB) change in quantized output value (Table 2.4). By nature, the ADC is exposed to band-limited analog signals and to digital signals with very fast rise times. Table 2.4 shows that even for a moderate ADC (10 bits) the LSB-equivalent voltage is approximately 10 mV. Contributions of impulsive digital noise, coupled by direct crosstalk between cables and induced because of voltage drops in common (shared) ground resistances, can easily exceed this value unless precautions are taken. Attention to guarding and to the shielding of signal cables is therefore important [2]. Analog signals should be carried in differential (balanced) form, preferably until they are near the ADC, using twisted-pair shielded cables. This procedure can markedly increase the probability of obtaining high-quality signals at the ADC input. Generally, cable shields should be connected to ground at only one end, preferably the transducer end, to prevent any current flowing in the shield. Shield current can induce voltages in the signal wires, thus reducing the effectiveness of the shield [3].

Analog-to-digital converters normally provide a digital ground connection and an analog (or signal) ground connection. Within the ADC, the comparator (or comparators in the case of the simultaneous ADC) is the component that bridges the gap between the analog and digital domains. The comparators have a common mode rejection capability, thus enabling the input side (analog) and the output side

**Table 2.4** LSB Signal Levels in ADCs (10 V Input Range

| Bits | LSB amplitude, mV |
|---|---|
| 4 | 625 |
| 6 | 156 |
| 8 | 39.1 |
| 10 | 9.77 |
| 12 | 2.44 |
| 14 | 0.61 |
| 16 | 0.15 |

(digital) to utilize separate ground references. The digital ground and the analog ground should be connected together at only one point in the system. To make this feasible and to prevent ground loops, isolated power supplies may be needed for remote signal-conditioning amplifiers.

## 6.2 Analog-to-Digital Converter Output Codes

A variety of output codes is available from commercial ADCs. The more common codes are straight (natural) binary, sign-magnitude, offset binary, and 2's complement. Binary-coded-decimal output can also be produced for applications in which numerical values are displayed for direct reading, as in a voltmeter [15].

The straight binary code is normally associated with ADCs that handle unipolar inputs and require no sign bit. Sign-magnitude, offset binary, and 2's complement codes are commonly used for bipolar signals; usually the most significant bit (MSB) of the code provides the polarity indicator (see Table 2.5). Of course, all that is required to convert a 0–10-V unipolar ADC into a ±10-V bipolar one is to halve the gain and shift the half-scale input point to zero volts. Thus, the MSB automatically becomes the sign bit. The absolute voltage resolution of the LSB is then twice as coarse and, compared to the maximum input (plus or minus), the fractional resolution similarly changes from $1/2^n$ to $1/2^{n-1}$, where $n$ is the number of bits in the unipolar ADC. The resulting output code is offset binary.

As seen in Table 2.5, complementing the MSB (leftmost bit) of the offset binary code results in the 2's complement code. Both codes are therefore normally

**Table 2.5** Common ADC Output Codes (3 Bits Plus Sign) for Bipolar Inputs (±10 V)

| Input voltage, V | Fractional input | Sign-magnitude | Offset binary | Two's complement |
|---|---|---|---|---|
| 8.75 | 7/8 | 0111 | 1111 | 0111 |
| 7.50 | 6/8 | 0110 | 1110 | 0110 |
| 6.25 | 5/8 | 0101 | 1101 | 0101 |
| 5.00 | 4/8 | 0100 | 1100 | 0100 |
| 3.75 | 3/8 | 0011 | 1011 | 0011 |
| 2.50 | 2/8 | 0010 | 1010 | 0010 |
| 1.25 | 1/8 | 0001 | 1001 | 0001 |
| 0 | +0 | 0000 | 1000 | 0000 |
| 0 | −0 | 1000 | 1000 | 0000 |
| −1.25 | −1/8 | 1001 | 0111 | 1111 |
| −2.50 | −2/8 | 1010 | 0110 | 1110 |
| −3.75 | −3/8 | 1011 | 0101 | 1101 |
| −5.00 | −4/8 | 1100 | 0100 | 1100 |
| −6.25 | −5/8 | 1101 | 0011 | 1011 |
| −7.50 | −6/8 | 1110 | 0010 | 1010 |
| −8.75 | −7/8 | 1111 | 0001 | 1001 |
| −10.00 | −8/8 | — | 0000 | 1000 |

available from the same ADC. Mathematically, the 2's complement is formed by complementing all the bits in a natural binary number and adding one to the complement. The 2's complement code can be considered as a set of positive and negative numbers that can readily be used in numerical data processing.

An important attribute of the 2's complement and the offset binary codes is the existence of a single representation for zero. In contrast, the sign-magnitude code is awkward, having two values for zero: zero plus and zero minus. Note also from Table 2.5 that an input of −10 V (or −8/8) results in a 2's complement number (1000) that is beyond the normal range of values associated with the fixed point negative fraction notation. For this reason, input values for 2's complement and offset binary are usually constrained to $\pm(2^{n-1})/2^n$ of full scale, where $n$ is the number of bits excluding the sign bit.

Sign-magnitude coded ADCs usually have external circuitry for sensing polarity and for taking the absolute value of an input before performing the analog-to-digital conversion using a unipolar ADC. Because the absolute value is taken, the voltage resolution and the fractional resolution ($1/2^n$) of the unipolar converter remain unchanged. However, an additional bit, the sign bit, is added to the output word.

## 7 CONCLUSION

In common with the design of most electronic systems, the acquisition and conditioning of analog signals for conversion to digital form should begin by examining the output signal requirements. The fundamental question to be answered is related to the quality needed in the digital data. A leading factor is the final signal-to-noise ratio to be achieved after any digital signal processing. Postprocessing of the data can reduce the random noise level through the use of averaging techniques. Aliased signal energy and noise or interference that is phase coherent in relation to the sampling frequency, however, cannot be reduced by averaging. Thus, a second important factor is a knowledge of the frequency spectrum of the input signal.

Based on the SNR requirements and the signal spectrum, Figure 2.14 (or an equivalent plot such as Figure 2.13) can be used to determine a suitable trade-off between the sampling frequency and the spectral rolloff needed to achieve an appropriately small level of aliased signal in the band 0 to $f_s/2$. In a good design, aliased energy will be smaller than the residual noise.

The level of random noise that is tolerable in the postprocessed signal can be used in Eq. (2.27) to determine an SNR target for the output of the signal conditioner. If an integrating ADC is to be used, its noise and interference reduction capabilities from Eq. (2.36) should be included at this point. By comparing the signal conditioner's target SNR to the signal-to-error ratio in Table 2.3, the minimum number ($n$) of bits in the ADC output can be determined. The SNR and SER values should, of course, take account of the peak-to-rms ratio of the input signal and of the noise in order to achieve the required confidence level in the sampled values. The sampling rate $f_s$ and the resolution ($n$ bits) determine the primary attributes of the ADC, multiplexer, and sample/hold required.

The SNR needed to achieve a given confidence level provides a comparison point for the quantitative level of noise and uncertainty from the system components. An RSS error analysis for the transducer, amplifier, filter, sample/hold, multiplexer, and ADC will indicate which components are major contributors to the total error. If the RSS error is higher than acceptable, then better components can be selected to reduce the major contributions to the error. Equally, less costly components can be substituted for those that add little to the RSS error. Techniques to reduce some errors may still be required to make the RSS error small enough, e.g., using periodic zero and full-scale calibration sequences to null low-frequency errors.

When the desired RSS error level has been achieved on paper, the more difficult task of hardware implementation begins.

## ACKNOWLEDGMENTS

The able and dedicated assistance of Elizabeth Vaughan in typing and proofreading the manuscript is gratefully acknowledged.

## REFERENCES

1. D. Sheingold, *Transducer Interfacing Handbook*, Analog Devices, Inc., Norwood, Massachusetts (1980).
2. R. Morrison, *Instrumentation Fundamentals and Applications*, Wiley, New York (1984).
3. H. W. Ott, *Noise Reduction Techniques in Electronic Systems*, Wiley, New York (1976).
4. *Analog Devices Databook*, Analog Devices, Inc., Norwood, Massachusetts (1984).
5. P. H. Garrett, *Analog I/O Design*, Reston, Reston, Virginia (1981).
6. A. B. Williams, *Electronic Filter Design Handbook*, McGraw-Hill, New York, (1981).
7. D. E. Johnson and J. N. Hilburn, *Rapid, Practical Design of Active Filters*, Wiley, New York (1975).
8. A. Ryan and T. Scranton, D-C amplifier noise revisited, *Analog Dialogue*, vol. 18, no. 1, pp. 3–10. Analog Devices, Inc., Norwood, Massachusetts (November 1, 1984).
9. P. F. Blackman, Introduction to sampling and Z-transforms, *Introduction to Digital Filtering*, (R. E. Bogner and A. G. Constantinides, eds.) Wiley, New York (1975).
10. R. Coates, Fourier Transform Methods, *Introduction to Digital Filtering*, (R. E. Bogner and A. G. Constantinides, eds.) Wiley, New York (1975).
11. D. Santucci, Maneuvering for top speed and high accuracy in data acquisition, *Electronics*, *48*(24): 115–119 (1975).
12. D. Santucci, Data acquisition can falter unless components are well understood, *Electronics*, *48*(23): 114–118 (1975).
13. E. Zuch, *Data Acquisition and Conversion Handbook*, Datel-Intersil, Mansfield, Massachusetts (1979).
14. C. L. Tang, A. S. Bennett, and D. J. Lawrence, Thermohaline intrusions in the frontal zones of a warm-core ring observed by batfish, *J. Geophys. Res.*, *90*: 8928–8942 (1985).

15. D. Sheingold, *Analog-Digital Conversion Notes*, Analog Devices, Inc., Norwood, Massachusetts (1977).
16. H. J. Blinchicoff and A. I. Zverev, *Filtering in the Time and Frequency Domains*, Wiley, New York (1976). P. 217.

# 3

# Digital Filters: Principles and Implementation

IOANNIS PITAS Department of Electrical Engineering, Aristotelian University of Thessalonica, Thessalonica, Greece

## 1 THE DIGITAL FILTER AS A DISCRETE SYSTEM

A digital filter is a discrete-time system $T$ whose input $x(n)$ and output $y(n)$ are discrete signals:

$$y(n) = T[x(n)] \tag{3.1}$$

A digital filter is said to be *linear* iff it satisfies the following relation:

$$T[ax_1(n) + bx_2(n)] = aT[x_1(n)] + bT[x_2(n)] \tag{3.2}$$

It is said to be *time-invariant*, if its internal parameters do not change with time:

$$y(n) = T[x(n)] \longrightarrow y(n-k) = T[x(n-k)] \tag{3.3}$$

A digital filter is causal if its response at a specific time is independent of subsequent values of the excitation.

In the rest of this chapter we shall consider only the linear time-invariant digital filters. Such filters are determined by their *impulse response* $h(n)$. The output $y(n)$

of a linear time-invariant digital filter is a *convolution* of its impulse response $h(n)$ with its input $x(n)$:

$$y(n) = x(n) * h(n) = \sum_{k=-\infty}^{\infty} h(k)x(n-k) \tag{3.4}$$

If the impulse response $h(n)$ is of finite duration, the filter is called a *finite impulse response* (FIR) digital filter. In this case Eq. (3.4) becomes

$$y(n) = \sum_{k=0}^{M} h(k)x(n-k) \tag{3.5}$$

If the impulse response $h(n)$ is of infinite duration, the filter is called an *infinite impulse response* (IIR) digital filter and it is described by the following difference equation:

$$\sum_{k=0}^{N} a_k y(n-k) = \sum_{k=0}^{M} b_k x(n-k) \tag{3.6}$$

where $a_k$, $k = 0, \ldots, N$, and $b_k$, $k = 0, \ldots, M$ are the IIR filter coefficients. The IIR and FIR digital filters are designed and implemented in different ways.

A digital filter is called *stable* iff any bounded excitation results in a bounded response, i.e.,

$$\forall n, \ |x(n)| < \infty \longrightarrow \forall n, \ |y(n)| < \infty \tag{3.7}$$

It can be proved [1] that a linear, time-invariant, and causal digital filter is stable iff

$$\sum_{k=0}^{\infty} |h(k)| < \infty \tag{3.8}$$

Equation (3.8) is always satisfied by FIR filters. Therefore they are always stable.

## 2 THE $Z$ TRANSFORM AND ITS INVERSE

### 2.1 The $Z$ Transform

A basic tool in the analysis of digital filters is the $Z$ transform. Its role in digital filter analysis is equivalent to that of the Laplace transform in analog filters. Its definition is the following:

$$X(z) = \sum_{n=-\infty}^{\infty} x(n)z^{-n} \tag{3.9}$$

where $z$ is a complex variable which defines the complex $z$ plane. The sum (3.9) does not necessarily converge over the entire $z$ plane. The *region of convergence* (ROC) of $X(z)$ depends on the sequence $x(n)$. The following cases may be examined separately:

1. *Finite duration sequence* $[x(n) = 0, n < n_1, n > n_2]$. In this case the sum

$$X(z) = \sum_{n=n_1}^{n_2} x(n)z^{-n} \tag{3.10}$$

converges for every $z$. Therefore the ROC is the entire $z$ plane.

2. *Right-sided sequence* $[x(n) = 0, n < n_1]$. In this case, if there exists $z_1$ such that

$$\sum_{n=n_1}^{\infty} |x(n)z_1^{-n}| < \infty \tag{3.11}$$

the following inequality is satisfied for every $|z| > |z_1|$:

$$\sum_{n=n_1}^{\infty} |x(n)z^{-n}| < \sum_{n=n_1}^{\infty} |x(n)z_1^{-1}| < \infty \tag{3.12}$$

Therefore the ROC is the exterior of a circle centered at $z = 0$ and having radius $|z_1|$.

3. *Left-sided sequence* $[x(n) = 0, n > n_2]$. The $Z$ transform has the following form:

$$X(z) = \sum_{n=-\infty}^{\infty} x(n)z^{-n} = \sum_{m=-n_2}^{\infty} x(-m)z^{m} \tag{3.13}$$

By substituting $z$ by $z^{-1}$ the result of the previous case can be applied. Therefore the ROC is the interior of a circle centered at $z = 0$ and having radius $|z_2|$.

4. *Two-sided sequence* $[x(n) \neq 0, \forall n]$. The $Z$ transform has the following form:

$$X(z) = \sum_{n=-\infty}^{\infty} x(n)z^{-n} = \sum_{n=0}^{\infty} x(n)z^{-n} + \sum_{n=-\infty}^{-1} x(n)z^{-n} \tag{3.14}$$

If the first term of $X(z)$ converges for $|z| > |z_1|$ and the second term converges for $|z| < |z_2|$, the ROC of $X(z)$ is the ring $z_1| < |z| < |z_2|$. The ROC for cases 2, 3, and 4 are shown in Figure 3.1.

The properties of the $Z$ transform are summarized in Table 3.1.

## 2.2 Inverse $Z$ Transform

The calculation of the inverse $Z$ transform is based on the Cauchy theorem, which states that:

$$\frac{1}{2\pi i}\oint_C z^{k-1}\,dz = \begin{cases} 1 & k=1 \\ 0 & k\neq 0 \end{cases} \tag{3.15}$$

**Table 3.1** Properties of the $Z$ Transform

| Sequence | Transform | ROC |
|---|---|---|
| $x[n]$ | $X(z)$ | $R_x$ |
| $x_1[n]$ | $X_1(z)$ | $R_1$ |
| $x_2[n]$ | $X_2(z)$ | $R_2$ |
| $ax_1[n] + bx_2[n]$ | $aX_1(z) + bX_2(z)$ | At least the intersection of $R_1$ and $R_2$ |
| $x[n-n_0]$ | $z^{-n_0}X(z)$ | $R_x$ except for the possible addition or deletion of the origin |
| $e^{j\Omega_0 n_x[n]}$ | $X(e^{-j\Omega_0}z)$ | $R_x$ |
| $z_0^n x[n]$ | $X\left(\frac{z}{z_0}\right)$ | $z_0 R_x$ |
| $\alpha^n x[n]$ | $X(\alpha^{-1}z)$ | Scaled version of $R_x$ |
| $x[-n]$ | $X(z^{-1})$ | Inverted $R_x$ |
| $w[n] = \begin{cases} x[r], & n = rk \\ 0, & n \neq rk \text{ for some } r \end{cases}$ | $X(z^k)$ | $R_x^{1/k}$ |
| $x_1[n] * x_2[n]$ | $X_1(z)X_2(z)$ | At least the intersection of $R_1$ and $R_2$ |
| $nx[n]$ | $-z\frac{dX(z)}{dz}$ | $R_x$ except for the possible addition or deletion of the origin |
| $\sum_{k=-\infty}^{n} x[k]$ | $\frac{1}{1-z^{-1}}X(z)$ | At least the intersection of $R_x$ and $\|z\| > 1$ |

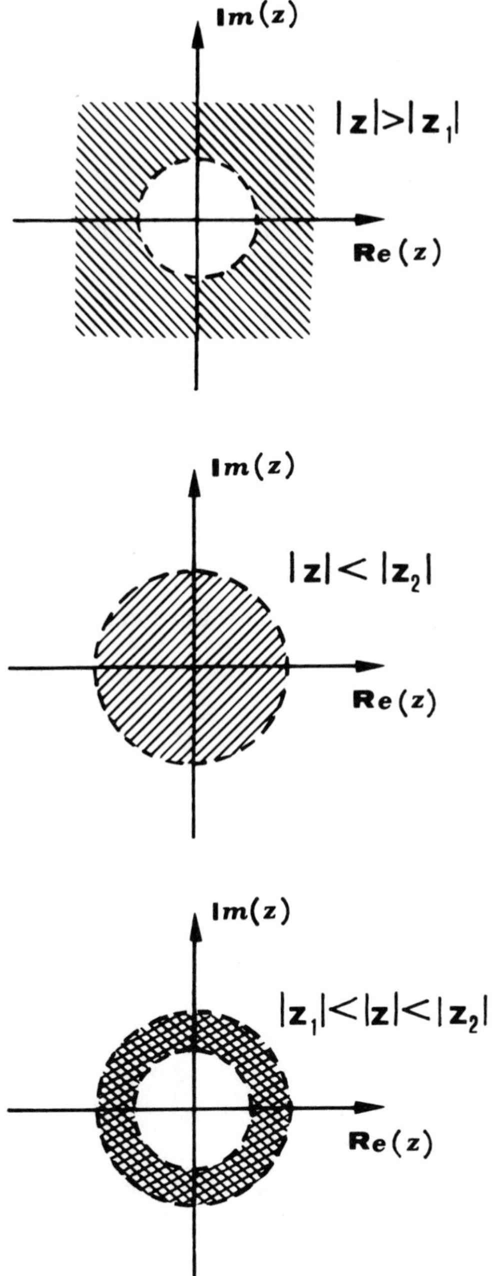


**Figure 3.1** Regions of convergence of the $Z$ transform.

where $C$ is a closed curve that circumscribes the origin $z = 0$. By integrating both sides of Eq. (3.9) we find that:

$$\frac{1}{2\pi i}\oint_C X(z)z^{k-1}\,dz = \frac{1}{2\pi i}\oint_C \sum_{n=-\infty}^{\infty} x(n)z^{-n+k-1}\,dz$$
$$= \sum_{n=-\infty}^{\infty} x(n)\frac{1}{2\pi i}\oint_C z^{-n+k-1}\,dz = x(k) \quad (3.16)$$

Thus, the inverse $Z$ transform is given by

$$x(n) = \frac{1}{2\pi i}\oint_C X(z)z^{n-1}\,dz \quad (3.17)$$

where $C$ is a closed curve containing the origin $z = 0$ and lies on the ROC of $X(z)$. The calculation of the inverse $Z$ transform by Eq. (3.17) is computationally cumbersome. Therefore one of the following methods is employed in practice.

1. *Use of residues.* If $X(z)$ is a rational function, its inverse transform is given by

$$x(n) = [\text{residues of } X(z)z^{n-1}\text{at the poles inside } C] \quad (3.18)$$

If $X(z)z^{n-1}$ is of the form $\Psi(z)/(z-z_0)^s$, the residue Res$[X(z)z^{n-1}$ at $z_0]$ is given by the formula

$$\text{Res}[X(z)z^{n-1} \text{ at } z_0] = \frac{1}{(s-1)!}\left[\frac{d^{s-1}\Psi(z)}{dz^{s-1}}\right]_{z-z_0} \quad (3.19)$$

If $s = 1$, Eq. (3.19) becomes

$$\text{Res}\left[X(z)z^{n-1} \text{ at } z_0\right] = \Psi(z_0) \quad (3.20)$$

2. *Use of power series.* If $X(z)$ is expandable in power series:

$$X(z) = \sum_{n=0}^{\infty} a_n z^{-n} \quad (3.21)$$

its inverse transform is simply given by

$$x(n) = a_n \quad (3.22)$$

3. *Use of partial fractions.* If $X(z)$ is expandable to a sum of fractions:

$$X(z) = \sum_{i=1}^{N} \frac{a_i}{1 - p_i z^{-1}}, \qquad |z| > \max_{1 \le i \le N} |p_i| \tag{3.23}$$

its inverse transform is given by

$$x(n) = \begin{cases} \sum_{i=1}^{N} a_i p_i^n u(n), & n \ge 0 \\ 0, & n < 0 \end{cases} \tag{3.24}$$

because the $Z$ transform of $a_i p_i^n$ is $a_i/(1 - p_i z^{-1})$, $(|z| \ge |p_i|)$.

The choice of the inverse $Z$ transform method depends on the mathematical form of $X(z)$.

## 3 TRANSFER FUNCTION OF A DIGITAL FILTER

As we have mentioned, the $Z$ transform is used as a basic tool for the analysis of digital filters. Its primary use is in the development of the transfer function of a digital filter. The transfer function of an FIR filter can be found by applying the $Z$ transform on both sides of Eq. (3.5) and using the time-shift property shown in Table 3.1:

$$H(z) = \frac{Y(z)}{X(z)} = \sum_{k=0}^{N} h(k) z^{-k} \tag{3.25}$$

The transfer function of an IIR filter is found by applying the same procedure on Eq. (3.6):

$$Y(z) \sum_{k=0}^{N} a_k z^{-k} = X(z) \sum_{k=0}^{M} b_k z^{-k} \tag{3.26}$$

Therefore $H(z)$ is given by

$$H(z) = \frac{Y(z)}{X(z)} = \frac{\sum_{k=0}^{N} b_k z^{-k}}{\sum_{k=0}^{N} a_k z^{-k}} \tag{3.27}$$

The poles and the zeros of $H(z)$ are found by factorizing the numerator and the denominator of $H(z)$:

$$H(z) = H\frac{\prod_{k=1}^{M}(1-c_k z^{-1})}{\prod_{k=1}^{N}(1-d_k z^{-1})} \tag{3.28}$$

The ROC of $H(z)$ is always bounded by the poles $d_k$, $k = 1, \ldots, N$ or infinity. For a given pole-zero pattern of $H(z)$ there exists only a limited number of ROCs consistent with the properties discussed in Section 2. Knowledge of both the pole-zero pattern and the ROC of $H(z)$ is required for a unique determination of the impulse response $h(n)$.

If a digital filter is stable, the unit circle $|z| = 1$ must lie in the ROC of $H(z)$. If the filter is causal, the ROC of $H(z)$ must be outside the circle defined by the outermost pole of $H(z)$. For the system that is both stable and causal, the ROC must include the unit circle and be outside the outermost pole. Therefore the poles of a stable and causal system must lie inside the unit circle. This is a necessary and sufficient condition for stability.

The direct check for stability is to find the poles of $H(z)$ and to check if all of them lie inside the unit circle. An alternative approach is to use Jury's criterion [2], described in Table 3.2. The elements $a_0, \ldots, a_N$ of the first and the second row are the coefficients of the denominator $D(z)$ of $H(z)$: $D(z) = \sum_{k=0}^{N} a_k z^{N-k}$. The elements of the third and fourth rows are computed as

$$c_i = \begin{vmatrix} a_0 & a_{N-i} \\ a_N & a_i \end{vmatrix}, \qquad i = 0, 1, \ldots, N-1 \tag{3.29}$$

and those of the fifth and sixth rows as

$$d_i = \begin{vmatrix} c_0 & c_{N-1-i} \\ c_{N-1} & c_i \end{vmatrix}, \qquad i = 0, 1, \ldots, N-2 \tag{3.30}$$

**Table 3.2** Jury's Array

| Row | Coefficients | | | |
|---|---|---|---|---|
| 1 | $a_0$ | $a_1$ | $\cdots$ | $a_N$ |
| 2 | $a_N$ | $a_{N-1}$ | $\cdots$ | $a_0$ |
| 3 | $c_0$ | $c_1$ | $\cdots$ | $c_{N-1}$ |
| 4 | $c_{N-1}$ | $c_{N-2}$ | $\cdots$ | $c_0$ |
| 5 | $d_0$ | $d_1$ | $\cdots$ | $d_{N-2}$ |
| 6 | $d_{N-2}$ | $d_{N-3}$ | $\cdots$ | $d_0$ |
| ...... | ............ | | | |
| $2N-3$ | $r_0$ | $r_1$ | $r_2$ | |

and so on, until $2N - 3$ rows are obtained. The filter $H(z)$ is stable if

$$
\begin{aligned}
&1.\ D(1) > 0\\
&2.\ (-1)^N D(-1) > 0\\
&3.\ a_0 > |a_N|\\
&\quad |c_0| > |c_{N-1}|\\
&\quad |d_0| > |d_{N-2}|\\
&\quad |r_0| > |r_2|
\end{aligned}
\tag{3.31}
$$

## 4 RELATION BETWEEN $Z$ TRANSFORM AND LAPLACE TRANSFORM

The sampling of the signal $x(t)$ is described by the following equation:

$$\hat{x}(t) = \sum_{n=-\infty}^{\infty} x(n)\delta(t - nT), \qquad \omega_s = \frac{2\pi}{T} \tag{3.32}$$

where $T$ is the sampling period and $x(n) = x(nT)$.

The Laplace transform of the sampled function $x(t)$ is given by

$$\mathcal{L}[\hat{x}(t)] = \sum_{n=-\infty}^{\infty} x(n)e^{-nTs} \tag{3.33}$$

By substituting

$$z = e^{Ts} = e^{\sigma T}e^{i\omega T} \tag{3.34}$$

we obtain the following relation between $X(z)$ and $\mathcal{L}[\hat{x}(t)]$:

$$X(z) = \mathcal{L}[\hat{x}(t)]_{z=e^{Ts}} \tag{3.35}$$

Equation (3.34) is a mapping from the $s$ plane to the $z$ plane, shown in Figure 3.2.

According to Eq. (3.34), we find that

$$|z| = e^{\sigma T}\begin{cases} < 1 & \text{for } \sigma < 0\\ = 1 & \text{for } \sigma = 0\\ > 1 & \text{for } \sigma > 0\end{cases} \tag{3.36}$$

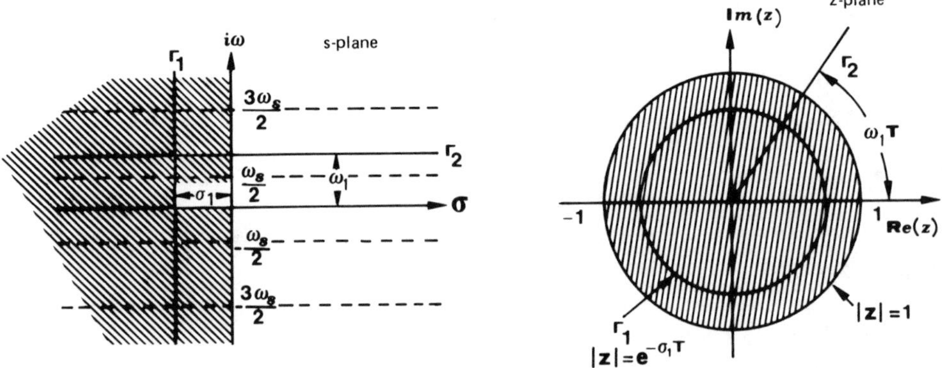


**Figure 3.2** Mapping from the $s$ plane to the $z$ plane.

Thus, the negative $s$ half-plane, the imaginary axis $i\omega$, and the positive $s$ half-plane are mapped onto the interior, the circumference, and the exterior of the unit circle $|z| = 1$, respectively. Straight lines parallel to the imaginary axis are mapped onto circles concentric to $|z| = 1$. Straight lines parallel to the real axis $\sigma$ are mapped onto radial lines. The segment $-i\omega_s/2 \leq i\omega \leq i\omega_s/2$ of the imaginary axis is mapped onto the circumference $|z| = 1$. The segments $i(\omega_s/2 + k\omega_s) \leq i\omega \leq i(\omega_s/2+k\omega_s)$ are mapped onto the unit circle too. This fact explains the periodicity of the frequency response of a discrete system.

## 5 DISCRETE-TIME FOURIER TRANSFORM AND DISCRETE FOURIER TRANSFORM

The $Z$ transform evaluated on the unit circle $z = e^{i\Omega}$ is given by

$$X(e^{i\Omega}) = \sum_{n=-\infty}^{\infty} x(n)e^{-i\Omega n} \tag{3.37}$$

$X(e^{i\Omega})$ is the discrete-time Fourier transform of $x(n)$ and $\Omega$ is the so-called *discrete frequency*. Its range is $-\pi \leq \Omega \leq \pi$ and its relation to the analog frequency $\omega$ is given by

$$\Omega = \omega T \tag{3.38}$$

The inverse discrete-time Fourier transform is

$$x(n) = \frac{1}{2\pi i} \oint_{z=e^{i\Omega}} X(e^{i\Omega})e^{i\Omega(n-1)}\, de^{i\Omega} = \frac{1}{2\pi} \int_{-\pi}^{\pi} X(e^{i\Omega})e^{i\Omega n}\, d\Omega \tag{3.39}$$

The properties of the discrete-time Fourier transform are summarized in Table 3.3.

The discrete Fourier transform (DFT) of a sequence $x(n)$ of finite extent $N$ is its discrete-time Fourier transform evaluated at frequencies $\Omega = 2\pi/N$, $0 \le k \le N-1$:

$$X(k) = X(\Omega)\Big|_{\Omega=(2\pi/N)k} = \sum_{n=0}^{N-1} x(n)e^{-i(2\pi nk/N)} = \sum_{n=0}^{N-1} x(n)W_N^{nk} \qquad (3.40)$$

$$W_N = e^{-i2\pi/N}$$

**Table 3.3** Properties of the Discrete-Time Fourier Transform

| Aperiodic signal | Fourier transform |
|---|---|
| $x[n]$ | $X(\Omega)$ periodic with period $2\pi$ |
| $y[n]$ | $Y(\Omega)$ periodic with period $2\pi$ |
| $ax[n] + by[n]$ | $aX(\Omega) + bY(\Omega)$ |
| $x[n - n_0]$ | $e^{-i\Omega n_0 X(\Omega)}$ |
| $e^{j\Omega_0 n}x[n]$ | $X(\Omega - \Omega_0)$ |
| $x^*[n]$ | $X^*(-\Omega)$ |
| $x[-n]$ | $X(-\Omega)$ |
| $x_{(k)}[n] = \begin{cases} x[n/k], & \text{if } n \text{ is a multiple of } k \\ 0, & \text{if } n \text{ is not a multiple of } k \end{cases}$ | $X(k\Omega)$ |
| $x[n] * y[n]$ | $X(\Omega)Y(\Omega)$ |
| $x[n]y[n]$ | $\frac{1}{2\pi}\int_{2\pi} X(\theta)Y(\Omega - \theta)\, d\theta$ |
| $x[n] - x[n-1]$ | $(1 - e^{-i\Omega})X(\Omega)$ |
| $\sum_{k=-\infty}^{n} x[k]$ | $\frac{1}{1 - e^{-i\Omega}}X(\Omega) + \pi X(0) \sum_{k=-\infty}^{+\infty} \delta(\Omega - 2\pi k)$ |
| $nx[n]$ | $i\frac{dX(\Omega)}{d\Omega}$ |
| $x[n]$ real | $X(\Omega) = X^*(-\Omega)$ |
| $x_e[n] = \text{ev}\{x[n]\}$, $[x[n]$ real$]$ | $\text{Re}\{X(\Omega)\}$ |
| $x_o[n] = \text{od}\{x[n]\}$, $[x[n]$ real$]$ | $i\,\text{Im}\{X(\Omega)\}$ |
| Parseval's relation for aperiodic signals $\sum_{n=-\infty}^{+\infty} \lvert x[n]\rvert^2 = \frac{1}{2\pi}\int_{2\pi} \lvert X(\Omega)\rvert^2\, d\Omega$ | |

The inverse DFT is given by the following relation:

$$x(n) = \frac{1}{N} \sum_{k=0}^{N-1} X(k) W_N^{-nk} \tag{3.41}$$

It can be shown that the transform pair Eqs. (3.40) and (3.41) can also be used for the exact representation of periodic sequences $x_p(n)$:

$$x_p(n) = x_p(n + kN), \qquad k \text{ integer}, N \text{ period} \tag{3.42}$$

If a sequence is neither periodic nor finite, it cannot be exactly represented by a DFT. In this case the sequence is truncated to a certain length $N$ and it is periodically repeated outside the domain $[0, N]$. The periodic sequence that results can be uniquely determined by a DFT.

## 6 DESIGN OF DIGITAL FILTERS

The design of a digital filter involves the following steps:

1. Specification of the desired properties of the filter.
2. Choice of the kind of filter (FIR or IIR).
3. Approximation of the filter specifications by a design method.
4. Approximation of the filter coefficients by finite-precision arithmetic.
5. Choice of the filter realization.
6. Simulation of the filter performance. If it is not satisfactory, we go back to step 3 and repeat the design procedure.

In this and in the subsequent section we shall analyze steps 1–3. The desired filter specification can be given either in analog frequency $\omega$ (whenever the filter is used for analog signal processing) or in discrete frequency $\Omega$. The analog frequency is easily transformed to discrete frequency, if the sampling frequency $\omega_s$ is known. The specifications of a low-pass digital filter are shown in Figure 3.3. Usually the passband cutoff frequency $\Omega_p$, the stopband cutoff frequency $\Omega_s$, the passband ripple $\delta_1$ (in dB), and the stopband attenuation $\delta_2$ (in dB) are specified. The specifications of high-pass and other digital filters are given in a similar way.

Different design methods are used for FIR and IIR filters. These methods will be discussed in the subsequent section and some hints for the choice between FIR and IIR filters will follow.

### 6.1 Design of IIR Digital Filters

There exist three approaches to the design of IIR filters:

1. Analog-to-digital filter transformation
2. Optimization techniques
3. Direct filter design in the time domain

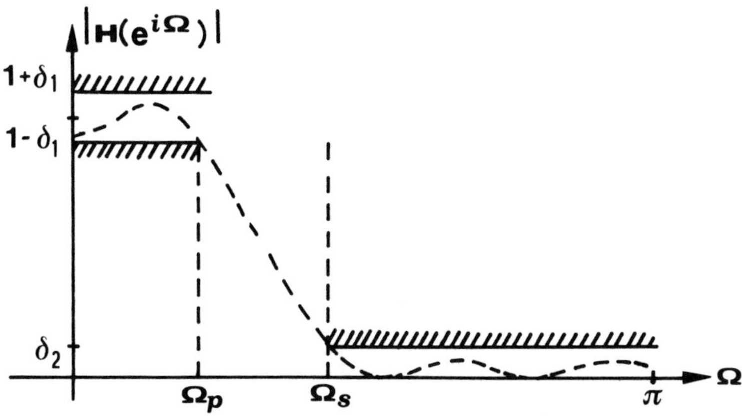


**Figure 3.3** Specifications of a digital low-pass filter.

The first is the most popular approach for the following reasons: (1) there exist very good analog design methods (e.g., Butterworth, Chebyshev, Bessel, and Elliptic filters), (2) there exist simple methods for analog-to-digital filter transformations, and (3) digital filters are used in many cases to substitute for existing analog filters.

Several methods are possible in all three of the above approaches. In the following we shall describe only the more popular transformation and optimization methods.

### Bilinear Transformation

The analog-to-digital filter transformation corresponds to a mapping from the $s$ plane onto the $z$ plane. There exist several such mappings [e.g., Eq. (3.34)]. Such a mapping is satisfactory if it is simple (possibly rational) and satisfies the following conditions [3]:

*Condition 1:* The imaginary axis $i\omega$ is mapped into the unit circle $|z| = 1$.
*Condition 2:* The left half-plane $s$ is mapped onto the interior of the unit circle of the $z$ plane.

The two conditions are illustrated in Figure 3.2. The first condition is needed to preserve the frequency characteristics of the analog filters. The second is needed to map stable analog filters to stable digital filters.

The bilinear transformation satisfies these conditions:

$$s = \frac{2}{T}\frac{1 - z^{-1}}{1 + z^{-1}}, \qquad z = \frac{1 + (T/2)s}{1 - (T/2)s} \tag{3.43}$$

The transfer function $H(z)$ of the designed digital filter is given by

$$H(z) = H_a(s)|_{s=(2/T)(1-z^{-1})/(1+z^{-1})}$$

where $H_a(s)$ is the transfer function of the analog prototype. However the mapping $\omega \rightarrow \Omega$ is nonlinear:

$$\Omega = 2\tan^{-1}\left(\frac{\omega T}{2}\right) \tag{3.44}$$

Therefore a warping in the characteristics of the digital filter is introduced, as shown in Figure 3.4. Having this fact in mind, we "prewarp" the characteristics of the prototype analog filter so that the digital filter has the desired specifications. The design procedure is illustrated by the following example.

**Example 1: Design of an Elliptic Bandstop Filter.** The transfer function

$$H_a(s) = \prod_{i=1}^{3} \frac{s^2 + \alpha_{0i}}{s^2 + \beta_{1i}s + \beta_{0i}} \tag{3.45a}$$

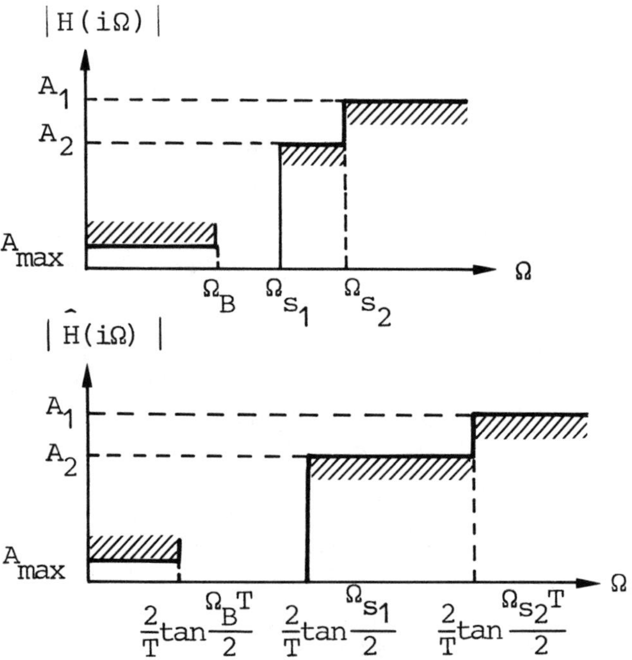


**Figure 3.4** Frequency warping introduced by the bilinear transformation.

represents an elliptic bandstop filter whose coefficients are given in Table 3.4 and whose response is shown in Figure 3.5. The digital filter which is obtained by the bilinear transformation is given by the following relation:

$$H(z) = \prod_{i=1}^{3} \frac{\alpha'_{2i} + \alpha'_{1i} z + \alpha'_{01} z^2}{\beta_{0i} + \beta_{1i} z + z^2} \tag{3.45b}$$

Its coefficients are given in Table 3.4 and its characteristics in Figure 3.5. The displacement due to the warping effect is evident.

## $L_p$ Norm Optimization Design

Let us suppose that the filter specifications are given as values of the desired filter response $H_d(e^{i\Omega})$ at frequencies $\Omega_i$, $i = 1, \ldots, M$, and that the filter transfer function $H(z)$ is given by

$$H(z) = A \prod_{k=1}^{K} \frac{1 + a_k z^{-1} + b_k z^{-2}}{1 + c_k z^{-1} + d_k z^{-2}} \tag{3.46}$$

The problem is to choose the $4K + 1$ unknown coefficients $\mathbf{X}^T = [A, a_1, b_1, c_1, d_1, \ldots, a_K, b_K, c_K, d_K]$ in such a way that the $L_p$ norm of the error of the magnitude function

$$E_p = \sum_{i=1}^{M} [H(e^{i\Omega_i}) - H_d(e^{i\Omega_i})]^p \, W(\Omega_i) \tag{3.47}$$

is minimized [$W(\Omega_i)$ is a weight function]. The optimization of $E_p$ with respect to $A$, $a_k$, $b_k$, $c_k$, $d_k$, $k = 1, \ldots, K$, leads to the solution of the following system of

**Table 3.4** Coefficients of Example 1

| $i$ | $\alpha_{0i}$ | $\beta_{0i}$ | $\beta_{1i}$ | |
|---|---|---|---|---|
| 1 | 6.250 | 6.250 | 2.618910 | |
| 2 | 8.013554 | $1.076433 \times 10$ | $3.843113 \times 10^{-1}$ | |
| 3 | 4.874554 | 3.628885 | $2.231394 \times 10^{-1}$ | |
| $j$ | $\alpha'_{0i}$ | $\alpha'_{1i}$ | $\beta'_{0i}$ | $\beta'_{1j}$ |
| 1 | $6.627508 \times 10^{-1}$ | $-3.141080 \times 10^{-1}$ | $3.255016 \times 10^{-1}$ | $-3.141080 \times 10^{-1}$ |
| 2 | $8.203382 \times 10^{-1}$ | $-1.915542 \times 10^{-1}$ | $8.893929 \times 10^{-1}$ | $5.716237 \times 10^{-2}$ |
| 3 | 1.036997 | $-7.266206 \times 10^{-1}$ | $9.0183366 \times 10^{-1}$ | $-8.987781 \times 10^{-1}$ |

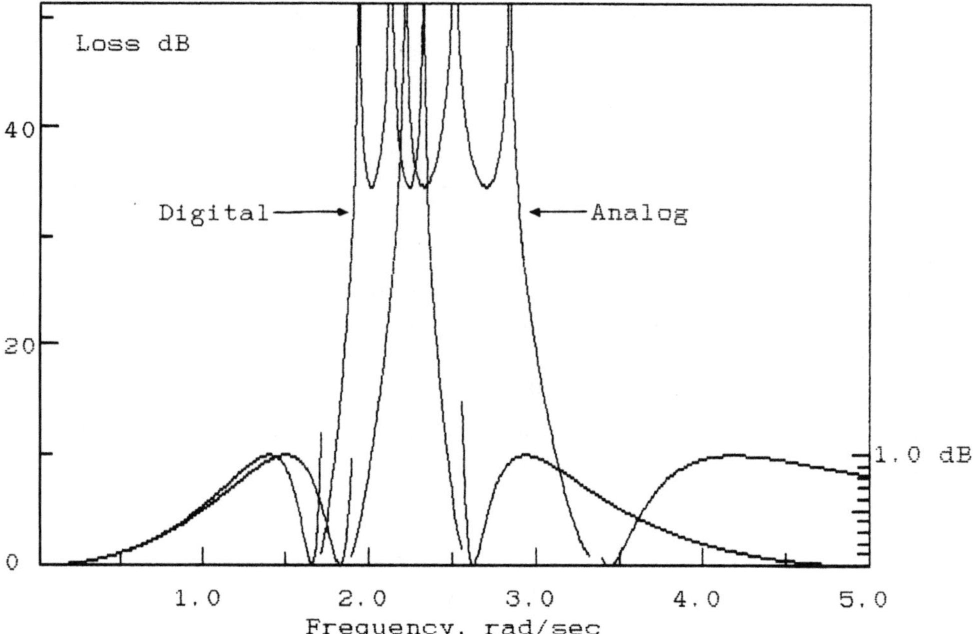


**Figure 3.5** Digital elliptic bandstop filter designed by bilinear transformation.

$4K + 1$ equations:

$$\frac{\partial E_p}{\partial X_i} = 0, \qquad i = 1, \ldots, 4K + 1 \tag{3.48}$$

by iterative optimization techniques (e.g., Fletcher–Powell). If $p = 2$ the mean square error is minimized. If $p \to \infty$ the $L_\infty$-norm is minimized [4]. In this case the designed filter is an equiripple one.

During the optimization design method, the coefficients $c_k$, $d_k$, vary without restriction. Therefore the designed filter may be unstable. This problem can be solved by two approaches:

1. Use constrained optimization techniques.
2. Use unconstrained optimization, but replace the poles of the final designed function that lie outside the unit circle by their reciprocals. Thus the designed filter magnitude remains unchanged and, at the same time, the filter becomes stable.

If the phase characteristics of the filter are of major importance, we can try to choose the coefficients $x_i$ of $X$ in such a way that the group delay:

$$\tau(\Omega) = -\frac{d}{d\Omega} \left\{ \arg[H(e^{i\Omega})] \right\} \tag{3.49}$$

of filter $H(e^{i\Omega})$ approximates the desired group delay $\tau_d(\Omega)$. This can be obtained by optimizing the following $L_p$ norm with respect to $X$:

$$E'_p = \sum_{i=1}^{M} [\tau(\Omega_i) - \tau_d(\Omega_i)]^p W(\Omega_i) \tag{3.50}$$

The optimization of $E'_p$ is performed by solving the following system of equations:

$$\frac{\partial E'_p}{\partial x_i} = 0, \qquad i = 1, \ldots, 4K + 1 \tag{3.51}$$

by using iterative techniques.

## Digital Filter Transformations

There are two ways to design a high-pass (or bandpass or bandstop) digital filter from a prototype analog low-pass filter:

1. Transformation of the analog low-pass filter to an analog high-pass filter and then transformation of this filter to a digital one.
2. Transformation of the analog low-pass filter to a digital low-pass filter and then transformation of this filter to a digital high-pass one.

A digital filter transformation method is needed for the second way. Such a transformation is a mapping $z \to Z$ of the form

$$Z^{-1} = G(Z^{-1}) \tag{3.52}$$

which must satisfy the following conditions:

1. $G(Z^{-1})$ is a rational function of $Z$.
2. The interior of the unit circle $|z| = 1$ is mapped onto the interior of the unit circle $|Z| = 1$.

The only function that satisfies these conditions is given by

$$G(Z^{-1}) = \pm \prod_{k=1}^{N} \frac{Z^{-1} - a_k}{1 - a_k Z^{-1}} \tag{3.53}$$

We usually choose $|a_k| < 1$ for stability. The simplest form of Eq. (3.53) that transforms a digital low-pass filter $H_1(z)$ having cutoff frequency $\theta_p$ to a digital low-pass filter $H_d(Z)$ having cutoff frequency $\Omega_p$ is the following:

$$z^{-1} = \frac{Z^{-1} - a}{1 - aZ^{-1}} \tag{3.54}$$

By substituting $z = e^{i\theta}$ and $Z = e^{i\Omega}$ we find that

$$\Omega = \arctan\left[\frac{(1-a^2)\sin\theta}{2a+(1+a^2)\cos\theta}\right] \tag{3.55}$$

which is a nonlinear relation. In this case $a$ is given by

$$a = \frac{\sin((\theta_p - \Omega_p)/2)}{\sin((\theta_p + \Omega_p)/2)} \tag{3.56}$$

The desired low-pass function $H_d(Z)$ is calculated from $H_1(z)$ as follows:

$$H_d(Z) = H_1(z)\,|_{z^{-1}=(Z^{-1}-a)/(1-a/Z^{-1})} \tag{3.57}$$

Various such transformations are shown in Table 3.5 [5].

## 6.2 Design of FIR Digital Filters

The transfer function of an FIR filter is given by Eq. (3.25) and its frequency response by

$$H(e^{i\Omega}) = \sum_{n=0}^{N-1} h(n)e^{-i\Omega n} \tag{3.58}$$

It can be proved that if the filter coefficients satisfy

$$h(n) = h(N-1-n) \tag{3.59}$$

the FIR filter has linear phase characteristics:

$$H(e^{i\Omega}) = \begin{cases} e^{-i\Omega(N-1)/2}\left\{h\left(\frac{N-1}{2}\right) + \sum_{n=0}^{(N-3)/2} 2h(n)\cos\left[\Omega\left(n-\frac{N-1}{2}\right)\right]\right\}, & N \text{ odd} \\ e^{-i\Omega(N-1)/2}\left\{\sum_{n=0}^{N/2-1} 2h(n)\cos\left[\Omega\left(n-\frac{N-1}{2}\right)\right]\right\}, & N \text{ even} \end{cases} \tag{3.60}$$

**Table 3.5** Transformations from a Low-Pass Digital Filter Prototype of Cutoff Frequency $\theta_p$

| Filter type | Transformation | Design Formulas |
|---|---|---|
| Low-pass | $z^{-1} = \dfrac{Z^{-1} - \alpha}{1 - \alpha Z^{-1}}$ | $\alpha = \dfrac{\sin\left(\frac{\theta_p - \omega_p}{2}\right)}{\sin\left(\frac{\theta_p + \omega_p}{2}\right)}$<br>$\omega_p$ = desired cutoff frequency |
| High-pass | $-\dfrac{Z^{-1} + \alpha}{1 - \alpha Z^{-1}}$ | $\alpha = -\dfrac{\cos\left(\frac{\omega_p + \theta_p}{2}\right)}{\cos\left(\frac{\omega_p - \theta_p}{2}\right)}$<br>$\omega_p$ = desired cutoff frequency |
| Bandpass | $\dfrac{Z^{-2} - \frac{2\alpha k}{k+1} Z^{-1} + \frac{k-1}{k+1}}{\frac{k-1}{k+1} Z^{-2} - \frac{2\alpha k}{k+1} Z^{-1} + 1}$ | $\alpha = \dfrac{\cos\left(\frac{\omega_2 + \omega_1}{2}\right)}{\cos\left(\frac{\omega_2 - \omega_1}{2}\right)}$<br>$k = \cot\left(\dfrac{\omega_2 - \omega_1}{2}\right) \tan \dfrac{\theta_p}{2}$<br>$\omega_2, \omega_1$ = desired upper and lower cutoff frequencies |
| Bandstop | $\dfrac{Z^{-2} - \frac{2\alpha}{1+k} Z^{-1} + \frac{1-k}{1+k}}{\frac{1-k}{1+k} Z^{-2} - \frac{2\alpha}{1+k} Z^{-1} + 1}$ | $\alpha = \dfrac{\cos\left(\frac{\omega_2 + \omega_1}{2}\right)}{\cos\left(\frac{\omega_2 - \omega_1}{2}\right)}$<br>$k = \tan\left(\dfrac{\omega_2 + \omega_1}{2}\right) \tan \dfrac{\theta_p}{2}$<br>$\omega_2, \omega_1$ = desired upper and lower cutoff frequencies |

which is equivalent to a delay by $(N - 1)/2$ samples. The ability of the FIR filters to have exactly linear phase characteristics is a major advantage over the IIR filters. Another important advantage is the fact that they are always stable and realizable. Therefore several techniques have been proposed for FIR filter design. We shall describe three of the most popular ones:

1. Design using windows
2. Frequency sampling design
3. Equiripple filter design

## Design Using Windows

If the desired impulse response $h_d(n)$, $-\infty < n < \infty$, is known, an FIR filter can easily be designed by truncating $h_d(n)$ with the aid of a *window* function

$w(n)$:

$$h(n) = h_d(n)w(n), \qquad w(n)\begin{cases} \neq 0, & 0 \leq n \leq N_1 \\ = 0 & \text{elsewhere} \end{cases} \tag{3.61}$$

The frequency response of $H(e^{i\Omega})$ of the FIR filter is given by

$$H(e^{i\Omega}) = \frac{1}{2\pi}\int_{-\pi}^{\pi} H_d(e^{i\theta})W(e^{i(\Omega-\theta)})\,d\theta \tag{3.62}$$

according to the convolution theorem described in Table 3.3.

$H_d(e^{i\Omega})$ and $W(e^{i\Omega})$ are the frequency responses of the ideal filter and of the window, respectively. An example of such a window is the rectangular window:

$$w(n) = \begin{cases} 1, & 0 \leq n \leq N-1 \\ 0 & \text{elsewhere} \end{cases}$$

Its frequency response is given by

$$W(e^{i\Omega}) = e^{-i\Omega(N-1)/2}\frac{\sin(\Omega N/2)}{\sin(\Omega/2)}$$

and is shown in Figure 3.6 for $N = 8$. Ideally, the frequency response should be a delta function. This can be obtained by increasing $N$. In this case the width of the main lobe $4\pi/N$ is decreasing. However, as $N$ increases, so does the computational effort required for realization of the FIR filter. Therefore the requirements for low computational complexity and close approximation of the ideal frequency response are contradictory. For a fixed $N$, which is a result of a compromise between the characteristics of the filter and its computational complexity, the filter characteristics are improved by choosing a window that has low sidelobes in $W(e^{i\Omega})$. This can be obtained by allowing $w(n)$ to approach 0 smoothly when $n$ approaches 0 or $N$. Several windows have been designed with this requirement in mind. The best known ones are included in Table 3.6. There exists no general rule for the choice of the optimal window. We usually apply several windows until we get the desired result. Despite this disadvantage, the simplicity of the window filter design method has rendered it very popular.

An example of the design of an FIR low-pass filter by using Blackman and Hamming windows is shown in Figure 3.7. The designed filter has cutoff frequency 2 rad/s and 21 taps.

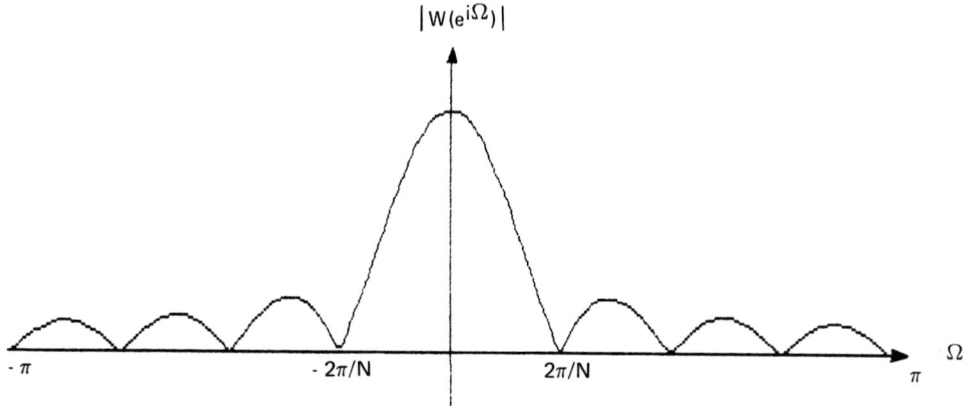


**Figure 3.6** Magnitude of the frequency response of a rectangular window for $N = 8$.

## Frequency Sampling Design

An FIR filter can be described in terms of its frequency samples:

$$\tilde{H}(k) = H(z)\,|_{z=e^{i(2\pi nk/N)}} = \sum_{n=0}^{N-1} h(n)e^{-i(2\pi kn/N)} \tag{3.63}$$

**Table 3.6.** Window Functions

| Window | $w(n)$ |
|---|---|
| Rectangular | $w(n) = 1, \qquad 0 \le n \le N-1$ |
| Bartlett | $w(n) = \begin{cases} \dfrac{2n}{N-1}, & 0 \le n \le \dfrac{N-1}{2} \\ 2 - \dfrac{2n}{N-1}, & \dfrac{N-1}{2} \le n \le N-1 \end{cases}$ |
| Hanning | $w(n) = \dfrac{1}{2}\left[1 - \cos\left(\dfrac{2\pi n}{N-1}\right)\right], \qquad 0 \le n \le N-1$ |
| Hamming | $w(n) = 0.54 - 0.46\cos\left(\dfrac{2\pi n}{N-1}\right), \qquad 0 \le n \le N-1$ |
| Blackman | $w(n) = 0.42 - 0.5\cos\left(\dfrac{2\pi n}{N-1}\right) + 0.08\cos\left(\dfrac{4\pi n}{N-1}\right), \qquad 0 \le n \le N-1$ |
| Kaiser | $w(n) = \dfrac{I_0\left\{\omega_\alpha\sqrt{\left(\frac{N-1}{2}\right)^2 - \left[n - \left(\frac{N-1}{2}\right)\right]^2}\right]}{I_0\left[\omega_\alpha\left(\frac{N-1}{2}\right)\right]}$ <br> $I_0(\cdot)$ = modified zeroth-order Bessel function |

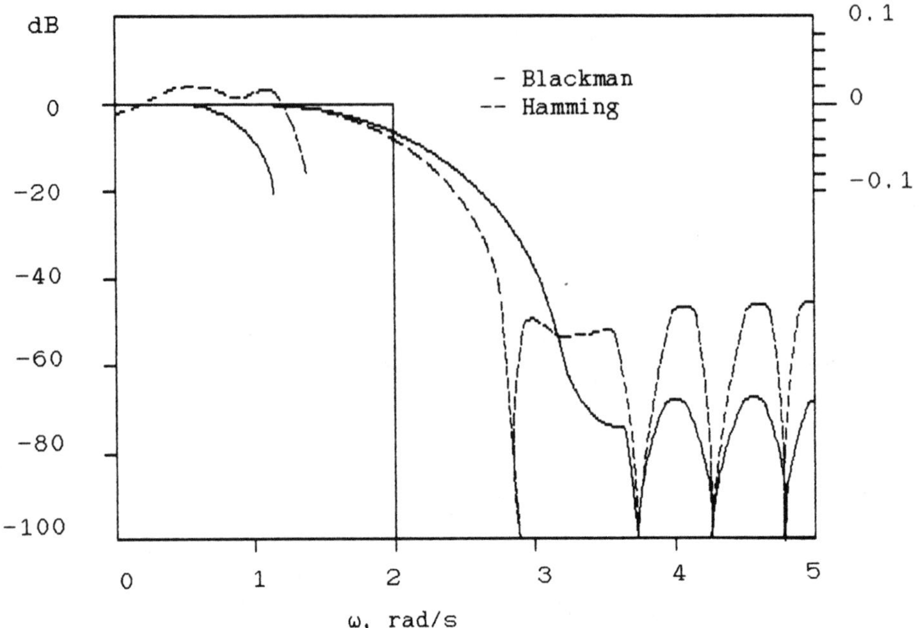


**Figure 3.7** Low-pass digital filter designed by using Blackman and Hamming windows.

In this case, its $z$ transform is given by the following relation [6]:

$$H(e^{i\Omega}) = \frac{e^{-i\Omega(N-1)/2}}{N} \sum_{k=0}^{N-1} \tilde{H}(k) e^{-i\pi k(1-1/N)} \frac{\sin[N(\Omega - 2\pi k/N)/2]}{\sin[(\Omega - 2\pi k/N)/2]} \tag{3.64}$$

A very simple, though unsatisfactory, design method is the following:

$$\tilde{H}(k) = H_d(e^{i2\pi k/N}), \qquad k = 0, 1, \ldots, N-1 \tag{3.65}$$

$$h(n) = \frac{1}{N} \sum_{k=0}^{N-1} \tilde{H}(k) e^{i2\pi kn/N}, \qquad n = 0, 1, \ldots, N-1$$

This method can be drastically optimized as follows. The frequency samples $\tilde{H}(k)$ in the passband and in the stopband are chosen as in Eq. (3.64):

$$\tilde{H}(k) = H_d(e^{i2\pi k/N}), \qquad k \text{ in filter passband or stopband} \tag{3.66}$$

The frequency samples in the transition band are chosen in such way that the following error is minimized:

$$E_p = \frac{1}{2\pi}\int_{-\pi}^{\pi} |H_d(e^{i\Omega}) - \tilde{H}(e^{i\Omega})|^p \, d\Omega \tag{3.67}$$

$H(e^{i\Omega})$ is a linear combination of the $\tilde{H}(k)$. Therefore simple optimization techniques can be used to minimize $E_p$ with respect to $\tilde{H}(k)$ of the transition band:

$$\frac{\partial E_p}{\partial \tilde{H}(k)} = 0 \tag{3.68}$$

## Equiripple Filter Design

The equiripple filter design minimizes the following error function:

$$\begin{aligned} E_{\max} &= \underset{\substack{0\leq\Omega\leq\Omega_p \\ \Omega_x\leq\Omega\leq\pi}}{\text{maximum}} |E(i\Omega)| \\ E(\Omega) &= W(\Omega)[H_d(e^{i\Omega}) - H(e^{i\Omega})] \end{aligned} \tag{3.69}$$

where $H_d(e^{i\Omega})$ is the desired filter response shown in Figure 3.3:

$$H_d(e^{i\Omega}) = \begin{cases} 1, & 0 \leq \Omega \leq \Omega_p \\ 0, & \Omega_s \leq \Omega \leq \pi \end{cases} \tag{3.70}$$

and $W(\Omega)$ is a weight function:

$$W(\Omega) = \begin{cases} 1/K, & 0 \leq \Omega \leq \Omega_p \\ 1, & \Omega_s \leq \Omega \leq \pi \end{cases}, \qquad K = \frac{\delta_1}{\delta_2} \tag{3.71}$$

where $\delta_1$, $\delta_2$ are the maximal errors in the passband and stopband respectively.

We shall limit our discussion to the design of linear phase FIR filters of the form

$$H(e^{i\Omega}) = \sum_{n=-M}^{M} h(n)e^{i\Omega n} = h(0) + \sum_{n=1}^{M} 2h(n)\cos\omega n, \qquad h(n) = h(-n) \tag{3.72}$$

There exist $M+1$ $h(n)$ coefficients to be determined. It is a generally impossible to specify $M$, $\delta_1$, $\delta_2$, $\Omega_p$ and $\Omega_s$ independently. Usually $M$, $\delta_1$, and $\delta_2$ are variable. We shall assume that $M$, $\Omega_p$, and $\Omega_s$ are fixed and we shall try to minimize Eq. (3.69) with the aid of the following theorem [7]:

**Alternation Theorem.** Let $F$ be any closed subset of the closed interval $0 \leq \Omega \leq \pi$. In order that $H(e^{i\Omega})$ of Eq. (3.72) is the unique best approximation on $F$ to $H_d(e^{i\Omega})$, it is necessary and sufficient that the error function $E(\Omega)$ exhibits on $F$ at least $M+2$ alternations: $E(\Omega_i) = -E(\Omega_{i-1}) = \cdots = \pm\|E\| = \max|E(\Omega)|$ with $\Omega_0 \leq \Omega_1 \leq \cdots \leq \Omega_{M+1}$ and $\Omega_i$ contained in $F$.

Thus the problem of optimizing $E_{\max}$ reduces to the problem of calculating the $M+2$ alternation frequencies $\Omega_i$. The frequencies $\Omega_i$ and the error $\rho = \max|E(\Omega)|$ are calculated by the REMEZ algorithm shown in Figure 3.8. Once $\Omega_i$, $i = 1, \ldots, M+2$ and $\rho$ are known, the filter coefficients $h(n)$ are the solutions of the system of the following $M+2$ equations:

$$W(\Omega_i)\left[H_d(e^{i\Omega_i}) - h(0) - \sum_{n=1}^{M} 2h(n)\cos(\Omega_i n)\right] = -(-1)^i \rho \qquad i = 0, 1, \ldots, M+1 \tag{3.73}$$

The linear system [Eq. (3.73)] can be solved either directly or by special techniques [7].

### 6.3 Comparison Between FIR and IIR Digital Filters

The main advantage of FIR filters over IIR filters is that they are always stable and realizable and that they can have exactly linear phase characteristics. For certain applications (e.g., in data transmission) constant group delay is mandatory. In these cases FIR filters are preferable, although it may sometimes be possible to find satisfactory IIR filters. The FIR filters are also suited for certain specific applications (e.g., numerical differentiation or integration, differential equation simulation). Their main disadvantage is that the required filter order is generally 5 to 10 times higher than that of an IIR filter. This results in high computational demands for their realization. This difficulty is alleviated by use of the fast Fourier transform realization. Furthermore, the modern very large scale integration (VLSI) digital signal processors have speeded up FIR filtering considerably.

The main advantage of IIR filters is their low order and their flexibility for realizing highly selective filters because of the placement of the filter poles anywhere inside the unit circle. Hence, for high-selectivity applications, where the delay characteristics are of secondary importance, the choice is an IIR filter. IIR filters are also suited for applications where the prescribed specifications can be met by conventional Butterworth, Chebyshev, or elliptic approximations.

## 7 REALIZATION OF DIGITAL FILTERS

The realization of digital filters is the process of converting a filter transfer function into a filter network. There exist several realization techniques. The choice of the best filter structure depends on the following criteria:

1. Number of additions and multiplications required
2. Number of adders and multipliers required for parallel implementation

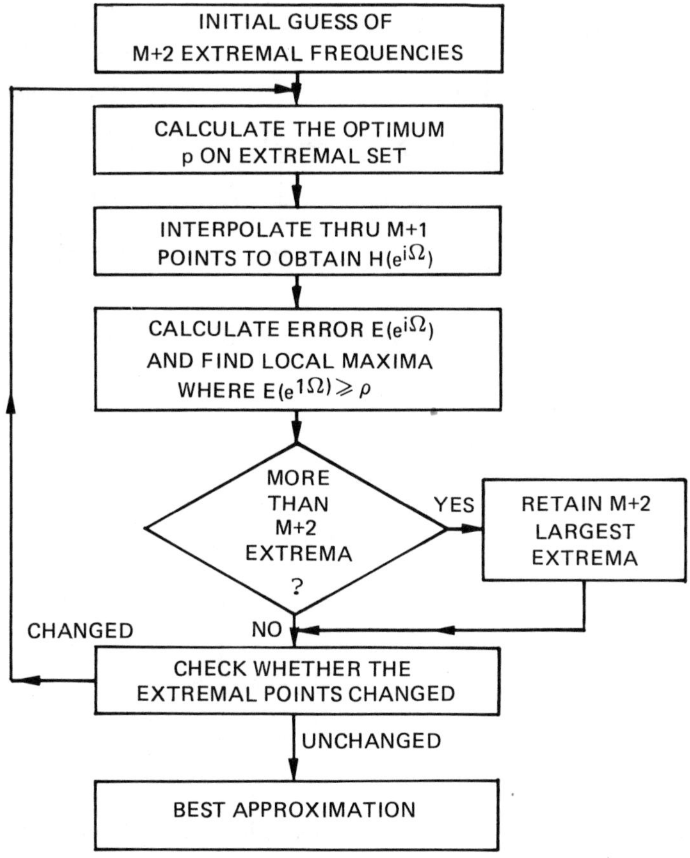


**Figure 3.8** REMEZ algorithm.

3. Number of registers and delay units
4. Throughput delay for serial and parallel computation
5. Error characteristics in finite arithmetic implementations

In the subsequent sections we shall discuss the realizations of both IIR and FIR digital filters.

## 7.1 IIR Filter Structures

An IIR digital filter is described by the difference equation (3.6). A simple realization of this filter (called direct structure 1) is based on Eq. (3.6) and is shown in Figure 3.9 for $N = M$ and $b_0 = 1$.

The transfer function $H(z)$ is given by Eq. (3.27), which can be written alternatively as follows:

$$H(z) = H_1(z)H_2(z), \qquad H_1(z) = \frac{1}{\sum_{k=0}^{N} a_k z^{-k}}, \qquad H_2(z) = \sum_{k=0}^{M} b_k z^{-k} \tag{3.74}$$

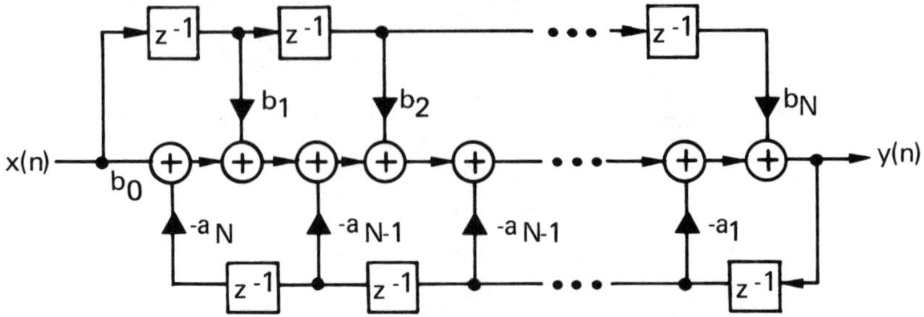


**Figure 3.9** Direct structure 1 of an IIR digital filter.

For $M = N$, Eq. (3.74) corresponds to the direct structure 2 shown in Fig. 3.10. Direct structure 2 has the theoretically minimal number of delay units $z^{-1}$.

Another realization stems from the factorization of $H(z)$ to second-order factors:

$$H(z) = a_0 \prod_{i=1}^{K} H_i(z), \qquad H_i(z) = \begin{cases} \dfrac{1 + a_{1i}z^{-1} + a_{2i}z^{-2}}{1 + b_{1i}z^{-1} + b_{2i}z^{-2}} \\[2ex] \dfrac{1 + a_{1i}z^{-1}}{1 + b_{1i}z^{-1}} \end{cases} \tag{3.75}$$

This is the so-called cascade realization and is shown in Figure 3.11.

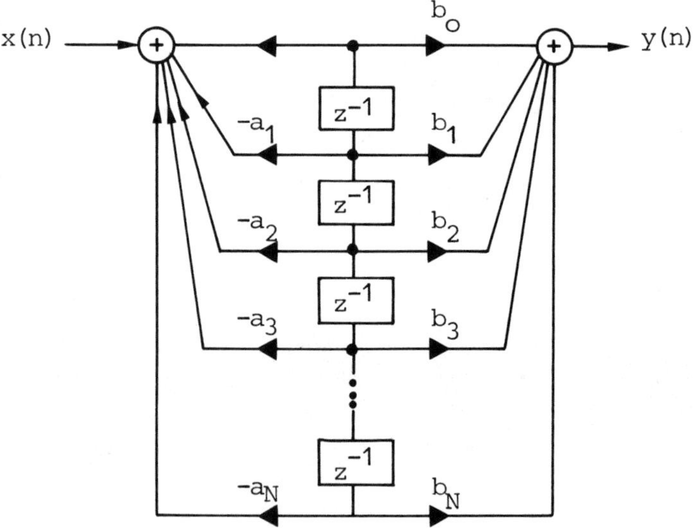


**Figure 3.10** Direct structure 2 of an IIR digital filter.

The transfer function $H(z)$ can be expanded into partial fractions:

$$H(z) = c + \sum_{i=1}^{K} H_i(z), \qquad H_i(z) = \begin{cases} \dfrac{a_i + b_i z^{-1}}{1 + c_i z^{-1} + d_i z^{-2}} \\ \dfrac{a_i}{1 + b_i z^{-1}} \end{cases} \tag{3.76}$$

Equation (3.76) corresponds to the parallel realization shown in Figure 3.12. The last realization to be considered here is the ladder realization. It is based on the continued fraction expansion of $H(z)$ [8]:

$$H(z) = \frac{N(z)}{D(z)}, \qquad D(z) = \cfrac{1}{c_1 z + \cfrac{1}{c_2 z + \cfrac{1}{\ddots \, \cfrac{1}{c_N z}}}} \tag{3.77}$$

$$N(z) = \sum_{i=1}^{N} d_i n_i(z)$$

where $n_i(z)$ are polynomials of degree $i-1$ having as coefficients expressions of $c_i$, $i = 1, \ldots, N$. The ladder realization is shown in Figure 3.13. Sometimes the continued fraction expansion of $D(z)$ is impossible. In this case there exists no ladder realization of the IIR filter [e.g., when $H(z)$ has poles on the imaginary axis of the $z$ plane].

## 7.2 FIR Filter Structures

The simplest possible realization of an FIR filter is shown in Figure 3.14. It is called the direct form. It is a special form of the direct structure of the IIR filters. The

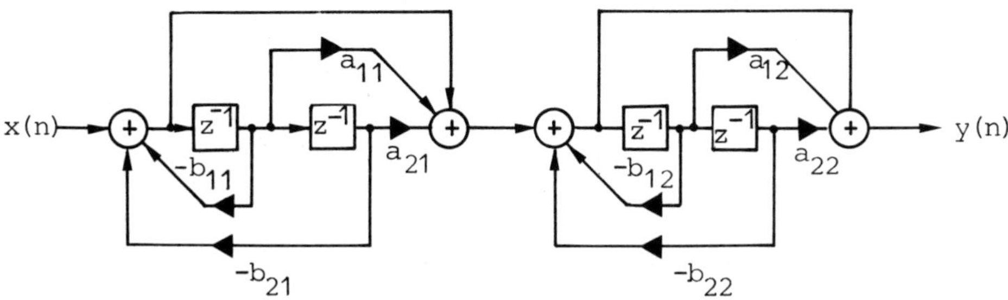


**Figure 3.11** Cascade IIR digital filter realization.

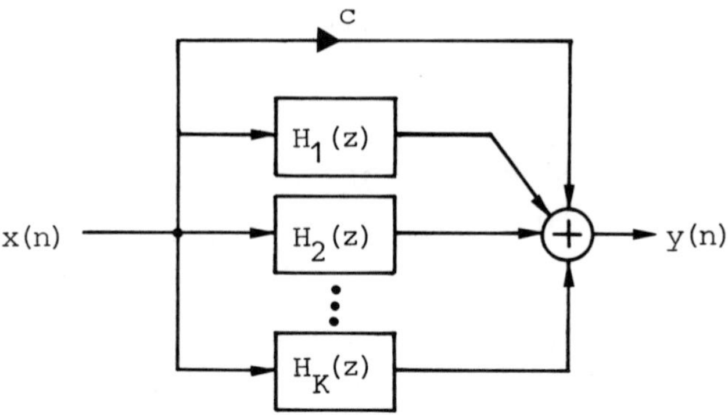


**Figure 3.12** Parallel IIR digital filter realization.

network of the direct structure can be transposed to obtain the transposed direct form shown in Figure 3.15.

A cascade realization, shown in Figure 3.16, can be obtained by factorizing $H(z)$:

$$H(z) = \prod_{i=1}^{[N/2]} (a_i + b_i z^{-1} + c_i z^{-2}) \tag{3.78}$$

Several realizations of the FIR filters can be obtained by using the Lagrange

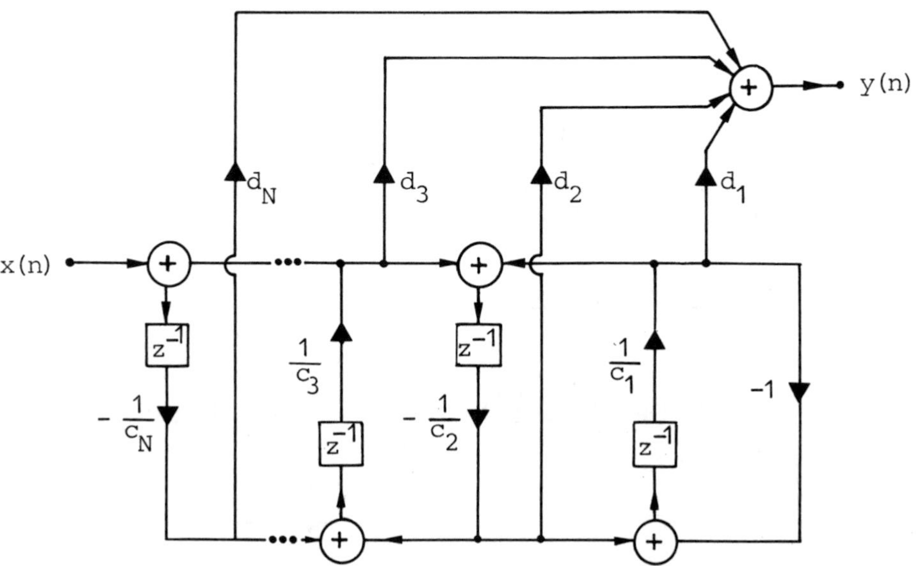


**Figure 3.13** Ladder IIR digital filter structure.

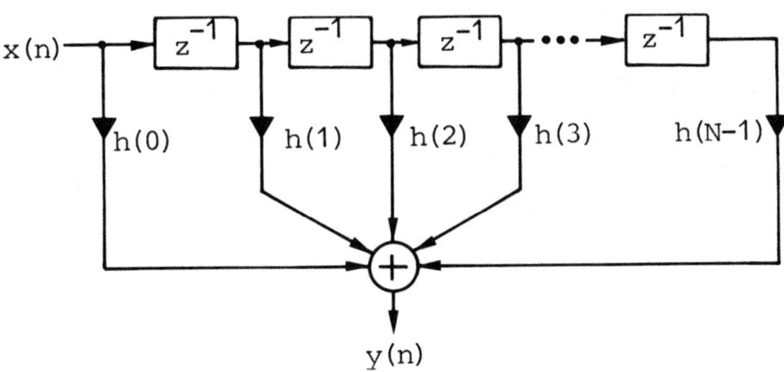


**Figure 3.14** Direct FIR digital filter structure.

interpolation formula for $H(z)$:

$$H(z) = \prod_{n=0}^{N-1}(1 - z^{-1}z_n)\sum_{m=0}^{N-1}\frac{A_m}{1 - z^{-1}z_m} \tag{3.79}$$

$$A_m = \frac{H(z_m)}{\prod_{\substack{n=0 \\ n\neq m}}^{N-1}(1 - z_n z_m^{-1})}$$

The Lagrange structure is shown in Figure 3.17.

An interesting special case of the Lagrange structure comes from the choice

$$z_n = e^{i2\pi n/N}, \qquad n = 0, 1, \ldots, N-1 \tag{3.80}$$

In this case Eq. (3.79) is equivalent to the frequency sampling formula [6]:

$$H(z) = \frac{1 - z^{-N}}{N}\sum_{n=0}^{N_1}\frac{H(e^{i2\pi n/N})}{1 - z^{-1}e^{i2\pi n/N}} \tag{3.81}$$

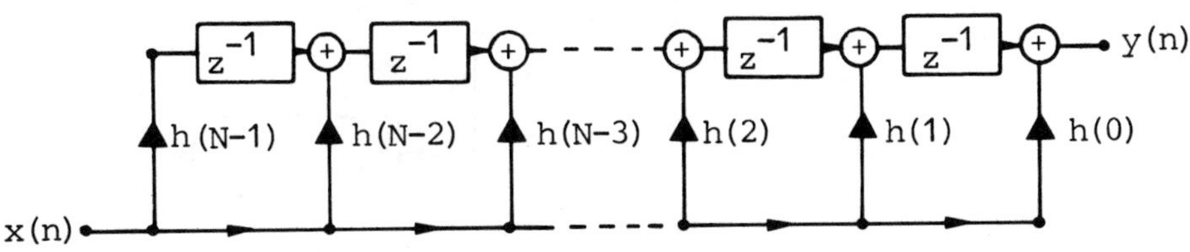


**Figure 3.15** Transposed direct FIR digital filter structure.

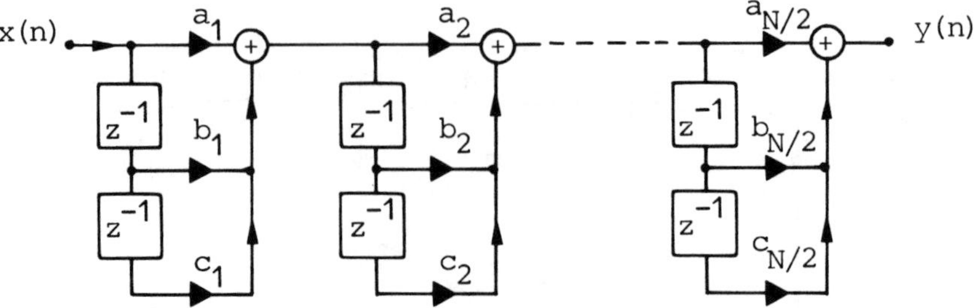


**Figure 3.16** Cascade FIR digital filter realization.

The frequency sampling structure is shown in Figure 3.18. This structure is closely related to the frequency sampling design method. Its advantage is that, in certain applications, various terms $H(e^{i2\pi n/N})$ are equal to zero and the corresponding paths in Figure 3.18 are eliminated.

Finally, a special structure for the realization of linear phase FIR filters satisfying Eq. (3.59) is shown in Figure 3.19, for odd order $N$.

### 7.3 Fast Calculation of Convolution and Correlation

A completely different realization of FIR filters is based on fast calculation of the convolution with the aid of the circular convolution property of the discrete Fourier transform:

$$y_p(n) = x_p(n) * h_p(n) \longleftrightarrow Y_p(k) = X_p(k)H_p(k) \tag{3.82}$$

where $y_p(n)$, $x_p(n)$, and $h_p(n)$ are periodic sequences and $Y_p(k)$, $X_p(k)$, and $H_p(k)$ their DFTs. If the sequence $h(n)$ is the impulse response of the FIR filter having length $N_1$ and $x(n)$ is the input sequence having length $N_2$, the filter response $y(n)$ has nonzero values in the interval $[0, L]$, $L = N_1 + N_2 - 1$. If we pad $x(n)$, $h(n)$

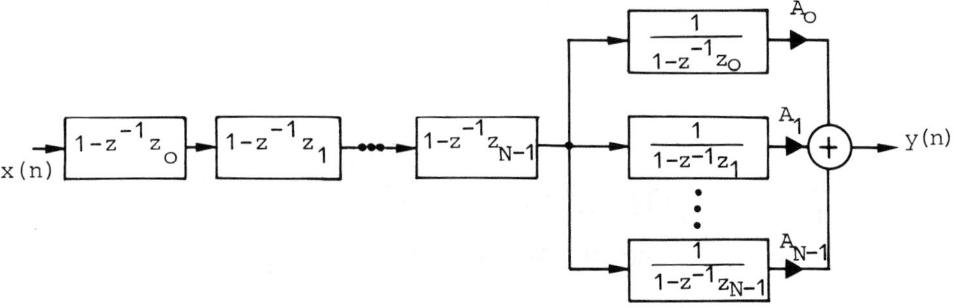


**Figure 3.17** Lagrange interpolation realization.

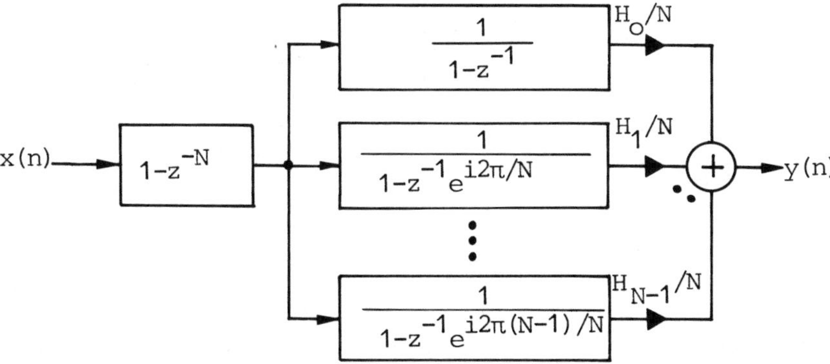


**Figure 3.18** Frequency sampling FIR digital filter realization.

with zeros:

$$
h_p(n) = \begin{cases} h(n) & 0 \le n \le N_1 - 1 \\ 0 & N_1 \le n \le L \end{cases}
$$
$$
x_p(n) = \begin{cases} x(n) & 0 \le n \le N_2 - 1 \\ 0 & N_2 \le n \le L \end{cases} \tag{3.83}
$$

and we assume that $x_p(n)$, $h_p(n)$ and $y_p(n)$ are extended periodically with period $L$, Eq. (3.82) can be used for the realization of the FIR digital filter. The corresponding structure is shown in Figure 3.20. Calculation of the DFTs and the inverse DFT shown in Figure 3.20 can be done very efficiently with the aid of the fast Fourier transform algorithms, described in Chapter 4.

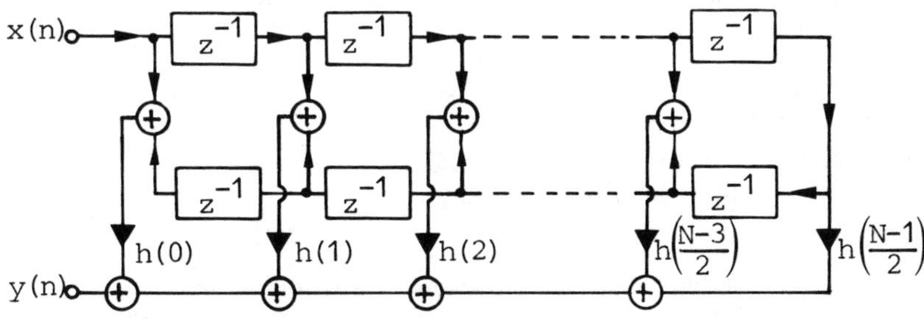


**Figure 3.19** Linear phase, odd-order FIR digital filter realization.

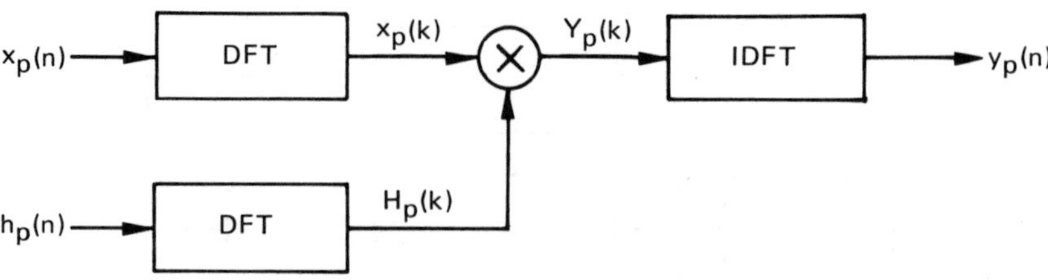


**Figure 3.20** Fast calculation of the cyclic convolution.

A similar method can be used for fast calculation of the cross-correlation of two discrete real sequences $x(n)$ and $h(n)$:

$$y(n) = \sum_{k=-\infty}^{\infty} x(k)h(n+k)$$

The discrete Fourier transform has the following circular correlation property:

$$y_p(n) = \sum_{k=0}^{N-1} x_p(k)h_p(n+k) \longleftrightarrow Y_p(k) = X_p^*(k)H_p(k) \tag{3.84}$$

Therefore the computation of the cross-correlation of two sequences $x(n)$, $h(n)$ of finite extent can be accomplished by padding them with zeros, as in Eq. (3.83), and by calculating $X_p^*(k)H_p(k)$ by a structure entirely similar to that of Figure 3.20.

## 8 IMPLEMENTATION OF DIGITAL FILTERS

Digital filters are essentially computation structures that can be implemented on any computer. However, in most applications, real-time processing is required. This requirement is met by the implementation of the digital filters on dedicated processors called digital signal processors. There exist the following possibilities for the implementation of a digital filter:

1. Dedicated hardware implementation
2. Bit-slice signal processors
3. VLSI signal processors
4. Parallel digital filter implementation
5. Digital filter implementation on a conventional microprocessor

The dedicated hardware implementation is the oldest one. It has the advantage of high speed and the disadvantages of development of the specialized hardware and limited flexibility. In contrast, the digital filter implementation on a microprocessor/microcomputer is very flexible and cheap, but it rarely meets the real-time processing requirements. Bit-slice signal processors are a compromise between the two previous solutions. They provide the speed required for real-time applications and they use slices from commercially available bit-slice families (e.g., the AM2900 family). A recent trend is digital filter implementation on VLSI signal processors. There are several commercial VLSI signal processor families (e.g., Texas Instruments TMS320, Phillips PCB 5010). They give a good compromise between cost and speed, as can be seen in Figure 3.21. However, the future of digital filter implementation is in massive parallel processing. The current technological level can already support parallel filter implementation. Parallel implementation using VLSI signal processors in efficient architectures will solve real-time processing requirements for computationally demanding applications (e.g., image processing).

In the following we shall give some techniques for digital filter implementation on specialized hardware, on VLSI signal processors, and on parallel machines.

## 8.1 Distributed Arithmetic Implementation

In this section we shall describe an implementation of the biquad digital filter:

$$y(n) = a_0x(n) + a_1x(n-1) + a_2x(n-2) - b_1y(n-1) - b_2y(n_2) \tag{3.85}$$

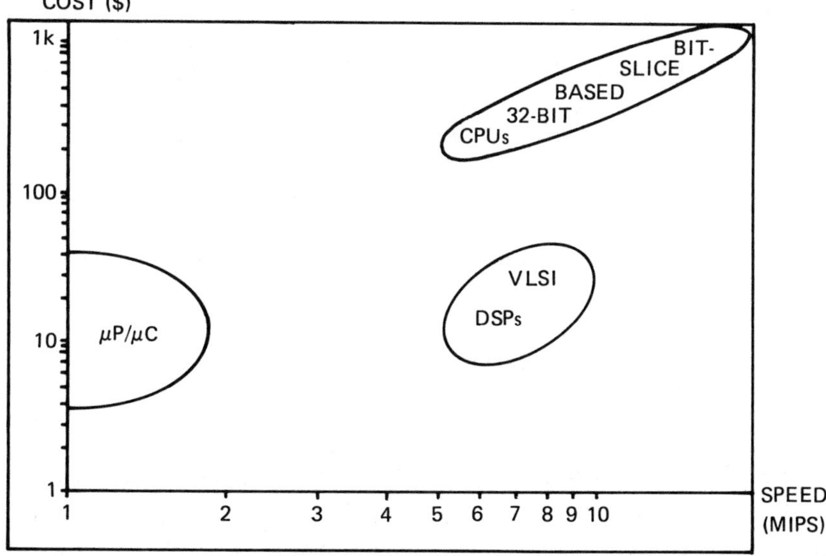


**Figure 3.21** Cost versus speed comparison of DSP families.

based on distributed arithmetic. If $y(n)$ and $x(n)$ are 2's complement numbers, they can be expressed as

$$x(n) = -x_0(n) + \sum_{i=1}^{L} x_i(n)2^{-i} \tag{3.86}$$

$$y(n) = -y_0(n) + \sum_{i=1}^{L} y_i(n)2^{-i}$$

where $x_i(n)$, $y_i(n)$, $i = 0, \ldots, L$ are the bits of $x(n)$, $y(n)$, respectively. In this case, Eq. (3.85) is rewritten as follows:

$$y(n) = \sum_{i=1}^{L} 2^{-i}F_i - F_0 \tag{3.87}$$

$$F_i = a_0 x_i(n) + a_1 x_i(n-1) - b_1 y_i(n-1) - b_2 y_i(n-2), \qquad i = 0, \ldots, L$$

The term $F_i$ can be computed using a look-up table having $x_i(n)$, $x_i(n-1)$, $y_i(n-1)$, $y_i(n-2)$ as inputs. The look-up table is implemented by a read-only-memory (ROM). A hardware implementation of Eq. (3.87) is shown in Figure 3.22 [9]. $R_1$, $R_2$, $R_3$, $R_4$, are $L+1$ bit shift-registers. The filter output is calculated after $L+1$ shifts. In each shift, $F_i$ is calculated and $2^{-i}F_i$ is added to $y(n)$.

The distributed-arithmetic implementation of Figure 3.21 can be used as a module for the cascade or parallel realization of higher-order digital filters. A comparison of various distributed-arithmetic implementations is presented in [10].

## 8.2 Digital Filter Implementation on VLSI Signal Processors

A basic operation in both FIR and IIR digital filters is the following:

$$a_i x(n-i) + R \longrightarrow R \tag{3.88}$$

where $a_i$ is a filter coefficient, $x(n-i)$ is a data sample, and $R$ is the content of a register. Therefore, a VLSI digital signal processor (DSP) must include a fast multiplier-accumulator unit (MAC) for the calculation of Eq. (3.88). The MAC included in the Texas Instruments TMS320 family is shown in Figure 3.23 [11].

The instruction set of a VLSI DSP usually includes a special instruction for the calculation of Eq. (3.88). By using this instruction repeatedly we can calculate the

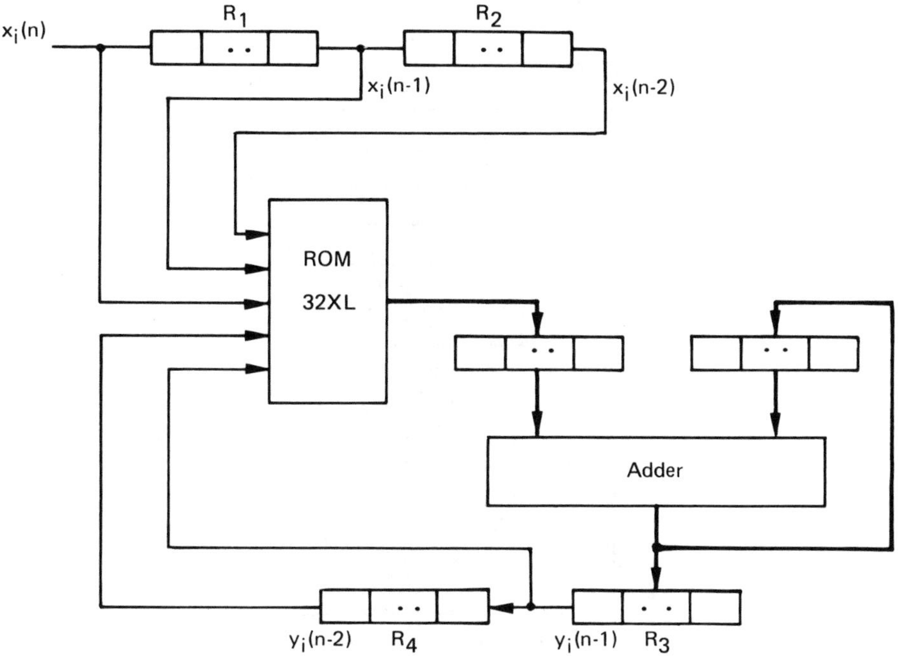


**Figure 3.22** Distributed-arithmetic realization of a digital biquad filter.

output $y(n)$ of the biquad filter [Eq. (3.85)] in five steps as follows:

$$\begin{aligned} a_0 x(n) &\longrightarrow R \\ a_1 x(n-1) + R &\longrightarrow R \\ a_2 x(n-2) + R &\longrightarrow R \\ -b_1 y(n-1) + R &\longrightarrow R \\ -b_2 y(n-2) + R &\longrightarrow R \end{aligned} \tag{3.89}$$

The currently available VLSI DSPs can calculate a digital biquad in $\sim 1\ \mu$s and an $N$-tap FIR filter in $0.1N\ \mu$s [12, 13].

In bit-slice DSPs, fast execution of Eq. (3.88) can be obtained by using a hardware multiplier that works in parallel with a bit-sliced arithmetic logic unit (ALU) [14] or by using a VLSI MAC [15].

## 8.3 Parallel Digital Filter Implementation

The development of parallel architecture and the drop in the price and power consumption of hardware components (adders, multipliers, MACs) has enabled the parallel implementation of digital filters. Several architectures are available for parallel filter implementation. An architecture commonly used in the cascade implementation of digital biquads is pipelining, which is shown in Figure 3.24.

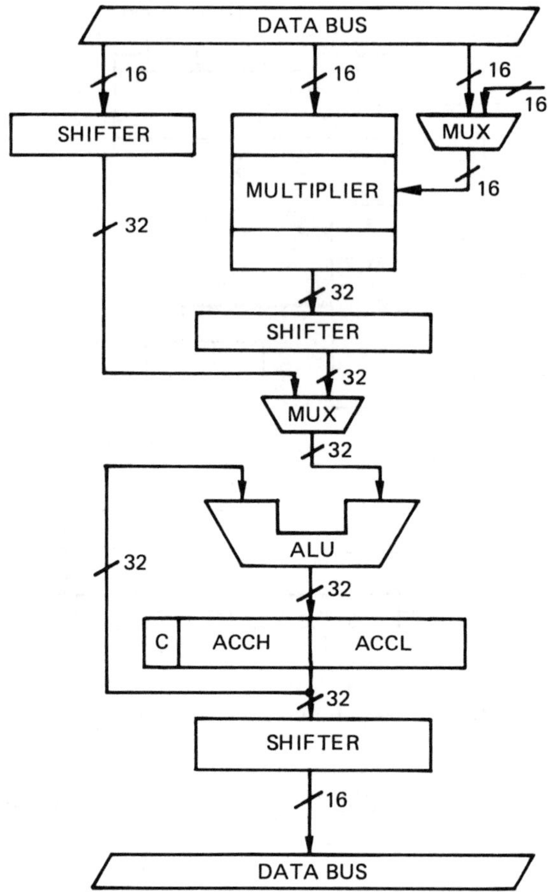


**Figure 3.23** Multiplier-accumulator unit of the TMS320 family.

Another structure, used especially in parallel executions of additions, is the tree structure [16]. The transposed direct structure of Figure 3.14 is also suitable for parallel calculation by using $N$ multipliers and $N - 1$ adders. A cascadable signal processor based on this architecture has already been constructed [17].

Each parallel structure has its advantages and its drawbacks. The two major criteria in the choice of a parallel filter structure are its hardware requirements (i.e., its cost) and its throughput delay. There have been several efforts to find a filter implementation, given the same hardware constraints, that uses the hardware resources optimally (i.e., that has the lowest possible throughput delay and the lowest idle time of the components). In fact, it is known that this problem is an NP-complete one. A novel technique [18, 19] proposes the use of operation research methods for the optimal parallel digital filter implementation under limited hardware resources. This technique considers additions, multiplications, and data transfers as "activities" that can be implemented in parallel by "workers" (adders, multipliers, buses). The following operation research methods are employed for the

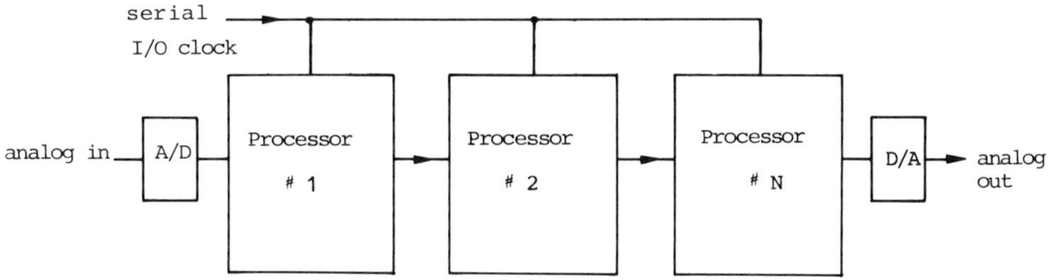


**Figure 3.24** Pipelined DSP architecture.

optimal filter implementation: (1) the critical path method (CPM) and (2) Project Planning Under Limited Resources (PPULR).

Such techniques, which take full advantage of the existing parallel architectures and VLSI components, can lead to digital filter implementations that fill most of the current industrial needs.

## ACKNOWLEDGMENTS

The author greatly appreciates the help of Prof. A. N. Venetsanopoulos and Prof. M. G. Strintzis in preparing the manuscript and thanks Prof. Antoniou for his permission to publish two examples from his book *Digital Filter Analysis and Design* (McGraw-Hill, 1979).

## REFERENCES

1. A. Antoniou, *Digital Filter Analysis and Design*, McGraw-Hill, New York (1979).
2. E. I. Jury, *Theory and Application of the z-Transform*, Wiley, New York (1964).
3. H. Y.-F. Lam, Analog and Digital Filters, Prentice-Hall, Englewood Cliffs, New Jersey (1964)
4. K. Steiglitz, Computer-aided design of recursive digital filters, *IEEE Trans. Audio Electroacoust.*, *AU-18*: (June 1970).
5. A. G. Constantinides, Spectral transformations for digital filters, *Proc. IEEE*, *117* (7): 1585–1590, (August 1970).
6. L. R. Rabiner and R. W. Schafer, Recursive and nonrecursive realizations of digital filters designed by frequency sampling techniques, *IEEE Trans. Audio Electroacoust.*, *19* (3): 200–207 (September 1971).
7. T. W. Parks and J. H. McClellan, Chebyshev approximation for nonrecursive digital filters with linear phase, *IEEE Trans. Circuit Theory*, *CT-19*: 189–194, (March, 1972).
8. S. K. Mitra and R. J. Sherwood, Digital ladder networks, *IEEE Trans. Audio Electroacoust.*, *AU-21*: 30–36 (February, 1973).
9. A. Peled and B. Liu, *Digital Signal Processing*, John Wiley, New York (1976).

10. M. Arjmand and R. A. Roberts, On comparing hardware implementations of fixed point digital filters, *IEEE Circuits Syst. Mag. 3* (2): 2–8 (1982).
11. Texas Instruments, *Signal Processing Products and Technology* (1983).
12. Phillips, *Introducing the SP 50 Family*, (1986).
13. Texas Instruments, *TMS320C25 Digital Signal Processor Description*, (1986).
14. H. J. Kolb, A signal processor using bit-slice elements for the audio frequency range, *Signal Process*, *2*: 339–346, (1980).
15. L. Guendel, A novel high speed Fourier-vector processor, *Signal Process. 9*: 107–120 (1985).
16. I. Pitas and A. N. Venetsanopoulos, Two-dimensional realization of digital filters by transform decomposition, *IEEE Trans. Circuits Syst.*, *CAS-32*, (10): 1029–1040, (October, 1985).
17. INMOS, IMS A100 cascadable signal processor description, (1987).
18. J. Zeman and G. Moschytz, A systematic approach to the design and speed comparison of signal processor architectures for digital filtering, *IEEE Trans. Acoust. Speech Signal Process. ASSP-31*: 1536 (December, 1983).
19. I. Pitas and M. G. Strintzis, An efficient and systematic technique for the parallel implementation of DFT algorithms, *Signal Processing III (Theories and Applications)* (I. T. Young et al., eds.) Elsevier, Amsterdam (1986).

## BIBLIOGRAPHY

A. V. Oppenheim and R. N. Schafer, *Digital Signal Processing*, Prentice-Hall, Englewood Cliffs, New Jersey (1975).

A. V. Oppenheim, A. S. Willsky, and I. T. Young, *Signals and Systems*, Prentice-Hall, Englewood Cliffs, New Jersey (1983).

L. R. Rabiner and B. Gold, *Theory and Application of Digital Signal Processing*, Prentice-Hall, Englewood Cliffs, New Jersey (1976).

# 4

# Discrete Fourier Transform and FFT

HOSSEIN HAKIMMASHHADI Electrical Engineering Department, Worcester Polytechnic Institute, Worcester, Massachusetts

## 1 INTRODUCTION

In applied science one is basically dealing with signals and systems. Signals contain information about variables of a phenomenon, and systems model the interaction of these signals. Usually these signals and their interaction are represented in the time domain, but this does not mean that progression versus time is always the best representation for their analysis. A number of different techniques can be used to map the time representation of signals and systems into another space and, as a result, make their analysis simpler. Among these techniques, Fourier analysis, which includes the Fourier series and Fourier transform, is one of the most powerful analytical tools and has application in the analysis of both continuous-time and discrete-time signals and systems.

Application of Fourier analysis for periodic signals results in the Fourier series representation. This representation decomposes the signal into a sum of weighted complex sinusoidal signals with frequencies that are integer multiples of the original signal fundamental frequency. For example, if $x_T(t)$ is a periodic signal with period $T$, Fourier series representation of the signal will be of the form

$$x_T(t) = \sum_{k=-\infty}^{\infty} a_k e^{jk\omega_0 t} \tag{4.1}$$

where $\omega_0$, which is equal to $2\pi/T$, is the fundamental frequency of $x_T(t)$ in radians per second. The $a_k$ coefficients are weighting factors and are in general complex numbers. If $x_T(t)$ is a real function, Eq. (4.1) can be manipulated to have the following alternative representation where all terms are real:

$$x_T(t) = a_0 + 2\sum_{k=1}^{\infty}[b_k \cos(k\omega_0 t) - c_k \sin(k\omega_0 t)] \tag{4.2}$$

In this equation $b_k$ and $c_k$ are real and imaginary parts of $a_k$, respectively. The set of Fourier series coefficients of $x_T(t)$, which is $\{a_k\}$, is called the spectrum of the signal and can be calculated by

$$a_k = \frac{1}{T}\int_{t_0}^{t_0+T} x_T(t)e^{-jk\omega_0 t}\,dt \tag{4.3}$$

where $t_0$ is an arbitrary real number.

If the signal under consideration is not periodic, it cannot be represented by a Fourier series. For this class of signals, however, there exists a representation called the Fourier transform representation which has similar properties. The Fourier transform representation of a nonperiodic signal is basically a decomposition of the signal into a continuum of weighted complex exponentials of the form $e^{j\omega t}$. This representation has the following mathematical form:

$$x(t) = \frac{1}{2\pi}\int_{-\infty}^{\infty} X(\omega)e^{j\omega t}\,d\omega \tag{4.4}$$

Here $X(\omega)$ is defined as the Fourier transform of $x(t)$ and can be derived by

$$X(\omega) = \int_{-\infty}^{\infty} x(t)e^{-j\omega t}\,dt \tag{4.5}$$

where $\omega$ takes any value in the continuum of frequencies.

Traditionally, Fourier analysis has been an analytical tool for continuous-time signals and systems. But rapid development of digital computers in the last few decades has challenged this tradition. Now in many applications of analog signal processing, it is preferable to sample the signal, process it by a discrete-time system, which is implemented on a digital computer, and then convert it back to an analog signal. The flexibility and advantages of this approach have resulted in the development of a set of powerful techniques for digital signal processing. Some of these techniques, such as discrete Fourier analysis, are actually discrete versions of the techniques which have been found to be useful in continuous-time signal processing.

For a discrete periodic signal $x_N(n)$ with period $N$, the discrete Fourier series pair is defined by

$$x_N[n] = \sum_{k=n_0}^{n_0+N-1} a_k e^{jk(2\pi/N)n} \tag{4.6}$$

$$a_k = \frac{1}{N} \sum_{n=n_1}^{n_1+N-1} x_N[n] e^{-jk(2\pi/N)n} \tag{4.7}$$

where $n_0$ and $n_1$ are arbitrary integer numbers [1–5]. Equation (4.6) shows clearly that the discrete Fourier series has only $N$ terms, as opposed to the continuous Fourier series, which can have an infinite number of terms. Also, there are at most $N$ distinct $a_k$, and the sequence $\{a_k\}$ is a periodic sequence with period $N$.

Nonperiodic discrete signals do not have a discrete Fourier series representation. For these signals, which in general are of infinite length, the discrete-time Fourier transform pair is defined as an alternative by

$$x[n] = \frac{1}{2\pi} \int_{\Omega_0}^{\Omega_0+2\pi} X(\Omega) e^{j\Omega n} \, d\Omega \tag{4.8}$$

$$X(\Omega) = \sum_{n=-\infty}^{\infty} x[n] e^{-j\Omega n} \tag{4.9}$$

where $\Omega_0$ is an arbitrary real number [1–5]. $X(\Omega)$ is called the discrete-time Fourier transform of $x[n]$ and is a continuous periodic function with a period of $2\pi$.

Since $x[n]$ is in general of infinite length and $X(\Omega)$ is a continuous function, the discrete-time Fourier transform does not lend itself to implementation on digital computers. To manage this difficulty, the discrete signal $x[n]$ is assumed to be nonzero only for $0 \leq n \leq N-1$, and a new transform called the discrete Fourier transform pair is defined by

$$x[n] = \frac{1}{N} \sum_{k=0}^{N-1} X[k] e^{jk(2\pi/N)n}, \qquad 0 \leq n \leq N-1 \tag{4.10}$$

$$X[k] = \sum_{n=0}^{N-1} x[n] e^{-jk(2\pi/N)n}, \qquad 0 \leq k \leq N-1 \tag{4.11}$$

Here $X[k]$, which is the discrete Fourier transform of $x[n]$. is also a discrete signal and is nonzero only for the range $0 \leq k \leq N-1$ [1–5].

This chapter is basically a discussion of the discrete Fourier transform and its efficient implementation on digital computers. Section 2 discusses the properties of the discrete Fourier transform, including its relation to the continuous-time Fourier transform and its application in digital signal processing. Section 3 reviews different algorithms for fast computation of the discrete Fourier transform and describes the details of the Cooley–Tukey and the split-radix algorithms. The conclusion of this chapter is presented in Section 4.

## 2 THE DISCRETE FOURIER TRANSFORM (DFT)

The mathematical expression for the discrete Fourier transform of a finite-length discrete signal was defined by Eq. (4.11) and the inverse discrete Fourier transform was defined by Eq. (4.10). These equations by themselves do not tell much about the usefulness of the DFT. Since the development of the DFT is based on the idea of the continuous-time Fourier transform, in order to evaluate the usefulness and limitations of the DFT for processing continuous signals, one has to understand the relation between the DFT and the continuous-time Fourier transform.

### 2.1 Relation Between the DFT and the Continuous-Time Fourier Transform (CFT)

The best way to understand the relation between the DFT and the CFT is to study the modifications of the Fourier transform of a continuous signal, as the signal goes through a series of mathematical operations to make it suitable for machine-based Fourier transform computation [6]. As an example consider the continuous signal

$$x(t) = \begin{cases} e^{-\alpha t}, & t > 0 \\ 0 & \text{otherwise} \end{cases} \tag{4.12}$$

which has a closed-form Fourier transform

$$X(\omega) = \int_{-\infty}^{\infty} x(t)e^{-j\omega t}\,dt = \frac{1}{\alpha + j\omega} \tag{4.13}$$

$x(t)$ and $|X(\omega)|$ are both illustrated in Figure 4.1a, where $\alpha$ is assumed to be a positive real number. Considering the fact that we are interested in using a computer to find the Fourier transform of $x(t)$, which in general does not have a closed-form Fourier transform, we have to convert the continuous-time representation of the signal to a discrete-time representation. Mathematically, this can be represented by multiplying $x(t)$ by the sampling function

$$s(t) = \sum_{n=-\infty}^{\infty} \delta(t - nT) \tag{4.14}$$

Figure 4.1b clearly shows that $s(t)$ is a periodic impulse train with a period $T$ and a weight of one. $T$ is the sampling period and $f_s = 1/T$ is the sampling frequency.

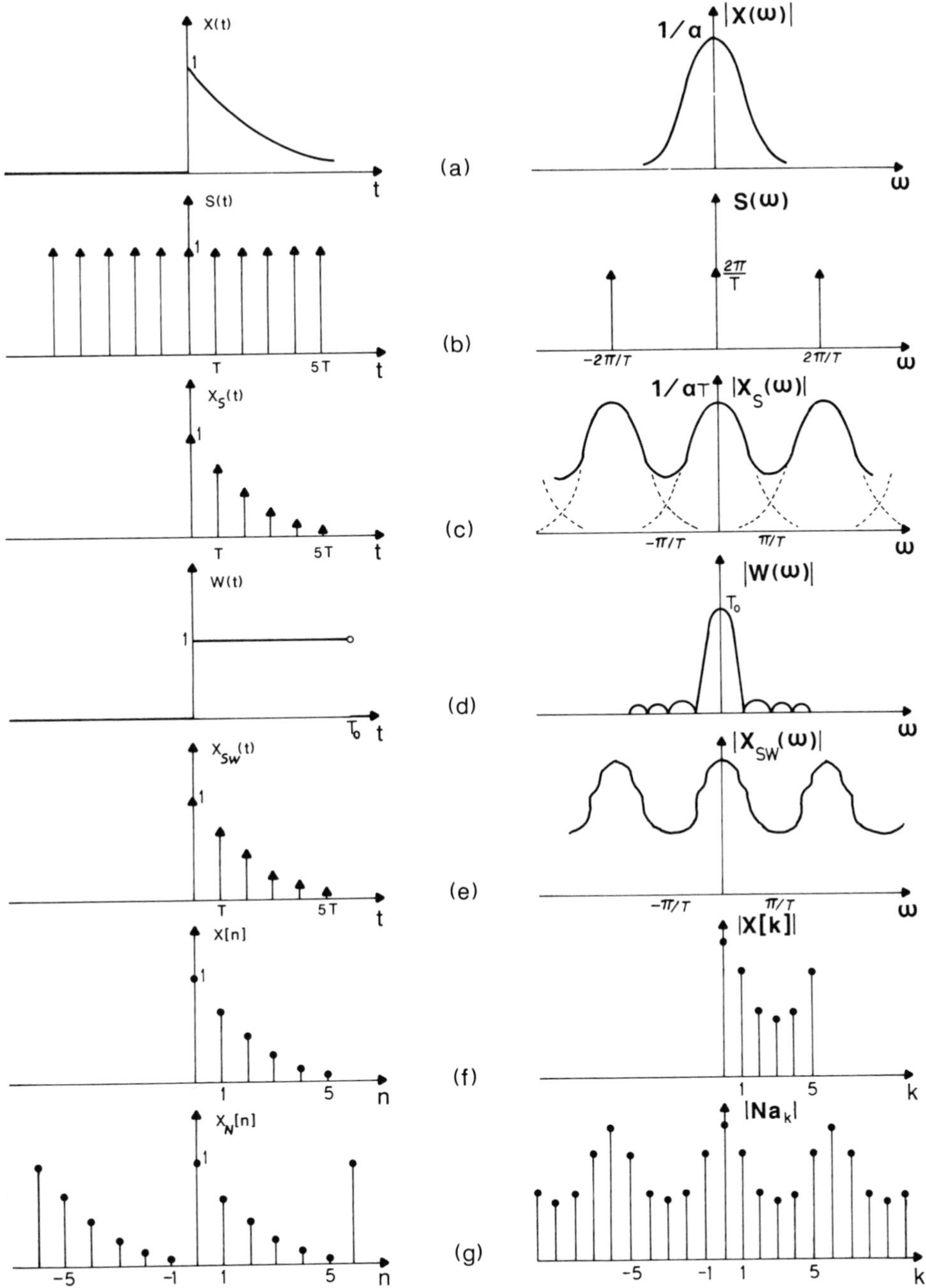


**Figure 4.1** Derivation of the discrete Fourier transform from the continuous-time Fourier transform.

The Fourier transform of $s(t)$ is

$$S(\omega) = \frac{2\pi}{T} \sum_{n=-\infty}^{\infty} \delta\left(\omega - n\frac{2\pi}{T}\right) \tag{4.15}$$

$S(\omega)$ is also a periodic impulse train, as shown in Figure 4.1b. But the weight of each impulse is $2\pi/T$ and the period is also $2\pi/T$. The result of multiplying $x(t)$ by $s(t)$ is

$$x_s(t) = x(t)s(t) = \sum_{n=-\infty}^{\infty} x(nT)\delta(t - nT) \tag{4.16}$$

$x_s(t)$, which is shown in Figure 4.1c, is an impulse train but is not necessarily a periodic function. The weight of the impulse at time $nT$ is actually the same as the amplitude of the signal $x(t)$ at that time. Since multiplication in the time domain corresponds to convolution in the frequency domain, $X_s(\omega)$, which is the Fourier transform of $x_s(t)$, can be found by convolving $X(\omega)$ and $S(\omega)$. Therefore

$$X_s(\omega) = \frac{1}{2\pi}[X(\omega) * S(\omega)] = \frac{1}{T} \sum_{n=-\infty}^{\infty} \left[X\left(\omega - n\frac{2\pi}{T}\right)\right] \tag{4.17}$$

As shown in Figure 4.1c, $X_s(\omega)$ is a periodic function with period $2\pi/T$. It is interesting to note that $X_s(\omega)$ is actually a sum of shifted and scaled versions of $X(\omega)$. If we neglect the scaling factor $1/T$, then for the range $|\omega| < \pi/T = 2\pi(f_s/2)$, $X_s(\omega)$ is similar to $X(\omega)$ but is not exactly the same. This is due to the fact that there is some overlap between adjacent shifted replicas of $X(\omega)$, and when these replicas are added, the high-frequency part of $X(\omega)$ is distorted. This effect is called aliasing. One way to decrease the distortion due to aliasing is to increase the sampling frequency. This forces the shifted replicas of $X(\omega)$, which are distanced by $2\pi/T$, to move away from each other, and as a result their overlap decreases. Since $X_s(\omega)$ and $X(\omega)$ are similar only for the range $|\omega| < 2\pi(f_s/2)$, $X_s(\omega)$ at its best can approximately represent the information in $X(\omega)$ only up to half of the sampling frequency. But if $x(t)$ is frequency band-limited or is filtered such that $X(\omega) = 0$ for $|\omega| > \omega_c$, and if the sampling frequency is greater than twice the highest frequency of $x(t)$, which means $2\pi/T > 2\omega_c$, then there will be no aliasing and one period of $X_s(\omega)$ contains exactly the same information as $X(\omega)$. In other words, under these ideal conditions, $x_s(t)$, which is the sampled version of $x(t)$, completely represents the original signal and there is no loss of information.

The sampled version of the original signal contains an infinite number of samples and is still not a suitable form for computer-based Fourier transform computation. This stems from the fact that it would take an infinite amount of time for a computer to acquire and process $x_s(t)$. This problem can be resolved by multiplying $x_s(t)$ by a time-limited function such as the rectangle window

$$W(t) = \begin{cases} 1, & 0 \le t < T_0 \\ 0 & \text{otherwise} \end{cases} \tag{4.18}$$

where $T_0$ is chosen such that there are only $N$ samples in the window, which means $T_0 = NT$. The Fourier transform of $W(t)$ is similar to a sinc function and is

$$W(\omega) = \int_{-\infty}^{\infty} W(t)e^{-j\omega t}\,dt = e^{-j\omega T_0/2}\left[2\frac{\sin(\omega T_0/2)}{\omega}\right] \tag{4.19}$$

$W(t)$ and $|W(\omega)|$ are both illustrated in Figure 4.1d. The result of multiplying $x_s(t)$ by $W(t)$ is

$$x_{sw}(t) = x_s(t)W(t) = \sum_{n=0}^{N-1} x(nT)\delta(t-nT) \tag{4.20}$$

As is seen in Figure 4.1e, $x_{sw}(t)$ is a sequence of $N$ impulses where the weight of the impulses at each particular time is the same as the amplitude of the original signal at that time. The Fourier transform of $x_{sw}(t)$ can be derived in two different ways. The first approach is to take the Fourier transform of $x_{sw}(t)$ directly, which results in

$$X_{sw}(\omega) = \int_{-\infty}^{\infty} x_{sw}(t)e^{-j\omega t}\,dt = \sum_{n=0}^{N-1} x(nT)e^{-j\omega nT} \tag{4.21}$$

Although this is a valid result, there is a drawback. By using this approach we cannot relate $X_{sw}(\omega)$ to $X_s(\omega)$ and the relation between the DFT and the CFT cannot be found. The second approach derives from the fact that since we know $x_{sw}(t)$ is product of $x_s(t)$ and $W(t)$, $X_{sw}(\omega)$ should be the convolution of $X_s(\omega)$ and $W(\omega)$; therefore

$$\begin{aligned} X_{sw}(\omega) &= \frac{1}{2\pi}[X_s(\omega) * W(\omega)] \\ &= \frac{1}{2\pi}\left\{\left[\frac{1}{T}\sum_{n=-\infty}^{\infty} X\left(\omega - n\frac{2\pi}{T}\right)\right] * \left[e^{-j\omega T_0/2}\left(2\frac{\sin(\omega T_0/2)}{\omega}\right)\right]\right\} \end{aligned} \tag{4.22}$$

$|X_{sw}(\omega)|$ is shown in Figure 4.1e. Although $|X_{sw}(\omega)|$ is still periodic and similar to $|X_s(\omega)|$, there are also some differences. Since $W(\omega)$ is not composed of impulses and has sidelobes, the result of convolving $X_s(\omega)$ with $W(\omega)$, $X_{sw}(\omega)$, is not exactly a replica of $X_s(\omega)$. The value of $X_{sw}(\omega)$ at a particular frequency $\omega_0$

is actually a weighted integral of $X_s(\omega_0)$ and its neighboring frequencies, and the weighting function is $W(\omega)$, which is centered at $\omega_0$. This distortion of $X_s(\omega)$, which is actually the result of truncating $x_s(t)$, is called leakage. The name is very appropriate because information at $X_{sw}(\omega_0)$ contains the information at $X_s(\omega_0)$ plus some leakage of information from other frequencies. The frequencies which are very close to $\omega_0$ will have the highest amount of leakage. The leakage effect can be decreased by increasing $T_0$. For very large values of $T_0$, $W(\omega)$ approaches an impulse function and $X_{sw}(\omega)$ approaches $X_s(\omega)$.

As a result of windowing $x_s(t)$, $x_{sw}(t)$ is generated, which has a finite number of impulses. Now the information in $x_{sw}(t)$ can be used to generate a discrete-time signal which is suitable for computer-based signal processing. This discrete-time signal is defined as

$$x[n] = \begin{cases} x(nT), & 0 \le n \le N-1 \\ 0 & \text{otherwise} \end{cases} \tag{4.23}$$

which can be also represented by

$$x[n] = \sum_{k=0}^{N-1} x(kT)\delta[n-k] \tag{4.24}$$

$x[n]$ is illustrated in Figure 4.1f. Since $x[n]$ is a finite-length discrete-time signal, it has a discrete Fourier transform $X[k]$ defined by

$$X[k] = \sum_{n=0}^{N-1} x[n]e^{-jk(2\pi/N)n} = \sum_{n=0}^{N-1} x(nT)e^{-jk(2\pi/N)n}, \qquad 0 \le k \le N-1 \tag{4.25}$$

But $N = T_0/T$, therefore $X[k]$ can be written as

$$X[k] = \sum_{n=0}^{N-1} x(nT)e^{-jk(2\pi/T_0)nT}, \qquad 0 \le k \le N-1 \tag{4.26}$$

By comparing Eqs. (4.26) and (4.21), one can conclude that $X[k]$ is actually the same as samples of $X_{sw}(\omega)$ at frequencies $k(2\pi/T_0)$, which means

$$X[k] = X_{sw}(k2\pi/T_0), \qquad 0 \le k \le N-1 \tag{4.27}$$

$|X[k]|$ is shown in Figure 4.1f. This is an interesting result because it suggests that the discrete Fourier transform of samples of a continuous signal is closely related to samples of the continuous-time Fourier transform of the signal. In fact, if we collect $N$ samples of a continuous signal such as $x(t)$, then by computing the $N$-point DFT of these $N$ samples, we can estimate $N$ samples of the CFT of $x(t)$. Accuracy of this estimation depends on the amounts of aliasing and leakage which occur during sampling and windowing. Fortunately, by controlling the sampling period $T$ and the window width $T_0$, one can control the aliasing and leakage effect. It should be also noted that of the $N$ samples computed by $X[k]$, the first half are estimates of samples of $X(\omega)/T$ for the frequency range $(0, f_s/2)$ and the second half are estimates of samples of $X(\omega)/T$ for the frequency range $(-f_s/2, 0)$. Therefore we have $N$ samples over a frequency range of $f_s$, which means that samples of the discrete Fourier transform are distanced by $f_s/N$. But $f_s/N = 1/TN = 1/T_0$, and this means that the frequency distance between successive points of the discrete Fourier transform is inversely proportional to the width of the time window $W(t)$.

There is also a second interpretation for the DFT of $x[n]$. In order to understand this interpretation, we first define a periodic discrete signal where each period of this signal is the same as $x[n]$. This new signal is

$$x_N[n] = x[n], \qquad 0 \leq n \leq N-1 \tag{4.28}$$

which is shown in Figure 4.1g. The discrete Fourier series coefficients of $x_N[n]$ can be computed by Eq. (4.7), which, for $n_1 = 0$, results in

$$a_k = \frac{1}{N} \sum_{n=0}^{N-1} x_N[n] e^{-jk(2\pi/N)n} \tag{4.29}$$

But for $0 \leq n \leq N-1$, $x_N[n] = x[n]$; therefore Eq. (4.29) can be written as

$$a_k = \frac{1}{N} \sum_{n=0}^{N-1} x[n] e^{-jk(2\pi/N)n} \tag{4.30}$$

Multiplying both sides by $N$ results in

$$Na_k = \sum_{n=0}^{N-1} x[n] e^{-jk(2\pi/N)n} \tag{4.31}$$

For $0 \leq k \leq N-1$, the right-hand side of Eq. (4.31) is actually the DFT of $x[n]$; therefore, one can conclude that

$$Na_k = X[k], \qquad 0 \leq k \leq N-1 \tag{4.32}$$

This equation states that the DFT of a finite-length discrete signal is proportional to discrete Fourier series coefficients of a periodic discrete signal, where each period is the same as the original finite-length discrete signal. $x_N[n]$ and the $\{|Na_k|\}$ sequences are both illustrated in Figure 4.1g.

## 2.2 Properties of the Discrete Fourier Transform

The DFT of a discrete signal $x[n]$ which has a length $N$ was defined by Eq. (4.11) as

$$X[k] = \sum_{n=0}^{N-1} x[n]e^{-jk(2\pi/N)n}, \qquad 0 \leq k \leq N-1 \tag{4.33}$$

As seen in this equation, $X[k]$ is also a discrete signal of length $N$. The correspondence between these two finite-length sequences $x[n]$ and $X[k]$ is unique, which means that once we have either of them, the other one is completely specified. $x[n]$ in terms of $X[k]$, which is called the inverse discrete Fourier transform of $X[k]$, has been defined by Eq. (4.10) and is

$$x[n] = \frac{1}{N}\sum_{k=0}^{N-1} X[k]e^{jk(2\pi/N)n}, \qquad 0 \leq n \leq N-1 \tag{4.34}$$

For the sake of simplicity of notation, usually Eqs. (4.33) and (4.34) are defined in terms of $W_N$, which is defined by

$$W_N = e^{-j(2\pi/N)} \tag{4.35}$$

Therefore the DFT pair is written as

$$X[k] = \sum_{n=0}^{N-1} x[n]W_N^{kn}, \qquad 0 \leq k \leq N-1 \tag{4.36}$$

$$x[n] = \frac{1}{N}\sum_{k=0}^{N-1} X[k]W_N^{-kn}, \qquad 0 \leq n \leq N-1 \tag{4.37}$$

In these equations $x[n]$ and $X[k]$ are finite-length sequences and they are nonzero only for $0 \leq n,\ k \leq N-1$, but it is interesting to note that for any integer $m$ we have

$$\begin{aligned} X[k+mN] &= \sum_{n=0}^{N-1} x[n]W_N^{(k+mN)n} = \sum_{n=0}^{N-1} x[n]W_N^{kn}W_N^{mnN} \\ &= \sum_{n=0}^{N-1} x[n]W_N^{kn}e^{-jmn(2\pi)} = \sum_{n=0}^{N-1} x[n]W_N^{kn} = X[k] \end{aligned} \tag{4.38}$$

and

$$x[n+mN] = \frac{1}{N}\sum_{k=0}^{N-1} X[k]W_N^{-(n+mN)k} = \frac{1}{N}\sum_{k=0}^{N-1} X[k]W_N^{-kn}W_N^{-mkN}$$
$$= \frac{1}{N}\sum_{k=0}^{N-1} X[k]W_N^{-kn}e^{jmk(2\pi)} = \frac{1}{N}\sum_{k=0}^{N-1} X[k]W_N^{-kn} = x[n] \quad (4.39)$$

This suggests that even though $x[n]$ and $X[k]$ are finite-length sequences, they can be interpreted as periodic sequences with a period of $N$. Now if we let Eqs. (4.36) and (4.37) be valid for all integers $k$ and $n$, then $X[k]$ and $x[n]$ on the left-hand side will be periodic sequences $X_N[k]$ and $x_N[n]$ and the DFT pair can be modified to

$$X_N[k] = \sum_{n=0}^{N-1} x[n]W_N^{kn} \qquad \text{for } all\ k \quad (4.40)$$

$$x_N[n] = \frac{1}{N}\sum_{k=0}^{N-1} X[k]W_N^{-kn} \qquad \text{for } all\ n \quad (4.41)$$

where

$$X_N[k] = X[k], \qquad 0 \le k \le N-1 \quad (4.42)$$

$$x_N[n] = x[n], \qquad 0 \le n \le N-1 \quad (4.43)$$

Since the summations in the DFT pair extend only over $0 \le n,\ k \le N-1$, Eqs. (4.42) and (4.43) can be used to modify Eqs. (4.40) and (4.41) to

$$X_N[k] = \sum_{n=0}^{N-1} x_N[n]W_N^{kn} \qquad \text{for } all\ k \quad (4.44)$$

$$x_N[n] = \frac{1}{N}\sum_{k=0}^{N-1} X_N[k]W_N^{-kn} \qquad \text{for } all\ n \quad (4.45)$$

These equations offer a new interpretation of the DFT pair because now the DFT pair can be interpreted as describing a correspondence between two periodic sequences $x_N[n]$ and $X_N[k]$ as defined by Eqs. (4.42) and (4.43). This new interpretation of the DFTs of finite-length sequences is helpful in understanding some of the properties of the DFT pair.

In the following, some of the properties of the DFT which are important for the application of digital signal processing techniques are summarized. Interested readers can find the proof of these properties in [1–5]. In describing the properties of the DFT, the double-headed arrow connecting two finite-length sequences $x[n]$ and $X[k]$, that is $x[n] \leftrightarrow X[k]$, is used to state that $x[n]$ and $X[k]$ establish a DFT pair. Also, two windowing sequences $R_N[n]$ and $R_N[k]$ are used which are defined by

$$R_N[n] = \begin{cases} 1, & 0 \leq n \leq N-1 \\ 0 & \text{otherwise} \end{cases} \tag{4.46}$$

$$R_N[k] = \begin{cases} 1, & 0 \leq k \leq N-1 \\ 0 & \text{otherwise} \end{cases} \tag{4.47}$$

1. *Linearity property*

$$\begin{aligned} &\text{if} \quad x_1[n] \longleftrightarrow X_1[k] \qquad \text{and} \qquad x_2[n] \longleftrightarrow X_2[k] \\ &\textit{then} \quad (ax_1[n] + bx_2[n]) \longleftrightarrow (aX_1[k] + bX_2[k]) \end{aligned} \tag{4.48}$$

2. *Circular shifting property*

$$\text{if} \quad x[n] \longleftrightarrow X[k]$$

$$\textit{then} \quad x_N[n-m]R_N[n] \longleftrightarrow W_N^{km}X[k] \tag{4.49}$$

$$\text{and} \quad W_N^{-nm}x[n] \longleftrightarrow X_N[k-m]R_N[k] \tag{4.50}$$

where $x_N[n]$ and $X_N[k]$ are the periodic counterparts of $x[n]$ and $X[k]$, respectively, which were defined by Eqs. (4.42) and (4.43).

3. *Circular convolution property*

$$\text{if} \quad x[n] \longleftrightarrow X[k] \qquad \text{and} \qquad y[n] \longleftrightarrow Y[k]$$

$$\begin{aligned} \textit{then} \quad &\left(\sum_{m=0}^{N-1} x_N[m]y_N[n-m]\right) R_N[n] \\ &= \left(\sum_{m=0}^{N-1} x_N[n-m]y_N[m]\right) R_N[n] \longleftrightarrow X[k]Y[k] \end{aligned} \tag{4.51}$$

$$\begin{aligned} \text{and} \quad x[n]y[n] &\longleftrightarrow \frac{1}{N}\left(\sum_{m=0}^{N-1} X_N[m]Y_N[k-m]\right) R_N[k] \\ &= \frac{1}{N}\left(\sum_{m=0}^{N-1} X_N[k-m]Y_N[m]\right) R_N[k] \end{aligned} \tag{4.52}$$

4. *Circular correlation property*

$$\text{if} \quad x[n] \longleftrightarrow X[k] \qquad \text{and} \qquad y[n] \longleftrightarrow Y[k]$$

$$\begin{aligned} \textit{then} \quad & \left(\sum_{m=0}^{N-1} x_N[m]y_N[n+m]\right) R_N[n] \\ & = \left(\sum_{m=0}^{N-1} x_N[m-n]y_N[m]\right) R_N[n] \longleftrightarrow (X_N[-k]R_N[k])Y[k] \end{aligned} \tag{4.53}$$

$$\begin{aligned} \text{and} \quad (x_N[-n]R_N[n])y[n] & \longleftrightarrow \frac{1}{N}\left(\sum_{m=0}^{N-1} X_N[m-k]Y_N[m]\right) R_N[k] \\ & = \frac{1}{N}\left(\sum_{m=0}^{N-1} X_N[m]Y_N[m+k]\right) R_N[k] \end{aligned} \tag{4.54}$$

5. *Conjugation property*

$$\text{if} \quad x[n] \longleftrightarrow X[k]$$

$$\textit{then} \quad x^*[n] \longleftrightarrow X_N^*[-k]R_N[k] \tag{4.55}$$

$$\text{and} \quad x_N^*[-n]R_N[n] \longleftrightarrow X^*[k] \tag{4.56}$$

6. *Reversion property*

$$\text{if} \quad x[n] \longleftrightarrow X[k]$$

$$\textit{then} \quad x_N[-n]R_N[n] \longleftrightarrow X_N[-k]R_N[k] \tag{4.57}$$

7. *Property of real and imaginary parts*

$$\text{if} \quad x[n] \longleftrightarrow X[k]$$

$$\textit{then} \quad \mathrm{Re}(x[n]) \longleftrightarrow X_{ep}[k] = \tfrac{1}{2}(X_N[k] + X_N^*[-k])R_N[k] \tag{4.58}$$

$$j\,\mathrm{Im}(x[n]) \longleftrightarrow X_{op}[k] = \tfrac{1}{2}(X_N[k] - X_N^*[-k])R_N[k] \tag{4.59}$$

$$x_{ep}[n] = \tfrac{1}{2}(x_N[n] + x_N^*[-n])R_N[n] \longleftrightarrow \mathrm{Re}(X[k]) \tag{4.60}$$

$$\text{and} \quad x_{op}[n] = \tfrac{1}{2}(x_N[n] - x_N^*[-n])R_N[n] \longleftrightarrow j\,\mathrm{Im}(X[k]) \tag{4.61}$$

8. *Parseval's property*

$$\text{if} \quad x[n] \longleftrightarrow X[k]$$

$$\textit{then} \quad \sum_{n=0}^{N-1} |x[n]|^2 = \frac{1}{N}\sum_{k=0}^{N-1} |X[k]|^2 \tag{4.62}$$

### 2.3 Application of the Discrete Fourier Transform

The DFT is a discrete-time signal transform whose properties are very similar to the properties of the continuous-time Fourier transform. Therefore, its application is also analogous to the application of the CFT. The DFT can be used to convert mathematical operations such as discrete convolution and correlation, which are widely used in digital signal processing, to mathematical operations such as vector multiplication which are simple in nature [Eqs. (4.51)–(4.54)]. It can also be used to extract the frequency content of a signal, which is sometimes essential in understanding and interpreting the information within the signal. These applications will become more practical when we see in the next section that there are algorithms for fast computation of the DFT.

## 3 THE FAST FOURIER TRANSFORM (FFT)

The discrete Fourier transform was defined in a previous section as

$$X[k] = \sum_{n=0}^{N-1} x[n] W_N^{kn}, \qquad 0 \leq k \leq N-1 \tag{4.63}$$

The summation in this equation involves $N$ terms and its computation requires $(N-1)$ complex additions. Also, computation of each term in the summation requires one complex multiplication; therefore computation of the $N$-point DFT of an $N$-point sequence requires $N(N-1)$ complex additions and $N^2$ complex multiplications. This is a heavy computational burden in many applications of the DFT, and it was exactly this computational burden which blocked the development of digital signal processing for some time. In 1965 Cooley and Tukey published a paper [7] introducing an algorithm which computes the DFT much faster than the direct computation of $X[k]$ as defined by Eq. (4.63). This was a breakthrough in the development of fast algorithms for computation of the DFT. Although this algorithm is not a transform, it was named the fast Fourier transform (FFT). Later it was realized that Good and Thomas had developed a different FFT algorithm a few years before Cooley and Tukey published their paper [8, 9]. Since 1965, there has been a great deal of effort to develop new FFT algorithms and improve the older ones [10–41]. This has resulted in a number of FFT algorithms, including the Good–Thomas algorithm [8, 9, 18], Cooley–Tukey algorithm [7, 10, 11, 12], Rader algorithm [13, 42], chirp $z$ algorithm [15, 17], mixed-radix algorithm [16], Winograd algorithm [19, 20, 22, 24], Rader–Brenner algorithm [21], prime factor algorithm [23, 32, 33], Brunn algorithm [25], radix-3 algorithm [27, 41], Walsh transform-based algorithm [28, 37], polynomial transform algorithm [29], radix-6 algorithm [31, 41], Preuss algorithm [34], dynamic programming-based algorithm [35], recursive cyclotomic factorization algorithm [36], split-radix algorithm [14, 39, 40], and radix-12 algorithm [41]. Interested readers can find comparative study of some of these algorithms in [26, 30, 38]. In this section we develop the details of the Cooley–Tukey algorithm and the split-radix algorithm and also describe the basics of the Winograd algorithm and the prime factor algorithm.

## 3.1 The Cooley–Tukey Algorithm

As discussed earlier, the direct computation of an $N$-point DFT requires $N^2$ complex multiplications and $N(N-1)$ complex additions. Cooley and Tukey showed that if $N$ is a composite number, the computation of the DFT as defined by Eq. (4.63) can be changed to a form which requires substantially less multiplications and additions [7]. In order to derive the Cooley–Tukey formulation of the FFT we assume that $N = N_1 N_2$, and express indices $k$ and $n$ in Eq. (4.63) as

$$n = n_2 N_1 + n_1, \qquad 0 \le n_1 \le N_1 - 1,\ 0 \le n_2 \le N_2 - 1 \tag{4.64}$$

$$k = k_1 N_2 + k_2, \qquad 0 \le k_1 \le N_1 - 1,\ 0 \le k_2 \le N_2 - 1 \tag{4.65}$$

Then one can write $X[k]$ as

$$X[k_1 N_2 + k_2] = \sum_{n_1=0}^{N_1-1} \sum_{n_2=0}^{N_2-1} x[n_2 N_1 + n_1] W_N^{(k_1 N_2 + k_2)(n_2 N_1 + n_1)} \tag{4.66}$$

But

$$\begin{aligned}
W_N^{(k_1 N_2+k_2)(n_2 N_1+n_1)} &= W_N^{k_1 n_2 N_2 N_1} \times W_N^{k_2 n_2 N_1} \times W_N^{k_2 n_1} \times W_N^{k_1 n_1 N_2} \\
&= [e^{-j2\pi/N_1 N_2}]^{k_1 n_2 N_2 N_1} \times [e^{-j2\pi/N_1 N_2}]^{k_2 n_2 N_1} \times [e^{-j2\pi/N}]^{k_2 n_1} \\
&\quad \times [e^{-j2\pi/N_1 N_2}]^{k_1 n_1 N_2} \\
&= [e^{-j k_1 n_2 (2\pi)}] \times [e^{-j2\pi/N_2}]^{k_2 n_2} \times [e^{-j2\pi/N}]^{k_2 n_1} \\
&\quad \times [e^{-j2\pi/N_1}]^{k_1 n_1} \\
&= W_{N_2}^{k_2 n_2} \times W_N^{k_2 n_1} \times W_{N_1}^{k_1 n_1}
\end{aligned} \tag{4.67}$$

Substituting this result into Eq. (4.66) yields

$$X[k_1 N_2 + k_2] = \sum_{n_1=0}^{N_1-1} \left[ \left( \sum_{n_2=0}^{N_2-1} x[n_2 N_1 + n_1] W_{N_2}^{k_2 n_2} \right) W_N^{k_2 n_1} \right] W_{N_1}^{k_1 n_1} \tag{4.68}$$

In this equation, for each fixed value of $n_1$, the inner sum is an $N_2$-point DFT, and for each fixed value of $k_2$, the outer sum is an $N_1$-point DFT. Moreover, for each fixed value of $k_2$ and $n_1$, the inner sum is multiplied by $W_N^{k_2 n_1}$, which is usually called the twiddle factor. In order to understand the computational requirements of Eq. (4.68), its computation is divided into the following five steps:

1. $x[n] = (x[n_2N_1], x[n_2N_1+1], x[n_2N_1+2], \ldots, x[n_2N_1+N_1-1])\,,$
$$0 \le n_2 \le N_2 - 1 \quad (4.69)$$

2. $X[n_1, k_2] = \sum_{n_2=0}^{N_2-1} x[n_2N_1 + n_1] W_{N_2}^{k_2 n_2},$
$$0 \le k_2 \le N_2 - 1,\ 0 \le n_1 \le N_1 - 1 \quad (4.70)$$

3. $X'[n_1, k_2] = X[n_1, k_2] W_N^{k_2 n_1}, \qquad 0 \le k_2 \le N_2 - 1,\ 0 \le n_1 \le N_1 - 1 \quad (4.71)$

4. $X[k_1N_2 + k_2] = \sum_{n_1=0}^{N_1-1} X'[n_1, k_2] W_{N_1}^{k_1 n_1},$
$$0 \le k_1 \le N_1 - 1,\ 0 \le k_2 \le N_2 - 1 \quad (4.72)$$

5. $X[k_1N_2 + k_2] = (X[k_2], X[N_2 + k_2], X[2N_2 + k_2], \ldots, X[(N_1 - 1)N_2 + k_2])\,,$
$$0 \le k_2 \le N_2 - 1 \quad (4.73)$$

These steps are shown graphically in Figure 4.2. The first step is a rearrangement of the input data sequence. The $x[n]$ sequence is rearranged into $N_1$ sequences of length $N_2$. It should be clear from Eq. (4.69) that the new sequences do not have the same order as the original sequence $x[n]$. In the second step the DFT of each sequence is computed. This means $N_1$ computations of an $N_2$-point DFT. The third step is just a multiplication which requires $N$ complex multiplications. The fourth step, as shown by Eq. (4.72), involves $N_2$ computations of an $N_1$-point DFT. The last step is redundant because the output of stage 4 is actually the DFT of $x[n]$ in the normal order.

If we neglect the time required for rearrangement of data in the first step, the computational requirement of Eq. (4.68) is equivalent to the computational requirement of $N_1$ $N_2$-point DFT computations plus $N_2$ $N_1$-point DFT computations plus $N$ complex multiplications. If we assume that $N_1$-point and $N_2$-point DFTs are done by direct computation, then the number of required complex multiplications $M_c(N)$ and the number of required complex additions $A_c(N)$ for computation of Eq. (4.68) are

$$\begin{aligned} M_c(N) &= N_1(N_2^2) + N_1N_2 + N_2(N_1^2) = N_1N_2(N_1 + N_2 + 1) \\ &= N(N_1 + N_2 + 1) \qquad (4.74) \\ A_c(N) &= N_1[N_2(N_2 - 1)] + N_2[N_1(N_1 - 1)] = N_1N_2(N_1 + N_2 - 2) \\ &= N(N_1 + N_2 - 2) \qquad (4.75) \end{aligned}$$

But for $N > 6$, $M_c(N)$ is less than $N^2$ and $A_c(n)$ is less than $N(N-1)$; therefore we can conclude that Eq. (4.68) represents a more efficient approach for computation of the DFT. Obviously, if $N_1$ and/or $N_2$ are also composite, then the required $N_1$-point DFT and/or $N_2$-point DFT can be computed in terms of smaller DFTs, which

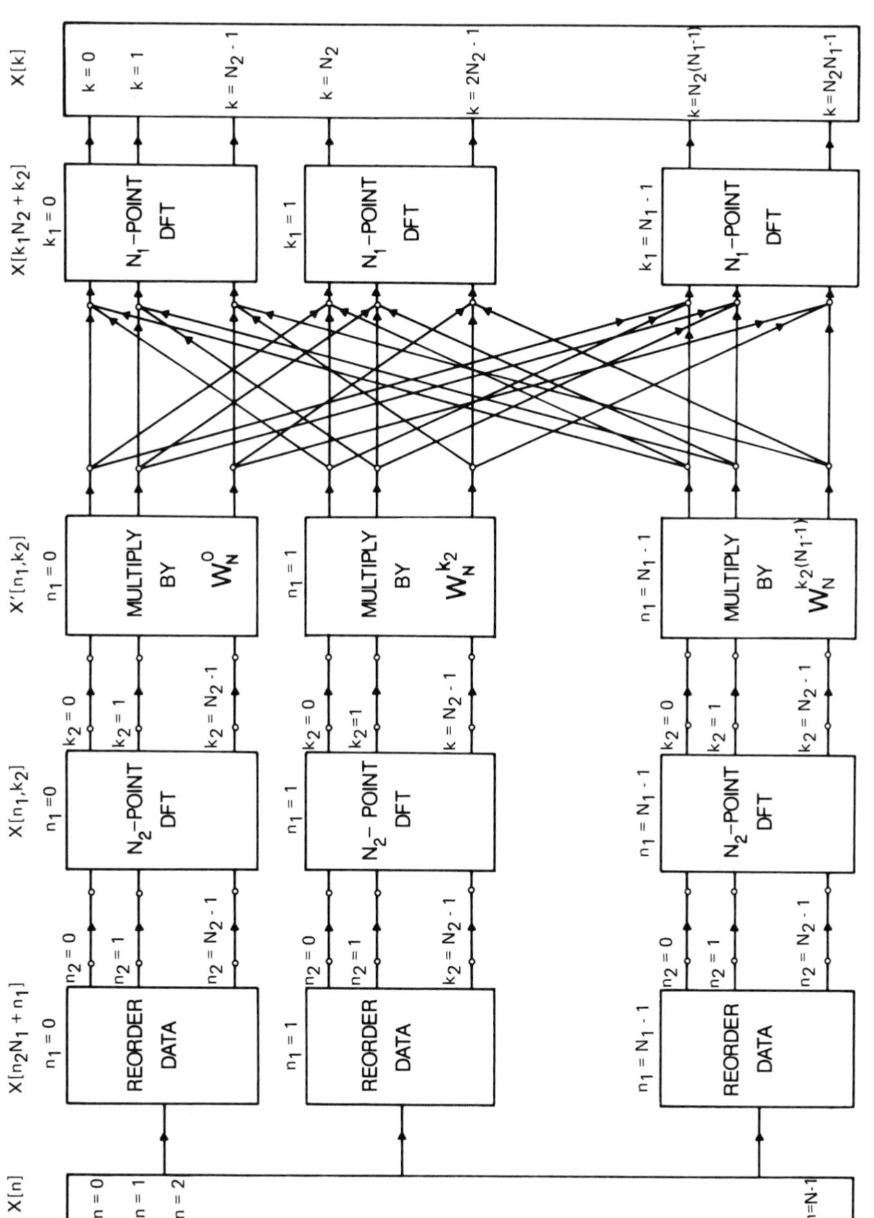


**Figure 4.2** Basic block diagram of the Cooley–Tukey algorithm, for an $N$-point DFT computation.

results in even more reduction in computational requirements. This reduction in computational requirements can be continued recursively until the point that small DFTs cannot be further broken down by the Cooley–Tukey technique. For example, if $N = N_1 N_2 N_3, \ldots, N_n$, then the Cooley–Tukey approach computes the $N$-point DFT by combining DFTs with length $N_1, N_2, N_3, \ldots,$ and $N_n$. The number of required complex multiplications and additions for this approach is approximately $N(N_1 + N_2 + N_3 + \cdots + N_n)$, which is much less than $N^2$ for large $N$. It is interesting to note that if $N_j = N_{j_1} N_{j_2}$ with $N_{j_1}, N_{j_2} > 1$, then $N_{j_1} + N_{j_2} < N_j$ unless $N_{j_1} = N_{j_2} = 2$, in which case $N_j = N_{j_1} + N_{j_2}$. From this argument one can easily conclude that the Cooley–Tukey algorithm is most efficient when $N$ is highly composite and all $N_j$ factors are primes. Moreover, it should also be noted that the need for a regular computational structure makes the cases where all $N_j$ factors are equal very attractive. Among all the different versions of the Cooley–Tukey algorithm, the radix-2 algorithm ($N_j = 2$) and the radix-4 algorithm ($N_j = 4$) are the most popular and widely used FFT algorithms.

### The Radix-2 Algorithm

The $N$-point Cooley–Tukey algorithm is referred to as a radix-2 algorithm when $N$ is a power of 2, that is, $N = 2^n$, and it is factored as $2 \times 2 \times 2 \times 2 \times \cdots \times 2$. In order to understand the implementation of the radix-2 algorithm, initially $N$ is factored as $2 \times N/2$, which means $N_1 = 2$ and $N_2 = N/2$.

Now Eqs. (4.64), (4.65), and (4.68) can be modified to

$$n = 2n_2 + n_1, \qquad 0 \le n_1 \le 1,\ 0 \le n_2 \le N/2 - 1 \tag{4.76}$$

$$k = k_1 N/2 + k_2, \qquad 0 \le k_1 \le 1,\ 0 \le k_2 \le N/2 - 1 \tag{4.77}$$

and

$$\begin{aligned} X[k_1 N/2 + k_2] &= \sum_{n_1=0}^{1} \left[ \left( \sum_{n_2=0}^{N/2-1} x[2n_2 + n_1] W_{N/2}^{k_2 n_2} \right) W_N^{k_2 n_1} \right] W_2^{k_1 n_1} \\ &= \sum_{n_2=0}^{N/2-1} x[2n_2] W_{N/2}^{k_2 n_2} + W_2^{k_1} + W_N^{k_2} \sum_{n_2=0}^{N/2-1} x[2n_2 + 1] W_{N/2}^{k_2 n_2} \\ &= \sum_{n_2=0}^{N/2-1} x[2n_2] W_{N/2}^{k_2 n_2} + (-1)^{k_1} W_N^{k_2} \sum_{n_2=0}^{N/2-1} x[2n_2 + 1] W_{N/2}^{k_2 n_2} \\ &= X_0'[k_2] + (-1)^{k_1} W_N^{k_2} X_1'[k_2] \end{aligned} \tag{4.78}$$

where

$$X_0'[k_2] = \sum_{n_2=0}^{N/2-1} x[2n_2] W_{N/2}^{k_2 n_2} \tag{4.79}$$

and

$$X_1'[k_2] = \sum_{n_2=0}^{N/2-1} x[2n_2+1]W_{N/2}^{k_2n_2} \tag{4.80}$$

Here $X_0'[k_2]$ is an $N/2$-point DFT of even-indexed samples of the original data sequence and $X_1'[k_2]$ is an $N/2$-point DFT of odd-indexed samples of the original data sequence. Therefore according to Eq. (4.78), the $N$-point DFT of the original data sequence can be expressed as a linear combination of two $N/2$-point DFTs. In order to better understand the implementation of the radix-2 algorithm, Eq. (4.78) is written as

$$X[k_2] = X_0'[k_2] + W_N^{k_2}X_1'[k_2], \qquad 0 \le k_2 \le N/2-1 \tag{4.81}$$

$$X[k_2+N/2] = X_0'[k_2] - W_N^{k_2}X_1'[k_2], \qquad 0 \le k_2 \le N/2-1 \tag{4.82}$$

and its graphical representation is shown in Figure 4.3. The convention used in the Figure 4.3 flow graph is 1) the output of each node is the sum of incoming signals, 2) each branch transmits a signal from one mode to another node, 3) if there is an integer number $m$ next to the arrowhead of a branch, and it is enclosed by parentheses, then the transmitted signal through that branch is multiplied by $W_N^m$, and 4) if there is a coefficient $c$ next to the arrowhead of a branch, and it is not enclosed by parentheses, then the transmitted signal through that branch

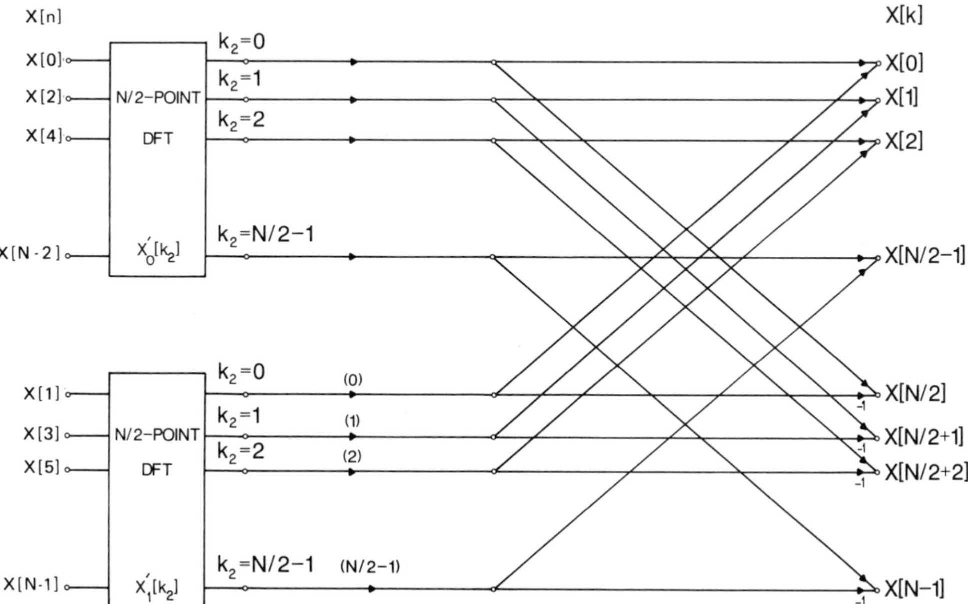


**Figure 4.3** Basic block diagram of the decimation-in-time radix-2 algorithm, for an $N$-point DFT computation.

is multiplied by $c$. Figure 4.3 indicates that the computational requirement of the radix-2 algorithm is equal to computational requirement of two $N/2$-point DFTs plus $N$ complex additions (subtraction is also counted as addition) and $N/2$ complex multiplications, which means

$$M_c(N) = 2M_c(N/2) + N/2 \tag{4.83}$$

$$A_c(N) = 2A_c(N/2) + N \tag{4.84}$$

But since $N/2 = 2 \times N/4$, the two $N/2$-point DFTs can be computed by four $N/4$-point DFTs at the cost of $N$ extra complex additions and $N/2$ extra complex multiplications. Therefore Eqs. (4.83) and (4.84) can be modified to

$$M_c(N) = 4M_c(N/4) + N/2 + N/2 \tag{4.85}$$

$$A_c(N) = 4A_c(N/4) + N + N \tag{4.86}$$

This procedure can be continued $(\log_2 N - 1)$ times, where the last stage requires $N/2$ two-point DFTs, and each stage requires $N$ complex additions and $N/2$ complex multiplications. Therefore, we have

$$M_c(N) = (N/2)\log_2 N \tag{4.87}$$

$$A_c(N) = N\log_2 N \tag{4.88}$$

Table 4.1 compares the number of required real multiplications $M_r(N)$ and real additions $A_r(N)$ for computation of the direct DFT, the radix-2 FFT algorithm, and the optimal radix-2 FFT algorithm. The calculation of computational requirements for the direct DFT and the radix-2 algorithm is based on Eqs. (4.63), (4.87), and (4.88) and the assumption that a complex multiplication requires four real

**Table 4.1** Number of Required Real Multiplications and Additions for Computation of Discrete Fourier Transform and Radix-2 FFT

| DFT Length $N$ | Direct DFT | | Radix-2 FFT | | Optimal radix-2 FFT | |
|---|---|---|---|---|---|---|
| | $M_r(N)$ | $A_r(N)$ | $M_r(N)$ | $A_r(N)$ | $M_r(N)$ | $A_r(N)$ |
| 8 | 256 | 240 | 48 | 72 | 4 | 52 |
| 16 | 1,024 | 992 | 128 | 192 | 24 | 152 |
| 32 | 4,096 | 4,032 | 320 | 480 | 88 | 408 |
| 64 | 16,384 | 16,256 | 768 | 1,152 | 264 | 1,032 |
| 128 | 65,536 | 65,280 | 1,792 | 2,688 | 712 | 2,504 |
| 256 | 262,144 | 261,632 | 4,096 | 6,144 | 1,800 | 5,896 |
| 512 | 1,048,576 | 1,047,552 | 9,216 | 13,824 | 4,360 | 13,576 |
| 1024 | 4,194,304 | 4,192,256 | 20,480 | 30,720 | 10,248 | 30,728 |
| 2048 | 16,777,216 | 16,773,120 | 45,056 | 67,584 | 23,560 | 68,616 |
| 4096 | 67,108,864 | 67,100,672 | 98,304 | 147,456 | 53,256 | 151,560 |

multiplications and two real additions, and a complex addition requires two real additions. As seen in this table, for large $N$, the radix-2 algorithm requires much less computation and as a result is much faster than the direct DFT. However, it should be mentioned that some of the required multiplications for the radix-2 algorithm, which are reflected in Eq. (4.87), are multiplication by $W_N^0$, $W_N^{kN/2}$, and $W_N^{kN/4}$, which are 1, $(-1)^k$, and $(-j)^k$ respectively. These operations are trivial and do not require actual multiplication. Moreover, some multiplications are multiplication by $W_N^{kN/8}$, which is $[(1-j)/\sqrt{2}]^k$ and requires less computation than a general complex multiplication. At the expense of a more complex computer program, one can handle these special cases separately and as a result enhance the computational speed of the radix-2 algorithm. The computational requirement for the optimal radix-2 algorithm, which is shown in Table 4.1 and can be calculated from Eqs. (4.89) and (4.90), is based on the assumptions that 1) all multiplications by 1, $(-1)^k$, and $(-j)^k$ are detected and avoided, 2) all multiplications by odd powers of $[(1-j)/\sqrt{2}]$ are detected and done by two real multiplications and two real additions, and 3) general complex multiplications are done by three real multiplications and three real additions [43–45].

$$M_r(N) = (N/2)(3\log_2 N - 10) + 8 \tag{4.89}$$

$$A_r(N) = (N/2)(7\log_2 N - 10) + 8 \tag{4.90}$$

The complete flow graph representation for a radix-2 implementation of a 16-point DFT is illustrated in Figure 4.4. As seen in this figure, the complete computation of this algorithm consists of four stages. In general, an $N$-point FFT is computed in $\log_2 N$ stages, which for $N = 16$ results in four stages. Each stage requires $N/2$ computations of modules of the form shown in Figure 4.5. These modules are called butterfly modules. In stage $(r+1)$ the inputs of each butterfly module are components of the $X^r(k)$ array which are multiplied by $W_N^m$ factors. It is important to note that the distance between two inputs of each butterfly is constant at each stage but is different at various stages. Actually, at stage $r$, this difference is $2^{r-1}$. The output array of stage $\log_2 N$ is $X[k]$, which is the desired DFT, and the input array of the first stage is the input data sequence. But it is important to note that while $X^n[k]$ is $X[k]$ in the normal order from 0 to $N-1$, $X^0[k]$ is not equal to $x[n]$ in the normal order. In order to get the DFT in normal order, $x[n]$ has to be rearranged according to a rule called bit reversal. This rule states that if $N = 2^n$ then

$$X^0[k] = X^0[(k_n k_{n-1}\cdots k_3 k_2 k_1)_{\mathrm{base2}}] = x[(k_1 k_2 k_3 \cdots k_{n-1} k_n)_{\mathrm{base2}}] \tag{4.91}$$

For example, for $N = 16$ and $k = 2$ we have

$$X^0[2] = X^0[(0010)_{\mathrm{base2}}] = x[(0100)_{\mathrm{base2}}] = x[4] \tag{4.92}$$

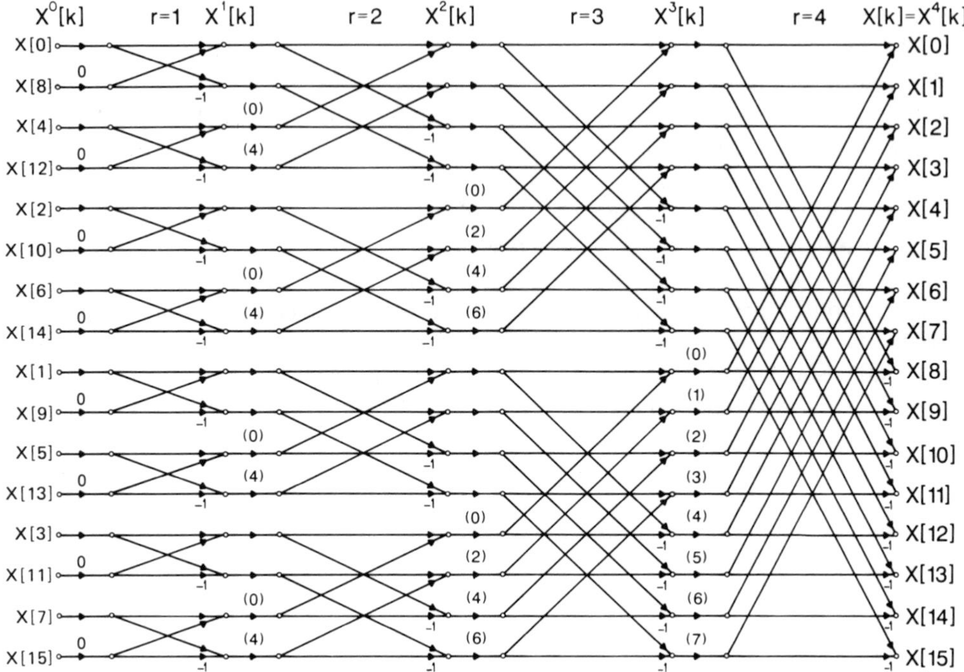


**Figure 4.4** Complete flow graph of the decimation-in-time radix-2 algorithm, for a 16-point DFT computation, with input in bit-reversed order and output in normal order.

and one can find the rest of $X^0[k]$ in the same way. Another property of the implementation in Figure 4.4 is that computation of the new array $X^{r+1}[k]$ from $X^r[k]$ does not require extra memory for storage because it can use the same memory location used for $X^r[k]$. This property, which is called "in-place computation" is evident in Figures 4.4 and 4.5. The inputs for the butterflies, $X^r[i]$ and $X^r[j]$ are used only for computation of $X^{r+1}[i]$ and $X^{r+1}[j]$, which have the same location in array $X^{r+1}[k]$ as the input samples have in the array $X^r[k]$. Therefore, as soon as $X^{r+1}[i]$ and $X^{r+1}[j]$ are computed, they can be stored in the same memory location

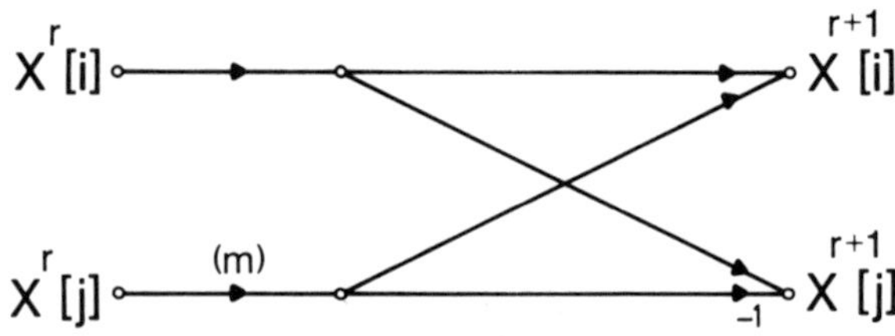


**Figure 4.5** Basic computational module of the decimation-in-time radix-2 algorithm, for an $N$-point DFT computation.

where $X^r[i]$ and $X^r[j]$ were stored. This destroys $X^r[i]$ and $X^r[j]$ but that is all right because this information is not required for the rest of the computation.

The implementation of Figure 4.4 was completely based on the development of the Cooley–Tukey approach for fast computation of the DFT, but, once obtained, the nodes of the flow graph can be rearranged in any arbitrary manner. This does not change the relation between input nodes $x[n]$ and output nodes $X[k]$ as long as the branches which connect nodes are not altered and the multiplier coefficient (transmittance) of each branch is not changed. Figure 4.6 shows an alternative form of Figure 4.4 where the input nodes $x[n]$ are in the normal order but the output nodes are bit reversed. This implementation also has the property of "in-place computation." Figure 4.7 is another alternative implementation where the input and output nodes are in normal order but the computation cannot be done in place and as a result requires $N$ more memory locations. Moreover, it does not have the regular pattern which is observed in Figures 4.4 and 4.6. All these alternative forms are equal from the point of view of computational requirements and their selection depends on the requirements of each application.

The Cooley–Tukey approach for computation of the DFT was based on the idea of decomposing the input data sequence into small sequences and then combining the DFTs of these small sequences to obtain the DFT of the original data sequence. In general, FFT algorithms which are based on decompositions of the input data sequence into small sequences are called decimation-in-time algorithms. Gentleman and Sande [10] proposed an alternative approach for computation of

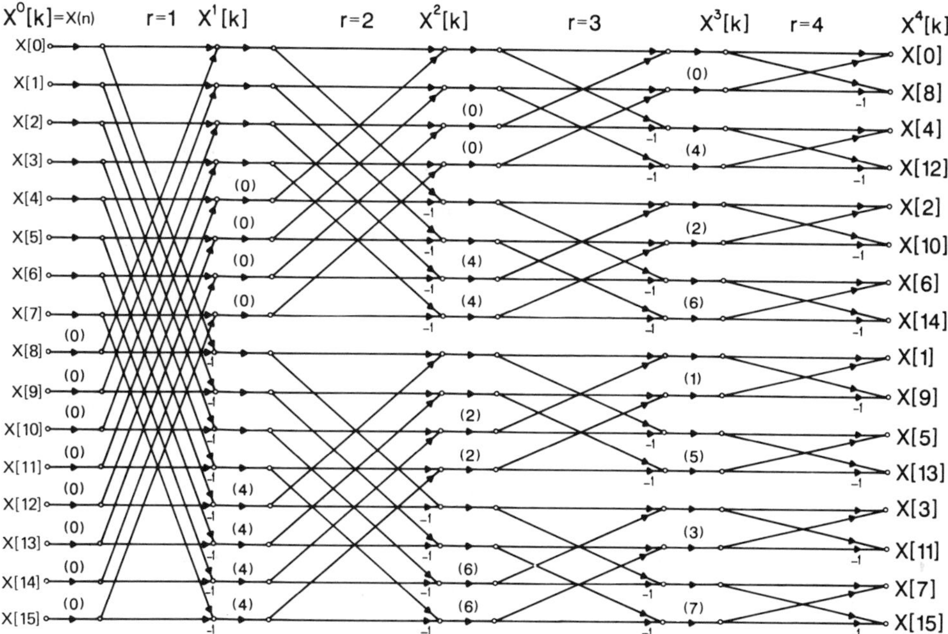


**Figure 4.6** Complete flow graph of the decimation-in-time radix-2 algorithm, for a 16-point DFT computation, with input in normal order and output in bit-reversed order.

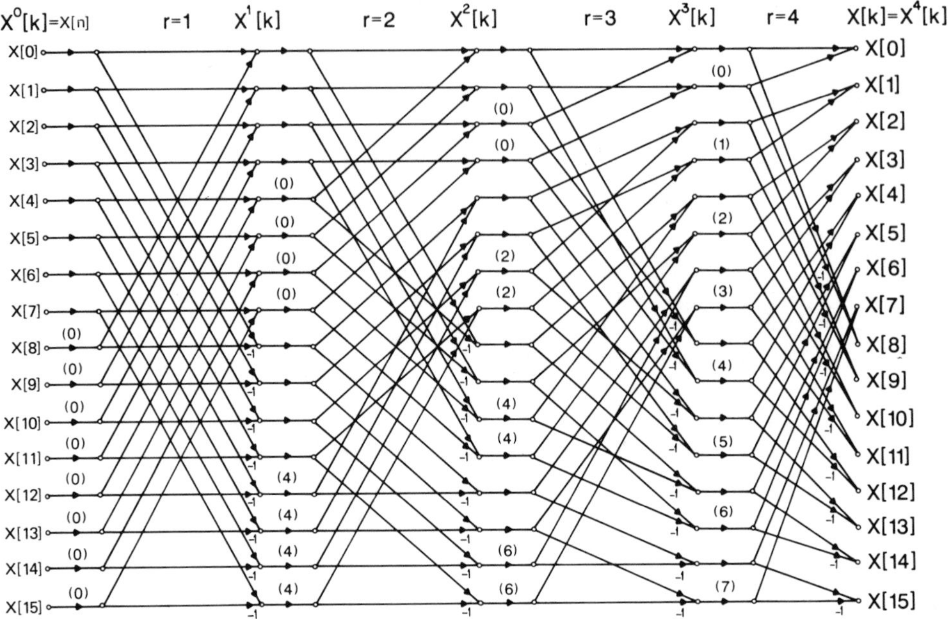


**Figure 4.7** Complete flow graph of the decimation-in-time radix-2 algorithm, for a 16-point DFT computation, with both input and output in normal order.

the DFT. Their idea was to decompose the DFT of the data sequence into smaller DFTs, but in a way that was different from the original Cooley–Tukey idea. Today FFT algorithms which are based on this idea are called decimation-in-frequency algorithms. In order to develop a decimation-in-frequency radix-2 algorithm, we use Eqs. (4.64), (4.65), and (4.68) and factor $N$ as $N/2 \times 2$; then we have

$$n = n_2 N/2 + n_1, \qquad 0 \le n_1 \le N/2 - 1,\ 0 \le n_2 \le 1 \tag{4.93}$$

$$k = 2k_1 + k_2, \qquad 0 \le k_1 \le N/2 - 1,\ 0 \le k_2 \le 1 \tag{4.94}$$

and

$$\begin{aligned} X[2k_1 + k_2] &= \sum_{n_1=0}^{N/2-1} \left[ \left( \sum_{n_2=0}^{1} x[n_2 N/2 + n_1] W_2^{k_2 n_2} \right) W_N^{k_2 n_1} \right] W_{N/2}^{k_1 n_1} \\ &= \sum_{n_1=0}^{N/2-1} \left[ \left( x[n_1] + (-1)^{k_2} x[N/2 + n_1] \right) W_N^{k_2 n_1} \right] W_{N/2}^{k_1 n_1} \end{aligned} \tag{4.95}$$

Now $X[2k_1 + k_2]$ can be divided into even-indexed and odd-indexed parts

$$X[2k_1] = \sum_{n_1=0}^{N/2-1} (x[n_1] + x[N/2 + n_1])\, W_{N/2}^{k_1 n_1} \tag{4.96}$$

$$X[2k_1 + 1] = \sum_{n_1=0}^{N/2-1} [(x[n_1] - x[N/2 + n_1])\, W_N^{n_1}]\, W_{N/2}^{k_1 n_1} \tag{4.97}$$

Implementation of Eqs. (4.96) and (4.97) is shown graphically in Figure 4.8. As this figure shows, this approach has converted an $N$-point DFT into two $N/2$-point DFTs at the expense of $N/2$ complex multiplications and $N$ complex additions. The same technique can now be applied to $N/2$-point DFTs, to convert them into four $N/4$-point DFTs at the expense of $N/2$ complex multiplications and $N$ complex additions. This technique can be applied successively until the DFTs are two-point DFTs. The total computational requirement will then be

$$M_c(N) = (N/2) \log_2 N \tag{4.98}$$

$$A_c(N) = N \log_2 N \tag{4.99}$$

This is exactly the same as the computational requirement for the decimation-in-time radix-2 algorithm. The flow graph for the computation of a 16-point DFT by a decimation-in-frequency radix-2 algorithm is shown in Figure 4.9 and the corresponding butterfly module is shown in Figure 4.10. As seen in these figures,

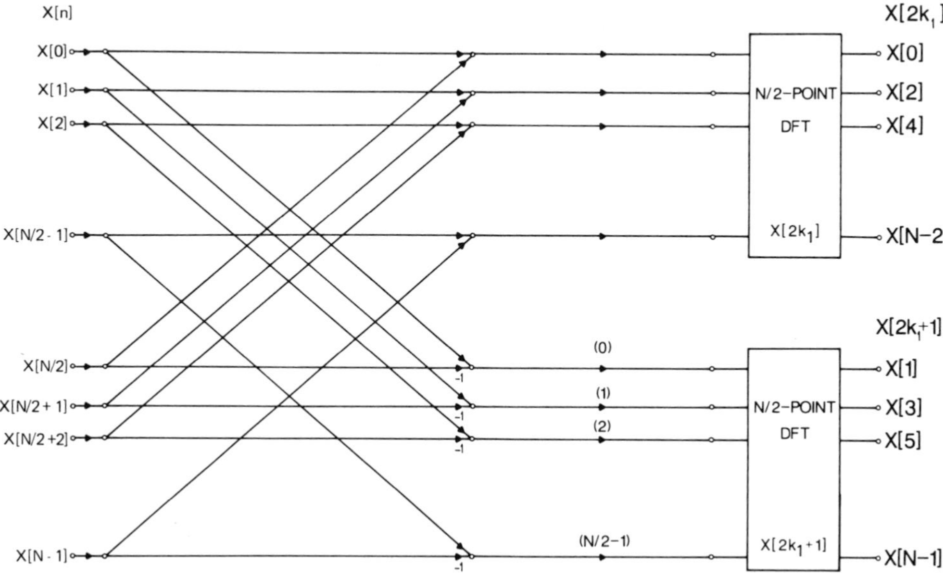


**Figure 4.8** Basic block diagram of the decimation-in-frequency radix-2 algorithm, for an $N$-point DFT computation.

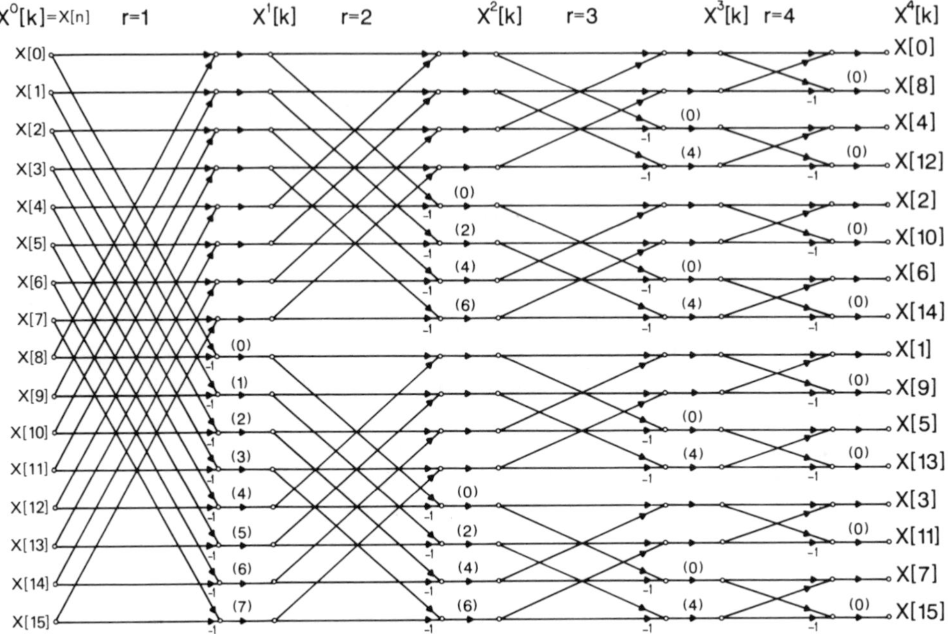


**Figure 4.9** Complete flow graph of the decimation-in-frequency radix-2 algorithm, for a 16-point DFT computation, with input in normal order and output in bit-reversed order.

the decimation-in-frequency radix-2 algorithm has a structure which is similar to the decimation-in-time radix-2 algorithm as shown in Figure 4.4. Actually, if we interchange the input and output nodes in Figure 4.9 and change the direction of all arrowheads, Figure 4.9 becomes an exact replica of Figure 4.4. In Figure 4.9 the input data sequence is in the normal order and the DFT sequence $X[k]$ is in bit-reversed order. As in the case with decimation-in-time algorithms, there can be many variations of the implementation in Figure 4.9 just by rearranging nodes of the flow graph. Figures 4.11 and 4.12 illustrate two of these alternatives which are more important than others. In Figure 4.11 the input data sequence is in bit-reversed order and the DFT sequence is in normal order, while in Figure 4.12 both

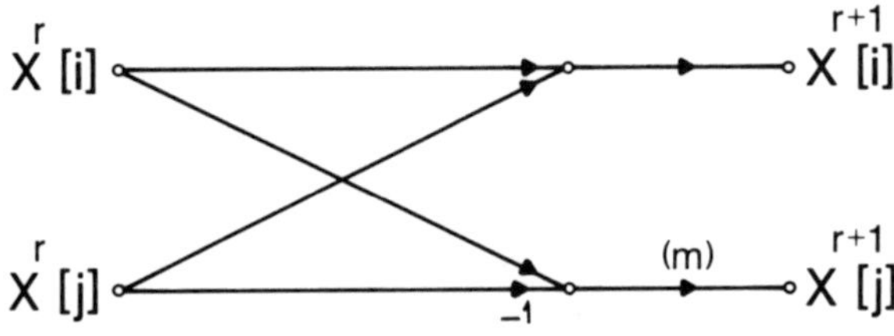


**Figure 4.10** Basic computational module of the decimation-in-frequency radix-2 algorithm, for an $N$-point DFT computation.

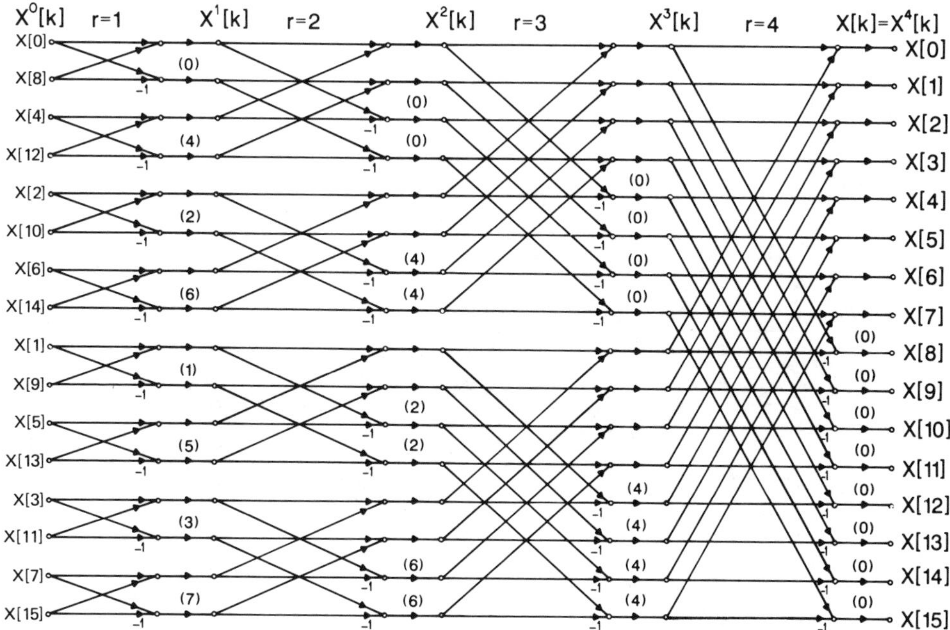


**Figure 4.11** Complete flow graph of the decimation-in-frequency radix-2 algorithm, for a 16-point DFT computation, with input in bit-reversed order and output in normal order.

input and $X[k]$ sequences are in normal order. As mentioned before, the selection of the appropriate implementation does actually depend on the requirements of each application.

## The Radix-4 Algorithm

The $N$-point Cooley–Tukey algorithm is referred to as the radix-4 algorithm when $N$ is a power of 4, that is, $N = 4^n$, and it is factored as $4 \times 4 \times 4 \times 4 \times \cdots \times 4$. In order to understand the implementation of the radix-4 algorithm, initially $N$ is factored as $4 \times N/4$, which means $N_1 = 4$ and $N_2 = N/4$. Now Eqs. (4.64), (4.65), and (4.68) can be written as

$$n = 4n_2 + n_1, \qquad 0 \le n_1 \le 3,\ 0 \le n_2 \le N/4 - 1 \tag{4.100}$$

$$k = k_1 N/4 + k_2, \qquad 0 \le k_1 \le 3,\ 0 \le k_2 \le N/4 - 1 \tag{4.101}$$

and

$$X[k_1 N/4 + k_2] = \sum_{n_1=0}^{3} \left[ \left( \sum_{n_2=0}^{N/4-1} x[4n_2 + n_1] W_{N/4}^{k_2 n_2} \right) W_N^{k_2 n_1} \right] W_4^{k_1 n_1}$$

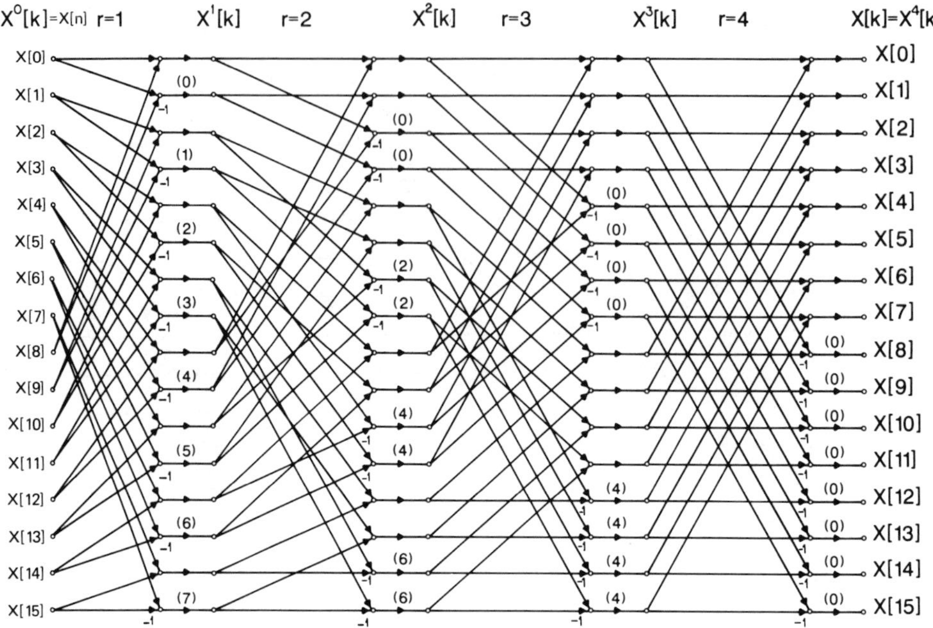


**Figure 4.12** Complete flow graph of the decimation-in-frequency radix-2 algorithm, for a 16-point DFT computation, with both input and output in normal order.

$$
\begin{aligned}
&= \sum_{n_2=0}^{N/4-1} x[4n_2] W_{N/4}^{k_2 n_2} + W_4^{k_1} W_N^{k_2} \sum_{n_2=0}^{N/4-1} x[4n_2+1] W_{N/4}^{k_2 n_2} \\
&\quad + W_4^{2k_1} W_N^{2k_2} \sum_{n_2=0}^{N/4-1} x[4n_2+2] W_{N/4}^{k_2 n_2} \\
&\quad + W_4^{3k_1} W_N^{3k_2} \sum_{n_2=0}^{N/4-1} x[4n_2+3] W_{N/4}^{k_2 n_2} \\
&= X_0'[k_2] + W_4^{k_1} W_N^{k_2} X_1'[k_2] + W_4^{2k_1} W_N^{2k_2} X_2'[k_2] \\
&\quad + W_4^{3k_1} W_N^{3k_2} X_3'[k_2]
\end{aligned}
\tag{4.102}
$$

where

$$
X_j'[k_2] = \sum_{n_2=0}^{N/4-1} x[4n_2+j] W_{N/4}^{k_2 n_2}, \qquad j = 0, 1, 2, 3 \tag{4.103}
$$

Equation (4.103) shows that $X_0'[k_2]$, $X_1'[k_2]$, $X_2'[k_2]$, and $X_3'[k_2]$ are $N/4$-point DFTs of the $N/4$-point sequences $x[4n_2]$, $x[4n_2+1]$, $x[4n_2+2]$, and $x[4n_2+3]$, respectively. Therefore, according to Eq. (4.102), the $N$-point DFT of the original data sequence can be expressed as a linear combination of four $N/4$-point DFTs. To better visualize the implementation of the radix-4 algorithm, Eq. (4.102) is written as

$$X[k_2] = X_0'[k_2] + W_N^{k_2} X_1'[k_2] + W_N^{2k_2} X_2'[k_2] + W_N^{3k_2} X_3'[k_2], \quad 0 \le k_2 \le N/4-1 \tag{4.104}$$

$$X[k_2+N/4] = X_0'[k_2] - jW_N^{k_2} X_1'[k_2] - W_N^{2k_2} X_2'[k_2] + jW_N^{3k_2} X_3'[k_2], \quad 0 \le k_2 \le N/4-1 \tag{4.105}$$

$$X[k_2+2N/4] = X_0'[k_2] - W_N^{k_2} X_1'[k_2] + W_N^{2k_2} X_2'[k_2] - W_N^{3k_2} X_3'[k_2], \quad 0 \le k_2 \le N/4-1 \tag{4.106}$$

$$X[k_2+3N/4] = X_0'[k_2] + jW_N^{k_2} X_1'[k_2] - W_N^{2k_2} X_2'[k_2] - jW_N^{3k_2} X_3'[k_2], \quad 0 \le k_2 \le N/4-1 \tag{4.107}$$

and its flow graph is shown in Figure 4.13. This figure indicates that the computational requirements of an $N$-point DFT is equal to the computational requirement of four $N/4$-point DFTs plus $3N/4$ complex multiplications and $3N$ complex addi-

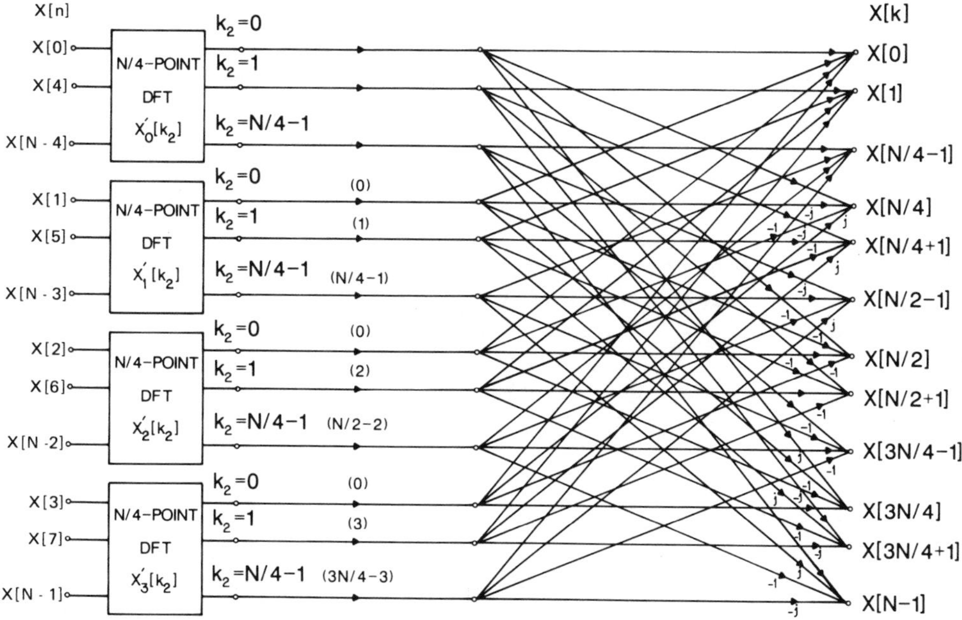


**Figure 4.13** Basic block diagram of the decimation-in-time radix-4 algorithm, for an $N$-point DFT computation.

tions, which means

$$M_c(N) = 4M_c(N/4) + 3N/4 \tag{4.108}$$

$$A_c(N) = 4A_c(N/4) + 3N \tag{4.109}$$

If we manipulate Eqs. (4.104)–(4.107) and write them down in the form

$$X[k_2] = \left(X_0'[k_2] + W_N^{2k_2} X_2'[k_2]\right) + \left(W_N^{k_2} X_1'[k_2] + W_N^{3k_2} X_3'[k_2]\right) \tag{4.110}$$

$$X[k_2 + N/4] = \left(X_0'[k_2] - W_N^{2k_2} X_2'[k_2]\right) - j\left(W_N^{k_2} X_1'[k_2] - W_N^{3k_2} X_3'[k_2]\right) \tag{4.111}$$

$$X[k_2 + 2N/4] = \left(X_0'[k_2] + W_N^{2k_2} X_2'[k_2]\right) - \left(W_N^{k_2} X_1'[k_2] + W_N^{3k_2} X_3'[k_2]\right) \tag{4.112}$$

$$X[k_2 + 3N/4] = \left(X_0'[k_2] - W_N^{2k_2} X_2'[k_2]\right) + j\left(W_N^{k_2} X_1'[k_2] - W_N^{3k_2} X_3'[k_2]\right) \tag{4.113}$$

the number of extra complex additions per stage is decreased from $3N$ to $2N$. This can be seen clearly from the butterfly module of the radix-4 algorithm, which is shown in Figure 4.14. Therefore we have

$$M_c(N) = 4M_c(N/4) + 3N/4 \tag{4.114}$$

$$A_c(N) = 4A_c(N/4) + 2N \tag{4.115}$$

Since $N/4 = 4^{n-1} = 4 \times N/16$, the four $N/4$-point DFTs can be then converted to 16 $N/16$-point DFTs at the expense of $3N/4$ extra complex multiplications and $2N$ extra complex additions. Therefore Eqs. (4.114) and (4.115) are modified to

$$M_c(N) = 16M_c(N/16) + 3N/4 + 3N/4 \tag{4.116}$$

$$A_c(N) = 16A_c(N/16) + 2N + 2N \tag{4.117}$$

This procedure can be continued $(\log_4 N - 1)$ times, where the last stage requires $N/4$ four-point DFTs and each stage requires $3N/4$ complex multiplications and $2N$ complex additions. As a result, we finally have

$$M_c(N) = (3N/4)\log_4 N = (3N/8)\log_2 N \tag{4.118}$$

$$A_c(N) = 2N\log_4 N = N\log_2 N \tag{4.119}$$

Comparison of these results with Eqs. (4.87) and (4.88) indicates that while the radix-4 algorithm requires the same number of additions as the radix-2 algorithm, it requires fewer multiplications, which means the radix-4 algorithm is more efficient than the radix-2 algorithm. This improvement is obtained at the expense of a little more complexity, which is involved in the program development of a radix-4 algorithm. As in the case of the radix-2 algorithm, multiplications by $(\pm 1)$, $(\pm j)$, and odd powers of

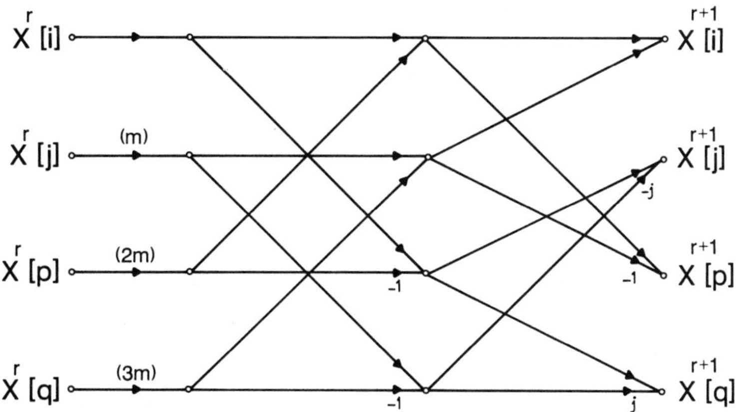


**Figure 4.14** Basic computational module of the decimation-in-time radix-4 algorithm, for an $N$-point DFT computation.

$W_8 = [(1 - j)\sqrt{2}]$ can be detected by the program and handled separately. If all this is done and complex multiplication is implemented by three real multiplications and three real additions, then the total number of real multiplications $M_r(N)$ and the total number of real additions $A_r(N)$ are

$$M_r(N) = (9N/8)\log_2 N - 43N/12 + 16/3 \tag{4.120}$$

$$A_r(N) = (25N/8)\log_2 N - 43N/12 + 16/3 \tag{4.121}$$

As seen in Figures 4.13 and 4.14, the computation of the radix-4 algorithm can also be done in place. Therefore, it needs only $N$ memory locations for storage of intermediate results. In the implementation shown in Figure 4.13, $X[k]$ is in normal order but the input data sequence is not in normal order. In a full implementation of the radix-4 algorithm which is obtained by successive application of the techniques in Figure 4.13, the input data sequence will be in digit-reversed order in base 4. This means that if $N = 4^n$, then $X^0[k]$, which is the rearranged array of the input sequence, is obtained by

$$X^0[k] = X^0[(k_n k_{n-1} \cdots k_3 k_2 k_1)_{\text{base4}}] = x[(k_1 k_2 k_3 \cdots k_{n-1} k_n)_{\text{base4}}] \tag{4.122}$$

Obviously, there are alternative forms of this implementation which can be obtained by rearranging nodes arbitrarily while keeping the connection between nodes and their transmittances fixed.

Our derivation of the radix-4 algorithm has been based on the decimation-in-time approach. The decimation-in-frequency approach can easily be derived by factoring $N$ as $N/4 \times 4$ and using Eqs. (4.64), (4.65), and (4.68). This results in

$$n = n_2 N/4 + n_1, \qquad 0 \le n_1 \le N/4 - 1,\ 0 \le n_2 \le 3 \tag{4.123}$$

$$k = 4k_1 + k_2, \qquad 0 \le k_1 \le N/4 - 1,\ 0 \le k_2 \le 3 \tag{4.124}$$

and

$$\begin{aligned} X[4k_1 + k_2] &= \sum_{n_1=0}^{N/4-1} \left[ \left( \sum_{n_2=0}^{3} x[n_2 N/4 + n_1] W_4^{k_2 n_2} \right) W_N^{k_2 n_1} \right] W_{N/4}^{k_1 n_1} \\ &= \sum_{n_1=0}^{N/4-1} \Big[ \Big( x[n_1] + W_4^{k_2} x[n_1 + N/4] + W_4^{2k_2} x[n_1 + 2N/4] \\ &\quad + W_4^{3k_2} x[n_1 + 3N/4] \Big) W_N^{k_2 n_1} \Big] W_{N/4}^{k_1 n_1} \end{aligned} \tag{4.125}$$

Equation (4.125) can be written in four parts for $k_2 = 0, 1, 2, 3$:

$$\begin{aligned} X[4k_1] = & \sum_{n_1=0}^{N/4-1} [(x[n_1] + x[n_1 + N/4] + x[n_1 + 2N/4] \\ & + x[n_1 + 3N/4])] W_{N/4}^{k_1 n_1} \end{aligned} \tag{4.126}$$

$$\begin{aligned} X[4k_1 + 1] = & \sum_{n_1=0}^{N/4-1} [(x[n_1] - jx[n_1 + N/4] - x[n_1 + 2N/4] \\ & + jx[n_1 + 3N/4]) W_N^{n_1}] W_{N/4}^{k_1 n_1} \end{aligned} \tag{4.127}$$

$$\begin{aligned} X[4k_1 + 2] = & \sum_{n_1=0}^{N/4-1} [(x[n_1] - x[n_1 + N/4] + x[n_1 + 2N/4] \\ & - x[n_1 + 3N/4]) W_N^{2n_1}] W_{N/4}^{k_1 n_1} \end{aligned} \tag{4.128}$$

$$\begin{aligned} X[4k_1 + 3] = & \sum_{n_1=0}^{N/4-1} [(x[n_1] + jx[n_1 + N/4] - x[n_1 + 2N/4] \\ & - jx[n_1 + 3N/4]) W_N^{3n_1}] W_{N/4}^{k_1 n_1} \end{aligned} \tag{4.129}$$

Implementation of Eqs. (4.126)–(4.129) is shown in Figure 4.15. As seen in this figure, the decimation-in-frequency approach is similar to the decimation-in-time approach; in the first step the $N$-point DFT is converted into four $N/4$-point DFTs at the expense of $3N/4$ complex multiplications and $2N$ complex additions. Successive application of Eqs. (4.126)–(4.129) will result in the decimation-in-frequency

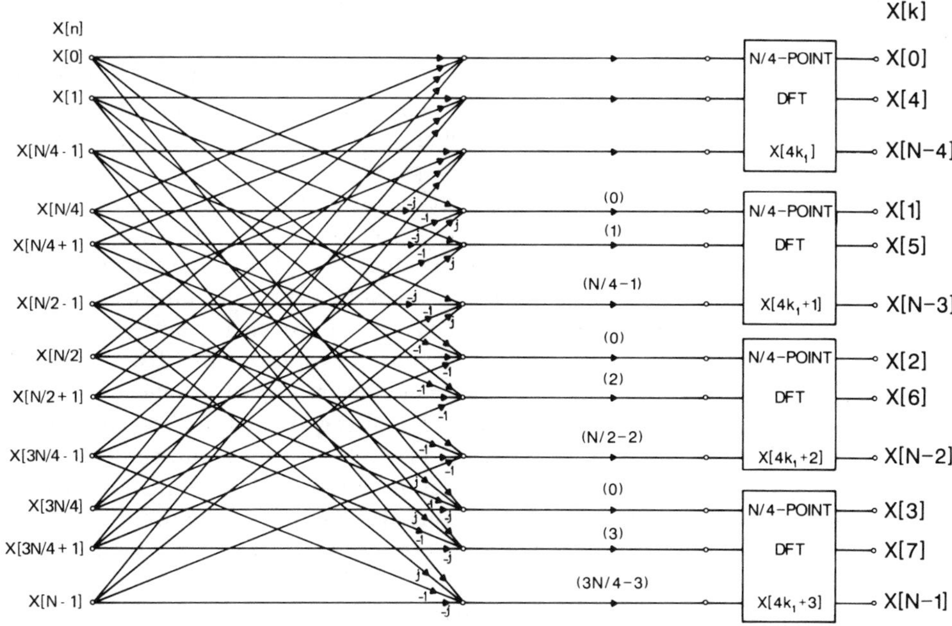


**Figure 4.15** Basic block diagram of the decimation-in-frequency radix-4 algorithm, for an $N$-point DFT computation.

radix-4 algorithm with computational requirements

$$M_c(N) = (3N/4)\log_4 N = (3N/8)\log_2 N \tag{4.130}$$

$$A_c(N) = (2N)\log_4 N = N\log_2 N \tag{4.131}$$

These are the same as the requirements for the decimation-in-time radix-4 algorithm. Basically there is not much difference between these two approaches, but one might find one of them more suitable for some specific applications.

## 3.2 The Winograd Algorithm

There are two distinct Winograd algorithms for fast computation of the discrete Fourier transform. The first is the Winograd short algorithm, which is feasible only for data sequences with a length less than approximately 20, and the second is the Winograd long algorithm, which has application for cases where the data sequence length is large. Because of the complexity of the Winograd algorithms, the details are not covered here, and only the general approach and properties of these algorithms are briefly discussed. For a more complete understanding of the Winograd algorithms, references [19, 20, 22, 24, 42–45] can be studied.

The Winograd short algorithm is based on two ideas:

1. Conversion of the DFT to a circular convolution

2. Fast computation of the circular convolution using the Chinese remainder theorem for polynomial products

The idea of converting the DFT to a circular convolution was initially suggested by Rader [13] for data sequences where the length $N$ is a prime integer. Later, Winograd [20] and McClellan and Rader [42] showed that the Rader approach can be extended to cases where $N$ is a power of a prime integer, that is $N = P^r$. The Rader algorithm did not receive much attention until 1975, when Winograd [19] developed a technique for fast computation of a circular convolution. Winograd represented the convolution operation as a product of two polynomials and used the Chinese remainder theorem to break the polynomial product into products of low-order polynomials. Winograd showed that his approach actually minimized the number of required multiplications for circular convolution. In 1976 Winograd integrated the Rader algorithm and his fast convolution algorithm and developed the Winograd short algorithm [20]. Computation of the Winograd short algorithm consists of a set of additions, followed by a set of multiplications, and then followed by another set of additions. This algorithms requires the minimum number of multiplications among all FFT algorithms, but the number of required additions grows rapidly as the data sequence length increases. Therefore, the algorithm is not recommended for $N$ greater than 20.

The Winograd long algorithm, which is feasible for large data sequences, is based on three ideas:

1. Decomposition of the large DFT into smaller DFTs
2. The Winograd short algorithm
3. Nesting of the short algorithms

Decomposition of the large DFT is done by the Good–Thomas indexing scheme [8, 9, 18]. Good and Thomas showed that if $N$ can be factored as $N = N_1 N_2$, where $N_1$ and $N_2$ are relatively prime integers, then the DFT of the sequence $x[n]$ can be decomposed as

$$X[k_1, k_2] = \sum_{n_1=0}^{N_1-1} \left( \sum_{n_2=0}^{N_2-1} x[n_1, n_2] W_{N_2}^{k_2 n_2} \right) W_{N_1}^{k_1 n_1} \tag{4.132}$$

where $[k_1, k_2]$ and $[n_1, n_2]$ are related to $k$ and $n$, respectively. This equation is similar to Eq. (4.68) without the twiddle factors, and its computation requires $N_1$ $N_2$-point DFT computations and $N_2$ $N_1$-point DFT computations. If $N_1$ or/and $N_2$ can be again factored as a product of relatively prime integers, then the Good–Thomas indexing scheme can again be used to decompose Eq. (4.132) into even smaller DFTs. This decomposition process can be continued until the DFTs are small enough to justify the application of Winograd short algorithms for their computation. The final step of the Winograd long algorithm is integration of all multiplications in short algorithms together by a nesting algorithm, which results in further reduction of the total number of required multiplications. In the final form, computation of the long algorithm consists of a set of additions called preweave additions, followed by a set of multiplications, and then followed by another set of additions called postweave additions.

**Table 4.2** Number of Required Real Multiplications for Computation of Discrete Fourier Transform by Different FFT Algorithms

| DFT Length $N$ | Cooley–Tukey algorithm Radix-2 | Radix-4 | Radix-8 | Winograd algorithm | Prime factor algorithm | Split-radix algorithm |
|---|---|---|---|---|---|---|
| 63 | — | — | — | 198 | 284 | — |
| 64 | 264 | 208 | 204 | — | — | 196 |
| 126 | — | — | — | 396 | 568 | — |
| 128 | 712 | — | — | — | — | 516 |
| 252 | — | — | — | 792 | 1,136 | — |
| 256 | 1,800 | 1392 | — | — | — | 1,284 |
| 504 | — | — | — | 1584 | 2,524 | — |
| 512 | 4,360 | — | 3204 | — | — | 3,076 |
| 1008 | — | — | — | 3564 | 5,804 | — |
| 1024 | 10,248 | 7856 | — | — | — | 7,172 |
| 2048 | 23,560 | — | — | — | — | 16,388 |
| 2520 | — | — | — | 9504 | 17,660 | — |

Tables 4.2 and 4.3 list the numbers of required real multiplications and additions for four of the more promising FFT algorithms. As seen in these tables, the Winograd algorithm requires the minimum number of multiplications, whereas the number of required additions is somewhat greater than for the other algorithms. Actually, the only advantage of Winograd algorithm is that it is optimal with respect to the number of required multiplications. This means that the algorithm has the potential to be the fastest FFT algorithm if the implementation is on a computer

**Table 4.3** Number of Required Real Additions for Computation of Discrete Fourier Transform by Different FFT Algorithms

| DFT Length $N$ | Cooley–Tukey algorithm Radix-2 | Radix-4 | Radix-8 | Winograd algorithm | Prime factor algorithm | Split-radix algorithm |
|---|---|---|---|---|---|---|
| 63 | — | — | — | 1394 | 1236 | — |
| 64 | 1032 | 976 | 972 | — | — | 964 |
| 126 | — | — | — | 3,040 | 2,724 | — |
| 128 | 2,504 | — | — | — | — | 2,308 |
| 252 | — | — | — | 6,584 | 5,952 | — |
| 256 | 5,896 | 5,488 | — | — | — | 5,380 |
| 504 | — | — | — | 14,428 | 13,164 | — |
| 512 | 13,576 | — | 12,420 | — | — | 12,292 |
| 1008 | — | — | — | 34,416 | 29,100 | — |
| 1024 | 30,728 | 28,336 | — | — | — | 27,652 |
| 2048 | 68,616 | — | — | — | — | 61,444 |
| 2520 | — | — | — | 99,068 | 82,956 | — |

where multiplication requires much more time than addition or data transfer. But it should also be noted that the Winograd algorithm has the following drawbacks:

1. It requires a great deal of data transfer.
2. In general, it requires more additions than other algorithms.
3. In general, it requires more program memory than other algorithms.
4. It is very complex and difficult to understand.
5. It is not the fastest FFT algorithm on modern computers where multiplication is not too costly.

## 3.3 The Prime Factor Algorithm

The prime factor algorithm is, to some extent, similar to the Winograd long algorithm. In the prime factor algorithm also, the large DFT is broken into smaller DFTs by the Good–Thomas indexing scheme and then the small DFTs are computed by the Winograd short algorithm. The difference is that the prime factor algorithm does not use the Winograd nesting technique but rather computes Eq. (4.132) by the conventional row-column technique. As a result, as shown in Table 4.2, the number of required multiplications is more than in the Winograd long algorithm. Although this is a drawback of the prime factor algorithm, the algorithm has the following advantages over the Winograd long algorithm:

1. It requires less data transfer.
2. It requires less program and data memory.
3. The total number of required multiplications and additions is less (Tables 4.2 and 4.3).
4. It is less complex and easier to understand.

It is because of these advantages that the prime factor algorithm is more attractive than the Winograd long algorithm. But in comparison with the split-radix algorithm, which is discussed next, the prime factor algorithm has the following drawbacks:

1. It requires more program memory.
2. The general-purpose version of it requires more data memory.
3. The total number of required multiplications and additions is greater (Tables 4.2 and 4.3).
4. It is more complex and more difficult to understand.

## 3.4 The Split-Radix Algorithm

The split-radix algorithm is essentially based on the ideas of the radix-2 and radix-4 algorithms [14, 39, 40], but it should be mentioned that it is different from the mixed-radix algorithm which was developed by Singleton [16]. In the mixed-radix algorithm, each stage of the algorithm is computed by a specific radix algorithm which is not necessarily the same for all stages. But in the split-radix algorithm a mixture of the radix-2 and radix-4 algorithms is used for all stages. The basic idea in the split-radix algorithm is to use the radix-2 algorithm for the even-indexed terms and the radix-4 algorithm for the odd-indexed terms. The decimation-in-frequency version of the split-radix algorithm can be obtained by using Eqs. (4.96), (4.127), and (4.129), which results in

$$X[2k_1] = \sum_{n_1=0}^{N/2-1} (x[n_1] + x[n_1 + N/2])W_{N/2}^{k_1 n_1}, \quad 0 \le k_1 \le N/2 - 1 \tag{4.133}$$

$$\begin{aligned} X[4k_1 + 1] = &\sum_{n_1=0}^{N/4-1} \{[(x[n_1] - x[n_1 + N/2]) - j(x[n_1 + N/4] \\ &- x[n_1 + 3N/4])]W_N^{n_1}\}W_{N/4}^{k_1 n_1}, \quad 0 \le k_1 \le N/4 - 1 \end{aligned} \tag{4.134}$$

$$\begin{aligned} X[4k_1 + 3] = &\sum_{n_1=0}^{N/4-1} \{[(x[n_1] - x[n_1 + N/2]) + j(x[n_1 + N/4] \\ &- x[n_1 + 3N/4])]W_N^{3n_1}\}W_{N/4}^{k_1 n_1}, \quad 0 \le k_1 \le N/4 - 1 \end{aligned} \tag{4.135}$$

These equations, which are for the first stage of the decimation-in-frequency split-radix algorithm, indicate that the $N$-point DFT is replaced by one $N/2$-point DFT and two $N/4$-point DFTs at the expense of $N/2$ complex multiplications. Figure 4.16 illustrates this decomposition and Figure 4.17 shows the computational

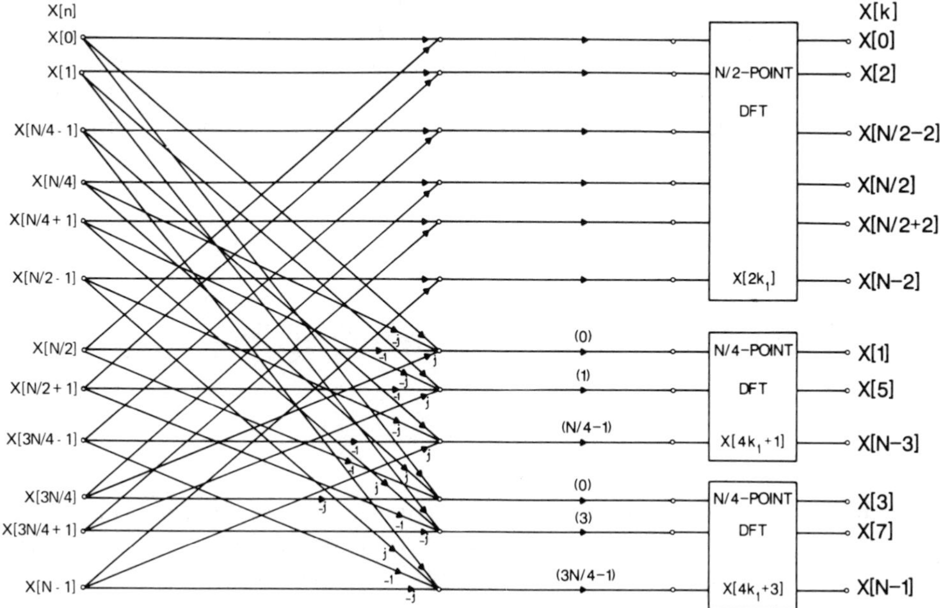


**Figure 4.16** Basic block diagram of the decimation-in-frequency split-radix algorithm, for an $N$-point DFT computation.

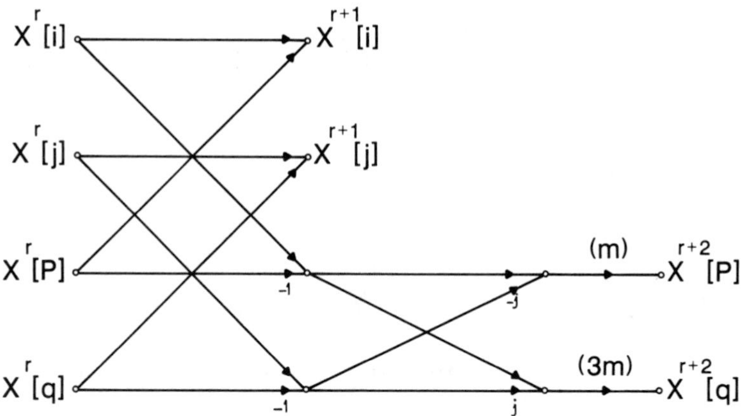


**Figure 4.17** Basic computational module of the decimation-in-frequency split-radix algorithm.

module of the split-radix algorithm. This decomposition is used repeatedly until the last stage, where two-point DFTs are encountered which are computed by radix-2 butterflies. Form this argument it should be clear that the split-radix algorithm is applicable where the data sequence length is a power of two, that is, $N = 2^n$.

As seen in Figure 4.16, from the required $N/2$ multiplications per stage, two are multiplications by $W_N^0$ and two are multiplications by odd powers of $W_8$. Therefore at each stage there are $(N/2 - 4)$ general complex multiplications, which can be done by $3(N/2 - 4)$ real multiplications and $3(N/2 - 4)$ real additions. The two multiplications by $W_N^0$ can be detected and avoided. The multiplications by odd powers of $W_8$ can also be detected and performed by two multiplications and two additions. Assuming that we utilize all these options, the total number of required real multiplications $M_r(N)$ will be

$$M_r(N) = M_r(N/2) + 2M_r(N/4) + 3(N/2 - 4) + 4 \tag{4.136}$$

and knowing that $M_r(2) = M_r(4) = 0$, the final result will be

$$M_r(N) = N(\log_2 N - 3) + 4 \tag{4.137}$$

The number of required real additions for computing the required multiplications is also $M_r(N)$. Moreover, there are $N \log_2 N$ complex additions in the complete flow graph, where each requires two real additions; therefore the total number of required real additions $A_r(N)$ will be

$$A_r(N) = 2N \log_2 N + M_r(N) = 3N(\log_2 N - 1) + 4 \tag{4.138}$$

$M_r(N)$ and $A_r(N)$ are listed in Tables 4.2 and 4.3. Although the number of required multiplications is more than that required by the Winograd algorithm or the prime factor algorithm, it is minimum among $2^n$ algorithms including various radix versions of the Cooley–Tukey algorithm. Moreover, in general, the total number of required multiplications and additions is less than in other algorithms, as shown in Tables 4.2 and 4.3. Therefore it has the potential to be the fastest FFT algorithm, if implemented on computers where multiplication is not too costly. It should also be noted that this is not the only promising feature of the split-radix algorithm, and in comparison with the Cooley–Tukey algorithm, the Winograd algorithm (WA), and the prime factor algorithm (PFA) it has other advantages such as:

1. It is more efficient than all Cooley–Tukey algorithms.
2. It requires less program memory than the WA and the PFA.
3. In general it requires less data memory than the WA and the PFA.
4. It has the flexibility of the radix-2 algorithm.
5. It has the possibility of "in-place implementation" for real and symmetric data with reduced arithmetic complexity.
6. It has a very regular structure which can easily be understood.

## 4 CONCLUSION

This chapter has provided a discussion of the discrete Fourier transform and its efficient implementation. The more promising FFT algorithms were found to be the Cooley–Tukey algorithm, the Winograd algorithm, the prime factor algorithm, and the split-radix algorithm. All of these algorithms offer features which are very attractive when the algorithm is implemented on certain classes of computers, but none of them is the best for implementation on all classes of computers [38]. The selection of the appropriate FFT algorithm for a specific application depends on the requirements of the algorithm; and many parameters such as a computational speed, memory requirements, complexity, generality, development time, ease of testing, and accuracy should be evaluated before the final selection. Having all this in mind, it is our conclusion that the split-radix algorithm seems to offer more promising features than other algorithms.

## REFERENCES


1. A. V. Oppenheim and R. W. Schafer, *Digital Signal Processing*, Prentice-Hall, Englewood Cliffs, New Jersey (1975).
2. L. R. Rabiner and B. Gold, *Theory and Application of Digital Signal Processing*, Prentice-Hall, Englewood Cliffs, New Jersey (1975).
3. A. Papoulis, *Signal Analysis*, McGraw-Hill, New York (1977).
4. A. V. Oppenheim, A. S. Willsky, and I. T. Young, *Signals and Systems*, Prentice-Hall, Englewood Cliffs, New Jersey (1983).
5. J. A. Cadzow, *Foundations of Digital Signal Processing and Data Analysis*, Macmillan, New York (1987).

6. E. O. Brigham, *The Fast Fourier Transform*, Prentice-Hall, Englewood Cliffs, New Jersey (1974).
7. J. W. Cooley and J. W. Tukey, An algorithm for the machine calculation of complex Fourier series, *Math. Comput.*, *19*: 297–301 (April 1965).
8. I. J. Good, The interaction algorithm and practical Fourier analysis, *J. R. Stat. Soc. Ser. B*, *20*: 361–372 (1958); Addendum, *ibid.* *22*: 372–375 (1960).
9. L. H. Thomas, Using a computer to solve problems in physics, *Applications of Digital Computers*, Ginn, Boston (1963).
10. W. M. Gentleman and G. Sande, "Fast Fourier Transform—for Fun and Profit," 1966 Fall Joint Computer Conf., *AFIPS Conf. Proc.*, vol. 29, pp. 563–578, Spartan Books, Washington, D. C. (1966).
11. J. W. Cooley, P. A. W. Lewis, and P. D. Welch, Historical notes on the fast Fourier transform, *IEEE Trans. Electroacoust.*, *AU-15*: 76–79 (June 1967).
12. G. D. Bergland, A fast Fourier transform algorithm using base 8 iterations, *Math. Comput.*, *22*: 275–279 (April 1968).
13. C. M. Rader, Discrete Fourier transforms when the number of data samples is prime, *Proc. IEEE*, *56*: 1107–1108 (June 1968).
14. R. Yavne, "An Economical Method for Calculating the Discrete Fourier Transform," 1968 Fall Joint Computer Conf. *AFIPS Conf. Proc.*, vol. 33, pp. 115–125 (1968).
15. L. R. Rabiner, R. W. Schafer, and C. M. Rader, The chirp $z$-transform algorithm, *IEEE Trans. Audio Electroacoust.*, *AU-17*: 86–92 (June 1969).
16. R. C. Singleton, An algorithm for computing the mixed radix fast Fourier transform, *IEEE Trans. Audio Electroacoust.*, *AU-17*: 93–103 (June 1969).
17. L. I. Bluestein, A linear filtering approach to the computation of discrete Fourier transform, *IEEE Trans. Audio Electroacoust.*, *AU-18*: 451–455 (December 1970).
18. I. J. Good, The relationship between two fast Fourier transforms, *IEEE Trans. Comput.*, *C-20*: 310–317 (March 1971).
19. S. Winograd, "The Effect of the Field of Constants on the Numbers of Multiplications," Proc. 16th Annu. Symp. on Foundations of Computer Science, pp. 1–2 (1975).
20. S. Winograd, On computing the discrete Fourier transform, *Proc. Natl. Acad. Sci. U.S.A.*, *73*: 1005–1006 (April 1976).
21. C. M. Rader and N. M. Brenner, A new principle for fast Fourier transformation, *IEEE Trans. Acoust. Speech Signal Process.*, *ASSP-24*: 264–266 (June 1976).
22. H. F. Silverman, An introduction to programming the Winograd Fourier transform algorithm, *IEEE Trans. Acoust. Speech Signal Process.*, *ASSP-25*: 152–165 (April 1977).
23. D. P. Kolba and T. W. Parks, A prime factor FFT algorithm using high speed convolution, *IEEE Trans. Acoust. Speech Signal Process.*, *ASSP-25*: 281–294 (August 1977).
24. S. Winograd, On computing the discrete Fourier transform, *Math. Comput.*, 32: 175–199 (January 1978).
25. G. Brunn, $z$-Transform DFT filters and FFT's, *IEEE Trans. Acoust. Speech Signal Process.*, *ASSP-26*: 56–63 (February 1978).

26. L. R. Morris, A comparative study of time efficient FFT and WFTA programs for general purpose computers, *IEEE Trans. Acoust. Speech Signal Process., ASSP-26*: 141–150 (April 1978).
27. E. Dubois and A. N. Venetsanopoulos, A new algorithm for the radix-3 FFT, *IEEE Trans. Acoust. Speech Signal Process., ASSP-26*: 222–225 (June 1978).
28. Y. Tadokoro and T. Higuchi, Discrete Fourier transform computation via the Walsh transform, *IEEE Trans. Acoust. Speech Signal Process., ASSP-26*: 236–240 (June 1978).
29. H. J. Nussbaumer and P. Quandalle, Fast computation of discrete Fourier transforms using polynomial transforms, *IEEE Trans. Acoust. Speech Signal Process., ASSP-27*: 169–181 (April 1979).
30. H. Nawab, J. H. McClellan, Bounds on the minimum numbers of data transfers in WFTA and FFT programs, *IEEE Trans. Acoust. Speech Signal Process., ASSP-27*: 394–398 (August 1979).
31. S. Prakash and V. V. Rao, A new radix-6 FFT algorithm, *IEEE Trans. Acoust. Speech Signal Process., ASSP-29*: 939–941 (August 1981).
32. C. S. Burrus and P. W. Eschenbacher, An in-place, in-order prime factor FFT algorithm, *IEEE Trans. Acoust. Speech Signal Process, ASSP-29*: 806–817 (August 1981).
33. S. Chu and C. S. Burrus, A prime factor algorithm using distributed arithmetic, *IEEE Trans. Acoust. Speech Signal Process., ASSP-30*: 217–227 (April 1982).
34. R. D. Preuss, Very fast computation of the radix-2 discrete Fourier transform, *IEEE Trans. Acoust. Speech Signal Process., ASSP-30*: 595–607 (August 1982).
35. H. W. Johnson and C. S. Burrus, The design of optimal DFT algorithm using dynamic programming, *IEEE Trans. Acoust. Speech Signal Process., ASSP-31*: 378–387 (April 1983).
36. J. B. Martens, Recursive cyclotomic factorization—a new algorithm for calculating the discrete Fourier transform, *IEEE Trans. Acoust. Speech Signal Process., ASSP-32*: 750–761 (August 1984).
37. Z. Wang, Fast algorithms for the discrete W transform and for the discrete Fourier transform, *IEEE Trans. Acoust. Speech Signal Process., ASSP-32*: 803–816 (August 1984).
38. M. A. Mehalic, P. L. Rustan, and G. P. Route, Effects of architecture implementation on DFT algorithm performance, *IEEE Trans. Acoust. Speech Signal Process., ASSP-33*: 684–693 (June 1985).
39. H. V. Sorensen, M. T. Heideman, and C. S. Burrus, On computing the split-radix FFT, *IEEE Trans. Acoust. Speech Signal Process., ASSP-34*: 152–156 (February 1986).
40. P. Duhamel, Implementation of split-radix FFT algorithms for complex, real, and real-symmetric data, *IEEE Trans. Acoust. Speech Signal Process., ASSP-34*: 285–295 (April 1986).
41. Y. Suzuki, T. Sone, and K. Kido, A new FFT algorithm of radix-3, 6, and 12, *IEEE Trans. Acoust. Speech Signal Process., ASSP-34*: 380–383 (April 1986).
42. J. H. McClellan and C. M. Rader, *Number Theory in Digital Signal Processing*, Prentice-Hall, Englewood Cliffs, New Jersey (1979).

43. H. J. Nussbaumer, *Fast Fourier Transform and Convolution Algorithms*, Springer-Verlag, New York (1982).
44. R. E. Blahut, *Fast Algorithms for Digital Signal Processing*, Addison-Wesley, Reading, Massachusetts (1985).
45. C. S. Burrus and T. W. Park, *DFT/FFT and Convolution Algorithms*, Wiley, New York (1985).

# 5

# High-Resolution Spectral Estimation Algorithms

C. H. CHEN Electrical and Computer Engineering Department, Southeastern Massachusetts University, North Dartmouth, Massachusetts

## 1 INTRODUCTION

Fourier spectral analysis with the use of fast Fourier transform (FFT) algorithms as discussed in Chapter 4 for discrete data sets has been the major tool for spectral estimation for the past 20 years. The availability of FFT software (even for cases where the number of data points is not a power of 2) and hardware is mainly responsible for such progress. Fourier-based spectral estimation includes two main nonparametric methods. The first is the periodogram method, which has the drawbacks of limited resolution and a variance of the spectral estimate that grows with increasing data length. Windowing (smoothing) techniques may be used to improve the situation. Welch's modified periodogram is among the best in the periodogram methods of spectral estimation. The assumption of independent segments, however, is not valid in practice. The second Fourier-based method is the Blackman–Tukey method. It takes the Fourier transform of the estimated autocorrelation function as the power spectrum. It is still the most popular nonparametric method, but again it suffers the drawback of limited resolution capability with short data length.

The resolution of the discrete Fourier transform is $1/N$, where $N$ is the number of data points. This is quite inadequate for small $N$. Modern spectral estimation techniques offer the possibility of enhanced spectral resolution and overall improvement in the spectrum for small to moderate $N$, with various amounts of added computational complexity. These are called collectively high-resolution spectral estimation techniques. A partial list of these power spectral density (PSD) estimation techniques (see, e.g., [1] includes the maximum likelihood spectral estimate, autoregressive PSD via the Yule–Walker approach, autoregressive PSD via the Burg algorithm, autoregressive PSD via the least-squares method, and Pisaranko's method. Most of the high-resolution spectral estimation methods are parametric, as parametric models are assumed for the signal process considered. A comparison of some of these methods is presented in this chapter.

## 2 THE MESA AND RELATED APPROACHES

For a finite record of short length, the autocorrelation values available are limited. An important parametric approach called maximum entropy spectral analysis (MESA) extrapolates the unknown autocorrelation values without zero padding. The maximum entropy method is thus least committal with respect to the unknown autocorrelation values, as its choice for these values is one that adds no information or entropy. Mathematically, this means setting the partial derivatives of the entropy rate with respect to the unknown autocorrelation values to zero, subject to the constraint that the autocorrelation values obtained from the estimated PSD are consistent with the known values. It turns out that the resulting spectral density is the same as that of an autoregressive (AR) process.

Among the autoregressive methods, Burg's procedure [2, 3] is one of the best-performing AR modeling techniques. A Levinson recursion constraint is imposed to give the method a data-adaptive capability. The technique provides superior resolution with respect to the FFT with short record lengths. The disadvantages of Burg's technique are bias in the location of spectral peaks, linesplitting for sine waves at high signal-to-noise ratio, and frequency shifting for sine waves at low signal-to-noise ratio.

Another procedure which makes use of the covariance method in AR modeling has been widely used in linear predictive coding of speech. It is computationally simpler than Burg's technique. Its resolution is inferior to that of the Burg method but is superior to the periodogram and Blackman–Tukey methods.

The autoregressive moving average (ARMA) method of spectral modeling is theoretically preferable to but harder to compute than AR models. The ARMA model requires fewer parameters than the AR model, but the AR model may still be preferred because of its simpler estimation algorithm.

Other maximum entropy-related spectral analysis procedures are the tapered Burg method, the nonlinear maximum entropy spectral analysis method [4], etc., which will be discussed in detail in this chapter.

## 3 BURG'S ALGORITHM

Given a stationary time series $x_1, x_2, \ldots, x_N$, the sample autocorrelation function $R_x(k)$ can be computed as

$$R_x(k) = \frac{1}{N} \sum_{i=1}^{N-|k|} x_i x_{i+k}^* \tag{5.1}$$

where the $*$ denotes the complex conjugate, as the observations in general are complex. The Fourier transform of $R_x(k)$ is an estimate of the power spectral density $S_x(f)$. Because $N$ is finite, only a finite number of autocorrelation values are known. The maximum entropy spectral estimate is one that maximizes the entropy rate

$$H = \frac{1}{2B} \int_{-B}^{B} \log S_x(f)\, df$$

with respect to the unknown $R_x(k)$ values. Here $B$ is the signal bandwidth corresponding to a sampling period of $T = 1/2B$. The resulting maximum entropy spectrum takes the form

$$\hat{S}_x(f) = \frac{P_m}{B|1 + \sum a_k \exp(-j2\pi f k T)|^2} \tag{5.2}$$

where $P_m$ is called the final prediction error. Equation (5.2) indicates that the maximum entropy spectrum is the same as that of an autoregressive model with coefficients $a_i$. Burg's algorithm estimates $a_i$ by using forward and backward recursion, as outlined in the following.

Let the linear prediction estimate of $x_t$ be given by $\hat{x}_t$.

$$\hat{x}_t = -\sum_{k=1}^{P} a_k x_{t-k}$$

for an $m$th-order AR model. The prediction error, i.e., the output of the prediction error filter, is $e_t = x_t - \hat{x}_t$. The final prediction error power is

$$P_p = \sum_{k=0}^{p} a_k R_x(-k), \qquad 1 \le p \le m \tag{5.3a}$$

with a recursive relation

$$P_p = P_{p-1}(1 - |r_p|^2) \tag{5.3b}$$

where $r_m$ is called the reflection coefficient. The Levinson recursion algorithm is given by

$$a_k^{(p)} = a_k^{(p-1)} + a_p^{(p)} a_{p-k}^{(p-1)*}, \qquad r_p = g_p^{(p)} \tag{5.4}$$

where the superscript denotes the order of the AR model considered. Define the forward and backward prediction errors as

$$e_{f,t}^{(p)} = \sum_{k=0}^{p} a_k^{(p)} x_{t-k} \tag{5.5a}$$

$$e_{b,t}^{(p)} = \sum_{k=0}^{p} a_{p-k}^{(p)*} x_{t+k-p} \tag{5.5b}$$

For the prediction error filter to be minimum phase, Burg has shown that the reflection coefficients must all be $|r_p| < 1$. This is also the condition for the filter to be stable. To guarantee the stability and thus minimum phase, Burg suggested minimizing the average prediction error power,

$$P_p = \frac{1}{2}(P_{f,p} + P_{b,p}) \tag{5.6}$$

where

$$P_{f,p} = (N-p)^{-1} \sum_{t=1}^{N-p} e_{f,t}^{(p)} e_{f,t}^{(p)*}$$

$$P_{b,p} = (N-p)^{-1} \sum_{t=1}^{N-p} e_{b,t}^{(p)} e_{b,t}^{(p)*}$$

Burg's algorithm starts with $p = 0$, for which

$$P_0 = R_x(0) = \frac{1}{N} \sum_{t=1}^{N} x_t x_t^*$$

and then computes $P_1$ from Eq. (5.6). The value of $r_1$ for which $P_1$ is a minimum is then determined by solving $\partial P_1 / \partial r_1 = 0$. Continuing this process for higher integer values of $m$, we can deduce a general expression for the reflection coefficient, given by

$$r_p = \frac{-2 \sum_{t=p+1}^{N} e_{f,t}^{(p-1)} e_{b,t-1}^{(p-1)*}}{\sum_{t=p+1}^{N} \left( |e_{f,t}^{(p-1)}|^2 + |e_{b,t-1}^{(p-1)}|^2 \right)}, \qquad p = 1, 2, \ldots \tag{5.7}$$

where the forward and backward prediction errors are given by Eqs. (5.5a) and (5.5b). It can be shown that $r_p$ given by Eq. (5.7) is always less than or equal to 1 in magnitude.

## 4 THE LEAST-SQUARES ALGORITHM

Burg's algorithm described in the preceding section is computationally efficient. Its shortcomings prompt the study of other high-resolution spectral estimation methods. The least-squares (LS) methods can provide better spectral estimates for short-sample harmonic processes than Burg's algorithm. The Levinson recursion in Burg's algorithm is sometimes not numerically stable because of rounding errors. The least-squares algorithms presented in this section, on the other hand, are numerically much more stable. Their required computational efforts are about the same as those of Burg's algorithm but they remove the line splitting and frequency shifting problems experienced in Burg's algorithm. The only problem with the least-squares methods is their possible instability when the poles of the prediction error filter fall outside the unit circle in the $Z$ plane. Empirically, it is determined that this situation rarely occurs.

Two least-squares methods are presented in this and the next section. The first method is due to Barrodale and Erickson [5] and is for real data, while the second method was developed by Marple [6] for a more general complex data case. The first method uses Cholesky's decomposition in determining the parameters, which has remarkable numerical stability. Define a forward prediction of $x_t$ as

$$\hat{x}_t = \sum_{k=1}^{p} \hat{a}_k x_{t-k} \qquad \text{for } t = p+1, p+2, \ldots, N \tag{5.8}$$

The LS criterion is to minimize the residual sum of squares

$$\sum_{t=p+1}^{N} \hat{e}_t^2$$

where $\hat{e}_t = x_t - \hat{x}_t$. Equivalently, an LS solution is sought for the AR scheme of order $p$, defined by

$$x_t = \sum_{k=1}^{p} \hat{a}_k x_{t-k} + \hat{e}_t \qquad \text{for } t = p+1, p+2, \ldots, N$$

Similarly, a backward prediction of $x_t$ is $\bar{x}_t$, where

$$\bar{x}_t = \sum_{k=1}^{p} \bar{a}_k x_{t+k} \qquad \text{for } t = 1, 2, \ldots, N-p \tag{5.9}$$

and its parameters $\bar{a}_k$ can be determined by minimizing

$$\sum_{t=1}^{N-p} \bar{e}_e^2$$

where $\bar{e}_t = x_t - \bar{x}_t$.

In vector and matrix notation, the forward prediction, Eq. (5.8), is described by the $(N-p) \times p$ matrix equation

$$\hat{X}\hat{a} = \hat{y} \tag{5.10}$$

where

$$\hat{X} = \begin{bmatrix} x_p & x_{p-1} & \cdots & x_1 \\ x_{p+1} & x_p & \cdots & x_2 \\ \vdots & & & \\ x_{N-1} & x_{N-2} & \cdots & x_{N-p} \end{bmatrix} \qquad \hat{a} = \begin{bmatrix} \hat{a}_1 \\ \hat{a}_2 \\ \vdots \\ \hat{a}_p \end{bmatrix} \qquad \text{and} \qquad \hat{y} = \begin{bmatrix} x_{p+1} \\ x_{p+2} \\ \vdots \\ x_N \end{bmatrix}$$

The error residual vector is $\hat{e} = \hat{y} - \hat{x}\hat{a}$ and the sum of error residuals is $S = S(\hat{a}) = \hat{e}^T\hat{e}$.

Assume $N > 2p$. A unique LS solution exists and is obtained by setting all partial derivatives of $S$ to zero, that is, $\partial S/\partial \hat{a}_k = 0$. It follows that $S$ is minimized when $\hat{X}^T\hat{e} = 0$ and the solution $\hat{a}_*$ must satisfy

$$\hat{x}^T(\hat{y} - \hat{x}\hat{a}_*) = 0$$
$$\hat{R}\hat{a}_* = \hat{S} \tag{5.11}$$

where $\hat{R} = \hat{X}^T\hat{X}$ and $\hat{S} = \hat{X}^T\hat{y}$.

Similarly, the backward prediction problem is described by the $(N-p)\times p$ matrix equation

$$\bar{X}\bar{a} = \bar{y} \tag{5.12}$$

where

$$\bar{X} = \begin{bmatrix} x_2 & x_3 & \cdots & x_{p+1} \\ x_3 & x_4 & \cdots & x_{p+2} \\ \vdots & & & \\ x_{N-p+1} & x_{N-p+2} & \cdots & x_N \end{bmatrix}, \qquad \bar{a} = \begin{bmatrix} a_1 \\ a_2 \\ \vdots \\ a_p \end{bmatrix} \qquad \text{and} \qquad \bar{y} = \begin{bmatrix} x_1 \\ x_2 \\ \vdots \\ x_{N-p} \end{bmatrix}$$

The LS solution $\bar{a}_*$ is given by the equation

$$\bar{R}\bar{a}_* = \bar{S} \tag{5.13}$$

where $\bar{R} = \bar{X}^T\bar{X}$ and $\bar{S} = \bar{X}^T\bar{y}$.

In Burg's algorithm, it is assumed that $x_t$ can be estimated by a weighted sum of $p$ previous observations and a weighted sum of $p$ future observations, using the same weight $a_k$ in both directions. Hence the maximum entropy spectral estimation problem is described by the $2(N-p) \times p$ matrix equation

$$Xa = y \tag{5.14}$$

where

$$X = \begin{bmatrix} \hat{x} \\ \cdots \\ \bar{x} \end{bmatrix}, \qquad a = \begin{bmatrix} a_1 \\ a_2 \\ \vdots \\ a_p \end{bmatrix}, \qquad y = \begin{bmatrix} \hat{y} \\ \cdots \\ \bar{y} \end{bmatrix}$$

The LS solution $a_*$ to Eq. (5.14) satisfies the $p \times p$ system of normal equations

$$Ra_* = S \tag{5.15}$$

where $R = X^TX$ and $S = X^Ty$. Equation (5.15) can be written as

$$(\hat{R} + \bar{R})a_* = \hat{S} + \bar{S} \tag{5.16}$$

The three solutions forward $\hat{a}_*$, backward $\bar{a}_*$, and forward-backward $a_*$ are not the same and are not linearly related.

Computationally, the algorithms for the three solutions allow any one of the three $(p+1) \times (p+1)$ systems of normal equations to be generated efficiently from the corresponding $p \times p$ system of normal equations, and the use of intermediate storage is avoided. Consider, for example, how $\hat{R}_3$ and $\hat{S}_3$ can be obtained from $\hat{R}_2$ and $\hat{S}_2$. Suppressing the elements above the diagonal (since any matrix for normal equations is symmetric), we have

$$\hat{R}_2 = \begin{bmatrix} \hat{r}^{(2)}_{1,1} & \\ \hat{r}^{(2)}_{2,1} & \hat{r}^{(2)}_{2,2} \end{bmatrix} = \begin{bmatrix} \sum_{t=1}^{N-2} x_{t+1}x_{t+1} & \\ \sum_{t=1}^{N-2} x_t x_{t+1} & \sum_{t=1}^{N-2} x_t x_k \end{bmatrix}$$

$$\hat{S}_2 = \left[ \hat{S}^{(2)}_{1,1} \quad S^{(2)}_2 \right]^T = \left[ \sum_{t=1}^{N-2} x_{t+1}x_{t+2} \quad \sum_{t=1}^{N-2} x_t x_{t+2} \right]^T$$

and

$$\hat{R}_3 = \begin{bmatrix} \hat{r}^{(3)}_{1,1} & & \\ \hat{r}^{(3)}_{2,1} & \hat{r}^{(3)}_{2,2} & \\ \hat{r}^{(3)}_{3,1} & \hat{r}^{(3)}_{3,2} & \hat{r}^{(3)}_{3,3} \end{bmatrix}$$

$$= \begin{bmatrix} \sum_{t=1}^{N-3} x_{t+2}x_{t+2} & & \\ \sum_{t=1}^{N-3} x_{t+1}x_{t+2} & \sum_{t=1}^{N-3} x_{t+1}x_{t+1} & \\ \sum_{t=1}^{N-3} x_{t+1}x_{t+2} & \sum_{t=1}^{N-3} x_t x_{t+1} & \sum_{t=1}^{N-3} x_t x_t \end{bmatrix}$$

$$\hat{S}_3 = \left[ \hat{S}^{(3)}_1, \quad \hat{S}^{(3)}_2, \quad \hat{S}^{(3)}_3 \right]^T$$

$$= \left[ \sum_{t=1}^{N-3} x_{t+2}x_{t+3} \quad \sum_{t=1}^{N-3} x_{t+1} \quad \sum_{t=1}^{N-3} x_t x_{t+3} \right]^T$$

It is noted that the leading $2 \times 2$ submatrix of $\hat{R}_3$ can be obtained from $\hat{R}_2$ by the equation

$$\hat{r}^{(3)}_{i,j} = \hat{r}^{(2)}_{i,j} - x_{3-i}x_{3-j} \qquad \text{for } 1 \leq j \leq i \leq 2$$

Equivalently, the leading $2 \times 2$ submatrix of $\hat{R}_3$ can be expressed as $\hat{R}_2 - \hat{\Delta}_2$, where the matrix $\hat{\Delta}_2 = [x_2 \ x_1]^T[x_2 \ x_1]$ represents a rank one modification of $\hat{R}_2$. Also the last row of $\hat{R}_3$ (other than its first element) can be obtained from the last row of $\hat{R}_2$ as follows:

$$\hat{r}^{(3)}_{3,j} = \hat{r}^{(2)}_{2,j-1} - x_{N-2}x_{N+1-j} \qquad \text{for } 2 \leq j \leq 3$$

and

$$\hat{r}^{(3)}_{3,1} = \hat{S}^{(2)}_2 - x_{N-2}x_N$$

Finally, $\hat{S}_3$ (other than its last element) can be obtained from $\hat{S}_2$ as follows:

$$\hat{S}^{(3)}_i = \hat{S}^{(2)}_i - x_{3-i}x_3 \qquad \text{for } 1 \leq i \leq 2$$

and

$$\hat{S}_3^{(3)} = \sum_{t=1}^{N_3} x_t x_{t+3}$$

Thus only one inner product, the element $\hat{S}_3^{(3)}$, must be calculated again when we derive the normal equations for $p = 3$ from the normal equations for $p = 2$; each of the remaining elements of $R_3$ and $\hat{S}_3$ is obtained at the cost of one operation.

In general, $\hat{R}_{p+1}$ and $\hat{S}_{p+1}$ can be obtained from $\hat{R}_p$ and $\hat{S}_p$, with only the element $\hat{S}_{p+1}^{(p+1)}$ requiring more than one operation. Similar statements can be made for backward prediction and forward-backward prediction solutions.

## 5 THE MARPLE LEAST-SQUARES ALGORITHM

The Marple least-squares algorithm estimate of the AR spectrum is applicable to complex samples and is computationally as efficient as the Burg algorithm. It is thus more effective than the algorithm of the preceding section. By generalizing Eq. (5.16) to the complex sample case and combining it with Eq. (5.3a), we have a $(p+1) \times (p+1)$ matrix equation

$$R_p a = e \tag{5.17}$$

where

$$R_p = \begin{bmatrix} r_p(0,0) & \cdots & r_p(0,p) \\ \vdots & & \\ r_p(p,0) & \cdots & r_p(p,p) \end{bmatrix}, \qquad a = \begin{bmatrix} 1 \\ a_1 \\ a_2 \\ \vdots \\ a_p \end{bmatrix} \qquad e = \begin{bmatrix} p_p \\ 0 \\ \vdots \\ 0 \end{bmatrix}$$

and

$$r_p(i,j) = \sum_{k=1}^{N-p} (x_{k+p-j} x^*_{k+p-i} + x^*_{k+j} x_{k+i})$$

Ulrych and Clayton [7] and Nuttall [8] independently proposed this least-squares approach for AR parameter estimation. Both found the LS estimate by computation of the $r_p(i,j)$ terms directly and then by solution of Eq. (5.17) for vector $a$ using matrix inversion. This would require on the order of $p^3$ computational operations, while Burg's algorithm requires on the order of $p^2$ computational operations.

Computational saving is available in the LS algorithm by using both the Hermitian symmetry $r_p(i,j) = r_p^*(j,i)$ and the Hermitian persymmetry $r_p(i,j) = r_p^*(p-i, p-j)$ of $R_p$. Note that $R_p$ does not have the Toeplitz symmetry $r_p(i,j) = r_p(i-j)$, but it has a special structure composed of the sum of two Toeplitz matrix products,

$$R_p = (T_p)^H T_p + (T_p^v)^H T_p^v \tag{5.18}$$

where $T_p$ is an $(N-p) \times (p+1)$ Toeplitz matrix of data samples,

$$T_p = \begin{bmatrix} x_{p+1} & x_p & \cdots x_1 \\ x_{p+2} & x_{p+1} & \cdots x_2 \\ \vdots & & \\ x_N & x_{N-1} & \cdots x_{N-p} \end{bmatrix}$$

with $T_p^v$ denoting the conjugated and reversed matrix

$$T_p^v = \begin{bmatrix} x_1^* & \cdots & x_{pH}^* \\ \vdots & & \\ x_{N-p}^* & \cdots & x_N^* \end{bmatrix}$$

and $H$ denoting the complex conjugate transpose operation. This special structure makes it possible to allow an algorithm of order $p^2$ operations to be developed. The basic procedure of the algorithm makes use of two additional prediction error terms that are time index-shifted variants of the prediction error defined earlier. Details are somewhat involved, but a detailed program is available in the software package MESA/IRL described in Section 8 and in Appendix A.

## 6 THE NONLINEAR MAXIMUM ENTROPY SPECTRAL ESTIMATION ALGORITHM

Another approach to high-resolution spectral estimation motivated by correcting the line-splitting problem in Burg's algorithm as suggested by Fougere [4] is to move the poles of the prediction error filter as close to the unit circle as possible while still maintaining stability. This requires that the reflection coefficients be less than 1 in magnitude. The other filter coefficients can be determined by using Levinson recursion. To enforce the condition, we can set

$$a_k^{(k)} = U \sin\theta_k \exp i\Phi_k \tag{5.19}$$

for complex samples, where $0 \le \theta_k \le \pi$, $\theta_k$ and $\Phi_k$ are both real, and $U$ is a positive constant slightly less than unity. A typical choice of $U$ is $1\text{–}10^{-6}$ or $1\text{–}10^{-5}$. For real samples, we can set

$$a_k^{(k)} = U \sin\theta_k \tag{5.20}$$

All $\theta_k$ and $\Phi_k$, for all $k = 1, 2, \ldots, p$, are independent variables. The prediction error $P_p$ taken as the average of the forward and backward error powers thus depends on $\theta_k$ for real samples and on $\Phi_k$ for complex samples, resulting in nonlinear operations for error minimization. For complex samples the nonlinear optimization starts with the expressions for variations in $\theta_k$, $\Phi_k$:

$$\theta_k = \theta_k^0 + \Delta\theta_k$$
$$\Phi_k = \Phi_k^0 + \Delta\Phi_k$$

Now expand the prediction errors $e_{f,t}$ and $e_{b,t}$ in a Taylor series about $\theta_k^0$ and $\Phi_k^0$ and retain only the first two terms:

$$e_{f,t} = e_{f,t}^0 + \sum_{k=1}^{p}\left(\frac{\partial e_{f,t}}{\partial\theta_k}\Delta\theta_k + \frac{\partial e_{f,t}}{\partial\Phi_k}\Delta\Phi_k\right)$$

and a similar expression for $e_{b,t}$. Then set $\partial P_p/\partial\Delta\theta_k = 0$ and $\partial P_p/\partial\Delta\Phi_k = 0$, where $k = 1, 2, \ldots, p$, to find the minimum error power. This results in $2p$ linear equations to solve for the corrections $\Delta\theta_k$ and $\Delta\Phi_k$ by standard matrix methods.

Another approach is to take the gradient $P_p$ with respect to the independent variables $\theta_k$ and $\Phi_k$ as given by

$$\nabla_{\dot{\theta}}\Phi_p^p = \sum_{k=1}^{p}\left(\frac{\partial P_p}{\partial\theta_k}\bar{\theta}_k + \frac{\partial P_p}{\partial\Phi_k}\bar{\Phi}_k\right)$$

where $\bar{\theta}_k$ and $\bar{\Phi}_k$ are unit vectors.

In practice, the gradient method is simple but unreliable. The rapid descent method of Fletcher and Powell [9] and improved Fletcher and Powell methods, as well as the conjugate gradient methods [10], perform much better even though more storage space may be required.

## 7 ORDER SELECTION

The order of a parametric model should depend on the characteristics of the data rather than on the spectral estimation method. In practice there is some dependence, however. For the AR model and maximum entropy spectral analysis (Burg's and nonlinear), Akaike's criterion is well accepted for order determination. Consider an AR model of order $p$ with a prediction error power $P_p$; the final prediction error is

$$(\mathrm{FPE})_p = \frac{N+p}{N-p} P_p \tag{5.21}$$

The order $p$ is selected such that $(\mathrm{FPE})_p$ is lowest. In most cases there is a small range of $p$ for $(\mathrm{FPE})_p$ to be minimum. Empirical evidence shows that such orders all provide suitable spectra. Another empirical result is that $p$ should be less than two-thirds of $N$.

For both Burg's and nonlinear maximum entropy spectral analysis, the prediction error filters are guaranteed to be stable. This is not the case in Marple's algorithm for AR model estimation. The stopping criterion we have employed is to use the maximum order beyond which the model becomes unstable, ill-conditioned, or a singular matrix arises. The order so selected has worked quite well.

Another procedure for order selection is to check the percentage decrease in the final prediction error with increased order. When the decrease falls below a certain threshold, no increase is made in the model order. It is noted that a large order requires considerably more computation. The final spectrum also depends critically on the order. Too large an order means overfitting the data, resulting in a worse spectrum. Thus careful selection of the order is indeed very important.

## 8 MICROCOMPUTER IMPLEMENTATION

Only a few years ago, single-channel maximum entropy spectral analysis required a large computer to perform accurate computation. Then microcomputers were shown to be just as effective [4]. Today almost all algorithms can be accurately implemented by powerful microcomputers, in FORTRAN and even in BASIC. Table 5.1 shows a comparison of software realizations of nonlinear maximum entropy spectral analysis. In fact, a microcomputer can offer almost as much accuracy as the CDC 6600 computer. We have developed a dedicated software package, MESA/IRL for high-resolution spectrum estimation using an IBM PCXT with an 8087 numeric processor and FORTRAN compiler. The software package, as mentioned in the Appendixes, includes implementation of 1) Burg's method, 2) the forward and backward least-squares method (the Marple algorithm), 3) the general least-squares method for real data only, 4) the original nonlinear maximum entropy spectral analysis method, due to Fougere, 5) an improved nonlinear method making use of an improved Fletcher algorithm [11], 6) a second improved nonlinear method that employs Broyden's function minimization and linear search [12], 7) the nonlinear method employing the conjugate gradient method of optimization, 8) the conventional fast Fourier transform method, 9) the multichannel maximum entropy method, and 10) the 2-D-maximum entropy method. This software package makes it possible to perform high-resolution spectral estimation even with the use of home computers. The data can be typed in or specified in functional form. It is noted that these high-resolution spectral estimation methods work particularly well with a small number of data points, say 100 or much less. If the number of data points is large, say 256 words or more, the fast Fourier transform method is adequate for

Table 5.1 Comparison of Software Realizations of Nonlinear Maximum Entropy Spectral Analysis[a]

| | Fougere's implementation | Previous implementation | Current implementation |
|---|---|---|---|
| Computer | CDC 6600 | PDP 11/45 | IBM PC/XT |
| Language | FORTRAN IV | FORTRAN IV | FORTRAN 77 |
| Precision | Single | Double | Double |
| Number of samples (max) | 50 complex | 25 complex | 1024 complex |
| Number of spectral points | 601 points | 251 points | 2500 points |
| Number of PEF coefficients | 10 weights | 25 weights | 100 weights (linear)<br>25 weights |
| Minimization procedures | FP<br>CG | FP | FP<br>Improved FP<br>CG<br>Broyden |
| Start of minimization | Burg's method | Burg's method | Burg's method<br>Least squares |
| Corrections | Splitting<br>Shifting | Splitting<br>Shifting | Splitting<br>Shifting<br>PEF instability |

[a]Abbreviations: FP, Fletcher and Powell; CG, conjugate gradient.

all practical purposes. Even for such a large number of data points, Burg's maximum entropy spectral analysis is still quite useful for data characterized by several resonant frequency components.

For a small number of data points, Marple's algorithm almost entirely corrects the spectrum splitting and shifting problem of Burg's method with comparable computational effort. The general least-squares method has good performance but takes longer and works for real data only. The second improved nonlinear method is the best-performing nonlinear maximum entropy spectrum estimation techniques and requires much less computation time than the original nonlinear method.

## 9 COMPUTER RESULTS

For various illustrative purposes, four examples of using the software packages are given in this section. The nonlinear spectrum is based on the second improved method and is also called the Broyden spectrum.

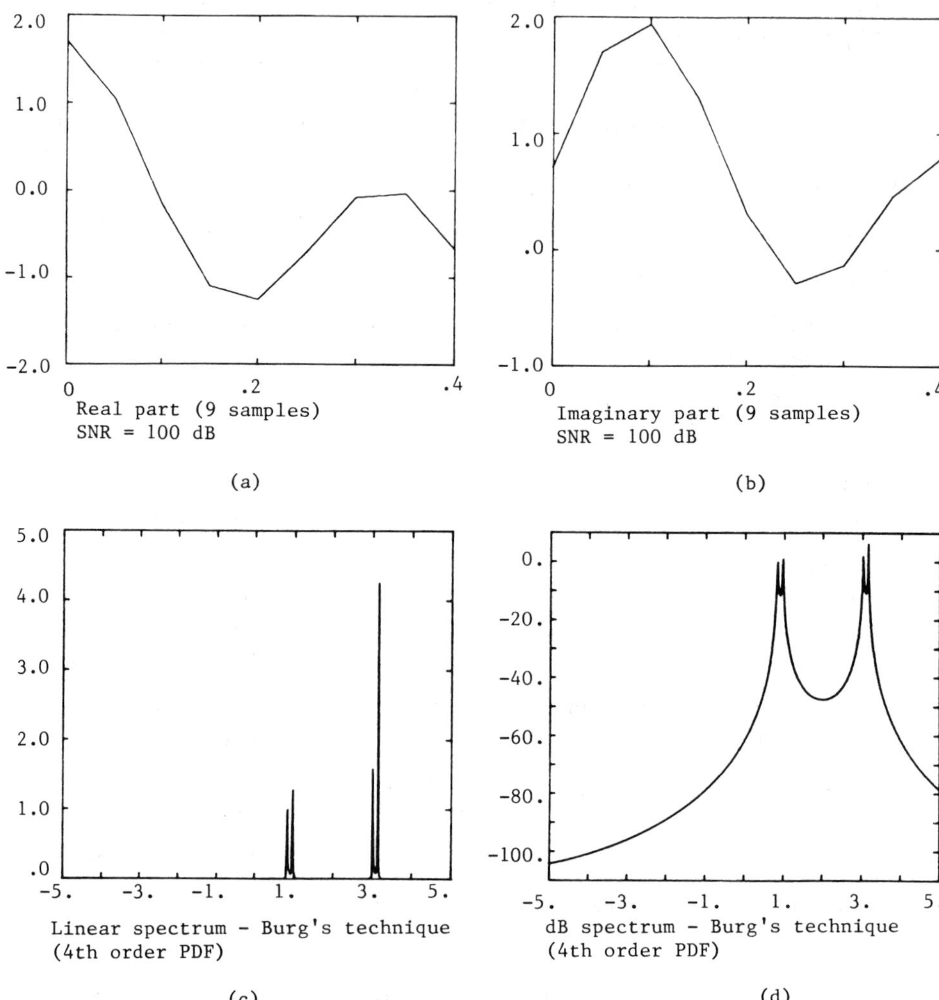


**Figure 5.1** (a and b) Plot of the waveform given by Eq. (5.22), (c and d) its Burg spectra, and its nonlinear spectra for (e and f) 30 iterations and (g and h) 80 iterations.

**Example 1.** Two complex sinusoids plus noise are given,

$$x(k) = \exp\left(\frac{j2\pi f_1 k}{20} + 45^\circ\right) + \exp\left(\frac{j2\pi f_2 k}{20}\right) + 0.00001n(k) \qquad \text{for } k = 0, 1, \ldots, 8 \tag{5.22}$$

where $f_1 = 1$ Hz, $f_2 = 3$ Hz, $n(k)$ is complex Gaussian noise of unit variance, and the sampling rate is 20 samples per second. A plot of this waveform is shown in Figures 5.1a and 5.1b. A fifth-order model was selected. Figures 5.1c and 5.1d show the Burg spectrum for the waveform in linear and logarithmic plots. The line-splitting phenomenon is quite evident here. In an attempt to show how the

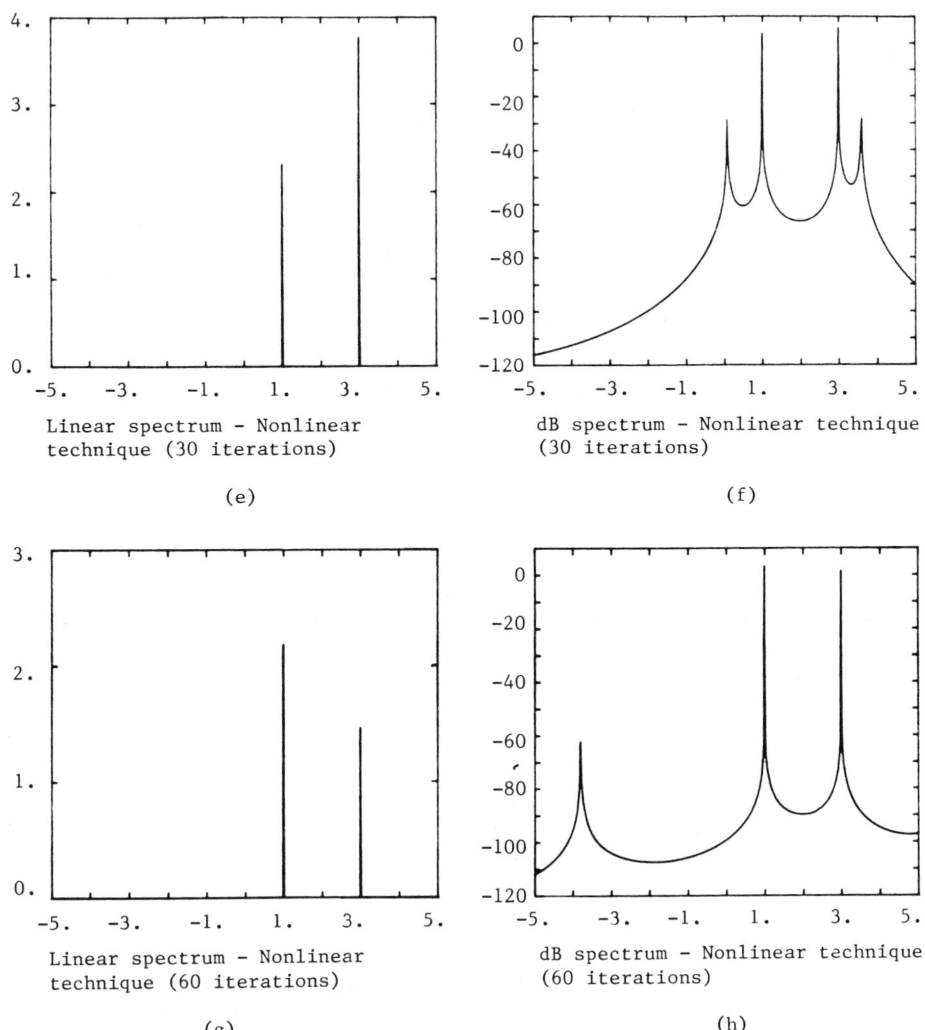


Linear spectrum - Nonlinear technique (30 iterations)

(e)

dB spectrum - Nonlinear technique (30 iterations)

(f)

Linear spectrum - Nonlinear technique (60 iterations)

(g)

dB spectrum - Nonlinear technique (60 iterations)

(h)

second improved nonlinear method corrects this problem, Figures 5.1e and 5.1f contain the nonlinear spectra when 30 iterations are used. For 80 iterations, the nonlinear spectra are shown in Figures 5.1g and 5.1h. It is noted that as the number of iterations increases, the "extraneous" peaks move farther away from the "true" peaks. Also, when compared with the true peaks, these extraneous peaks carry insignificant power.

**Example 2.** The complex waveform considered here is given by

$$x(k) = \exp(j2\pi f_1 k) + \exp(j2\pi f_2 k) + 0.05n(k) \qquad \text{for } k = 0, 1, \ldots, 14 \qquad (5.23)$$

where $f_1 = 0.44$ Hz, $f_2 = 0.46$ Hz, $n(k)$ is complex Gaussian noise, and the sampling rate is 1 sample per second. A plot of this waveform is given in Figures 5.2a and 5.2b.

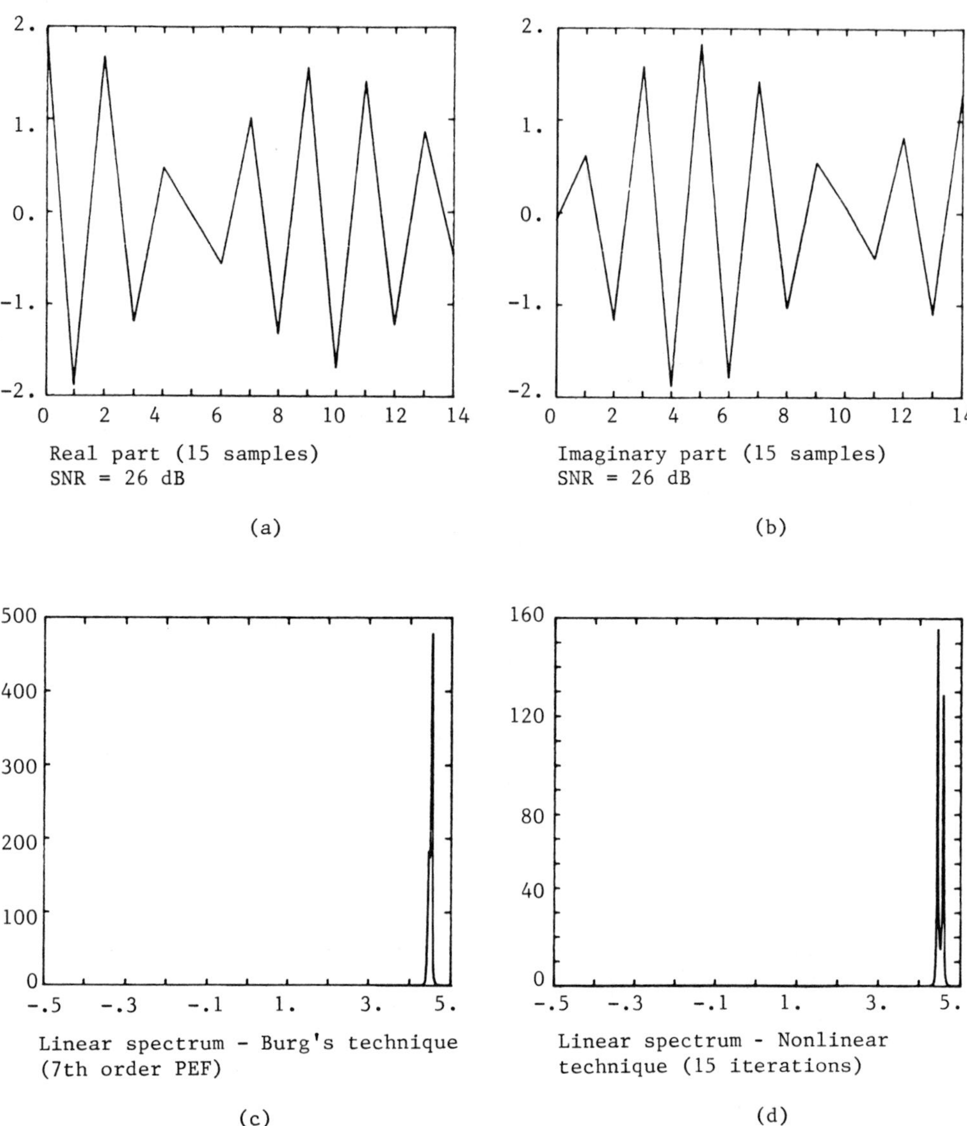


**Figure 5.2** (a and b) Plot of the waveform given by Eq. (5.23); (c and d) its Burg spectrum and its nonlinear spectrum, respectively, on linear plots.

An eighth-order model was selected to test the Burg and nonlinear methods for the signal given by Eq. (5.23) at the 26-dB SNR level. Figure 5.2c shows the Burg spectrum while Figure 5.2d contains the nonlinear spectrum with 15 iterations. The nonlinear method can resolve the two spectral peaks much better than the Burg method.

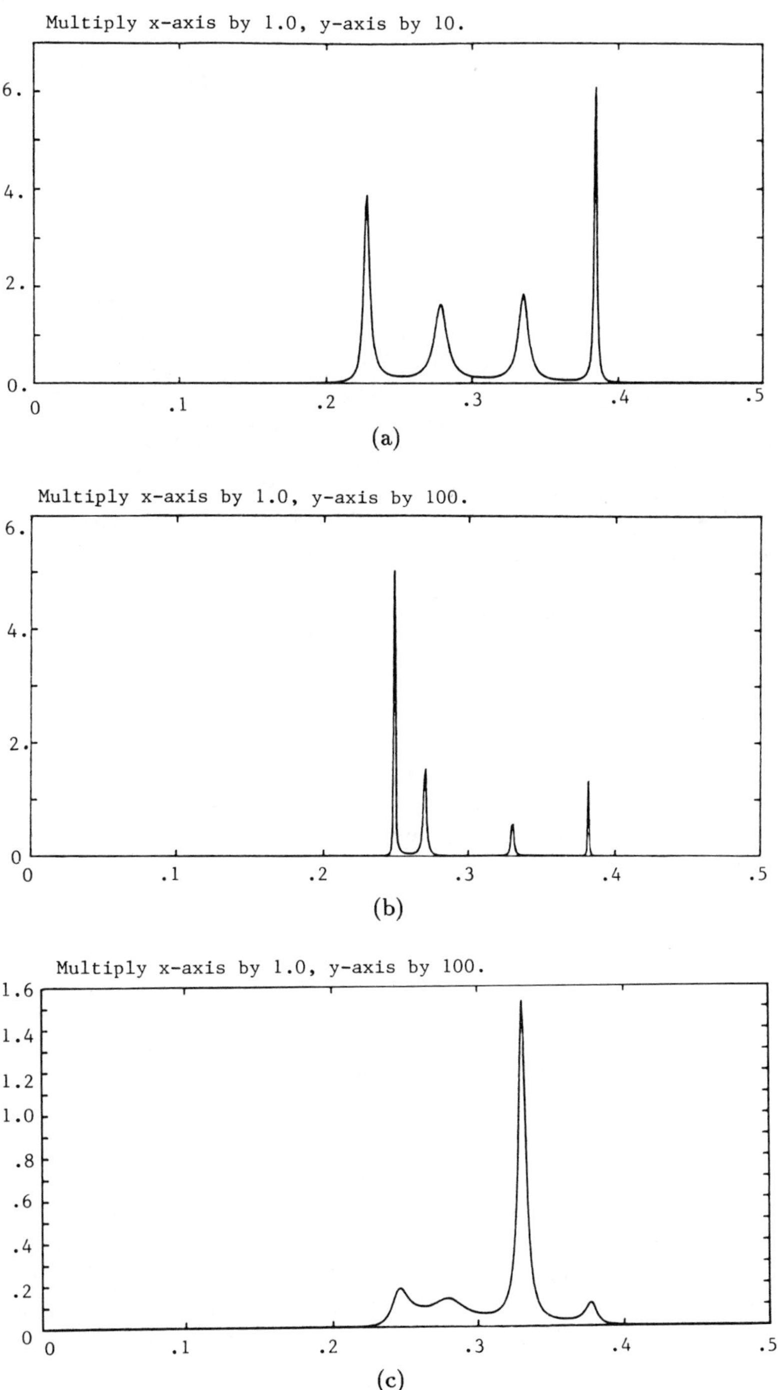


**Figure 5.3** For the waveform given by Eq. (5.24): (a) the Burg spectrum, (b) the Broyden spectrum, and (c) the spectrum of Marple's least-squares algorithm, all on linear plots.

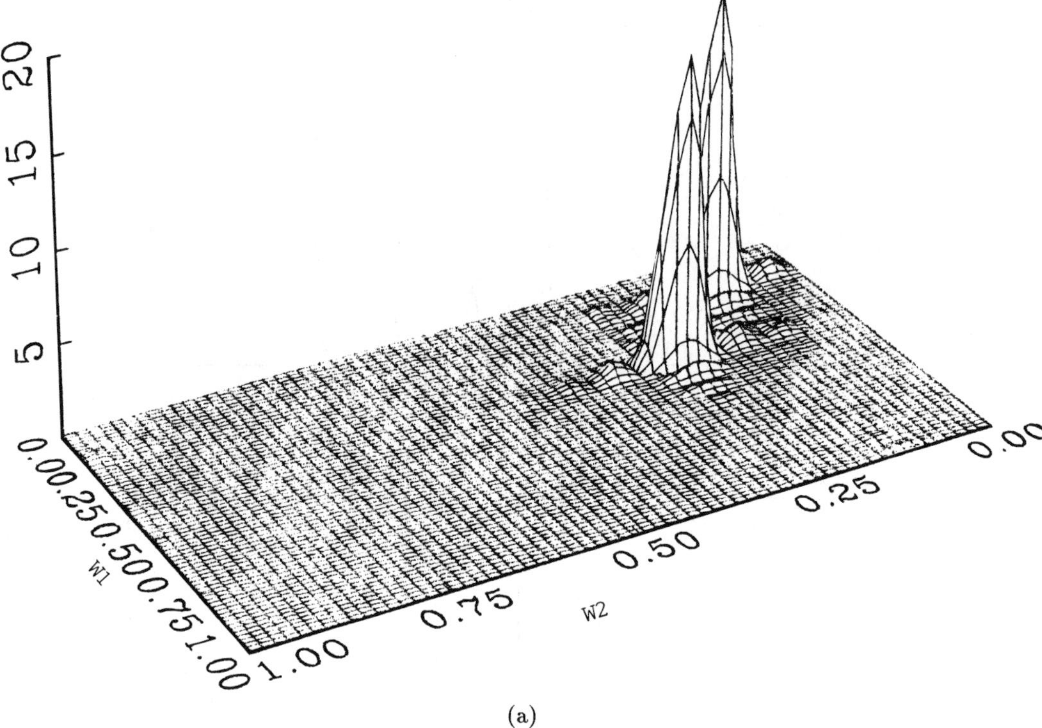


(a)

**Figure 5.4** 2-D power spectrum of the two 2-D sine waves of size 128 × 128 using (a) 2-D FFT, (b) 2-D maximum entropy method (MEM), and (c) 2-D hybrid FFT-Burg method.

**Example 3.** The waveform considered here is given by

$$x(k) = \sum_{i=1}^{4} \cos(2\pi f_i k) + 0.1n(k) \qquad \text{for } k = 0, 1, \ldots, 21 \tag{5.24}$$

where $f_1 = 0.24$ Hz, $f_2 = 0.28$ Hz, $f_3 = 0.34$ Hz, $f_4 = 0.38$ Hz, $n(k)$ is white noise, and the sampling rate is 1 sample per second. A 14th order model was selected to test the Burg's method, the Broyden method with 25 iterations, and the Marple algorithm, as shown respectively in Figures 5.3a, 5.3b, and 5.3c, all in linear plots. Although the four spectral peaks are all detected, the Burg method has significant frequency shifting (bias); the Broyden method resolves the spectral peaks much better than the Marple algorithm.

**Example 4.** The two two-dimensional (2-D) sine waves of frequency $w_{11}/2\pi = 0.125$, $w_{12}/2\pi = 0.125$ and frequency $w_{21}/2\pi = 0.375$, $w_{22}/2\pi = 0.25$, as described in [4, p. 78], are used to illustrate the three methods of 2-D power spectral analysis. The 2-D FFT, 2-D maximum entropy spectral analysis of [4], and 2-D hybrid FFT-

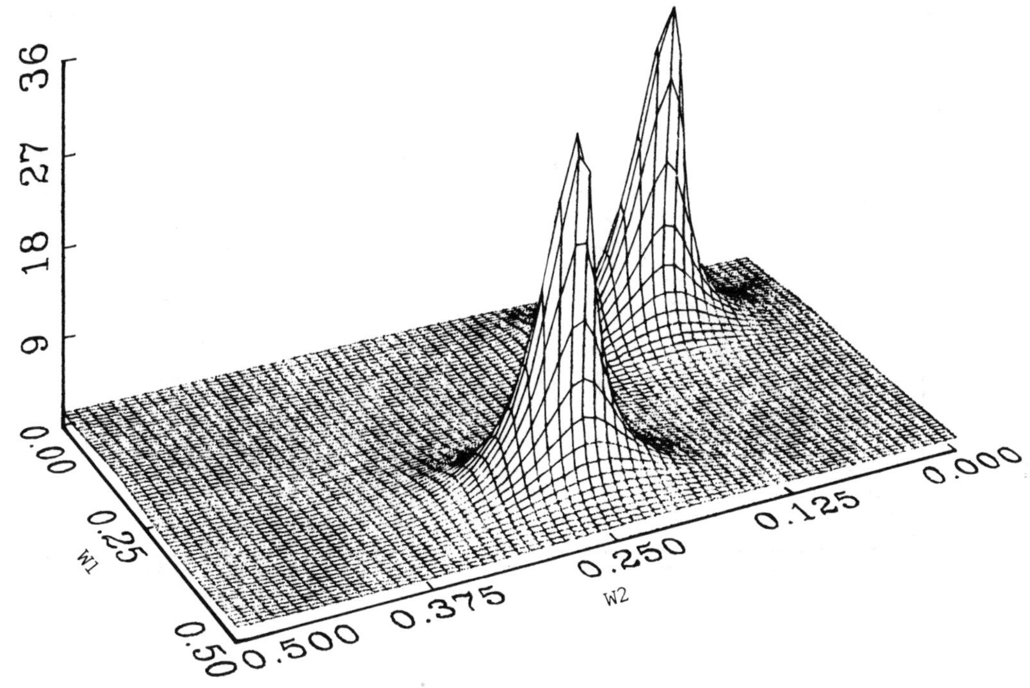
0.00
9
18
27
36
0.00
0.25
W1
0.50
0.500
0.375
0.250
W2
0.125
0.000

(b)

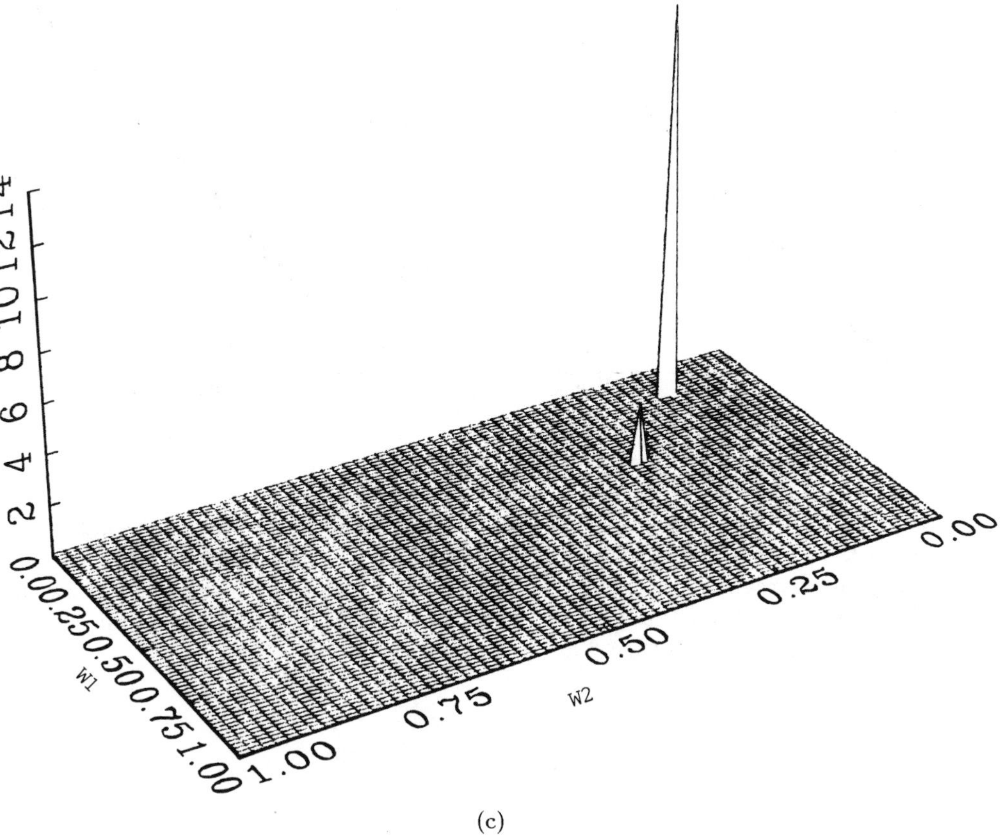

0.00
2
4
6
8
10
12
14
0.00
0.25
0.50
0.75
1.00
W1
1.00
0.75
0.50
W2
0.25
0.00

(c)

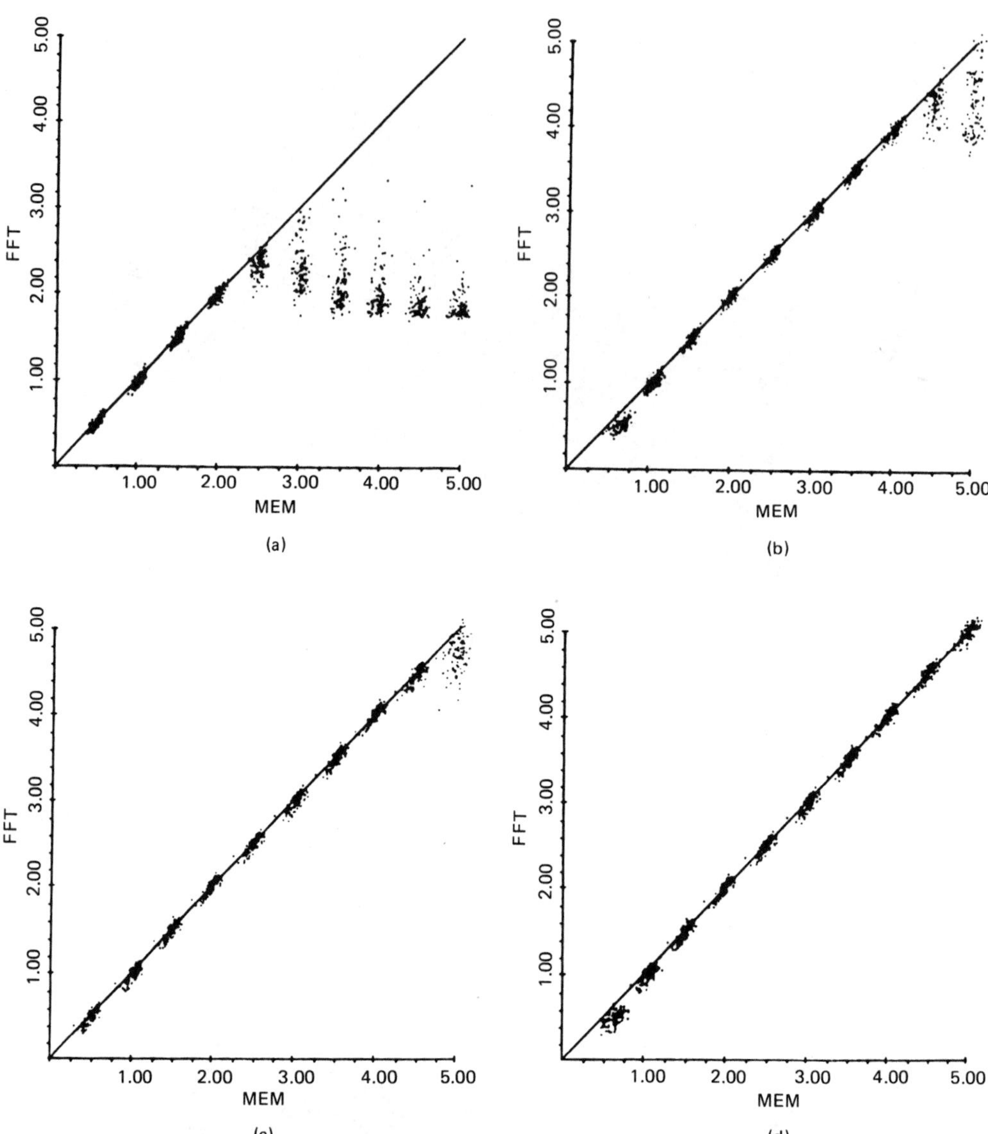


**Figure 5.5** Comparison of FFT and MEM analysis for red noise data. Each point is the observed FFT index (slope) versus the MEM index. There are 100 independent realizations of the red noise for each of the 10 slopes. (a) Raw data; (b) data are end-matched; (c) data are windowed; (d) data are end-matched and windowed. (Courtesy of Dr. Paul F. Fougere.)

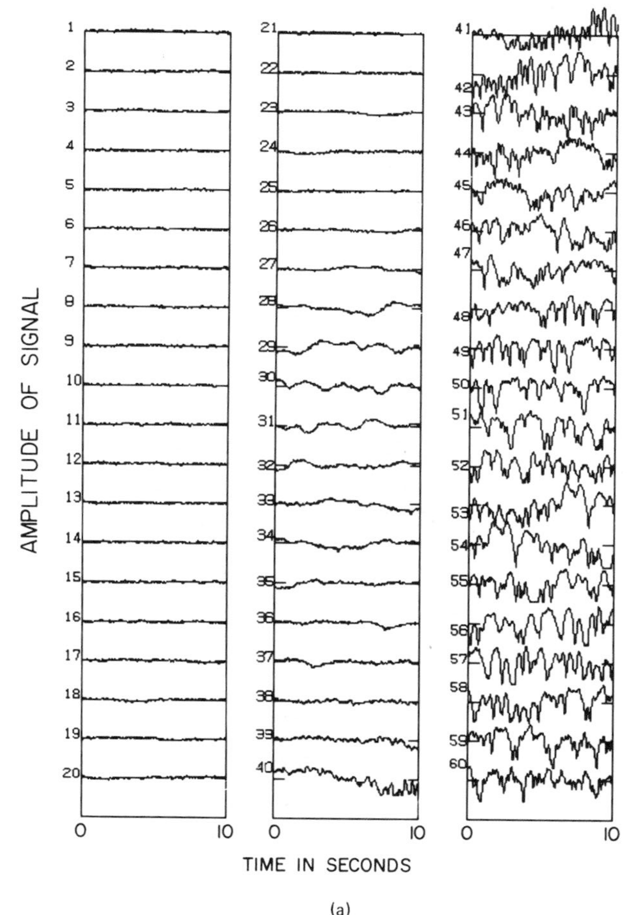


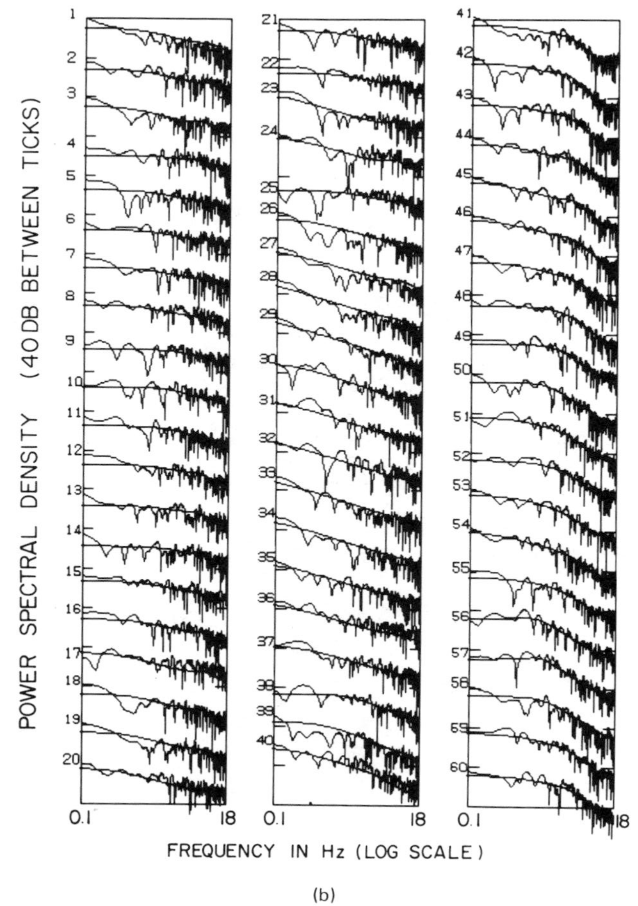


**Figure 5.6** (a) Satellite data records; (b) corresponding FFT periodogram spectra with the Burg spectrum superimposed. (Courtesy of Dr. Paul F. Fougere.)

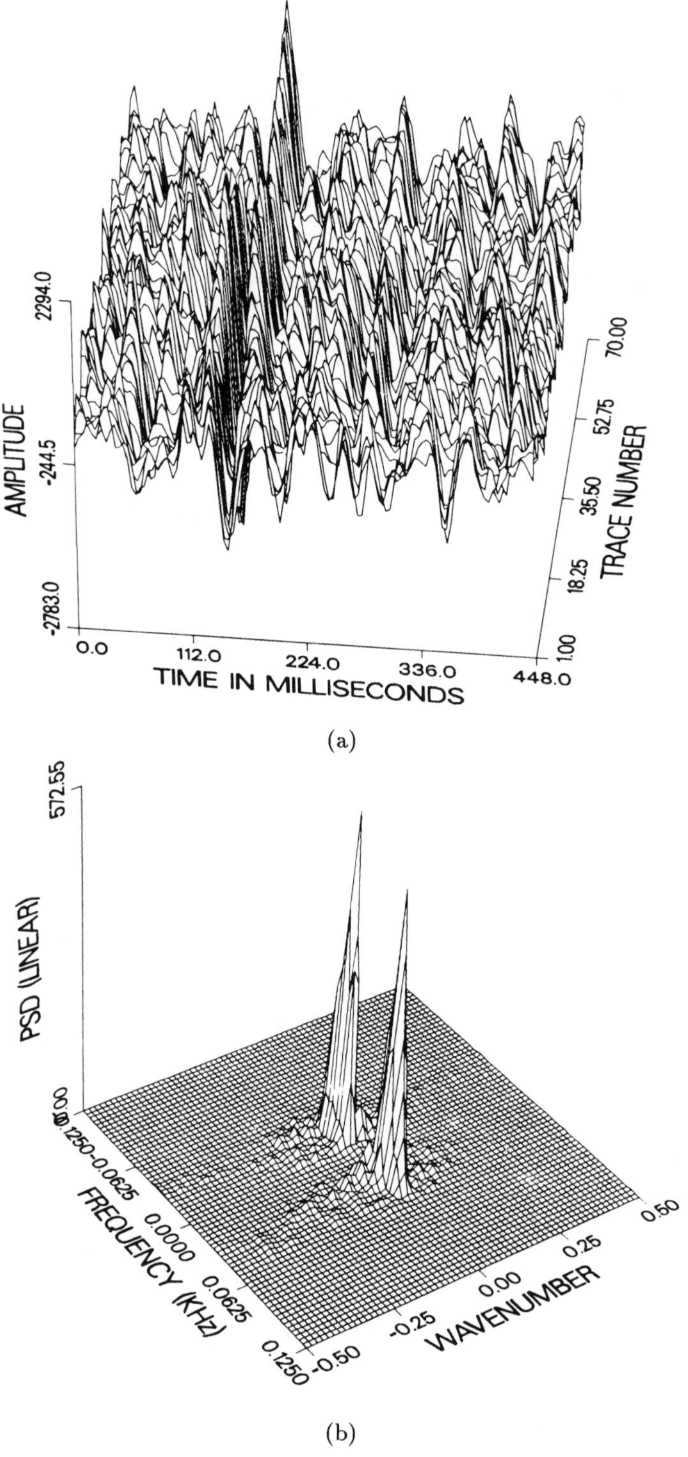


(a)

(b)

**Figure 5.7** (a) Section of seismic exploration data, and the corresponding (b) 2-D FFT power spectrum, (c) 2-D FFT-Burg spectrum, and (d) 2-D FFT-modified covariance spectrum, based on a 64 × 64 data set.

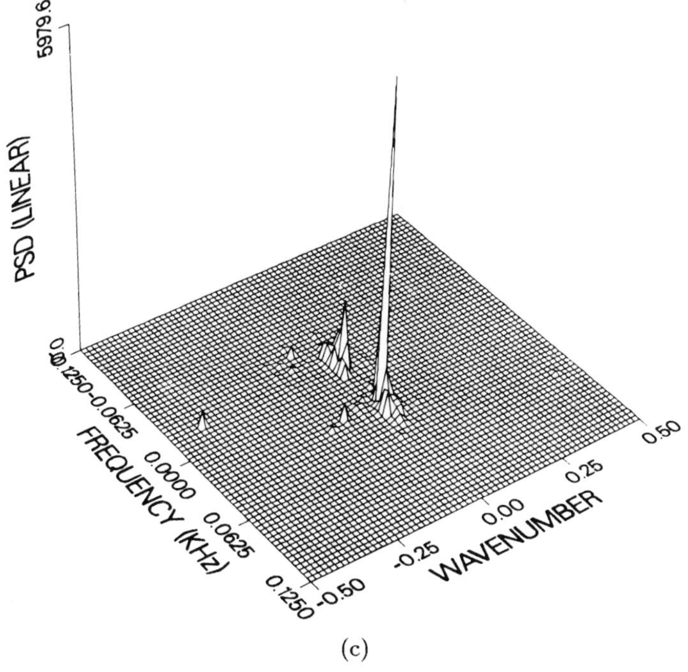
5979.6
PSD (LINEAR)
0.00
-0.1250
-0.0625
0.0000
0.0625
0.1250
FREQUENCY (KHz)
-0.50
-0.25
0.00
0.25
0.50
WAVENUMBER

(c)

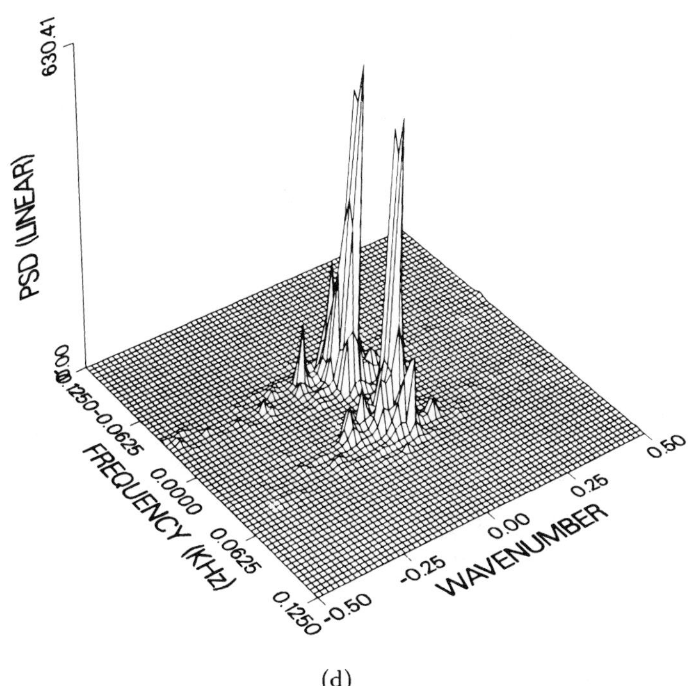
630.41
PSD (LINEAR)
0.00
-0.1250
-0.0625
0.0000
0.0625
0.1250
FREQUENCY (KHz)
-0.50
-0.25
0.00
0.25
0.50
WAVENUMBER

(d)

Burg spectrum are shown in Figures 5,4a, 5.4b, and 5.4c, respectively. The 2-D hybrid method employs FFT in one dimension and the results are used in the other dimension to calculate the Burg coefficients, and then the Fourier transform of the coefficients padded by zeros is taken to obtain the 2-D power spectrum. The hybrid method provides much sharper spectral peaks with less computation time than the 2-D maximum entropy spectrum. The 2-D maximum entropy spectrum is much smoother with no sidelobes and is thus better than the 2-D FFT power spectrum, which has significant and undesirable sidelobes.

Because the 2-D hybrid spectral analysis method is new, a complete FORTRAN program listing is presented in Appendix B.

## 10 APPLICATION EXAMPLES

High-resolution spectral analysis has now been applied in almost every area dealing with signal analysis [13]. In this section only a small number of examples are presented. Red noise is typically experienced in geophysical satellite data [14]. Figure 5.5 shows a comparison of FFT and MEM analysis with the observed FFT (periodogram) index plotted versus the MEM index. Here the index is the negative value of the shape ($-2p$) in a log-log plot of the red noise spectrum:

$$s(f) = f^{-2p}$$

There are 100 independent realizations of the power law process for each of the 10 indices 0.5, 1.0, ..., 5.0. In every case the same time series was used as input to both the MEM and FFT (periodogram). Figure 5.5A is for the raw data. The remaining figures are for (b) matched data, (c) windowed data, and (d) end-matched and windowed data. The MEM method is consistently better in (a), (b), and (c) and is equivalent to end-matched and windowed data in the spectral representation.

For the second example, a set of amplitude scintillation data from the MARISAT satellite was digitized at 36 Hz and examined in spectrum. Figure 5.6a shows the 60 overlapped data batches. Figure 5.6b shows for each trace of Figure 5.6a the Burg maximum entropy spectrum (order 4) in smooth curves and the spectrum of end-matched windowed data in noisy curves. The Burg spectrum in this case appears as an envelope of the periodogram. More detailed harmonic content can be obtained from the Burg spectrum if a higher-order prediction error filter is employed.

As a third application example, we consider a set of real seismic exploration data. Seventy traces of such data with 113 samples/trace are shown in Figure 5.7a. The 2-D FFT plot, which takes the FFT in both time and spatial dimensions, is shown in Figure 5.7b where the wave number is the frequency in the spatial domain. Figure 5.7c is the 2-D FFT-Burg spectrum with the 15th-order AR model in the Burg spectrum computation. Figure 5.7d is the 2-D FFT-modified covariance spectrum in which the modified covariance method is a realization of Marple's algorithm. The latter has better resolution than the other two in showing the direction of incoming seismic waves.

The results of the 2-D FFT and 2-D FFT-Burg methods are similar, however.

## REFERENCES

1. S. M. Kay and S. L. Marple, Jr., Spectrum analysis—a modern perspective, *Proc. IEEE*, *69* (11): 1380–1419 (1981).
2. J. P. Burg, Maximum entropy spectral analysis, Ph.D. Dissertation, Stanford University, Stanford, California (1975).
3. J. P. Burg, "A New Analysis Technique for Time Series Analysis," presented at the NATO Advanced Study Institute on Signal Processing with Emphasis on Underwater Acoustics, Enschede, Netherlands. (August, 1968)
4. C. H. Chen, *Nonlinear Maximum Entropy Spectral Analysis Methods for Signal Recognition*, Research Studies Press, a division of John Wiley & Sons, Chichester, England (1982).
5. I. Barrodale and R. E. Erickson, Algorithm for least-squares linear prediction and maximum entropy spectral analysis—Part I: Theory; Part II: Fortran program, *Geophysics*, *45*: 420–446 (March 1980).
6. L. Marple, A new autoregressive spectrum analysis algorithm, *IEEE Trans. Acoust. Speech Signal Process.*, *ASSP-28*, (4): 441-454 (1980).
7. T. J. Ulrych and R. W. Clayton, Time series modeling and maximum entropy, *Phys. Earth Planet. Inter.*, *12*: 188–200 (August 1976).
8. A. H. Nuttall, Spectral Analysis of a Univariate Process with Bad Data Points, via Maximum Entropy and Linear Predictive Techniques, Technical Report 5303, Naval Underwater Systems Center, New London, Connecticut (1976).
9. R. Fletcher and M. J. D. Powell, A rapid descent method for minimization, *Comput. J.*, *5* (2): 163–168 (1963).
10. R. Fletcher and C. M. Reeves, Function minimization by conjugate gradients, *Comput. J.*, *7* (2): 149–154 (1964).
11. C. H. Chen and A. H. Costa, Interactive Maximum Entropy Spectral Analysis on Personal Computers, Technical Report SMU-ECE-TR-8, Southeastern Massachusetts University, North Dartmouth, Massachusetts (1986).
12. K. Fielding, Function minimization and linear search, algorithm 387, *Commun. ACM*, *13* (8): 509–510 (1970).
13. C. H. Chen, *Digital Waveform Processing and Recognition*, CRC Press, Boca Raton, Florida (1982).
14. P. F. Fougere, On the accuracy of spectrum analysis of red noise processes using maximum entropy and periodogram methods: Simulation studies and application to geophysical data, *J. Geophys. Res.* (1985).

# 6

# Design Methodologies for Systolic Arrays: Mapping Algorithms to Architecture

S. Y. KUNG,* S. N. JEAN, S. C. LO,† and P. S. LEWIS‡ Department of Electrical Engineering, University of Southern California, Los Angeles, California

## 1 INTRODUCTION

A systolic system is a network of processors which rhythmically compute and pass data through the system. It features the important properties of modularity, regularity, local interconnection, a high degree of pipelining, and highly synchronized multiprocessing. It is also scalable architecturally; i.e., the size of the array may be indefinitely extended, as long as the system synchronization can be maintained. Hence systolic arrays are very amenable to implementation with very large scale integration (VLSI).

Due to the fast progress of VLSI technology, algorithm-oriented array architectures appear to be effective, feasible, and economic. In particular, a large class of regular computations, especially those for signal and image processing, can be efficiently implemented on systolic arrays. These trends have necessitated a systematic methodology of mapping computations onto systolic arrays.

---

This research was supported in part by the National Science Foundation under Grant ECS-82-13358, by the Semiconductor Research Corporation under the USC SRC program, and by the Innovative Science and Technology Office of the Strategic Defense Initiative Organization and was administered through the Office of Naval Research under Contract No. N00014-85-K-0469 and N00014-85-K-0599.

Present affilations:
*Department of Electrical Engineering, Princeton University, Princeton, New Jersey
†Microelectronic Center, Hughes Aircraft Company, Newport Beach, California
‡Work performed while on leave from the Electronics Division of the Los Alamos National Laboratory, Los Alamos, New Mexico

To facilitate the mapping, these regular computations can be represented in some computational graphs. In this chapter we introduce 1) a *canonical mapping* methodology for mapping homogeneous computational graphs onto systolic arrays, 2) a *generalized mapping* methodology for mapping heterogeneous computational graphs onto systolic arrays [1], 3) *algorithm matching* techniques to ensure the efficient execution of algorithms on a given array, and 4) applicational domains of VLSI array processors.

The methodology proposed in this chapter is an extension of graph-based mapping techniques investigated by many other authors [2–7]. These previous efforts all share as a common basis the work on uniform recurrence equations (UREs) by Karp et al. [8]. Their approaches have recently been further popularized by Rao [9]. The basic idea of UREs is to express the algorithm as a set of assignment statements, ranging over a set of index points. In FORTRAN this would correspond to nested DO loops, with the assignment statements as the loop body and the set of index points the range of the loop variables. For example, a sample FORTRAN program is shown here.

```
   DO 10 I = 1, 10
   DO 10 J = 1,100
   A(I,J)=A(I,J−1)+C(I,J)
   B(I,J)=A(I,J)+B(I−1,J−1)*C(I,J)
10 CONTINUE
```

Here the index space corresponds to the points $(I, J)$ for $1 \leq I \leq 10$ and $1 \leq J \leq 100$. There exists a unique indexed occurrence (on the left-hand side of the assignment statements) of the variables $A$, $B$, and $C$ at each index point. The expressions on the right-hand side are considered to be *primitive operations* which are not decomposed any further.* The total computation corresponds to computing values for $A(I, J)$ and $B(I, J)$ at each point in the index space. Since each of these indexed variables is assigned only a single value, we call this an *indexed, single assignment* form.

Computations at each point depend on $A(I, J-1)$, and $B(I-1, J-1)$. Hence, node $(I, J)$ is dependent on the results of nodes $(I, J-1)$ and $(I-1, J-1)$.† To explicitly express the data dependences, a directed graph, termed a *dependence graph* (DG), can be embedded in the index space. The nodes in a DG correspond to the index points. The (dependence) *edges* represent the data dependencies. An index point has incoming edges from all points it is directly dependent on. To facilitate the concurrent computation, this graph is assumed to be acyclic.

In the following, we shall briefly illustrate the ideas of DG and the key ideas of single assignment, data dependences, and locality through a simple example.

---

*In this case, and in the examples later in this chapter, the primitive operations are at the instruction-set level. However, these methods can be applied at other levels, such as the logic or register transfer levels.

†An operation $\alpha$ is data-dependent on another operation $\beta$ if one of the inputs to $\alpha$ is an output of $\beta$. Operations which are independent of each other may be executed in parallel.

**Example: DG for a Matrix-Vector Multiplication.** Consider matrix-vector multiplication, $\mathbf{c} = \mathbf{Ab}$, which can be written as

$$c_i = \sum_{k=1}^{N} a_{ik} \times b_k$$

If we assume that multiply/accumulate is the primitive operation, then we can write a simple recursive formula for this computation, uniquely identifying each operation.

For $i, k$ from 1 to $N$

$$c_i^k \longleftarrow c_i^{k-1} + \{a_{ik} \times b_k\}$$

with initial condition $c_i^0 = 0$ and final result $c_i = c_i^N$. This is a *single assignment* form, since each indexed variable is assigned a unique value.

The parallel implementation of an algorithm in this form is quite intuitive. Each element of $\mathbf{c}$ can be computed recursively in $N$ serial steps. Since the computation of each element is independent of the others, all can be done in parallel. This can be seen graphically by drawing the DG as shown in Figure 6.1a. In this graph, each indexed variable $c_i^k$ resides in an index point. Into each point are directed edges corresponding to the data needed to compute $c_i^k$. Examining this DG, we see that the dependences on $c_i^{k-1}$ and $a_{ik}$ are local, while the dependence on $b_k$ is global.*

In implementation, global dependences may imply the need for global communication. Some algorithms, like the fast Fourier transform, require global data dependencies. Others, like matrix-vector multiplication, can be written in a *locally recursive* form in which all dependencies are local. This can be done by introducing a *transmittent variable*, used to distribute the $b_k$ values without broadcasting. The resulting recursions, treating super- and subscripts equivalently, are

For $i, k$ from 1 to $N$

$$b(i,k) \longleftarrow b(i-1,k)$$
$$c(i,k) \longleftarrow c(i,k-1) + a(i,k) \times b(i,k)$$

The inputs are $a(i,k) = a_{i,j}$, $b(0,k) = b_i$, and $c(i,0) = 0$. The output is $c_i = c(i,N)$. The above recursion lead *immediately to a locally recursive algorithm* for matrix-vector multiplication.† The DG for this is shown in Figure 6.1b.

Note that the DG shown in Figure 6.1b is quite homogeneous. In fact, it is a shift-invariant DG. A DG is *shift-invariant* if the dependence arcs corresponding to *all* nodes in the index space are independent of their positions.

---

*The length of local dependence edges is independent of the problem size $N$, while the length of global edges is not.

†A locally recursive algorithm is an algorithm whose corresponding DG has only local dependences [1].

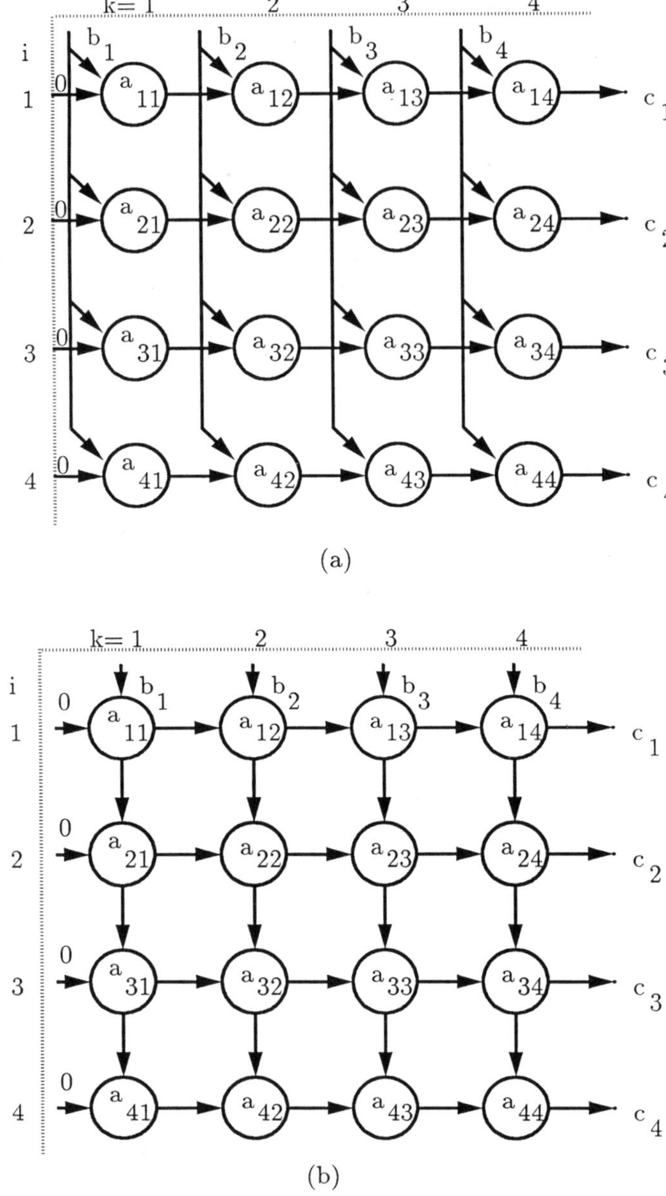


**Figure 6.1** DG for matrix-vector multiplication ($N = 4$): (a) global version; (b) locally recursive version.

Formally, this means that for index vectors $\mathbf{i}_1$, $\mathbf{i}_2$, and $\mathbf{j}$, if a variable at $\mathbf{i}_1$ depends on a variable at $\mathbf{i}_1 - \mathbf{j}$, then a variable at $\mathbf{i}_2$ will depend on a variable at $\mathbf{i}_2 - \mathbf{j}$. Note that the node functions can be different and input/output (I/O) to the border nodes is exempted from the shift-invariant requirement. Shift-invariance of a DG is a very basic assumption for the canonical mapping methodology.

## 2 CANONICAL MAPPING METHODOLOGY

For a shift-invariant DG, a *canonical mapping methodology* may be used to map the algorithm to a systolic array. This mapping is divided into *three stages*, each based on an appropriate canonical form.

*Stage 1:* *Local DG design.* In the first stage, a local DG is developed to perform the computation. Since the structure of a DG greatly affects the final array design, either a very "good" DG is derived directly or further modifications of the DG are required in order to achieve a better design.

*Stage 2:* *Signal flow graph (SFG) design.* In the second stage, the nodes of the DG are mapped onto an abstract processor array architecture, which is modeled by a *signal flow graph* (SFG). This mapping, involving both the assignment of DG nodes to processors and their scheduling in time, is guided by the structure of the DG. Since the DG treated is shift-invariant, linear mapping is used in this stage.

*Stage 3:* *Systolic array processor design.* The third stage addresses the issues of implementing the SFG via a systolic array. In this case the necessary transformations involve temporal localization, which may not be unique. We can find "optimal" transformations to minimize various objective functions, such as the pipelining period (which will be explained later) or the total number of delay elements in the array.

The progression through these stages is not strictly top-down. Factors at one stage may prompt a designer to backtrack to a previous stage. In general, any design effort will involve iterations through this cycle.

### 2.1 Stage 1: Local DG Design

The question of how to develop a local DG is not easy to answer. It is much like trying to describe how to invent an algorithm. However, given a general approach to the problem, such as a serial algorithm or a set of recursive equations, there are several steps to go through in deriving a local DG.

1. Identify the individual primitive computations.
2. Define an index space structure that naturally fits these computations. Associate the intermediate result of each computation with a unique indexed variable, therefore generating the single assignment form. From this, a set of basic (possibly nonlocal) dependence relations can be determined. This generates the indexed single assignment form.
3. Localize the DG so that only local dependences exist.

The first two steps are rather intuitive and it is difficult to give specific techniques that are helpful. In the last (localization) stage, however, several design options can be used if the nonlocal dependences are of the *broadcast* type; i.e., there is a single data *source* and multiple data *receivers*. Since all these receiver nodes are dependent on the source data, this set of receiver nodes constitute a *broadcast contour*. Localized data dependences can be derived from the broadcast contour by using a *transmittent variable*, which propagates data without being modified. Design flexibilities can be obtained by the following techniques to deal with localization.

1. If there is no restriction on where the source data are generated, the distribution of the transmittent variable can be very flexible. For example, the direction of propagation can be reversed.
2. If a chain of dependences represents a series of associative operations, like the chain of partial sums in the matrix-vector example, then the direction of these dependences may be reversed as long as no other computations are dependent on any of the intermediate results. If the data source of a broadcast contour is the output of some associative operations, then reversing the order of associative operations can change the position where the source data are generated. Thus some flexibility in revising the distribution along a broadcast contour can be achieved by this technique.
3. Sometimes the DG, which is localized using the previous techniques, can be further modified by permuting nodes on a broadcast contour without *greatly* increasing the complexity of the DG (since the dependences on the broadcast contour can still be easily localized). With this technique, the localized DG may possess some preferable properties in the design via mapping, e.g., shift-invariance in some direction or simpler I/O control.

**Example: DG for Convolution Algorithm.** The problem of convolution is defined as follows: Given two sequences $u(j)$ and $w(j)$, $j = 0, 1, \ldots, N-1$, the convolution of the two sequences is

$$y(j) = \sum_{k=0}^{j} u(k)w(j-k)$$

or

$$y_j = \sum_{k=0}^{j} u_k w_{j-k}$$

where $j = 0, 1, \ldots, 2N-2$.

The derivation of a DG for convolution is similar to the matrix-vector multiplication case. The first step of deriving the recursive equation is to introduce a recursive variable $y_j^k$. Then the convolution equation can be written in terms of a recursive form:

$$y_j^k = y_j^{k-1} + u_k \cdot w_{j-k} \tag{6.1}$$

where $k = 0, 1, \ldots, j$ when $j = 0, 1, \ldots, N-1$ and $k = j - N + 1, jN + 2, \ldots, N-1$ when $j = N, N+1, \ldots, 2N-2$.

Note that Eq. (6.1) is already in a single assignment form; therefore, the DG can be readily sketched, as shown in Figure 6.2a. Equation (6.1) is an expression with global data dependences and it is therefore not a locally recursive algorithm.

By replacing the broadcast contours by local arcs, the global DG can easily be converted to a localized version as shown in Figure 6.2b. The localized DG has a

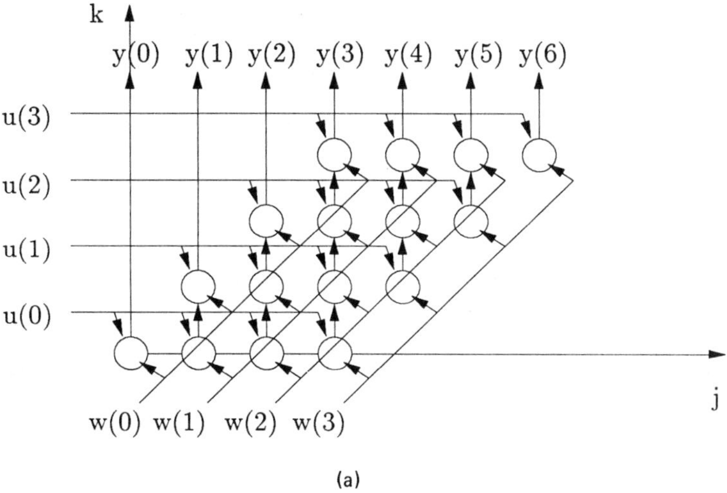


(a)

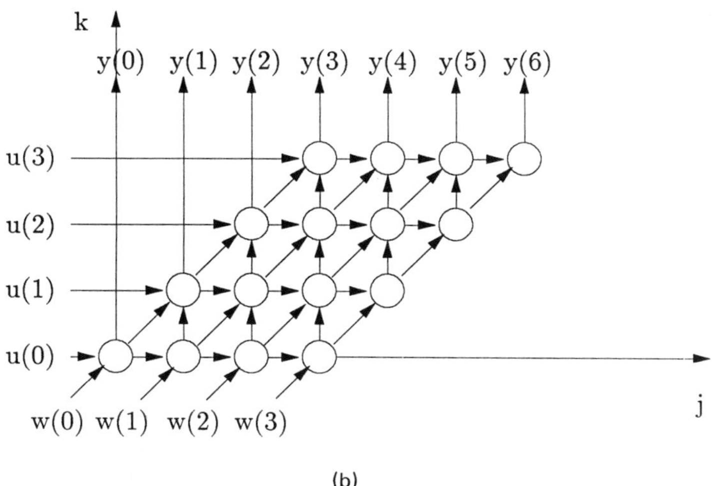


(b)

**Figure 6.2** DG for convolution; (a) global; (b) localized.

corresponding locally recursive algorithm, as follows:

$$y_j^k = y_j^{k-1} + u_j^k \cdot w_j^k, \qquad y_j^0 = 0$$

$$u_j^k = u_{j-1}^k, \qquad u_0^k = u_k$$

$$w_j^k = w_{j-1}^{k-1}, \qquad w_{j-k}^0 = w_{j-k}$$

where $k = 0, 1, \ldots, j$ when $j = 0, 1, \ldots, N-1$ and $k = j - N + 1, j - N + 2, \ldots,$ $N - 1$ when $j = N, N + 1, \ldots, 2N - 2$.

**Example: DG for Autoregressive (AR) Filtering.** To illustrate the general procedure involved in the localization of a recursive algorithm, we present a more sophisticated example of AR filtering. An AR filtering algorithm is described by the following difference equation: For $j = 1, 2, \ldots,$

$$y(j) = \sum_{k=1}^{N} a_k y(j-k) + u(j)$$

Note that the output of an AR filter is an indefinite-length sequence even if the input sequence $u(j)$ has a finite duration. To derive a local DG for the AR algorithm, one might follow the formal procedure given earlier. To simplify the illustration, here we take advantage of the derivation used for the convolution algorithm and propose in a similar fashion a preliminary DG for the AR algorithm as shown in Figure 6.3a. Here it is assumed that $N = 4$. Note that the key differences are: 1) the data $y(c)$, where $c = 0, 1, 2, \ldots,$ are assumed available without showing where they come from, and 2) the DG does not indicate the direction of data flow along the vertical lines because both directions are viable (cf. Figure 6.3b), according to the associative operation principle. Consequently, the input data $u(i)$ may be input either from the top or the bottom of the vertical lines.

Recall that Figure 6.3a does not indicate exactly how $y(c)$ is produced and utilized. To see how this may be produced, let us verify for example the operations in the fifth column ($j = 4$):

$$y(4) = u(4) + a_4 y(0) + a_3 y(1) + a_2 y(2) + a_1 y(3) \tag{6.2}$$

*Intermediate and external variables.* We observe that in the DG there are two types of broadcast contours. One comprises index points of a constant $k$-index, transmitting the *broadcast data* $a_k$. The other comprises the index points defined by $j - k$ = constant, transmitting the *broadcast data* $y(j - k)$. A variable in a set of recursive equations is said to be an *intermediate variable* if it appears on both the right-hand side and the left-hand side of these recursive equations. Otherwise, the variable is termed an *external variable.* Thus $y(j - k)$ is considered an intermediate variable while $a_k$ is an external variable.

Just as in the procedure for localization of the convolution DG discussed previously, the broadcast data $a_k$, which is an external variable, can be localized simply by replacing a global arc by many local arcs. However, the broadcast data $y(j-k)$, which is an intermediate variable, requires more thought to achieve a best localized DG.

Recall that in the preliminary DG we have not decided where the data $y(c)$ should be produced. In fact, it depends on where the produced data will be utilized. Note that the data $y(c)$ may be injected at any of the nodes on the broadcast contour $j - k = c$. However, it can be shown that the bottom node $(c + 1, 1)$ is the most suitable place to inject the data.* For example, if $c = 4$, then the data $y(4)$ will be injected at node (5,1). Referring to Eq. (6.2), the data $y(4)$ can be produced either at the top node (4,4) or the bottom node (4,1) of the vertical line ($j = 4$). The latter is of course closer to the node (5,1) where the produced data will be utilized.

In general, there are two possible positions where the intermediate variable $y(c)$ may be produced, which correspond to two different designs proposed below:

• *Spiral communication approach*: If $y(c)$ is produced at the top nodes of the vertical lines, then the recursive algorithm for AR filtering is

$$y_j^k = y_j^{k-1} + a_k \cdot y_{j-k}^N$$
$$\text{Input:} \quad y_j^0 = u(j); \qquad \text{output:} \quad y(j) = y_j^N \tag{6.3}$$

To verify Eq. (6.3), let us again display the node activities (from bottom up) along the vertical line for $j = 4$,

$$\begin{aligned} y_4^1 &= y_4^0 + a_1 y_3^4 \\ y_4^2 &= y_4^1 + a_2 y_2^4 = y_4^0 + a_1 y_3^4 + a_2 y_2^4 \\ y_4^3 &= y_4^2 + a_3 y_1^4 = y_4^0 + a_1 y_3^4 + a_2 y_2^4 + a_3 y_1^4 \\ y_4^4 &= y_4^3 + a_4 y_0^4 = y_4^0 + a_1 y_3^4 + a_2 y_2^4 + a_3 y_1^4 + a_4 y_0^4 \end{aligned}$$

The result produced at the top node is

$$y(4) = y_4^4 = u(4) + a_1 y(3) + a_2 y(2) + a_3 y(1) + a_4 y(0)$$

This shows that in general $y(c)$ is produced at node $(c, N)$ ($y_c^N = y_c^{N-1} + a_N y_{c-N}^N$) [since $y(c)$ depends on $y_c^N$]. A spiral communication arc will be required to feed the results of node $(c, N)$ to node $(c + 1, 1)$. This leads to a DG with *spiral communication* (see Figure 6.3c).

• *Local communication approach*: Note that the operation in Eq. (6.3b). Therefore, an alternative recursive equation (via a reversed direction) is also possible. This corresponds to the case where the data $y(c)$ is produced at the bottom nodes of the vertical lines. Correspondingly, the recursive algorithm now becomes

$$y_j^k = y_j^{k+1} + a_k \cdot y_{j-k}^1$$
$$\text{Input:} \quad y_j^{N+1} = u(j), \qquad \text{output:} \quad y(j) = y_j^1 \tag{6.4}$$

Thus $y(c)$ is produced at node $(c, 1)$ and only a *local communication* arc is required to send the data to node $(c + 1, 1)$. The corresponding DG is shown in

*In order to obtain a DG which has some permissible linear schedules (to be discussed in a moment), it turns out that node $(c + 1, 1)$ is the only choice.

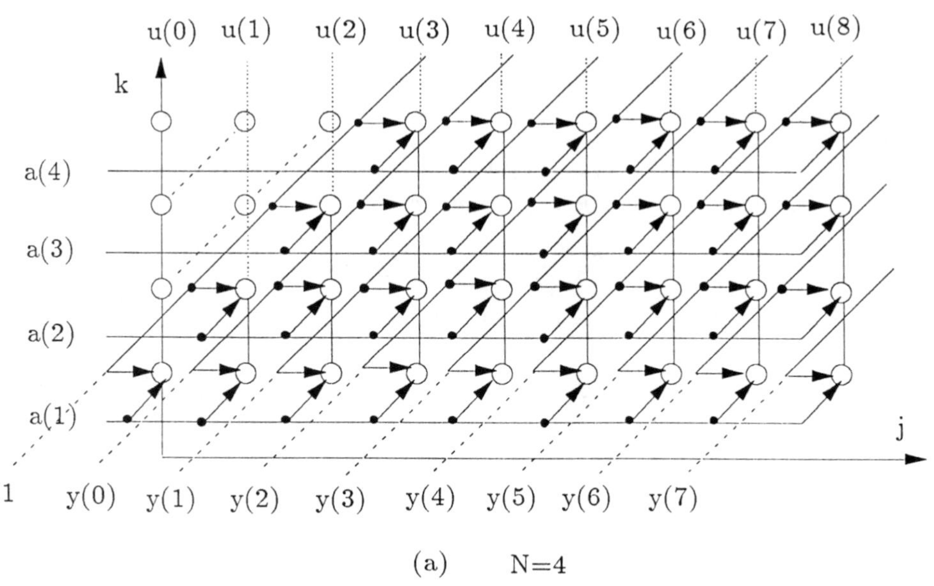


(a) N=4

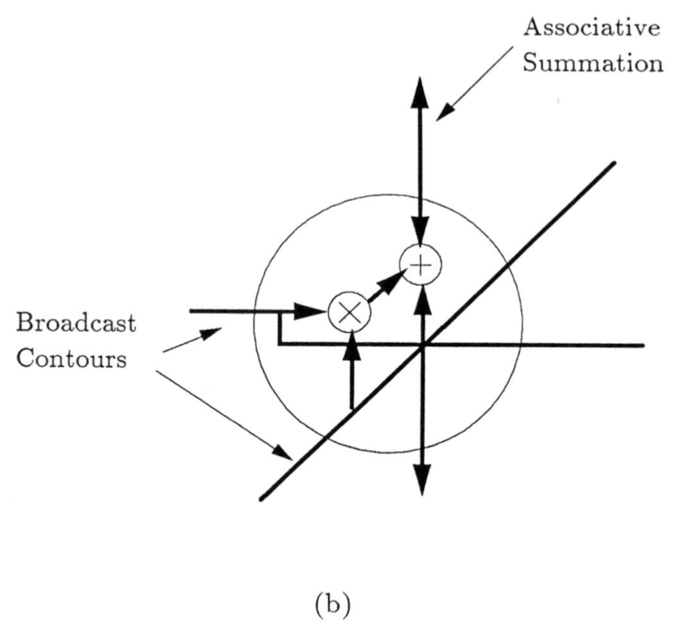


(b)

**Figure 6.3** (a) Preliminary (global) DG for AR filtering; (b) a detailed node; (c) a spiral DG; (d) a localized DG.

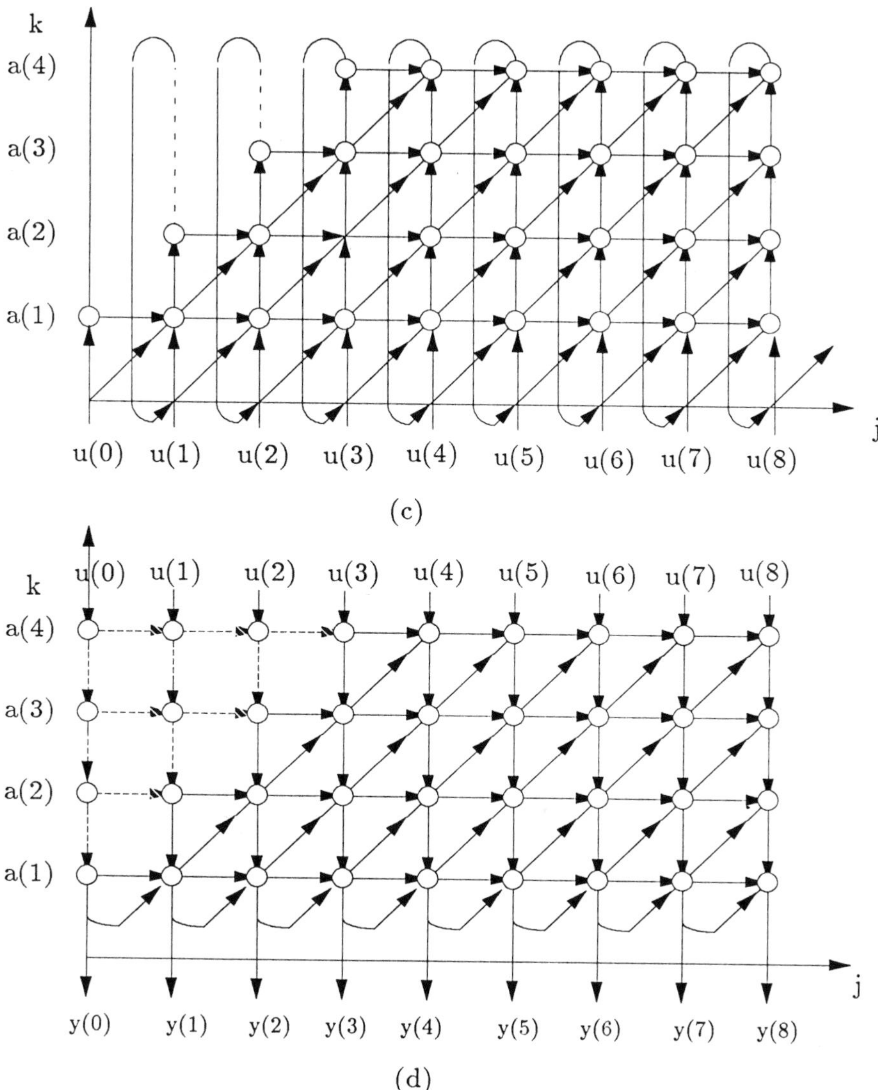


Figure 6.3d, which is completely localized. Obviously, this localized DG is preferable to the spiral version. The locally recursive algorithm for this DG can now be obtained:

$$y_j^k = y_j^{k+1} + a_j^k \cdot B_j^k$$

$$a_j^k = a_{j-1}^k, \qquad a_{-1}^k = a(k)$$

$$B_j^k = B_{j-1}^{k-1}, \qquad B_j^0 = y_j^1$$

$$y_j^{N+1} = u(j) \qquad \text{and} \qquad y(j) = y_j^1$$

where $j = 0, 1, 2, \ldots, \infty$, $k = N, N-1, N-2, \ldots, 1$. Note that, because the variable $y_j^k$ has already been used in the $j = c$ contour to denote the data involved in the iterative summation process, a new variable $B_j^k$ is introduced to propagate the data $y(c)$ along the broadcast contour $j - k = c$.

## 2.2 Stage 2: Signal Flow Graph Design

**Definition: Signal Flow Graph (SFG).** A *signal flow graph* is a directed graph consisting of *nodes* and *edges* weighted with *edge delays.* The nodes model computations and the edges communications. Time is explicitly modeled by the delays, since node computations and edge communications are assumed to take zero time. An SFG must have a nonnegative number of delays on every edge. A systolic array differs from an SFG only in that there should be positive number of delays on every edge. An example of an SFG for matrix-vector multiplication is shown in Figure 6.4.

A complete SFG description includes both functional and structural description parts. The functional description defines the behavior within the nodes, while the structural description specifies the interconnection (edges and delays) between the nodes. The simplest type of SFG is one which is both structurally and functionally time-invariant. This type of SFG will be termed a *time-invariant* (TI) SFG: Another, less restrictive type of SFG is a *structurally time-invariant* (STI) SFG. Its structure is constant, but its node functions may vary over time.*

**SFG Design.** We introduce the SFG to serve as an abstract representation of a processor array. Hence, the SFG can model the implementation, in time and space, of a computation described by a local DG. To obtain this implementation we *map* the DG onto the SFG. This mapping includes *assigning* the nodes in the index space to particular nodes of the SFG and *scheduling* the order in which these nodes are to be computed. Methods of achieving this mapping are the subject of

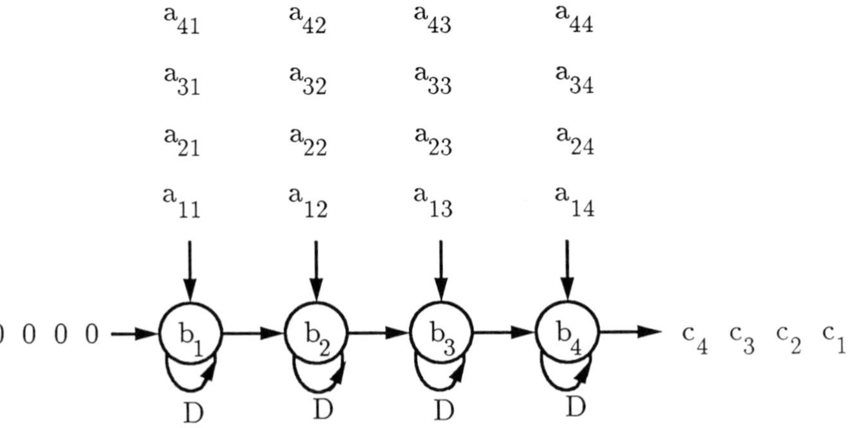


**Figure 6.4** SFG for matrix vector multiplication.

*A functionally time-invariant SFG can also be defined in this fashion.

the next section. Because of the more abstract nature and fewer timing constraints of the SFG, it is easier to map a DG onto an SFG than it is to map the DG directly to hardware. The SFG representation so obtained can then be *retimed* to achieve a systolic array. Basically, the SFG represents an intermediate level which allows many of the structural and timing issues to be treated separately from each other.

## Mapping from a DG to an SFG (Graphical Approach)

To implement a DG, a projection from the index space onto an array processor can be used. The meaning of projection is to assign the operations of all nodes along a line (corresponding to the projection direction $\vec{d}$) to a single processor. See Figure 6.5a.

The projection should be accompanied by a specific schedule, which defines the sequence of the operations in *all* the processing elements (PEs). That is, we need to specify at what time each node of the DG is executed. Only linear schedules will be adopted in the canonical mapping methodology. A linear schedule is a linear mapping in which all nodes on an *equitemporal hyperplane* are mapped to the same integer and thus are executed at the same time. The set of integers then corresponds

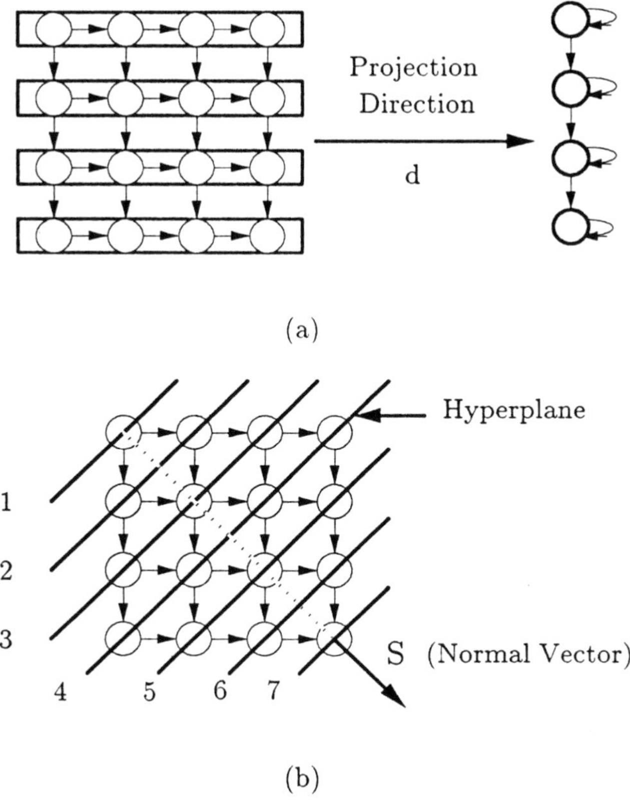


**Figure 6.5** Illustration of (a) a linear projection with projection vector $\vec{d}$ and (b) a linear schedule $\vec{s}$ and its hyperplanes.

to a set of parallel hyperplanes. Mathematically, a linear schedule $L(\mathbf{i})$, denoted by a column vector $\vec{s}$, maps a node index $\mathbf{i}$ to an integer $L(\mathbf{i}) = \vec{s}^T\mathbf{i}$. See Figure 6.5b.

To ensure effective pipelining, the scheduling vector $\vec{s}$, which shows the normal direction of the equitemporal hyperplanes, must satisfy the following conditions:*

1. $$\vec{s}^T\vec{e} \geq 0, \text{ for any dependence arc } \vec{e} \tag{6.5}$$
2. $$\vec{s}^T\vec{d} > 0. \tag{6.6}$$

A linear schedule whose scheduling vector $\vec{s}$ satisfies these two conditions is called a *permissible linear schedule.* Using a projection combined with a permissible linear schedule, an SFG can then be produced where the schedule or time dimension is represented by the edge delays and the processors in the processor space by the indexed nodes.

**Example: SFGs for Convolution.** The two-dimensional (2-D) index space of convolution, as shown in Figure 6.2b, may be decomposed into a direct sum of a 1-D *processor space* and a 1-D *schedule space.* With the projection direction ($\vec{d}$) $[1\ 1]^T$ and the scheduling vector ($\vec{s}$) $[0\ 1]^T$, the SFG obtained is shown in Figure 6.6a. Note that all the outputs $y(n)$ are obtained from the boundary processor. This is achieved by a technique explained in Section 3.3. With $\vec{d} = [1\ 0]^T$ and $\vec{s} = [1\ 0]^T$,† another SFG can be obtained (see Figure 6.6b)).

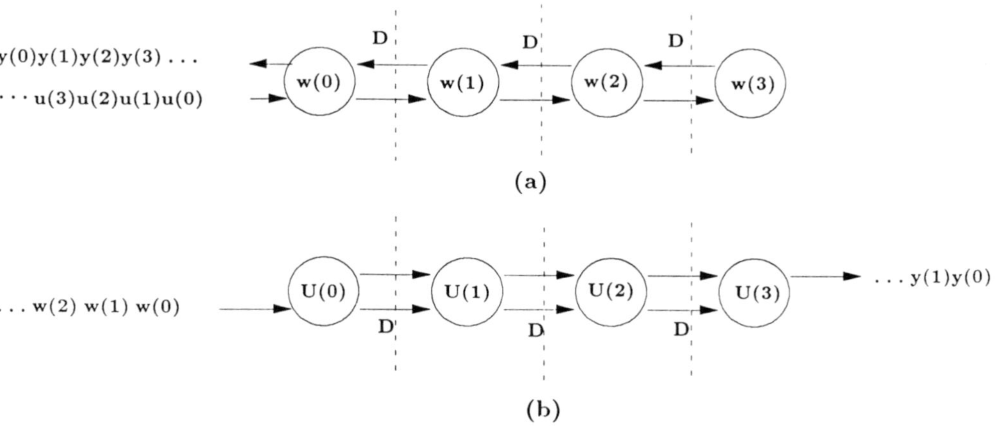


**Figure 6.6** Two SFGs for convolution: (a) $\vec{d} = [1\ 1]^T$ and $\vec{s} = [0\ 1]^T$; (b) $\vec{d} = [1\ 0]^T$ and $\vec{s} = [1\ 0]^T$.

---

*These conditions can be checked by inspection. In fact, these conditions are true if and only if there exists a hyperplane whose normal vector $\vec{s}$ is not orthogonal to the direction $\vec{d}$, and all if the dependence edges are pointing in the same direction across the hyperplane.

†A schedule is called *default schedule* if $\vec{s} = \vec{d}$.

### SFG Projection Procedure (Algebraic Approach)

The above graph-based projection procedure may be formally described in terms of algebraic transformations. Given a DG of dimension $n$, a projection vector $\vec{d}$, and a permissible linear schedule $\vec{s}$, then an SFG may be derived based on the following mappings:

1. *Node mapping.* This mapping assigns the node activities in the DG to processors. The index set of nodes of the SFG are represented by the mapping

$$\mathbf{P}: R \longrightarrow I^{n-1}$$

where $R$ is the index set of the nodes of the DG and $I^{n-1}$ is the Cartesian product of $(n-1)$ integers. The mapping of a computation $\mathbf{c}$ in the DG onto a node $\mathbf{n}$ in the SFG is found by

$$\mathbf{n} = \mathbf{P}^T \mathbf{c}$$

where the *processor basis* $\mathbf{P}$, denoted by an $n \times (n-1)$ matrix, is *orthogonal* to $\vec{d}$. Mathematically,

$$\mathbf{P}^T \vec{d} = \mathbf{0}$$

2. *Arc mapping.* This maps the arcs of the DG to the edges of the SFG. The edge $\vec{e}$ into each node of the SFG and the number of delays $D(\vec{e})$ on it are derived from the dependence $\vec{e}$ at each point in the (shift-invariant) DG by

$$\begin{bmatrix} D(\vec{e}) \\ \cdots \\ \vec{e} \end{bmatrix} = \begin{bmatrix} \vec{s}^T \\ \cdots \\ \mathbf{P}^T \end{bmatrix} [\vec{e}]$$

3. *I/O mapping.* The SFG node position $\mathbf{n}$ and time $t(\mathbf{c})$ of an input of the DG computation $\mathbf{c}$ is derived as

$$\begin{bmatrix} t(\mathbf{c}) \\ \cdots \\ \mathbf{n} \end{bmatrix} = \begin{bmatrix} \vec{s}^T \\ \cdots \\ \mathbf{P}^T \end{bmatrix} [\mathbf{c}]$$

A similar mapping applies to output nodes.

**Remark:** The elements of $\vec{s}$, $\vec{d}$, and **P** are integers. Clearly, it is desirable to have these vectors (matrices) represented by the smallest integers whenever possible. To avoid confusion, from now on the elements of $\vec{s}$ are restricted to be coprime. That is, *the greatest common divisor of all the elements of* $\vec{s}$ *is 1.* The elements of $\vec{d}$ are also assumed to be coprime, as are the elements of each column vector of **P**.

## 2.3 Stage 3: Systolic Array Processor Design

Many issues are involved in the implementation of algorithms described by SFGs, and a number of previous works have focused on them [10, 11]. Note that the SFG obtained from a local DG in stage 2 is always spatially localized but not necessarily *temporally localized.* That is, some edges of the SFG contain no delay. An STI SFG is easily implemented as a systolic array once it is temporally localized. Temporal localization involves a *retiming* or *systolization* transformation of the SFG to an equivalent SFG (that performs the same computation) which has no zero delay edges. A number of techniques exist to perform this transformation [12, 13]. A systolic array can then be constructed with a PE corresponding to each SFG node and interconnections corresponding to the SFG edges.

**Cut-set Systolization.** The cut-set systolization procedure is based on two simple rules:

*Rule 1:* *Time-scaling.* All delays **D** may be scaled by a single positive integer $\alpha$,* that is, **D** can be replace by $\alpha\mathbf{D}'$. Correspondingly, the input and output rates also have to be scaled by a factor of $\alpha$ (with respect to the new time unit $\mathbf{D}'$). The time-scaling factor (or, equivalently, the slowdown factor) $\alpha$ is determined by the slowest cycle in the SFG array [11].

*Rule 2:* *Delay-transfer.* Given any cut-set of the SFG,† which partitions the graph into two components, we can group the edges of the cut-set into *inbound edges* and *outbound edges*, as shown in Figure 6.7, depending on the directions assigned to the edges. Rule 2 allows advancing $k(\mathbf{D}')$ time units on all the outbound edges and delaying $k$ time units on the inbound edges, or vice versa. It is clear that for a (time-invariant) SFG the general system behavior is not affected, because the effects of lags and advances cancel each other in the overall timing. Note that the input-input and input-output timing relationships also remain exactly the same only if they are located on the same side of the cut-set. Otherwise, they should be adjusted by a lag of $+k$ time units or an advance of $-k$ time units. If more than one cut-set is involved and if the input and output are separated by more than one cut-set, then such adjustment factors should be accumulated.

The following theorem guarantees that the basic cut-set rules are sufficient for the systolization procedure. Its proof can be found in [11].

---

*$\alpha$ is also known as the *pipelining period* of the SFG.

†A cut-set in an SFG is a minimal set of edges which partitions the SFG into two parts.

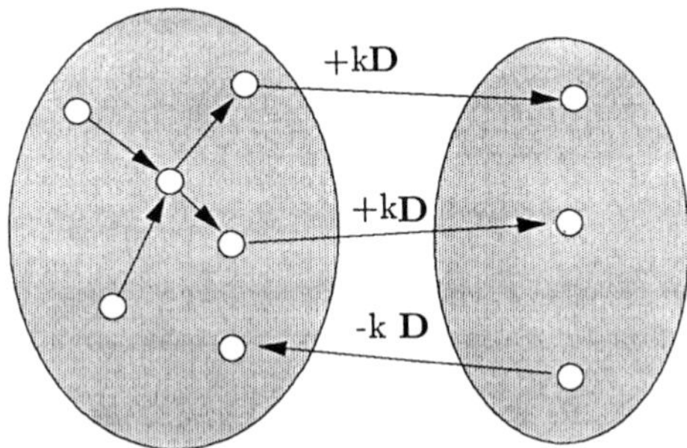


**Figure 6.7** Illustrations of delay-transfer rule. Notice that the computation performed by a component is not changed if we advance the inputs and delay the outputs by the same number of time units.

**Theorem:** *All computable SFGs can be made temporally local by following the cut-set systolization rules. Consequently, a spatially local and regular SFG array is always systolizable.*

**Example: Systolization of SFGs for Convolution.** To systolize the SFG in Figure 6.6a, rule 1 is applied with $\alpha = 2$ and then the rule 2 is applied. The resulting systolic array is shown in Figure 6.8a. The systolized version of the SFG in Figure 6.6b is obtained by applying rule 2 only and is shown in Figure 6.8b.

## 3 GENERALIZED MAPPING METHODOLOGY

There is a large class of problems that can be solved by algorithms that are *regular*, but not regular enough to be expressed as UREs. These algorithms can still be expressed in indexed single-assignment format, but the dependence edges and node functions will vary over the index space. Processor array implementations can sometimes still be found by linear mappings, but only in certain directions. In order to handle these cases, we have developed a *generalized mapping methodology* to address a wider class of problems. This mapping methodology, together with canonical mapping, has been used to unify array designs for numerous signal processing algorithms [11, 14] and has served as the basis for the development of new systolic architectures for the algebraic path problem [15, 16].*

### 3.1 Characterizing DGs Based on Projection Directions

In UREs the homogeneity of the DG has been utilized to reduce the complexity of the mapping process and obtain a time-invariant regular array implementation. Unlike UREs, local DGs are not necessarily homogeneous (or totally shift-invariant).

*The algebraic path problem is a general class of problems that includes transitive closure, shortest path, and matrix inversion as special cases.

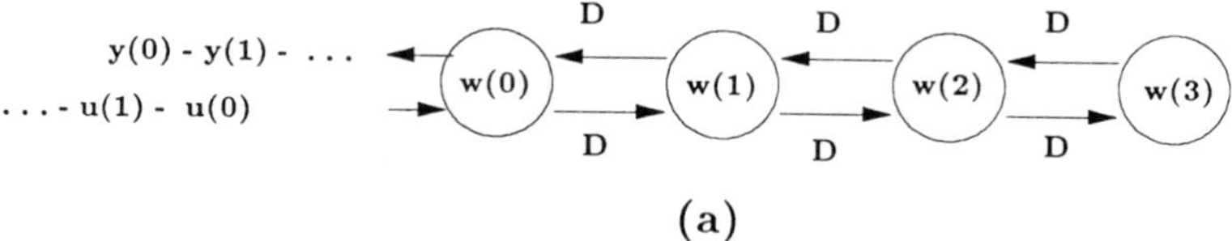


(a)

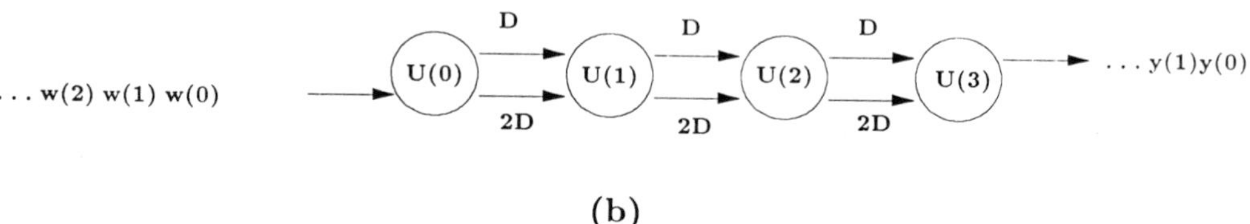


(b)

**Figure 6.8** Two systolic arrays for convolution: (a) systolized version of the SFG in Figure 6.6a; (b) systolized version of the SFG in Figure 6.6b.

In order to examine the property of the SFGs derived from mapping in a certain projection direction, let us focus on direction-specific properties of the DG. This leads to the following extended classification of algorithms based on their shift-invariance characteristics.

Extended classification based on shift-invariance property:

1. *Shift-invariance.* If the dependence arcs corresponding to *all* nodes in the index space do not change with respect to the node positions, then the DG is said to be shift-invariant (SI). An example is given in Figure 6.9a.
2. *Directional shift-invariance.* Given a direction $\vec{d}$, if the dependence arcs corresponding to the nodes along $\vec{d}$ remain invariant with respect to the node position, then the DG is said to be *directionally shift-invariant* (DSI) along $\vec{d}$. Note that SI is equivalent to DSI in all directions. An example of DSI is given in Figure 6.9b.
3. *Pseudo DSI.* Given a projection direction $\vec{d}$, if the projected components of the dependence arcs (which are obtained by applying the projection along $\vec{d}$) are invariant for the nodes along $\vec{d}$, then the DG is said to be *pseudo DSI* (PDSI) along $\vec{d}$. Note that the dependence arcs along $\vec{d}$ may have global or even opposite components. The components will, however, not affect the projected arcs, since they correspond to zero components after projection. An example is given in Figure 6.9c.

Recall that an STI SFG has static interconnection; i.e., the interconnection arcs are invariant with time. (However, by definition, in an STI SFG the functions performed within the nodes may be time-varying.) An SI DG is mapped into an STI SFG by projection in any direction. From a broader perspective, an STI SFG can be obtained as long as the corresponding projections of the dependency arcs are DSI. In other words, an STI SFG can be obtained by projecting in any direction that is at least PDSI. These properties may be exemplified by the Gauss–Jordan elimination algorithm discussed below.

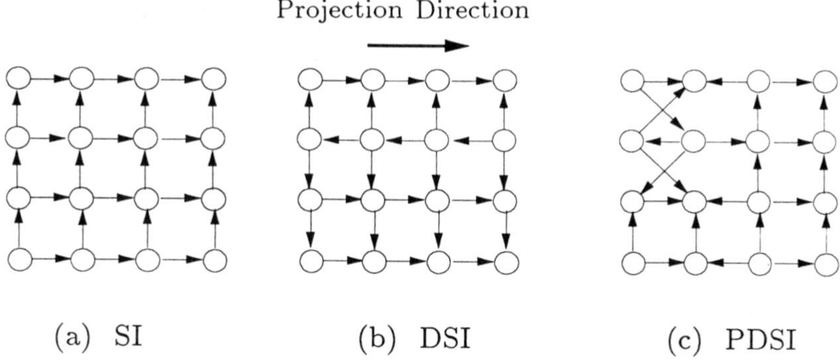


**Figure 6.9** Illustration of (a) SI, (b) DSI, and (c) PDSI projections.

**Example: Gauss–Jordan Elimination.** Given an $N \times N$ nonsingular matrix $\mathbf{A}$, the inverse of $\mathbf{A}$ can be found by applying elementary row operations on $\mathbf{A}$ and successively reducing $\mathbf{A}$ to an identity matrix $\mathbf{I}$. Then the product of all basic row operations applied on $\mathbf{A}$ is the inverse of $\mathbf{A}$, that is, $\mathbf{A}^{-1}\mathbf{A} = \mathbf{I}$. If this same set of row operations is applied to the identity matrix, the result will be $\mathbf{A}^{-1}$. This is the basic idea of the Gauss–Jordan elimination method, which is very similar to the Gaussian elimination only the entries below the diagonal are eliminated, so the resulting matrix is an upper triangular matrix; while in Gauss–Jordan elimination the entries both below and above the diagonal are eliminated and eventually the matrix is reduced to an identity. (We assume that no pivoting is needed.) Mathematically, Gauss–Jordan elimination starts initially with an augmented matrix $\mathbf{X}^0 = [\mathbf{A} \mid \mathbf{I}]$ and finally ends with $\mathbf{X}^N = [\mathbf{I} \mid \mathbf{A}^{-1}]$. The recursive equation is described in the following form:

For $k$ from 1 to $N$, $i$ from 1 to $N$, $j$ from $k$ to $2N$

$$x_{ij}^{k} \longleftarrow \begin{cases} x_{ij}^{k-1} \cdot (x_{kk}^{k-1})^{-1} & \text{if } i = k \\ x_{ij}^{k-1} - x_{ik}^{k-1} \cdot (x_{kk}^{k-1})^{-1} \cdot x_{kj}^{k-1} & \text{otherwise} \end{cases}$$

where $\mathbf{X}^0 = [\mathbf{A} \mid \mathbf{I}]$ and $\mathbf{X}^N = [\mathbf{I} \mid \mathbf{A}^{-1}]$.

Suppose that for the purpose of solving a linear system, we want to compute $\mathbf{A}^{-1}\mathbf{b}$. We can do it simply by initializing $\mathbf{X}^0 = [\mathbf{A} \mid \mathbf{b}]$ instead of $[\mathbf{A} \mid \mathbf{I}]$, and the final result $\mathbf{X}^N$ will be $[\mathbf{I} \mid \mathbf{A}^{-1}\mathbf{b}]$.

*DG (stage 1 design).* The preceding dependences can be localized by adding propagating (transmittent) variables. One is the row transmittent variable $r$, which propagates data to all the nodes in a row. The other is the column transmittent variable $c$, which propagates data to all the nodes in a column.

For $k$ from 1 to $N$, $i$ from 1 to $N$, $j$ from $k$ to $2N$

$$r(i,j,k) \longleftarrow \begin{cases} x(i,j,k-1) & \text{if } j = k \\ r(i,j-1,k) & \text{if } j > k \end{cases}$$

$$c(i,j,k) \longleftarrow \begin{cases} 1 & \text{if } i = k \text{ and } j = k \\ x(i,j,k) & \text{if } i = k \text{ and } j > k \\ c(i+1,j,k) & \text{if } i < k \\ c(i-1,j,k) & \text{if } i > k \end{cases}$$

$$x(i,j,k) \longleftarrow \begin{cases} x(i,j,k-1) \cdot r(i,j,k)^{-1} & \text{if } i = k \\ x(i,j,k-1) - r(i,j,k) \cdot c(i,j,k) & \text{otherwise} \end{cases}$$

where input $\mathbf{X}^0 = [\mathbf{A} \mid \mathbf{I}]$ and output $\mathbf{X}^N - [\mathbf{I} \mid \mathbf{A}^{-1}]$.

The 3-D DG is shown in Figure 6.10, with each $ij$ plane drawn separately. [Not drawn are the additional dependence lines in the $k$ direction that run from points $(i,j,k)$ to $(i,j,k+1)$.] In each plane the points that serve as the *source* of the row and column are marked as dark nodes.

Starting from this DG, we present three different projections, each having different properties, and the resulting SFGs.

*Case 1:* *Projection in the $[1\ 1\ 1]^T$ Direction—A DSI projection.* The $[1\ 1\ 1]^T$ is a DSI direction in the DG. If we project the DG along this direction and use the recursion schedule in the $k$ direction, the resulting SFG will be a TI SFG as shown in Figure 6.11a. The function of each node is fixed and the structure of the SFG does not change with time.

*Case 2:* *Projection in the $j$ Direction—a DSI projection.* The projection direction in the $j$ axis is a DSI direction, and the resulting SFG array is an STI SFG, which can be scheduled linearly. The SFG obtained from this projection is shown in Figure 6.11b, in which a default schedule is used.

*Case 3:* *Projection in the i-Direction—a PDSI projection.* The projection direction in the $i$ axis is a PDSI direction. The schedule we are using is still the default schedule. However, due to the PDSI projection, the SFG does not have a linear schedule. This SFG is shown in Figure 6.11c.

Note that if the projection direction chosen is the $k$ axis, then the projection does not exhibit either DSI or PDSI properties. Therefore, the corresponding SFG is no longer regular.

**Remark:** Matrix multiplication, LU decomposition, and Gauss–Jordan elimination all share a common recursive formulation:

$$x_{ij}^{k} = x_{ij}^{k-1} + x_{ik}^{k-1} \cdot (x_{kk}^{k-1})^{*} \cdot x_{kj}^{k-1} \tag{6.7}$$

where $+$, $\cdot$, and $^*$ denote algebraic operators to be specified by the application. This generalized formulation covers a broad and useful application domain. In addition to the matrix operations mentioned above, it also covers transitive closure, shortest-path problems, and many others. The option of adopting different projection directions as well as some reindexing schemes allows many varieties of SFGs to be derived. It is essential for achieving an optimal systolic array.

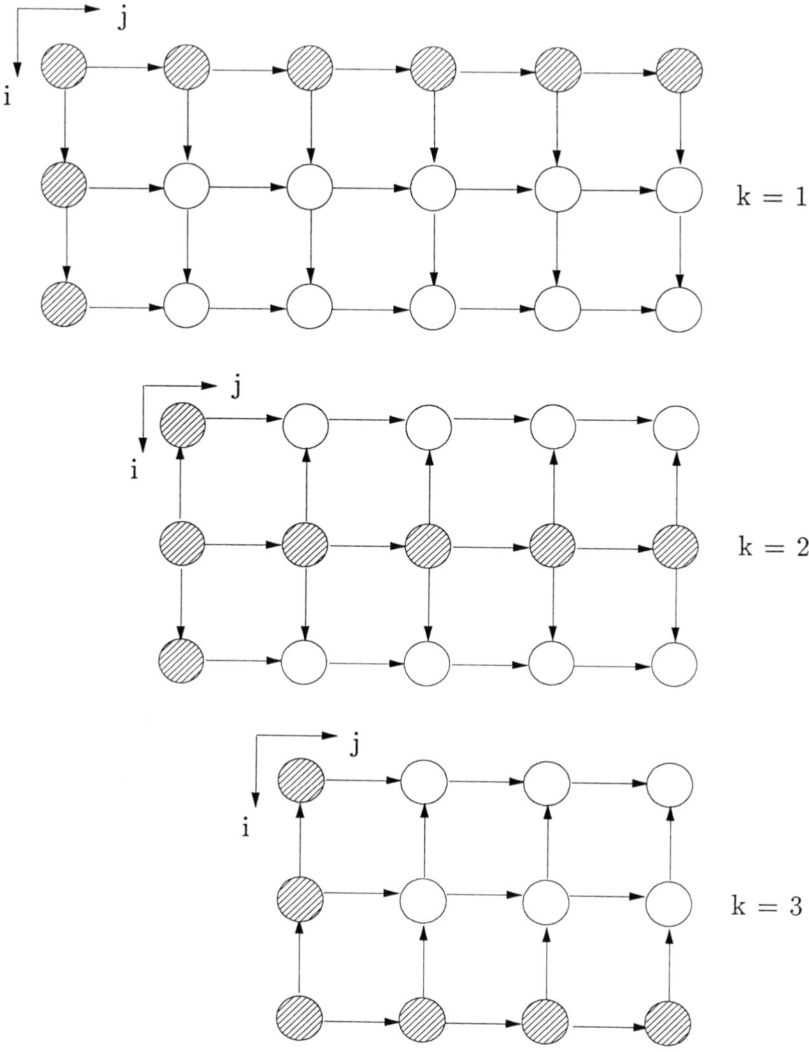


**Figure 6.10** DG for Gauss–Jordan Elimination ($N = 3$).

## 3.2 Nonlinear Schedule and Nonlinear Assignment

### Piecewise Linear Schedule

If the projection vector $\vec{s}$ is in a DSI direction, there are two kinds of mappings: linear schedule mapping (LS-DSI) and nonlinear schedule mapping (NLS-DSI). An LS-DSI mapping maps a DG to an SFG with a linear schedule $\vec{s}$, which satisfies the following conditions:

1. $\vec{s}^T \vec{e} \geq 0$, for any dependence arc $\vec{e}$.
2. $\vec{s}^T \vec{d} > 0$.

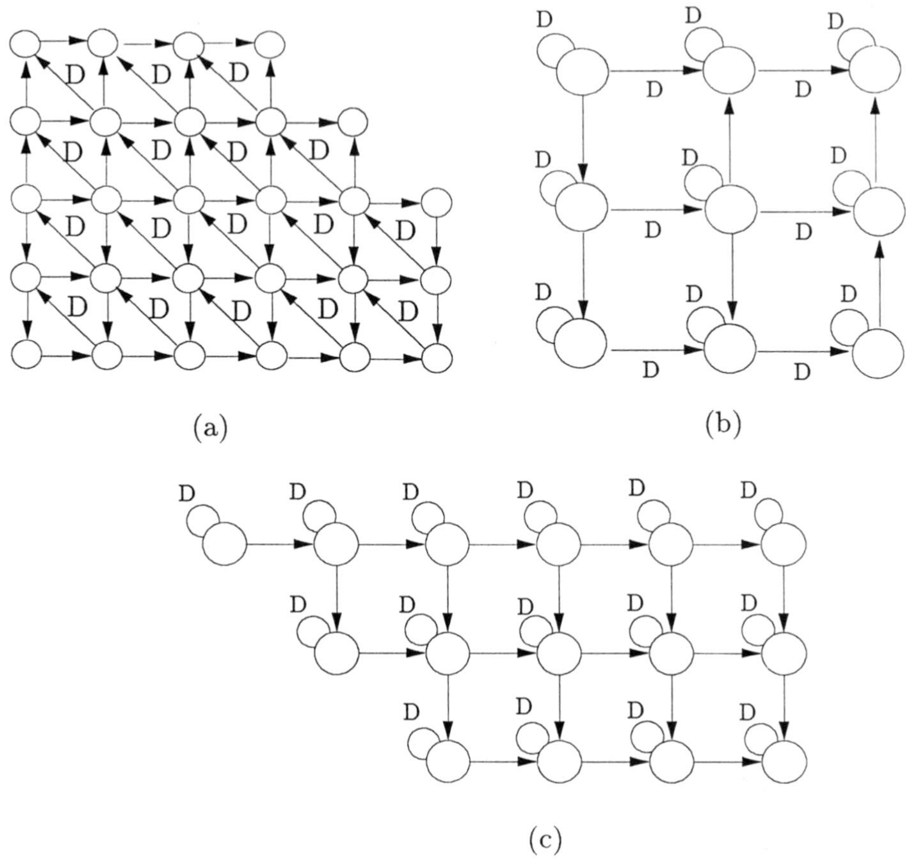


**Figure 6.11** (a) TI SFG derived by projection in the $[1\ 1\ 1]^T$ direction; (b) an STI SFG derived by projection in the $j$ direction; (c) an STI SFG derived by projection in the $i$ direction.

An example is given in Figure 6.12a. A mapping other than an LS-DSI mapping is called an NLS-DSI mapping.

Usually, an NLS-DSI mapping can be viewed as a combination of many relaxed LS-DSI mappings, where condition 2 is replaced by

$$\vec{s}^T \vec{d} \neq 0$$

In this case, such a mapping is called a *piecewise linear schedule* (PWLS). An example is given in Figure 6.12b.

Since a PWLS-DSI mapping can be viewed as a combination of many LS-DSI mappings, the corresponding SFG should require only simple control. To prove that the circuit design will be regular and modular, we note that the schedules are linear on both sides of a partitioning "interface" plane (see Figure 6.13). Actually, a linear schedule may be used for each of the LS-DSI

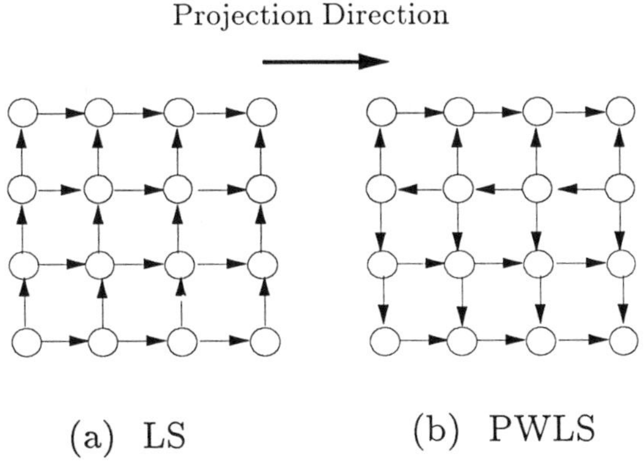


**Figure 6.12** (a) An LS-DSI DG mapping; (b) a PWLS-DSI DG mapping.

DG regions and each region may be projected to an STI SFG with simple control. What is worth noting is the design of interface modules required for handling data crossing the interface plane. There are two possible types of relationships between the two neighboring linear schedules. The first possibility is that both the linear scheduling time indices increase along the $\vec{d}$ direction. The second is that the two linear schedules progress in opposite directions with respect to $\vec{d}$. Correspondingly, there are two possible interface models. For the first case, for the purpose of adjusting and balancing the speeds of data processing rates, a buffer (FIFO: first in, first-out) suffices to provide a proper interface between the two regions. On the other hand, for the second case, a stack (LIFO: last-in, first-out) suffices to handle the interface. Note that all the FIFOs and LIFOs mentioned require only simple control. A PWLS-DSI DG to SFG mapping and its corresponding SFG with FIFOs and LIFOs are shown in Figure 6.13.

Projections in non-DSI directions generally allow only nonlinear schedules. However, the projected SFGs may have other advantages, such as fewer PEs, faster pipelining, or higher utilization of the array. If an advantageous trade-off can be reached, a nonlinear schedule mapping may become preferred.

## Nonlinear Assignment

By *nonlinear assignment* we mean that multiple node *not necessarily along a straight line* are assigned to a PE. Nonlinear assignment to map a DG to an SFG usually incurs the expense of somewhat sophisticated control. In certain special circumstances, a nonlinear assignment mapping may offer some unique flexibility and advantages.

**Example: Consecutive Matrix-Vector Multiplication.** By using a nonlinear assignment, a more flexible design can be devised and a broader range of algorithms can be covered. One such example is an algorithm represented by *cascaded DG*, in which the algorithm comprises a group of DGs connected in cas-

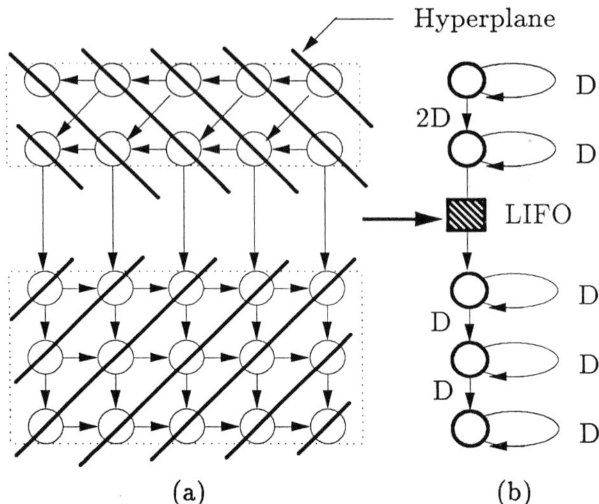


**Figure 6.13** (a) A PWLS-DSI DG; (b) its corresponding SFG with FIFOs and LIFOs.

cade. A simple example is shown in Figure 6.14, where three DGs for matrix-vector multiplication are cascaded to compute $\mathbf{e} = \mathbf{ABCd}$. Here, $\mathbf{A}$, $\mathbf{B}$, and $\mathbf{C}$ are $3 \times 3$ matrices and both $\mathbf{e}$ and $\mathbf{d}$ are $3 \times 1$ vectors.

For this cascaded DG, all the DGs involved are the same. A nonlinear assignment can be applied to this cascaded DG (see Figure 6.14). In this case, the nonlinear mapping allows handling of the three DGs together in one piece rather than separately. This will ease the data reformatting and increase the pipelining rate.

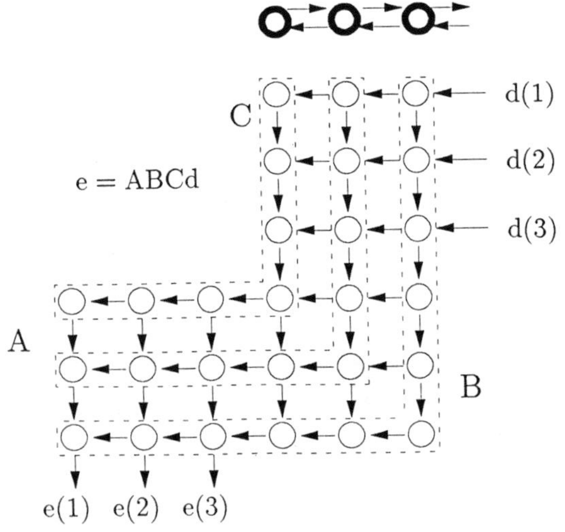


**Figure 6.14** Cascaded DG for matrix-vector multiplication.

## 3.3 Mapping to Arrays Without Interior I/O

In many cases projecting a DG onto an SFG will cause the resulting I/O to fall on interior nodes of the SFG array, resulting in a large number of I/O ports. For example, Figure 6.15a shows a two-dimensional $N \times N$ DG in which inputs occur at the top and left sides and outputs occur at the bottom and right sides. Although all I/O for this DG occurs on the boundaries, any projection of this DG onto a linear array will result in I/O on the array's interior nodes. This is illustrated in Figure 6.15b, which shows a vertical projection onto a linear SFG array.

To circumvent this problem, it is possible to *extend* the index space of the DG so that all I/O occurs at points that will be mapped to the boundary nodes of the SFG. This extension corresponds to defining the communication scheme to transport input (output) data from (to) the boundary nodes of the SFG to (from) the interior nodes, where it is needed (produced). Extending index space to eliminate interior I/O was proposed in [9] for restricted projection directions. Here this concept is generalized to deal with arbitrary projection directions. Given a DG with a projection direction $\vec{d}$, the procedure of index space extension is as follows:

1. *Decide the I/O border of the DG.* Since the structure of the SFG is determined by $\vec{d}$, the boundary nodes of the SFG are well defined and the intended I/O

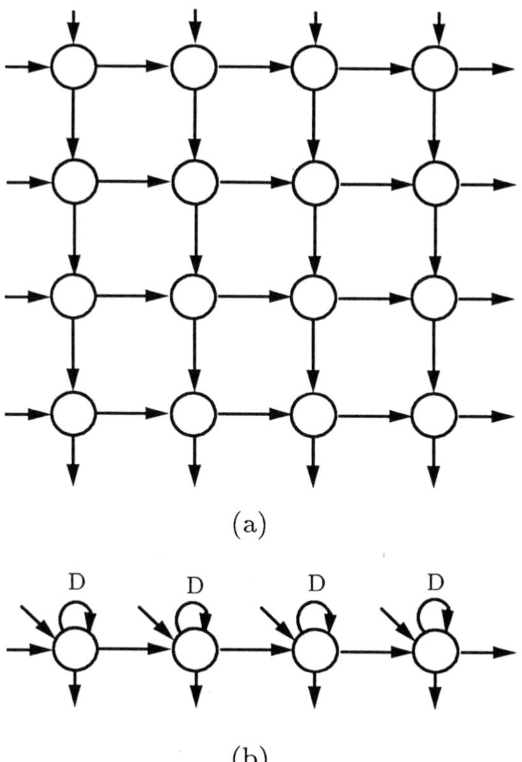


**Figure 6.15** (a) Dependence graph; (b) SFG from vertical projection.

positions in the SFG are known. Some borderlines (or planes), which will be projected to the boundary nodes of the SFG, may be decided in the DG. Figure 6.16a shows borderlines for the previous example with $\vec{d} = [1\ 1]^T$ and Figure 6.16b shows borderlines with $\vec{d} = [0\ 1]^T$.

2. *Extend the DG with some NOP (no operation) nodes.* The purpose of these NOP nodes is to transmit the input (output) data from (to) the borderlines to (from) the destination nodes. *To simplify the control of the resulting array, no extra dependence edge directions (other than those of the original DG) should be used for these NOP nodes.* The insertion of NOP nodes for the previous example with $\vec{d} = [1\ 1]^T$ and its resulting SFG (using the default schedule) is shown in Figure 6.17. Note that only a single-dependence edge direction is used for each NOP node. The insertion of NOP nodes for the previous example with $\vec{d} = [0\ 1]^T$ and its resulting SFG (using the schedule $\vec{s} = [1\ 1]^T$) is shown in Figure 6.17. Note that a single-dependence edge direction for each NOP node is not enough for this projection direction and thus two directions are used.

**Comment:** The detailed I/O operation of the array in Figure 6.17 can be stated as follows. The data are first loaded from boundary PEs one by one. Once the destination PE is reached, the computation starts. Note that this is different from the concept of *preloading of data*, where the computation is started only after *all* the input data reach their destination PEs. Similarly, the output data will go toward the boundary PEs once they are produced without waiting until *all* the output data are produced. In this array each boundary PE handles both input and output.

The detailed I/O operation of the array in Figure 6.18 is as follows. Each PE takes its required input data, lets the other data pass through, and starts to do computation after all the data with other PEs as destination pass through. Once the computation is started, the succeeding data are treated as normal incoming

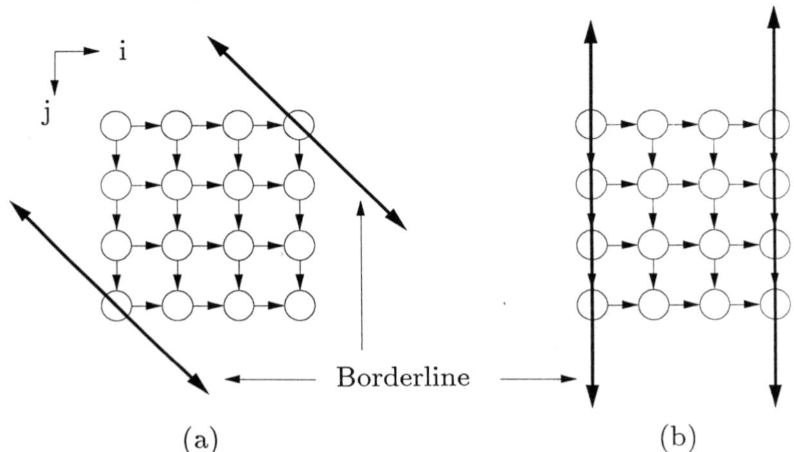


**Figure 6.16** (a) Borderlines with $\vec{d} = [1\ 1]^T$; (b) borderlines with $\vec{d} = [0\ 1]^T$.

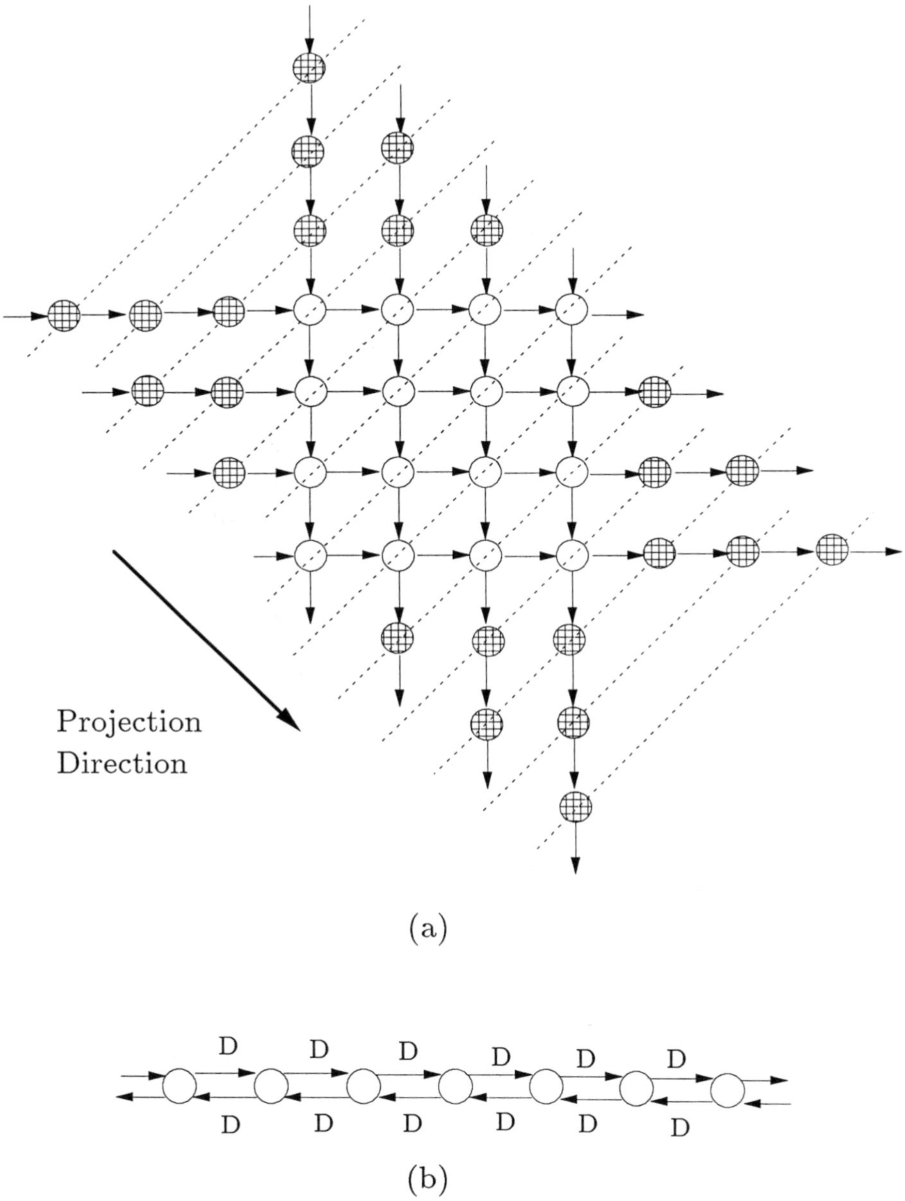

Projection
Direction

(a)

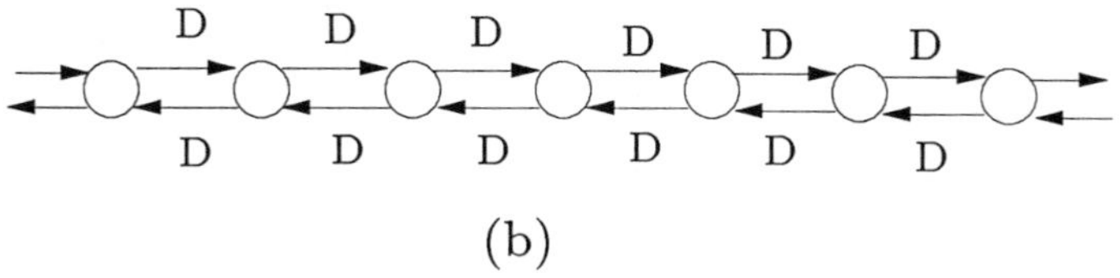


(b)

**Figure 6.17** (a) Index space extension with $\vec{d} = [1\ 1]^T$; (b) resulting SFG.

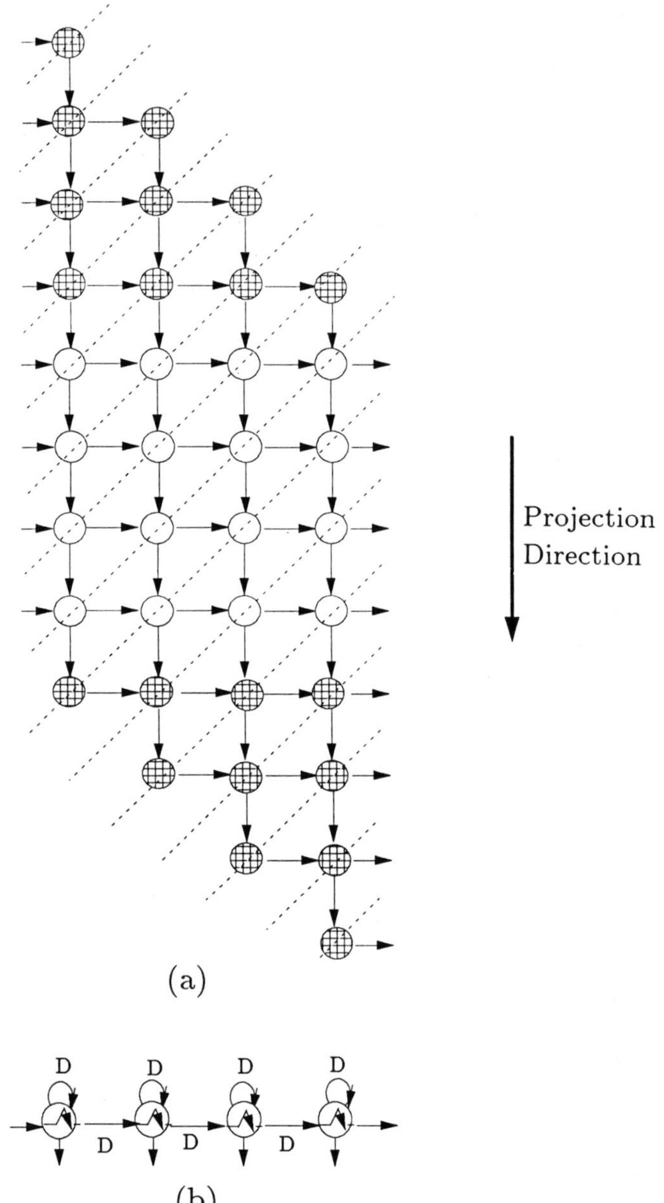


**Figure 6.18** (a) Index space extension with $\vec{d} = [0\ 1]^T$; (b) resulting SFG.

data (no more passing through unless the data are transmittent). In this array the input and output are handled by different boundary PEs.

Note that both arrays (Figure 6.17b and Figure 6.18b) require some control in each PE to indicate whether to take the incoming data and do computation or to let the data pass through. This kind of control mechanism is also required in the implementation using preloading of data or dumping of output data after all computation is done. But the implementation of preloading is easier since the whole array is under the same process at one time. This is a trade-off between speed and hardware.

## 3.4 Multiprojection

In general, it is possible to map an $N$-dimensional DG directly onto an $(N-k)$-dimensional SFG ($k = 1, 2, \ldots, N-1$) [7, 9, 17]. In the previous sections we proposed a methodology that maps an $N$-dimensional DG to an $(N-1)$-dimensional SFG. In principle, the method can be applied $k$ times and thus reduces the dimension of the array to $N-k$. More elaborately, a similar projection method can be used to map an $(N-1)$-dimensional SFG into an $(N-2)$-dimensional SFG, and so on. This scheme is called the *multiprojection* method.

To facilitate the discussion on the multiprojection technique, we introduce the notion of *DSI subspace*, which extends the previous notion of DSI direction. Given a DG, a $k$-dimensional subspace is a DSI subspace if and only if the DG is DSI in any direction in the $k$-dimensional subspace. [Equivalently, a $k$-dimensional subspace is a *DSI subspace* of a DG if and only if the DG is DSI in $k$ independent vectors (directions) in the subspace.] To simplify the discussion, let us assume that the DG is $N$-dimensional, the DSI subspace is 2-D and is on the $ij$ plane, and, after one projection (say in the $i$ direction), the resulting SFG if $(N-1)$-dimensional. The question is *how to further project the SFG to an (N - 2)-dimensional SFG.*

The potential difficulties of this mapping are 1) the presence of delay edges in the $(N-1)$-dimensional SFG and 2) the possibilities of loops or cycles in an SFG, although no loops or cycles can exist in a DG. Therefore, mapping this SFG to an $(N-2)$-dimensional SFG will require additional care.

To handle the cycle problem, an *instance graph* (at a certain time $t = t_0$) is defined as *the SFG with all the delay edges removed.* Hence the *instance graph* has no loops or cycles. According to the SFG schedule, all the nodes in the instance graph are executed simultaneously at $t = t_0$. An activity instance (at $t_0$) can be defined as the nodes represented by the instance graph at $t = t_0$. According to the SFG schedule, there will be no overlap in time between two consecutive activity instances (one at $t = t_0$ and the next instance at $t = t_0 + D$).

Recall that a mapping methodology should consist of a projection part and a schedule part.

- As to the projection part, since the assumption of $ij$ DSI subspace implies that the instance graph is itself DSI in the $j$ direction, the same projection method previously proposed may be applied in the $j$ direction. *The original SFG (with delay edges included) is now projected along the $\vec{d}$ (i.e., $j$) direction.*

- The schedule part is somewhat more complicated. First, it is always possible to find a valid SFG schedule vector $\vec{s}$ for the instance graph, since there is no loop or cycle in an instance graph. To project the $(N-1)$-dimensional instance graph to an $(N-2)$-dimensional graph, it is necessary to create a new type of delay, denoted by $\tau$. Note that the ***global*** delay $D$ and *local* delay $\tau$ are intimately related. The relationship depends on the constraints imposed by both the processor availability and the data dependency (i.e., data availability). ***The original delay edges with $\beta D$ map to an edge bearing delay weight $\beta D + \vec{s} \cdot \vec{e}\tau$ (see Figure 6.19b)).***

Now let us more closely examine the relationship between $D$ and $\tau$. First, to ensure *processor availability*, an activity instance must have adequate time to be complete before the next activity starts. So the following condition is necessary:

(a) $D \geq \tau + (M-1)(\vec{s} \cdot \vec{d})\tau$

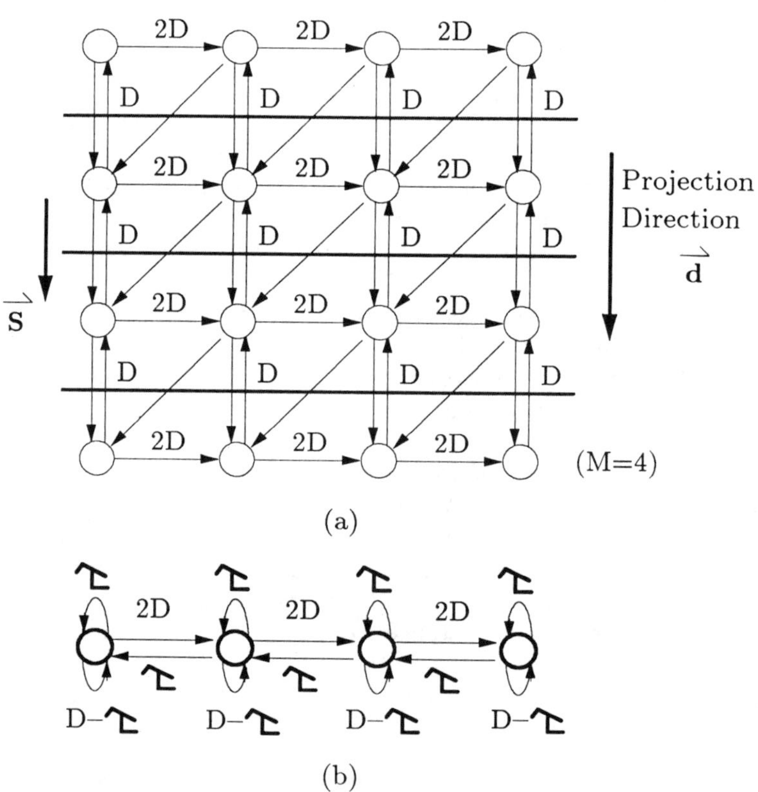


**Figure 6.19** (a) Original SFG; (b) further projected SFG.

where $M$ is the maximal number of nodes along the $\vec{d}$ direction in the instance graph (see Figure 6.19a). Note that this condition also guarantees that there is no time overlap between two activity instances.

Second, to ensure *data availability*, conditions (b) and (c) are necessary for all edges $\vec{e}$:

(b) $\beta D + (\vec{s} \cdot \vec{e})\tau \geq 0$

(c) $\beta D + (\vec{s} \cdot \vec{e})\tau \geq \tau$ for at least one edge in every cycle

Condition (b) is necessary in order to satisfy the causality condition of the SFG. Condition (c) is necessary to ensure that every cycle in the new SFG has at least one delay element ($\tau$) in the cycle. Note that these two conditions are necessary conditions for a graph to be qualified as an SFG.

Conditions (a), (b), and (c) together guarantee both the processor and data availability for the new SFG. Therefore, the mapping is complete. Note that condition (a) is very often the dominant constraint. In fact, condition (a) would be sufficient to ensure conditions (b) and (c) whenever a locally interconnected SFG is considered.

Obviously, this mapping methodology may be directly applied to map from an $(N-2)$-dimensional SFG to an $(N-3)$-dimensional SFG. The method can be applied $k$ times (i.e., using $k$ steps of simple projection) to reduce the dimension of the array to $N-k$.

**Example: Band Matrix Multiplication.** To demonstrate the procedure, a 2-D SFG for *band matrix multiplication* and its corresponding projected SFG are shown in Figures 6.20a and 6.20b. In Figure 6.20b, note that the matrix **B** is input in parallel and will lead to I/O bandwidth of order $M$, which is usually not desirable. Since there is only one data in $M$ time steps for each parallel input edge, these data can be interleaved and input from the boundary PE, as shown in Figure 6.20c. An alternative way to obtain this SFG is to extend the index space (see Section 3.3) of the SFG and overlap the execution of the extended index space (for I/O) with the next instance graph (for computation). By a similar technique, the output directly from interior nodes (cf. Figure 6.20c) may also be avoided.

## 4 MATCHING ALGORITHMS TO SYSTOLIC ARRAYS

The issue addressed in previous sections has been *mapping* an algorithm onto a dedicated array design. The main question in algorithm mapping is: *How is an array processor design dependent on the algorithm under consideration?* The key parameters of the array, such as the number of PEs or communication links or the memory size, are assumed to be very flexible and without any constraints. A closely related question is: *How is a set of algorithms best programmed and executed on a specific array processor?* This is referred to as the *algorithm matching* problem.

From a designer's point of view, matching is basically a "constrained mapping." In other words, it is to fit algorithms to an array, based on the existing constraints

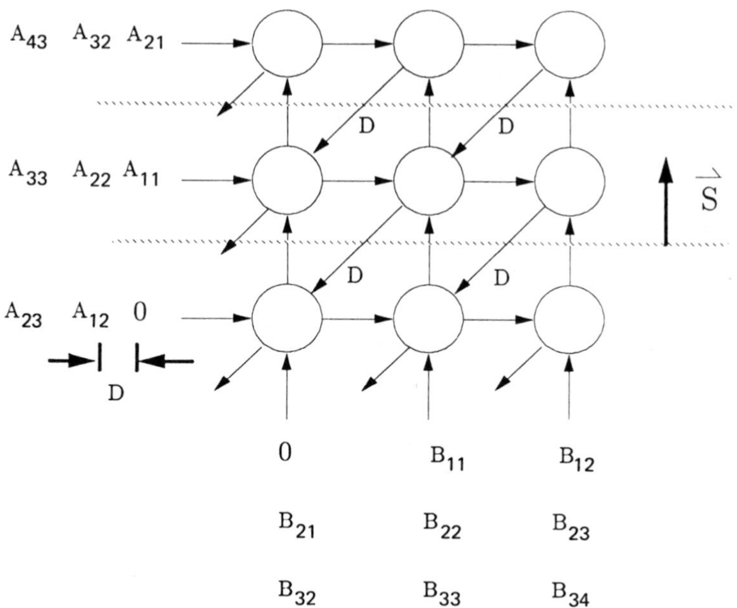


(a)

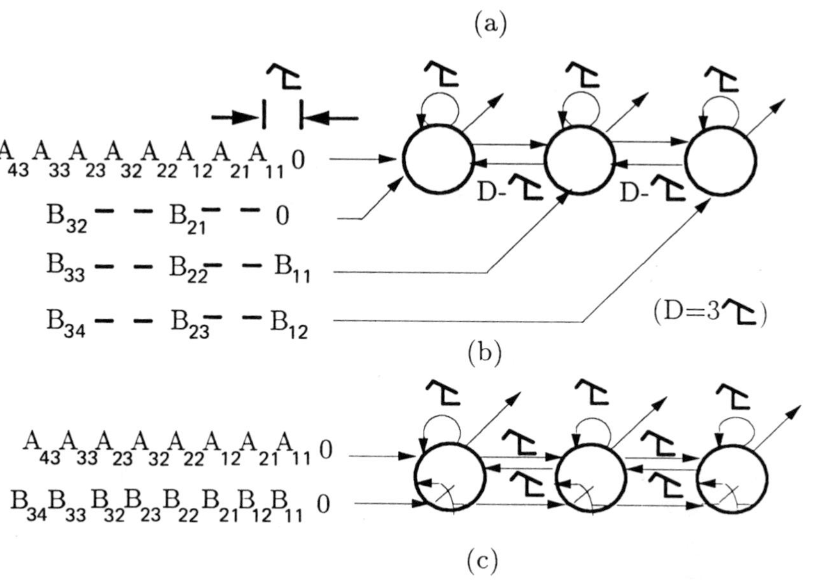


(b)



(c)

**Figure 6.20** Multiplication on band matrix multiplication: (a) A 2-D SFG and a "good" schedule; (b) the 1-D projected SFG; (c) the final SFG using $\mathbf{D} = 3\tau$.

on array architecture, such as PE number, communication, and (global or local) memory size. If constraints are on the communication links, some special methods will be necessary to solve the problem. When the constraints are on the number of PEs or the size of memory, the matching problem becomes a partitioning issue or multiprojection issue as explained in Section 3.4.

## 4.1 Matching Algorithms to Fixed Array Structures

In the literature, there has been very little study of a general methodology for matching algorithms with certain structures. Here we discuss two case study examples. One is about matching local DGs to mesh arrays and the other about matching mesh algorithms to hypercube structures.

### Matching Local DGs to Mesh Arrays

When the SFG is defined on a lattice space, local DGs require only local communication edges. Note that some of these local edges may be diagonal or antidiagonal edges. We now show that all these edges can be implemented on a mesh array. The basic idea is to replace, say, a *southeastern* diagonal edge (cf. Figure 6.21a) by one *south* edge and one *east* edge (cf. Figure 6.21b). However, in general, a mesh architecture can support only a single channel communication each way. Therefore, the newly created south edge will have to share the same channel with the existent south edge by a time-sharing scheme. For example, a possible time-sharing schedule is represented by the two delay parameters on the arcs shown in Figure 6.21c. In fact, they indicate different buffer sizes used for the two different edges shown in Figure 6.21b which share the same channel. The schedule for time-sharing is not necessarily unique and it may greatly affect the processing speed performance. Therefore, a good scheduling scheme is very desirable. Fortunately, a systematic DG modification procedure may be devised, based on which a channel time-sharing schedule may easily be determined. The basic idea is to modify the DG so that the DG arcs originally projected to diagonal edges may be replaced by some new arcs which are projected to edges implementable by mesh connections. This objective may be accomplished in three steps: 1) insert some extra layers of nodes in the DG, 2) "bend" the original undesirable arcs in the DG into some segments, with the "inserted" nodes as intermediate nodes, and 3) make sure each segment may be projected onto a channel in the mesh connections.

For the scheduling of the modified DG, there are at least two approaches. A straightforward scheme is to assign the "inserted" nodes the same amount of time as assigned to the "original" nodes; then the slowdown overhead will be severe. Another scheme is to assign only communication time, which is usually shorter than the computation time of the original nodes, to the inserted nodes (since no data processing occurs), and thus PEs execute different nodes at different speeds. In this case, the resulting array will have higher PE utilization efficiency at the expense of a more complicated control. It can be shown that under such a matching scheme, or any other matching scheme, the PE utilization efficiency will be hampered. This is an unavoidable cost when the algorithm does not directly fit the architecture. In short, with different degrees of overhead, all the local DGs may eventually be

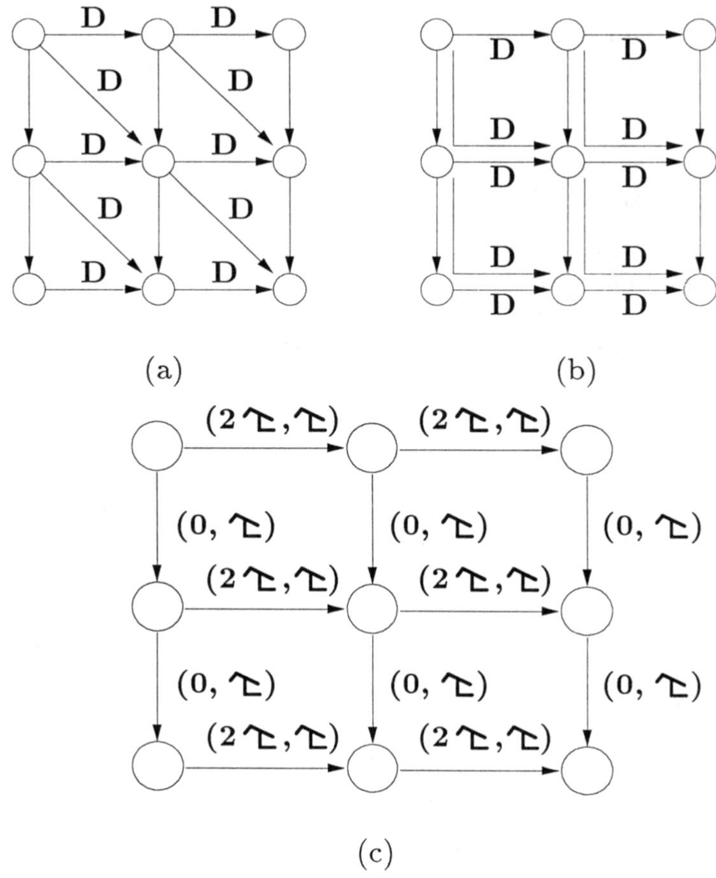


**Figure 6.21** (a) SFG with (southeastern) diagonal edges. (b) SFG with one south edge and one east edge to replace the southeastern diagonal edge in (a). (c) SFG whose edges are time-shared. Each edge is labeled with a delay vector to indicate the different buffer sizes for different data sharing the same edge. The first element shows the size of the buffer for the data originally using the edge. The second element shows the size of the buffer for the data "routing through" the edge.

matched to mesh arrays. This is an important step in the task of matching a local DG to a hypercube structure. This will be our next topic.

## Matching Mesh Algorithms to Hypercube Structures

Our aim here is to demonstrate that a mesh algorithm (of any dimension) may be embedded in a hypercube [18]. Combining this with the previous observation, we may thus conclude that all local DGs may be naturally matched to a hypercube array.

An $n$-cube is a hypercube consisting of $2^n$ nodes. The nodes of an $n$-cube can be labeled by $n$-bit binary indices, from 0 to $2^n - 1$. *By the definition of hypernetwork, two processors are directly linked if and only if their binary index representations differ only by 1 bit.* For example, if $n = 3$, then the eight nodes can be represented as the vertices of a three-dimensional cube as shown in Figure 6.22. To see how to embed meshes in a hypercube, let us consider a two-dimensional mesh of size $8 \times 4$ as shown in Figure 6.23 and examine how it might be embedded into a 5-cube. (Note that $2^5 = 32$.) By using a 3-bit gray code $(b_1 b_2 b_3)$ for labeling the horizontal dimension and a 2-bit gray code $(c_1 c_2)$ for the vertical dimension, each node in the mesh may be labeled as a 5-bit representation $(b_2 b_2 b_3 c_1 c_3)$ [18]. Due to the gray coding, the corresponding indices between any two neighboring nodes in the mesh differ by exactly 1 bit. By the definition of *hypernetwork*, their corresponding nodes in the hypercube are thus directly connected. Thus, the embedding of this $8 \times 4$ mesh into a 5-cube is complete. In a similar manner, any algorithm executable in a mesh can easily be matched to (and executed by) a hypercube computer.

Note that each PE of an $n$-cube connects to $n$ other PEs and each PE of a mesh connects only to four neighboring PEs. In our method, the communication ability of a hypercube is not fully utilized. How to further enhance the utilization is not clear to us yet.

## 4.2 Partitioning

The partitioning process is basically ***mapping computations of a larger problem to an array processor of a smaller size.*** It is a basic requirement in many practical system designs, since no matter what special-purpose computing hardware is available, there is a computation too large for it [19]. To take into account the partitioning problem, the following factors should be considered [3]:

1. Minimize the amount of overhead in external hardware and external communication caused by partitioning.
2. Minimize additional delays caused by the partitioning process.
3. Minimize control overheads and any increase in the complexity of PEs.
4. Maximize the flexibility and extensibility to cope with variable problem sizes.

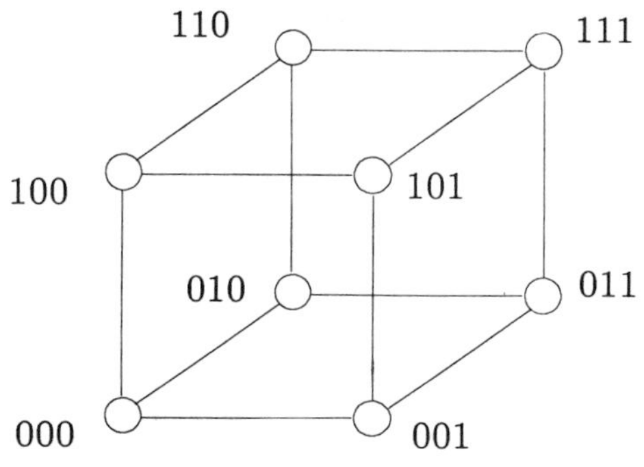


3-cube

**Figure 6.22** Three-dimensional view of the 3-cube.

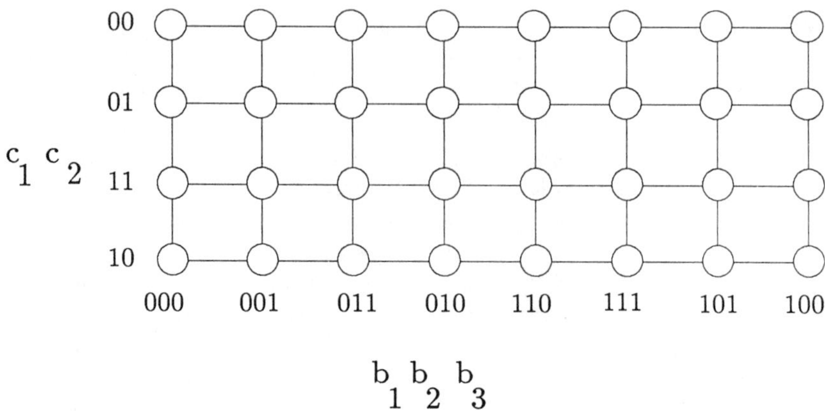


**Figure 6.23** Two-dimensional gray code for an $8 \times 4$ mesh.

There have been several studies of this problem [3, 9, 19–21]. In our treatment of the partitioning problem, we consider the partitioning design as a special option in the stage 2 design, i.e., mapping the DG onto an SFG array of a specified size. This size should reflect what is realistically and economically affordable to build in hardware. In general, the mapping scheme (including both node assignment and scheduling) will be much more complicated than the regular projection methods discussed earlier.

To preserve the regularity of the mapping, two schemes are commonly adopted to map the DG onto an SFG, namely the *locally sequential globally parallel* (LSGP) method and the *locally parallel globally sequential* (LPGS) method. In both schemes, the DG is partitioned into many *blocks* (or bands in [3]). Each block consists of some neighboring nodes of the DG.

**Example: Matrix-Vector Multiplication Partitioning.** Assume that we want to compute a matrix-vector multiplication $\mathbf{Ab} = \mathbf{c}$, where $\mathbf{A}$ is $6 \times 6$ and $\mathbf{b}$ and $\mathbf{c}$ are $6 \times 1$. The DG of this problem is shown in Figure 6.24a. Figure 6.24b shows the DG is partitioned into three blocks. Figure 6.24c shows another partition where two blocks are obtained.

## LSGP Scheme

In the LSGP scheme, one block is mapped to one PE. The number of blocks is equal to the number of PEs in the array. Each PE sequentially executes the nodes of the block assigned to it. This mapping should still preserve the regularity of the array. To ease the sequential computation, local memory within each PE is needed. Note that this partitioning scheme will result in the growth of local memory size with the computation size. As long as the local memory is large enough for the computation under consideration, the LSGP approach is quite appealing. For example, if we have only a linear array of three PEs to solve the previous matrix-vector multiplication problem, the result of the LSGP partitioning is as shown in Figure 6.25.

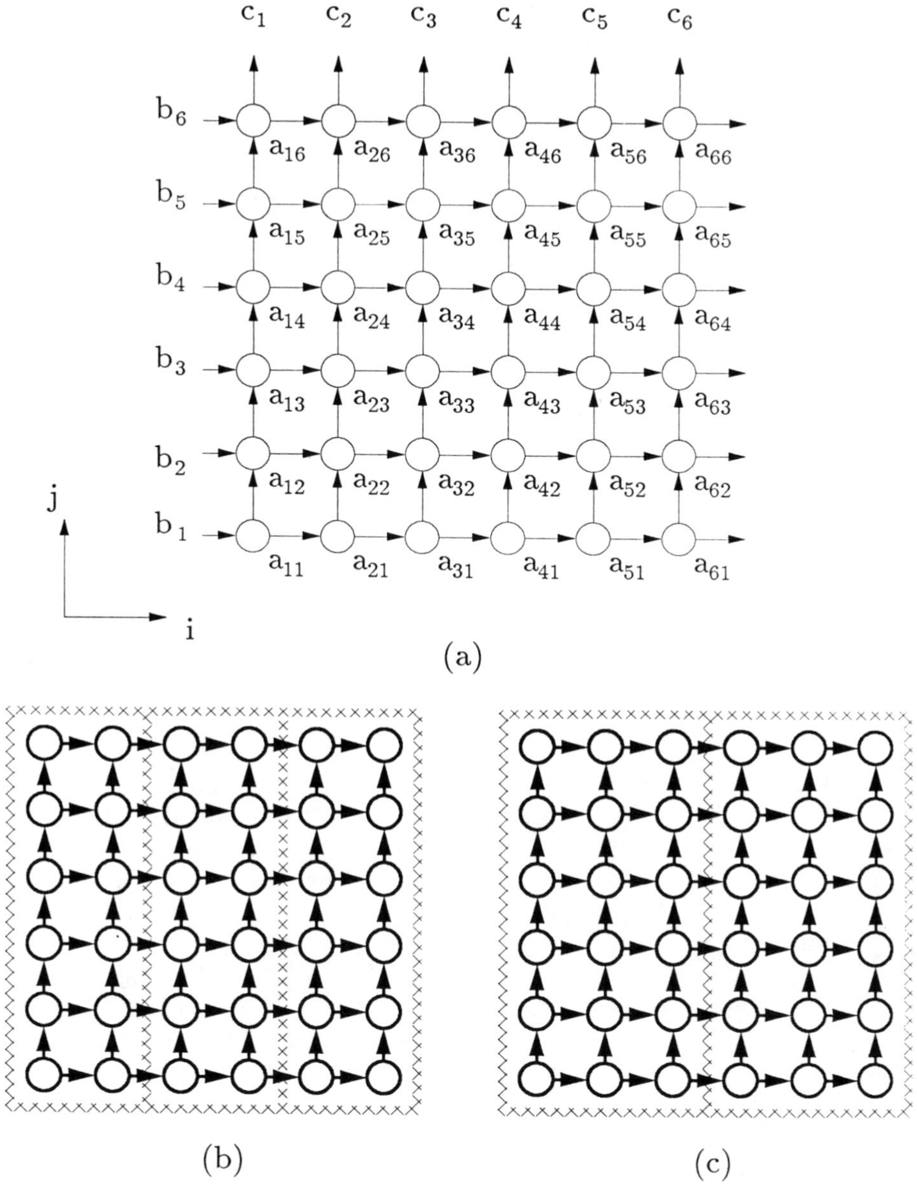


**Figure 6.24** (a) DG for a matrix-vector multiplication; (b) DG being partitioned into three blocks; (c) DG being partitioned into two blocks.

The preceding discussion concentrates only on the assignments (spatial mapping) of a DG to a processor array. The scheduling part has not been specified. In other words, we want to specify at what time and at which processor a computation in the DG takes place. In the matrix-vector multiplication, a linear schedule

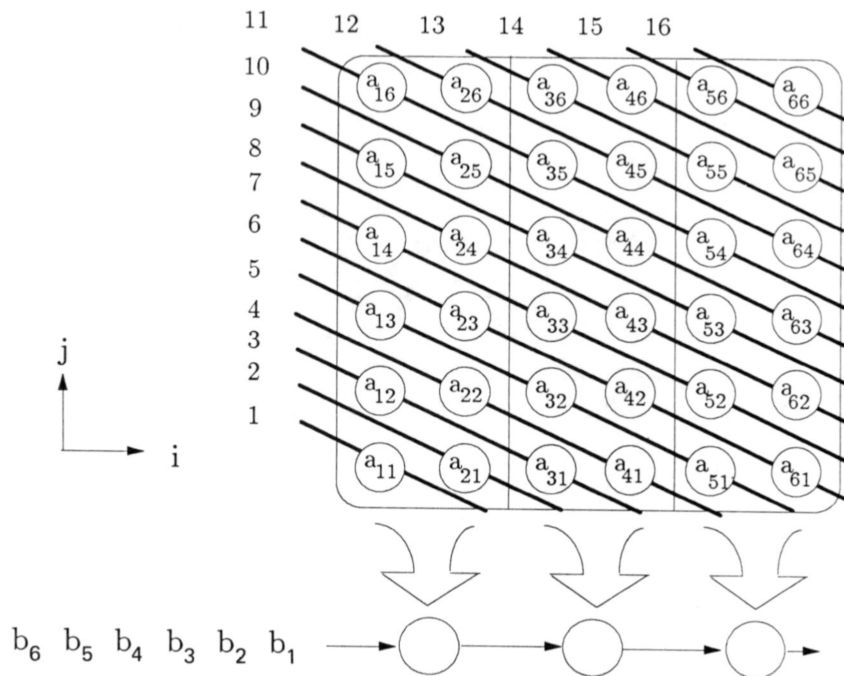


**Figure 6.25** LSGP partitioning of matrix-vector multiplication.

can be found to achieve the shortest computation time under the extra constraints on the *processor availability*; i.e., nodes within each block should be executed sequentially (see Figure 6.25). Such an "optimal" linear schedule, in general, exists for a 1-D array; i.e., the corresponding DG (not necessarily 2-D) is partitioned into blocks along one single direction.* This is, however, not the case if the array is of higher dimensionality. An example is given in Figure 6.26, which shows the LSGP partitioning and scheduling of a 3-D DG. Note that only one layer of the DG is shown and the DG is partitioned into nine blocks ($3 \times 3$), each consisting of four nodes ($2 \times 2$) in one layer of the DG. In total, three sets of blocks may be formed by combining three blocks along the horizontal direction into one set. The nodes within each set are linearly scheduled. The equitemporal hyperplanes of neighboring sets are related in a "piecewise linear" fashion as shown in the figure. In general, if the dimensionality of the array is higher than one, a *piecewise linear schedule* is necessary to achieve the shortest computation time.

## LPGS Scheme

In the LPGS scheme, the block size is chosen to match the array size; i.e., one block can be mapped to one array. All nodes within one block are processed by

*The scheduling is such that 1) the nodes assigned to the same PE are executed in the shortest pipelining period, and 2) the succeeding PE can start can start as soon as possible.

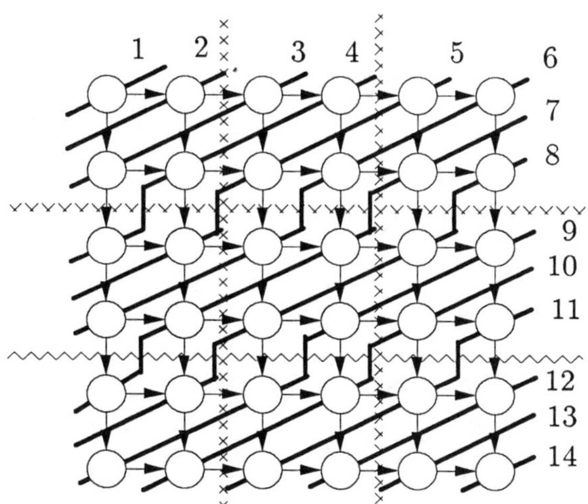


**Figure 6.26** PWLS in an LSGP scheme. Note that only one layer of the 3-D DG is shown and the DG is partitioned into nine blocks (3 × 3), each consisting of four nodes (2 × 2) in one layer of the DG.

the array concurrently—thus locally parallel. The array processes one block after another in a sequential manner—thus globally sequentially. Hence the name LPGS is used for this scheme. In this scheme, local memory size in the PE can be kept constant, independent of the size of the computation. All intermediate data can be stored in certain buffers outside the processor array. Usually, simple FIFO buffers are adopted to store and recirculate the intermediate data efficiently. The result of the LPGS scheme for matrix-vector multiplication is shown in Figure 6.27.

Note that the previous example can be handled by a simple scheme because there is no *reverse data dependence* for the chosen blocks; i.e., the interblock dependences have the good property that blocks can be executed one by one. An example with reverse data dependence for the chosen blocks is shown in Figure 6.28. Thus, in general, the choice of blocks depends on the data dependences. A procedure for choosing the blocks can be found in [3]. The basic idea is to partition an $n$-dimensional DG into blocks by $k$ sets of parallel hyperplanes, where $k$ is the dimensionality of the array. Usually $k$ is assumed to be $n - 1$. Each set of hyperplanes can be represented by a vector normal to all the hyperplanes in that set. Thus a partition is to find linearly independent vectors $\vec{b}_i$ $(i = 1, 2, \ldots, k)$ such that there is no reverse data dependence. Or, equivalently,

$$\vec{b}_i^T \cdot \vec{e} > 0 \qquad \text{for all dependence } \vec{e} \text{ and } i = 1, 2, \ldots, k$$

As to the scheduling part of the LPGS scheme, a global scheduling vector $\vec{s}$, normal to a set of equitemporal hyperplanes, should be decided. Since $\vec{s}$ should not

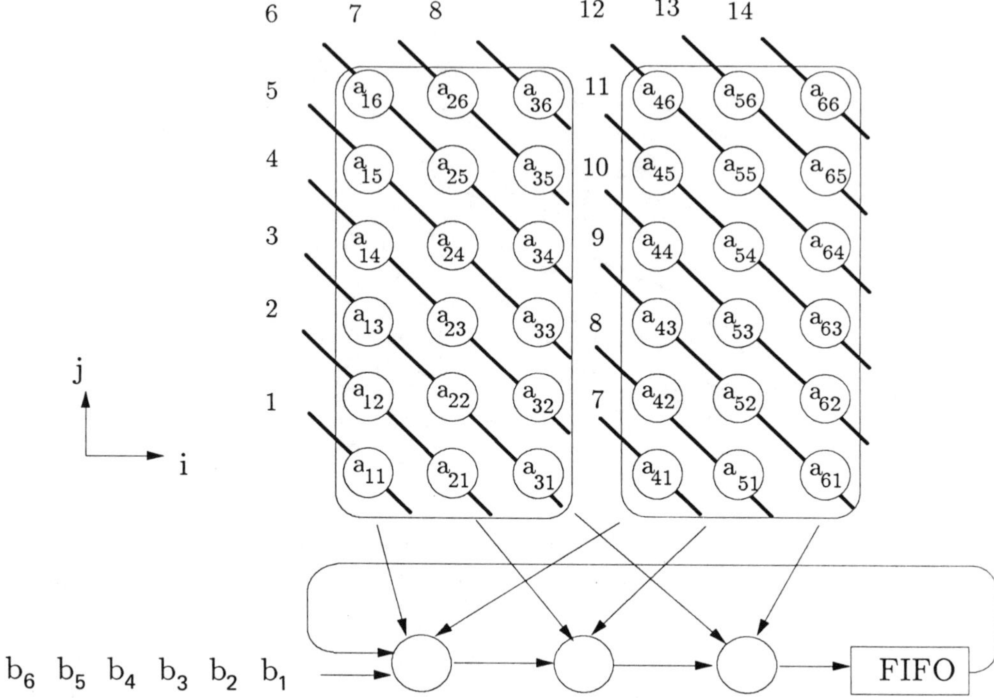


**Figure 6.27** LPGS partitioning of matrix-vector multiplication.

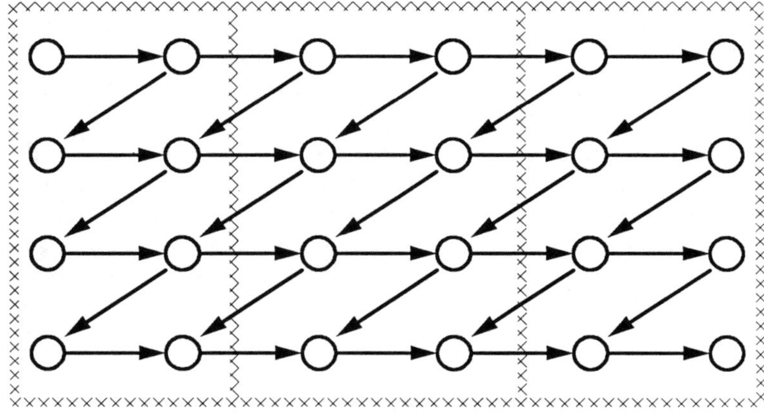

**Figure 6.28** A partition of DG with reverse data dependence.

violate data dependences, the constraints become

$$\vec{s}^T \cdot \vec{e} > 0 \qquad \text{for all dependence } \vec{e}$$

Furthermore, if the projection direction $\vec{d}$ is assumed to be orthogonal to all the $\vec{b}_i$ (this is assumed in [3] to provide an efficient algorithm), then $\{\vec{s}, \vec{b}_i, i = 1, 2, \ldots, k\}$ should be linearly independent. Otherwise, some nodes on an equitemporal hyperplane will be assigned to the same PE.

When using the LPGS scheme, some FIFOs are required to feed the *output* of the array back to the array. By noting that some *output* of the array may have the same value as the input if the output variable is transmittent, some FIFOs are actually not necessary. A $2 \times 2$ array for $4 \times 4$ matrix multiplication without using FIFOs is shown in Figure 6.29. Thus, *transmittent variables should be utilized to reduce the number of FIFOs in LPGS schemes.*

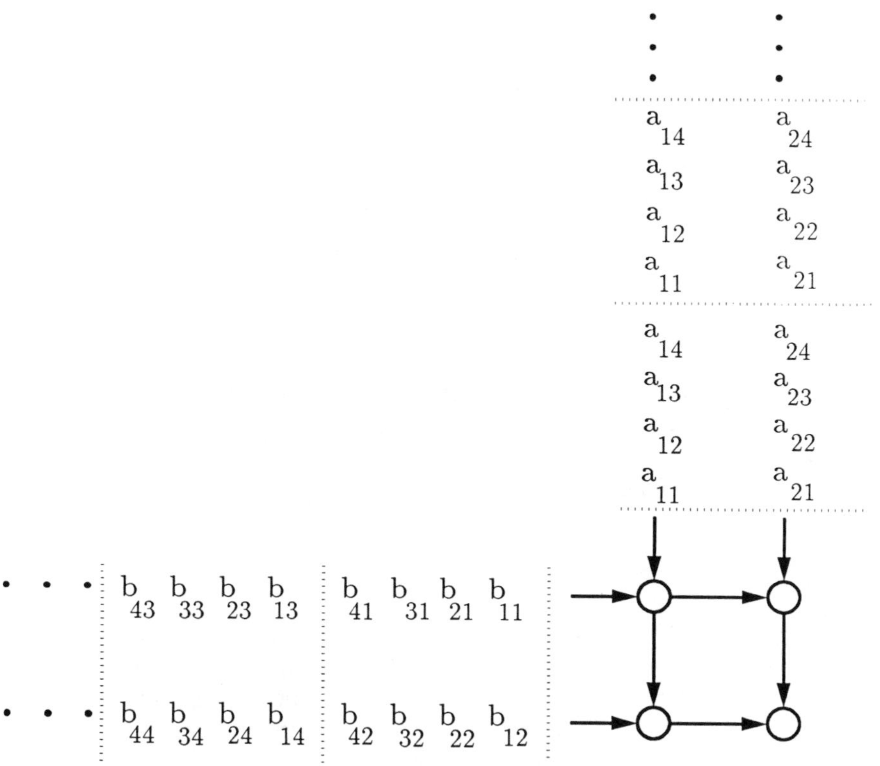


**Figure 6.29** A $2 \times 2$ array for $4 \times 4$ matrix multiplication without using FIFOs.

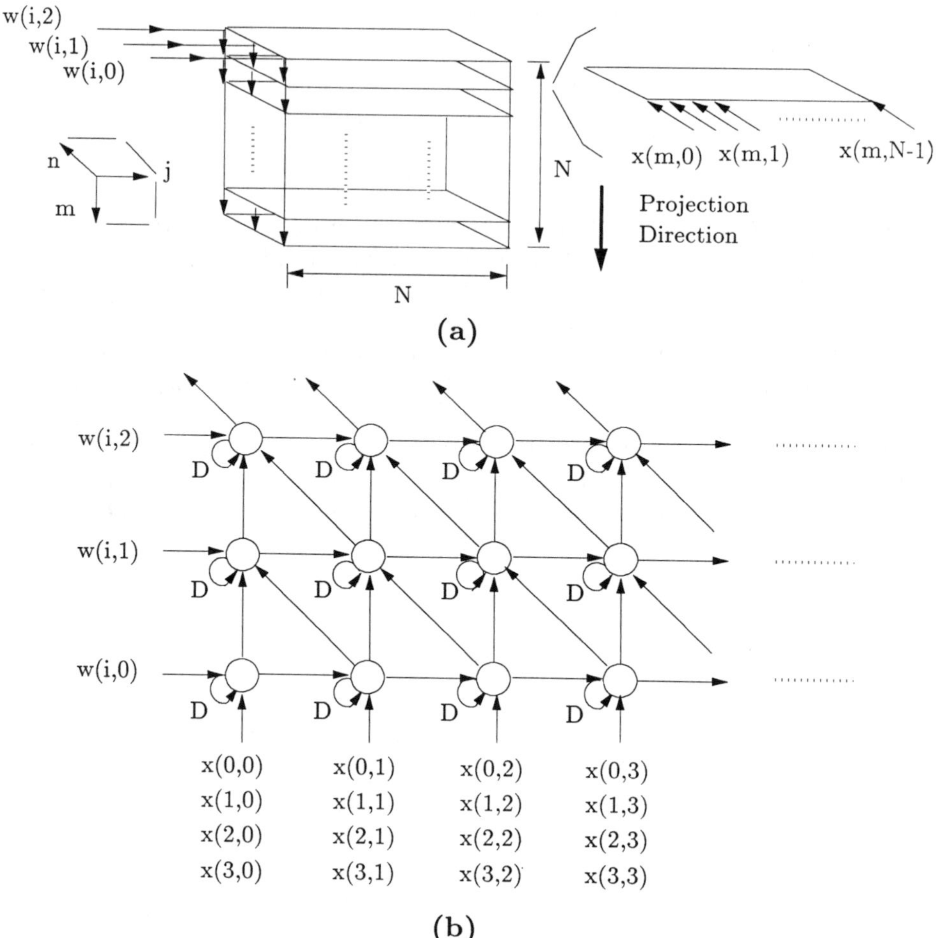


**Figure 6.30** (a) A 3-D DG for the image ($N \times N$) convolved with one row of the window pattern ($1 \times 3$). The 4-D DG may be constructed by summing up the outputs of three such 3-D DGs. (b) A 2-D SFG obtained by projecting the 3-D DG along the $m$ direction. (c) The 3-D SFG obtained by summing up the outputs of three 2-D SFGs. (d) The 2-D SFG obtained by adopting multiprojection.

## 5 SIGNAL/IMAGE PROCESSING APPLICATIONS

The applicational domain of VLSI array processors covers image processing, computer vision, nuclear physics, structural analysis, speech, sonar, radar, seismic studies, weather, astronomy, medical signal processing applications, and so on. A successful array processor design requires an understanding of the signal and image formation process, the algorithm class involved, and the specifications of the intended application system. For example, in a real-time vision processing system, it is possible for more than $2.5 \times 10^8$ pixels to be processed every second. If each pixel requires 10 operations, a

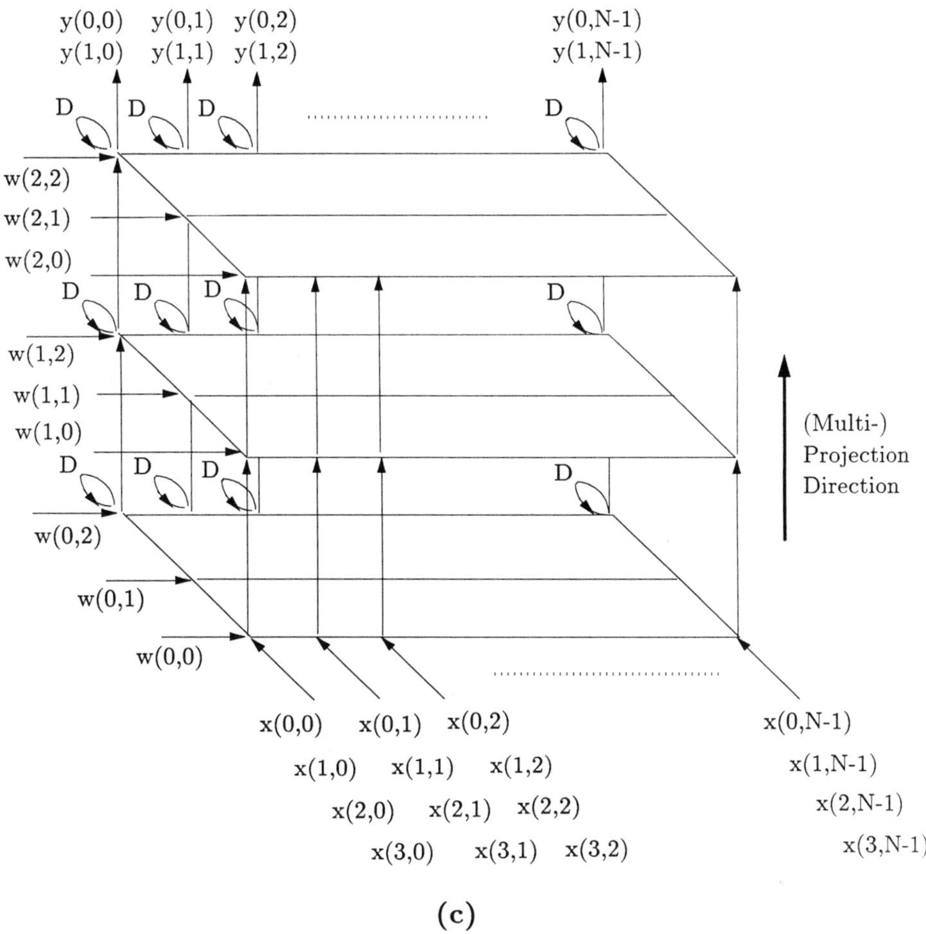


(c)

speed of 2500 million operations per seconds (MOPS) is needed. It is apparent that special-purpose parallel processing architectures are indispensable.

Digital signal and image processing encompasses a wide variety of mathematical and algorithmic techniques. Two dominating aspects of signal and image processing requirements are enormous throughput rates and huge amounts of data and memory. On the other hand, most signal and image processing algorithms are dominated by transform techniques, convolution/correlation filtering, and some key linear algebraic methods. These algorithms have common properties such as regularity, recursiveness, and locality, which can be naturally exploited in array processor design.

**Example: Edge Detection by Means of 2-D Convolution.** Two dimensional convolution is the most common method of image edge detection. Assume that image **x** is to be edge-detected; the 2-D convolution involves convolving a small window pattern **w** with the image **x**. This is done by moving the window pattern **w** over the image. At each point the convolution is com-

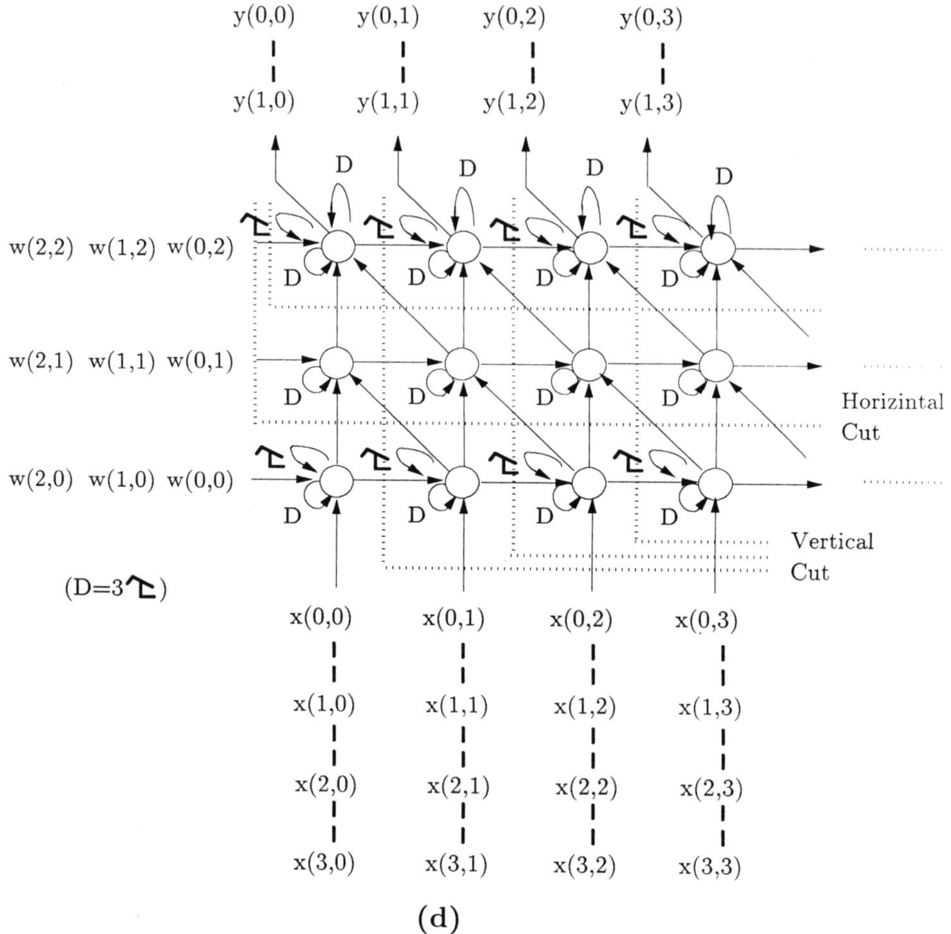

(d)

puted:

$$y(m,n) = \sum_i \sum_j w(i,j)x(m-i,n-j) \tag{6.8}$$

For 2-D convolution we define a $k_1 \times k_2$ matrix as a window. Since the size of an image may typically vary from 256 × 256 up to 8K × 8K pixels, we immediately see the necessity of extremely fast operation. For real-time applications, where the image is updated 30 to 60 times per second, the speed requirements are even more critical. In the following the systolic design of the 2-D convolution is described.

1. *DG design* Without loss of generality, assume that the window size is $3 \times 3$ and the image size is $N \times N$. The Eq. (6.8) can be rewritten as

$$y(m,n) = \sum_{i=0}^{2}\sum_{j=0}^{2} w(i,j)x(m-i,n-j)$$

Since there are four indices, $i$, $j$, $m$, $n$, in this equation, the DG for this algorithm is 4-D and the SFGs obtained via projection are 3-D. Because of the difficulties in drawing a 4-D DG, a 3-D DG for the image ($N \times N$) convolved with one row of the window pattern ($1 \times 3$) is shown in Figure 6.30a. Note that each 2-D ($3 \times N$) layer of the 3-D DG is the same as that shown in Figure 6.2b for the convolution of two 1-D signals. The 4-D DG may be constructed by summing up the outputs of three such 3-D DGs. Mathematically, this is equivalent to saying

$$y(m,n) = \sum_{j=0}^{2} w(0,j)x(m,n-j) + \sum_{j=0}^{2} w(1,j)xj(m-1,n-j)$$
$$+ \sum_{j=0}^{2} w(2,j)x(mj-2,n-j) \tag{6.9}$$

Note that each term on the right-hand side of Eq. (6.9) is represented by a 3-D DG.

2. *SFG design* From the 4-D DG, we can generate 3-D SFGs. Here we consider only a 3-D SFG with (approximate) size $3\times3\times N$. This 3-D SFG may be obtained by first projecting each 3-D SFG to one 2-D SFG (of size $3 \times N$) (see Figure 6.30b) and then combining these three 2-D SFGs by summing up the outputs (see Figure 6.30c). If 2-D SFGs are preferred, then multiprojection can be adopted to project the 3-D SFG to 2-D SFGs. Figures 6.30d shows a 2-D SFG with size $3 \times N$. Different projection directions can also be applied if desired. For example, it is apparent that a $3 \times 3$ SFG can also be obtained.

3. *Systolic array design* The SFGs shown in Figures 6.30c and 6.30d can easily be systolized by applying cut-set procedure. For example, the 2-D SFG shown in Figure 6.30d can be systolized by using rule 2 (delay-transfer) only. First, two delays are added to each edge of horizontal "cuts" (see Figure 6.30d). Then, applying vertical cuts, one delay can be transferred from each diagonal edge to the horizontal edges.

## 6 CONCLUSION

This chapter has presented a framework for the mapping of regular algorithms onto systolic arrays. The mapping methodology uses a three-stage design approach. An algorithm is first expressed in terms of a DG. The DG is then mapped to an abstract SFG array, which is in turn implemented by a systolic array. Depending on the regularity of the DGs, either a canonical or a generalized mapping method may be adopted. Because total algorithm regularity is not required, the generalized mapping is more general than existing techniques, but it retains many of their advantages. Furthermore, the generalized mapping methodology has served in many cases to unify and relate different systolic approaches to the same problem.

Based on the above mapping methodologies, algorithm matching techniques have been proposed in this chapter. These techniques map algorithms to an architecture subjected to specific constraints. The problems of different communication constraints and different array size constraints are treated. Some applicational domains of VLSI array processors are mentioned in the last section, and a 2-D convolution example is used to show the application of systolic arrays in image processing.

In this chapter, there has been no automatic generation of "optimal designs." The reason is that we believe people do not design an array processor by choosing a single optimality criterion. Different criteria, e.g., number of PEs, computation time, pipelining period, block pipelining period, and I/O, can produce drastically different optimal designs. Usually, depending on the applications, tradeoffs will be necessary in choosing a desired design. Moreover, in this chapter we do not concentrate on a dependence matrix, which is a simple matrix specifying the dependences of a shift-invariant DG, and then do mathematical transformations. The reason is that, in our experience, many good designs can be produced only if the DG is used and more insight into the problems is obtained. That is also why we usually try to modify the DG if a design is not satisfactory. Of course, it is still very desirable if an "optimal" design can really be obtained without human interaction via a fast algorithm, e.g., using a dependence matrix. The real challenge in developing design methodologies and notations for systolic arrays is to make them general enough to handle the problems of interest, yet specific enough to clearly exhibit the properties deemed important. We believe that the techniques proposed in this chapter achieve this balance.

## REFERENCES


1. S. Y. Kung, P. S. Lewis, and S. N Jean, "Canonic and Generalized Mapping from Algorithms to Arrays—a Graph Based Methodology," Proc. Hawaii Int. Conf. System Sciences, pp. 124–133 (January 1987).
2. D. I. Moldovan, On the design of algorithms for VLSI systolic arrays, *Proc. IEEE*, *71*(1), (1983).
3. D. I Moldovan and J. A. B. Fortes, Partitioning and mapping of algorithms into fixed size systolic arrays, *IEEE Trans. Comput.* *35*(1): 1–12 (1986).
4. P. Quinton, "Automatic Synthesis of Systolic Arrays from Uniform Recurrent Equations," Proc. 11th Annu. Symp. Computer Architecture, pp. 208–214 (1984).
5. W. L. Miranker, Space-time representations of computational structures, *Computing*, (1984).
6. P. R. Cappello and K. Steiglitz, "Unifying VLSI array designs with geometric transformations," International Conference on Parallel Processing, 1983.

7. Yiwan Wong and J. M. Delosme, Optimal systolic implementations of $n$-dimensional recurrences. ICCD pp. 618–621, (1985).
8. R. M. Karp, R. E. Miller, and S. Winograd, The organization of computations for uniform recurrence equations, *J. ACM*, *14*(3): 563–590 (1967).
9. S. K Rao, Regular iterative algorithms and their implementations on processor arrays, Ph.D. thesis, Stanford University, Stanford, California (1985).
10. S. Y. Kung, on supercomputing with systolic/wavefront array processors, *Proc. IEEE*, *72*(7); (1984).
11. S. Y. Kung, *VLSI Array Processors*, Prentice-Hall, Englewood Cliffs, New Jersey (1988).
12. S. Y. Kung, S. C. Lo, and J. Annevelink, Temporal localization and systolization of signal flow graph (SFG) computing networks, *SPIE*, McGregor & Werner, San Diego, (1984).
13. C. E. Leiserson, F. M Rose, and J. B. Saxe, "Optimizing Synchronous Circuitry by Retiming," Proc. Caltech VLSI Conf., Pasadena (1983).
14. S. Y. Kung, J. N. Hwang, and S. C. Lo, "Mapping Digital Signal Processing Algorithms onto VLSI Systolic/Wavefront Arrays," Proc. 12th Annu. Asilomar Conf. Signals, Systems and Computers (November 1986).
15. S. Y. Kung, P. S. Lewis, and S. C. Lo, "On Optimally Mapping Algorithms to Systolic Arrays with Application to the Transitive Closure Problem, Proc. 1986 IEEE Int. Symp. Circuits and Systems, 1316–1322 (1986).
16. P. S. Lewis and S. Y. Kung, "Dependence Graph Based Design of Systolic Arrays for the Algebraic Path Problems," Proc. 12th Annu. Asilomar Conf. Signals, Systems, and Computers (November 1986).
17. I. R. Saal, A Linear Systolic Array for Computation Matrices, Technical Report, Institute on Network and Circuit Theory, Technical University of Munich (1986).
18. T. F. Chan and Y. Saad, Multigrid algorithms on the hypercube multiprocessor, *IEEE Trans. Comput.*, pp. 969–977 (1986).
19. D. Heller, Partitioning big matrices for small systolic arrays, *VLSI and Modern Signal Processing*, Prentice-Hall, Englewood Cliffs, New Jersey (1984).
20. K. Hwang and Y. H. Cheng, Partitioned matrix algorithms for VLSI arithmetic systems, *IEEE Trans. Comput.*, *C-31*(12): 1215–1224 (1982).
21. K. Jainandunsing, Optimal Partitioning Schemes for Wavefront/Systolic Array Processors, Technical Report, Delft University of Technology, Delft, The Netherlands (1986).

# 7

# VLSI Signal Processors

HASSAN M. AHMED Department of Electrical, Computer and Systems Engineering, Boston University, Boston, Massachusetts

## 1 INTRODUCTION

Digital signal processing (DSP) is a well-studied discipline. It has long been recognized that processing signals in a digital rather than analog domain offers significant performance advantages; however, owing to the computational complexity of DSP algorithms, cost-effective digital realizations have not always been at hand. The preceding 7 to 10 years have witnessed widespread application of DSP techniques, primarily because of

1. The availability of commercial DSP microprocessors
2. Advances in LSI and VLSI technology
3. Advances in design methodology making custom VLSI design a generally available capability

Advances in integrated circuit (IC) technology have made possible microprocessors of ever increasing complexity whose architectures are tailored to DSP algorithms [1–6]. The emergence of such a general-purpose cost-effective DSP capability has sparked many commercial uses of DSP technology. Furthermore, the emergence of application-specific integrated circuits (ASICs), due to advances in design methodology, has sparked many custom DSP efforts tailored to a specific problem or a class of specific problems (e.g., see [7–10]). Custom circuits allow even more cost-effective (or performance-intensive) solutions than general-purpose processors because they need not waste chip area on architectural capability that is extraneous to the specific problem being solved. In contrast, general-purpose processors must include sufficient architectural features in anticipation of being employed for a broad range of applications.

In this chapter, we discuss the evolution of DSP IC architectures (or VLSI signal processors). We begin in Section 2 by reviewing some of the constraints imposed by VLSI technology. Section 3 concentrates on broadly defining the needs of DSP applications. The architectures of general-purpose processors are described in Section 4. Finally, in Section 5, we discuss special-purpose architectures. It would be impossible to review special-purpose processors in detail owing to their architectural diversity, so we restrict our attention to a specific multiprocessor structure known as the systolic array.

## 2 THE VLSI ENVIRONMENT

VLSI is a well-known acronym for "very large scale integration." It is distinguished from MSI (medium-scale integration) and LSI (large-scale integration) in several ways. Although the spirit of the distinction is well understood, the exact definition of VLSI remains an issue of (inconsequential) debate.

The most frequent definition of VLSI states that any chip having more than 100,000 devices is a VLSI chip. The number of devices necessary for a circuit to be stamped "VLSI" is not sharply defined and varies over several orders of magnitude (some establish the cutoff at 1,000,000 devices, for example).

Perhaps a better definition of VLSI emerges when a communications constraint is considered. Whenever a large number of devices are integrated on a single IC, the number of interconnections is also large. As the integration level increases, the interconnect becomes the dominant user of chip area and such chips are defined as VLSI chips. It is clear from this definition that VLSI circuits demand a different design approach than MSI and LSI circuits. Whereas the latter are designed using classical methodologies that attempt to minimize the gate count or device count of a circuit, VLSI chips should ideally be designed with a calculus that minimizes interconnections.

The definition of VLSI aside, there are several constraints in the IC world that dictate architectural (and hence performance) trade-offs in the design of VLSI signal processors. The key constraint is chip yield, which determines chip cost and hence establishes the cost effectiveness of the DSP approach. Figure 7.1 shows the relationship between yield and chip (or die) area. As die area increases, the yield falls dramatically. In order to fully appreciate the impact of die area on cost, it is important to understand how chips are manufactured.

Integrated circuits are built on silicon (usually) wafers of fixed diameter, typically 4 to 6 inches. The cost of a processed wafer is usually constant for a given process and facility; hence the more die that can be built on a single wafer, the lower the die cost (and hence chip cost). A larger die area reduces the total number of chips per wafer (referred to as possible die per wafer of PDPW). This is the first effect of die area.

Of the possible die, only a fraction will actually be functional once the wafer has been processed. This fraction, referred to as yield, decreases with increasing chip area as shown in Figure 7.1. Yield is the second area effect. Generally, as a

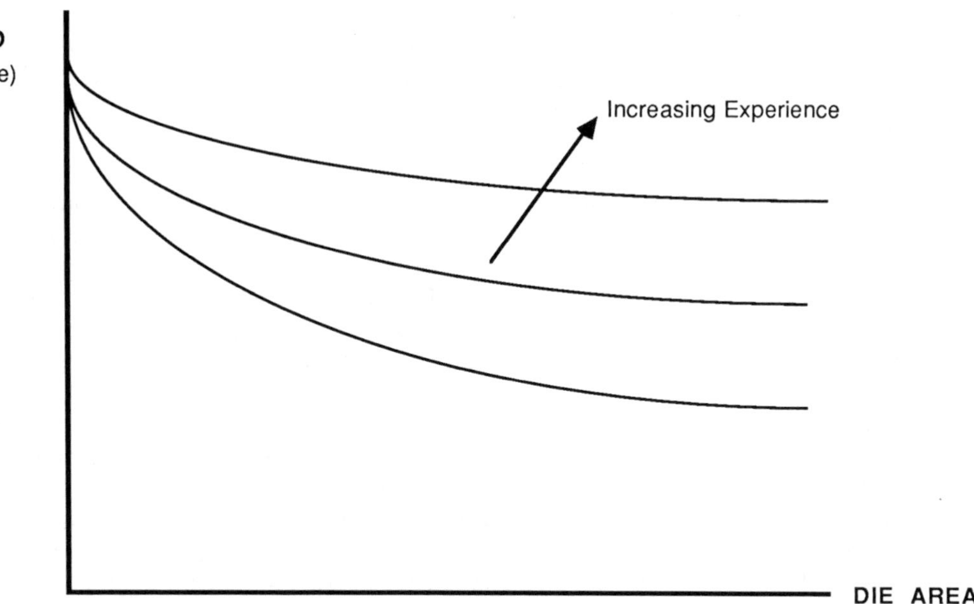


**Figure 7.1** VLSI yield curve.

processing facility gains experience with a particular process, the yield improves (see Figure 7.1); however, its inverse dependence on chip area remains.

DSP applications are especially sensitive to the yield phenomenon since their speed requirements dictate large amounts of fast hardware. This results in large chips with a relatively low layout density. Technological advances lead to increasingly dense layout rules as well as higher device speeds, thereby allowing reductions in DSP circuit area; however, the fact remains that the yield curve imposes limits on circuit size which can 1) compromise the chip architecture in favor of reduced area and 2) lead to unattractive system partitioning if the DSP problem must be divided among several chips. Much of the evolution of DSP microcomputers has been limited by yield considerations, as we will discover in a later section.

## 3 DSP SYSTEM ARCHITECTURES

DSP applications mandate several system requirements that affect the architecture of DSP chips. These requirements can be broadly stated as follows:

1. Real-time operation. Most applications are in real time, being presented with a constant stream of input samples at the sample rate.
2. High throughout. DSP algorithms are quite computationally complex, requiring real-time computation rates typically several orders of magnitude above the sample rates. For example, voiceband data communication sample rates are typically 8–10 kHz, but the algorithms may require 2–3 million operations per second [1] to operate in real time.

3. Limited area. Generally, real-time DSP applications are allotted limited area due to the application. For example, a sophisticated target tracking system for a cruise missile has very high throughout needs but must still fit inside the head of the missile! Limited area is also a requirement at the chip level, as the yield arguments of Section 2 indicated.

Items 2 and 3 obviously pose conflicting requirements. High throughput is generally achieved with increased hardware and lowered circuit density, both of which adversely affect area. Most DSP systems involve architectural trade-offs to balance these conflicting requirements, as the next few sections will demonstrate. Of course, VLSI has been the key component allowing higher levels of performance for a given area. Without integration, few DSP algorithms would have found widespread practical application.

The art of designing DSP system architectures (whether for a chip or a board) involves establishing the aforementioned trade-off between throughput and area. Area could refer to the total board area of the system, which may in turn limit the total number of chips into which the problem may be partitioned. When our view is restricted to a specific chip, area refers to the die area. The chip architect's task is to exploit both process technology and problem structure to yield a high throughout per area. Exploiting technology involves clever circuit design and ever-shrinking device geometries to achieve greater speed and/or reduced area. Exploiting problem structure involves constructing architectures that are well suited to the problem or the class of problems being considered. Clever DSP architectures often provide a better throughput enhancement than simply clever circuit designs (e.g., [9–11]). Architecture, rather than circuit design, is the primary focus of this chapter.

DSP architectures can be broadly classified as general purpose or special purpose. By general purpose, we mean that they are suited to a broad class of DSP problems and applications. Typically, general-purpose DSPs are programmable (e.g., DSP microcomputers such as the TMS 320 [2]), thereby allowing the users to program their algorithm of choice. Generally, the architecture and instruction set are fixed, so certain problems are more efficiently solved than others; however, this drawback is outweighed by the flexibility of use. In contrast, special-purpose architectures are specific to a single application or a few related applications. Special-purpose structures offer the best throughput/area for a specific DSP problem, since every aspect of the architecture can be tuned to the problem. The digital convolver of [7] is an example of a special-purpose device. It achieves a much higher throughput/area than would have been possible with a general-purpose architecture.

DSP architectures can also be classified as centralized or decentralized. The former implies the existence of a single powerful resource (such as a multiplier/accumulator) which is time-shared to realized the DSP algorithm. Commercial DSP processors (e.g., [6]) are examples of a centralized architecture. In contrast, decentralized architectures imply a multiplicity of less powerful (or even more powerful) resources to achieve the required throughput. The DSP task must be segmented into pieces which map onto each of the resources as well as onto the communication paths between the resources. We will consider a special type of decentralized architecture known as the systolic array in Section 5.

By way of summary, Figure 7.2 gives a sample classification of architecture along the lines described above. Note that commercial DSP microcomputers are centralized, general-purpose architectures. As such, they offer a high degree of flexibility but a limited throughput/area.

## 4 GENERAL-PURPOSE DSP ARCHITECTURES

This section will discuss the design of general-purpose digital signal processors. We will restrict our attention to microprocessors ($\mu$Ps) and use the term microprocessor and microcomputer interchangeably. When a distinction between a processor and a computer is required, it will be clear from the context.

Our approach will be to begin by discussing some typical DSP applications suited to DSP $\mu$Ps. We will then suggest an ideal $\mu$P structure for these applications. Less ideal structures used in commercial DSP microprocessors will be outlined and their evolution toward the ideal structure will be discussed. Sources of overhead tasks

| | GENERAL PURPOSE | SPECIAL PURPOSE |
|---|---|---|
| CENTRALIZED | DSP microprocessors and microcomputers [1 - 6] | e.g. ADPCM Transcoder [8]<br>Speech Analysis Chip [9] |
| DECENTRALIZED | e.g. Arrays of General Purpose Processors<br>Array Processors | e.g. Systolic Arrays [19 - 22]<br>Digital Convolver [7] |

**Figure 7.2** Sample architecture classification.

that impair DSP throughput will be described and architectural solutions to reduce this overhead will be discussed.

## 4.1 Typical DSP $\mu$P Applications

Commercial DSP microprocessors are mostly employed in voiceband applications such as speech analysis and synthesis, data communications, and voiceband echo cancellation, to mention a few. Higher-bandwidth applications will be discussed in Section 5 together with special-purpose architectures.

Figure 7.3 presents a simplified view of the speech analysis/synthesis problem. From the digital samples of a band-limited speech signal, it is required to repeatedly calculate the signal's short-term spectral characteristics. The speech signal is then represented solely by these spectral parameters and can be reproduced (synthesized) by exciting a time-varying filter having spectral characteristics equal to those of the speech signal. Speech analysis is the process of finding the spectral parameters of the speech signal from its samples. It is generally accomplished by whitening the speech samples with an adaptive filter [12]. Notice that both the analysis and synthesis tasks are based on adaptive and fixed filtering.

Adaptive filtering also plays a major role in data communications applications, such as the high-speed modem of Figure 7.4. A complete explanation of this application is beyond the scope of this chapter (see [1] instead); however, we note once again that fixed and adaptive filters are the key ingredients.

## 4.2 DSP $\mu$P Architectures

The architectures of commercial DSPs have been driven by

1. The pervasiveness of fixed and adaptive digital filters and the need to execute them efficiently
2. A desire for programmability for widespread application
3. Die size limitations for cost effectiveness

Commercial DSPs are microprocessors (or microcomputers) having a centralized general-purpose architecture. The components of a $\mu$P are typically:

1. Arithmetic unit
2. Program memory for instruction store
3. Data memory for data store
4. Program sequencer for program flow control
5. Data address generator for data structure maintenance

We will examine each of these functions in light of the needs of DSP algorithms, thereby providing insight into DSP $\mu$P architectures. It is important to note that a DSP chip need not embody all of the above functions; e.g., the processor of [4] does not have any on-chip program or data memory.

### Program and Data Memory

Traditional von Neumann architectures are based on the equivalence of data and instructions. They maintain a single address space and therefore a single memory

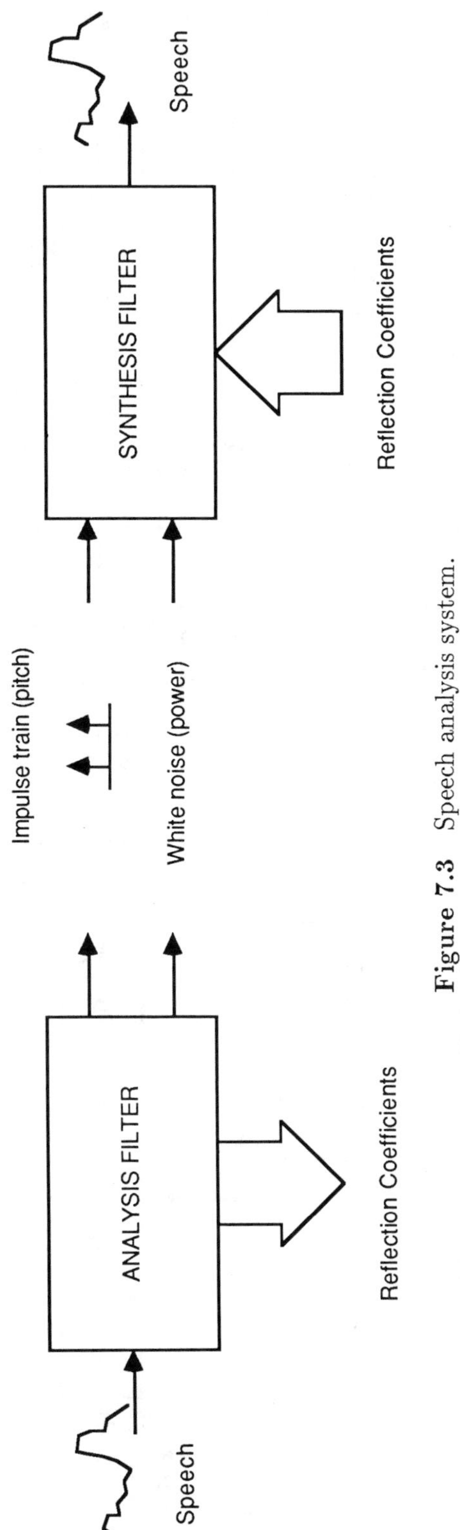


**Figure 7.3** Speech analysis system.

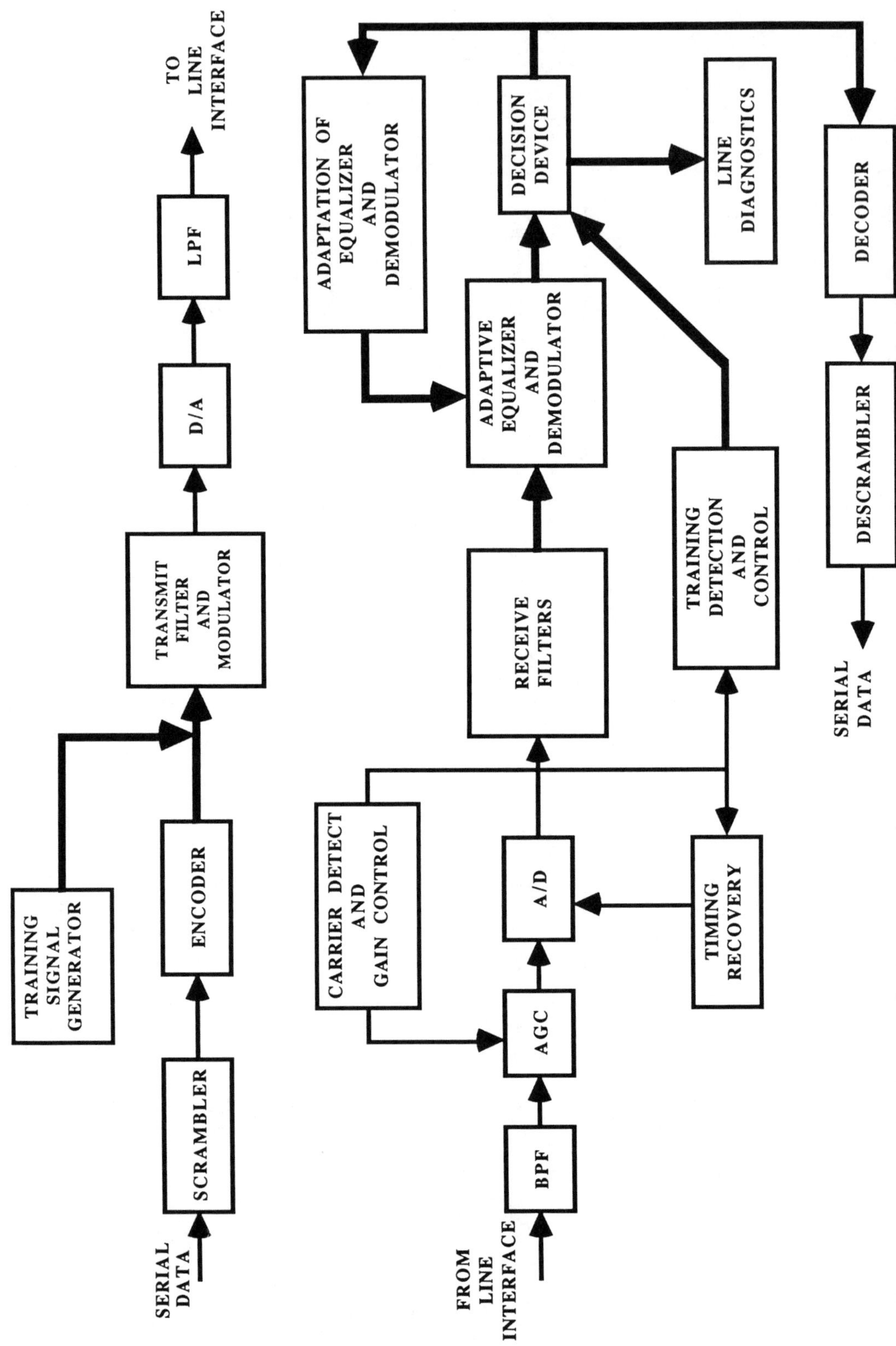
SERIAL DATA
SCRAMBLER
ENCODER
TRAINING SIGNAL GENERATOR
TRANSMIT FILTER AND MODULATOR
D/A
LPF
TO LINE INTERFACE
FROM LINE INTERFACE
BPF
AGC
A/D
CARRIER DETECT AND GAIN CONTROL
TIMING RECOVERY
RECEIVE FILTERS
TRAINING DETECTION AND CONTROL
ADAPTIVE EQUALIZER AND DEMODULATOR
ADAPTATION OF EQUALIZER AND DEMODULATOR
DECISION DEVICE
LINE DIAGNOSTICS
DECODER
DESCRAMBLER
SERIAL DATA

area, in which both instructions and data are stored. The speed of such architectures may be limited by the need to fetch instructions and their associated data in sequence. Most general-purpose microprocessors have a von Neumann architecture.

In contrast, present DSP microcomputers have a Harvard (or modified Harvard) architecture. This architecture is distinguished by the fact that data and memory occupy separate address spaces and reside in separate memories. Since the program memory and data memory buses are separate, instructions can be fetched concurrently with their associated operands, thereby improving the DSP's throughput. The penalty for this improvement lies in the need for separate program and data memory address and data buses, separate address generation units for each memory, and increased instruction width to exercise control over these separate memory subsystems. Since buses occupy chip area, as do wider program stores, the throughput improvement offered by the Harvard architecture comes at the expense of increased chip area. This trade-off is nonetheless imperative to achieve the desired throughput, and we will assume a Harvard architecture for the remainder of this section.

## Arithmetic Facility

The arithmetic unit (AU) and its associated bus structure to and from data memory is one of the most important elements of a DSP microprocessor. AU architectures are driven by the computations that are prevalent in the target DSP algorithms. We have already noted that many DSP $\mu$P applications are based on fixed and adaptive filtering. In fact, the AU design of most DSP $\mu$Ps has been driven by the structure of the transversal filter shown in Figure 7.5 and defined by

$$y(k) = \sum_{i=1}^{N} c(i)x(k-i)$$

where $y(k)$ is the output of the filter at time $k$, $x(k)$ the filter input at time $k$, $c(i)$ the $i$th filter coefficient, and $N$ the order of the filter.

Adaptive filtering has also been a major determinant of AU bus structure design, particularly the many forms of the least mean square (LMS) coefficient update [13]. More recently, DSP, $\mu$Ps, (e.g., [6]) can also efficiently execute the various recursive least squares (RLS) updates. Details of these algorithms can be found in [13].

Returning to the transversal filter output equation, we see that the output is calculated as a weighted sum of past inputs. Its computation involves $N$ multiply and accumulate operations, which we can represent with the pseudocode of Figure 7.6. It should be apparent now why all DSP $\mu$Ps to date have AUs centered around the ability to rapidly multiply and accumulate.

Now we know the key ingredient of the AU. However, it is equally important to have a bus structure around the AU that does not reduce throughput by constraining communications to and from the AU. The three-bus structure of Figure 7.7 mated with a triple-port RAM for data memory is ideal in this sense. Recognizing that multiplication is a dyadic operation, the structure provides two operand buses to the multiplier: effectively one bus for $c(i)$ and the other for $x(k-i)$ in our transversal filter example. The third bus is provided for writing the result,

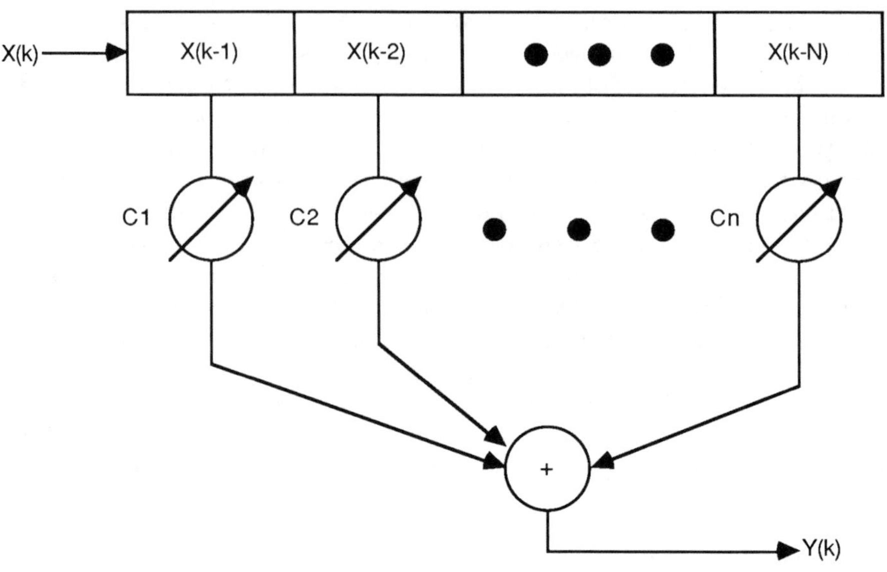


**Figure 7.5** Transversal filter.

$y(k)$, back to memory. Only one bus is necessary here since the kernel operation produces a single result.

The structure of Figure 7.7 provides maximal throughput on a uniprocessor; i.e., one tap of the filter is computed per cycle (assuming a single-cycle multiplier). However, the structure is very area-intensive in many ways:

1. There are three buses which take up area.
2. Up to three addresses must be supplied simultaneously, which widens the program memory and substantially increases the control complexity as noted earlier. We will discuss address generation separately.
3. A triple-port memory is very area-intensive since it requires triplication of the address decode and read/write circuits.

The evolution of DSP $\mu$P architectures has been such as to compromise the above structure to achieve reasonable die sizes. As process technology advanced and denser circuit layouts became possible, fewer compromises had to be made. The very early commercial DSP $\mu$Ps, such as the TMS 32010 [2], are based on

```
MPY     C(1), X(K-1)     ; Calculate 1st tap and store in ACC
MAC     C(2), X(K-2)     ; Calculate and accumulate 2nd tap
MAC     C(3), X(K-3)     ; 3rd tap
.
.
.
MAC     C(N), X(K-N)     ; ACC now contains Y(K)
```

**Figure 7.6** Transversal filter pseudocode.

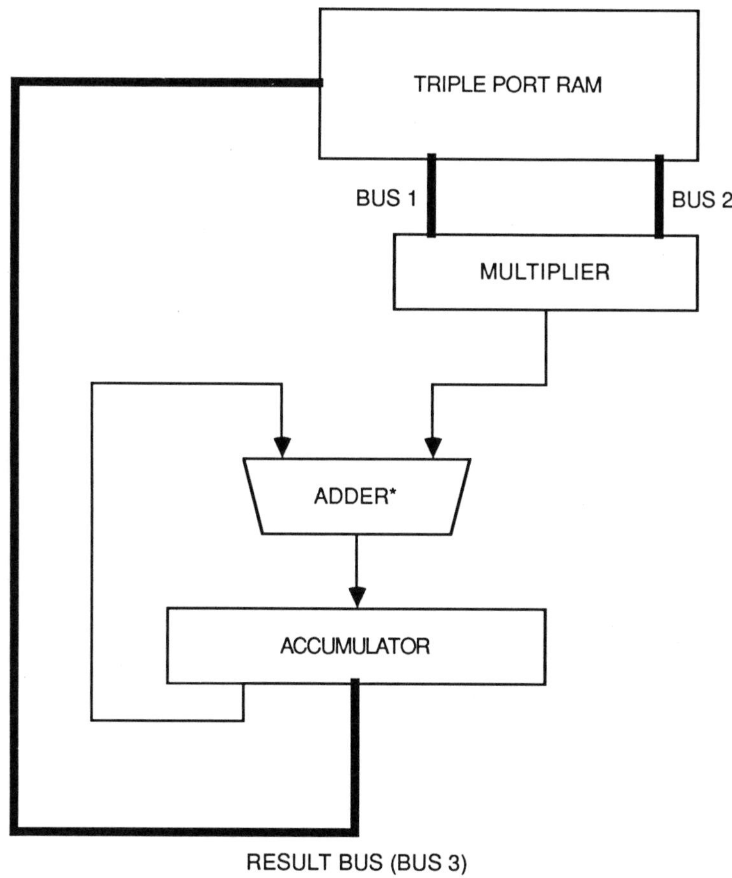


**Figure 7.7** Three-bus structure.

a single-bus structure, depicted schematically in Figure 7.8a. This arrangement eliminates many of the area-hungry features of the "ideal" structure, but at a throughput penalty. Now each tap of the transversal filter requires at least two cycles to compute since the two multiplier operands must be transferred in sequence across the single bus. A temporary holding register is used to store the first operand while the second is being fetched. Figure 7.8b shows the pseudocode for the transversal filter implemented on a single AU bus structure and a single cycle-multiplier.

Although the TMS 320 has a single-cycle multiplier, the bus limits all multiplications to two cycles. Recognizing this, the DSP of [1], which also has a single-bus structure, employs a pipelined two-cycle multiplier (which requires less area than a single-cycle multiplier). It was noted that in a long sequence of multiplications, such as with the transversal filter, the present multiplication can be

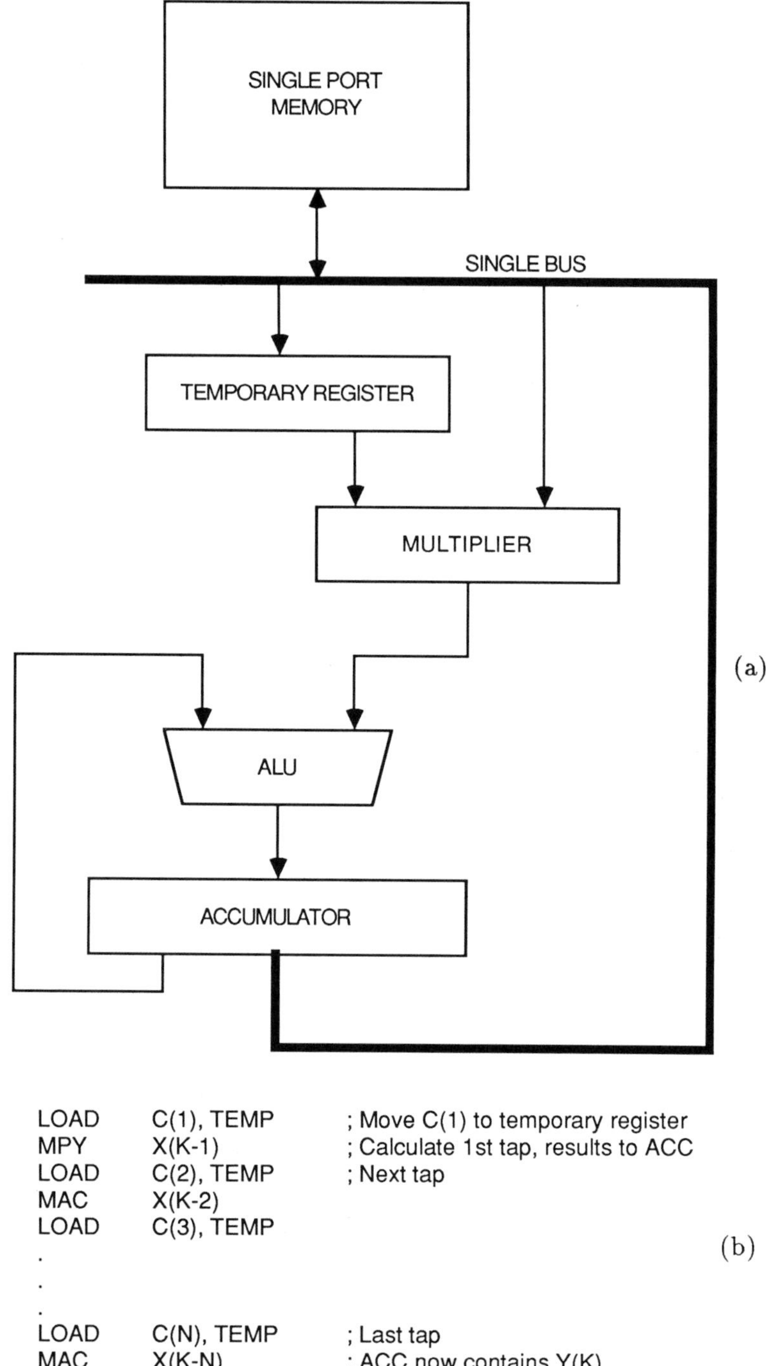


**Figure 7.8** (a) Single-bus structure; (b) Transversal filter pseudocode.

completed in the two cycles that are required to fetch the next two multiplier operands. The present multiplication finishes by the time the operands for the next multiplication are in place. This approach introduces a two-cycle pipeline latency. Therefore, single multiplications require four cycles rather than two to complete; however, a sequence of $M$ multiplications requires $2M + 2$ cycles. For large $M$, as is the case for DSP operations such as filters, this approaches the performance of the unpipelined structure ($2M$ cycles) while consuming less chip area.

Returning to the structure of Figure 7.7, we note that an obvious area-saving simplification would be to eliminate the third bus from the accumulator and reduce it to the dual-bus arrangement of Figure 7.9. This eliminates one bus and reduces the data memory and its associated control (such as address generation) to only two ports. In DSP applications, the throughput penalty for this modification is minor, because the third bus is used much less frequently than either of the other two. We can see this readily from the transversal filter output equation for $y(k)$. The output $y(k)$ is available only after $N$ multiplications and additions have been computed. Therefore, while the multiplier operand buses are used every cycle to

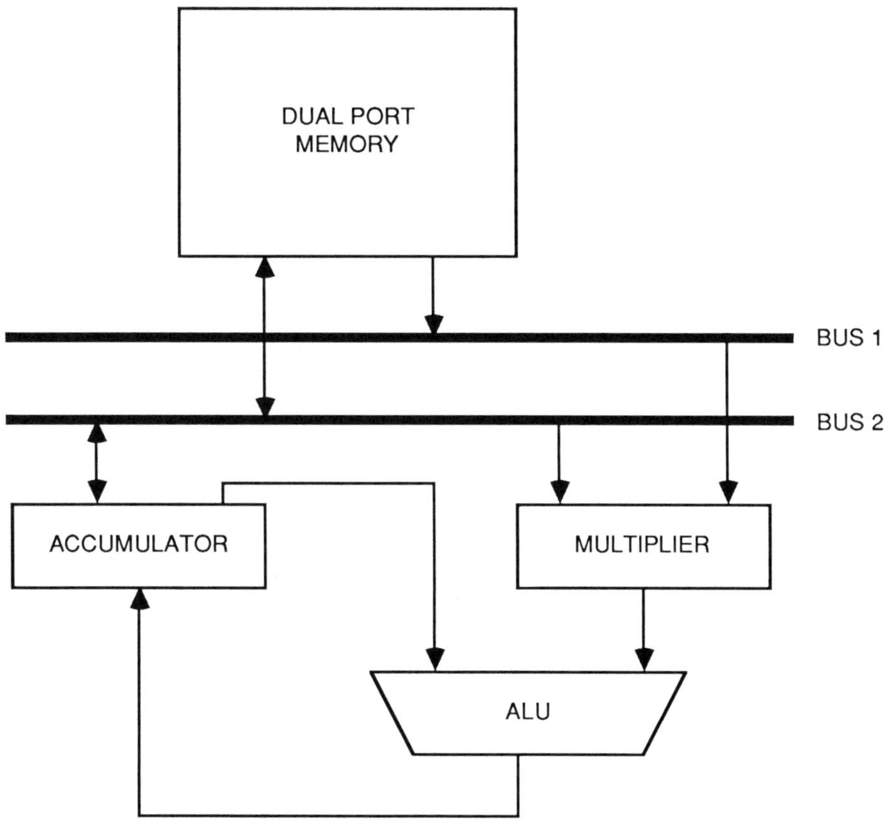


**Figure 7.9** Dual-bus arrangement.

compute a single filter output, the third bus is used at most once every $N$ cycles to store the result. For large filters (large $N$), the performance of the two-bus architecture approaches that of the "ideal" structure, although at a substantially reduced chip area.

The third-generation DSP $\mu$Ps, such as the Motorola DSP56000 [6], taking advantage of denser layout geometries than were available to first-generation processors, utilize the two-bus structure (although the processors have more than two buses, only two are used for data memory/AU communications). Figure 7.10a shows the DSP56000 chip architecture with the two AU buses highlighted. Once again, the AU is a single-cycle multiplier-accumulator. The filter pseudocode for the 56000 is shown in Figure 7.10b. It requires one cycle per filter tap, which is the same as the performance of the triple-bus architecture.

We end our discussion of AU architectures by noting a variant of the two-bus structure found in the ADSP 2100 [4]. Depicted in Figure 7.11, this chip has three separate arithmetic units, a multiplier-accumulator (MAC), an arithmetic logic unit (ALU), and a shifter, all arranged in parallel on the two operand buses which feed each AU. However, a third result bus, R-BUS, also exists. This bus is akin to the third bus of our ideal structure; however, it is not used for communicating with data memory. Rather it serves as an alternate operand bus to each of the three AUs, thereby allowing the output of one AU to be used as an operand in another AU without having to go through the data memory (alternatively, the input registers associated with each AU may be considered an extension of data memory). Whereas the structures that we discussed previously always have the adder placed at the output of the multiplier (to achieve a multiply-accumulate), the ADSP 2100 structure offers the benefit of being able to dynamically alter the relative placement of the three AUs based on how the R-BUS is used. Therefore, the ADSP 2100 can calculate

$$(a+b)*c+d \qquad \text{or} \qquad (a+b*c)+d$$

with equal ease.

As a final point, we have noted that all commercial DSP $\mu$Ps are based on a multiply-accumulate capability owing to the prevalence of that operation in the transversal filter and other DSP algorithms. Alternative arithmetic units for DSP have been explored in the literature; see [11] and [14] for examples.

## Data Address Generation

The previous section illustrated a performance/area trade-off for the AU section of a DSP $\mu$P. Other components of the microprocessor are equally important for high levels of performance. In particular, data address generation and data structure maintenance represent overhead tasks that can severely hamper throughput without the appropriate hardware support. This point is best illustrated by example. We will assume a dual-bus architecture with a dual-port memory. Since there are two address ports, we assume that there are two data address generators to feed these ports. The usual addressing modes common to microprocessors are assumed

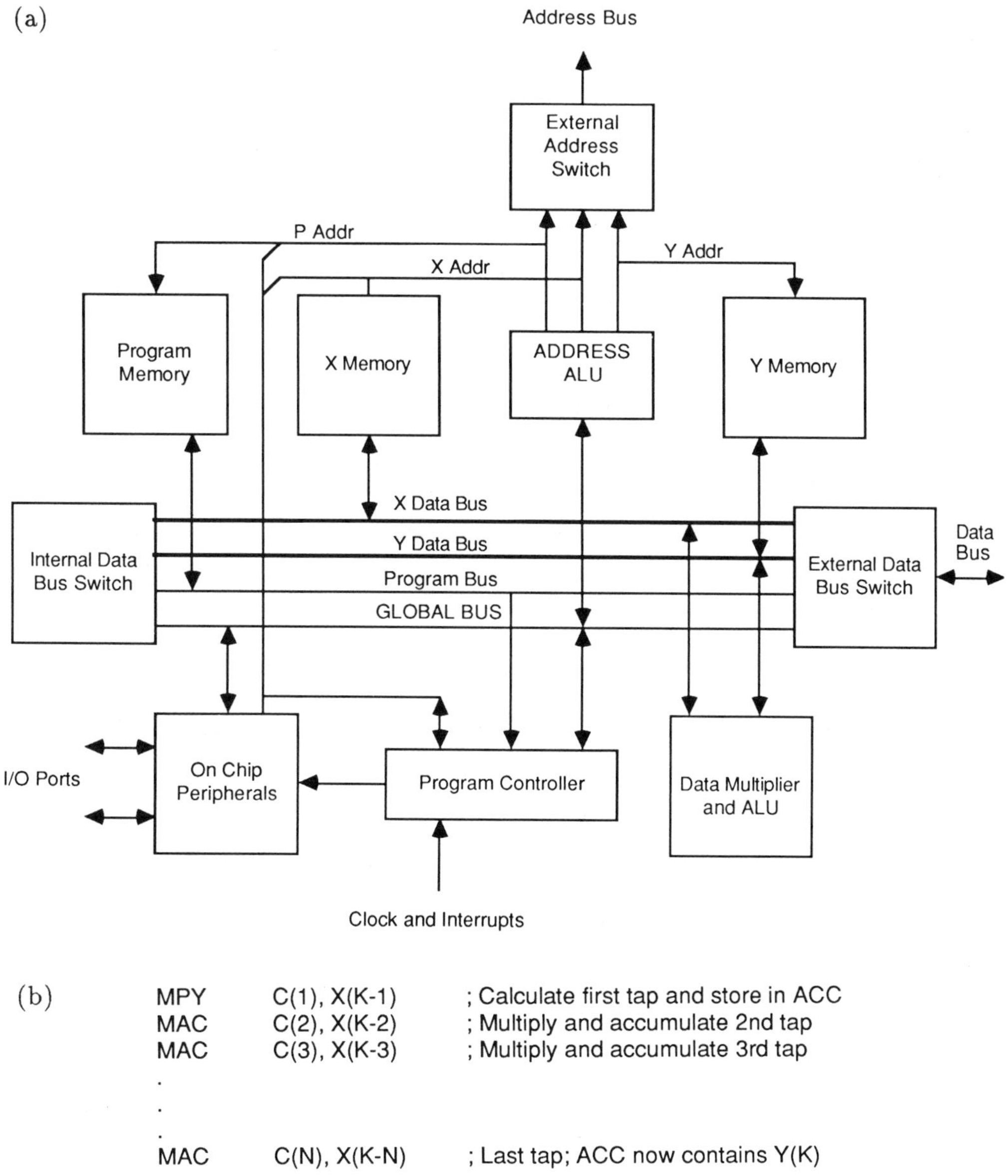


**Figure 7.10** (a) Motorola DSP56000; (b) transversal filter pseudocode.

to be supported, i.e., direct, immediate, indirect, and indirect with post increment, and decrement. While these modes are often sufficient in general-purpose $\mu$P applications, DSP tasks require additional addressing modes for efficient operation. This is the point we now attempt to illustrate by example.

Consider again the transversal filter of Figure 7.5. Imagine that the coefficients $c(i)$ are stored in data memory as an array. The array is accessed using indirect addressing with a pointer (presumably a register in the address generator) into the

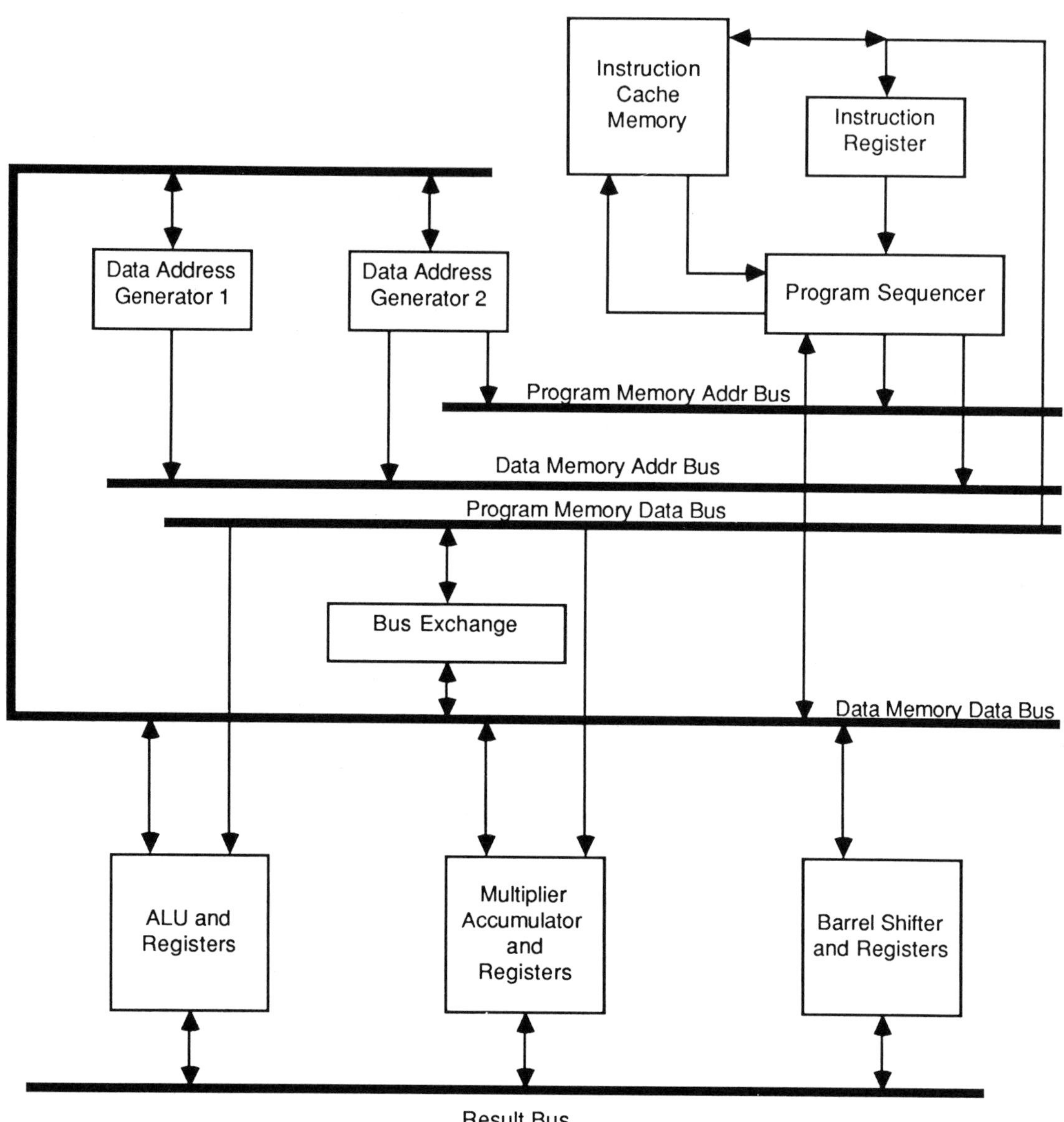


**Figure 7.11** ADSP 2100 internal architecture.

array. The filter delay line containing the input samples is best realized in data memory as a circular buffer. Such a buffer is an array of length $N$ with a floating starting entry; i.e., any of the array elements can be considered to be the head (or tail) of the buffer. A pointer is used to designate the head of the buffer. The pointer is therefore the location in the array at which the newest input sample, $x(k)$, will be stored. When a new sample arrives, it is placed in the array and the pointer is incremented. The next sample will be stored at this new location addressed by the pointer. Once the pointer reaches the end of the array, it must be reset to the starting address of the array in data memory, thus forming a circular buffer. By using such a buffer, we avoid the need to physically shift the delay line entries every time a new sample arrives.

It is clear that without any address generation capability, valuable processing time is spent in updating the pointer in software, i.e., in maintaining the data structure. Even with the traditional indirect addressing with postincrement modes mentioned above, a circular buffer cannot be implemented transparently since processor cycles must be used to check for the end-of-buffer condition. Figure 7.12 illustrates the pseudocode for the transversal filter when postincrement is the only pointer modification mode available. Notice the substantial overhead in checking for end of buffer at each sample—overhead that can readily be eliminated with a modulo addressing mode.

Modulo addressing essentially provides hardware support for checking for the end of buffer and resetting the pointer. This occurs transparently with the DSP AU operation; therefore no time overhead is incurred. Modulo addressing may be thought of as an additional post-modification mode of a pointer used for indirect addressing. Its operation is as follows: The length of the circular buffer is placed in a modulo register, M, contained in the data address generation circuitry. A pointer update modulo M performs the operation:

$$P \longleftarrow [(P - B + 1) \bmod M] + B$$

where $P$ is the pointer being updated and $B$ the base address of the circular buffer. We repeat that modulo addressing allows the realization of data structures common to DSP without compromising throughput, but at the expense of chip area. Modulo addressing provides a valuable performance/area trade-off and appears in several high-performance DSPs including those in [4] and [6].

```
; Let P be an index register into the coefficient array, C(i)
; Let Q be an index register into the data array, X(k)
; Let (P)+ denote indirect addressing with post increment
; Let XMAX be the address of the final element in X(k)
; Let XMIN be the address of the first element in X(k)

        CLEAR   ACC             ; clear accumulator
        MAC     (P)+, (Q)+      ; accumulate C(i)*X(K-i)
        SLE     Q, XMAX         ; skip next instruction if Q≤ XMAX
        LOAD    Q, XMIN         ; reset Q to start of array
        .
        .                       ; repeat above code for each
        .                       ; filter tap
        .
        .
        MAC     (P)+, (Q)+      ; final tap. ACC now contains Y(k)
```

**Figure 7.12** Transversal filter code without modulo addressing.

We have already discussed one addressing mode that is uncommon to general-purpose $\mu$Ps but provides a major performance benefit to DSP $\mu$Ps. There are others, such as pre- or postincrement by $K$, a programmable constant, which is useful for decimation filters [15]. More details can be found in, for example, [16].

## Program Flow Control

Another major source of overhead in programmable machines is the management of program flow. Consider a simple example. Recall that the samples of some input signal to the DSP (e.g., speech samples) are provided at a fixed rate known as the sample rate. Suppose that every time a sample is available, the $\mu$P is interrupted. It reads the sample and resumes its operation. The interrupt service routine would appear as:

```
READ sample AND MOVE TO DATA MEMORY
RTI                                          ; return from interrupt
```

Only one instruction of the two-instruction procedure is performing useful work. The RTI is merely a 50% overhead associated with the interrupt. There are several hardware techniques for avoiding this overhead and enhancing the performance of the DSP $\mu$P. For example, the terminal address of the service routine could be stored in a register. Upon interrupt, the program sequencer begins monitoring the program address. When the terminal address is encountered, a hardware return from interrupt is executed in parallel with the last instruction of the service routine. Therefore the RTI is never programmed and the overhead associated with fetching, decoding, and executing that instruction is never incurred. Such techniques again trade off area for performance. They have been utilized in DSP $\mu$Ps (e.g., [1]) in an effort to maintain high utilization of the AU of the DSP for the DSP algorithm.

Perhaps the biggest savings in program flow control overhead can be realized by providing hardware support for looping. Because looping is such a common program construct and the overhead associated with it is so severe, nearly all DSP $\mu$Ps now offer a hardware looping capability. The source of looping overhead stems from the need to increment the loop counter, check for its expiry, and conditionally branch back to the start of the loop body. These operations must be performed every time through the loop and can represent several machine cycles when no hardware support is provided and the looping construct must be programmed. In light of our transversal filter example, the loop body is simply one instruction, namely the multiply-accumulate. A software loop control would amount to at least 50% overhead, but in general would be higher.

Hardware looping eliminates the throughput loss, but with increased chip area. There are several techniques for hardware looping; however, most are based on a variant of the basic technique of [1]. The latter approach utilizes the hardware of Figure 7.13 and operates as follows. The loop register contains the address of the last instruction in the loop body, while the loop counter contains the number of times the loop is to be executed. Prior to entering the loop, the loop counter is loaded with the desired number of iterations and the current program address is pushed onto a stack (a stack together with several loop

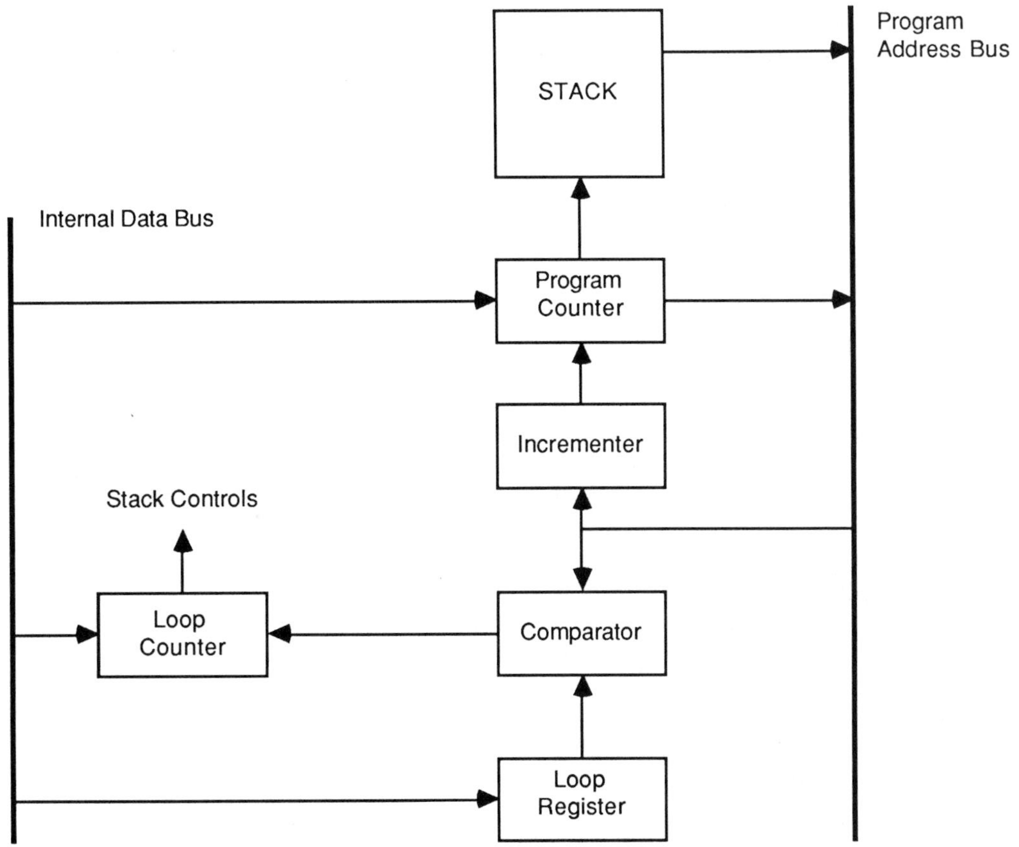


**Figure 7.13** DO loop hardware architecture.

counters and registers provides a convenient method for nested loops). The program counter is monitored. Whenever the last instruction in the loop is encountered, the loop counter is decremented. If it has not expired, the loop start address is read from the stack and loaded into the program counter and the loop repeats. With hardware support, these operations are performed in parallel with the AU operation, therefore introducing no temporal overhead. When the loop counter does expire, the stack is purged but the program counter is not loaded; therefore the instruction following the loop is executed. It is clear that such a looping facility requires one or two machine cycles of overhead to set up the loop; however, no additional overhead is required during loop execution. For a large number of iterations, the setup overhead becomes inconsequential.

The previous three sections have illustrated several sources of throughput overhead in the three major components of any DSP microprocessor, namely the arithmetic unit, the address generator, and the program sequencer. In all three cases, we have provided architectural solutions that eliminate or substantially reduce the overhead at the expense of increased chip area. These architectural trade-offs have guided the development of all present DSP $\mu$Ps. Although there are other sources of overhead, we believe that we have covered the most important ones, which appear to be universally recognized in the third-generation DSP processors available commercially.

## 5 SPECIAL-PURPOSE DSP ARCHITECTURES

While general-purpose DSP $\mu$Ps offer much design flexibility, they often provide insufficient throughput for many important signal processing applications. Image processing is an application demanding much higher throughput than offered by DSP $\mu$Ps. Although many such processors can be interconnected to yield the desired throughput, specialized architectures suited to the image processing algorithms provide much more efficient realizations, as indicated by the chip set of [7]. Many architectural features appropriate to the algorithms are employed by the chip set. In particular, the convolver chip uses small barrel shifters rather than full array multipliers, since in most cases the convolver coefficients are powers of two. Immediately, substantial area savings are realized. The reader is referred to [7] for additional details on the chip set.

Another example of the inadequacy of general-purpose DSPs can be found in the coding of speech signals via a recently recommended Adaptive Differential Pulse Code Modulation (ADPCM) algorithm [17]. One channel of full duplex transcoding as defined by the algorithm requires two general-purpose processors of considerable capability [18]. However, a special-purpose architecture tailored to the algorithm can perform *four* such full duplex channels in about the same die area as a *single* processor [8]. Although ADPCM is a voiceband application, the DSP $\mu$P falls short in delivered throughput as many of its features are redundant or cumbersome for the algorithm. The custom chip, however, eliminates these redundant features and utilizes the chip area for other architectural capability to which the ADPCM algorithm is closely matched. For example, it replaces the ever-present multiplier with a barrel shifter which is more effective for the implementation.

A severe case of the inadequacy of general-purpose processors occurs in spectral estimation, which is discussed in Chapter 5 of this volume. It is noted there that a common requirement of spectral estimation algorithms is to factor an $N$th-order correlation matrix derived from signal samples. The factorization is typically an orthogonal decomposition, that is,

$$R = QU$$

where, $R$ is the correlation matrix, $Q$ an orthogonal matrix, and $U$ an upper triangular matrix, or a Cholesky factorization, that is,

$$R = LL^T$$

where $L$ is a lower triangular factor and $L^T$ is the transpose of $L$.

Consider a real-time spectral estimation problem in which a frame of $N$ samples is collected at a sample rate of $D$ samples per second. During each frame, $N$

samples are collected and the correlation matrix is formed and then factored. It is clear that at most $N/D$ seconds are available for factoring the matrix $R$ before the matrix from the next frame of samples becomes available.

It is instructive to observe the relationship between the available computation time $N/D$ and the required computation as a function of the frame size $N$. The factorization requires $O(N^3)$ operations in general. However, when the underlying process is stationary, its correlation matrix assumes a Toeplitz structure and can be factored in $O(N^2)$ operations. Figure 7.14 illustrates these relationships. Notice that while the computation requirements grow quadratically or cubically, the available computation time (frame period) grows only linearly. For real-time operation, the frame size $N$ is severely limited because of this mismatch. A general-purpose uniprocessor architecture provides insufficient throughput/area to deal effectively with this problem. Ideally, we would like to discover an architecture that allows us to linearize the required computation time as a function time of $N$, thereby matching it to the available computation time. This will remove the frame size restriction while maintaining real-time operation. We will use this example to illustrate a special-purpose DSP architecture for this problem.

## 5.1 Systolic Arrays

The processing time required to factor $R$ can be linearized with a special-purpose multiprocessor architecture. Consider, for example, the Toeplitz case, where $O(N^2)$ operations are required. Intuitively, this could be accomplished in $O(N)$ time steps

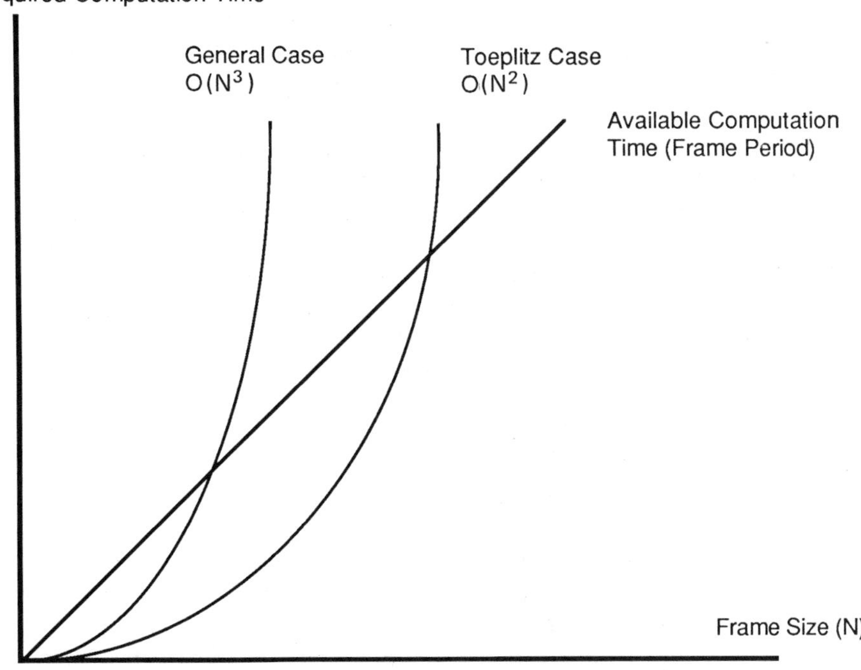


**Figure 7.14** Required computation time for matrix factorization.

if $O(N)$ processors could be used simultaneously to each perform a portion of the calculation. For the most effective silicon usage, each processor would have a special-purpose architecture suited to computing only the tasks associated with that processor. Effectively, the problem would be solved with a special-purpose decentralized architecture, rather than a general-purpose centralized one. (Of course, a general-purpose decentralized architecture could also be employed, but with reduced efficiency.)

We will develop a specialized decentralized multiprocessor known as a "systolic array." This structure is particularly attractive in the VLSI environment because it has the added architectural constraint that processors within the array may communicate only with their nearest neighbors. Hence, all communications paths are short, thus limiting the amount of chip area consumed by interconnections. Another benefit of short communications paths is their relatively high bandwidth, which prevents communications delay from becoming a constraint in the performance of the array.

The remainder of this section will be devoted to developing the systolic array for the orthogonal decomposition of a general matrix $R$. There is insufficient space to describe also the Cholesky factorization of a Toeplitz matrix. The interested reader is referred to [19]. Systolic arrays have been studied for a variety of problems. A small sampling of the literature is [20–22].

## 5.2 A Systolic Array for Orthogonal Decomposition

Orthogonal decomposition is most commonly achieved by the Givens method [23]. Somewhat akin to the familiar technique of Gaussian elimination, the Givens method operates on pairs of rows or columns of the matrix $R$ to reduce it to upper triangular form. However, unlike Gaussian elimination, the elementary row operations are chosen to be plane rotations. Figure 7.15 defines the algorithm more completely. It clearly requires $O(N^3)$ operations, that is, $O(N^3)$ time steps on a uniprocessor.

Recall that our goal is to develop an array structure requiring $O(N)$ time steps, perhaps using $O(N^2)$ processors. As an interim step we will develop a linear array of $O(N)$ processors requiring $O(N^2)$ time steps. Such an array is shown in Figure 7.16. It consists of $N$ ($N = 6$ here) processors, each having one associated first-in, first-out (FIFO) memory for intermediate storage, initialized as shown. Processors can communicate only with their neighbors, either directly with the right neighbor or via the FIFO to the left neighbor. The leftmost processor calculates the angles for the plane rotations that are used in the algorithm (see Figure 7.15) and passes these angles to the right. The row operations are computed in the remaining processors as the angles are received in sequence from the left. The results appear skewed in time at the output of the processors in the array. Figure 7.17 shows the operation of each processor in the array at successive time steps. We will term Figure 7.17 the "activity chart" of the array.

The activity chart of the linear array immediately suggests the triangular array of Figure 7.18, which can compute the factorization of $R$ in $O(N)$ time steps. The triangular array is shown for $N = 4$ and consists of three linear arrays "stacked" one above the other. Each array operates in the manner described above, with the leftmost processor calculating the rotation angles and passing them to the right.

```
; Let A=[a_ij] be the matrix to be factored
; Let B=[b_i] be a vector in the equation Ax=b
; Here is the Givens  Algorithm pseudocode

    for r=1 to n begin;

        for k=r+1 to n begin;

            theta_rk = -tan^-1(a_kr/a_rr) ;        Find rotation angle
            a_rr=√(a^2_rr+ a^2_kr) ;
            cos=cos(theta_rk); sine=sin(theta_rk);

            for j=r+1 to n begin;
```

$$\begin{bmatrix} a_{rj} \\ a_{kj} \end{bmatrix} = \begin{bmatrix} \cos & -\text{sine} \\ \text{sine} & \cos \end{bmatrix} \begin{bmatrix} a_{rj} \\ a_{kj} \end{bmatrix} \quad \text{; Elementary row operation ; on rows 'k' and 'r'}$$

```
            end;
```

$$\begin{bmatrix} b_r \\ b_k \end{bmatrix} = \begin{bmatrix} \cos & -\text{sine} \\ \text{sine} & \cos \end{bmatrix} \begin{bmatrix} b_r \\ b_k \end{bmatrix} \quad \text{; Row operation on b}$$

```
        end;

    end;
```

**Figure 7.15** Givens algorithm.

In this case, each linear array implements one level of the $k$ loop in Figure 7.15. Therefore, we have developed an array of processors which can compute the factorization of $R$ in $O(N)$ time steps. The array requires $O(N^2)$ processors, each of which must compute a plane rotation (the boundary processors calculate an arctan function to get the rotation angles). This array represents a special-purpose architecture since the array interconnections and data movement are specific to the Givens algorithm. Furthermore, the processors in the array perform operations specific to that algorithm as well. In fact the array is also useful for performing Gaussian elimination (a touch of generality). The details appear in [21]. However, we note that the array interconnections and data movements remain the same; only the individual processor operations change to reflect the new algorithm.

## 5.3 Processor Architecture

In Section 5.2 we developed an efficient array architecture for a specific problem. A key aspect of that architecture was the ability of the individual processors in the array to perform plane rotations. Once again, a special-purpose architecture for this calculation must be developed for maximum throughput. There are many

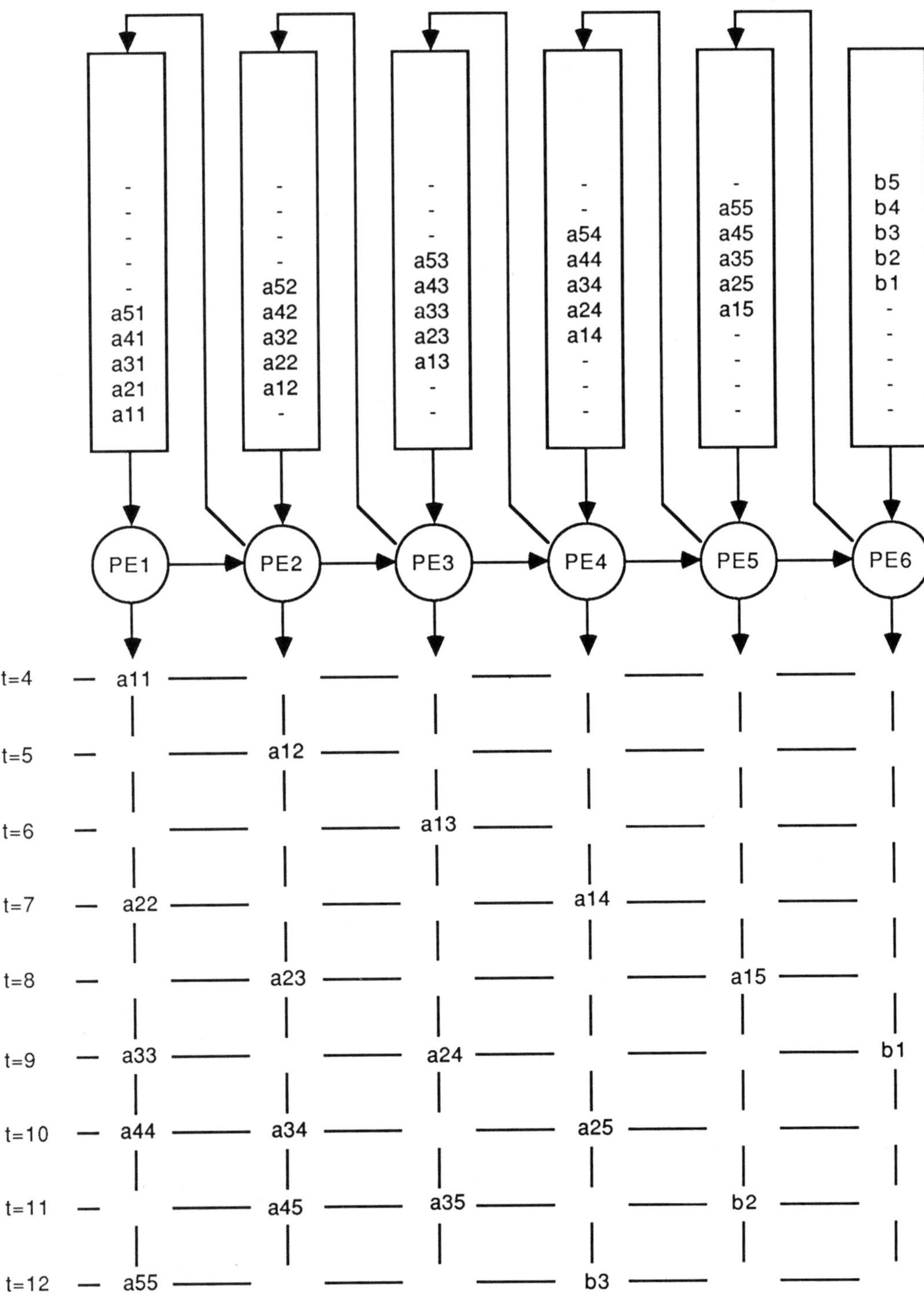


**Figure 7.16** Fully pipelined Givens method on a linear array.

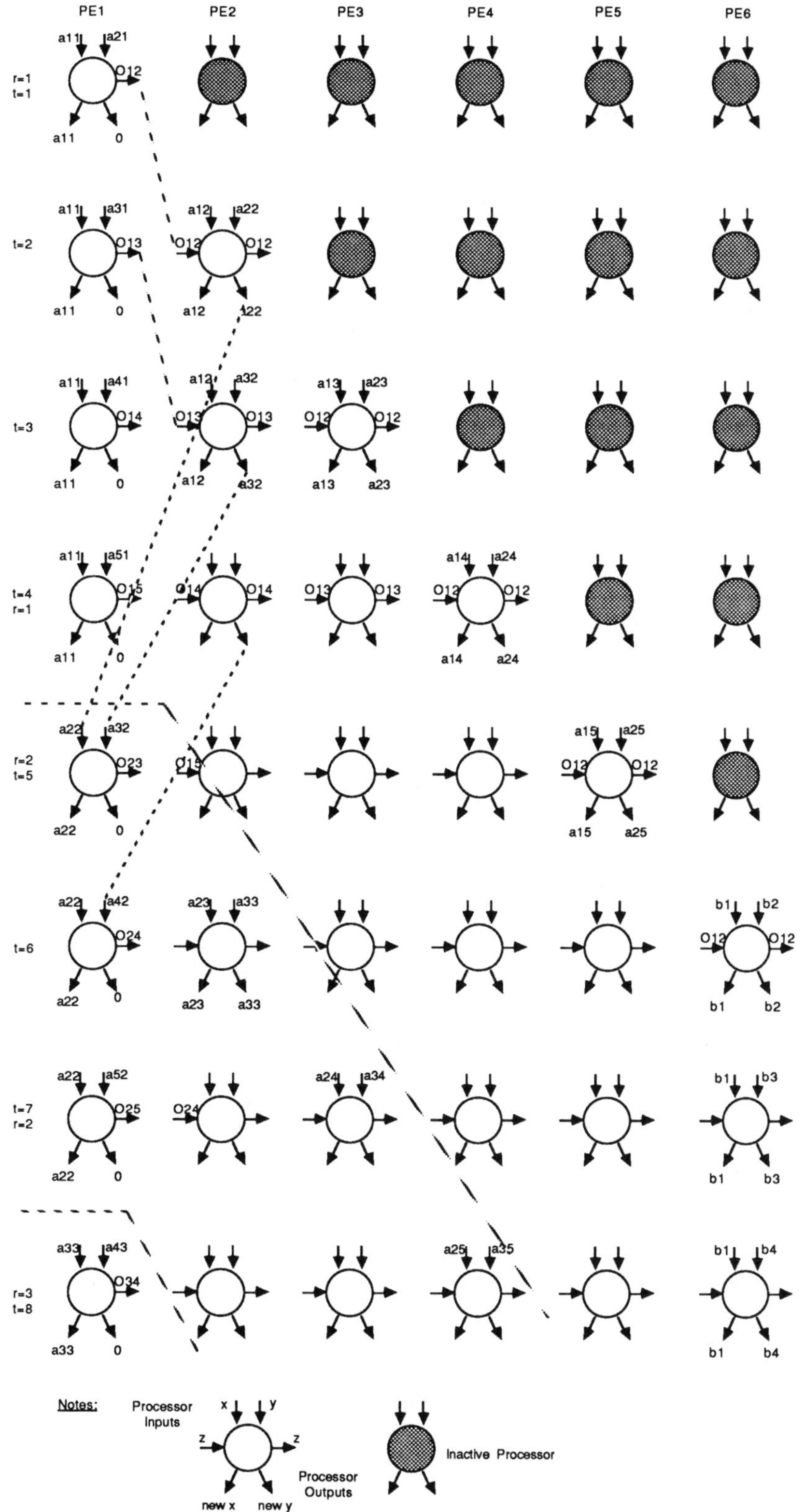


**Figure 7.17** Linear array activity chart.

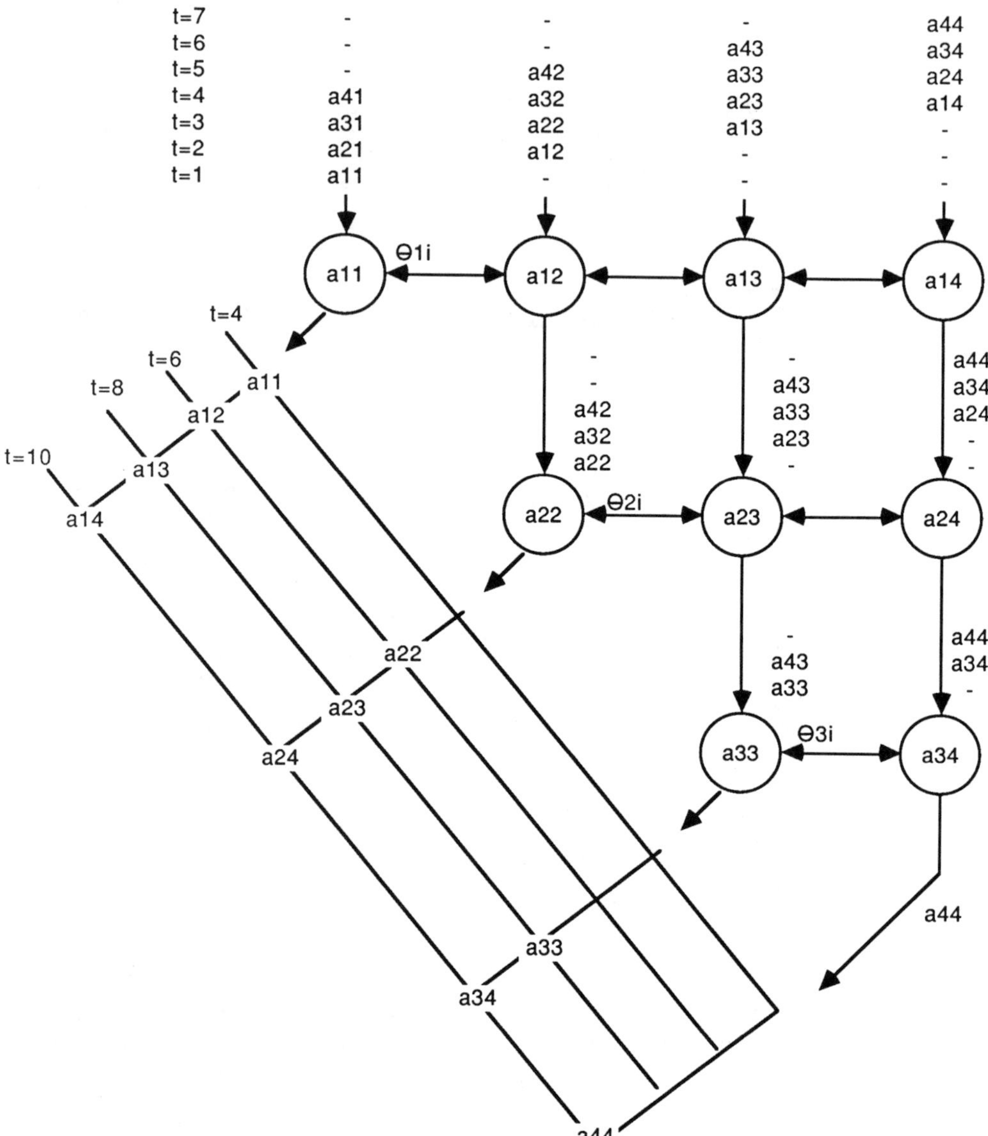


**Figure 7.18** A two-dimensional array for the Givens algorithm.

approaches to calculating rotations. One especially effective method is a numerical technique known as the CORDIC (Coordinate Rotation Digital Computer) algorithm. Efficient processor architectures for the CORDIC algorithm are given in [24]. A side benefit of using processors based on the CORDIC algorithm is that they are also then capable of performing the kernel operations of Gaussian elimination (see [21] for details). Therefore one array would be capable of efficiently executing two factorization algorithms in spite of its special-purpose nature.

## 6 CONCLUSIONS

This chapter has considered several aspects of VLSI signal processing architectures. We noted that efficient DSP architectures are the result of exploiting both process technology and algorithm structure to achieve a high throughput/area.

General-purpose programmable architectures offer much flexibility of use, but sacrifice performance for flexibility. Commercial DSPs are examples of general-purpose centralized architectures. Their arithmetic units are based on the celebrated transversal filter algorithm, hence the need to rapidly compute multiply-accumulate operations. A three-bus structure is ideal for the AU owing to the dyadic operations common to DSP algorithms. DSP $\mu$P architectures have evolved to dual-bus structures of comparable efficiency to the three-bus architecture. Their evolution has also recognized the need for hardware support for program flow control and data structure maintenance, since these chores otherwise represent substantial overhead and severely limit the DSP $\mu$P's performance.

Special-purpose DSP architectures are tailored to the needs of a specific application and therefore provide the best throughput/area at the price of reduced flexibility of use. We noted several diverse instances of the performance advantages of special-purpose architectures over the general-purpose programmable approach. The systolic array is an especially attractive special-purpose architecture for VLSI since it offers the benefits of multiprocessing in a structure suitable for integration. We developed one- and two-dimensional systolic arrays with individual processors based on the CORDIC algorithm for the Givens factorization algorithm. Although the arrays were special purpose, we noted that they were sufficiently flexible to implement Gaussian elimination as well.

## REFERENCES


1. S. U. Qureshi and H. M. Ahmed, A custom VLSI chip set for digital signal processing in high speed voiceband modems, *IEEE J. Selected Areas Commun. SAC-4* (1986), pp. 81–91.
2. S. S. Magar, "A Microcomputer with DSP Capability," ISSCC82 Conference Digest, San Francisco (1982), pp. 32–33.
3. Meerbergen et al., "An 8 MIPS CMOS Digital Signal Processor," Proc. ISSCC86, Anaheim, California (February 1986).
4. J. Roesgen and S. Tung, Moving memory off chip, DSP microprocessor squeezes on more computational power, *Electron. Des.*, *34* (4) (1986), pp. 131–143.
5. Kawakami et al., "A 32b Floating Point CMOS Digital Signal Processor," Proc. ISSCC86, Anaheim, California (February, 1986).
6. J. Bates, "Motorola's DSP56000: A Fourth Generation Digital Signal Processor," Proc. Electro86, Boston (May 1986).
7. P. Ruetz and R. W. Broderson, "A Realtime Image Processing Chip Set," Proc. ISSCC86, Anaheim, California (February 1986).
8. M. Song and M. Narasimha, "A VLSI for 32kbps ADPCM Transcoding," Proc. 1985 Custom Integrated Circuits Conference, Portland, Oregon (May, 1985).

9. H. M. Ahmed et al., " A VLSI Speech Analysis Chip Set Based on Square-Root Normalized Ladder Forms," Proc. 1981 ICASSP, Atlanta, Georgia (March 1981).
10. Galinski et al., "A Processor for Graph Search Operations," Proc. ISSCC87, New York (February 1987).
11. H. M. Ahmed, Alternative arithmetic unit architectures for VLSI digital signal processors, *VLSI and Modern Signal Processing* (S. Y. Kung, H. Whitehouse, T. Kailath, eds.) Prentice-Hall, Englewood Cliffs, New Jersey (1985).
12. J. Markel and A. Gray, *Linear Prediction of Speech*, Springer-Verlag, New York (1976).
13. M. Honig and D. Messerschmitt, *Adaptive Filters: Structure, Algorithms and Applications*, Kluwer, Boston (1984).
14. M. Soderstrand, W. Jenkins, G. Jullien, and F. Taylor, eds., *Residue Number System Arithmetic: Modern Applications in Digital Signal Processing*, IEEE Press, New York (1986).
15. R. Crochiere and L. Rabiner, *Multirate Digital Signal Processing*, Prentice-Hall, Englewood Cliffs, New Jersey (1983).
16. Motorola Inc., *DSP56000 Digital Signal Processor Users Manual*, Phoenix, Arizona (1986).
17. International Telegraph and Telephone Consultative Committee (CCITT) Recommendation G.721, International Telecommunications Union, Geneva.
18. T. Matsumura et al., Implementation of a 32bps ADPCM codec using a general purpose digital signal processor, *IEEE J. Selected Areas Commun. SAC-4*: (1986).
19. H. Ahmed and M. Morf, VLSI array architectures for matrix factorization, *Outils et Modèles Mathématiques pour l'Automatique, l'Analyse de Systèmes et le Traitement du Signal*, Editions du CNRS, Paris (1982).
20. H. T. Kung and C. E. Leiserson, Systolic array (for VLSI), *Sparse Matrix Proceedings*, Society for Industrial and Applied Mathematics, Philadelphia (1979).
21. H. Ahmed, J. M. Delosme, and M. Morf, Highly concurrent computing structures for digital signal processing and matrix arithmetic, *IEEE Comput. Highly Concurrent Syst.*, special issue (January, 1982).
22. S. Y. Kung, On supercomputing with systolic wavefront array processors, *Proc. IEEE, 72*(7): 867–884 (1984).
23. J. Stoer and R. Bulirsch, *Introduction to Numerical Analysis*, Springer-Verlag, New York (1980).
24. H. M. Ahmed, *Signal processing algorithms and architectures*, Ph.D. Dissertation, Stanford University, Stanford, California (1982).

# 8

# VLSI Signal Processing Systems

EARL E. SWARTZLANDER, JR.* Digital Processing Laboratory, TRW Electronic Systems Group, Redondo Beach, California

In the last two decades, integrated circuit technology has evolved from primitive devices with a few gates per package to the current level of very large scale integration (VLSI). Although circuit capability has grown by several orders of magnitude since the first integrated circuits, it is likely that significant further growth will occur in the future.

VLSI devices are, in fact, small systems which differ greatly from earlier technologies in how they are designed and used. Effective VLSI architecture development requires coordination of the technological constraints, internal functional structure, and external interfaces. The results of successful architecture development are chip designs that are producible, efficient, and applicable to many signal processing systems. References [1–3] are recommended sources for further details on signal processing with VLSI that could not be included here because of space limitations.

## 1 INTRODUCTION

Signal processing has become practical with the advent of VLSI. It is a major application of VLSI and provides much of the justification for continued VLSI technology development. One explanation for the emergence of ever-increasing requirements is that the historical growth in computing capability leads potential users to expect continually increasing levels of performance. Without the assumption of growth in capability, many potential users would concentrate on refinement of existing systems and much of the impetus for technology development would be lost. Economic

Present affiliation:

*TRW Defense Systems Group, Redondo Beach, California

pressure would push the technology, but to a much lesser extent than is currently observed.

The challenge of VLSI in this context is to develop chip and system architectures that facilitate the development of signal processing. In spite of the five order of magnitude improvement in circuit performance over the past two decades, chip architecture remains a problem area. To understand the constraints and limitations, it is necessary to be aware of the characteristics of VLSI circuit technology; these will be examined (from an architectural perspective) in Section 2.

Section 3 examines the VLSI circuit architectures which have been developed in response to previous signal processing requirements. Examination of the design of a multiplier-accumulator in Section 4 illustrates the VLSI architecture design process in detail. Section 5 examines a signal processor systems architecture, the next higher level of design.

Finally, Section 6 describes the development of VLSI components to implement a specific signal processing system. Beyond the issues involved in developing a specialized signal processor, this example demonstrates the process of VLSI architecture development, including algorithm and technology selection.

## 2 CIRCUIT TECHNOLOGY

In this section, the evolution of circuit technology will be examined. To provide the appropriate perspective, the history of integrated circuit technology is surveyed. A brief examination of chip complexity metrics and current technology limitations provides the guidance necessary for chip and system architecture design.

### 2.1 History of Circuit Technology

Integrated circuit technology has improved dramatically over the past couple of decades, as shown in Table 8.1. The initial integrated circuits were built using what is now called small-scale integration (SSI). SSI circuits consist of a few unconnected gates or simple flip-flops. These functions duplicate vacuum tube (and transistor) logic modules that were proven by use in the large computer of the 1940 to 1960 time frame. A significant characteristic of the SSI devices is that all logical variables are externalized (i.e., connected to package pins). There were two main reasons for this: 1) implementation technology was such that complex functions (with many

**Table 8.1** Circuit Technology Evolution

| Technology | Date | Gates per chip | Functions |
|---|---|---|---|
| SSI | 1963–1967 | 1–30 | Gates, flip-flops |
| MSI | 1967–1971 | 10–300 | Counters, multiplexers, adders |
| LSI | 1971–1976 | 100–3,000 | Microprocessors, bit slice ALUs |
| VLSI | 1976–1985 | 1,000–30,000 | Multipliers, A/D converters |
| WSI | 1986– | 10,000– | Signal processors, computers |

internal gates) were not producible, and 2) the chip designers wanted to develop "universal" circuits where the user could connect devices in the optimum way to meet the application demands.

With the transition to medium-scale integration (MSI), more complex functions such as 4-bit binary, decimal, and duodecimal (i.e., divided by 12) counters and 4-bit ripple carry adders were developed. In addition to the increase in complexity, a significant change between SSI and MSI was that the number of externalized logic signals had to be reduced. This requirement constrained the selection of functions for MSI implementation to those with a high degree of internal connectivity and relatively modest requirements for external connectivity. A typical example is a 4-bit counter: it consists of four flip-flops arranged to permit division by 2 and eight with lines available to allow cascading to provide division by 16. This MSI circuit has an average of only 2 leads/flip-flop versus 11 leads/flip-flop for the SSI circuit (in both cases power and ground pins are disregarded).

LSI devices were customs chips specially optimized for specific applications (as opposed to general-purpose functional building blocks suitable for a wide range of applications). The normal approach was to take an existing system design that had been implemented using SSI and MSI, and create a unique custom LSI chip design which contained all the specialized interfaces required for compatibility with the earlier design. This approach was rationalized in two ways: 1) if the LSI chip was not producible, the "old" SSI/MSI design could be used, and 2) the LSI design would replace a maximum number of existing chips.

With the increasing cost and complexity of VLSI design, this highly customized approach has fallen into disfavor. Currently, generic designs are developed that stress generality and flexibility even though a few "extra" interface chips may be required. The resulting generic VLSI designs may then be used in a wide variety of systems.

With the move to even higher levels of integration, where hundreds or thousands of gates are placed on chips with 16 to 64 input/output (I/O) pins, the technique of duplicating tube or transistor modules became a clearly unsatisfactory method of chip architecture development. The microprocessor was the critical architectural breakthrough! With the development of the "single-chip" microprocessor, a complete functional entity (i.e., CPU, register set, control sequencer, etc.) was implemented on a single chip, thus eliminating many package pins. Beyond the obvious system simplification, this resulted in higher speed (since fewer signals had to go off chip) and better reliability (due to the reduction in the numbers of chips and connections) and opened many new vistas for computing.

## 2.2 Chip Complexity Metrics

As is becoming widely recognized, the utility of an integrated circuit depends on the product of speed and density. This is because most systems may be implemented with a simple but very fast processor or with a multiplicity of proportionately slower processors.

A chart showing the attainable limits of chip complexity and gate performance for gallium arsenide (GaAs), silicon emitter coupled logic (ECL), and CMOS technologies is shown in Figure 8.1. These speeds and densities are indicative of devices

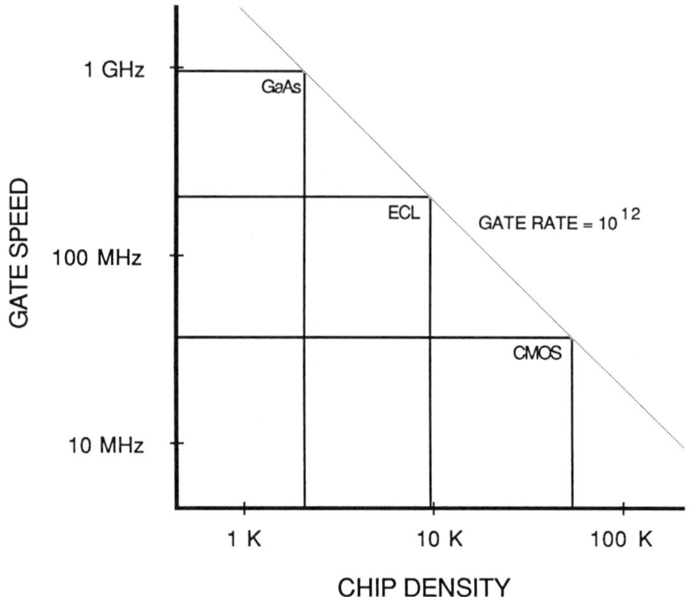


**Figure 8.1** VLSI gate-rate chart.

that were commercially available in 1986. The speeds refer to the maximum clock rates of typical chips and the densities to the number of equivalent two-input gates available on standard chips. This chart also shows a line corresponding to a "gate rate" of $10^{12}$. The gate rate directly measures the chip performance by computing the product of chip complexity (i.e., the gate count) times gate speed. This measure is equivalent to the functional throughput rate (FTR) used by various military programs to evaluate chip technology [4]. From the chart, it is clear that gate rate levels of approximately $10^{12}$ transitions per second are achievable over a range of roughly two orders of magnitude in speed (i.e., 10 MHz to 1 GHz) by selection of the proper technology.

## 2.3 Current Limitations

The current most significant constraint on the development of high-complexity VLSI chips is the problem of packaging. This is due in part to the prevalence of the dual-in-line package (DIP), which has leads in two rows for insertion into printed circuit boards or wire-wrap socket panels. The growth in size of the largest commercially available DIP has been approximately linear with time from 16 pins in the mid-1960s to 64 pins in the mid-1980s. This trend suggests that packages will not become substantially larger in the foreseeable future. Part of the explanation for this is the difficulty in constructing and using the large packages, which are fragile and must be treated with considerable care to avoid breaking the package or damaging the pins. These constraints are expected to continue, posing a significant challenge in device and system architecture, as the exponentially increasing chip complexity must communicate through, at most, about 100 package pins. It is possible to use pin grid array or surface-mounted packages to increase the contacts per package by about a factor or two, but this increases the packaging complexity (and cost). Surface-mounted packages are not amenable to low-cost breadboard fabrication.

An obvious solution to the need for more gates per chip subject to the pin-out limitations is to employ architectures that reduce the number of chip outputs. Any reduction in the number and speed of outputs also reduces the chip power dissipation. Lithographic scaling of devices has reduced the power dissipation per device, but the number of devices per chip has increased at a higher rate so that power management remains a significant VLSI problem. Power dissipation can be reduced by minimizing the number and speed of device output signals.

An intuitive relationship between the module pin count $P$ and the number of logic blocks $B$ was developed by E. F. Rent in 1960. The relation was based on analysis of the IBM 1400 series of computers and is referred to as Rent's rule:

$$P = 4.17B^{.65}$$

Subsequent analysis and examination of additional systems [5] has confirmed this relation with minor variation in multiplicative constant and exponent. Rent's rule does not model pin requirements accurately for high levels of integration. For example, a 2000-gate VLSI circuit would require over 500 pins if Rent's rule was accurate. The explanation for the disparity is that when a complete function is implemented, many interconnections are eliminated. For example, for the ultimate system on a chip, inputs from sensors are required, all processing is done on chip, and a single output is generated. Thus as the level of integration continues to increase, the pin requirements should begin to decrease.

A final problem area is the large amount of time and effort required to design an advanced high-complexity custom chip [6]. Considering advanced VLSI circuits (e.g., with 50,000 or more gates), several hundred thousand design decisions are required to optimize the design. At the very least, such large circuits must be partitioned into relatively independent functional entities so that several designers can be working concurrently. A second approach involves use of cellular logic structures where a small number of cell designs are replicated in one or two dimensions to realize a complex structure.

## 2.4 The Design Challenge

Thus, the challenge posed by signal processing system designers to the developers of VLSI devices is to develop chips using highly structured logic design (to simplify design and test), clean functional partitioning (to simplify chip usage and to minimize pin count), and minimum output pins (to reduce power and to simplify packaging). When VLSI device architectures are developed that satisfy these requirements, signal processing will become practical for a wide variety of applications. At this point, it is appropriate to consider current VLSI architectural concepts.

## 3 VLSI ARCHITECTURE

In designing VLSI circuits several constraints arise, as shown in Section 2. There is a real need to develop improved chip design approaches which will reduce the system design time and effort. Effective use of the new generation of integrated circuit technology demands careful attention to the chip-level functional partitioning. These VLSI architecture issues will be examined in this section and examples of current VLSI chips will be presented to illustrate some of the compromises implicit in the development of real chips.

### 3.1 Hierarchical Chip Design and Layout

The importance of efficient VLSI design techniques is best understood by examining the design cost. For typical circuits, the design cost averages $100.00/gate [6] when advanced computer-aided design (CAD) techniques are used. At this rate, the design of 100,000-gate signal processing circuits would cost tens of millions of dollars and require that hundreds of designers cooperate efficiently. The design problem cannot be isolated to a single horizontal level (i.e., logical, physical, fabrication, etc.) but instead requires a vertical view of the integrated circuit as a system. This approach was used and extended in the design of the Intel iAPX-432 microcomputer chip set [7].

The basic design approach is shown in Figure 8.2. The design is viewed as hierarchical decomposition from the architecture through the logical, physical, and mask design. At each level in the hierarchy, the data base and design tools are different, but the basic design activity flow remains constant as shown in Figure 8.3. Although a detailed examination of the actual design process is beyond the scope of this chapter, a few architectural-level design guidelines will be reviewed.

The basic architectural requirements are clean functional partitioning, regular structures, and minimal chip output. Functional partitioning refers to the design of system-level building blocks. It is desirable to provide a complete function (i.e., multiplier, adder, filter, memory, etc.) on a single chip or to provide interfaces allowing several chips to be concatenated to provide the desired function. Functional design emphasizes the development and implementation of generic functions as opposed to absolute logic minimization. The intent is that the resulting chip will be useful in many applications even though it may include some features that are not required in each application.

Design simplification has become increasingly important with the growth in chip complexity. It is desirable to employ cellular chip structures where a carefully designed cell is replicated in one or two dimensions to produce the complete functions. Cellular designs are generally somewhat less efficient than "custom-crafted circuits" in terms of silicon usage, but the saving in design time is currently the overwhelming consideration. Cellular designs are directly applicable to arithmetic circuits because of the regularity of human pencil-and-paper arithmetic algorithms. Similarly, memory circuits are inherently highly regular. The main problem area in this regard is "random logic" as used to implement control units. Microprogramming [8, 9] offers an approach to "regularizing" random logic,

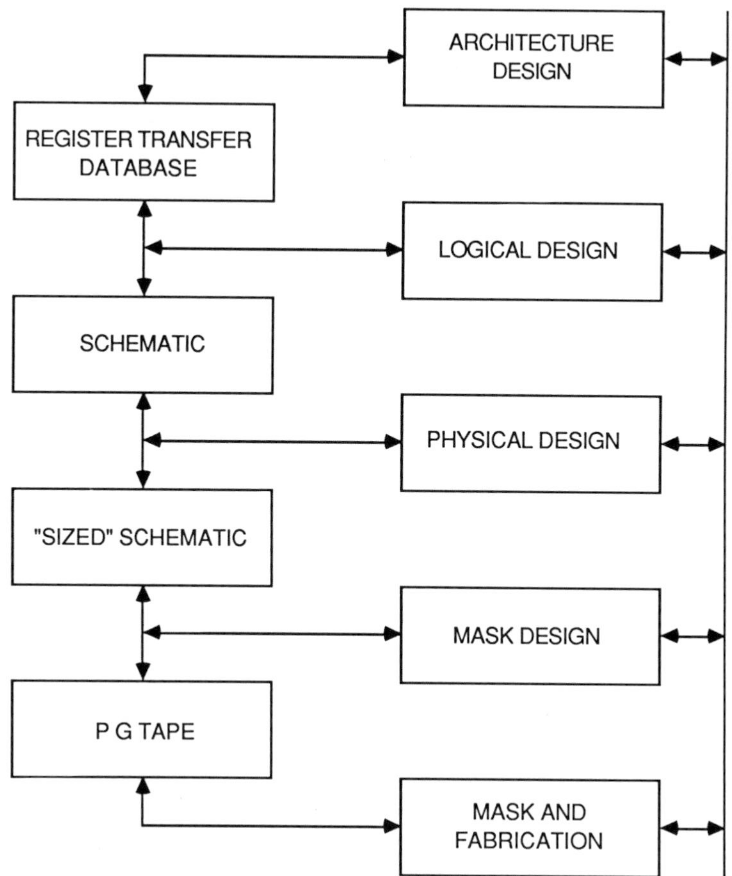


**Figure 8.2** Hierarchical design decomposition. (After Ref. [7].)

but often involves a performance penalty. As highly automated design procedures based on gate array chips become available, some of the need for regularity in chip structures may be mitigated. At present, however, gate arrays achieve relatively low levels of gate density (i.e., gates per chip) in comparison with custom designs.

The third architectural requirement is chip output minimization. The importance of this was discussed in Section 2, where the packaging issues were examined. Another consideration is the chip area required for output drivers and bonding pads, which can be as much as 30% of the total chip area for simple chips with high interface requirements. Although it is commonly thought that the number of I/O pins is a monotonically increasing function of chip complexity, this does not apply at the VLSI complexity levels. Specifically, if a complete function is implemented on a single chip, the output requirements are generally less than for the implementation of subfunctions.

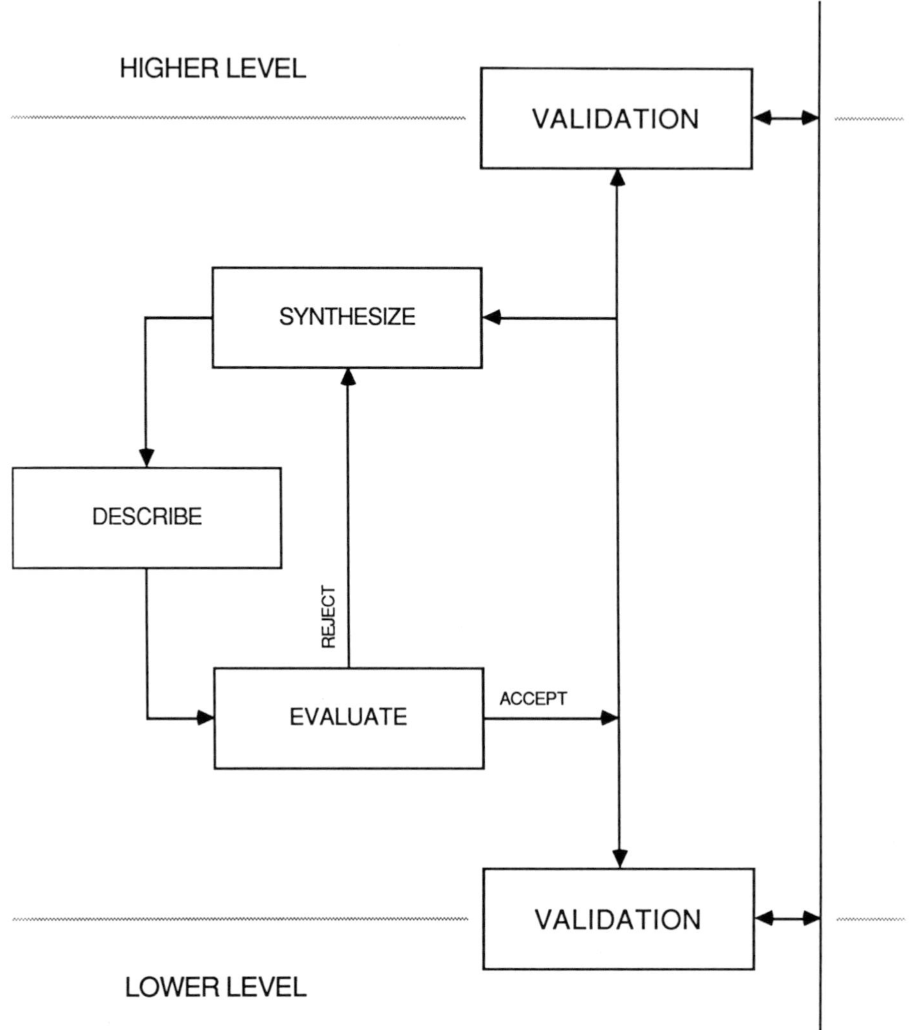


**Figure 8.3** Chip design activity flow. (After Ref. [7].)

### 3.2 Use of Metrics

The gate rate metric presented in Section 2 is useful in the initial partitioning of a system. When a technology is selected, the logic speed indicates the degree of parallelism required to implement the system. The system logic is partitioned into building blocks based on the attainable chip complexity.

### 3.3 Example VLSI Designs

A few examples of current VLSI circuit designs will serve to illustrate the types of functions which are presently being implemented and will further clarify some of the architectural issues. The circuits are building-block circuits suitable for a variety of applications.

## Multipliers

One of the first commercial VLSI devices was the 16 by 16 parallel multiplier that was first sold in 1976. The basic functional architecture is shown in Figure 8.4. The two operands are loaded into input latches, the arithmetic function (in this case multiplication) is performed asynchronously, and the result is loaded into an output register. Use of three-state drivers on the output simplifies interfacing in a variety of applications as shown in Figure 8.5. In single-port operation (typically used in general-purpose data processors), a single bus supplies operands to both data inputs of the arithmetic function by paralleling the data A and B chip inputs and time-phasing the input register clocks. Both halves of the product are communicated on the same bus, if desired, since there is a three-state output. Alternatively, the multiplier can be used in a three-port pipelined mode, as is often required for signal processing applications. The output drivers are continuously enabled and all three registers are clocked simultaneously. On each cycle, two operands are loaded into the input register and the previous result is clocked into the output register. This provides maximum speed, yet requires only simple control (a common signal clock for the three registers) since all data transfers are synchronous. Multipliers for 8-, 12-, and 16-bit word sizes are currently available, with speeds on the order of 10 MHz and power levels of 1 to 3 watts. For the 16-bit multiplier, the use of a 64-pin package precludes use of separate pins for each of the four 16-bit data ports; instead, the least significant half of the product (LSP) is multiplexed with the B input port. In data processing applications (where the least significant half of the product is frequently required), the multiplier is connected with all three data ports tied to a system data bus, so much multiplexing incurs no penalty. In signal processing the least significant half of the product is generally not used, so no penalty is incurred.

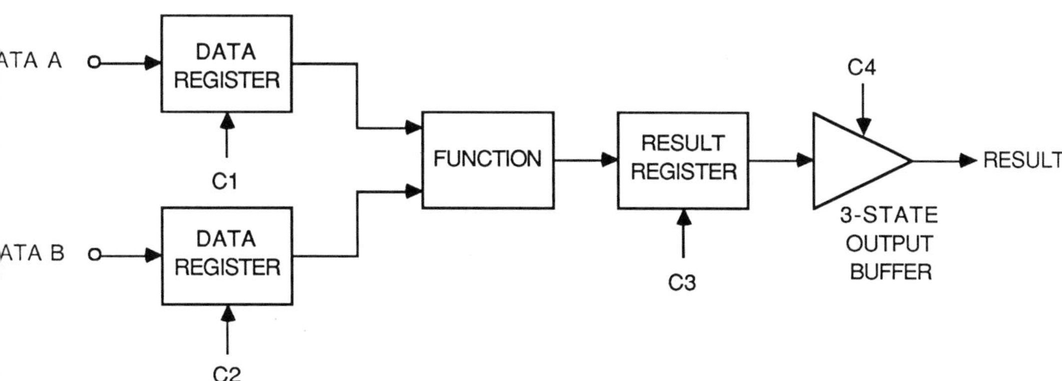


**Figure 8.4** Arithmetic function architecture.

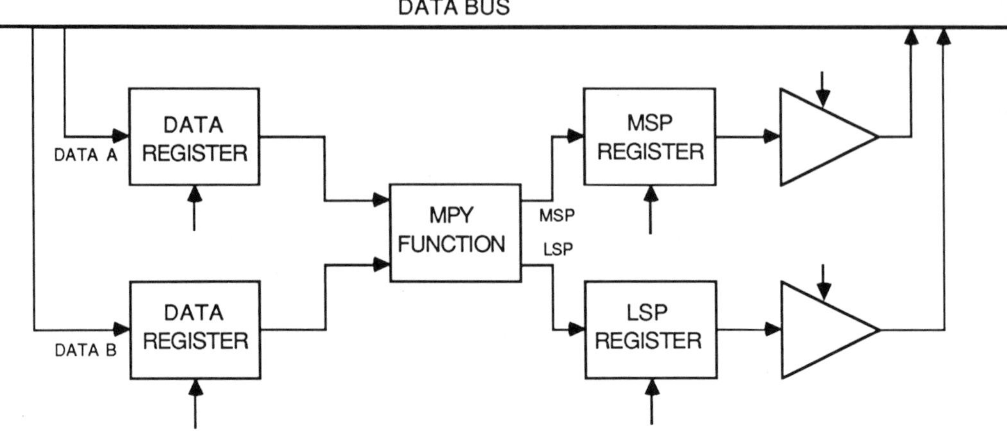


(A) One-Port (Data Bus) Operation

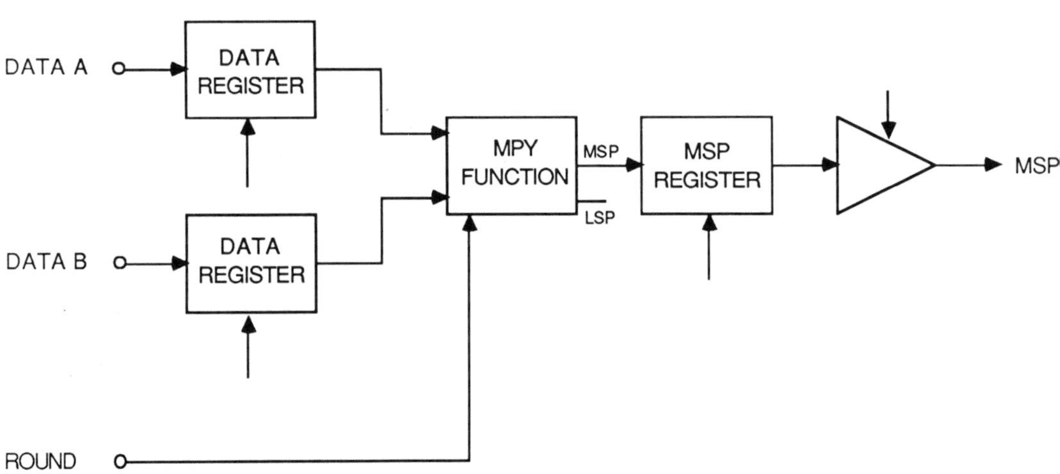


(B) Three-Port Operation

**Figure 8.5** Multiplier interface options.

## Four-Port Memory

The second example VLSI device is the four-port memory shown in Figure 8.6. It consists of a K-bit-wide "slice" of $2^N$ memory locations with four data ports (two read ports and two write ports). Each read port comprises an $N$-bit address input and an output enable control which activates the three-state output driver. Similarly, the write ports consist of $N$-bit address and K-bit data inputs and write enable strobe signals.

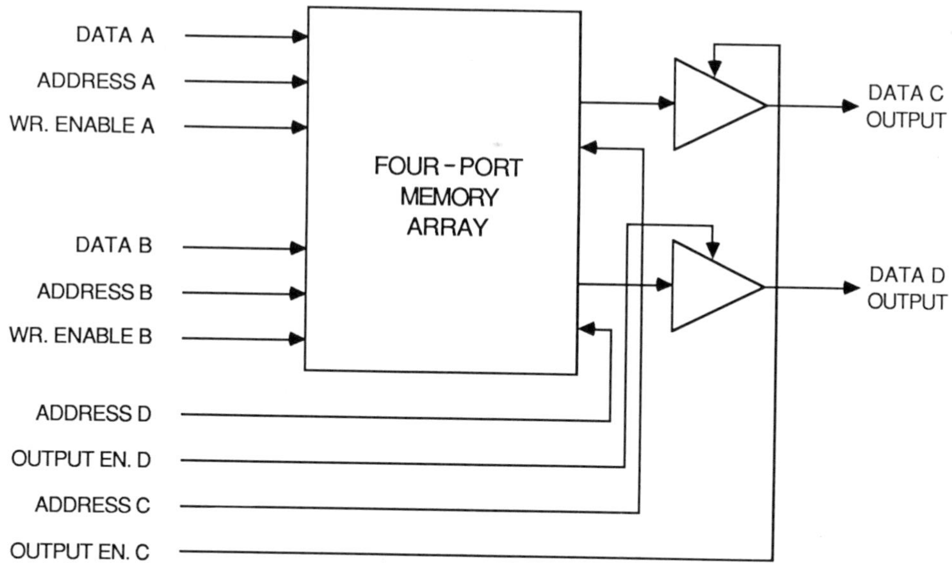


**Figure 8.6** Four-port memory.

A signal processor can be constructed quite efficiently using the multiport memory. Figure 8.7 shows a basic signal processor implemented with a four-port memory, an arithmetic element, and a microprogrammed controller. Arithmetic elements can be implemented using multiplier (described previously) and commercially available MSI/LSI components (carry look-ahead adders, etc.) to operate at

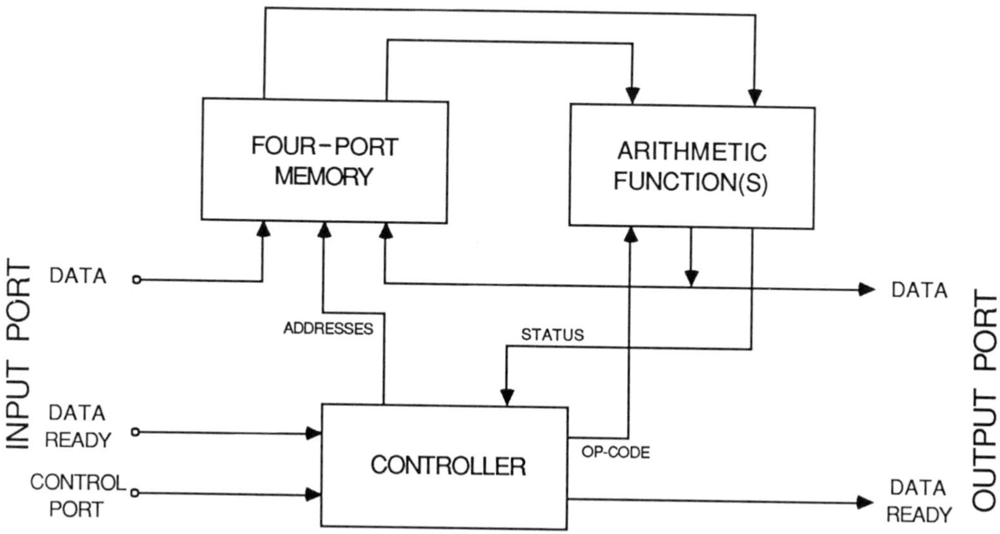


**Figure 8.7** Signal processor implemented with the four-port memory.

pipeline rates of 10 MHz (i.e., accepting a new pair of operands and generating a result every 100 ns) This necessitates a memory that supplies two operands and stores at least one operand each 100 ns, which can be performed efficiently with the four-port memory operating at a 10-MHz cycle.

This architecture recognizes the dyadic nature of signal processing operations (i.e., two operands are accessed, a computation performed, and the result returned to memory) and the relative infrequency of data-dependent jump operations in signal processing.

## 4 MULTIPLIER-ACCUMULATOR DESIGN EXAMPLE

To clarify the issues in chip-level architecture and approaches toward their solution, the design of a multiterm multiplier-adder will be examined in detail. The multiplier-adder is used for the implementation of finite impulse response (FIR) filters for digital signal processing and the evaluation of inner products for image processing applications like high-speed computed tomography; however, this analysis is directly applicable to many other systems elements.

### 4.1 Performance Requirements

The basic circuit structure which will be implemented is a multiterm inner product processor, as shown in Figure 8.8. To form an inner product of two vectors of length $M$, the first $K$ elements of the vectors are multiplied and summed; the next $K$ elements are multiplied, summed, and added to the previous sum, etc. After repeating this process $M/K$ times, the inner product has been computed [10]. The obvious implementation uses $K$ multipliers and a $K-1$ element adder tree to multiply and sum the terms. An additional adder and latch accumulate the sums. At each stage in the adder tree, adders of sufficient width are used to avoid rounding or truncating the data. The accumulator width sets an upper limit on the number of vector elements which can be multiplied without the possibility of overflow; however, this width can easily be made large enough to satisfy practical requirements (i.e., with 10 growth bits, 1024-term inner products may be computed without overflow).

Three distinct arithmetic implementation approaches will be considered: 1) a modular array which serves as a building block for constructing multipliers and adders, 2) a merged arithmetic multiterm multiplier-adder, and 3) a commercial VLSI multiplier-accumulator. All three approaches use conventional two's complement arithmetic, but because of different assumptions about the available technology, markedly different designs have been developed.

### 4.2 Modular Array

The modular array is based on an ultra-high-speed two's complement arithmetic circuit which is used for addition, subtraction, and multiplication. Such a device implemented with advanced silicon ECL, gallium arsenide, or Josephson junction technology can operate at projected clock rates in excess of 500 MHz by pipelining, although the latency (the time from when the operands are entered until the result

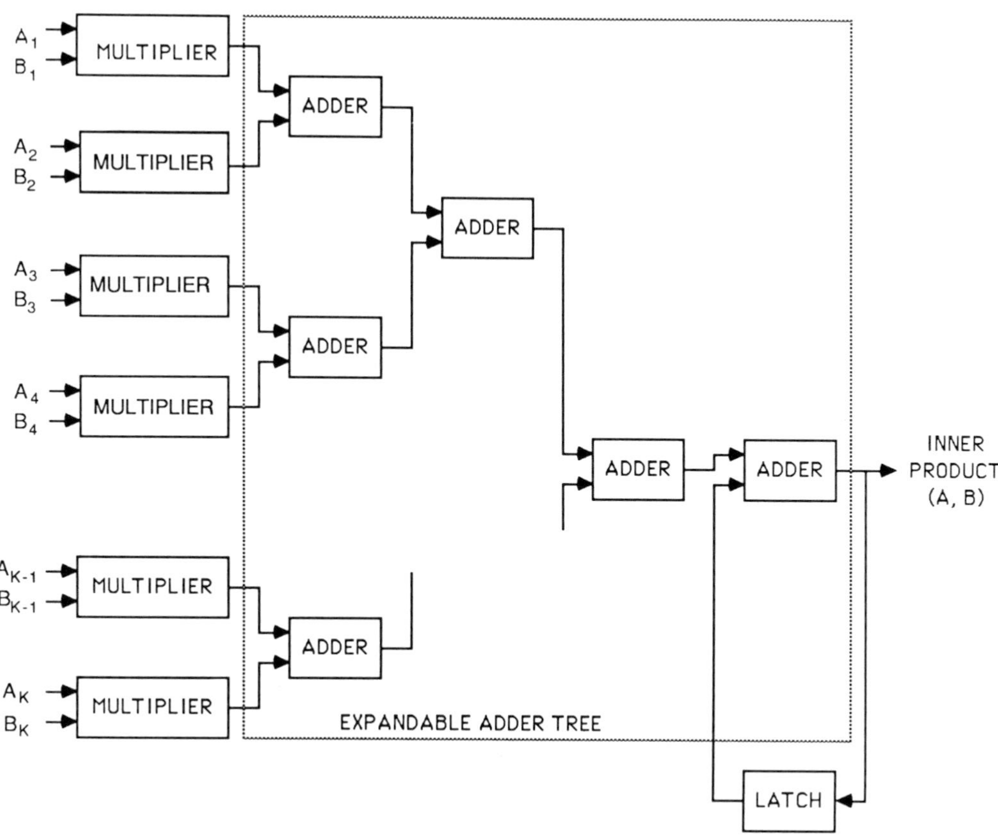


**Figure 8.8** Multiterm inner product processor.

is available) may be many tens of clock cycles. In the multiterm inner product processor (as well as many other signal processing applications) large blocks of data are processed without data-dependent branch instructions, so that achieving high processing rates is of greater utility than minimizing the latency.

The circuit is a carry-save implementation of Booth's multiplication algorithm [11] (Booth's paper has been reprinted in [12]), which has been widely accepted for both software and hardware multiplication applications. This algorithm is implemented for pipelining using a carry-save approach for the addition/subtraction of the multiplicand to the partial product. For example, a 16 by 16 multiplier requires a 32-stage pipeline; the first 16 stages of the pipeline form two intermediate results (i.e., a sum word and a carry word) which sum to the product via the carry-save process; the final 16 stages add these two words. Since the pipeline clock interval need be only long enough to permit execution of a carry-save addition operation between successive clock cycles, pipeline rates in excess of 500 MHz can be attained with advanced oxide isolated ECL process technology. Gallium arsenide technology, which is currently emerging from the laboratory into commercial production, should allow gigahertz clock rates. Booth's algorithm was selected for possible implementation as a modular array because a single VLSI circuit design can be used to multiply (by using the complete multiplier configuration) or to add/subtract by using only the carry-save adder array.

In Booth's algorithm, adjacent bits in the multiplier operand are examined in overlapping pairs, and a decision is made to add or subtract the multiplicand from the accumulated sum of partial products or to retain the previous sum. Two adjacent multiplier bits are inspected at a time; if both multiplier bits are the same (either 00 or 11), the partial product is shifted one position to the right. If the multiplier bit pattern is 10, the multiplicand is subtracted from the partial product; if the bit pattern is 01, the multiplicand is added to the partial product. This procedure is repeated $N$ times for an $N$-bit multiplier. The requirement for physical realization of this algorithm is an array of full adders and a small amount of peripheral logic, as described in [10]. Algebraic subtraction is performed, when required, by adding the one's complement, and then by adding a 1-bit at the least significant bit position.

The carry-save adder eliminates the usual requirement for horizontal carry propagation within a row of the full adder matrix. This results in an array with maximum pipeline rate, since only a single full add operation must be completed within each clock period. The integrated circuit implementing this modified form of Booth's algorithm consists of three sections: a triangular array of (multiplier) pipelining latches, a rectangular array of (multiplicand) pipelining latches, and an array of full adders supported by a small amount of ancillary logic. The two arrays of latches have all input and output leads externalized for independent use as delay elements. The triangular latch array appropriately delays the multiplier bits; this corresponds to the delay of the multiplicand and partial product in the carry-save adder circuit. Including pipeline latches at every level in the array ensures a 500-MHz operating speed.

The sizing of the modular array has been set at 8 by 4, which requires 56 signal pins and can be accommodated by a 64-pin package. Further increases in the array size (e.g., to 16 by 8 bits) would require packages with more than 100 pins.

The modular arithmetic array may be used to implement a variety of operations, which include addition, subtraction, and multiplication. The simplest designs generate the results on a "time-skewed" basis; that is, the resulting bits are not all present simultaneously at the output, but emerge from the unit sequentially from the least significant to more significant bit positions as they are formed. For example, the least significant bit is available one clock period after the operands are presented to the device inputs. The next most significant bit is available one clock period later, and so on. If necessary, the skewed result can be deskewed by using the triangular latch arrays.

Table 8.2 summarizes the number of modular arrays required for the implementation of the various arithmetic circuits. The complexity level ranges from 2 devices for realization of an 8-bit adder to 30 devices for a 40-bit adder. All entries in this table include deskewing logic.

A pipelined adder with a 40-state latency (e.g., the output is available 40 clock periods after the operands enter the adder) may appear unsuitable for use in computing inner products since the inner product requires serial accumulation. It is possible to achieve highly efficient operation by interleaving the computations for $N$ ($> 40$) distinct inner products and computing a new component for each of the inner products every $N$ clock cycles. This process is shown for FIR filtering in

**Table 8.2** Modular Array Count to Implement Various Arithmetic Functions

| Function | Number of 4 by 8 arrays |
|---|---|
| 8-bit adder | 2 |
| 16-bit adder | 6 |
| 24-bit adder | 12 |
| 32-bit adder | 20 |
| 40-bit adder | 30 |
| 8-bit multiplier | 6 |
| 12-bit multiplier | 12 |
| 16-bit multiplier | 20 |

Figure 8.9. For each input data, $H_i$, $N$ filter kernel values are accessed and multiplied to form $N$ intermediate products, which become available after the multiplier latency time. These products are then summed in the accumulator, adding to each of the $N$ partial results which are in progress. During each cycle, one of the FIR filter outputs will be completed and another initiated in its place.

To implement the inner product processor, a multiplier and an accumulator are used. Assuming 16-bit operands and a 40-bit result, which allows summation of 256 full-precision products without overflow, a total of 50 modular arrays are required

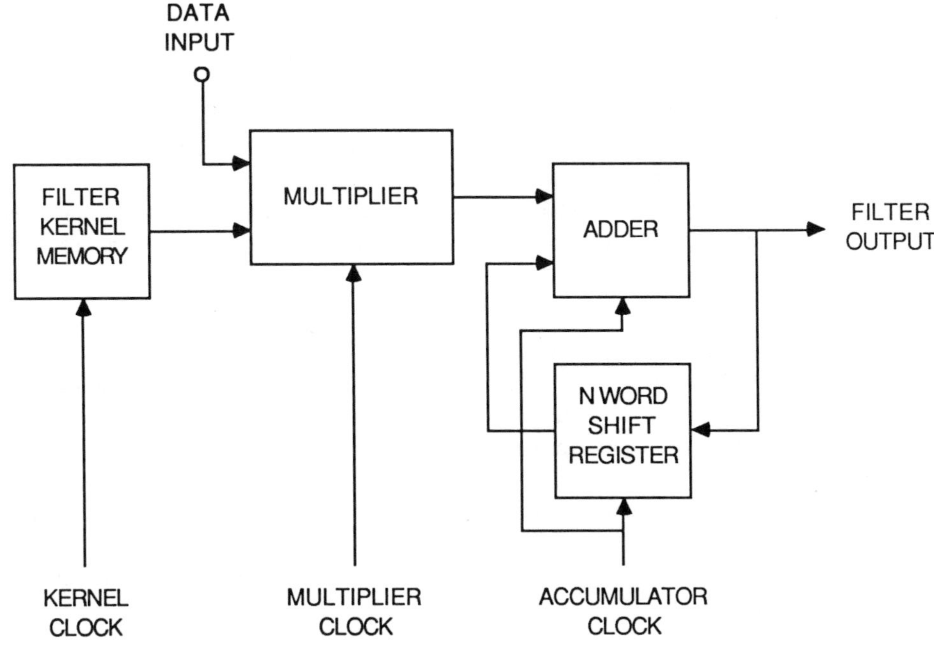


**Figure 8.9** Modular array FIR filter.

(20 for the multiplier and 30 for the adder). Allowing for control logic and clock buffers, the convolver implementation complexity increases to approximately 70 integrated circuits, with a computational capacity of at least 500 million multiply-add operations/second. Dividing the throughput by the chip count gives a useful metric which is evaluated as 7 million multiply-add operations/second per integrated circuit for this approach. It should be noted that the modular arithmetic array has not been developed but that its complexity is well within the current state of the art.

By adopting an unconventional approach to the development of a logical array, which is realized via three separate (interacting) iterative functions on a single integrated circuit, a circuit design results which can be used efficiently as either an adder/subtractor or a multiplier. This circuit is well suited to signal processing applications in which its relatively long pipeline length and attendant latency is not a limitation. In contrast, this level of latency would not be acceptable for general-purpose data processing applications (because of the frequent needs to execute data-dependent branch instructions).

## 4.3 Merged Arithmetic Multiterm Multiplier-Adder

Merged arithmetic results from recognition that an optimal realization of a composite arithmetic function such as a multiterm inner product does not require distinct arithmetic operators (i.e., adders, subtractors, and multipliers), but instead may generate the output bits comprising the function directly from the operand bits [13]. To simplify the discussion, the algorithm will be described for sign/magnitude numbers. This design is easily modified to provide direct two's complement operation by inclusion of correction terms as for two's complement multiplication [14]. The sequence of operations required to compute an inner product of two $M$-element (each of $I$ bits) vectors via merged arithmetic involves three basic steps: first, the bit product matrix is generated with an array of $I^2M$ AND gates; then the matrix is reduced by counting the 1-bits in each column and by performing carry processing to produce a two-row matrix; and finally, the two rows are summed in a carry look-ahead adder to generate the desired inner product.

The algorithm used for the second step, matrix height reduction, is the key to achieving an efficient design. Dadda's heuristic minimization procedure [15], which results in minimum-complexity designs, is based on use of full adders as the counters:

1. Let $d_1 = 2$ and $d_j = [3d_{j-1}/2]$ where $[x]$ denotes the largest integer less than or equal to $x$. Find the largest $j$ such that at least one column of the bit matrix has more than $d_j$ elements.
2. Use full or half adders as required to achieve a reduced matrix with no column containing more than $d_j$ elements. Note that only columns with more than $d_j$ elements (or those which receive carries from less significant columns) are reduced.
3. Repeat step 2 with $j = j - 1$ until a matrix with only two rows is generated (i.e., $j = 1$).

In implementing step 2, each full adder accepts three inputs from a given column and produces a sum bit which remains in that column and a carry bit which moves to the next significant column. Thus, each use of a full adder reduces the number of elements in the composite partial product matrix by one. Similarly, half adders take in two elements from a column and produce two outputs: the sum in the original column and the carry in the next more significant column.

Figure 8.10 shows the computation of a two-term inner product (i.e., $A \times B + C \times D$) for 8-bit operand precision. At the top, the two 8 by 8 trapezoidal bit product matrices are formed: one for $A \times B$ and one for $C \times D$. The highest column is that in the middle of the figure with a height of 16. Since the $d_j$ sequence is 2, 3, 4, 6, 9, 13, 19, ..., the first matrix reduction is to a matrix (matrix II) where each column has 13 or fewer elements. A total of six full adders are used to effect the reduction; these are shown by connecting the outputs of the adders with a line as they appear in matrix II. Thus, a full adder is used to reduce three of the entries in the seventh column from the right (which has a total of 14 entries) in the composite bit product (matrix) to a sum bit in column 7 of matrix II and a carry into column 8. The seventh column of matrix II now has 12 entries (the 11 "unreduced" entries and the sum bit) and satisfies the desired constraint. Columns 8 and 9 each require two full adders to satisfy the constraint, and column 10 requires a single full adder. Note that column 10 in the composite bit product matrix had only 12 entries but that the two carries that resulted from reducing column 9 would have caused violation of the constraint. The reduction process continues with successive column height limits of 9 (matrix II), 6 (matrix IV), etc. Half adders are used in forming columns 7 to 10 of matrix IV. They are shown as two outputs connected by a line (like full adders) except that the connecting line is "crossed." Half adders are also used to form columns 5 and 6 of matrix V and column 3 of matrix VI. All of the numbered matrices in Figure 8.10 have been drawn with the dots pushed to the highest possible row of each matrix (with the sole exception of the third column of matrix V) to simplify checking each column's height.

In comparison to the direct methods exemplified by the modular array, the merged arithmetic implementation generates all of the bit products "en masse" and reduces them to a two-row result matrix using a single (i.e., merged) reduction network. For the two-term 8-bit merged inner product element example, the two bit-product matrices are reduced through six stages of adders to produce a pair of intermediate operands which are summed with a single carry look-ahead adder.

Comparison of the merged approach with conventional practice [13] demonstrates that, for this example, merged arithmetic requires 20 more adder modules (i.e., full adders) but two fewer 15-bit carry look-ahead adders. Since a 15-bit carry look-ahead adder is significantly more complex than 20 adder modules, the merged approach is considerably simpler than the conventional implementation. In terms of timing, the merged approach incurs two additional levels of full adder delays at a saving of one 15-bit carry look-ahead delay. Examination of the gate-level timing indicates that the merged approach is slightly faster.

From the viewpoint of VLSI technology implementation, the complexity reduction achieved through merged arithmetic is even greater than might be expected from gate count considerations. Since carry look-ahead adders require interconnection topologies which are difficult to realize in VLSI circuits (because of the large number of signal crossovers), the saving of $2M-2$ ($M$ at the multiplier outputs and $M-2$ in the adder tree) carry look-ahead adders in an $M$-term convolver is quite significant. An expandable two-term convolver building block forms the sum of the

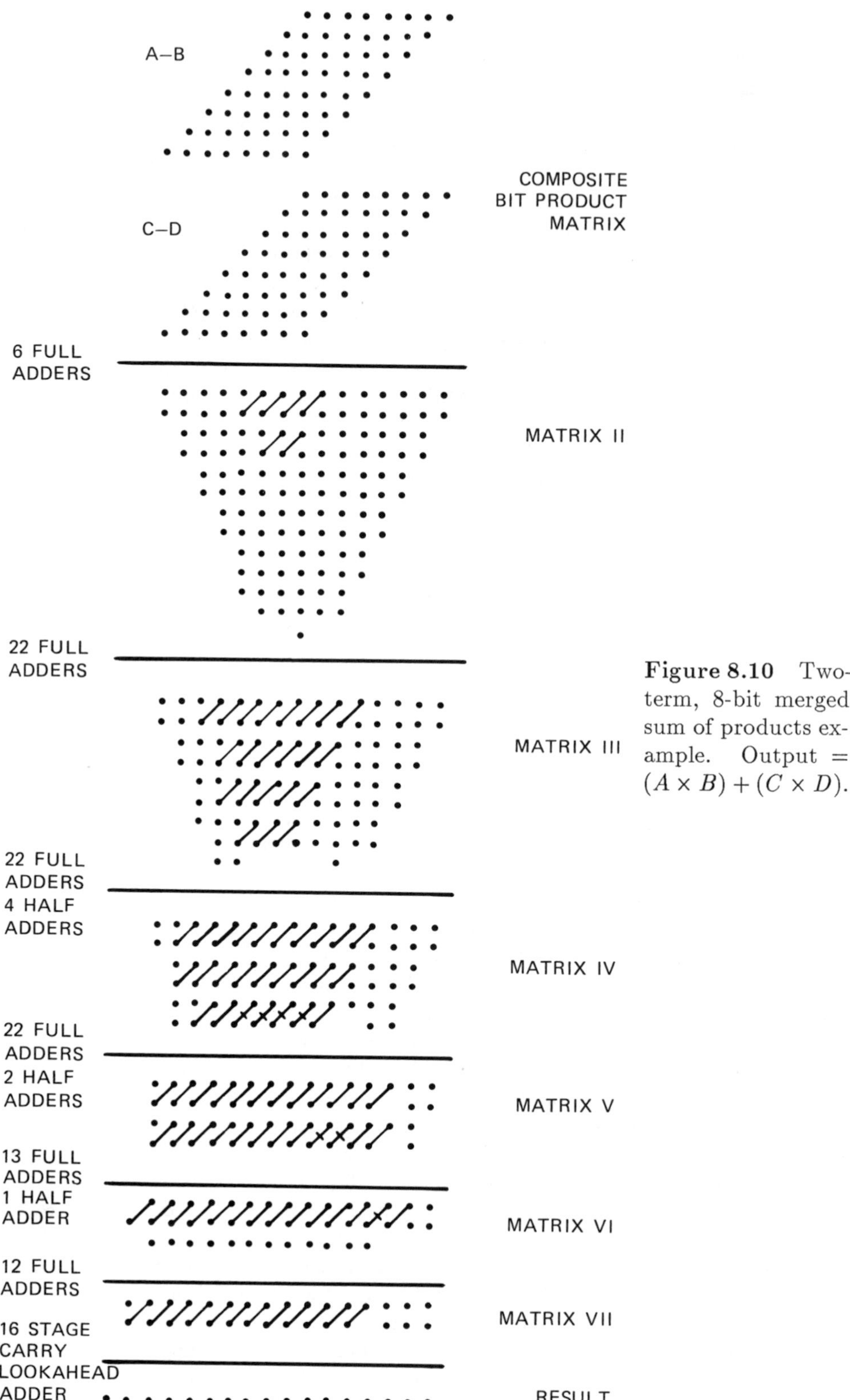


**Figure 8.10** Two-term, 8-bit merged sum of products example. Output = $(A \times B) + (C \times D)$.

three terms (the two products and the expansion input). The expansion input is used to convolve vectors of arbitrary length by repeatedly computing pairs of terms which are summed with other partial results.

Within current VLSI processing constraints, the implementation of a two-term convolution block is feasible as a single VLSI circuit. If implemented with a high-density bipolar technology, throughput of the merged arithmetic two-term multiplier-adder is estimated conservatively at 30 million operations/second. This is equivalent to 60 million multiply-add operations/second per integrated circuit. It is a structure which lacks the desired regularity for VLSI design at the present time, but development of sophisticated computer-aided design techniques may allow the design and production of this device in the near future.

The large number of package pins required for the signals is an apparent problem with this circuit. A total of 144 signals are required for the two-term convolver (two pairs of 16-bit inputs, a 40-bit expansion input, and a 40-bit result). In fact, in most applications of large inner products (FIR filters, convolution, etc.), the vectors either are elements of a time sequence or are fixed kernels. In the first instance, a single input port is required with shift registers to propagate the data from one multiplier to the other. In the second, the same shift register structure is used, but the kernel is loaded only at start-up or when a new kernel is required. This reduces the input pin count to 32 (excluding the expansion input). The expansion input and output still require 80 pins. A potential solution is to establish a communications protocol at the integrated circuit level which allows transfer of input data operands into the VLSI circuits in either a bit-serial or a byte-serial stream.

## 4.4 Commercial VLSI Multiplier-Accumulator

The third approach is to use a commercial multiplier-accumulator. This is exemplified by a 16 by 16 parallel multiplier-accumulator, which has been commercially available since 1978. The basic structure is shown in Figure 8.11. This multiplier-accumulator consists of two 16-bit input registers connected to a combination multiplier array. The double-precision multiplier output drives a 35-bit adder and an accumulator register. Three-state output drivers allow reading and loading the accumulator register. A variety of control signals are required: selection of two's complements or unsigned arithmetic, accumulator commands (i.e., add/subtract, accumulator or pass-through, and preload), three-state driver enables, and clocks for the input registers and the output register.

This design shows the impact of package pin limitations. Since 67 signal pins would be required to accommodate the two 16-bit inputs and the 35-bit output word, compromise was required to use a standard 64-pin package. Specifically, the least significant 16 product bits are multiplexed with the $Y_{IN}$ input on a bidirectional port. This reduces the number of data lines to 51, and a 64-pin DIP is used.

The typical delay time for a commercial 16 by 16 multiplier-accumulator is 115 ns, so a multiply rate in excess of 8 million multiplies/second and an equivalent accumulation rate are achieved. This is comparable to the throughput/chip count of the modular array which assumed use of advanced oxide isolated ECL technology.

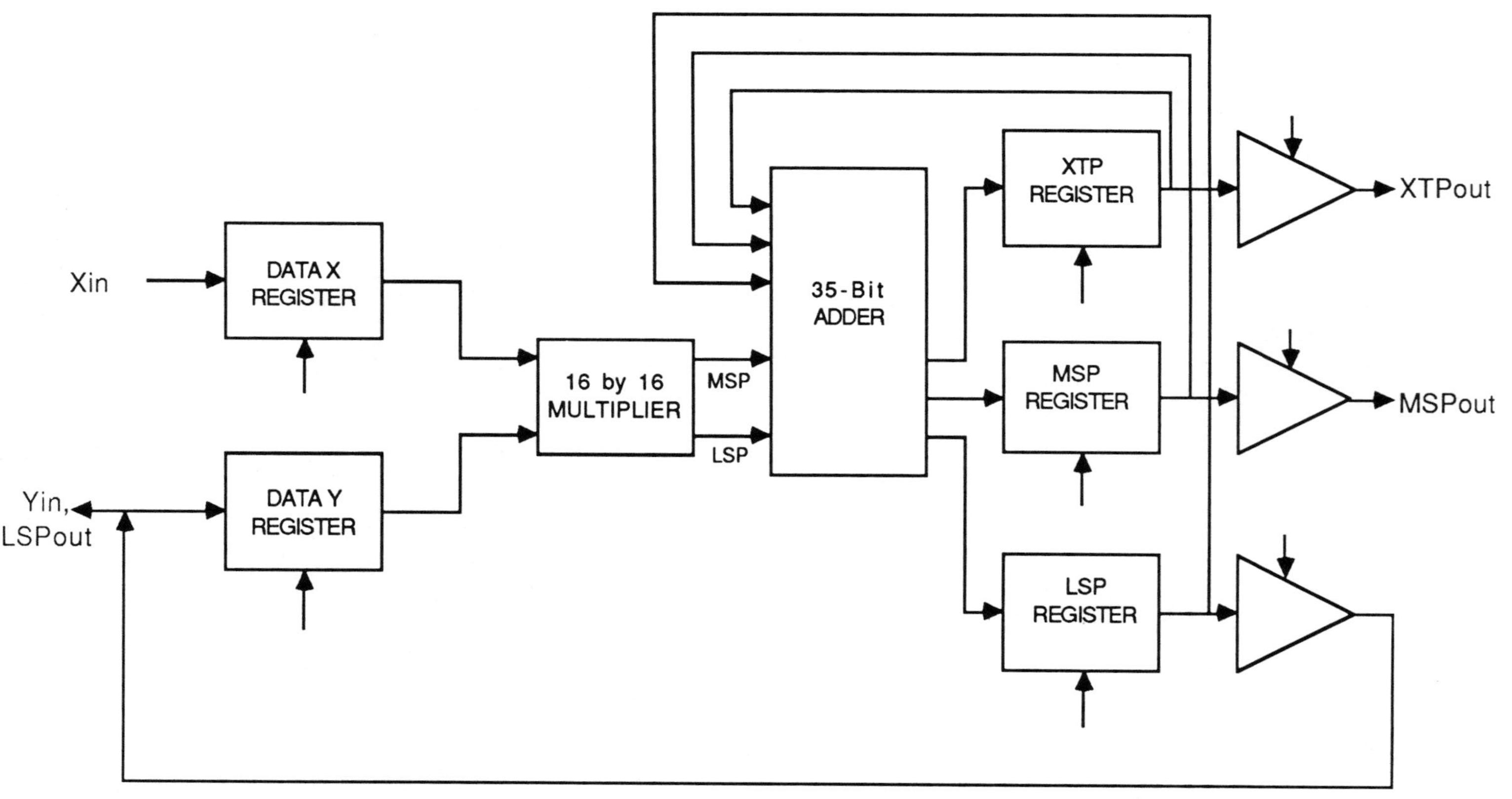


**Figure 8.11** Multiplier-accumulator.

### 4.5 Comparison of Implementations

Formal comparisons of the three implementations presented here for multiterm multiplication-addition must consider the state of the art of VLSI circuit technology, which affects both circuit speed and circuit density. Since the three implementations considered involve three technologies, the merit is best evaluated by comparing the performance and complexity. For this application, the performance is defined as the functional throughput (i.e., the number of multiply-add operations performed per second). A variety of complexity measures have been used for specific applications, but one of the most useful is the number of integrated circuits required to implement the design. This measure corresponds well with more conventional criteria (e.g., cost, size, power, gate count, and LSI chip area), is easily computed, and is useful in preliminary feasibility assessments [16]. A large numerical value of the figure of merit implies increased computation per integrated circuit or, alternatively, fewer circuits to achieve equivalent computation rates.

As shown in Table 8.3, the merged arithmetic element achieves a figure of merit of 80, which is an order of magnitude greater than that of the modular array and the commercial multiplier-accumulator. The use of high-density moderate-speed VLSI technology is an order of magnitude more efficient than the ultra-high-speed ECL commercial bipolar technology. This finding corroborates one of the major motivations for the development of VLSI technology: increases in circuit density greatly reduce partitioning problems by executing a complete function within a single circuit package. This comparison may also be viewed as confirmation that, for this application, moderate-performance technology provides higher utility than a technology which is at the leading edge of the current state of the art. The advantage of the merged arithmetic approach with respect to the commercial multiplier-accumulator is attributed partly to the growth in technology capability over the past few years and partly to the advantages of a custom chip design.

A final consideration which is related to the high functional density of these mechanizations is the practical problem of packaging, given the limitations of present integrated circuit packages. A potential solution is to establish a communications protocol at the integrated circuit level which allows transfer of input data operands into the VLSI circuits in either a bit-serial or a byte-serial stream.

## 5 SIGNAL PROCESSOR ARCHITECTURE

Over the four decades since the development of the first stored-program computers there has been much evolution in systems architecture, in part because of the increasing complexity of the requirements. In this section, signal processor archi-

**Table 8.3** Comparison of FIR Filter Implementation

| Approach | Terms per second | Complexity | Figure of merit |
|---|---|---|---|
| Modular array | 500,000,000 | 70 chips | 7 |
| Merged arithmetic | 80,000,000 | 1 chip | 80 |
| Commercial multiplier-accumulator | 8,000,000 | 1 chip | 8 |

tectures are surveyed to identify the basic approaches available to the systems architect. One of the major design decisions in developing a signal processing system is the choice of processor architecture.

## 5.1 Computer Architectures

The first computers, developed in the late 1940s and early 1950s, were designed with what has come to be called the von Neumann architecture, which is shown in Figure 8.12. It uses a central processing unit (CPU) coupled to a read/write memory. The memory stores both the program instruction sequence and data values. During operation, the CPU reads the first word from memory and executes the appropriate instruction. Then the next word from memory is read, executed, and the process continues until the program is completed. The significant aspect of this architecture is that a single memory is used for both the program and the data; accordingly, a program may access the memory and modify instructions to change the program. It is this stored program concept which represents the attraction as well as the disadvantage of the von Neumann architecture: the attraction is the ease of program modification, while the disadvantage is that all operations performed within the machine, either arithmetic, control, or input/output, require memory access. Therefore, the machine performance is directly constrained by the

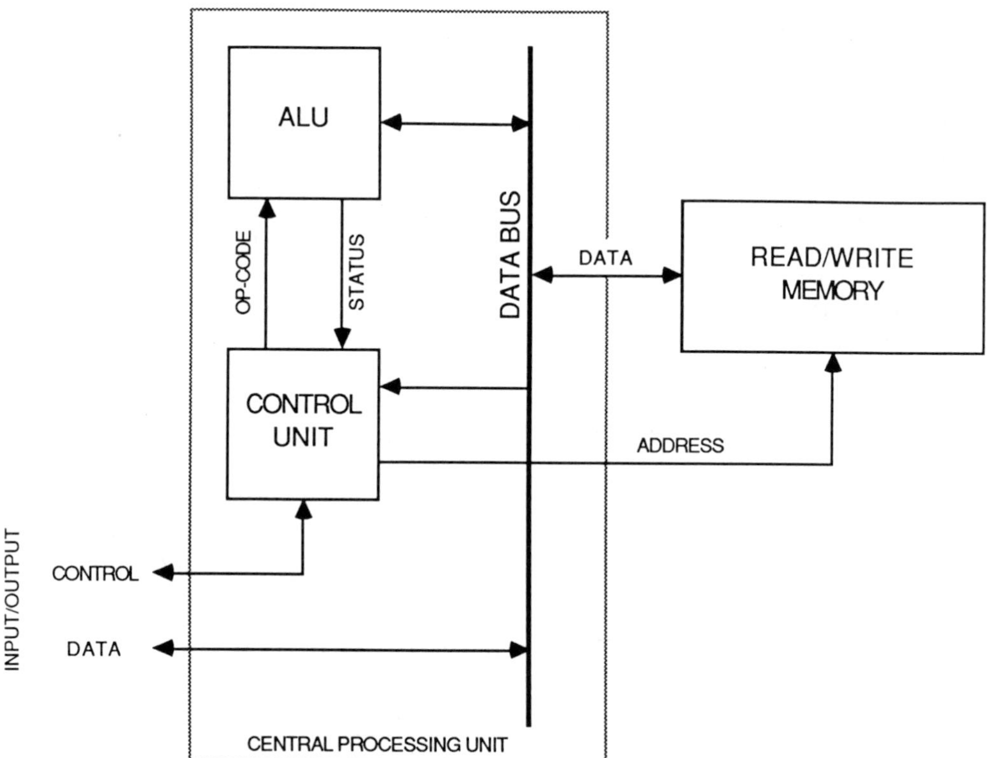


**Figure 8.12** Von Neumann processor architecture.

memory performance. Since three or more memory access operations are required to execute a single instruction, achieving high performance is very difficult. Part of the explanation for the preeminence of the von Neumann architecture is that it is relatively low in cost to implement: only a single read/write memory is required to implement the entire storage for the system. This consideration was, of course, far more critical in the early days of computing, when the memory cost often dominated the cost of a computer.

Another major class of architectures is the Harvard architecture shown in Figure 8.13. Here, two separate memories are used; one for data and another for the program. The data memory, like the memory used in the von Neumann architecture, is a read/write memory, while the program memory may be either a read-only memory (ROM) or a read/write memory. Generally, read/write memories are used early in the machine development cycle, and a ROM is used when the design is more mature (i.e., stable). This architecture achieves a degree of concurrency in its operation: in a single cycle, an operand is accessed from the data memory and used in a calculation by the arithmetic/logic unit (ALU). Simultaneously, the program memory and control unit are accessing the next instruction and computing the addresses for the next data access. The processors built using this architecture are generally tailored to reflect the requirements of the system; thus they are relatively inflexible and not suitable for other processing applications.

In contrasting the von Neumann and Harvard architectures, the von Neumann architecture is clearly more flexible but offers lower performance. This is, in fact, a recurring theme in architecture design: flexibility and generality are obtained only by sacrificing performance.

To achieve the performance required for signal processing, ultra-high-speed logic is required. Candidate technologies for this approach include ECL, CMOS, and gallium arsenide. ECL is currently in commercial production, but effort is required

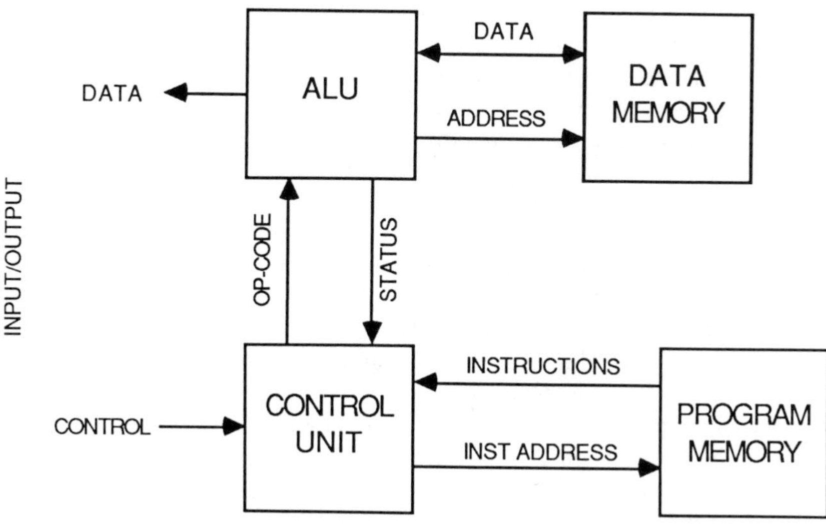


**Figure 8.13** Harvard processor architecture.

to increase the speeds and density to levels necessary for signal processing systems. With each of the other technologies, subanosecond logic delays are currently attainable on an experimental basis, and significant efforts are being devoted to achieve commercial producibility. Severe problems currently exist in the areas of circuit density, thermal management, packaging, and propagation of ultra-high-speed logic signals between the integrated circuits. It is appropriate to anticipate progress in all of these areas, but in view of the level of improvement required to achieve practical systems, it appears that at least several years of additional development will be necessary.

When ultra-high-speed logic is used, a system is implemented with fewer integrated circuits, thereby improving the system reliability. This occurs because the reliability of a system is generally proportional to the number of individual components and particularly to the number of interconnects between these devices. Provided that proper design, fabrication, and test techniques are employed, the mean time to failure is not significantly affected by increases in the system clock rate. Hence, a well-designed system which achieves a given throughput by an exploitation of high-speed device technology will demonstrate improved reliability in comparison to a system employing slower clock rates and a larger number of integrated circuits to achieve the same throughput.

### 5.2 Array Processors

The second category of systems architecture is the array processor. These are general-purpose (host) computers with special-purpose adjunct processors that are used to provide high performance for restricted classes of processing. Since the host computers are conventional single computers, attention is focused here on the special-purpose adjuncts. The three primary types of array processors in common use are parallel processors [17], pipeline processors [18], and tailored processors. Generally, parallel processors consist of a single specialized control unit that issues commands to an array of computational units that operate in parallel. Such processors often use a highly regular interconnection concept, allowing limited communication between the computational units; in the Illiac IV, a computational unit is connected to its four nearest neighbors (right, left, up, down) and data routing between nonadjoining units requires a series of transfer steps during which no computation occurs.

As noted elsewhere [19], parallel processors can achieve substantial speed increases over single computers while using moderate-speed logic, but must be tailored to the problem. Commercial parallel processors are designed to be highly flexible so that they may be used for a variety of applications at a sacrifice of efficiency for any specific application. Pipeline processors also achieve high computational throughput with moderate device speeds. Like the parallel processors, they pose significant software development problems.

The three main types of tailored array processors are the hardwired processor, programmable signal processor, and multilevel processor. The early hardwired array processors were direct digital emulations of previous analog processors that used fixed data routing and could not be modified without extensive redesign. Hardwired processors are finely tuned to implement specific algorithms (matrix inversion, correlation, etc.). By creating a custom design for each application, the hardwired processor achieves a higher level of efficiency than other array processors and minimizes the parts count to perform a given function.

In the past two decades, the programmable signal processor (PSP) concept has evolved. One to several specialized arithmetic units and an assortment of I/O channels are operated under stored-program control to provide an extremely flexible system. Although a single PSP design can be used with different software to implement signal processors for a variety of applications, most PSP designs have special features to optimize their use within a single application area (radar, signal processing, sonar beam forming, speech recognition, image enhancement, etc.) Programmability provides the flexibility to implement multiple operating modes within the single application area. Because of the desire to achieve efficiency with the PSP design, hardware is added to achieve a high level of data path flexibility. As a result, PSP hardware is much more complex than a hardwired processor to do a given job. Also, it is becoming well known that the wide-word microcode used by early PSP designs is quite expensive to design, code, debug, modify, and maintain. The PSP is an attractive approach where a high degree of flexibility is required, as in the early phase of system development, where the software flexibility of the PSP allows correcting design errors and optimizing the algorithms.

An alternative approach to signal processing architecture is the multilevel processor. This approach uses a family of processing modules that are interconnected as required to optimally implement a specific application. A typical multilevel processor architecture is shown in Figure 8.14. A control microprocessor generates a stream of "high-level" commands to a number of specialized processing mod-

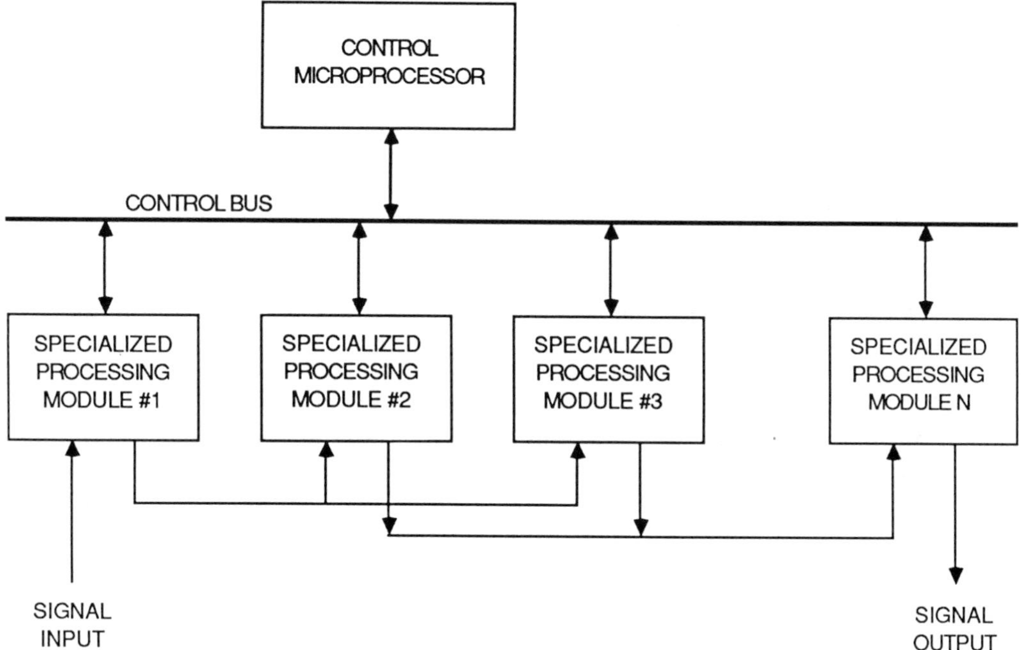


**Figure 8.14** Multilevel processor architecture.

ules. The modules are customized to implement frequently used functions (fast Fourier transformation, digital filtration, correlation, matrix inversion, etc.). In operation, the control microprocessor issues a command to one of the specialized processor modules, which then begins processing; the control processor can now generate the command for another processor, etc. Since the first processor may require from 0.1 to 10 ms to perform the high-level task that it was commanded to do, and since there are seldom more than a dozen such processors in a system, the control microprocessor issues fewer than 100,000 commands/second. Typical rates are in the range 1000 to 10,000 commands/second. Software is used in this system at two different levels: 1) high-level software in the control microprocessor to generate the stream of commands for the specialized processing modules, and 2) wide-word microcode used in the specialized processing modules to provide a limited degree of flexibility. The high-level microcode for the control microprocessor does not need to be as efficient as PSP software because it is not "in line" with the data processing; therefore, it is easier (and less expensive) to write and maintain. On the other hand, the microcode for the processing modules is written only at design time as the modules are being developed. As a result, it is developed by the module design team and can be optimized for the specific applications.

Each one of these tailored array processor architectures is optimal for specific types of applications. Generally, signal processing applications with well-established fixed algorithms are best suited for realization with the hardwired processor. The PSP approach is best where a high degree of flexibility is required, as in the early development stages where algorithms have not yet been finalized or where a processor must be reconfigured to accommodate greatly changed operations modes. The multilevel processor offers a compromise, with more flexibility than hardwired processors and better implementation efficiency than PSP-based systems. Part of the reason for the improvement and implementation efficiency is that the multilevel processor uses specialized interconnection between the processing modules as required by the specific problem. This specialization of the interconnection avoids much of the complexity of the PSP.

For these three architectures, it is important to understand the interaction with the technology. The hardwired approach, which was developed initially with SSI technology, uses technology most efficiently in the sense that all gates are used. As greater levels of integration were achieved, the need to take full advantage of every gate on every chip became less important and the desire for flexibility led to the creation of the programmable signal processor. To develop optimized hardwired processors with current (LSI or VLSI) technology means that new custom chips must be developed for every processor (and modified for every major modification to existing processors). At the other extreme, the PSP (because it achieves a high level of data path flexibility) does not use LSI or VLSI efficiently. This is because packaging constraints limit the interconnection flexibility. The middle ground is more attractive than either of these extremes at VLSI and higher levels of integration. For example, the specialized processing modules of the multilevel processor may be shared by many systems in much the same way as standard software subroutines are used for a variety of scientific applications with general-purpose computers. Thus, an investment in optimization can be amortized across a number of projects. Furthermore, because the specialized processing modules perform complete primitive operations, the input/output rates are relatively low and great interconnection flexibility is not required.

As the cost of software development has come to dominate most computational applications, array processors are recognized as useful for many (but not all) applications. All array processors are susceptible to failures in their centralized control units. Such failures may disable a complete system by disrupting the function of only a small portion of the system.

# 6 DEVELOPMENT OF A VLSI-BASED SIGNAL PROCESSOR

This section describes recent progress in the implementation of a high-speed signal processing system with commercial and semicustom VLSI circuits. The specific application is to develop fast Fourier transform (FFT) and inverse FFT processors. The processors employ the radix 4 pipeline FFT algorithm to achieve data rates of up to 40 million samples per second (MSPS) with modest 10-MHz clock rates. The interstage reordering is performed by delay commutators realized with semicustom VLSI, while the arithmetic is performed by commercial floating-point adders and multipliers. This section describes the development process and explains the pipeline FFT implementation.

## 6.1 Signal Processor Development Approach

The first step is determining the appropriate circuit technology. This is done by considering application environmental constraints (temperature, radiation, etc.) expected computational characteristics (throughput, arithmetic, etc.), and likely algorithm types (digital filtering, maximum entropy estimation, etc.). The characteristics of the selected circuit technology directly affect the algorithm design and determine appropriate architectures. For example, fast limited-complexity technologies such as GaAs are most effective in implementing simple serial architectures which perform recursive algorithms, while technologies that achieve higher complexity such as CMOS are most effective when implementing parallel or systolic architectures.

The next step is to perform an initial (high-level) design. This identifies areas where a better understanding is required, such as the arithmetic rounding characteristics. The initial design is simulated to resolve uncertainties about the operation of the algorithm.

Successive iterations of the technology selection, algorithm design, and processor architecture serve to refine the design at ever-increasing levels of detail. On completion of this process, a chip-level hardware design has been developed, software (if any) has been coded and debugged, and the algorithm execution has been extensively simulated (taking into account the arithmetic characteristics of the hardware design). When the system is constructed, the simulation serves as a reference for component checkout, debugging, and system integration.

## 6.2 FFT Processor Implementation

Although the Cooley–Tukey FFT algorithm developed in 1965 made it possible to apply digital techniques to many signal processing applications, many others (described in other chapters of this book) require computational performance that exceeds the present state of the art. Current data acquisition technology generates input data streams at rates of 25–50 MSPS, which can be processed only with special-purpose signal processors. Signal processing systems require many diverse functions: transformation, time and frequency domain processing, and general-purpose computation. Work is under way to produce a growing family of building-block modules to facilitate the development of such systems on a semicustom basis. The result is the ability to quickly develop high-performance signal processing systems for a wide variety of algorithms. The use of predesigned and precharacterized modules reduces cost, development time, and most important, risk.

The initial set of signal processing modules includes a data acquisition module, building-block elements that are replicated to realize pipeline FFT and inverse FFT modules, a frequency domain filter module, a power spectral density computational module, and an output interface module [20]. The modules all have separate data and control interfaces. The data interfaces satisfy a common interface protocol so that modules can be connected together to form architectures to match the data flow of each specific application. Separation of the data and control is a contemporary realization of the Harvard architecture described in the previous section, which uses the separate data and instruction memories to eliminate the "von Neumann bottleneck." In this context, the separation of data and control allows the (simple) data interfaces to operate at high speed while the more flexible (and complex) control interfaces operate at a lower rate.

The FFT processor implemented here uses the radix 4 pipeline algorithm developed at Lincoln Laboratory [21]. With this algorithm, four complex data pass in parallel through a pipeline network comprising computational elements and delay commutators as shown in Figure 8.15. Data rates of 40 MSPS are achieved using 10-MHz clock rates since the radix 4 architecture processes four data streams concurrently. An important feature of this algorithm and architecture is that only two types of elements are used: computational elements and delay commutators. Only minor changes are required to implement forward and inverse transforms of lengths that are powers of 4. The changes involve varying the number of stages connected in series, changing the counter sequence and step size on the computational elements, and changing the length of the delays on the delay commutator. In this realization 22-bit floating-point arithmetic is performed with single-chip adder-subtractors, and multipliers. The delay commutator reorders the data between computational stages as required for the FFT algorithm.

## 6.3 The Delay Commutator Circuit

Careful examination of the FFT module design revealed that much of the complexity was due to the delay commutator element. Initial complexity estimates were 80 commercial integrated circuits for the computational element and 180 circuits for the delay commutator. Given the high complexity of the commercial implementation of the delay commutator, alternative approaches were examined. A $B$-bit-wide

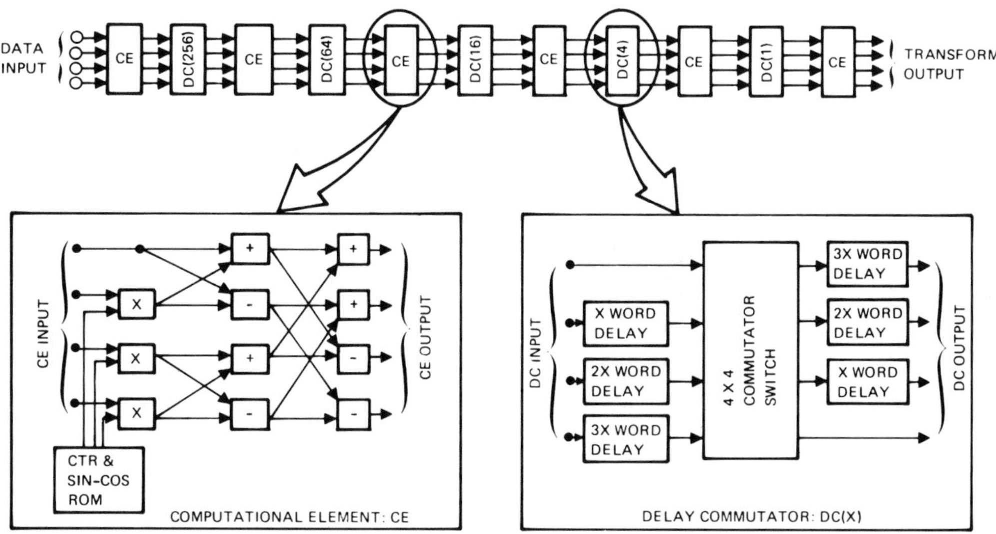


**Figure 8.15** Radix 4 pipeline FFT architecture.

data slice of a delay commutator that can be programmed for $X = 1, 4, 16, 64$, and 256 requires approximately $400(B+1)$ logic gates and $3072B$ shift register stages. Since a shift register stage is comparable in complexity to three random logic gates, this reduces to $400 + 9616B$ gates. Table 8.4 compares a commercial implementation with three VLSI versions (based on gate arrays, standard cells, and custom technology). For the VLSI implementations the maximum achievable bit slice widths are currently limited to 1, 4, and 10, respectively. With such width limitations 44, 11, or 5 delay commutator circuits would be required for the complex pairs of 22-bit data. Given the desire to minimize system complexity and avoid an expensive custom VLSI development activity, the standard cell approach was selected. The resulting delay commutator circuit is a 4-bit-wide slice that uses programmable-length shift registers and a 4 by 4 switch as shown in Figure 8.16. Data enter through shift registers with taps and multiplexers to set the delay at 1, 4, 16, 64, or 256 ($= X$) in the uppermost input register and multiples of $2X$ and $3X$ in the middle and lower registers, respectively. Four 4:1 multiplexers implement the commutator function under the control of the programmable rate counter. The final 2-bit counter/decoder that controls the multiplexer setting can be reset and

**Table 8.4.** Comparison of Delay Commutator Implementations

| Approach | Gates/chip | Width | Chips/stage | Relative development cost |
|---|---|---|---|---|
| Commercial | 0 | 0 | 179 | 1 |
| Gate array | 10,000 | 1 | 44 | 2 |
| Standard cell | 40,000 | 4 | 11 | 3 |
| Custom | 100,000 | 10 | 5 | 10 |

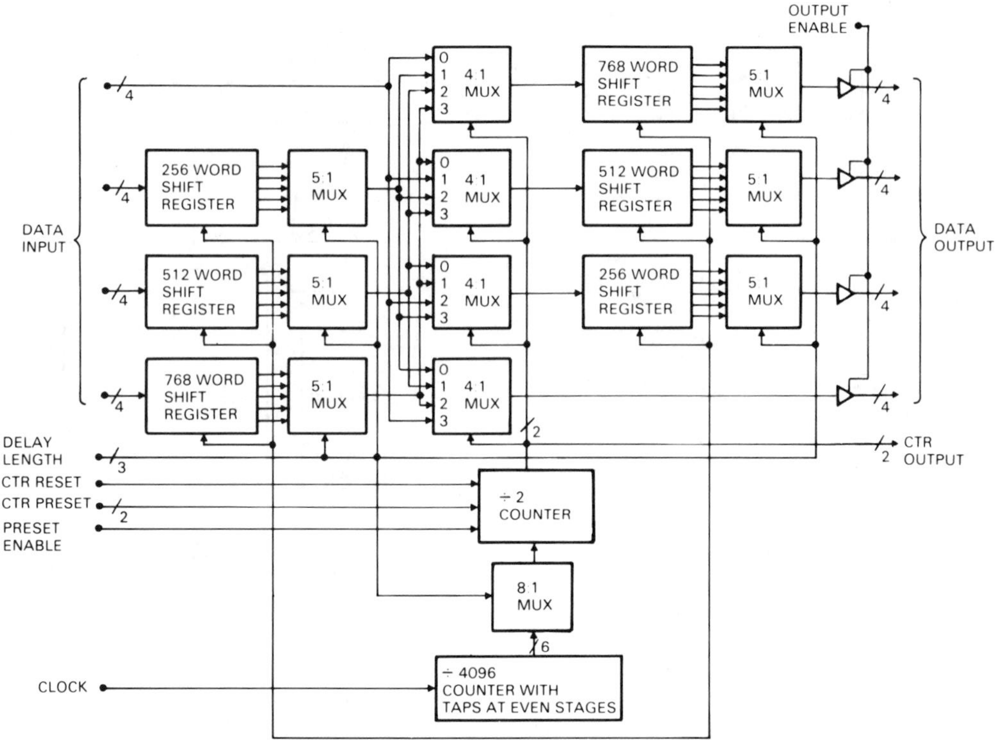


**Figure 8.16** Delay commutator block diagram.

held to disable the commutator switch function. Data from the 4:1 multiplexers are output through programmable-length shift registers that are similar to the input registers.

Operation of the delay commutator to reorder data is shown in Figure 8.17, where the data flow for a 64-point transform is shown [22]. The input data (with a spacing of 16) are applied to a radix 4 butterfly computational element producing output data with a spacing of 16. The reordering necessary to produce a data spacing of 4 is accomplished with delays of 0, 4, 8, and 12; commutation at a rate of one-fourth the data rate; and delays of 12, 8, 4, and 0. The data (with a spacing of 4) are applied to a second radix 4 butterfly computational element. The resulting data are reordered by use of delays of 0, 1, 2, and 3; commutation at a rate equal to the data rate; and delays of 3, 2, 1, and 0. The final data have a spacing of 1, as required for the final radix 4 butterfly computational element.

This circuit was designed and implemented with standard cell CMOS technology. The chip contains 12,288 shift register stages and about 2000 gates of random logic, for a total complexity of 108,000 transistors [23]. At a clock rate of 10 MHz, the power dissipation is under 1/2 watt. The 340 mil by 376 mil chip is packaged in a 48-pin dual-in-line ceramic package. The chip is shown in Figure 8.18. Each of the bit slices is constructed with input registers in a column, switching logic in a second "random logic" column, and output register in a column. The four nearly identical

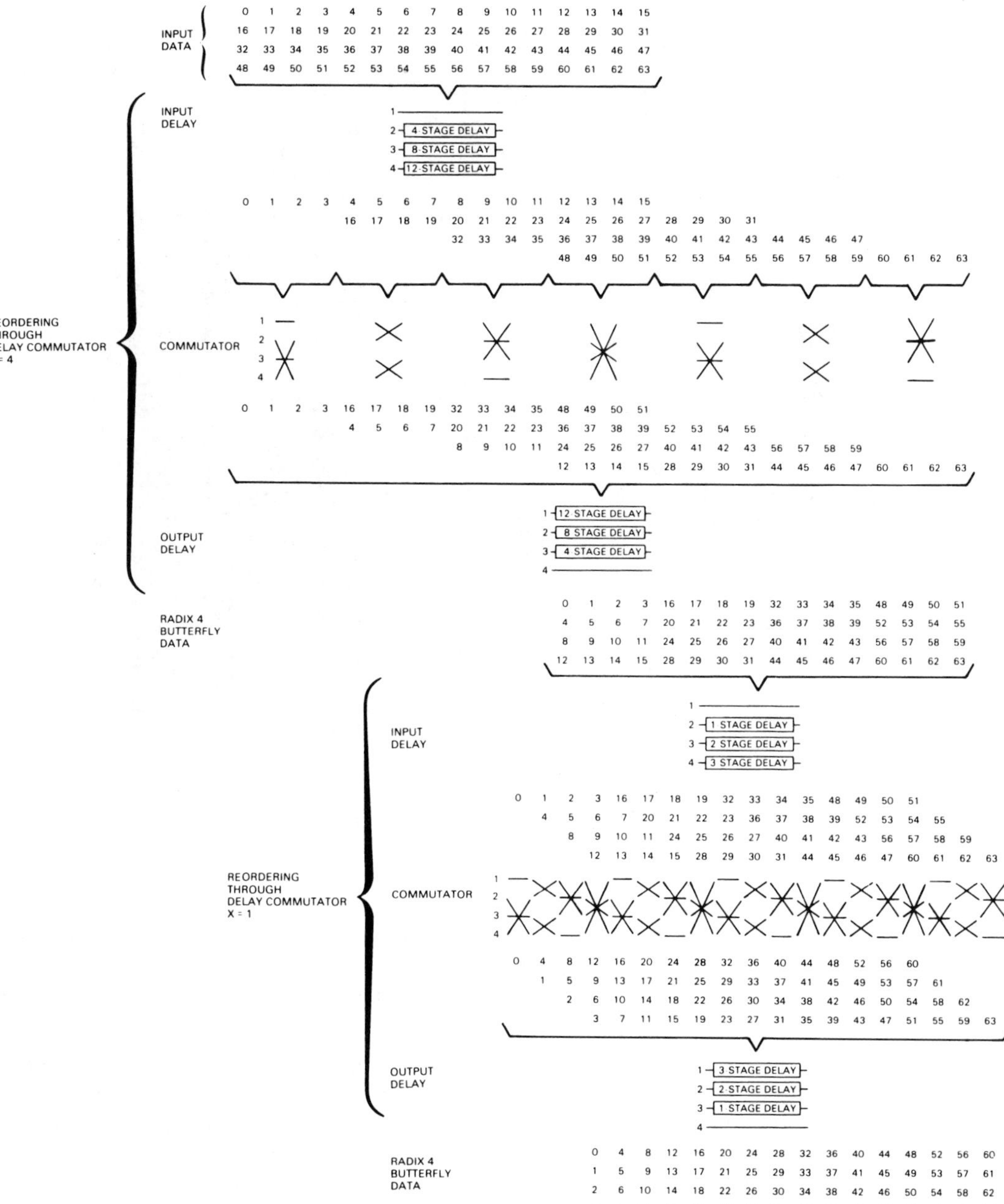


**Figure 8.17** Data patterns for a 64-point FFT. (After Ref. [22], p. 661.)

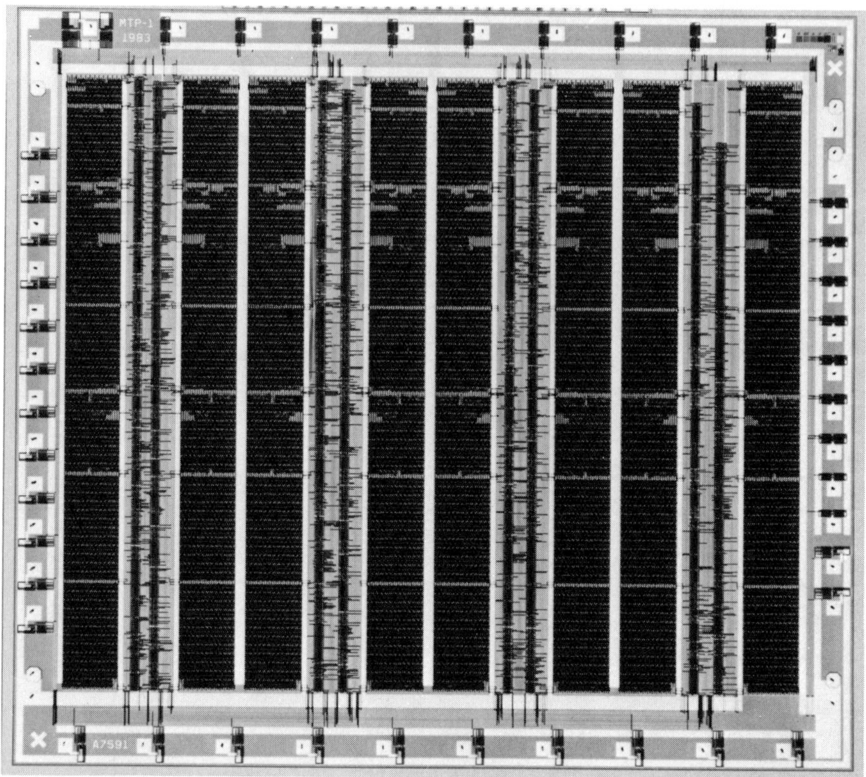


**Figure 8.18** Delay commutator circuit photograph.

slices are about four times as tall as they are wide, producing a roughly square chip when they are properly stacked. There is minor variation in the random logic of each bit slice to account for sharing of the counters, decoders, clock drivers, etc.

Development of the delay commutator chip reduces the complexity of the 40-MHz 4096-point FFT from 1375 commercial integrated circuits to 546 circuits (of which 66 are delay commutator chips). This is a 60% complexity reduction achieved through the development of a single semicustom chip. Such a reduction greatly improves system reliability since connections between circuits represent the dominant failure mechanism in modern systems [24]. With 60% fewer circuits (and a corresponding reduction in the number of interconnections) the reliability is greatly improved.

## 6.4 Arithmetic Realization

A critical design decision in signal processing concerns the arithmetic implementation. There are three somewhat contradictory requirements: 1) high speed to accommodate high signal bandwidths, 2) high precision to minimize computational error, and 3) wide dynamic range to avoid overflow. A wide variety of techniques are employed in signal processing because of the different relative importance of these requirements for specific applications. At the highest speeds, analog techniques are often employed. In audio and geophysical applications, where high accuracy is needed but speed is less critical, minicomputers can perform high-precision operations in firmware. In most digital signal processing applications, both accuracy and speed are important. Many users have compromised on 16-bit fixed-point arithmetic. Although fast, fixed-point arithmetic can lead to overflow errors or loss of precision unless complex data-dependent scaling is provided. Floating-point arithmetic has not been used because of size, cost, and speed limitations of available hardware. Recently, single-chip adders and multipliers using the 22-bit floating-point format (16-bit fraction, 6-bit exponent) have been developed. This section describes the use of these components to perform the arithmetic required for the radix 4 butterfly computational element shown in the lower left of Figure 8.15. The 22-bit format, with a 16-bit two's complement fraction and a 6-bit two's complement exponent, is a reasonable compromise among performance, speed, and size. Although single-chip 32-bit floating-point devices are commercially available, for a given technology the 22-bit format will always produce chips that are simpler and as a result cheaper and faster, with adequate dynamic range and precision for most applications. The 32-bit arithmetic is useful in scientific computation when inverting matrices, evaluating eigenvectors, etc., but these operations are usually performed at much lower rates than those required for signal processing, where input data are often limited in precision to 14 bits or less.

The chief advantage of floating point is increased dynamic range. As shown in Table 8.5, 22-bit floating-point arithmetic provides 96 dB of precision (i.e., equivalent to 16-bit fixed-point arithmetic) over a dynamic range of 476 dB [25]. Although this dynamic range is less than the 1686 dB provided by 32-bit floating-point arithmetic, it is more than adequate for most signal processors. The 16-bit fixed-point format has a dynamic range of 96 dB; the 22-bit floating-16-bit precision is available over a dynamic range of 380 dB.

## Adder

Under user control, the 22-bit adder performs floating-point addition, accumulation, and conversions between fixed- and floating-point formats. Rounding and scaling (i.e., divide by 2) are also selectable if desired. For proper operation and maximum accuracy, nonzero floating-point operands are normalized.

The adder performs the three component operations of floating-point addition: denormalization (exponent alignment), addition, and renormalization. The first

**Table 8.5** Comparison of Arithmetic Implementations

| Approach | Dynamic range, dB | Precision, dB |
|---|---|---|
| 12-bit fixed point | 72 | 72 |
| 16-bit fixed point | 96 | 96 |
| 22-bit fixed point | 132 | 132 |
| 22-bit floating point | 476 | 96 |
| 32-bit floating point | 1686 | 144 |

and last steps are hardware-intensive, involving shifting of the fraction and compensatory incrementing of the exponent. In the addition mode, the adder first selects the addend with the smaller exponent. It denormalizes this operand by shifting it rightward by the difference between the exponents of the two addends. The denormalized fraction is added to the other addend's unshifted fraction, and their sum passes to the renormalizing section, along with the larger addend's exponent. The sum is normalized by shifting its fraction leftward until the sign bit differs from the next bit, and the exponent is decremented by the number of bit positions of this shift.

Subtraction is identical to addition, except that the fraction of the subtrahend is complemented before the addition is performed. In standard two's complement fashion, the bits are inverted and a "hot one" is introduced at the adder's LSB carry-in position. Fixed-to-floating-point conversion and normalization of floating-point numbers is performed by left shifting the fraction as necessary to eliminate redundant leading zeros or ones while decrementing the exponent to compensate.

### Multiplier

The floating-point multiplier is basically a 16-bit two's complement fixed-point multiplier, a 6-bit adder, and a simple normalizer. It does not require an operand conditioner, such as the adder's denormalizer. Furthermore, the output conditioning requirements are minimal: if the input operands are normalized, then the product is at most one shift left or right from normalization. Hence, the 16-bit half-barrel shifter of the adder is replaced by a small 16-bit, three-position multiplexer. The only "communication" between fraction and exponent occurs in the final product normalization step, where the exponent must be incremented or decremented to compensate for any shift in the fraction. With the normalizer defeated, the chip performs 16-bit two's complement, fixed-point multiplications.

This section shows the high payoff of synergistic use of commercial and semicustom integrated circuits. Specifically, a 40-MHz pipeline FFT processor implementing 22-bit floating-point arithmetic has been developed. The processor complexity was decreased by 60% through the development of a standard cell VLSI delay commutator circuit. The arithmetic is performed with single-chip adders and multipliers that use a 22-bit floating-point format. The processor is simpler and correspondingly lower in power, size, and cost than the designs using 32-bit floating-point arithmetic. Similar improvements can be achieved in a wide variety of signal processing systems by carefully tailoring the algorithms, processor architecture, arithmetic precision, and technology selection.

## 7 CONCLUSIONS

This chapter has addressed the major issues of VLSI technology in the context of signal processing systems development. The critical issues include basic technology constraints, VLSI design methodologies, and systems architecture. Examples to illustrate the concepts include current VLSI chip functional designs, a case study of a multiterm multiplier-accumulator, and a high-performance FFT processor development.

Although the growth in circuit technology has solved many system problems in the past, technology constraints in chip design, packaging, and interfacing remains as serious issues. In the future, these issues will doubtless be solved through technology innovation, but other problems will arise. Through the use of advanced technology, system costs will continue their historic decline.

Device architecture for VLSI requires striking a delicate balance between extremely efficient custom chips optimized for use in a single system and generic integrated functions designed for broad applicability to many systems. The design issues require development of cellular arrays to permit design of very complex chips with reasonable amounts of time and personnel. It is necessary to minimize the number of chip I/O signals (especially the number of outputs). Such I/O reduction minimizes the chip area required for bonding pads, reduces package size and cost, and decreases output driver power requirements. It requires great care, however, to ensure that the application flexibility of the chip is not compromised.

As the level of integration has grown, the distinction between chip architecture and system architecture has begun to dissolve. Crucial issues remain in the design of distributed systems, especially in the areas of control, fault detection/localization, and implementation.

Examination of the development of an advanced FFT signal processing system illustrates the compromises implicit in systems optimization. Standard cell technology was used to design a delay commutator for FFT interstage data reordering. This approach is more expensive than gate arrays, but produces a greatly reduced processor chip count. Full custom design technology would have further reduced the chip count, but would have cost much more. A similar compromise led to the selection of 22-bit floating-point arithmetic of 16-bit fixed-point arithmetic (with dynamic range limitations) or 32-bit floating-point arithmetic (with greater precision than justified by the data).

## REFERENCES


1. E. E. Swartzlander, Jr., *VLSI Signal Processing Systems*, Kluwer, Boston (1986).
2. P. R. Capello, et al., eds., *VLSI Signal Processing*, IEEE Press, New York (1984).
3. S.-Y. Kung, et al., eds., *VLSI Signal Processing, II*, IEEE Press, New York (1986).
4. L. W. Sumney, VLSI with a vengeance, *IEEE Spectrum, 17*: 24–27 (April 1980).
5. B.S. Landman and R. L. Russo, On a pin versus block relationship for partitions of logic graphs, *IEEE Trans. Comput., C-20*: 1469–1479 (1971).
6. A. L. Robinson, Are VLSI microcircuits too hard to design? *Science, 209*: 258–262 (1980).
7. W. W. Lattin, J. A. Bayliss, D. A. Budde, F. R. Rattner, and W. S. Richardson, A methodology for chip design, *Lambda, 1*: 34–44 (Second Quarter, 1981).
8. M. V. Wilkes, The growth of interest in microprogramming: a literature survey, *Comput. Surv., 1*: 139–145 (1969).

9. A. K. Agrawala and T. G. Rauscher, *Foundations of Microprogramming*, Academic Press, New York (1976).
10. E. E. Swartzlander, Jr., B. K. Gilbert, and I. S. Reed, Inner product computers, *IEEE Trans. Comput.*, *C-27*: 21–31 (1978).
11. A. D. Booth, A signed binary multiplication technique, *Q. J. Mech. Appl. Math.*, *4*: 236–240 (1951). This paper is reprinted in [12].
12. E. E. Swartzlander, Jr., ed., *Computer Arithmetic*, Dowden, Hutchinson & Ross, Stroudsburg, Pennsylvania (1980).
13. E. E. Swartzlander, Jr., ed., Merged Arithmetic, *IEEE Trans. Comput.*, *C-29*: 946–950 (1980).
14. C. R. Baugh and B. A. Wooley, A two's complement parallel array multiplication algorithm, *IEEE Trans. Comput.*, *C-22*: 1045–1047 (1971). This paper is reprinted in [12].
15. L. Dadda, Some schemes for parallel multipliers, *Alta Freq.*, *34*: 349–356 (1965). This paper is reprinted in [12].
16. E. E. Swartzlander, Jr., and B. K. Gilbert, Arithmetic for ultra-high-speed tomography, *IEEE Trans. Comput.*, *C-29*: 341–353 (1980).
17. K. Hwang and F. A. Briggs, *Computer Architecture and Parallel Processing*, McGraw-Hill, New York (1984).
18. P. M. Kogge, *The Architecture of Pipelined Computers*, McGraw-Hill, New York (1981).
19. D. J. Kuck, *The Structure of Computers and Computations*, vol. 1, Wiley, New York (1978).
20. E. E. Swartzlander, Jr., L. S. Lome, and G. Hallnor, Digital signal processing with VLSI technology, *Proc. IEEE Conf. Acoust. Speech, Signal Process.*, 951–954 (1983).
21. B. Gold and T. Bially, Parallelism in fast Fourier transform hardware, *IEEE Trans. Audio Electron.*, *AU-21*: 5–16 (1973).
22. L. R. Rabiner and B. Gold, *Theory and Applications of Digital Signal Processing*, Prentice-Hall, Englewood Cliffs, New Jersey, p. 611 (1975).
23. E. E. Swartzlander, Jr., W. K. W. Young, and S. J. Joseph, A radix 4 delay commutator for fast Fourier transform processor implementation, *IEEE J. Solid-State Circuits*, *SC-19*: 702–709 (1984).
24. G. W. Preston, The very large scale integrated circuit, *Am. Sci.*, *71*: 466–472 (1983).
25. J. A. Eldon and C. Robertson, A floating point format for signal processing, *Proc. IEEE Int. Conf. Acoust. Speech, Signal Process*, 717–720 (1982).

# 9

# Syntactic Signal Processing

HSI-HO LIU* Department of Electrical and Computer Engineering, University of Miami, Coral Gables, Florida

## 1 INTRODUCTION

The many different techniques used to solve pattern recognition problems may be grouped into two general approaches: the decision-theoretic (or discriminant) approach and the syntactic (or structural) approach. In the decision-theoretic approach, a set of characteristic measurements, called features, are extracted from the patterns; the recognition of each pattern (assignment to a pattern class) is usually made by partitioning the feature space. In some signal analysis problems, the structural information that describes each pattern is important, and the recognition process includes not only the capability of assigning the pattern to a particular class (to classify it), but also the capacity to describe structures of the pattern that make it eligible for the assignment. In order to represent the hierarchical (treelike) structural information for each pattern—that is, the pattern described in terms of simpler subpatterns, each simpler subpattern again described in terms of even simpler subpatterns, and so on—the syntactic or structural approach has been proposed [1, 2]. This approach draws an analogy between the (hierarchical, or treelike) structure of a pattern and the syntax or grammar of languages. A simple block diagram of a typical syntactic pattern recognition system is shown in Figure 9.1. In this chapter we will outline the theory of formal grammars, string-to-string distance computation, and error-correcting parsing, and describe some applications of syntactic pattern recognition to signal analysis problems.

Present affiliation:

*Department of Electrical and Biomedical Engineering, Vanderbilt University, Nashville, Tennessee

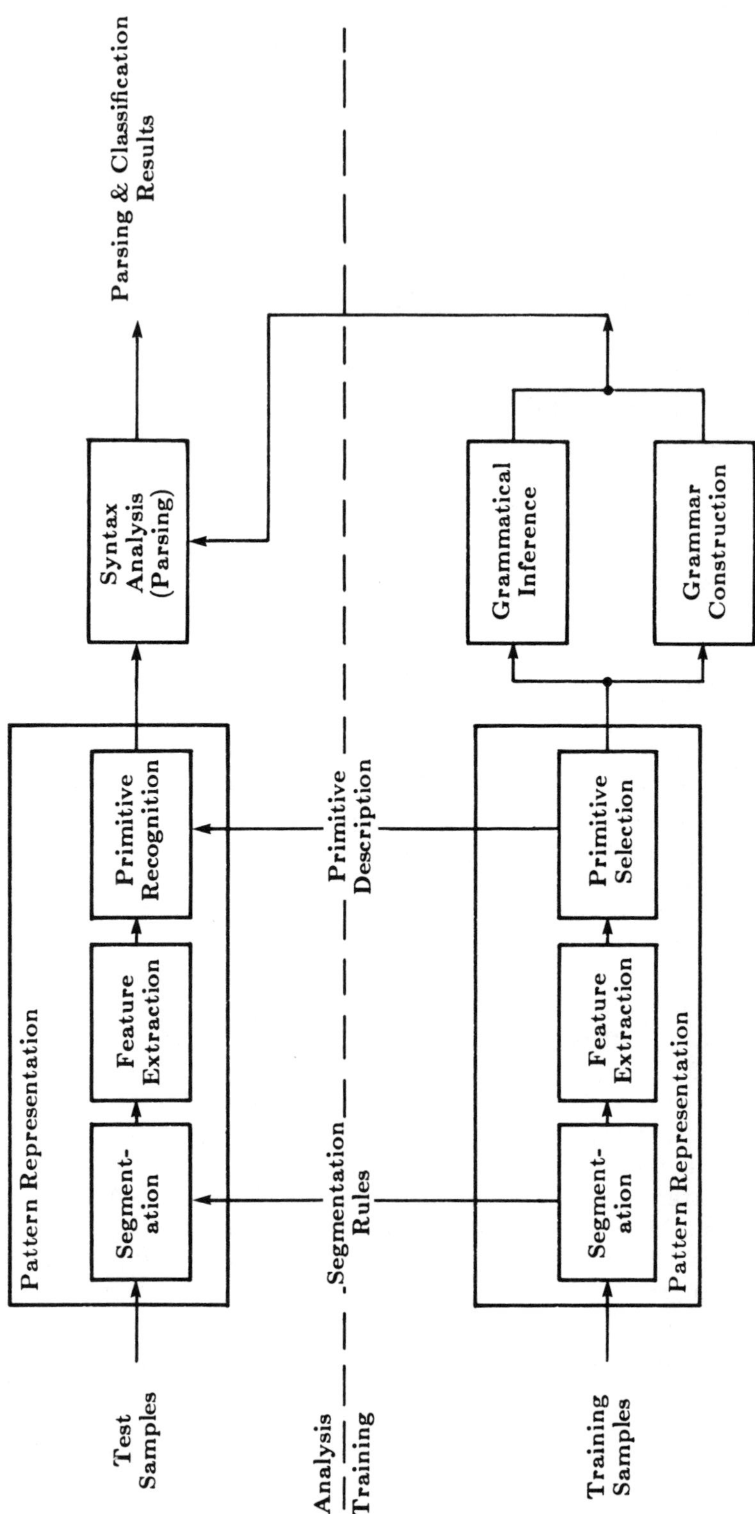


Figure 9.1 Block diagram of a typical syntactic pattern recognition system.

## 2 PATTERN REPRESENTATION AND GRAMMARS

In the syntactic approach, patterns are specified as being built up out of subpatterns by various methods of composition, just as phrases and sentences are built up by concatenating words, and words are built up by concatenating characters. Evidently, for this approach to be advantageous, the simplest subpatterns selected, called pattern primitives, should be much easier to recognize than the patterns themselves. The language that provides the structural description of patterns in terms of a set of pattern primitives and their composition operations is sometimes called the pattern description language. The rules governing the composition of primitives into patterns are usually specified by the so-called grammar of the pattern description language. Since a signal waveform has a one-dimensional structure, it is more natural to represent it by a string of primitives, although other types of representation (e.g., tree) have been proposed.

### 2.1 Primitive Selection

The pattern primitives should serve as basic pattern elements in describing the structural relations and they should be easily extractable, usually by nonsyntactic methods. To convert a signal waveform into a string of primitives, typically we need to perform three steps: pattern segmentation, feature selection, and primitive recognition. For some cases, such as electrocardiogram (ECG) and carotid pulse wave analysis, every single peak and valley are significant; therefore, these waveforms are segmented according to their shape. For others, like electroencephalogram (EEG) and seismic wave analysis, a single peak or valley does not contain significant information, especially when the signal-to-noise ratio is low; consequently, they are usually segmented by length, either a fixed or a variable length. A variable-length segmentation is more efficient and accurate in representation, but it is usually very difficult and time-consuming to find an appropriate segmentation. A fixed-length segmentation, on the other hand, is much easier to implement. If the length is properly selected, it will be adequate to represent the original waveform without losing much information. There is a compromise between the representation accuracy and analysis efficiency. With shorter segmentation, the representation is more accurate, but the analysis becomes more inefficient because the string length is longer and the computation time is proportional to string length. Another problem that should be considered is the noise. If the segmentation is too short, it will be very sensitive to noise.

Any linear or nonlinear mapping of the original measurements, in either the time domain or the frequency domain, can be used as features as long as they can characterize the signal segment. Features can be selected subjectively, as by zero-crossing count and log energy, or we can use general features such as $K - L$ expansion. The selection of primitives varies largely in digital signal analysis. Line segments from a linear approximation of signals have been used in ECG analysis, and parabolas and line segments have been used in carotid pulse wave analysis to describe the shape of the signal waveform. When the shape of the signal waveform is not important, other types of primitives must be selected. For example, in spoken word recognition, silence interval, stable zone, and lines are used as primitives. In EEG analysis, a group of seven primitives has been specified and a linear classifier is used to recognize the testing segments. When there is no predefined number of primitives, a clustering procedure can classify any number of signal segments into a certain number of clusters such that the classification minimizes some criterion function. After the signal waveform has been converted to a string of primitives, the theory of formal languages [3] can be applied to define pattern grammars.

## 2.2 String Grammars

A phrase structure grammar is a 4-tuple $G = (V_N, V_T, P, S)$ where

1. $V_N$ is a finite set of nonterminal symbols (or variables).
2. $V_T$ is a finite set of terminal symbols, $V = V_N \cup V_T$, $V_N \cap V_T = \varnothing$.
3. $P$ is a finite set of rewrite rules or productions denoted by $\alpha \rightarrow \beta$, where $\alpha$ and $\beta$ are strings over $V$ and $\alpha$ involves at least one symbol of $V_N$.
4. $S \in V_N$ is the start symbol of a sentence.

The following notation is frequently used:

1. $V^*$ is the set of all finite-length strings of symbols in a finite set of symbols $V$, including $\lambda$, the string of length 0. $V^+ = V^* - \{\lambda\}$.
2. If $x$ is a string, $x^n$ is $x$ written $n$ times.
3. $|x|$ is the length of the string $x$, or the number of symbols in the string $x$.
4. $\eta \underset{G}{\Longrightarrow} \gamma$, or a string $\eta$ directly generates or derives another string $\gamma$ if $\eta = \omega_1 \alpha \omega_2$, $\gamma = \omega_1 \beta \omega_2$, and $\alpha \rightarrow \beta$ is a member of $P$. Usually, if it is clear which grammar $G$ is involved, the $G$ under $\Rightarrow$ can be omitted.
5. $\eta \overset{k}{\Longrightarrow} \gamma$, or a string $\eta$ generates or derives another string $\gamma$ if there exists a sequence of strings $\alpha_1$, $\alpha_2$, ..., $\alpha_k$ such that $\eta = \alpha_1$, $\gamma = \alpha_k$, $\alpha_i \Rightarrow \alpha_{i+1}$, $i = 1, 2, \ldots, n-1$. The sequence of strings $\alpha_1$, $\alpha_2$, ..., $\alpha_k$ is called a derivation of $\gamma$ from $\eta$. $\eta \overset{*}{\Longrightarrow} \gamma$ if and only if $\eta \overset{i}{\Longrightarrow} \gamma$ for some $i \geq 0$.

The language generated by a grammar $G$, denoted $L(G)$, is the set of sentences generated by $G$. A formal grammar $G$ is said to be ambiguous if there is a string $x \in L(G)$ that has more than one derivation.

**Example.** An example of a grammar is $G = (V_N, V_T, P, S)$, where $V_N = \{S, A\}$, $V_T = \{a, b\}$ and $P$:

$$S \longrightarrow A$$

$$A \longrightarrow aAb$$

$$A \longrightarrow \lambda$$

and the language generated by grammar $G$ is

$$L(G) = \{a^n b^n \mid n \geq 0\}$$

Chomsky divided the phrase structure grammars into four types according to the forms of the productions. In type 0 (unrestricted) grammars, there is no restriction on the productions, which may have any strings on either the right or the left of the substitution arrow. Production of type 1 (context-sensitive) grammars are restricted to the form

$$\alpha A \beta \longrightarrow \alpha \gamma \beta$$

where $A \in V_N$, $\alpha, \beta, \gamma \in V^*$, and $\gamma \neq \lambda$. Productions of type 2 (context-free) grammars are of the form

$$A \longrightarrow B$$

where $A \in V_N$ and $\beta \in V^+$. Productions of type 3 (finite-state or regular) grammars are of the form

$$A \longrightarrow aB$$

or

$$A \longrightarrow b$$

where $A, B \in V_N$ and $a, b \in V_T$.

### 2.3 Stochastic Grammars

In practical applications, the languages describing patterns of different classes may overlap one another. In other words, a sentence describing a pattern may be generated by two or more pattern grammars. Stochastic languages have been suggested for those situations [1]. For every string $x$ in a language $L$ is a probability $p(x)$ can be assigned such that $0 < p(x) \leq 1$ and $\sum_{x \in L} p(x) = 1$.

A stochastic phrase structure grammar (or simply stochastic grammar) is a 4-tuple $G_S = (V_N, V_T, P_S, S)$, where $V_N$ and $V_T$ are finite sets of nonterminals and terminals; $S \in V_N$ is the start symbol; $P_S$ is a finite set of stochastic productions, each of which is of the form

$$\alpha_i \xrightarrow{p_{ij}} \beta_{ij}, \qquad j = 1, \ldots, n_i, \; i = 1, \ldots, k$$

where

$$\alpha_i \in (V_N \cup V_T)^* V_N (V_N \cup V_T)^*, \qquad \beta_{ij} \in (V_N \cup V_T)^*$$

and $p_{ij}$ is the probability associated with the application of this stochastic production,

$$0 < p_{ij} \leq 1, \qquad \sum_{j=1}^{n_i} p_{ij} = 1$$

Suppose that $\alpha \xrightarrow{p_{ij}} \beta_{ij}$ is in $P_S$. Then the string $\xi = \gamma_1 \alpha_i \gamma_2$ may be replaced by $\eta = \gamma_1 \beta_{ij} \gamma_2$ with probability $p_{ij}$. We denote this derivation by

$$\xi \overset{P_{ij}}{\Longrightarrow} \eta$$

and we say that $\xi$ directly generates $\eta$ with probability $p_{ij}$. If there exists a sequence of strings $\omega_1, \ldots, \omega_{n+1}$ such that

$$\xi = \omega_1, \quad \eta = \omega_{n+1}, \quad \omega_i \overset{p_i}{\Longrightarrow} \omega_{i+1}, \qquad i = 1, \ldots, n$$

then we say that $\xi$ generates $\eta$ with probability $p = \prod_{i=1}^{n} p_i$ and denote this derivation by

$$\xi \overset{p}{\underset{*}{\Longrightarrow}} \eta$$

**Example.** Consider the stochastic finite-state grammar $G_S = (V_N, V_T, P_s, S)$ where $V_N = \{S, A, B\}$, $V_T = \{0, 1\}$, and $P_S$:

$$S \xrightarrow{1} A, \qquad B \xrightarrow{0.3} 0$$
$$A \xrightarrow{0.8} 0B, \qquad B \xrightarrow{0.7} 1S$$
$$A \xrightarrow{0.2} 1$$

A typical derivation, for example, would be

$$S \Longrightarrow 1A \Longrightarrow 10B \Longrightarrow 100, \qquad p(100) = 1 \times 0.8 \times 0.3 = 0.24$$

The stochastic language generated by $G_S$, $L(G_S)$ is as follows:

| String generated, $x$ | $p(x)$ |
|---|---|
| 11 | 0.2 |
| 100 | 0.24 |
| $(101)^n$ 11 | $1.2 \times (0.56)^n$ |
| $(101)^n$ 100 | $0.24 \times (0.56)^n$ |

It is noted that

$$\sum_{x \in L(G_s)} p(x) = 0.2 + 0.24 + \sum_{n=1}^{\infty} (0.2 + 0.24)(0.56)^n = 1$$

## 3 PARSING AND RECOGNITION

After each primitive within the pattern is identified, the recognition process is accomplished by performing a syntax analysis, or parsing [3], of the sentence (string of primitives) describing the given pattern to determine whether or not it is syntactically (or grammatically) correct with respect to the specified grammar. In the meantime, the syntax analysis also produces a structural description of the sentence representing the given pattern (usually in the form of a tree structure).

### 3.1 Parsing for Context-Free Grammars

The following parsing algorithms are applicable to the entire class of context-free languages. The full backtracking algorithms simulate nondeterministic parsers, which require linear space but may take exponential time as a function of the length of the string to be parsed. The tabular methods take space $n^2$ and time $n^3$. Earley's algorithm requires time $n^2$ whenever the context-free grammar is unambiguous.

#### Backtracking Algorithms

**Top-Down Parsing.** The top-down parsing algorithm can be thought of as building the parse tree by trial and error from the root (top) and proceeding downward to the leaves. The procedure begins with a tree containing one node labeled $S$. That node is the initial active node. We then perform the following steps recursively:

1. If the active node is labeled by a nonterminal, say $A$, then choose the first alternate, say $X_1 \cdots X_k$, for $A$ and create $k$ direct descendants for $A$ labeled $X_1, X_2, \ldots, X_k$. Make $X_1$ the active node. If $k = 0$, then make the node immediately to the right of $A$ active.
2. If the active node is labeled by a terminal, say $a$, then compare the current input symbol with $a$. If they match, then make the node immediately to the right of $a$ active and move the input pointer one symbol to the right. If $a$ does not match the current input symbol, go back to the node where the previous production was applied, adjust the input pointer if necessary, and try the next alternate. If no alternate is possible, go back to the next previous node, and so forth.

If the grammar is not left-recursive, we shall either successfully find a parse or exhaust all possibilities and report that the input string is not accepted by the grammar.

**Bottom-Up Parsing.** Bottom-up parsing, which in a sense is opposite to top-down parsing, starts with the leaves and attempts to build the tree upward toward the root. This method can be viewed as considering all possible sequences of moves of a nondeterministic right parser for a grammar. A move consists of scanning the string on top of the pushdown list to see if there is a right side of a production that matches the symbols on top of the list. If more than one production is possible, we order the possible reductions in some arbitrary manner and apply the first.

If no reduction is possible, we shift the next input symbol onto the pushdown list and proceed as before. We shall always attempt to make a reduction before shifting. If we come to the end of the string and no reduction is possible, we backtrack to the last move at which we made a reduction. If another reduction was possible at that point, we try that. The parsing is successful if we can reach the root. In order to avoid certain pitfalls we shall rule out grammars with cycles, i.e., derivations of the form $A \stackrel{+}{\Longrightarrow} A$ for some nonterminal $A$ and $\lambda$-productions.

## Tabular Parsing Methods

**The Cocke-Younger-Kasami Parsing Algorithm.** The CYK algorithm requires that the context-free grammar (CFG) be in Chomsky normal form; i.e., all the productions in $P$ are of the form $A \rightarrow BC$ with $A$, $B$, and $C$ in $V_N$, or $A \rightarrow a$ with $a \in V_T$, with no $\lambda$-production.

*Cocke-Younger-Kasami Parsing Algorithm*

*Input.* A Chomsky normal form CFG $G = (V_N, V_T, P, S)$ with no $\lambda$-production and an input string $w = a_1 a_2 \cdots a_n$ in $V_T^+$.

*Output.* The parse table $T$ for $w$ such that $t_{ij}$ contains $A$ if and only if $A \stackrel{+}{\Longrightarrow} a_i a_{i+1} \cdots a_{i+j-1}$.

*Method.*

1. Set $t_{i1} = \{A \mid A \rightarrow a_i \text{ is in } P\}$ for each $i$. After this step, if $t_{ij}$ contains $A$, then clearly $A \stackrel{+}{\Longrightarrow} a_i$.
2. Assume that $t_{ij}$ has been computed for all $i$, $1 \leq i \leq n$, and all $j'$, $1 \leq j' \leq j$. Set $t_{ij} = \{A \mid \text{for some } k,\ 1 \leq k < j,\ A \rightarrow BC, \text{ in } P, B \text{ is in } t_{ik} \text{ and } C \text{ is in } t_{i+k,j-k}\}$. Since $1 \leq k < j$, both $k$k and $j - k$ are less than $j$. Thus, both $t_{ik}$ and $t_{i+k,j-k}$ are computed before $t_{ij}$ is computed. After this step, if $t_{ij}$ contains $A$, then

$$A \Longrightarrow BC \stackrel{+}{\Longrightarrow} a_i \cdots a_{i+k-1} C \Longrightarrow a_i \cdots a_{i+k-1} a_{i+k} \cdots a_{i+j-1}$$

3. Repeat step 2 until $t_{ij}$ is known for all $a \leq i \leq n$ and $1 \leq j \leq n - i + 1$.

The string $w$ is in $L(G)$ if $S$ is in $t_{1n}$.

**Example.** Consider the context-free grammar $G = (V_N, V_T, P, S)$, where $V_N = \{S, A\}$, $V_T = \{a, b\}$, and $P$:

$$S \longrightarrow AA, \qquad S \longrightarrow AS, \qquad S \longrightarrow b$$

$$A \longrightarrow SA, \qquad A \longrightarrow AS, \qquad A \longrightarrow a$$

Let $w$ = *abaab*. Applying the CYK parsing algorithm, we obtain the following parse table $T$. Since $S$ is in $t_{15}$, *abaab* is in $L(G)$.

| $j \uparrow$ | | | | | |
|---|---|---|---|---|---|
| 5 | $A,S$ | | | | |
| 4 | $A,S$ | $A,S$ | | | |
| 3 | $A,S$ | $S$ | $A,S$ | | |
| 2 | $A,S$ | $A$ | $S$ | $A,S$ | |
| 1 | $A$ | $S$ | $A$ | $A$ | $S$ |
| | 1 | 2 | 3 | 4 | 5 |

$i \rightarrow$

**Earley's Parsing Algorithm.** Earley's algorithm [4] is a top-down parser that considers all possible parses simultaneously. This characteristic cuts down on duplication of effort and also avoids the left-recursion problem.

*Earley's Parsing Algorithm*

*Input.* A context-free grammar $G = (V_N, V_T, P, S)$ and an input string $w = a_1 a_2 \cdots a_n$.

*Output.* The parse list $I_0, I_1, \ldots, I_n$ for $w$.

*Method.* First, construct $I_0$ as follows:

1. If $S \rightarrow \alpha$ is in $P$, add item $[S \rightarrow \cdot\alpha, 0]$ to $I_0$.

Perform steps 2 and 3 until no new items can be added to $I_0$.

2. If $[B \rightarrow \gamma\cdot, 0]$ is on $I_0$, add $[A \rightarrow \alpha B \cdot \beta, 0]$ for all $[A \rightarrow \alpha \cdot B\beta, 0]$ on $I_0$.
3. Suppose that $[A \rightarrow \alpha \cdot B\beta, 0]$ is an item in $I_0$. Add to $I_0$, for all productions in $P$ of the form $B \rightarrow \gamma$, the item $[B \rightarrow \cdot\gamma, 0]$.

Construct $I_j$ from $I_0, I_1, \ldots, I_{j-1}$.

4. For each $[B \rightarrow \alpha \cdot a\beta, i]$ in $I_{j-1}$ such that $a = a_j$, add $[B \rightarrow \alpha a \cdot \beta, i]$ to $I_j$.

Perform steps 5 and 6 until no new items can be added.

5. Let $[A \rightarrow \alpha\cdot, i]$ be an item in $I_j$. Examine $I_i$ for items of the form $[B \rightarrow \alpha \cdot A\beta, k]$. For each one found, add $[B \rightarrow \alpha A \cdot \beta, k]$ to $I_j$.
6. Let $[A \rightarrow \alpha \cdot B\beta, i]$ be an item in $I_j$. For all $B \rightarrow \gamma$ in $P$, add $[B \rightarrow \cdot\gamma, j]$ to $I_j$.

The string $w$ is in $L(G)$ if and only if there is some item of the form $[S \rightarrow \alpha\cdot, 0]$ in $I_n$. After a string is accepted, its parse (or derivation tree) can easily be extracted.

**Example.** Consider the context-free grammar $G = (V_N, V_T, P, S)$, where $V_N = \{S, T, F\}$, $V_T = \{a, +, *, (, )\}$, and $P$:

$$S \longrightarrow S + T, \qquad S \longrightarrow T$$

$$T \longrightarrow T * F, \qquad T \longrightarrow F$$

$$F \longrightarrow (S), \qquad F \longrightarrow a$$

Let $w = a * a$. Applying Earley's parsing algorithm, we obtain the following parse lists for $w$:

| $I_0$ | $I_1$ | $I_2$ | $I_3$ |
|---|---|---|---|
| $[S \to \cdot S + T, 0]$ | $[F \to a\cdot, 0]$ | $[T \to T * \cdot F, 0]$ | $[F \to a\cdot, 2]$ |
| $[S \to \cdot T, 0]$ | $[T \to F\cdot, 0]$ | $[F \to \cdot(S), 2]$ | $[T \to T * F\cdot, 0]$ |
| $[T \to \cdot T * F, 0]$ | $[S \to T\cdot, 0]$ | $[F \to \cdot a, 2]$ | $[S \to T\cdot, 0]$ |
| $[T \to \cdot F, 0]$ | $[T \to T \cdot *F, 0]$ | | $[T \to T \cdot *F, 0]$ |
| $[F \to \cdot(S), 0]$ | $[S \to S \cdot +F, 0]$ | | $[S \to S \cdot +T, 0]$ |
| $[F \to \cdot a, 0]$ | | | |

Since $[S \to T\cdot, 0]$ is in $I_3$, $a * a$ is in $L(G)$.

## 3.2 Error-Correcting Parsing

### Similarity Measure for String Patterns

String similarity measures can be defined in terms of two different concepts: the distance concept for nonstochastic models and the likelihood concept for stochastic models. Consider two strings $x = a_1 a_2 \cdots a_n$ and $y = b_1 b_2 \cdots b_n$; the string similarity measure between $x$ and $y$ is defined as the distance or probability that string $y$ is transformed from string $x$. One string distance definition is called the Levenshtein distance. The Levenshtein distance between strings $x$ and $y$, $x, y \in V_T^*$, denoted as $d^L(x, y)$, is defined as the smallest number of transformations required to derive string $y$ from string $x$, and the transformations include insertion, deletion, and substitution of terminal symbols [5].

For any two strings $x, y \in V_T^*$, we can define a sequence of transformations $J = \{T_1, T_2, \ldots, T_n\}$, $n \geq 0$, $T_i \in \{T_S, T_D, T_I\}$ for $1 \leq i \leq n$, such that $y \in J(x)$. The transformations $T_S$, $T_D$, and $T_I$ are defined as follows:

1. Substitution transformation, $T_S$

$$\omega_1 a \omega_2 \mid \overset{T_S}{\text{—}} \omega_1 b \omega_2 \qquad \text{for all } a, b \in V_T,\ a \neq b$$

2. Deletion transformation, $T_D$

$$\omega_1 a \omega_2 \mid \overset{T_D}{\text{—}} \omega_1 \omega_2 \qquad \text{for all } a \in V_T$$

3. Insertion transformation, $T_I$

$$\omega_1 \omega_2 \mid \overset{T_I}{\text{—}} \omega_1 a \omega_2 \qquad \text{for all } a \in V_T$$

where $\omega_1, \omega_2 \in V_T^*$.

The Levenshtein distance $d^L(x,y)$ is defined as

$$d^L(x,y) = \min_j \{k_j + m_j + n_j\}$$

where $k_j$, $m_j$, and $n_j$ are respectively the numbers of substitution, deletion, and insertion transformations in $J$. Since all the insertion, deletion, and substitution transformations are counted equally, the Levenshtein distance is symmetric.

Computation of the Levenshtein distance can be implemented by a dynamic programming technique on a grid matrix as shown in Figure 9.2. The partial distance $\delta[i,j]$, which denotes the minimum distance from points (0,0) to $(i,j)$, can be computed from the partial distances $\delta[i,j-1]$, $\delta[i-1,j-1]$, and $\delta[i-1,j]$ as shown in Figure 9.3. The total distance is simply $\delta[n,m]$, where $n$ is the length of the reference string and $m$ is the length of the test string.

*Algorithm—Levenshtein Distance Between Two Strings*
*Input.* Two strings $x = a_1a_2\cdots a_n$, $y = b_1b_2\cdots b_m$.
*Output.* The Levenshtein distance $d^L(x,y)$.
*Method.*

1. $\delta[0,0] = 0$.
2. Do $i = 1, n$.

$$\delta[i,0] = \delta[i-1,0] + 1$$

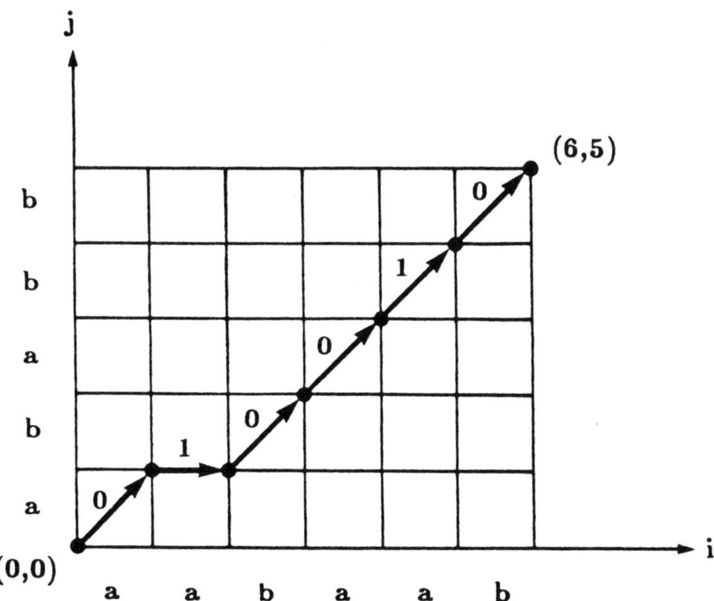


**Figure 9.2** The transformation from string *aabaab* to *ababb*. The distance $d^L(aabaab, ababb) = 2$.

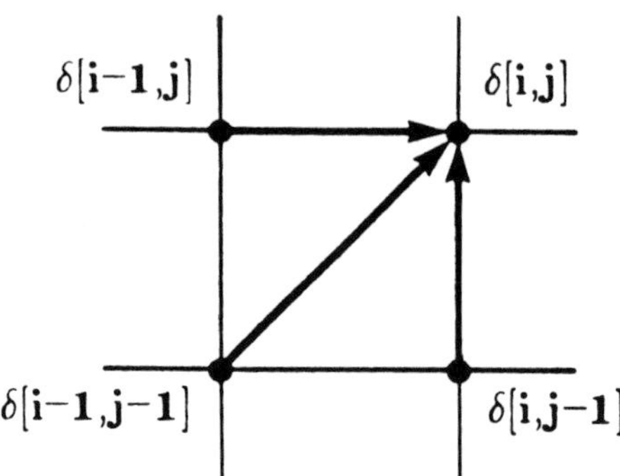


**Figure 9.3** The partial distance $\delta[i,j]$ computed from $\delta[i,j-1]$, $\delta[i-1,j-1]$, and $\delta[i-1,j]$

Do $j = 1, m$.

$$\delta[0,j] = \delta[0,j-1] + 1$$

3. Do $i = 1, n$; do $j = 1, m$.

$$\begin{aligned} e_1 &= \delta[i-1,j-1] + 1 \text{ if } a_i \neq b_j;\ \ e_i - \delta[i-1,j-1] \text{ if } a_i = b_j \\ e_2 &= \delta[i-1,j] + 1 \\ e_3 &= \delta[i,j-1] + 1 \\ \delta[i,j] &= \min(e_1, e_2, e_3) \end{aligned}$$

4. $d^L(x,y) = \delta[n,m]$.

Since the minimum distance is unlikely to occur in some area of the grid matrix—for example, the upper left and lower right corner—a global path constraint can be imposed to save computation time.

The Levenshtein distance appears not to be powerful enough for many pattern recognition applications. A weighted Levenshtein distance has been proposed where different weights are associated with different transformations: If $\sigma$, $\gamma$, and $\rho$ are the costs of substitution, deletion, and insertion, respectively, then we can define the weighted Levenshtein as

$$d^{WL}(x,y) = \min_j \{\sigma \cdot k_j + \gamma \cdot m_j + \rho \cdot n_j\}$$

where $k_j$, $m_j$, and $n_j$ are the numbers of error transformations as defined before. We can assign different weights to both transformations and terminals. For example, if $S(a,b)$ is the cost of substituting $b$ for $a$, $a, b \in V_T$ and $S(a,a) = 0$, $D(a)$ is the cost of deleting $a$, and $I(a)$ is the cost of inserting $a$, then the weighted distance becomes

$$d^{WL} = \min_{j} \left\{ \sum S_j(a,b) + \sum D_j(a) + \sum I_j(a) \right\}$$

where $a, b \in V_T$ and $J$ is the sequence of transformations used to derive string $y$ from $x$.

The string measures discussed so far are all absolute distance; i.e., the string length is not considered. A normalized distance, which is defined as absolute distance divided by string length, has been proposed. When absolute distances are equal, the normalized distances tend to favor longer strings. For example, consider two pairs of strings $(x_1, y_1)$ and $(x_2, y_2)$,

$$x_1 = aaabbbcccddd$$
$$y_1 = acabbbcccdbd$$
$$x_2 = ad$$
$$y_2 = cb$$

The distance between $x_i$ and $y_i$ is two (substitution errors). The distance between $x_2$ and $y_2$ is also two. However, when taking the string length into consideration, $x_1$ is more similar to $y_1$ than $x_2$ to $y_2$.

## Minimum-Distance Error-Correcting Parsing

An error-correcting parser has been proposed in the areas of compiler design and syntactic pattern recognition. When a conventional parser fails to parse a string, it will terminate and reject the string. An error-correcting parser produces the same results as a conventional one when the string is syntactically correct. However, it also generates a parse for the string even when it has minor syntax errors. In real pattern recognition problems, conventional parsers usually fail to work because of noise and incomplete information. The error-correcting parsing algorithms can be classified into two categories; one uses a minimum-distance criterion for the nonstochastic model and the other uses a maximum-likelihood criterion for the stochastic model. We shall discuss only minimum-distance error-correcting parsing for context-free grammars.

Aho and Peterson [5] have shown a minimum-distance error-correcting parsing algorithm using the Levenshtein distance. They first transform the original grammar into an expanded grammar that includes all the possible error productions. Then they modify Earley's parsing algorithm so that the number of error productions used is stored in the item list. The productions of the expanded grammar, $P'$, are constructed from $P$ as follows:

1. For each production in $P$, replace all terminals $a \in V_T$ by a new nonterminal $E_a$ and add these productions to $P'$.
2. Add to $P'$ the productions

a. $S' \to S$
b. $S' \to SH$
c. $H \to HI$
d. $H \to I$

3. For each $a \in V_T$, add to $P'$ the productions
   a. $E_a \to a$
   b. $E_a \to b$ for all $b$ in $V_T$, $b \neq a$
   c. $E_a \to Ha$
   d. $I \to a$
   e. $E_a \to \lambda$, $\lambda$ is the empty string.

In step 3, the productions $E_a \to b$, $I \to a$, and $E_a \to \lambda$ are called terminal error productions. The production $E_a \to b$ introduces a substitution error, $I \to a$ introduces an insertion error, and $E_a \to \lambda$ introduces a deletion error. For the Levenshtein distance, a constant weight 1 is associated with each of these productions. It can also handle the weighted distance in a similar way with, of course, variable weights.

*Minimum-Distance Error-Correcting Parsing Algorithm*

*Input.* An expanded grammar $G' = (V'_N, V'_T, P', S')$, and an input string $y = b_1 b_2 \cdots b_m$ in $V_T^*$.

*Output.* $I_0$, $I_1$, ..., $I_m$, the parse list for $y$, and $d^W L(x,y)$, where $x$ is the minimum-distance correction of $y$.

*Method.*

1. Set $j = 0$. Then add $[E \to \cdot S', 0, 0]$ to $I_j$.
2. If $[A \to \alpha \cdot B\beta, i, \xi]$ is in $I_j$, and $B \to \gamma, \eta$ is a production rule in $P'$, then add item $[B \to \cdot\gamma, j0]$ to $I_j$.
3. If $[A \to \alpha\cdot, i, \xi]$ is in $I_j$, and $[B \to \beta \cdot A\gamma, k, \xi]$ is in $I_i$, and if no item of the form $[B \to \beta A \cdot \gamma, k, \varphi]$ can be found in $I_j$, then add an item $[B \to \beta A \cdot \gamma, k, \eta + \xi + \zeta]$ to $I_j$, where $\zeta$ is the weight associated with production $A \to \alpha$. If $[B \to \beta A \cdot \gamma, k, \varphi]$ is already in $I_j$, then replace $\varphi$ by $\eta + \xi + \zeta$ if $\varphi > \eta + \xi + \zeta$.
4. If $j = m$ go to step 6; otherwise, $j = j + 1$.
5. For each item in $I_{j-1}$ of the form $[A \to \alpha \cdot b_j \beta, i, \xi]$, add item $[A \to \alpha b_j \cdot \beta, i\xi]$ to $I_j$ and go to step 2.
6. If item $[E \to S'\cdot, 0, \xi]$ is in $I_m$, then $d^{WL}(x,y) = \xi$, where $x$ is the minimum-distance correction of $y$.

The right parse of the input string can be constructed from the parse list.

### 3.3 Cluster Analysis and Nearest-Neighbor Rule

In syntactic pattern recognition of one-dimensional signals, a pattern is usually represented by a string of pattern primitives. Using the similarity measure between two strings, the conventional clustering methods, such as the minimum spanning tree and the method of cluster centers ($K$-means), can be extended to syntactic patterns. If a complete description of the pattern structure is needed, we should use parsing (or error-correcting parsing). On the other hand, if only the classification of the pattern is required, we can use the nearest-neighbor classification rule for its advantage in computation speed. In the following sections we show some of the clustering algorithms and the nearest-neighbor classification rule for syntactic patterns.

## Clustering Algorithms

The first algorithm requires a preset threshold $t$, and the cluster center ($K$-means) algorithm needs a predetermined number of clusters $K$. In most practical cases the application of these algorithms will require experimenting with various values of $t$ and $K$ as well as different choices of starting configurations.

### A Simple Cluster-Seeking Algorithm.

*Simple Cluster-Seeking Algorithm*
*Input.* A set of samples $X = \{x_1, x_2, \ldots, x_n\}$ and a parameter, or threshold, $t$.
*Output.* A partition of $X$ into $m$ clusters, $C_1$, $C_2$, ..., $C_m$.
*Method.*

1. Assign $x_1$ to $C_1$, $j = 1$, $m = 1$.
2. Increase $j$ by one. If $D = \min_l d(x_L^i, x_j)$ is the minimum, $1 \leq i \leq m$, and
   a. $D \leq t$, then assign $x_j$ to $C_i$;
   b. $D > t$, then initiate a new cluster for $x_j$, and increase $m$ by one.
3. Repeat step 2 until all the elements of $X$ have been put in a cluster.

### Minimum Spanning Tree.

*Minimum Spanning Tree Algorithm*
*Input.* A set of sentences (string) $X = \{x_1, x_2, \ldots, x_n\}$.
*Output.* The minimum spanning tree of $X$.
*Method.*

1. Assume that there are $n$ nodes.
2. Compute distances $d(s_i, x_j)$ for all $i, j$. Let $d(x_i, x_j)$ be the length of the arc connecting nodes $i$ and $j$ and denoted as $d(i, j)$.
3. List all arcs $(i, j)$ in the order of increasing $d(i, j)$.
4. Put the first arc $(p, q)$ on the list into list $A$.
5. Put the next arc on the list into $A$, except if a circuit can be found with the arcs already in $A$.
6. If all nodes are connected, stop; otherwise, go to step 5.

**Cluster Center ($K$-Means) Method.** Suppose we define a $\beta$-metric for a sentence $x_j^i$ in cluster $C_i$ as follows:

$$\beta_j^i = \frac{1}{n_i} \sum_{l=1}^{n_i} d(x_j^i, x_l^i)$$

Then $x_j^i$ is the cluster center of $C_i$ if $\beta_j^i = \min_l \{\beta_l^i \mid 1 < l < n_i\}$; $x_j^i$ is also called the representation of $C_i$, denoted as $A_i$.

*The Cluster-Center Algorithm*
*Input.* A sample set $X = \{x_1, x_2, \ldots, x_n\}$.

*Output.* A partition of $X$ into $K$ clusters.
*Method.*

1. Let $K$ elements of $X$, chosen at random, be the center of the $K$ clusters. Let them be called $A_1$, $A_2$, ..., $A_K$.
2. For all $i$, $x_i \in X$ is assigned to cluster $j$ iff $d(A_j, x_i)$ is minimum.
3. For all $j$, a new mean $A_j$ is computed. $A_j$ is the new center of cluster $j$.
4. If no $A_j$ has changed, stop; otherwise, go to step 2.

Although no general proof of convergence exists for this algorithm, it can be expected to yield acceptable results in most practical cases.

### Nearest-Neighbor Classification Rule

Let $C_1$ and $C_2$ be two pattern classes, represented by sentences $X_1 = \{x_1^1, x_2^1, \ldots, x_{n_1}^1\}$ and $X_2 = \{x_1^2, x_2^2, \ldots, x_{n2}^2\}$. For an unknown syntactic pattern $y$, decide that $y$ is in the same class as $C_1$ if

$$\min_j d(x_j^1) < \min_l d(x_1^2, y)$$

and $y$ is in class $C_2$ if

$$\min_j d(x_j^1, y) > \min d(x_l^2, y)$$

To determine $\min_j d(x_j^i, y)$, for some $i$, the distance between $y$ and every element in the set $X$ must be computed individually. The Levenshtein distance algorithm can be used to compute the distance between two strings.

The nearest-neighbor classification rule can easily be extended to the $K$-nearest-neighbor rule. Let $\tilde{X}_i = \{\tilde{x}_1^i, \tilde{x}_2^i, \ldots, \tilde{x}_n^i\}$ be a reordered set of $X_i$ such that $d(\tilde{x}_j^i, y) \leq d(\tilde{x}_l^i, y)$ iff $j < l$, for all $1 \leq j, l \leq n_i$; then

$$\text{decide} \quad y \in \begin{matrix} C_1 \\ C_2 \end{matrix} \quad \text{if} \quad \sum_{j=1}^{K} \frac{1}{K} d(\tilde{x}_j^1, y) \begin{matrix} < \\ > \end{matrix} \sum_{j=1}^{K} \frac{1}{K} d(\tilde{x}_j^2, y)$$

## 4 SYNTACTIC METHODS IN SIGNAL ANALYSIS

Applications of syntactic pattern recognition to digital signal processing have attracted increasing attention and made significant progress in the past decade. We will show some examples in the areas of biomedical waveform analysis, geophysical signal analysis, and speech recognition.

### 4.1 Biomedical Signal Analysis

Most biomedical waveforms, such as ECG and carotid pulse wave, are generated by certain organs of the body whose functions are well understood; therefore, it is possible to construct a grammar for these waveforms based on their functions. Horowitz [6] has applied the syntactic approach to peak recognition in ECG waveforms. A piecewise linear approximation is first performed on an input waveform, which is then encoded as a string of primitives or terminal symbols. A deterministic context-free grammar is constructed to generate sentences describing various waveforms. The proposed method has been successfully applied to peak recognition in electrocardiograms. An example of a waveform and its structure representation is given in Figure 9.4, where $p$ denotes positive slope, $n$ denotes negative slope, and 0 denotes zero slope. A deterministic context-free grammar $G$ was constructed that recognizes positive and negative peaks (if any) in a waveform represented by a string $w$:

$$G = (V_N, V_T, P, \{W\})$$

where

$$V_N = \{W, \langle\text{wave}\rangle, \langle\text{peak}^+\rangle, \langle\text{peak}^-\rangle, \langle\text{pos}_1\rangle, \langle\text{neg}_1\rangle, \langle\text{pos}_2\rangle, \langle\text{neg}_2\rangle, \langle\text{zero}\rangle\}$$

$$V_T = \{p, n, 0\}$$

and $P$:

$$
\begin{aligned}
W &\longrightarrow \langle\text{zero}\rangle\langle\text{wave}\rangle\langle\text{zero}\rangle \\
W &\longrightarrow \langle\text{zero}\rangle\langle\text{wave}\rangle \\
W &\longrightarrow \langle\text{wave}\rangle\langle\text{zero}\rangle \\
W &\longrightarrow \langle\text{wave}\rangle \mid \langle\text{zero}\rangle \\
\langle\text{wave}\rangle &\longrightarrow \langle\text{peak}^+\rangle \mid \langle\text{peak}^-\rangle \\
\langle\text{wave}\rangle &\longrightarrow \langle\text{pos}_1\rangle \mid \langle\text{neg}_1\rangle \\
\langle\text{peak}^+\rangle &\longrightarrow \langle\text{peak}^-\rangle\langle\text{zero}\rangle\langle\text{neg}_1\rangle \\
\langle\text{peak}^+\rangle &\longrightarrow \langle\text{peak}^-\rangle\langle\text{neg}_1\rangle \\
\langle\text{peak}^+\rangle &\longrightarrow \langle\text{pos}_1\rangle\langle\text{zero}\rangle\langle\text{neg}_1\rangle \\
\langle\text{peak}^+\rangle &\longrightarrow \langle\text{pos}_1\rangle\langle\text{neg}_1\rangle \\
\langle\text{peak}^-\rangle &\longrightarrow \langle\text{peak}^+\rangle\langle\text{zero}\rangle\langle\text{pos}_1\rangle \\
\langle\text{peak}^-\rangle &\longrightarrow \langle\text{peak}^+\rangle\langle\text{pos}_1\rangle \\
\langle\text{peak}^-\rangle &\longrightarrow \langle\text{neg}_1\rangle\langle\text{zero}\rangle\langle\text{pos}_1\rangle \\
\langle\text{peak}^-\rangle &\longrightarrow \langle\text{neg}_1\rangle\langle\text{pos}_1\rangle \\
\langle\text{pos}_1\rangle &\longrightarrow \langle\text{pos}_1\rangle p \mid \langle\text{pos}_2\rangle p \mid p \\
\langle\text{neg}_1\rangle &\longrightarrow \langle\text{neg}_1\rangle n \mid \langle\text{neg}_2\rangle n \mid n
\end{aligned}
$$

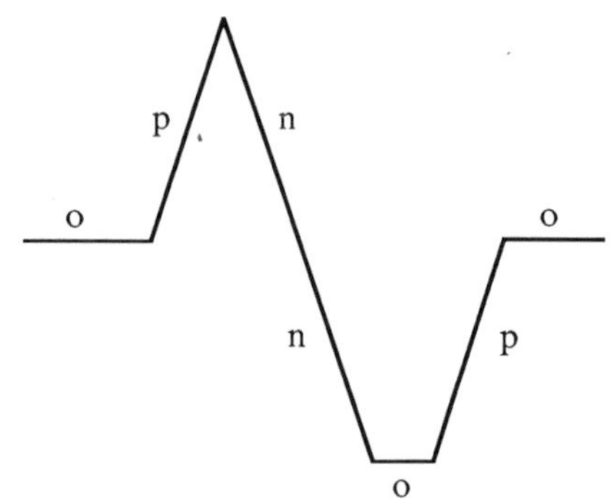


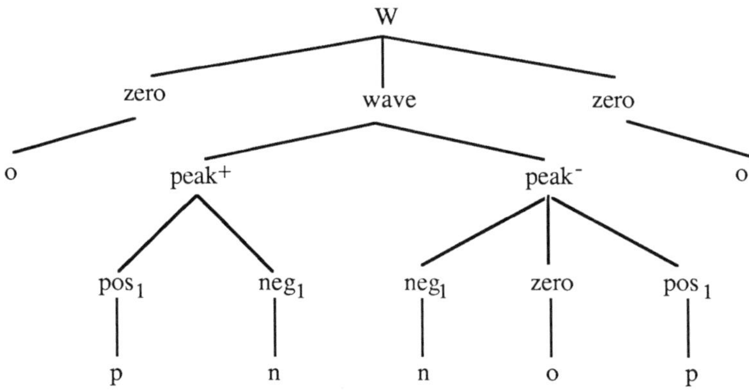


**Figure 9.4** Waveform and its structure representation. (From Fu [1].)

$$\langle \text{pos}_2 \rangle \longrightarrow \langle \text{pos}_1 \rangle 0 \mid \langle \text{pos}_2 \rangle 0$$
$$\langle \text{neg}_2 \rangle \longrightarrow \langle \text{neg}_1 \rangle 0 \mid \langle \text{neg}_2 \rangle 0$$
$$\langle \text{zero} \rangle \longrightarrow \langle \text{zero0} \rangle \mid 0$$

When parsing a specific $w$, a positive peak is recognized if and only if a section of $w$ is completely reduced by some production to the nonterminal $\langle \text{peak}^+ \rangle$. The same relation holds between a negative peak and the nonterminal $\langle \text{peak}^- \rangle$.

Stockman et al. [7] have applied syntactic pattern recognition to the analysis of carotid pulse waves. Specifically, they used a top-down and non-left-right parser to perform the syntax analysis. Segmentation and primitive recognition are essentially guided by the syntax of the waveforms. A typical segment of a carotid pulse wave and its structural description are shown in Figure 9.5. The primitives selected are linear and parabolic waveform segments. A context-free grammar was constructed for the carotid pulse wave. Top-down parsing is used to identify the most plausible pattern structure that matches the input waveform data. Straight-line primitives are extracted by scanning the data interval in either direction for a least-squares

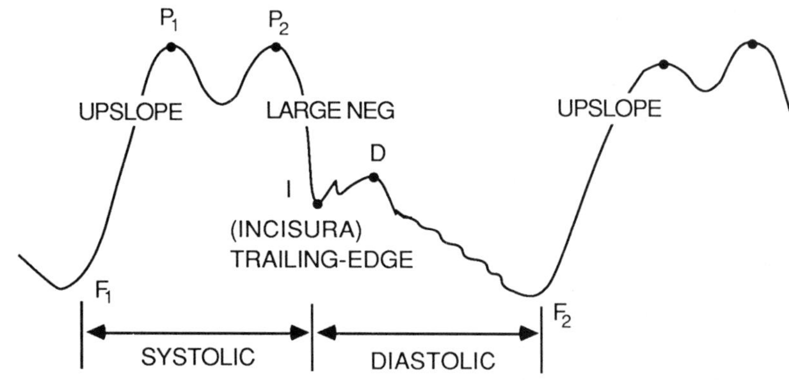


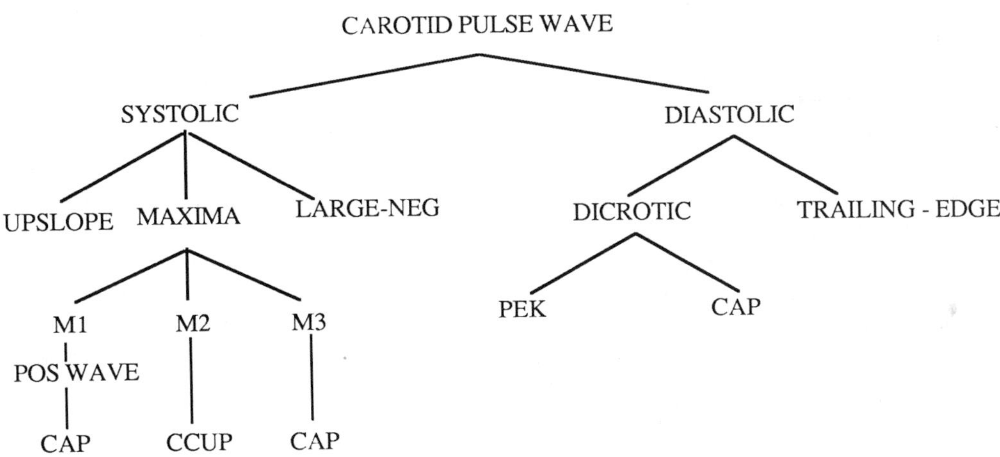


**Figure 9.5** Typical carotid pulse wave and its structure representation. (From Fu [1].)

straight-line fit of the minimum interval that has satisfactory parameters and error tolerance. The segment is then grown from the end where root-mean-square error accumulation is smaller until one of the parametric or interval constraints fails to be satisfied.

Giese et al. [8] applied a syntactic method to the analysis of electroencephalograms. A typical EEG pattern consists of 100-second four-channel waveforms. Each 100-second waveform was segmented (or sampled) into 1-second EEG segments, and each 1-second epoch or segment was considered a primitive. Seven different classes of primitives were identified. Recognition of each primitive was accomplished from 17 features by a linear classifier (or a maximum-likelihood classifier with Gaussian class density functions and equal covariance matrix assumptions). A preparse contextual analysis was used to reduce the misclassifications of primitive. Specifically, epochs or primitives, classified as slow (S or SL), were examined with respect to their spatial neighbors. An isolated slow epoch that occurred at the same time that other channels produced an artifactual epoch (A or AL) was implicated as a false slow epoch and reclassified as an artifactual primitive. An EEG grammar was constructed, and the parser used was a simple bottom-up parser without backtrack capabilities. It deals with ambiguities in the grammar by assigning implicit precedences to each production based on its relative position in the file.

## 4.2 Geophysical Signal Analysis

Structural information on the waveform has played an important role in some geophysical signal analysis problems. Liu and Fu [9] applied syntactic approaches to seismic discrimination, i.e., classification of underground nuclear explosions and earthquakes based on the seismic P-waves. They used two different approaches. In the first method, a pattern representation procedure converts the seismic waveforms into strings of primitives. The string-to-string distances between the test sample and all the training samples are computed and then the nearest-neighbor classification rule is applied. The second method consists of pattern recognition, automatic grammatical inference, and error-correcting parsing. The pattern representation procedure includes three steps: pattern segmentation, feature extraction, and primitive recognition. Each seismic record was normalized to 1200 sample points and divided into 20 segments (i.e., 60 points in each segment). Zero-crossing count and log energy of each segment were used as features, and based on these features a clustering method selected 13 primitives. A typical example from each class and their strings of primitives are shown in Figure 9.6. The automatic grammatical inference procedure infers a finite-state (regular) grammar from a finite set of training samples. An error-correcting parser was used which can accept erroneous and noisy patterns.

## 4.3 Speech Signal Analysis

The syntactic approach has been applied to the recognition of sounds, words, and continuous speech. We summarize the method used by De Mori [10] for recognition of spoken digits. Zero crossings are used to characterize each segment of the incoming speech signal. A careful analysis of the sequences of zero-crossing intervals from many spoken words indicated that intervals of 20-ms length provide meaningful short-time statistics of the intervals. Inspection of these statistics leads to the conclusion that the zero-crossing intervals in each 20-ms segment can be classified into a few groups. The numbers of zero-crossing intervals classified into the groups during a segmentation interval can be used as features of the speech segment. Consequently, the segmentation of each pattern (speech signal) is based on 20-ms time intervals and the range of zero-crossing intervals of the outputs of the low-pass filter (LPF) and high-pass filter (HPF). Each spoken word can be represented pictorially on a two-dimensional plane. The primitives (terminals) consist of silence interval (SI), stable zone (SZ), and lines (LN), nonstationary portions of the acoustics waveform approximated by straight-line segments. A context-free grammar was constructed for the 10 spoken digits, and a bottom-up parser implemented in terms of two pushdown transducers is used for analysis.

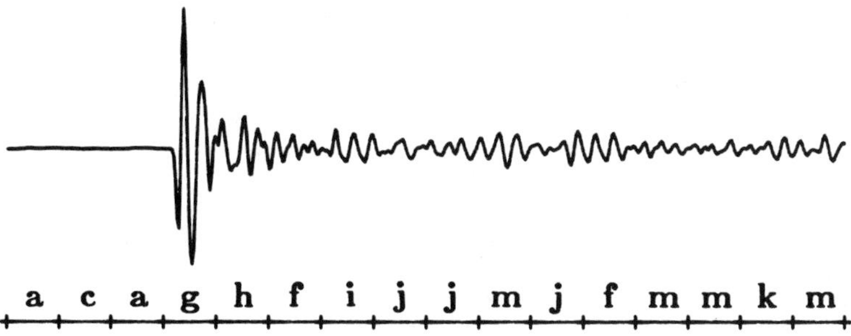


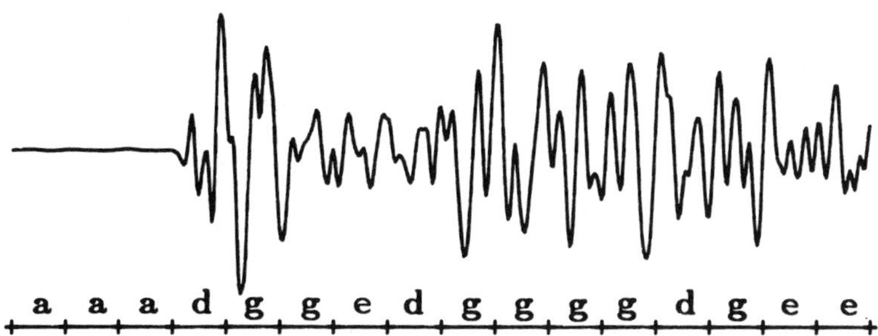


**Figure 9.6** Typical example of an explosion seismic P-wave (top), an earthquake seismic P-wave (bottom), and their corresponding string representations.

## 5 CONCLUSION

Syntactic pattern recognition has been successfully applied to some signal analysis problems. The syntactic approach is most effective when the waveforms have a clear and well-understood structure. Because of noise, missing information, and segmentation error, error-correcting parsing and/or stochastic grammars are usually required. The future direction in signal analysis may be a combination of the semantic and syntactic approaches using, for example, an attributed grammar.

## REFERENCES

1. K. S. Fu, *Syntactic Pattern Recognition and Applications*, Prentice-Hall, Englewood Cliffs, New Jersey (1982).
2. T. Pavlidis, *Structural Pattern Recognition*, Springer-Verlag, New York (1977).
3. A. V. Aho and J. D. Ullman, *The Theory of Parsing, Translation and Compiling*, vol. 1, Prentice-Hall, Englewood Cliffs, New Jersey (1972).

4. J. Earley, An efficient context-free parsing algorithm, *Commun. ACM*, *13*: 94–102 (1970).
5. A. V. Aho and T. G. Peterson, A minimum-distance error-correcting parser for context-free languages, *SIAM J. Comput.*, *4*: 305–312 (1972).
6. S. L. Horowitz, Peak recognition in waveforms, in *Syntactic Pattern Recognition Applications*, ed. K. S. Fu, Springer-Verlag, New York (1977).
7. G. Stockman, L. Kanal, and M. C. Kyle, Structural pattern recognition of carotid pulse waves using a general waveform parsing system, *Commun. ACM*, *19*: 688–695 (1976).
8. D. A. Giese, J. R. Bourne, and J. W. Ward, Syntactic analysis of the electroencephalogram, *IEEE Trans. Syst. Man Cybern.*, *SMC-9*(8): 429–435 (1979).
9. H. H. Liu and K. S. Fu, An application of syntactic pattern recognition to seismic discrimination, *IEEE Trans. Geosci. Remote Sensing*, *GE-21*(2): 125–132 (1983).
10. R. De Mori, Syntactic recognition of speech patterns, in *Syntactic Pattern Recognition Applications*, ed. K. S. Fu, Springer-Verlag, New York (1977).

# 10

# Algorithms and Architectures for Machine Vision

WOLF-EKKEHARD BLANZ, DRAGUTIN PETKOVIĆ, and JORGE L. C. SANZ
Computer Science Department, IBM Almaden Research Center, San Jose, California

## 1 INTRODUCTION

It is well known that we obtain most of our information through vision. Yet vision is probably one of the least understood sensorial processes. Research work is progressing toward the understanding of human vision as well as the use of machines to emulate it for practical purposes. We will consider the second area and refer to it as machine vision, and the question of how closely a specific solution emulates intelligent human behavior or just solves a given task without trying to emulate human vision will be of less concern for our discussions. Communication between real-world environment and machine vision systems is provided by some kind of sensing mechanism for input, followed by analog and/or digital computation. The final output may be an image prepared for human visual evaluation or some statement about the image content such as relevant measurements or recognition and location of objects. Machine vision is a highly multidisciplinary field, interacting closely with optics and sensing, image processing and analysis, computer architectures, pattern recognition, artificial intelligence, robotics, and data base management. Often the term computer vision is used to denote the processing and analysis of the vision data by digital computers, which thus is a subset of machine vision.

Applications of machine vision are found in various fields such as manufacturing, biomedicine, and defense. The manufacturing applications include qualitative and quantitative inspection, position sensing, alignment, measurement, robot guidance, and recognition. In biomedicine typical applications are in cytology, X-ray diagnosis, and reconstructing the images for computer tomograpy (CT), and magnetic resonance imaging (MRI). Scientific applications include areas such as analysis of bubble chambers, analysis of fringes, analysis of satellite images for agricultural, geological, and meteorological studies, and enhancement of sensor outputs for easier human interpretation. Defense applications include surveillance, reconnaissance, target recognition and tracking, and guidance of autonomous vehicles. Recent surveys (e.g., [1–5] give a thorough review of many industrial and defense applications.

The word "image" may denote any two-dimensional (2-D) array of digitized sensor data, not only intensities measured in the visible part of the spectrum of light but also other data, such as intensity of X-rays, ultrasound, radar, or other signals. It actually represents the projection of the 3-D world onto a 2-D array, and this imposes severe problems in applications where 3-D information is essential, such as robotics vision. Still, many applications, including satellite image interpretation and printed circuit (PC) board inspection, can be satisfied using the 2-D representation.

The technologies involved in machine vision are optics, sensing and image capture, algorithms, architectures, and application development environments (ADEs). Proper integration of all of these technologies is needed for a successful solution of practical problems. For example, smart sensing can often significantly simplify and improve results of the subsequent image processing and analysis. In addition, even a correct algorithm, if it is not executed sufficiently fast, becomes useless for practical purposes. Furthermore, the lack of skilled people makes it important to have application development environments where nonexperts can quickly develop solutions to new problems. These factors are especially true in the manufacturing industry, which is one of the major sources of investment in machine vision.

Within these technologies we distinguish image processing, image analysis, and image formation. The first two take a two-dimensional input and create either two-dimensional output, in the case of image processing, or symbolic descriptions of the two-dimensional data in the case of image analysis. Applications of the third group include diverse reconstruction problems, such as those related to X-rays, nuclear magnetic resonance (NMR), and synthetic aperture radar. Image understanding and image pattern recognition are closely related to image analysis. Although some authors make a distinction, we will consider them synonymous.

The requirements in all fields of application involve computations on large amounts of data. Today's standard images have about 0.25 Mbyte and often must be processed in short times, say 1/30 of a second for real-time applications to a few seconds for "near real time," to produce the output image or symbolic data. Image analysis tasks are further complicated by the fact that our understanding of vision is limited, in terms of both understanding human vision and trying to emulate it using digital or optical processing.

It is hard to cover all the aspects of machine vision in one book chapter. Therefore, we do not cover sensing, related optical technology, and optical computing. We review numerous digital algorithms and architectures, with the intention of presenting the whole process of digital image analysis. We put more emphasis on 2-D problems, since they represent a more mature part of machine vision. For 3-D problems, which are especially important in robotics applications and are receiving much attention in the research community, the reader is referred to proper references.

In Section 2 we analyze digital image algorithms in terms of functional and architectural considerations. A review of basic algorithms for solving (usually 2-D) image analysis tasks is given, followed by a review of basic control strategies used for their integration. In Section 3 we present an overview of machine vision architectures.

## 2 ALGORITHMS FOR MACHINE VISION

Let us assume that the information from an appropriate sensor is digitized and stored in a two-dimensional array. The elements or "pixels" of the image matrix somehow describe the intensity of the input signal being projected from the 3-D world onto a 2-D sensor plane. We assume that each pixel can take a value, which we will refer to as its gray value, out of a certain range, irrespective of what it describes physically. Today, images of 512 rows of 512 pixels per row with each pixel represented by 8 bits are standard, with larger images used for special applications. Positional information is implicitly encoded in the address of the pixel within the two-dimensional image array.

A detailed and functional categorization of image analysis [6, 7] reflects differences between 2-D and 3-D problems (see Figure 10.1). As mentioned before, 2-D sensor planes provide only the projection of a 3-D environment. Without some a priori knowledge it is impossible to obtain 3-D structures of the scene unambiguously. However, many problems like microelectronic inspection, satellite image analysis, and PC board inspection can be treated as 2-D problems and are typically referred to as $2\frac{1}{2}$-D problems, indicating that the extension of the third dimension is much less than of the other two.

The basic functional steps in 2-D image analysis are

1. Image preprocessing: used to reduce noise, compensate or correct some distortions in geometry or light distribution, and enhance certain desired features such as edges
2. Image segmentation: used to partition the image into regions of interest such as edges, or blobs corresponding to objects of interest, or to classify pixels or smaller image regions with different groups as in satellite image analysis
3. Feature extraction: the process of extracting certain symbolic or geometric measurements from the images, such as geometric features of segmented objects
4. Recognition/decision making: the process of recognizing the objects in the image, verifying inspection specifications, etc.

In the case of 3-D vision the steps are slightly different. There the key problems are to compute surface orientation at each image point, to describe (match) the surfaces with preestablished models, and to compute object motion parameters. For further information see [8–11].

The first two steps are typically referred to as low-level processing. All operators act on so-called iconic image data, given as an array of numbers. The output is of the same structure, thus preserving the same positional encoding. Typical low-level operations are point and neighborhood transformations, where the output

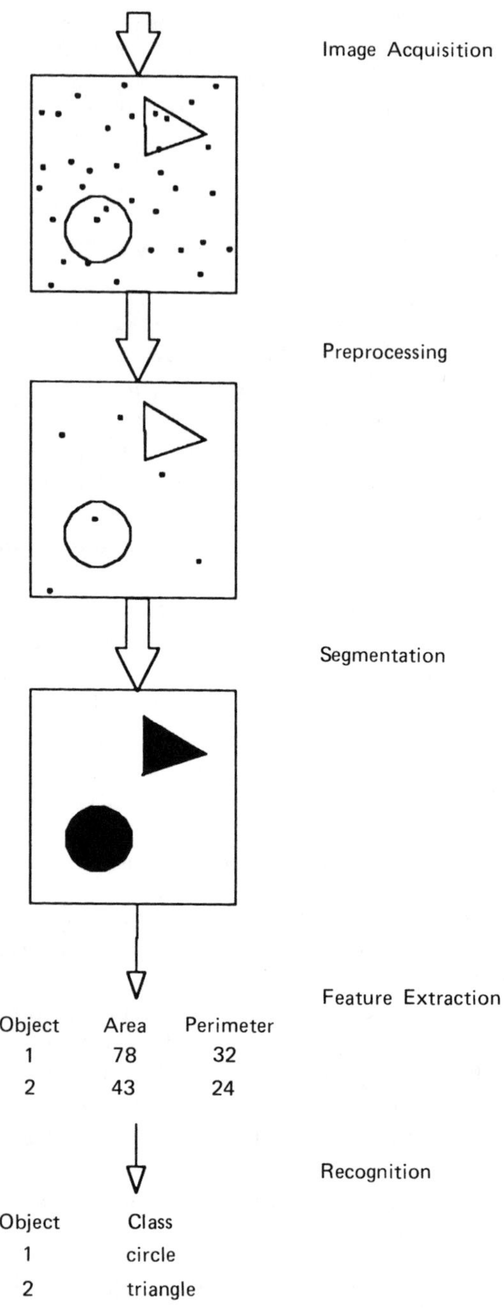


**Figure 10.1** Simplified diagram of 2-D image analysis algorithms.

pixel value is a function of the input pixel value or of this value and the pixel values of a certain neighborhood, respectively.

Step 3 in our categorization is called mid-level vision. It has an image (matrix) data structure as input and symbolic data structures as output. Relevant information is encoded explicitly in the symbolic structures, and iconic information is lost. This process usually consists of some form of feature extraction. Unfortunately, vision theory has still not provided a general solution to how to find systematically entities or features that must be measured to solve a given task. In addition, the complex nature of this problem makes it questionable whether a systematic solution will ever be found. Furthermore, because of the different data representations of the input and output, these algorithms pose serious challenges for the supporting architectures.

The last step, high-level vision, has symbolic data as both input and output. This step usually consists of pattern recognition, or artificial intelligence (AI) related processes, where objects are recognized and a final decision about these objects is made. Although many standard techniques are available (see [12–17]), none of them can be viewed as general. In addition, research is needed both in pattern recognition and in AI that specifically addresses vision problems. Data structures and operations in high-level image analysis such as list operations, pointer manipulations, and traversing tree structures are irregular and much different from those in low- and midlevel vision. Because of the reduced amount of data there is no urgent need for special architectures at this level for most applications, and, in practice, most of these processes are today implemented in general von Neumann-type architectures.

In the following sections we elaborate on some image analysis algorithms, referring to their low-, mid-, and high-level categorization and means of integrating them using various control strategies.

## 2.1 Low-Level Algorithms

The task of low-level algorithms is to process the input image and transform it into another image which better fulfills the needs of subsequent processing steps. This is the field where we find the greatest variety of algorithms in the literature [18–21].

### Point Operators

The first and simplest operators in this group are point operators. These operators create new images in which the gray level of a pixel is a function only of the gray level of the pixel at the same location in the input image:

$$g_{\text{new}}(x, y) = f(g_{\text{old}}(x, y))$$

Typical examples are thresholding and histogram equalization.

Thresholding is a basic segmentation operation, partitioning the image into pixels of interest (the objects) and the background. This operation assumes that object and background are separable by their gray values; i.e., there exists a certain threshold gray value $g_t$ such that all object pixels are greater than $g_t$ and all other pixels are smaller or vice versa. Thus a two-valued or binary image with gray

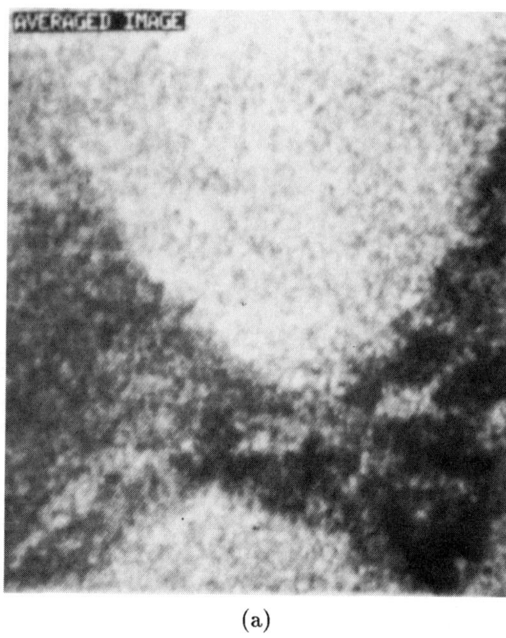


(a)

**Figure 10.2** Thresholding process: (a) original (averaged to reduce noise); (b) gray level histogram; (c) thresholded original—threshold set at the valley of the histogram.

values $g_1$ and $g_2$ can be produced by applying a threshold $g_t$ such that

$$g_{\text{new}} = \begin{cases} g_1 & \text{if } g_{\text{old}} > g_t \\ g_2 & \text{otherwise} \end{cases}$$

The assumption that gray level distributions of object and background do not overlap is often not valid, and the more the distributions overlap, the greater the segmentation error. Even if they do not overlap, it is not easy to find the proper threshold. This problem is typically solved by putting it at a valley in a bimodal gray value histogram, thus minimizing the error which is given by the amount of overlap of the partial histograms of object and background [22]. An example of the thresholding process is shown in Figure 10.2. In spite of its problems, thresholding is widely applied, especially in applications where the contrast between object and background is high. In industrial applications, it may be possible to adjust the illumination and background color to minimize the overlap. In all these cases thresholding is a recommended and inexpensive technique for producing binary images for further shape analysis. In more textured or outdoor scenes, and in almost all biological applications, more sophisticated procedures are necessary.

Point operators are often implemented as table lookups. A table is generated in which the original gray value of a pixel serves as an address in the table and the

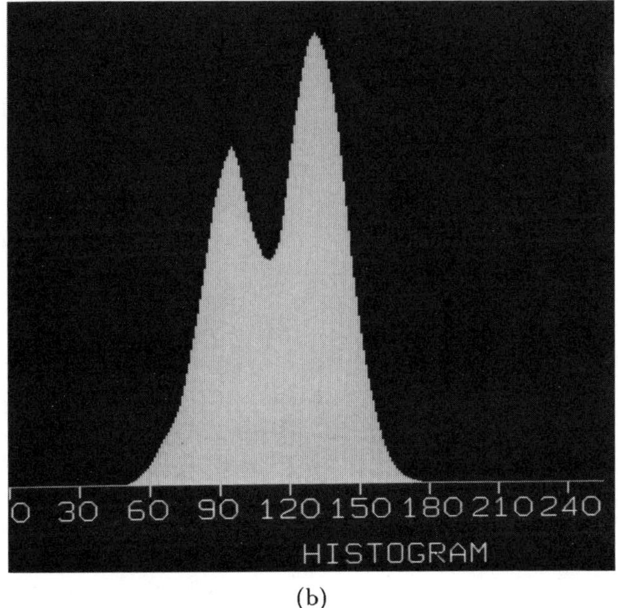


(b)

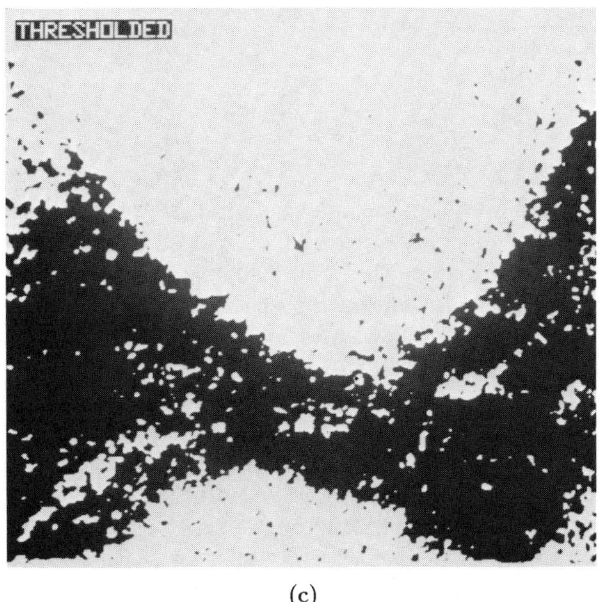


(c)

new gray value of the pixel is the content of the table at this address:

$$g_{\text{new}} = t(g_{\text{old}}(x, y))$$

Lookup-table operations are used to implement thresholding, correct sensor gray level characteristics, produce pseudo-color images. For 8-bit images a 256-element table is necessary, and arbitrarily complex operations can be applied efficiently by computing the table entries once and applying the table to one or many images.

More sophisticated lookup-table operations are those where the entries in the lookup-table are computed from image data. A typical example is histogram equalization, which yields an output image with a transformed (usually reduced) gray level range. The intervals of the old gray level range producing one unique new gray level are chosen such that the gray level histogram of the new image is as uniform as possible [23]. The histogram of an image which is typically presented as a bar graph diagram is defined as the vector of gray level frequencies

$$h(i) = \sum_{x,y} \delta_{i,g(x,y)}$$

where the sum is over all pixels of the image and $\delta$ is the Kronecker delta. The limits $l_j$ of the gray level interval of the original image which is transformed in one new gray value are given as

$$l_0 = 0$$

$$\sum_{i=l_{j-1}}^{l_j} \leq \frac{N}{K-1}$$

if $N$ is the number of pixels in the image and $K$ is the number of transformed gray levels. Histogram equalizations are typically used to scale data over the range of displayable levels or to reduce the gray value range of images while keeping or enhancing essential detail information for further analysis. Figure 10.3 shows an original image on the left-hand side and the result of a histogram equalization on the right.

Operations which are similar to point operations but involve two or more images are also useful in a variety of situations. For example, shading corrections to make digital images have a uniform spatial response to illumination involve subtraction and multiplication of two images [24]. Also, addition and lookup tables are the basis for efficient operations arising in Hough transform [25], coloring masks [26] and reconstruction in computed tomography [27], among many others [28].

In the early stages of automated visual inspection, especially for pattern inspection of printed circuit and printed wiring boards, image subtraction methods were attempted. The idea was to apply simple exclusive OR operations between a reference image ("golden sample") and the inspected image on a pixel-by-pixel basis. The advantage is that this is easy to implement in specialized hardware, thus enabling high pixel rates. However, image subtraction has disadvantages: 1) variations in parts are usually sufficient to create numerous false alarms after subtracting the images; 2) prior to the subtraction, a precise alignment of the two images is necessary, and this requires extensive processing or expensive mechanics;

(a) (b)

**Figure 10.3** Histogram equalization: (a) original $256 \times 512$ image; (b) histogram equalization to eight gray levels.

and 3) it is difficult to express allowed tolerances in terms of feature dimensions (i.e., inspection specifications).

## Image Filters

Another important group of low-level algorithms are image filters. We distinguish filters in the spatial domain and filters in the Fourier domain. The purpose of these filters is manyfold; for example, they can be used to reduce noise, enhance features, and compensate for deficiencies in sensors. The Fourier transform in two dimensions in the continuous case is defined as an extension of the well-known one-dimensional Fourier transform as

$$F(u,v) = \int_{-\infty}^{+\infty} \int_{-\infty}^{+\infty} f(x,y) e^{-i(ux+vy)}\, dx\, dy$$

where $x$ and $y$ are the coordinates in the spatial domain and $u$ and $v$ the coordinates in the Fourier domain. For practical purposes we are more interested in the discrete Fourier transform applied to sampled and digitized signals, which is similarly defined as

$$F(k,l) = \frac{1}{MN} \sum_{m=0}^{M-1} \sum_{n=0}^{N-1} f(m,n) \exp\left[-i2\pi\left(\frac{km}{M} + \frac{ln}{N}\right)\right]$$

Here $n$ and $m$ are the column and row indices of the spatial image matrix, $k$ and $l$ the indices of the matrix in the Fourier domain, and $N$ and $M$ the numbers of columns and rows, respectively. As for one-dimensional signal processing, there exist algorithms for fast Fourier transformations [19]. Filters in the Fourier domain, although of great theoretical interest, are not often used in practice. One reason is that they are time-consuming when carried out in software or expensive when done in hardware. Another reason is that many useful filter operations are not linear. An exception is in reconstruction problems, such as X-ray tomography, NMR tomography, and synthetic aperture radar [29, 30].

Since many operation in the Fourier domain are well known and understood from one-dimensional signal processing, there is a strong interest in the 2-D signal processing community in applying them for images as well, and therefore approximations of the frequency filters are used in the spatial domain. The realization of those filters is achieved by so-called local neighborhood or window operations. The impulse response $k(x,y)$ of the desired filter is computed in a certain region, typically a window of $3 \times 3$ to $15 \times 15$ pixels, and convolved with the input image $g_{\text{old}}(x,y)$ to produce the output image $g_{\text{new}}(x,y)$:

$$g_{\text{new}}(x,y) = \sum_{x'y' \in W(x,y)} g_{\text{old}}(x',y') * k(x'-x, y'-y)$$

where $W(x,y)$ is the window around the pixel $g_{\text{old}}(x,y)$.

This operation is repeated for each pixel. The border of the image must be handled as a special case, since there the neighborhood is not fully defined. Typical border handlings are extension of the image pixel values to the borders, or generating slightly smaller output images.

Figure 10.4 shows a typical example of low-pass and high-pass filtering.

Common filter kernels that are applied as window operators are edge operators which enhance the discontinuities in image pixel values. Two typical examples are the Roberts and the Laplacian operator [18]. The Roberts operator is equivalent to convolving the images with

$$\begin{matrix} 0 & 1 \\ -1 & 0 \end{matrix} \quad \text{and} \quad \begin{matrix} 1 & 0 \\ 0 & -1 \end{matrix}$$

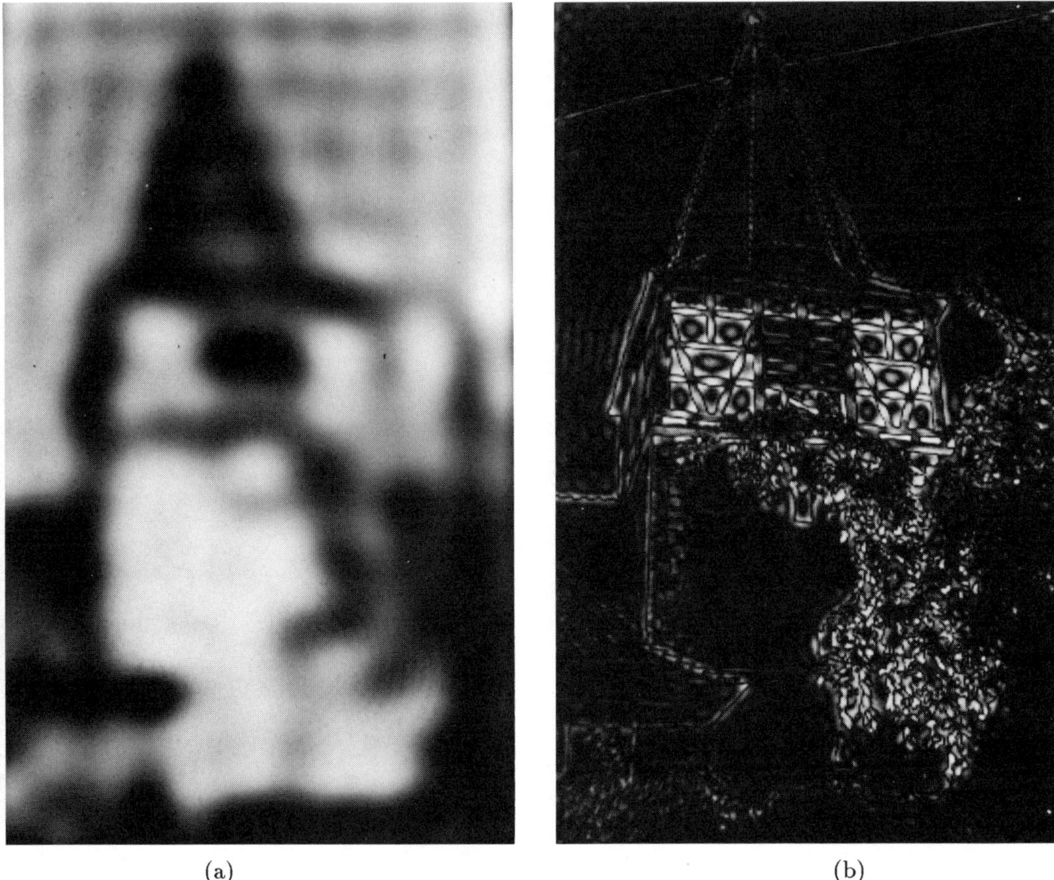

(a) (b)

**Figure 10.4** (a) Low pass filtered (original from Figure 10.3); (b) high-pass filtered.

and taking the maximum of the two responses. One approximation of the Laplacian operator consists of convolving the image with

$$\begin{matrix} 0 & 1/4 & 0 \\ 1/4 & -1 & 1/4 \\ 0 & 1/4 & 0 \end{matrix}$$

Since the Laplacian always gives two responses for each edge (one being positive and one negative on the bottom and the top of a gray level slope, respectively) we can produce the absolute value, take only the positive parts, and clip all negatives to zero if we want only positive responses, or use the zero crossings. Figure 10.5 shows results obtained with the Roberts and Laplacian operators.

Another well-known group of window-operations are the nonlinear filters. Here, the pixels in a window are used to calculate a new gray value by a nonlinear

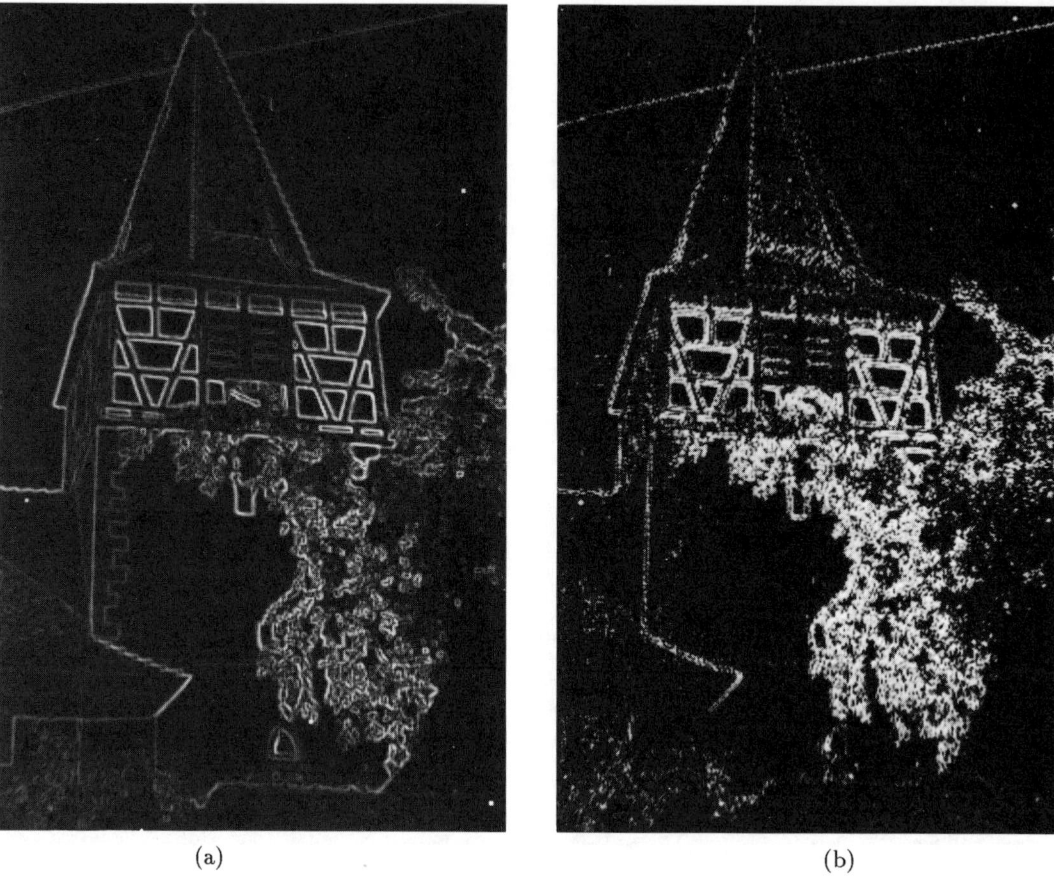


**Figure 10.5** (a) Roberts gradient operator applied to original in Figure 10.3; (b) Laplacian operator.

transformation. Three examples are the so-called rank order operations: local minimum, local maximum, and median. The first assigns the smallest gray value in the window to the new pixel; the second assigns the largest gray value; and the median assigns the gray level such that there are equal numbers of pixels with lower and higher gray levels in the window. Figure 10.6 shows results for the three filters using a window size of $5 \times 5$ pixels.

The median filter can be used for noise suppression where, in contrast to a low-pass filter, edges are not blurred although small noise structures are completely suppressed. Local minimum and local maximum are used mainly to suppress small bright or dark particles, respectively. If applied to binary images, these operations lead to the so-called morphological operators [31, 32]. The basic operators in morphology are expansion or dilation and contraction or erosion applied to binary images, as illustrated in Figure 10.7. In fact, it can be shown that local minimum and local maximum are obtained by generalizing the dilation and erosion operators [33]. Repetitive erosion of binary images, preserving local connectivity (sometimes called a thinning operation), produces a pattern skeleton. These skele-

(a)

(b)

(c)

**Figure 10.6** (a) Local minimum applied to original in Figure 10.3; (b) local maximum; (c) median filter.

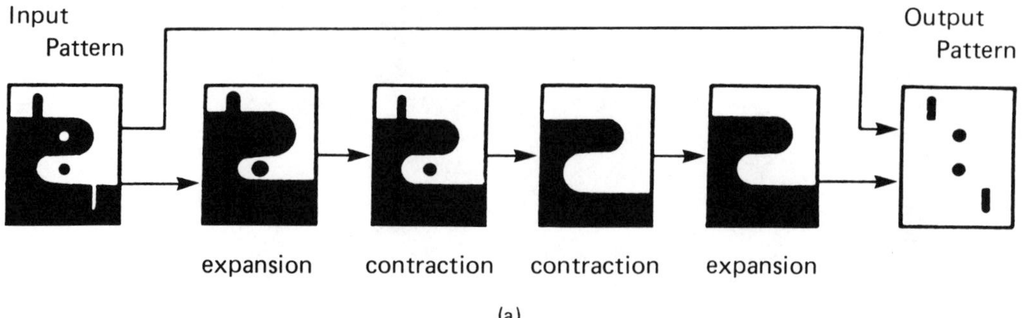


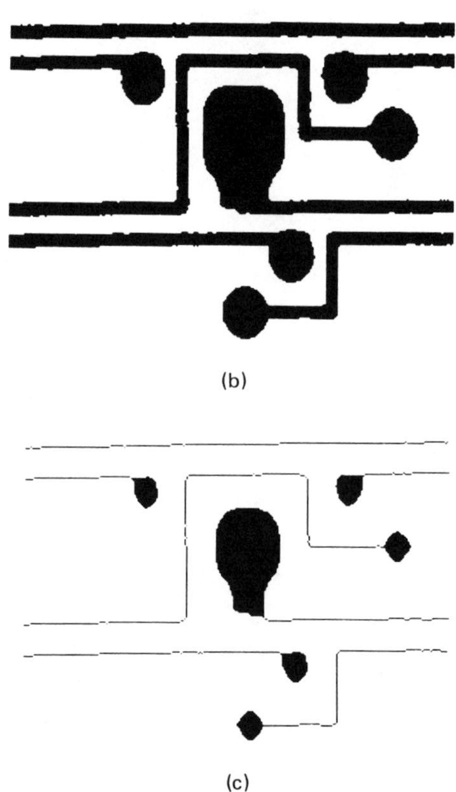


**Figure 10.7** (a) Contraction-expansion principle in morphology; (b) original binary image; (c) processed with morphic operators.

tons are then used for subsequent image analysis tasks. Some distance checking measurements can be made, for example, by applying a certain number of erosion steps. If this process eliminates a feature, its width was less than the number of basic (3 × 3) erosion steps. Morphology-based approaches have been used for PC board inspection. Morphology operators can be cascaded in various ways and are a useful building block in certain applications.

Morphological operators can be implemented in special hardware for real-time applications, and systems have been built to facilitate their computation [34, 35]. They are most often implemented as a pipeline of nonlinear neighborhood operators using lookup tables to obtain the output value. Extensions to gray level morphology have also been attempted [36].

## Segmentation

There is a great variety of image segmentation algorithms [22, 37]. Two basic approaches are to look for regions of change (like edges) or for regions of uniformity in some sense (region growing or pixel clustering, texture classification or pixel classification, etc.) [38].

There is also a great variety of edge detection techniques, and some of the better-known edge operators were already discussed in an earlier section. When used for object segmentation purposes, all edge detection techniques have one inherent difficulty in common: it cannot be guaranteed that all border points of an object will be detected as edge points, thus forming a closed boundary. On the other hand, border points are not necessarily the only edge points detected in a textured scene. Therefore, if edge detection techniques are used for object segmentation, special efforts are necessary to trace and combine the edges to form boundaries of objects [39, 42].

Besides techniques based on edge detectors and simple gray level thresholding, there exist some other techniques. One group of them extracts regions where all pixels have certain features in common which distinguish them from pixels of other regions. These techniques resemble cluster analysis techniques, and there are procedures which start with small regions and progressively combine them into larger regions with similar characteristics [41, 42] and others which begin with large regions and progressively divide them so that the new sub-regions have a better homogeneity than the larger region [43].

Another method is to describe pixels by one or more features and apply common pattern recognition techniques such as supervised or unsupervised classification to those features. Pixel features may be derived from multispectral data, where each channel at the location of a pixel contributes its gray value as a feature [21], or from local neighborhood information derived by certain combinations of the gray values of the neighbors of the pixel in a certain window, thus describing the texture to which the pixel belongs [44, 45]. This approach has recently been successfully applied to specific industrial inspection problems [46]. Pixels described by feature vectors can be classified with different regions either by unsupervised classification, i.e., feature clustering [47, 48], or by supervised classification techniques, which are essentially feature thresholding techniques where the thresholds are obtained by error minimization on a training set. Of course, there exist other methods with global thresholding [49, 50] where one global threshold is applied to the entire image to discriminate between two different regions. If the gray level of the pixels is used as the only feature, these procedures become the thresholding technique mentioned earlier.

A common problem of segmentation algorithms using unsupervised classification techniques is that the generated segments do not necessarily correspond to the objects needed by later processing steps. This problem can be solved by trainable algorithms which can be systematically adapted to produce segments that closely correspond to desired objects [51]. This is, of course, possible only if gray level or texture features sufficiently discriminate between the objects of interest and the rest of the scene. In all other cases low-level algorithms produce edges or subsegments with common characteristics but object recognition must be done in subsequent higher levels of analysis.

There is a whole class of geometric operations which we cannot address in detail because of space limitations. They include simple translations, rotations, scale changes, conform transformations, and distortion compensations. Their purposes are manyfold, and sometimes they are tailored for a specific application. For more details, see, for example, Ref. [21].

## 2.2 Midlevel Algorithms

The task of midlevel image processing is to provide abstract symbolic descriptions of scenes, starting from the image data provided by low-level processing. Thus midlevel processing performs a considerable data reduction while keeping the relevant information that is to be used in a subsequent recognition task. This symbolic representation usually consists of measurements or features corresponding to certain objects, most often provided by the low-level image segmentation. For example, in some applications we need basic geometric features, like area, perimeter, and enclosing box for objects in the image, while in others we need textural information about the objects of interest.

General pattern recognition and AI theory have not yet provided a formal methodology of "feature invention," e.g., a process for determining what to measure from the signal in order to solve a recognition task. This is even more complicated in vision. However, a few general guidelines for feature invention exist: the features should have high discriminatory power and be robust in the presence of noise, economical to extract, and invariant to the expected transformations that objects in the image might undergo. They should also correspond as much as possible to the measurements or attributes sought in the final image analysis task. On the other hand, the process of "feature selection," which denotes the selection of the best subset or combination from a given set of features, has been widely explored (e.g., [12, 14, 52]).

We distinguish two types of features: Features of type I are variables or quantizations of variables which can assume numeric values, like pixel gray level or object area. Features of type II encode nonnumerical, usually relational, measurements, like "left-off" or "inside," which are common in vision. The feature type influences the recognition procedure in high-level algorithms, and in practical applications we usually need a mixture of both feature types.

A simple form of feature extraction that can be found in some commercially available vision boards is the collection of a set of $x, y$ coordinates of image pixels having some predefined properties. This can be used, for instance, to collect edge point coordinates from an image in which edge pixels have been segmented.

Another well-known operation is component labeling and geometric feature extraction (sometimes called component analysis) [53–55]. Assume that we have a set of segmented objects in the image (see Figures 10.1 and 10.8). Each object consists of a set of 8-connected pixels, meaning that each object pixel is connected to another pixel of the same object by at least one of its 8 immediate neighbors. These objects may correspond to manufacturing defects or biological objects. The process of component analysis assigns a unique label to each object and computes its basic geometric features of area, perimeter, center of mass, and enclosing box; see Figure 10.8. To reduce computational costs, 4-connectedness may be used instead, meaning that the unique label will be assigned to the pixels being connected at least in the north, east, west, or south direction.

By extending these features, we can compute gray level features for each object, including minimal, maximal, and mean gray level, where gray levels are taken from the original full-range image [56]. Component analysis can be done recursively for objects spanning multiple image frames [57] and also can include other types of useful geometric features (for an example see [58]).

For more complex tasks, gray level features alone are not sufficient to provide discriminatory power for subsequent high-level processes, and textural features may be used. There are many techniques for measuring textures of images [59], and they may be applied to full images or only portions of images, such as already segmented objects. A paradigm of a complete system which analyzes uterine cells and is amenable to special feasible architectures is demonstrated in [60]. Given an architecture to derive a certain set of features, this approach shows how to use inexpensive nonlinear filter techniques (local minimum, local maximum, and median filters) to obtain a new set of linearly independent textural features for biological objects.

One of the most common tasks in industrial machine vision is that of robust and precise line and curve detection. The most common algorithms applied in this case are the Hough transform [61] and its generalizations. In its original form, the Hough transform yields information on slopes and locations of lines (or curves) in the image and thus may be considered a form of feature extraction. The reason for the robustness of this transform is that it is global, and minor local distortions have little or no effect. A line in $x, y$ coordinates can be represented as

$$\rho = x^* \cos\theta + y^* \sin\theta$$

where $\rho$ is the length of the normal to the line from the origin, and $\theta$ is the angle of that normal; see Figure 10.9.

If each $x, y$ point that is considered to be a line point (for example, that has a high gradient) adds its contribution to corresponding buckets in $(\rho, \theta)$ parameter space, peaks are created which correspond to the lines with high evidence. These peaks will correspond to collinear points, not necessarily uninterrupted lines, but in most applications this information is sufficient for line detection and location. The Hough transform has two distinct parts: accumulating the bucket counts, and searching for peaks in the parameter space. Both operations are time-consuming. However, Sanz

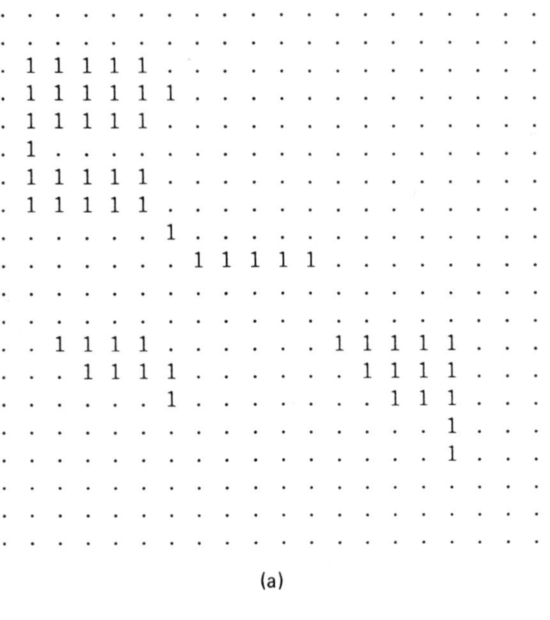

(a)

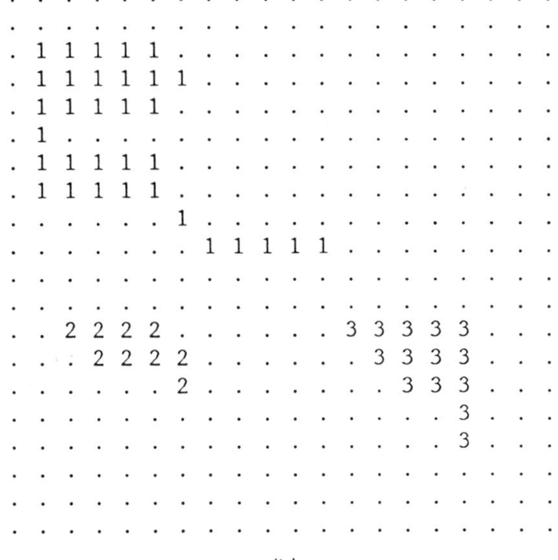

(b)

| LABEL | AREA | PERIMETER | ΔX | ΔY |
|---|---|---|---|---|
| 1 | 33 | 23 | 11 | 8 |
| 2 | 9 | 9 | 5 | 3 |
| 3 | 14 | 12 | 5 | 5 |

(c)

**Figure 10.8** Component labeling and feature extraction: (a) original binary image; (b) labeled image; (c) extracted features.

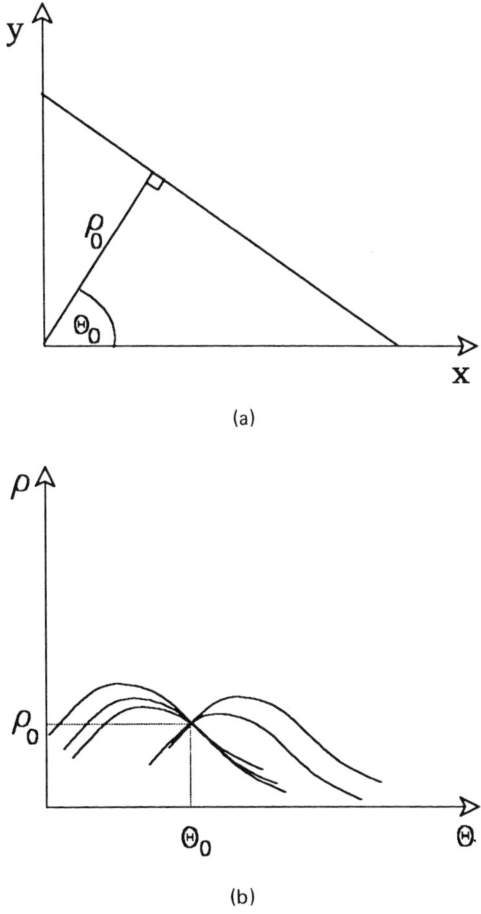


**Figure 10.9** Hough transform: (a) image space; (b) parameter space.

et al. [28] describes an architecture for the computation of Hough space parameters which enables the Hough transform to find widespread use in applied machine vision (for an example see [56]). The analysis in the parameter space, namely the search for peaks, can be greatly simplified using a priori knowledge, such as expected edge location and orientation.

Often we need to describe shapes of image objects in more detail than is provided by basic geometric features of area, perimeter, moments, and enclosing box. Although the problem of shape analysis, like many vision problems, has not been completely solved, several approaches are available. Extensive information on shape analysis can be found in [62]. Shapes of complex boundary curves can be described by graph-based techniques or Fourier descriptors [63, 64], the latter using the Fourier coefficients of parametrized boundary curves as features to describe the shape of an object. A related approach is to keep access to the iconic image data and to use information from the primal sketch [65], which is generated from zero crossing of the Gaussian-filtered image followed by a Laplacian operator. In this method the images are described by the interrelation of lines and edges in the primal sketches, producing graphlike structures that can be further processed by recognition algorithms.

Another method for shape analysis is the line adjacency graph (LAG) introduced by Pavlidis [66], where runs of the same gray level in subsequent image raster lines are converted to a special data structure. In these graphs, connected sets of boundary pixels of different objects in the scene are easily obtained. Other procedures describe chains of curves with [67, 68] or without [69] intersection.

Since midlevel vision algorithms operate on large image arrays, mapping them to an architecture for efficient computation is of extreme importance. This mapping is complicated by the fact that the output data are not in image form, but in symbolic form. Also, the process of feature extraction is application-dependent, requiring programmability and flexibility of the supporting architectures.

## 2.3 High-Level Algorithms

The task of high-level image analysis algorithms is to interpret the symbolic data provided by the midlevel algorithms. Typically, this means to locate and recognize objects in a scene. Recognition processes are often required even in the case of simple object measurement, where, for example, proper object edges must be selected and used in measurement computation.

Most general techniques described in this section are extensively analyzed in the nonvision literature. Consequently, we give a brief overview and point to some specific vision-related problems [6, 7, 70].

Classical pattern recognition offers two basic approaches: decision-theoretic [12–14], and syntactic [17, 71]. Their selection depends, among other things, on the application at hand, on the type of the extracted features, and often on the experience and preference of the designer. In the decision-theoretic approach, a pattern is represented as a set of features and the class to which the pattern belongs is decided on the basis of a similarity measure between the feature set (feature vector) of the pattern and the feature representation of the known classes. There are various similarity or distance measurements, such as maximum-likelihood measures and discriminant functions. The basic methodology is to establish training and test data bases and to define a similarity measure. Then the most discriminatory and useful features are selected. After that, the system must be tested on the test data set to estimate the recognition accuracy. The training can be supervised, where we know to which classes the feature vectors of the training set belong, or unsupervised, where the classes are defined as clusters in the feature space. Further, the design of these classifiers can be parametric, where we know the type of the feature probability density distributions and look only for its parameters (e.g., mean and variance for a Gaussian distribution), or nonparametric, where we base our design only on the feature statistics that are experimentally obtained (e.g., histograms). Obviously, classifiers of this type work best with features of type I.

There are many strategies for estimating the eventual recognition performance of the system. Parameters which describe this performance are false negative rate (ratio of missed to total number of samples), false positive or false alarm rate (ratio of erroneously detected objects to total number of samples), and misclassification rate (ratio of incorrectly classified to total number of samples). False positive and false negative rates are not independent and the receiver operating characteristic (ROC) [72, 73] illustrates their relationship. ROC analysis is especially useful in practice, since it can provide an optimum operating point with respect to false negative, false positive, and related costs of those errors. See Figure 10.10.

One strategy for evaluating the performance rates of classifiers is to train on half of the available data, test on the other half, reverse the process, and average the results. If fewer data are available, a viable strategy is to train on all but one sample and test on the remaining object, to repeat the process while always keeping another sample out of the training set, and to average the results. There are many other approaches, and details can be found in Sklansky and Wassel [13].

Since there is no systematic way to derive the optimal preprocessing steps or the optimal feature set, it often happens that results of a first investigation are unsatisfactory. In these cases the designer must improve the previous image analysis steps, invent new features, and improve the preprocessing and sensing; in other words, the intelligent behavior, expertise, and intuition of the designer are still needed to optimize the entire image analysis system.

A disadvantage of the decision-theoretic approach is its difficulty in dealing with a large number of classes and with spatial arrangements which are inherent and crucial in vision. This has led to syntactic approaches, where the pattern is represented as a string, tree, or graph of pattern primitives together with their relations.

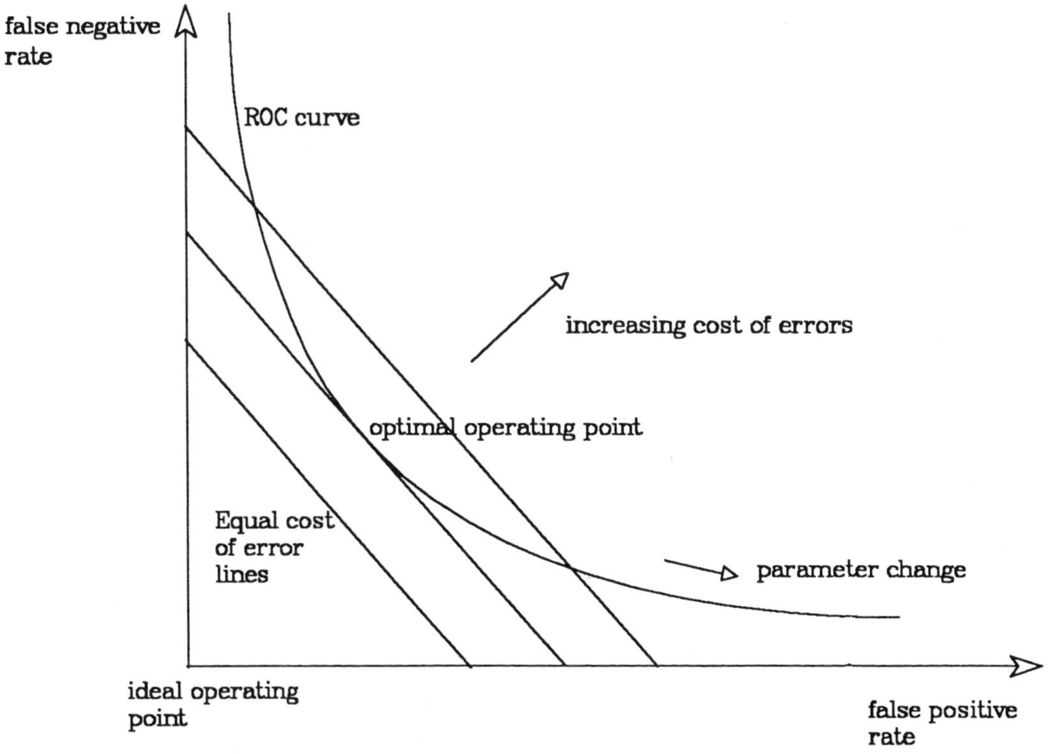


**Figure 10.10** Receiver operating characteristics (ROC) analysis.

The decision making is then transformed to a parsing procedure where certain strings are recognized by a grammar. These classifiers work best with features of type II. The disadvantage of the syntactic approach is its noise sensitivity, which is addressed by using error-correcting graph matching procedures, distance measures between strings, etc.

AI-based techniques [15], most notably expert systems, represent new attempts in solving image analysis problems. While previous techniques belong to a class of numeric and deterministic algorithms, AI techniques involve symbolic computing and nondeterministic algorithms, such as search, sorting, selection, and pattern matching. Therefore, they represent radically new approaches which still need to mature and which require novel architectures. One of the key notions here is knowledge representation. Methods for doing this include rules, semantic nets, frames, and predicate logic [15].

Expert systems are systems that operate using conveniently (i.e., close to natural language) encoded expert knowledge about the problem, usually in the form of IF-THEN-ELSE rules, and general inference procedures (for example, backward chaining, i.e., hypothesis testing, or forward chaining, i.e., concluding from facts) to arrive at a solution. Expert systems offer separation of knowledge and inference procedures, increased flexibility compared to "hard-coded" programs, and a self-explanation capability. However, they have low efficiency and do not incorporate self-learning processes; i.e., any new knowledge to be added to the system stems from knowledge acquisition from a human expert. Other AI techniques, namely heuristic search techniques [15], and formalizations such as semantic networks (labeled directed graphs where nodes represent the objects and links represent the relations among objects) [74] are useful in cases where structural models are matched with extracted features. This is especially the case in 3-D vision. These approaches also work best when operating on type II features. A variety of high-level languages geared toward AI paradigms have been developed, most notably LISP and Prolog.

One of the major bottlenecks in developing an expert systems is knowledge acquisition, i.e., encoding the knowledge into rules. In general, it is never clear whether a set of rules incorporated in a knowledge base is complete and unambiguous, and syntactic and AI-based approaches must be rigorously tested and recognition performance statistics, similar to those for decision-theoretic approaches, obtained. For vision, there is still discussion of whether it is possible to encode visual (i.e., spatial) knowledge using simple IF-THEN-ELSE rules [75]. A new theory of spatial reasoning is definitely needed.

One image analysis task where rule-based expert systems fit well is the final process of matching the results of the object detection and classification procedures with inspection specifications to determine whether an inspected part is good or bad. If the inspection specifications are well defined and the inspection system provides all necessary measurements, the rule-based approach is straightforward. The flexibility of rule-based systems matches well with the fact that inspection specifications can often change. In general, visual inspection applications are often characterized by complex, context-dependent rules that must be applied to the list of defects. A general and efficient rule-based approach for specification verification, applied to disk head inspection, is presented in Petkovic and Hinkle [76].

A "cookbook" of high-level approaches showing which should be used does not exist, and the solution is often application-dependent. In any real application we find a mixture of numeric and symbolic data representations, requiring traditional pattern recognition (often closer to low-level algorithms) and AI-based approaches (especially at the higher level of analysis), as well as a mixture of type I and type II features.

One of the main problems with all recognition systems is to establish confidence in the system's recognition performance on future samples. There is no firm theory, but there are general rules about training and testing strategies for the decision-theoretic approach. For example one rule states that the number of training samples must be much higher (at least 10 times) than the number of features. However, many application developments start with a limited number of training samples, but with reasonably good pattern models, especially in industrial machine vision. For syntactic and AI-based approaches there are no rules, although clearly the training data and/or models should be as representative of the real situation as possible.

For this reason the analysis of scenes with limited variety, as in industrial applications, offers much better hope of success than the analysis of, e.g., outdoor scenes. The former offer well-defined problems, better models, sufficient training data, a smaller number of possible classes, and fewer rules to encode.

Another problem in high-level machine vision is that of mapping algorithms onto specific architectures for their efficient execution. Often, high-level algorithms take a small fraction of the overall execution time, but sometimes their efficiency must be considered. The numeric and deterministic nature of traditional pattern recognition approaches makes them easier to map to special architectures, usually of von Neumann type with possible pipeline or array processors for speedup. Symbolic computing, however, imposes different requirements such as large memories with dynamic allocation, nondeterministic algorithms, and variable sizes of messages [77]. In general, high-level algorithms are application-dependent, imposing demands on programmability and flexibility of the supporting architectures.

## 2.4 Control Strategies in Machine Vision

Given the variety of machine vision algorithms, the natural question to ask is how to integrate them to solve a specific problem. This is addressed by control strategies in vision algorithms [78] and by methodologies for practical development of machine vision applications [73, 79].

If the algorithms are arranged from low level to high level, the control is said to be bottom-up. One typical arrangement of algorithms is preprocessing, segmentation, feature extraction, and recognition, as explained before. Here the control is derived by the designer and explicitly (procedurally) encoded in the analysis program. A problem may occur if the segmentation and feature extraction do not produce satisfactory results; the high-level recognition task therefore fails, and there is no feedback to correct the segmentation process. Still, for a variety of tasks where the scene content is restricted, which is the case in industrial machine vision, this algorithm arrangement, given proper algorithm design and sensing, gives satisfactory results.

The top-down approach starts with a model of the objects that are expected in the image. This model is then decomposed into lower model components and their existence is checked. The problem here is that much unnecessary searching may be done if a model assumption is wrong.

To avoid the problems described above, a combination of bottom-up and top-down approaches can be used. The global features are first extracted in a bottom-up approach and are used to select only the most likely candidate models, and these guide the subsequent top-down analysis. The last two approaches are called model-based vision [80].

Another strategy is feedback processing, where each algorithm is corrected depending on the result of its application to the data. For example, depending on the results of a thresholding operation, the parameters can be corrected by slightly changing them and repeating the operation until satisfactory results are obtained, although it may be difficult to quantify what constitutes "satisfactory" results. Unfortunately, there is no objective quality criterion for most processing steps to provide the necessary feedback data.

An important issue in designing control strategies is whether to implement them in a procedural or declarative way. The procedural way means implementation by explicit encoding of all algorithm steps. This strategy is perhaps the least elegant, but most current working systems are constructed this way. Declarative methods use constructs such as rule-based systems, frames, semantic nets, and blackboard systems [15]. In blackboard systems, the analysis modules read from and write data to the blackboard memory. Each module may be designed to detect and characterize shape, texture, edges, etc. The blackboard is analyzed by other modules, which make the final decision and synchronize the whole process. Although blackboard systems offer good modularity, enabling independent addition of various processing modules, the convergence of the process cannot always be guaranteed. In addition, contradictions may occur if various modules provide conflicting results for the same objects. For interesting applications see Nagao [78] and Nazif and Levine [81]. In most cases, the choice of the proper strategy will depend on the application at hand and the availability of software and hardware support.

We now address the practical issue of how to develop a typical industrial machine vision application. Today, this development takes much time and requires hard-to-find experts. Another task that remains is to speed up the application development and open this technology to nonexperts. The success of a machine vision solution depends on the observation of a few basic rules: 1) careful analysis of application requirements to find out if an automated vision-based solution is feasible; 2) design of the proper sensing and image capture to satisfy further digital processing; 3) investigation of machine vision algorithms, having possible architectures in mind; 4) small- and large-scale feasibility studies that provide accuracy analysis (false negative, false positive, and misclassification statistics on sufficient test data bases, measurement accuracy, and repeatability); and 5) on obtaining satisfactory results, final design of the architecture and the whole system, followed by final system testing. Note that the algorithm development actually resembles feedback processing, where the feedback information is supplied by the designer.

When we say "having possible architectures in mind" we mean that, although the algorithms are not known in advance, from the outset designers should try to exploit those that are likely to be mappable onto a feasible architecture because many machine vision applications require high processing speeds not achieved by general von Neumann architectures. Reversing the application development methodology and imposing a specific architecture on a problem typically produces suboptimal results.

It is important to add that, because of the limited theoretical knowledge about vision, it is usually not possible to analytically prove or derive the final system performance. Therefore, realistic and extensive experimentation is the ultimate judge of system performance. While there are many performance measures relevant to image processing, we still lack the "engineering" methodology for performance measurements in image analysis systems (see for example [73, 82]). The most important parameters are accuracy, repeatability, reliability, execution speed, cost, false negative rate, and false positive rate. Since extensive experimental work must be done to validate machine vision algorithms, proper architectures are also critical for research and development environments. The interrelation between algorithms and architectures is one of the keys to successful application of machine vision to real-life problems. More understanding of vision, a formal theory of how to do this algorithm-architecture mapping, and software and hardware tools to speed up the application development and architecture design process are necessary.

## 3 MACHINE VISION ARCHITECTURES

Under architectures for machine vision we denote the hardware structures for fast execution of image-oriented algorithms. Most machine vision applications involve high-speed processing of a wide variety of algorithms. In some problems, each individual step of the solving methodology must run at video-rate speeds to meet the required performance. In this section, a survey of some machine vision architectures is presented. We analyze relevant characteristics of machine vision algorithms that strongly influence the selection and applicability of a particular architecture.

A common goal of machine vision systems is the derivation of symbolic information from the pictorial data. This task goes beyond the transformations used in image processing problems, as it requires data structures other than the simple iconic representations encountered in low-level image processing. These structures involve linked and circular lists, trees, and other general graphs. The multiplicity of data structures, the variety of operations which are performed on the data, and the real-time requirements make architectures and parallel computing research challenges in the vision field.

Among the problems facing the designer of machine vision architectures is that of selecting and/or building the proper hardware for efficiently implementing image-to-image, image-to-symbols, and purely symbolic operations [83]. Because of the nature of these operations, underlying architectural requirements are rather different. This makes it difficult to use a single architecture to successfully implement these transformations. In particular, we still do not have a formal mechanism for mapping algorithms onto optimal architectures. Some attempts have been made [83–85]. Many factors, such as processor power, granularity, topology, fault tolerance, and cost, must be considered in evaluating the applicability of architectures to machine vision. In addition, relevant algorithm properties (inherent parallelism, memory requirements, etc.) must be taken into account to formalize this matching. In the end, no unequivocal ranking can be achieved without experimental results [83].

As opposed to trying to map algorithms onto architectures, there is another approach which consists of designing-application-specific architectures [86]. With the advances of VLSI technology and related design tools, this seems to be a more and more attractive methodology. VLSI technology also facilitates cost-effective fabrications of complex general-purpose processors and parallel fine-grained machines. Both VLSI trends are expected to yield new architectures in the upcoming years.

Any manipulation of images involves a huge amount of data. Typical image sizes range from 512 × 512 × 8 bits in industrial applications, to 4096 × 4096 × 8 bits for several spectral bands in Landsat satellite images. Obviously, most standard von Neumann architectures are not capable of processing this amount of data at high speeds. An important factor limiting the use of von Neumann architectures is their inability to exploit the high regularity of image data and image operators. While this argument is true for low-level operations on iconic data, it is still a research topic for other representations such as trees and other graphs. These factors have led to active research and development on parallel vision architectures. Great success has been obtained for image-to-image operations, for which a large number of architectures have been built. In this section, we first classify machine vision architectures into certain groups and give some examples of functioning machines. We discuss briefly hardware building blocks and future research trends in the field.

### 3.1 Classification of Architectures for Machine Vision

Machine vision architectures can be classified according to different parameters, depending on whether the emphasis is on system control, processor communication, granularity of the architecture, or flow of data [87–90]. A traditional classification proposed by Flynn [91] gives particular emphasis to control and flow of data, while other authors have considered parameters such as word length and the number of words which are processed in parallel [92].

A common classification involving architecture control is the following.

1. *Single instruction, single data stream (SISD).* This group includes the standard von Neumann processor, where operations are performed sequentially on each data item. Examples include conventional microprocessors (8086, 68000, etc.) and the reduced instruction set computers (RISC). This architecture is not well suited for low- and midlevel vision. It is most appropriately used in high-level algorithms, mainly because of the ease of programmability, wide range of commercial offerings, and low prices. With the improvement in speed of these processors, it is likely that they will continue to be used successfully for high-level vision tasks. Some special architectures are needed, however, for some AI applications and rule-based systems [77, 93–96]. Since computers for AI applications are also related to nonvision problems, we will devote slightly less attention to them.

2. *Single instruction, multiple data stream (SIMD).* (See Figure 10.11.) In these systems, each operation is performed on many data items in parallel. Instructions are broadcast from a common controller to each processor in the architecture. There are several modes of SIMD processing, differing in memory addressing and communication. In general, massively parallel systems where the architecture grain is tailored to the size of the problem (e.g., the image size for low-level operations) require that memory addressing also be accomplished in a SIMD mode. This implies that all processors address the same position in their local memory, this address also being broadcast by the central controller.

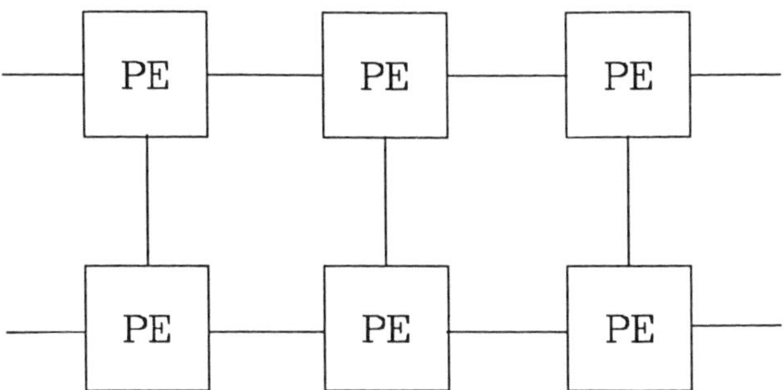


**Figure 10.11** SIMD Architecture: processors arranged in a mesh array.

SIMD communication requires that each processor choose the same physical port to communicate to neighboring processors. This distinction makes certain trivial operations more difficult to execute.

Although SIMD instruction processing simplifies controllability and synchronization for fine-grained systems, SIMD memory addressing greatly complicates the programing of midlevel algorithms and the efficient manipulation of data structures. SIMD memory addressing is needed as a consequence of a trade-off between chip pin count and the number of on-chip processors.

There is another trade-off between the number of processors and processor computing power in the design of architectures. A machine with hundreds of thousands of processors must be composed of bit-serial processing nodes. For up to about 1000 processors integrated in one system, cost considerations allow for 16-bit and even 32-bit processing nodes. The issue of architecture-grain versus computing efficiency of individual processing elements is undergoing new thinking and different proposals are advocated in the image processing community. Former designers of fine arrays today favor coarse-grained systems [97].

Examples of SIMD machines include the Illiac III [98], CLIP [99], MPP [100], GAPP [101], DAP [102], and the Connection Machine [103].

In the image processing community, most SIMD machines have been explored only for image processing applications involving low-level or pixel-oriented transformations. The suitability of this architecture to window-based operators is obvious, and fine-grained mesh array architectures have optimal performance for neighborhood-based operations because the complexity of the algorithms scales only with the window size. Local operations offer an optimal improvement over serial machines. In particular, the ratio asymptotic complexity in a serial machine to number of processors in the array is theoretically exactly the complexity of neighborhood-based operations, i.e., the number of pixels in the window. In practical terms, this statement remains valid only within some bounds. For example, the mesh processors are bit-serial but many local operators need full 8-bit or 16-bit manipulations. In addition, the arithmetic and logic units cannot build a complete logic function with nine inputs; therefore not all nonlinear $3 \times 3$ morphology operators can be computed using only ALUs.

It is important to recognize that these architectures also offer advantages for certain mid- and high-level operations, a point which is often overlooked. Operations such as histogramming or the Hough transform yield significant improvements over serial machines. Even connected component labeling is feasible with low demand on processing element hardware resources (e.g., amount of local memory), as recently shown in Cypher and Sang [104]. Asymptotically, all these midlevel operators have time complexity $O(N)$ for $N \times N$ mesh arrays. In general, much research is ongoing in the area of parallel algorithms for SIMD machines and models. The output of some of this work is reflected in the existence of machines which support parallel versions of LISP and the manipulation of noniconic data structures [103]. The treatment of vision algorithms in different parallel SIMD computers could well be the topic of a complete textbook, but we will not elaborate on these issues any further.

3. *Multiple instruction, single data stream (MISD).* (See Figure 10.12.) These architectures are also called pipelines, because multiple processing stages are cascaded to perform many operations on the same data stream. This terminology concerning "single data" is somewhat misleading because several different streams of data can actually exist in concurrent pipelines. In some cases, the output of a stage can be fed recursively to the same stage again. These multiple-stream machines are usually referred to as "parallel pipelines." "Single data" refers to the unique method of data flow in the architecture, which in this case is in raster or line-by-line mode. In this manner, data flow through each stage of the pipeline where a certain operation is applied to all pixels within a certain window.

The definition of a pipeline reveals its major limitations: algorithms must be constrained to raster processing. Although this is an advantage for several problems, and some operators can be efficiently implemented by processing pixels in a raster format, it is a great disadvantage for many other algorithms which are not pipelinable. In particular, rich data structures arising in most mid- and high-level algorithms cannot be manipulated efficiently.

Pipelines can be classified as homogeneous or heterogeneous. Homogeneous pipelines are those in which all stages are identical; examples are morphology machines [35, 105], Radon transform machines [27], and the WARP processor [106]. Heterogeneous pipelines include stages for performing different specific functions, such as histogramming, convolution boards, and component labeling.

Examples of current machines include the Environmental Research Institute in Michigan (ERIM) Cyto-computer [34, 105], the Carnegie-Mellon University (CMU) WARP processor [106], and the Jet Propulsion Laboratory (JPL) programmable feature extractor [107]. Pipeline architectures actually constitute the majority of commercial offerings because of their compatibility with serial video signal processing, relative ease of programming, and reasonable I/O capabilities. Their disadvantage is the limited speed improvement proportional to the number of pipeline stages. They are well suited to raster-based algorithms that find application in

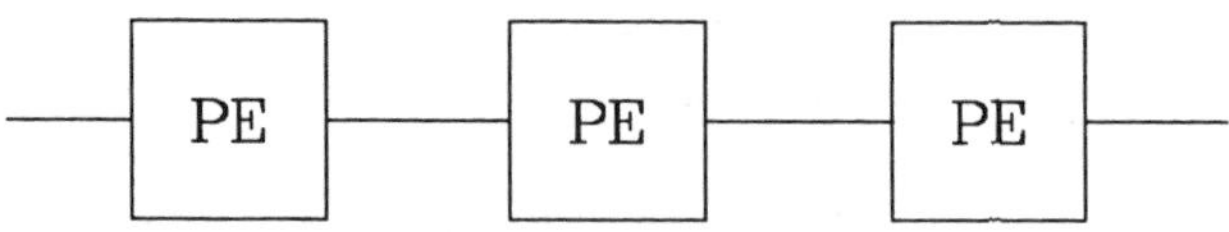


**Figure 10.12** MISD architecture (pipeline)

window-based operations and, in particular, in point-operators. A few midlevel algorithms such as two-pass component labeling and feature extraction [56] can be implemented by processing accessory information and intermediate data structures in conventional general-purpose microprocessors. MISD machines constitute an appropriate approach to many algorithm-specific hardware implementations. The main reasons for this success are their cost effectiveness and simple interconnection, which allows systematic plug-in of new VLSI-based stages that also use raster-format data.

4. *Multiple instruction, multiple data stream (MIMD).* (See Figure 10.13.) These architectures consist of several processors that apply different instructions to different data streams; i.e., they are real multiprocessors. Each processor has its own program memory and instruction-flow control. MIMD architectures potentially offer the highest speed improvement and versatility, but they still have many obstacles to conquer. Programming of these systems is not yet well understood because completely decentralized control presents synchronization problems, which is an area of active research. Nevertheless, commercial systems are appearing: "Cosmic Cube" [108], Flexible Computing Flex [86], the NCUBE machine [109], and specifically for machine vision, the ZMOB [110] and PICAP [111]. These architectures are best suited for high-level operations, where each processor analyzes a certain portion of the data (or image) or matches different sets of rules.

High-level vision algorithms can be of a numeric (traditional pattern recognition) or symbolic (AI-based) nature. Numeric problems are successfully solved using traditional computing machines (usually von Neumann architectures with some number-crunching facilities), but computers for AI are still in the research phase. AI machines pose new requirements including large memories, dynamic allocation, and interactive I/O, in order to support symbolic operations such as search, pattern matching, selection, and sorting. AI architectures are analyzed for these and nonvision algorithms in [77, 96].

Among the large variety of other machine vision architectures (see [88]), systems that allow partitionable processor control are of particular interest. These systems consist of sets of processors operating in SIMD mode under different controllers. The reconfigurable machine PASM, where processors and controllers can be dynamically reconfigured [112], has been proposed and a prototype built. These systems are called multiple SIMD or SIMD-MIMD machines.

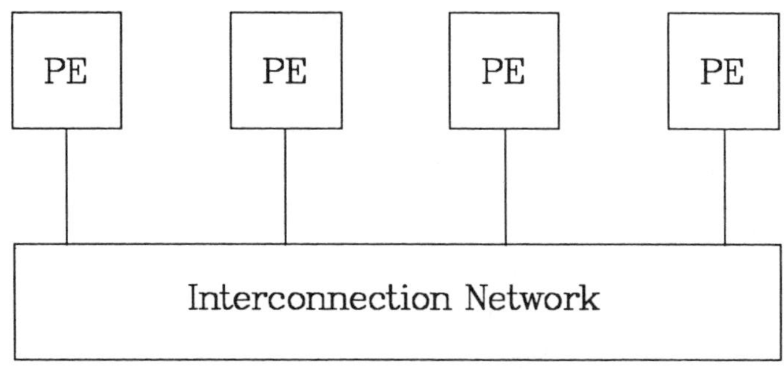


**Figure 10.13** MIMD architecture

Other systems and proposed architectures involve multiple control and different interprocessor communication. Most notable is an architecture suited to pyramid data structures [113]. This approach offers many interesting capabilities, such as ease of internode communication and efficient processing of multiresolution data. This results in simple image processing at different levels of abstraction and/or resolution. These factors are important to consider for both low- and midlevel vision (edge detection, segmentation, feature extraction). Paradigms for pyramid-machine algorithms are further described in Tanimoto [114].

However, pyramid machines which have been attempted in hardware present a number of constraints. First, the fine-grained nature of these architectures restricts the controllability to a SIMD mode of operation. At most, separate controllers could be provided for each layer of the pyramid. On the other hand, the desired property of increasing processor communication bandwidth toward the apex has not yet been realized [115–117]. Studies on pyramids seem to indicate that low-level segmentation could be accomplished only in the presence of augmented pyramids by interleaving the nodes at different levels of the machine structure [118]. These findings appear to be in agreement with the results reported in Reeves [119], where it was shown that the additional hardware of conventional pyramid schemes offers little improvement in performance over planar mesh array architectures. In addition, interleaving pyramids and fat pyramids are more complex than the conventional pyramidal structure and hence pose even more challenges to hardware implementation.

No single type of architecture is the most appropriate when we consider the whole spectrum of operations (low, mid, and high level). Therefore, the best combination depends on the particular machine vision application at hand, cost considerations, speed, and flexibility of the system.

## 3.2 Hardware Building Blocks for Machine Vision Architectures

With the rapid advance of hardware technology and the ever-increasing demand for signal processing power at low cost, there has been a proliferation of hardware building blocks and technologies available to the designer of machine vision systems. These building blocks may be divided into two categories: general-purpose components and special signal processing devices. The first category consists of various bit-slice processors, analog-to-digital (A/D) converters, and semicustom digital devices (gate arrays, programmable array logics (PALs), etc.). The latter category includes special signal processing microprocessors, image processing elements (cascadable convolvers etc.), and special function units (like the Bell Laboratories moment chip [123], the AMI, fast Fourier transform (FFT) chip, and the American Micro Devices (AMD) AM7970 data compression chip).

In addition, improvements in VLSI technology and design tools (silicon compilers and computer-aided design tools) not only will result in repackaging of some older architectures but also will open up many opportunities for mapping application-specific algorithms onto VLSI in a relatively short time. The impact of VLSI is not only in improving the performance/cost ratio but also in opening up completely new avenues to explore in the design of architectures [86, 120–122]. VLSI encourages different approaches to architecture design, favoring concurrency, proximity of memory and processors, regularity in data and control steps, regular and localized connections between processors, simple operations in many processors, minimization of I/O communications, etc. Image processing is well suited for architectures of this type. In fact, the major gains in VLSI-based solutions may be a result of the use of algorithms well-suited to this technology, rather than the increase of the speed and density of circuits.

The proliferation of low-cost and powerful workstations that can be hosts for specific machine vision architectures is helping to spread the use of machine vision systems. They have considerably improved speed and memory capabilities, although technological limits have not yet been reached. This development, in turn, has resulted in a rich variety of vendor offerings of modular machine vision hardware-digitizers, convolvers, display controllers, and FFT boards that are easily attachable to those workstations.

## 3.3 Future Trends in Machine Vision Architectures

We will try to point out some factors that will influence the future development of machine vision architectures and comment on some open issues. There are two lines of future research: VLSI-oriented activities and parallel architectures. VLSI and very high scale integration circuit (VHSIC) technology and tools will have a major impact on improving the cost/performance ratio that is so critical in machine vision applications. Significant progress has been made in manufacturing chips for specific tasks, and some attempts have been made to formalize the mapping of machine vision algorithms onto architectures [83–85, 120]. However, we still lack a formal mechanism for this mapping.

Improved programming tools are also needed, as our ability to put more and more elements on one chip already surpasses our ability to program them. Thus, one of the major needs is novel programming tools for specific VLSI architectures. As these tools become available, there will be greater availability of image/signal processing building blocks of both the specific and more general programmable types.

Powerful vision architectures will probably look more and more similar to general parallel processing machines. On the other hand, vision applications rarely justify the high cost of these systems. However, the availability of parallel computers for research purposes will permit better understanding of new algorithmic paradigms for computer vision.

Many different issues must be considered in parallel computing for machine vision. The problem of finding efficient algorithms for solving key low-, mid-, and high-level vision tasks is receiving considerable public attention. Optimal and efficient algorithms for different models of parallel computing, involving a variety of interconnections and architecture grains, are being sought. The availability of VLSI technology will allow the design of more sophisticated processing nodes and processor interconnections. Ideally, both of these trends—the conceptual understanding of algorithms and new architects' favorite designs—will converge at some point in the near future. Last, but not least, with the advent of more parallel systems, problems involving fault tolerance [124] will have to be resolved.

It is likely that by the end of the decade we will see many new massively parallel systems fully implemented in hardware. Some of them will have a quite general scope and will find applicability in research projects including machine vision. Many other systems will be tailored to solve restricted classes of vision problems. Because of the cost reduction for digital technology, these systems will be widely available for manufacturing automation and other industrial applications.

## ACKNOWLEDGMENTS

We gratefully acknowledge the excellent help of Wayne Niblack.

## REFERENCES

1. Special Issue on Machine Perception, *Computer*, *13*(5): (1980).
2. N. J. Zimmerman, G. J. R. Van Boven, and A. Oosterlink, "Overview of Industrial Vision Systems," Proc. Workshop on Industrial Applications of Image Analysis, Antwerp, pp. 193–231 (October 1983).
3. R. Chin et al. Automated visual inspection: a survey, *IEEE Trans. Pattern Anal. Machine Intell.*, *4*(6): 557–573 (1982).
4. J. L. C. Sanz and A. K. Jain, Machine vision techniques for inspection of printed wiring boards and thick-film circuits, *J. Opt. Soc. Am. 9*: 1465–1482 (1986).
5. J. L. C. Sanz, J. Apffel, W. Sander, and A. K. Jain, Industrial machine vision, *Encyclopedia of Robotics*, Wiley, New York, (in press).
6. K. S. Fu et al., Pattern recognition and computer vision, *IEEE Comput. 17*(10): 274–282 (October 1984).
7. A. Rosenfeld, Image analysis: Problems, progress and prospects, *Pattern Recogn.*, *17*(1): 3–12 (1984).
8. P. Besl and R. Jain, 3-D object recognition, *ACM Comput. Surv.*, *17*(1): 75–145 (March 1985).
9. R. A. Jarvis, A perspective on range finding techniques for computer vision, *IEEE Trans. Patt. Anal. Machine Intell.*, *PAMI*, *5*(2): 122–139 (1983).
10. D. Nitzan, Development of intelligent robots: Achievements and issues, *IEEE Robot. Autom. 1*(1): 2–13 (1985).
11. K. L. Boyer and A. C. Kak, Color-encoded structured light for rapid active imaging, *IEEE Trans. Pattern Anal. and Machine Intell.*, *PAMI-9*(1): 19–28 (1987).
12. R. O. Duda and P. E. Hart, *Pattern Classification and Scene Analysis*, Wiley, New York (1973).
13. J. Sklansky and G. Wassel, *Pattern Classifiers and Trainable Machines*, Springer-Verlag, New York (1981).
14. K. Fukunaga, *Statistical Pattern Recognition*, Wiley, New York (1972).
15. A. Barr and E. Feigenbaum, *The Handbook of Artificial Intelligence*, vol. I, William Kaufman, Los Altos, California (1981).
16. F. Harary, *Graph Theory*, Addison-Wesley, Reading, Massachusetts (1969).
17. K. S. Fu, *Syntactic Pattern Recognition and Applications*, Prentice-Hall, Englewood Cliffs, New Jersey (1982).
18. A. Rosenfeld and A. Kak, *Digital Picture Processing*, 3rd ed., Academic Press, Orlando, Florida (1982).

19. W. Pratt, *Digital Image Processing*, Wiley, New York (1978).
20. D. Ballard, *Computer Vision*, Prentice-Hall, Englewood Cliffs, New Jersey (1982).
21. W. Niblack, *An Introduction to Digital Image Processing*, Prentice-Hall, Englewood Cliffs, New Jersey (1986).
22. K. S. Fu and J. K. Mu, A survey on image segmentation, *Pattern Recogn.*, *13*: 3–16 (1981).
23. E. L. Hall, Almost uniform distribution for computer image enhancement, *IEEE Trans. Comput.*, *C-23*: 207–208 (1974).
24. J. L. C. Sanz, F. Merkle, and K. Y. Wong, Automated Digital Visual Inspection with Darkfield microscopy, *J. Op. Soc. Am.* *2*(9): 1857–1862 (1985).
25. J. L. C. Sanz and E. B. Hinkle, Computing projections of digital images in pipeline architectures, *IEEE Trans. Acoust. Speech Signal Process.*, *55*(2): 198–207 (1987).
26. J. L. C. Sanz, I. Dinstein, and D. Petkovic, Computing multicolor polygonal masks in pipeline architectures and its application to visual inspection, *Commun. ACM*, *30*(4): 318–329 (1987).
27. E. B. Hinkle, J. L. C. Sanz, A. K. Jain, and D. Petkovic, PPPE: A new life for projection based image processing, *J. Parallel Distributed Comput.*, *30*(4): 45–78 (February 1987).
28. J. L. C. Sanz, E. B. Hinkle, and A. K. Jain, *Radon and Projection Transform-Based Machine Vision: Algorithms, Architectures and Industrial Applications*, Springer-Verlag, New York, (in press).
29. D. Munson, F. D. O'Brien, and K. W. Jenkins, A tomographic formulation of spot-light synthetic aperture radar, *Proc. IEEE*, *71*: 917–925 (1983).
30. D. Munson and J. L. C. Sanz, Image reconstruction from frequency-offset Fourier data, *Proc. IEEE*, *72*: 661–669 (1984).
31. J. Serra, *Image Analysis and Mathematical Morphology*, Academic Press, New York (1982).
32. Special Section on Mathematical Morphology, Computer Vision, Graphics and Image Processing, *35*(3): (1986).
33. S. R. Sternberg, "Grayscale Morphology," *Comput. Vision, Graphics, Image Process.*, *35*: 333–335 (1986).
34. R. M. Lougheed and D. L. McCubberey, "The Cytocomputer: A Practical Pipelined Image Processor," Proc. 7th Ann. Int. Symp. Computer Architecture, La Boule, France pp. 271–277, May 6–8 (1980).
35. M. J. Kimmel, R. S. Jaffe, J. R. Mandeville, and M. A. Lavin, "MITE: Morphic Image Transform Engine: An Architecture for Reconfigurable Pipelines of Neighborhood Processors," IEEE Workshop on Computer Architecture for Pattern Analysis and Image Database Management, Miami Beach, Florida (1985).
36. S. R. Sternberg, "A Morphological Approach to Finished Surface Inspection," IEEE Proc. Int. Conf. on Robotics and Automation, IEEE, New York (1985).
37. R. M. Haralick and L. B. Shapiro, Survey of image segmentation techniques, *Comput. Vision, Graphics, Image Process.*, *29*: 100–132 (1985).
38. R. Pavlidis, "A Critical Survey of Image Analysis Methods," Proc. Int. Conf. Pattern Recognition, Paris (October 1986).

39. M. Lineberry, Image segmentation by edge tracing, *SPIE Proceedings, 359*: 361–368 (1982).
40. J. M. S. Prewitt, Object enhancement and extraction *Picture Processing and Psychopictures* (B. S. Lipkin and A. Rosenfeld, eds.), Academic Press, New York, 100–135 (1970).
41. J. L. Muerle and D. C. Allen, Experimental evaluation of techniques for automatic segmentation of objects in a complex scene, *Pictorial Pattern Recognition* (C. C. Cheng et al., eds.), Thompson, Washington, D.C., 3–13, (1968).
42. J. N. Gupta and P. A. Wintz, "Computer Processing Algorithms for Locating Boundaries in Digital Pictures," Proc. Int. Joint Conf. Pattern Recognition, Copenhagen, Denmark, 155–156 (1974).
43. A. Klinger and C. R. Dyer, Experiments on picture representation using regular decomposition, *Comput. Graphics Image Process., 4*: 68–105 (1975).
44. N. Ahuja, A. Rosenfeld, and R. Haralick, Neighbor gray levels as features in Pixel classification, *Pattern Recogn., 12*: 251–260 (1980).
45. W. E. Blanz and E. R. Reinhardt, "General Approach to Image Segmentation," Proc. 6th Int. Conf. Pattern Recognition, Munich, 188–191 (1982).
46. W. E. Blanz, J. L. C. Sanz, and E. B. Hinkle, Image analysis methods for visual inspection of solder balls integrated circuit manufacturing, *IEEE Trans. Robot. Autom.* (in press).
47. R. M. Haralick and I. Dinstein, A spatial clustering procedure for multi-image data, *IEEE Trans. Circuits Syst., CAS-22*: 440–450 (1975).
48. J. N. Gupta and P. A. Wintz, A boundary finding algorithm and its applications, *IEEE Trans. Circuits Syst., CAS-22*: 351–362 (1975).
49. W. Doyle, Operations useful for similarity-invariant pattern recognition, *ACM, 9*: 259–267 (1962).
50. J. M. S. Prewitt and M. L. Mendelsohn, The analysis of cell images, *Trans. N. Y. Acad. Sci. 128*: 1035–1053 (1966).
51. W. E. Blanz, J. L. C. Sanz, and D. Petkovic, "Control-Free Low Level Image Segmentation: Theory, Architecture, and Experimentation," Proc. 1st Int. Conf. Computer Vision, London, (July 1987).
52. W. E. Blanz, Feature Selection and Polynomial Classifiers for Industrial Decision Analysis, IBM Research Report RJ 5242 (1986).
53. A. K. Agraval and A. V. Kulkarni, A sequential approach to the extraction of shape features, *Computer Graphics Image Process., 6*: 538–557 (1977).
54. A. Rosenfeld and J. Pfaltz, Sequential operations in digital picture processing, *Assoc. Comput. Mach. 14*(4): 471–494 (1966).
55. F. Veillon, One pass computation of morphological and geometrical properties of objects in digital pictures, *Signal Process. 1*: 175–189 (1979).
56. D. Petkovic, J. L. C. Sanz, K. Mohiuddin, M. D. Flickner, E. B. Hinkle, C. Cox, and K. Y. Wong, "An Experimental System for Disk Head Inspection," Proc. 8th Int. Conf. Pattern Recognition, Paris (October 1986).
57. D. Petkovic and K. Mohiuddin, "Combining Component Features from Multiple Frame Images,"' IEEE Workshop on Computer Architecture for Pattern Analysis and Image Database Management," Miami Beach, Florida (November 1985).
58. D. Petkovic, J. L. C. Sanz, and I. Dinstein, "On Application of Multi-Colored Polygonal Masks and Their Computation in Image Processing Pipeline Ar-

chitectures," IEEE Workshop on Computer Architecture for Pattern Analysis and Image Database Management, Miami Beach, Florida (November 1985).

59. L. Van Gool, P. Dewaele, and A. Osterlink, Texture analysis anno 1983, *Computer Vision Graphics Image Process.*, *29*: 336–357 (1985).
60. E. R. Reinhardt, W. E. Blanz, R. Erhardt, W. Greiner, W. Kringler, R. Lenz, I. Schimpf, W. Schlipf, P. Schwarzmann, G. Straessle, and W. H. Bloss, "Automated Classification of Cytological Specimen Based on Multistage Pattern Recognition," Proc. 6th Int. Conf. Pattern Recognition, Munich (October 19–22, 1982).
61. R. O. Duda and P. E. Hart, Use of the Hough transform to detect lines and curves in pictures, *Commun. ACM*, *15*(1): 11–15 (1972).
62. Special Issue on Shape Analysis, *IEEE Trans. Pattern Anal. Machine Intell.*, (January 1986).
63. G. H. Granlund, Fourier preprocessing for hand printed character recognition, *IEEE Trans. Comput.*, *C-21*: 195–201 (1972).
64. C. T. Zahn and R. Z. Roskies, Fourier descriptors for plane closed curves, *IEEE Trans. Comput.*, *C-21*: 269–281 (1972).
65. D. Marr, Visual information processing: The structure and creation of visual representations, *Philos. Trans. R. Soc. London Ser.*, *B-290*: 199–218 (1980).
66. T. Pavlidis, Segmentation of pictures and maps through functional approximation, *Comput. Graphics Image Process.*, *1*: 360–372 (1972).
67. R. L. T. Cederberg, Chain-link coding and segmentation for raster scan devices, *Comput. Graphics Image Process.*, *10*: 224–234 (1979).
68. D. L. Milgram, Constructing trees for region description, *Computer Graphics and Image Process.*, *11*: 88–99 (1979).
69. P. T. Speck, "Automated Recognition of Line Structures on Noisy Raster Images Applied to Electron Micrographs of DNA," Proc. 5th Int. Conf. Pattern Recognition, Miami, Florida, 604–609 (1980).
70. A. Rosenfeld, Image pattern recognition, *Proc. IEEE*, *69*(5): 596–605 (1981).
71. Special Issue on Advances in Syntactic Pattern Recognition, *Pattern Recogn.*, *19*(4): 249–342 (1986).
72. C. E. Metz, Applications of ROC analysis in diagnostic image evaluation, *The Physics of Medical Imaging: Recording Systems, Measurements and Techniques* (A. G. Haus, ed.), American Association of Physicists in Medicine, New York, 546–572 (1979).
73. J. A. Swets and R. M. Pickett, *Evaluation of Diagnostic Devices in Clinical Medicine*, Academic Press, New York, (1982).
74. D. Norman and D. Rumlhart, *Exploration in Cognition*, Freeman, San Francisco (1975).
75. A. Rosenfeld, Expert vision systems: Some issues, *Comput. Vision Graphics Image Process.*, *34*: 99–117 (1986).
76. D. Petkovic and E. B. Hinkle, A rule-based system for verifying engineering specifications in industrial visual inspection applications, *IEEE Trans. Anal. Machine Intell.*, *PAMI-9*(2): 306–311 (March 1987).
77. K. Hwang, J. Ghish, and R. Chowkwanyun, Computer architectures for artificial intelligence processing, *IEEE Comput.* *20*(1): 19–27 (January 1987).
78. M. Nagao, Control strategies in pattern analysis, *Pattern Recogn.*, *17*(1): 45–56 (1984).

79. B. G. Batchelor, "A Laboratory-Based Approach for Designing Automated Inspection Systems," IEEE Workshop on Industrial Applications of Machine Vision, pp. 80–86, May 3–5 (1982).
80. T. O. Binford, Survey of model-based vision, *Int. J. Robot. Res.*, *1*(1): 18–64 (1982).
81. A. Nazif and M. Levine, Low level image segmentation: an expert system, *IEEE Trans. Pattern Recogn. PAMI-6*: 555–557 (1984).
82. I. E. Abdou and N. J. Dusaussoy, "Survey of Image Quality Measurements," IEEE-ACM Joint Conference, Dallas, Texas (November 1986).
83. T. J. Fountain, "Array Architectures for Iconic and Symbolic Image Processing," Proc. 8th Int. Conf. Pattern Recognition, Paris, (October 27–31, 1986).
84. L. Jamieson, H. J. Siegel, E. J. Delp, and A. Wonston, "The Mapping of Parallel Algorithms to Reconfigurable Parallel Architectures," ARO Workshop on Future Directions in Computer Architecture and Software, Charleston, South Carolina, (May 1986).
85. S. Yalamancili and J. K. Aggarwal, Analysis of a model for parallel image processing, *Pattern Recogn.*, *18*(1): 1–16 (1985).
86. P. Ruetz and R. Brodersen, "A Custom Chip Set for Real-Time Image Processing," ICASSP Conf., San Jose, California, (1986).
87. T. J. Fountain, A survey of bit-serial processor circuits, *Computing Structures for Image Processing*, (M. J. B. Duff, ed.), Academic Press, London (1983).
88. P. E. Danielson and S. Levialdi, Computer architectures for pictorial information processing, *IEEE Comput. 19*(11): 53–67 (November 1981).
89. A. P. Reeves, Survey: Parallel computer architectures for image processing, *Comput. Vision Graphics Image Process.*, *25*: 68–88 (1984).
90. K. Hwang, (guest ed.), Computer architectures for image processing, *IEEE Comput. 16*(1): 10–12 (January 1983).
91. M. Flynn, Very high computing systems, *Proc. IEEE*, *54*: 1901–1909 (1966).
92. T. Feng, "Some Characteristics of Associative-Parallel Processing," Proc. Sagamore Computer Conf., Syracuse University, pp. 5–16 (1972).
93. D. Kibler and J. Conery, "Parallelism in AI Programs," 9th Int. Joint Conf. Artificial Intelligence, Los Angeles (1985).
94. H. Tanaka, "A parallel inference machine," *IEEE Comput.*, *19*(5): 48–54 (1986).
95. R. Douglass, A qualitative assessment of parallelism in expert systems, *IEEE Software 2*(3): 70–81 (May 1985).
96. B. Wah and G. Li, "Computers for Artificial Intelligence Applications," IEEE Computer Society Press (May 1986).
97. T. J. Fountain, Plans for the CLIP7 chip, *Integrated Technology for Parallel Image Processing*, (S. Levialdi, ed.), Academic Press, New York, pp. 199–214 (1985).
98. B. H. McCormick, The Illinois pattern recognition computer—ILLIAC III, *IEEE Trans. Comput. EC-12*(5): 791–813 (December 1963).

99. M. J. B. Duff, "CLIP4: A Large Scale Integrated Circuit Array Parallel Processor," 3rd Int. Joint Conf. Pattern Recognition, Coronado, California, pp. 728–732 (1976).
100. L. W. Fung, A massively parallel processing computer, *High-Speed Computer and Algorithm Organization* (D. J. Kuck et al., eds.), Academic Press, New York, pp. 203–204 (1977).
101. T. Davis, Systolic array chip matches the pace of high-speed processing, *Electron. Des. 32*(2): 207–218 (October 31, 1984).
102. D. J. Hunt, The ICL DAP and its application for image processing, *Languages and Architectures for Image Processing*, (M. J. B. Duff and S. Levialdi, eds.) Academic Press, London, (1981).
103. D. Hillis, *The Connection Machine*, MIT Press, Cambridge, Massachusetts (1985).
104. R. Cypher and J. L. C. Sanz, SIMD Mesh Array Algorithms for Image Component Labeling, IBM Technical Report, Almaden Research Center, San Jose, California (February 1987).
105. S. R. Sternberg, "Parallel Architectures for Image Processing," 3rd Int. IEEE COMPSAC, Chicago, pp. 712–717 (1979).
106. T. Gross, H. T. Kung, M. Lam, and J. Webb, "Warp as a Machine for Low Level Vision," Proc. IEEE Int. Conf. Robotics and Automation, St. Louis, Missouri (March 1985).
107. D. B. Genery and B. Wilcox, "A Pipelined Processor for Low Level Vision," Proc. IEEE CVPR Conf., San Francisco, pp. 608–613, June 19–23 (1985).
108. C. Seitz, The cosmic cube, *Commun. ACM, 28*(1): 22–33 (1985).
109. NCUBE Corp. Product Report, NCUBE Corp. Headquarters, Beaverton, Oregon (1986).
110. C. Rieger, "ZMOB: Doing It in Parallel," IEEE Comput. Soc. Workshop on Computer Architecture for Pattern Analysis and Image Database Management, Hot Springs, Virginia, pp. 119–124 (November 11–13, 1981).
111. D. Antonson et al. "PICAP—A System Approach to Image Processing," IEEE Comput. Soc. Workshop on Computer Architecture for Pattern Analysis and Image Database Management, Hot Springs, Virginia (November 11–13, 1981).
112. H. J. Siegel et al., PASM: A partitionable SIMD/MIMD system for image processing and pattern recognition, *IEEE Trans. Comput., C-30*(12): 934–949 (1981).
113. L. Uhr, Pyramid multi-computer structures, and augmented pyramids, *Computing Structures for Image Processing*, (M. J. B. Duff, ed.), Academic Press, London (1983).
114. S. Tanimoto, Paradigms for pyramid machine algorithms, *Pyramidal Systems for Computer Vision*, (V. Cantoni and S. Levialdi, eds.), NATO Series, vol. 25, Springer-Verlag, New York, pp. 173–194 (1987).
115. V. Cantoni, I. P. Hierarchical systems: Architectural features, *Pyramidal Systems for Computer Vision*, (V. Cantoni and S. Levialdi, eds.), NATO Series, vol. 25, Springer-Verlag, New York, pp. 21–39 (1987).
116. A. Merigot, P. Clermonst, J. Mehat, F. Devos, and B. Zavidovique, A pyramidal system for image processing, *Pyramidal Systems for Computer Vision*

(V. Cantoni and S. Levialdi, eds.), NATO Series, vol. 24, Springer-Verlag, New York, pp. 109–124 (1987).

117. D. Schaefer and P. Ho, Counting on the GAM pyramid, *Pyramidal Systems for Computer Vision*, (V. Cantoni and S. Levialdi, eds.), NATO Series, vol. 25, Springer-Verlag, New York, pp. 125–131 (1987).

118. W. Grosky and R. Jain, A pyramid-based approach to segmentation applied to region matching, *IEEE Trans. Pattern Anal. Machine Intell., PAMI-8*(5): 639–650 (1986).

119. A. P. Reeves, Pyramid algorithms on processor arrays, *Pyramidal Systems for Computer Vision*, (V. Cantoni and S. Levialdi, eds.) NATO Series, vol. 25, Springer-Verlag, New York, pp. 195–213 (1987).

120. R. J. Offen, ed., *VLSI Image Processing*, McGraw-Hill, New York (1985).

121. Proc. IEEE Comput. Soc. Workshop on Computer Architecture for Pattern Analysis and Image Data Base Management, Miami Beach, Florida, (November 1985).

122. H. W. Carter, Computer-aided design of integrated circuits, *Computer 19*(4): 19–36 (April 1986).

123. M. Hatamian, A real-time two-dimensional moment generating algorithm and its single chip implementation, *IEEE Trans. Acoust. Speech, Signal Process., ASSP-34*(3): 546–552 (1986).

124. A. P. Reeves, Fault tolerance in highly parallel mesh connected processors, *Computing Structures for Image Processing*, (M. J. B. Duff, ed.), Academic Press, London (1983).

125. N. Ayache, O. D. Faugeras, HYPER: A new approach for the recognition and positioning of two-dimensional objects, *IEEE Trans. Pattern Anal. Machine Intell., PAMI-8*(1): 54–55 (1986).

126. V. Cantoni, C. Guerra, and S. Levialdi, Towards an evaluation of an image processing system, *Computing Structures for Image Processing*, (M. J. B. Duff, ed.), Academic Press, London (1983).

127. B. A. Wandell, The synthesis and analysis of color images, *IEEE Trans. Pattern Anal. Machine Intell., PAMI-9*(1): 2–13 (1987).

# 11

# Image Processing in Robotics

GIOVANNI GARIBOTTO Central Research Department, ELSAG S.p.A., Genoa-Sestri Ponente, Italy

## 1 INTRODUCTION

Image processing represents an essential tool in providing industrial robotic systems with the fundamental sense of vision. A great deal of research effort has been spent to reach such an ambitious goal. However, because of the wide spectrum of problems encountered in practice, most of the available solutions are still heavily conditioned by application constraints. In the following sections a synthetic classification of industrial applications is used in an attempt to stress the most significant trends and processing needs in the industrial field.

In the past decade, computer vision has quickly evolved as a field of research, with contributions from digital signal processing, computer science, and artificial intelligence. In fact, there has been increasing interest in the use of flexible and autonomous systems to perform complex tasks of industrial object manipulation with minimal external supervision and control.

The full process of scene analysis can be schematically divided into three different processing levels. Low-level techniques are mainly data-driven and provide the basic preprocessing results (noise removal, contrast enhancement, smoothing, etc.) as well as the extraction of relevant features in the scene. Intermediate levels are supposed to take advantage of some a priori knowledge (mainly geometric or statistical) to perform preliminary segmentation and image analysis (shape detection and recognition) to obtain three-dimensional (3-D) maps and object representations. High-level methods should perform scene interpretation using different kinds of knowledge, both generic and specific, related to the particular task at hand (knowledge-based systems).

Image processing tools are essential for low and intermediate levels, and some of the most commonly used algorithms for linear, nonlinear, and adaptive filtering will be reviewed in Section 3. Section 4 is intended to refer to some specific subtasks of computer vision which make extensive use of image processing tools, as in edge detection, region segmentation, stereo correspondence, and motion field estimation. The problem of image analysis and recognition will be discussed briefly with some reference to the literature, since this topic is outside the scope of this chapter. Finally, some critical comments are given to underline the main trends of research and the most promising areas of development in the industrial field.

## 2 SHORT REVIEW OF ROBOTIC APPLICATIONS

Machine vision systems are widely used in many different fields. In the following only industrial robotic applications will be considered, even though most of the techniques presented are also widely used in the analysis of aerial scenes, in computer-assisted medical image diagnosis, etc.

A major area of industrial application is in visual inspection of manufactured parts, from very simple tasks of completeness checking to more complex detection of small defects, size and shape measurements, etc.

Industrial automation has proved quite efficient both in reducing repetitive human tasks and in achieving reliable quantitative measurements at high speed. Most of the currently implemented solutions are often simplified by the use of controlled lighting conditions, either to obtain very high contrast pictures (backlighting) or to emphasize the visibility of surface defects. A more general and flexible system should be able to handle both 2-D and 3-D scenes, variable contrast conditions, and color and texture properties with high efficiency in terms of processing power.

Industrial applications can be roughly classified into three main categories: inspection, robot manipulation, and mobile robot guidance and control.

### 2.1 Visual Inspection

Many machine vision applications are in the inspection of integrated circuits and electronic printed circuit boards, usually by comparing the recorded images against reference defect-free samples. Major problems are due to differences in alignment and orientation as well as to size and scale factors. A simple and widely used approach [1] consists of determining object orientation by analysis of the histogram of the edges detected in the image. 3-D vision tasks are becoming more and more important in the measurement of surfaces (especially in the automobile industry) as an alternative to mechanical techniques using contacting probes.

### 2.2 Robot Manipulator

Another important area of application is the handling of industrial pieces by a robot manipulator either to feed a production line or to perform assembly tasks. An often cited vision task consists of the identification, recognition, and grasping of industrial parts from a heap or a bin of other similar or different objects [2]. The main difficulties in this case are encountered in the recognition on the basis of partial information only, due to overlapping and shadow effects. Many simplifications are currently used in practice, such as marking objects with patches of different shapes or colors [3].

### 2.3 Autonomous Moving Robot

Since the beginning of the 1980s there has been increasing activity in the development of mobile vehicles able to move almost autonomously in selected environments. Besides their military applications, these machines should be used to accomplish particular tasks in hostile environments (fire fighting and nuclear power plants) and to give some autonomy to flexible loading/unloading in factory automation.

Future applications may also be oriented toward performing more conventional tasks, such as office cleaning or document delivery, but it is still questionable how necessary and cost effective mobile robots will be in practical situations.

An active area of research is intended to provide a mobile robot with vision capabilities for autonomous guidance and navigation, where the model of the route and the surrounding environment are reasonably predictable. Traffic monitoring is another area of increasing research and application. Both situations rely heavily on monocular viewing and geometric perspective constraints to obtain the required 3-D information from the scene.

## 3 FUNDAMENTAL TOOLS OF IMAGE PROCESSING

Image processing techniques are currently used in a very broad range of applications, such as telecommunications, geophysics, medical imaging, and industrial robotics. Significant progress has been achieved in technology for image acquisition, data storage, and parallel processing as well as in the development of advanced algorithms, strongly reflecting the variety of these applications. The main objective of this section is to provide a brief review of basic digital image processing techniques that are frequently used in robot vision applications.

### 3.1 Histogram Processing

Gray-scale modification is the simplest image processing technique used to modify the image dynamic range and contrast. A suitable transformation $l_o = T(l_i)$ relates an input level $l_i$ to an output gray level $l_o$. Since we are dealing with discrete intensity values $(l_i, l_o)$, such a relation can be implemented by use of a lookup table working at TV rate on the output video signal of any available display unit.

Beside its use to compensate for nonlinearities of the video system, histogram processing is often used in the automatic selection of image thresholds and in contrast enhancement to emphasize the detection of low-contrast details (as shown in Figure 11.1). The histogram of an image is defined as the frequency of occurrence of its gray levels $l$, as shown in Figure 11.2.

Adaptive thresholding has proved to be profitable when lighting conditions are appropriate and the observed scene is particularly simple (a few objects with an

**Figure 11.1** Image of an industrial object (above) and result of contrast enhancement (below).

almost uniform gray level). Otherwise, the binary image obtained may exhibit irregular and/or wrong edge contours, and more accurate segmentation procedures must be used, as shown in Section 4.3.

In the histogram of Figure 11.2 most of the image intensities are clustered around a few values and the available dynamic range (8 bits) is not optimally used, for display purposes.

Histogram equalization is used to obtain an almost flat distribution of levels to maximize the average information per pixel. To achieve this result, as well as any other desired histogram transformation, it is necessary first to compute the cumulative function $C(l)$ through the following recursive relation:

$$C(l) = \sum_{k=0}^{l} h(k) = C(l-1) + h(l) \tag{11.1}$$

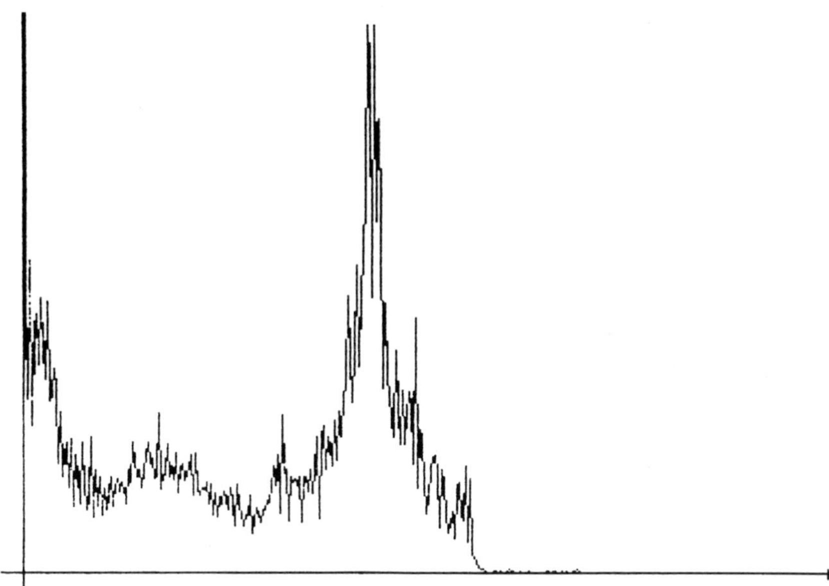

**Figure 11.2** Gray level histogram of image in Figure 11.1a.

The output gray level is obtained as

$$l_o = l_{\min} + \frac{(l_{\max} - l_{\min})C(l_i)}{\sum_{k=0}^{N} h(k)} \tag{11.2}$$

where $l_{\min}$ and $l_{\max}$ allow control of the dynamic range of the processed image, which is quantized over $N = 2^b$ levels. Unfortunately, the discrete nature of the available data prevents an exact implementation, since a few gray levels in the input image are usually quite densely populated and there is no unique way to split them between neighboring levels to obtain the desired output histogram.

In any case, image histogram equalization would be a valid solution only under the assumption of a stationary process, where the computed global histogram exhibits the same behavior as local histograms for all subimages greater than a minimum size. This is not the case for most pictorial images and in particular for industrial scenes, so local properties must be included in the process.

Ideally, it would be necessary to perform histogram equalization within a local window centered on the current pixel, with unacceptable computational costs. Suboptimal solutions are obtained by processing disjoint picture windows, but the result is often degraded by an unsatisfactory mosaicking effect. Two more efficient techniques are available to solve this problem. In the first case, bilinear interpolation is performed on the cumulative transfer functions obtained from nonoverlapping neighboring windows [4]. In the second case [5], adaptive equalization is performed on the difference image component according to a Gaussian model for the local histogram. An example of this processing is shown in Figure 11.1, where some local features in the scene are made more visible. This filtering technique is particularly promising in feature enhancement for printed circuit board analysis and the detection of faults or defects in automatic visual inspection.

## 3.2 Digital Linear Filtering

Many textbooks are concerned with digital linear filtering in one and/or more dimensions [6], with particular emphasis on space-invariant linear systems, which are fully described by their impulse response (and/or frequency response). Multidimensional systems have been a major field of research in the past decade, with the most significant contributions made in digital television geophysics and acoustic imaging. In the following the main properties of 2-D space-invariant linear filters are briefly discussed.

Let us suppose $h(m,n)$ to be the impulse response of the filter and $H(u,v)$ (complex frequency response) to be its Fourier transform. Using the same notation for an output signal $x(m,n)$ $[X(u,v)]$ and an output signal $y(m,n)$ $[Y(u,v)]$, the input/output (I/O) relation of the linear filter $h(m,n)$ can be written as

$$y(m,n) = x(m,n) \otimes h(m,n) = \sum_k \sum_l h(k,l)x(m-k,n-l) \tag{11.3}$$

which represents 2-D convolution, and the summation limits depend on the extent of the impulse response (either finite or infinite). In the 2-D frequency domain $(u,v)$ this convolution corresponds to the product of the complex functions

$$Y(u,v) = X(u,v)H(u,v) \tag{11.4}$$

where the filter mask $H(u,v)$ can affect both the amplitude and phase of the signal.

A 2-D filter is called "separable" when its impulse response $h(m,n)$ can be expressed as a product of two 1-D sequences, say $f(m)$ and $g(m)$:

$$h(m,n) = f(m)g(n) \tag{11.5}$$

This filter implementation is obtained by two 1-D convolutions along the two orthogonal axes $(m,n)$ in a sequential order (it is irrelevant which operates first). In the edge detection section this favorable property will be used extensively to achieve significant savings in the computation.

Another relevant classification of 2-D systems is according to the extent of their impulse response, so that we can have finite impulse response (FIR) filters if $h(m,n)$ is nonzero only within a finite support $(-M \leq m \leq M, -N \leq n \leq N)$. Otherwise the linear system is called an infinite impulse response (IIR), and its frequency response can be expressed as the ratio of two functions:

$$H(u,v) = \frac{A(u,v)}{B(u,v)} \tag{11.6}$$

Using the same notation as before, the I/O relation is given by the difference equation

$$\sum_k \sum_l b(k,l)y(m-k, n-l) = \sum_r \sum_s a(r,s)x(m-r, n-s) \tag{11.7}$$

The region of support for the coefficients $[a(\ ), b(\ )]$ determines an additional level of classification (first quadrant, symmetric or asymmetric half-planes as shown in Figure 11.3), provided that stability constraints are satisfied.

In fact, a linear system is stable (bounded input/bounded output) iff the impulse response is absolutely summable:

$$\sum_{m=-\infty}^{\infty} \sum_{n=-\infty}^{\infty} |h(m,n)| < \infty \tag{11.8}$$

A thorough analysis of multidimensional system stability may be found in [7] and efficient mapping techniques have been proposed in many papers to perform approximate tests on different domains, such as looking for the continuity of the phase function, through analysis of the cepstrum, etc. Moreover, some of these mapping techniques can be used to deal with general masks such as asymmetric half-plane filters [8], which is not possible using an algebraic approach.

Reliable optimization techniques are also available in 2-D filter design, with stability constraints, according to arbitrarily shaped frequency functions using either nonrecursive or recursive structures [9].

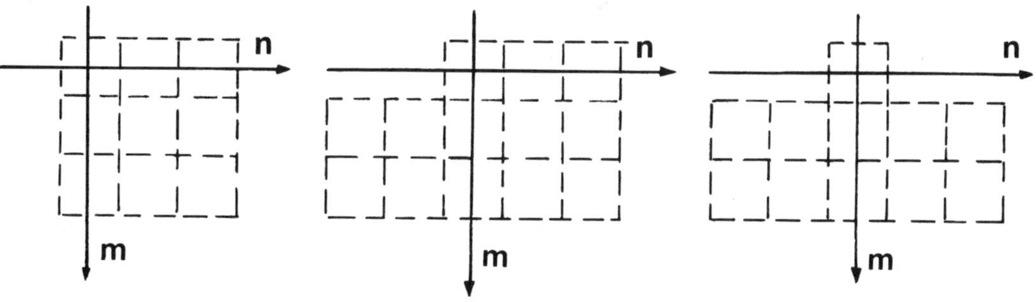


**Figure 11.3** Regions of support for recursive 2-D filter masks: (a) first quadrant; (b) asymmetric half-plane; (c) symmetric half-plane.

Linear filtering has also been proposed for some 3-D applications: in geophysical wave migration, in smoothing biomedical tomograms and cerebral structures [10], and in motion analysis using differential techniques.

As a general comment, digital linear filtering has proved to be more appropriate in problems of physical science that can be described by partial derivative models (inverse scattering, wave propagation, optimal filtering for noise removal). On the other hand, most image analysis problems require ad hoc solutions, including nonlinearities, space-variant properties, and some heuristic constraints (as in edge and line detection), with poor requirements for highly flexible and general 2-D filter design techniques.

In fact, one of the most common specifications in computer vision involves Gaussian filtering and derivatives of Gaussian functions (noise smoothing and edge detection). For that reason these filter specifications will be discussed in more details in the following.

A circularly symmetric 2-D Gaussian function

$$G(x, y, \sigma) = k^2 \exp\left(-\frac{x^2 + y^2}{2\sigma^2}\right) \qquad k = \frac{1}{\sqrt{2\pi}\sigma} \tag{11.9}$$

is separable and can be written as

$$G(x, y, \sigma) = G(x, \sigma)G(y, \sigma) \tag{11.10}$$

Henceforth, 2-D Gaussian convolutions can be implemented by separable 1-D filters which are required to approximate the symmetric 1-D impulse response

$$G(x, \sigma) = k \exp\left(-\frac{x^2}{2\sigma^2}\right) \tag{11.11}$$

When using an FIR implementation, the filter length $N$ is usually selected so that it includes most of the Gaussian signal (i.e., 99%). This means that the filter size depends on the standard deviation $\sigma$ as well as the number of multiplications per sample.

Recursive implementation can be performed with a lower number of multiplications and the filter design can be performed using well-known nonlinear optimization methods [6].

Of course, because of causal constraints, the impulse response $h(n)$ is defined only for positive samples $n \geq 0$.

To achieve a zero-phase symmetric response, either a cascade or a parallel implementation should be used, and the resulting frequency response (real function) will be the square of the amplitude or twice that value, respectively.

The major drawback of this approach is the necessity to design a new set of filter coefficients for any of the selected values of the Gaussian standard deviation.

A parametric solution is obtained using the following scheme [11]. First the target Gaussian impulse response is approximated by an exponentially damped cosine function:

$$F(x) = \exp(-\alpha x)\cos(\omega x), \qquad x \geq 0 \tag{11.12}$$

Using standard $Z$-transform methods, the corresponding recursive structure is obtained as a second-order filter:

$$y(n) = x(n) + a_1 x(n-1) - b_1 y(n-1) - b_2 y(n-2) \tag{11.13}$$

with coefficients

$$b_1 = -2\exp(-\alpha)\cos\omega, \qquad b_2 = -\exp(-2\alpha)$$

Furthermore, to improve the Gaussian approximation the coefficient $a_1$ is corrected as

$$a_1 = \exp\left(-\frac{1}{2\sigma^2}\right) + b_1$$

In this approach the cosine term $\omega$ must be selected to avoid significant oscillations of the impulse response and the exponential term $\alpha$ is inversely proportional to the Gaussian standard deviation $\sigma$:

$$\alpha = \frac{p}{\sigma}$$

The proportionality factor $p$ can be obtained by putting additional constraints in the frequency domain, and the filter coefficients are parametrically computed for any arbitrary choice of the standard deviation $\sigma$.

An example is shown in Figure 11.4, where the shape of the approximated impulse response is compared with the reference Gaussian function.

### 3.3 Nonlinear Filtering

Quite often linear filtering techniques are not sufficient to accomplish particular image processing tasks and ad hoc nonlinear solutions are quite common in machine vision systems. Examples are found in homomorphic filtering for picture enhancement, statistical differencing for contrast improvement, removal of cloud effects in satellite images, etc. In the following we refer mainly to two classes of filters which are often used in the robotic field, namely adaptive Kalman filtering and nonlinear median smoothing.

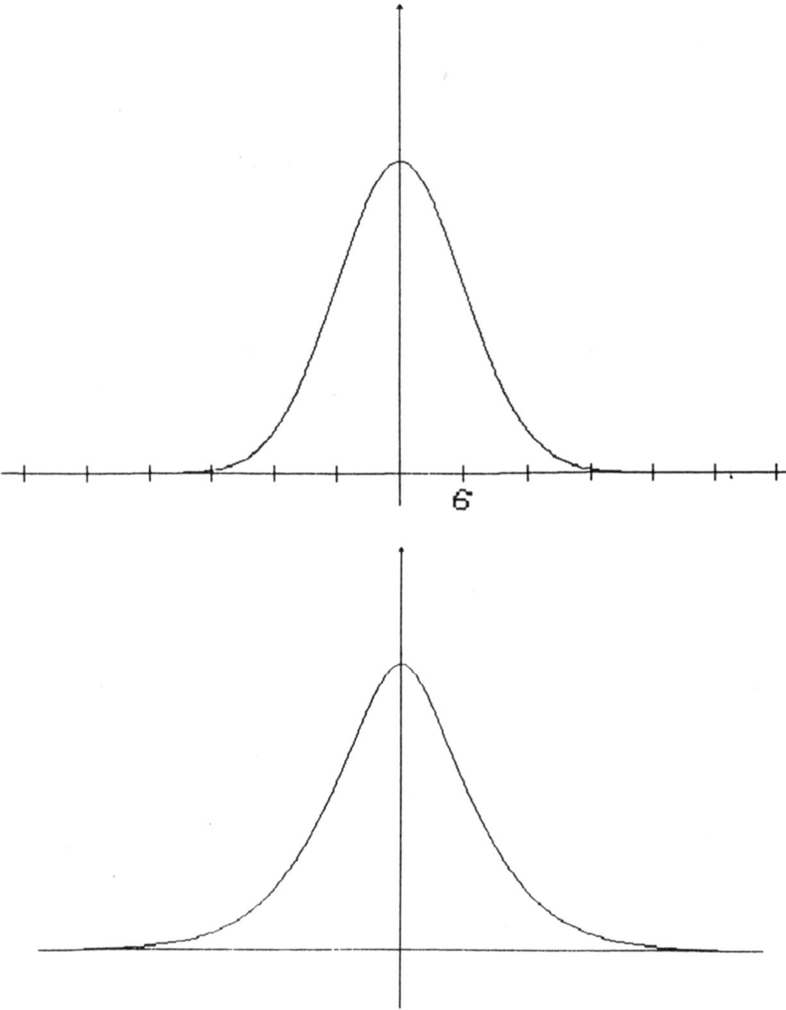


**Figure 11.4** Example of approximation of a Gaussian function: ideal specifications (above); impulse response of a recursive filter (below).

### Kalman Filter

This adaptive linear filtering technique is used mainly in digital communications for linear prediction to obtain the mean square estimator based on $n$ past values of the signal. A state-space model of the linear discrete time observed system is required in the hypothesis of an input random Gaussian process. In the simplest case it is supposed to provide the best estimate in the least-squares sense, given a set of measures affected by errors.

Besides such classical applications, this processing tool is also used in some signal processing environments where the dynamics of the signal are given by a nonlinear equation (extended Kalman filtering). This is particularly true for navigation systems and in 3-D nonlinear estimation, where the observed signal may be affected by errors and it is necessary to reduce data ambiguity and uncertainty.

In this case the nonlinear relations are linearized around some a priori estimate to achieve an equivalent set of linear equations describing the state space system so that ordinary Kalman filter equations are used. Some examples of this approach are given in [12], where the linearized Kalman filter is used to reduce uncertainty in stereo matching and motion analysis to obtain a reliable 3-D model of the environment for a mobile robot.

It is worth noting that these filters are suboptimal, while the Kalman filter for linear systems is optimal. The initial error in the measurements is particularly critical [13].

### Median Filtering

Median filtering is a nonlinear process which is quite useful in reducing impulsive and salt-and-pepper noise as well in performing edge-preserving smoothing. Its implementation requires a moving window sliding across the image, and the current central pixel is replaced by the median intensity value of the samples enclosed in this local neighborhood. To compute this median value, a sorting procedure must be performed on the intensity values to achieve a monotonic ordering of their amplitudes. An alternative solution consists of computing the local cumulative function up to the half value of the sample number. This histogram evaluation is advantageous only when using a window size greater than $5 \times 5$.

Improved results are obtained using recursive estimation of the median value for consecutive pixels. The search for the new value of the median $M_p$ starts from the previous one $M_{p-1}$, which is increased or decreased according to the number of new samples added to or subtracted from the histogram in a fast updating procedure [14].

An example of median filtering is shown in Figure 11.5, for two industrial objects. This nonlinear smoothing tends to provide more homogeneous regions, which can be particularly useful in image segmentation, without affecting the main discontinuities with respect to the background. On the other hand, when using large processing windows it is possible to lose some details such as the internal holes as well as to round the main corners of the objects.

## 4 SCENE ANALYSIS IN ROBOTIC APPLICATIONS

This section introduces some typical machine vision techniques, which make use to some extent of the previously discussed processing tools. As already mentioned, it is not the purpose of this discussion to investigate problems of object recognition and scene understanding; we will concentrate mainly on subtasks of computer vision that are aimed at extracting the most relevant information from the observed scene. The first subsection will recall some of the most commonly used shape description tools, in terms of edge features and area descriptors. Then boundary detection, image segmentation, and 3-D range computation will be briefly summarized.

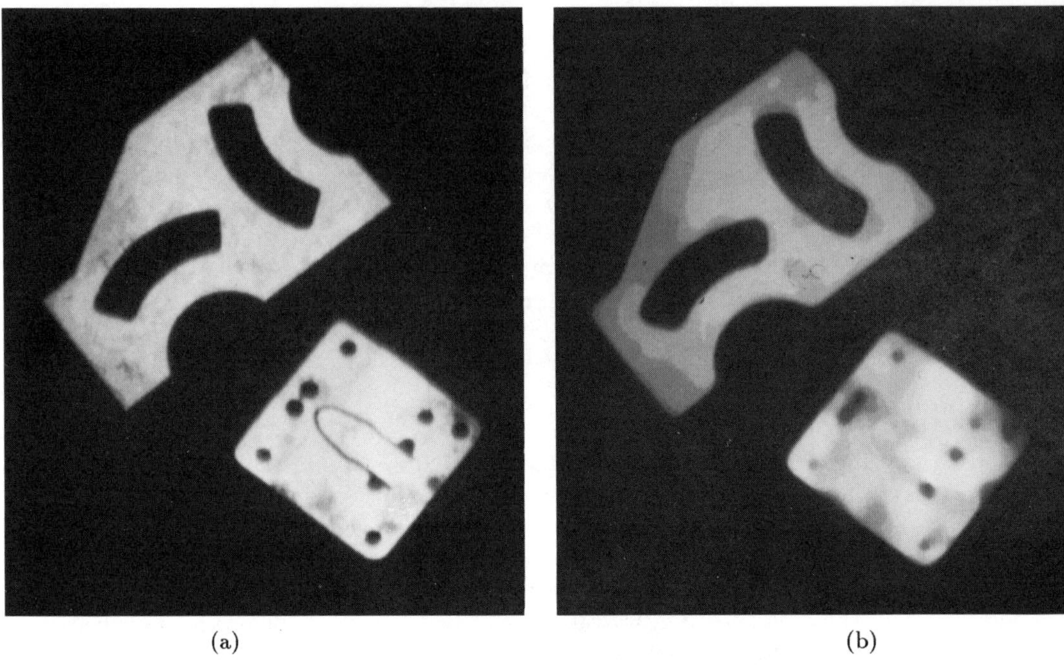


**Figure 11.5** Example of median filtering: (a) original image; (b) result obtained using a fairly large filter mask, 9 × 9. Some small details are smoothed out even if the external boundary discontinuities are unaffected.

## 4.1 Shape Analysis

A good shape representation should allow object recognition irrespective of minor noise effects in data acquisition and possible distortion in image presentation (scale factor and rotation). The recorded object will be described at the symbolic level as a set of relations between the individual shapes that compose it. They can be either regular geometric forms (circles, squares, etc.) or arbitrarily shaped contours of industrial components. As far as planar 2-D objects are concerned, such primitive forms will be described mainly in terms of edge boundaries derived from suitable edge detection or region segmentation procedures. The resulting region adjacency graph turns out to be an extremely powerful tool in scene description, and it can be analyzed using a syntactic approach [15] where a large set of complex patterns can be described efficiently using a small set of primitives with structural rules. The topic of shape analysis is discussed extensively in textbooks on computer vision such as [16], to which the interested reader is referred for more detailed information.

Most industrial machine vision systems are limited to operating with binary images, obtained by thresholding the digitized input gray level picture. This approach allows high-speed hardware implementation using low-cost configurations. Storage requirements are extremely low, using efficient run-length coding techniques which do not prevent easy computation of relevant geometric features such as area, position, and orientation [17]. On the other hand, a fairly high object-to-background contrast is required and the observed scene should be essentially flat or two-dimensional, since only the silhouette of the object is detected by thresholding.

More complex 3-D scenes often require sophisticated segmentation procedures to distinguish individual objects as well as shadows and overlaps. Most of the difficulties arise from the fact that only partial contour descriptions are available. A quite common approach consists of formulating hypotheses on the presence of a certain shape based on a local contour matching, followed by a verification procedure to propagate this hypothesis to other segments in the scene [2].

## Line Description

A digital boundary is efficiently described as an ordered list of coordinates. Significant savings can be obtained by storing the starting point and the incremental changes for the successive samples in a chain coding scheme. In this way an integer code is assigned to each of the neighbors of a pixel $P$, with eight values as shown in the following table:

| | | |
|---|---|---|
| 3 | 2 | 1 |
| 4 | $P$ | 0 |
| 5 | 6 | 7 |

so that a 3-bit code is sufficient for each element in the chain for a neighborhood of $3 \times 3$ samples of pixel P using counterclockwise scanning.

A further step consists of forming a symbolic description of the contour as a sequence of primitives such as angles, straight lines, and curvilinear arcs. Polygonal approximation represents a suboptimal solution where the contours are described in terms of linear segments. Such a description is quite suitable in a highly structured environment such as an indoor scene but may be not fully appropriate when significant curvature boundaries are present in the recorded images.

Iterative end point fitting [18] is a very simple technique for doing the job. It consists of connecting two selected end points by a straight line and searching for the intermediate point which is farthest from this line. If this distance is beyond a suitable threshold, the curve is segmented in two parts and the previous process is iterated.

Higher-order interpolation (spline functions) provides more accurate boundary description at the expense of increasing both the computation time and the amount of recorded data.

A common requirement in robotic applications is a stable contour segmentation and description in terms of corners, inflection points, segments, and circular arcs. To achieve this goal, the contour function $f(x, y) = 0$ must be expressed in a more appropriate way, for example, in terms of the angle $\theta$ of the local tangent to the curve as a function of the arc length $s$. A more efficient description is obtained as

$$\varphi(t) = \frac{\theta L t}{2\pi} + t \tag{11.14}$$

where $t = (2\pi/L)s$ is defined in the range 0–$2\pi$ and $L$ is the total length of the curve. This function turns out to be invariant to translation, rotation, and scaling of the contour. The curvature function can also be used, but it is quite ill-behaved for curves with sharp corners.

Another useful representation is given by the polar description, which is defined as the angle $\beta(s)$ formed by the line joining a point along the curve sample with its centroid. This periodic function is commonly used in shape recognition since its correlation with the reference functions provides not only a reliable measure of classification but also the amount of rotation.

Another approach to line shape recognition is to extract a reduced subset of discriminant features from these contour functions, to be compared against reference prototypes, or to use suitable mapping transformations such as the Hough transform [19].

A common approach consists of expanding the function through an orthogonal basis (such as the Fourier series) and using only low-order coefficients as descriptors [20].

### Area Descriptors

Quite simple and robust shape descriptions can be obtained from area and perimeter estimates. The ratio of the area to the squared value of the perimeter is a measure that is invariant with the size, position, and orientation of a shape.

Qualitative shape measures are based on the topology of the figure, including the number of connected components and holes. Quantitative descriptors can be obtained again by expansion in basis functions such as the 2-D Fourier series or by using moments, which are defined as

$$M_{pq} = \sum_{(x,y)\in R} x^p y^q \tag{11.15}$$

where $M_{pq}$ is the moment of order $pq$ and pixels $(x, y)$ are considered inside or on the boundary of the region $R$ to be described. For instance, the zero-order moment $M_{00}$ describes the area and the first-order moments $M_{0,1}$ and $M_{1,0}$ give the position of the centroid. Using suitable combinations of these moments it is possible to obtain invariance to position, scale, and also rotation and to achieve global measures that are quite useful in scene recognition and classification [21].

### 3-D Object Representation

Recognition of 3-D objects in an observed scene is a far more complicated task when only partial information from a single point of view is available. This is mainly due to perspective distortion and occlusion effects, which can be only partially reduced by additional range information using passive stereo arrangements or multiple views (fixed or moving configurations). In most general robotic applications 2-D boundary or region descriptions are not sufficient and a true 3-D representation of the scene is required, for both recognition purposes and action planning.

Shape description of 3-D objects can be obtained by representing either the outer surface or the enclosing volume. Surface computer-aided design (CAD) models are available as combinations of various high-order functions, mainly for design and accurate display proposes, but it is very difficult to match this kind of representation against the observed intensity data. Perspective constraints are quite promising, since regular shape distortions can easily be predicted from monocular view, such as for triangles and circles or ellipses [22].

When sparse range data are available, from active or passive stereo, 3-D interpolation is required to achieve a surface description of the scene. A powerful tool for this purpose is the Delaunay triangulation, which has been used for both points and segment feature interpolation [23].

## 4.2 Edge Detection

Most of the processing techniques in scene analysis are based on edge features, which are used for different purposes (segmentation, recognition, stereo matching, motion analysis). An efficient edge detection scheme should satisfy the following requirements: robustness against noise (spurious edges and low signal-to-noise conditions) and accurate and stable localization in the spatial domain. Many different alternatives are currently used, depending on the context of the application, and many review papers and textbooks summarize this topic of research [24, 25]. Basic methods can be classified in two main categories: local (linear and nonlinear) operators, and partial derivatives (gradient or Laplacian) of some regularization function (Gaussian smoothing).

### Local Operators

The simplest and most common class of edge detection techniques is based on the use of local difference masks, and candidate edge samples should exhibit sufficiently high response to these filters. In this category we find linear matched filtering with simple low-size masks corresponding to all possible local orientations of an edge.

Commonly used convolution masks are the discrete gradient approximations of the horizontal component $x$ and vertical component $y$,

$$\begin{matrix} -1 & 1 \end{matrix} \qquad \begin{matrix} 1 \\ -1 \end{matrix}$$

or the popular $2 \times 2$ Roberts cross operator (particularly simple but extremely sensitive to noise)

$$\begin{matrix} 0 & 1 \\ -1 & 0 \end{matrix} \qquad \begin{matrix} 1 & 0 \\ 0 & -1 \end{matrix}$$

and the Sobel operator [24]

$$\begin{matrix} -1 & 0 & 1 \\ -2 & 0 & 2 \\ -1 & 0 & 1 \end{matrix} \qquad \begin{matrix} 1 & 2 & 1 \\ 0 & 0 & 0 \\ -1 & -2 & -1 \end{matrix}$$

(a)

**Figure 11.6** Example of edge detection: (a) original image; (b) fixed threshold of the gradient magnitude; (c) hysteresis thresholding.

In all these cases the gradient magnitude is finally computed as

$$|\nabla_{x,y}| = (\nabla_x^2 + \nabla_y^2)^{1/2} \tag{11.16}$$

or, quite often, approximated by

$$|\widehat{\nabla}_{x,y}| = |\nabla_x| + |\nabla_y| \tag{11.17}$$

to simplify hardware implementation. A discrete version of the isotropic Laplacian operator is the $3 \times 3$ convolution mask

$$\begin{matrix} 0 & -1 & 0 \\ -1 & 4 & -1 \\ 0 & -1 & 0 \end{matrix}$$

Many different processing masks can be found in the literature, using different filter sizes and weights, in the attempt to achieve template matching, which has been extensively used for line and curve detection.

(b)

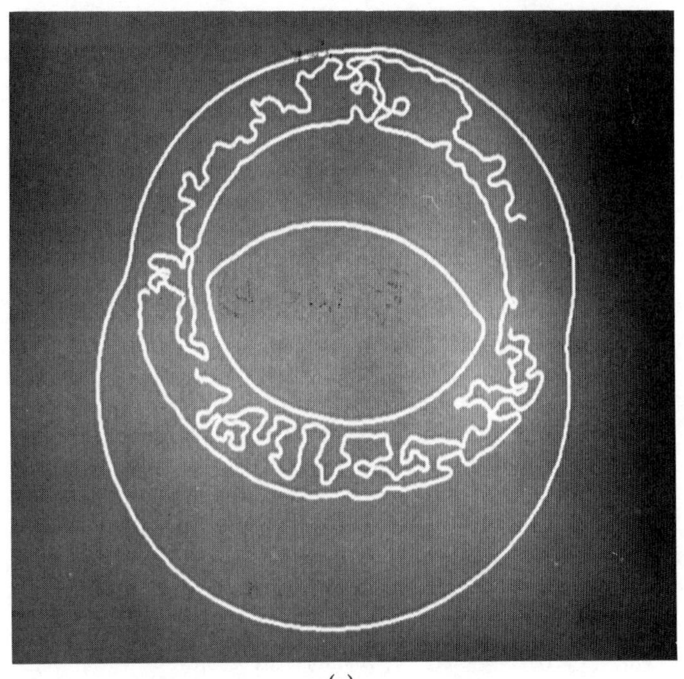

(c)

## Optimal Edge Detection

Since all recorded images are in some way affected by noise, any differentiation process must be coupled with smoothing operations to regularize such an ill-posed problem in order to achieve the following goals: maximize the signal-to-noise ratio, increase accuracy in edge localization, and avoid multiple responses to a single edge in the scene. According to these three criteria, an optimal edge detector of finite size has been proposed for 1-D signals [11]. An infinite impulse response implementation has recently been proposed [26] using second-order recursive filters.

In 2-D computer vision the most popular edge detection scheme is based on the use of a Laplacian operator with Gaussian filtering as an optimal regularization function, since it minimizes the bandwidth product in frequency and space. Edges are then detected at the zero crossings of the Laplacian of the Gaussian operator [27]. This approach always provides closed contours, but because of noise or complex texture, some of them are irregular and do not correspond to real edges in the scene. A quite common solution is to keep only the edge points that exhibit a reasonably high gradient magnitude above a selected threshold (see Figure 11.6). In this way many disconnected variable-length contours will appear, and it is often necessary to perform a further edge-linking process to come up with a more compact description of the edge features.

An alternative solution is obtained by hysteresis thresholding with two different values, $T_1 < T_2$. Edge samples having a gradient magnitude above threshold $T_2$ are selected as candidate starting points. Contiguous edge elements are then connected, provided they exhibit a local gradient above $T_1$. In this way only significant edge contours are retained with increased connectivity (Figure 11.6c).

Improved results in edge localization can be achieved by using directional gradient techniques. In fact, the two schemes are equivalent in one dimension, where the maximum of the gradient corresponds to the zero crossing of its derivative; but this is not always true in two dimensions.

Let us consider the ideal 2-D edge $e(x,y)$ depicted in Figure 11.7, where $(x,y)$ are referred to the object boundary and direction $x$ is always normal to the contour, with angle $\theta(X,Y)$ with respect to the orthogonal reference system $(X,Y)$. Then

$$\nabla^2 e(x,y) = \frac{\partial}{\partial_x}|\underline{\nabla}_x e(x,y)| + |\underline{\nabla}_x e(x,y)|\frac{\partial\theta}{\partial y} \tag{11.18}$$

The results obtained by the isotropic Laplacian operator would correspond to the derivative of the gradient function only for locally straight segments, where $\partial\theta/\partial y = 0$. Otherwise, the zeros of this function (zero crossings) will no longer coincide with gradient maxima, and the difference is due to the second term in Eq. (11.18), which represents an estimate of the local curvature of the contour. To remove this problem a simplified approach has been proposed [11] using first derivatives (local maxima estimation) along the gradient direction, which, on the average, yields better localization of the edges in the image.

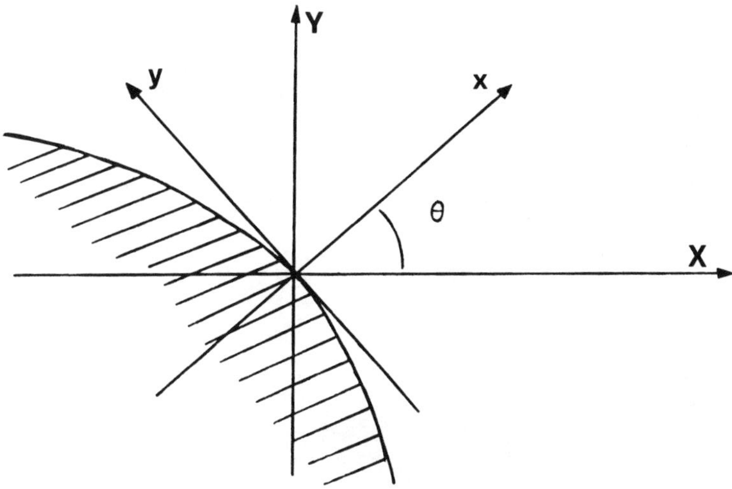


**Figure 11.7** Ideal edge in 2-D; the coordinate system $(x, y)$ is oriented along the edge with respect to the reference system $(X, Y)$.

## 4.3 Image Segmentation

The purpose of this process is to select meaningful regions in the recorded image which are homogeneous in some property, such as intensity, color, or gradients. Scene segmentation can also be performed using range data or reconstruction of the motion field to detect different objects characterized by different motions. Thus, the general topic of scene segmentation is vast and this goal cannot easily be distinguished from scene understanding, being often constrained to some knowledge about the environment. In this sense, image segmentation corresponds to clustering, that is, grouping a set of pixels in the picture based on more or less local properties of a feature vector. In fact, it is possible to use either pixel-based attributes or neighbors' properties such as texture. 2-D region segmentation techniques can be classified in two main categories, region splitting and region growing, or a combination of both.

### Region Splitting

Suitable classification schemes are used to split large regions into smaller ones, and the process is repeated iteratively to obtain a selected level of uniformity. The simplest technique is thresholding, and it is usually performed on image histograms obtained from intensity values or from the output of any appropriate linear or nonlinear process. Thresholding is well suited for simple scenes to detect uniform regions against flat background, as in character recognition, where threshold selection can be optimized using a priori statistics on gray level distribution. Quite often the statistical model consists of a linear combination of Gaussian distributions.

A well-known image segmentor has been proposed [28] in which many different attributes of the image are histogrammed (color hue and saturation, intensity, etc.). This recursive segmentation consists of achieving a first clustering based on a cer-

(a)

**Figure 11.8** Example of image segmentation: (a) original image with significant brightness gradient; (b) optimal thresholding cannot produce a satisfactory segmentation and the resulting binary image presents noisy irregular boundaries; (c) result of gray level segmentation, using a region-growing technique.

tain property. Further nonlinear constraints are used to check for compactness of the segmented regions, to avoid isolated or small groups of pixels. The same thresholding criterion is then repeated on the previously selected regions to improve local detection and localization.

Since most images exhibit space-variant properties, adaptive techniques are often required, to operate on small neighbors. Multiresolution schemes using pyramids of images, where low-resolution segmentation is used as a plan for finer segmentation, are also quite efficient. Split and merge procedures have also been proposed, to achieve a picture tree describing relations between regions at different levels of resolution [18]. Quad-trees are obtained when the pixels at a given level are split into four pixels at a lower level. These segmentation procedures have been used mainly in cartography and processing of satellite images, but they have proved to be quite sensitive to orientation (translation and rotation), which determines dramatic changes in the symbolic tree description.

## Region Growing

To segment an image into regions that satisfy a similarity criterion, it is possible to start at a point "seed" within a region and grow it by grouping all neighbors which

(b)

(c)

possess a similar property. In this scheme neighboring pixels whose attribute values are within a fixed predefined range are grouped together to form the first clusters. Then neighboring clusters are examined and eventually merged together based on suitable properties and relations. The criterion for region merging may be that an average value (intensity, gradient) is within a threshold and spatial constraints (adjacency, inclusion) are satisfied, or that the local contrast along the common boundary is below a selected value, etc.

The performance of such a region grower is strongly affected by suitable choice of the different thresholds, which should be determined adaptively, possibly including some a priori knowledge of the scene to be processed. Successful results are obtained in segmentation of simple industrial scenes, without complex textured objects, which would generate a very large number of candidate clusters.

An example of segmentation is shown in Figure 11.8 [29], where the first region-growing is performed on the output of the Laplacian of the Gaussian, so that the number of clusters at the first level is conditioned mainly by the choice of the standard deviation $\sigma$ of the Gaussian smoothing.

In spite of the simple example considered here, it is quite evident that a conventional optimal thresholding process would be unable to discriminate correctly between object and background, due to nonuniformity of the contrast in the picture.

## 4.4 3-D Stereo Techniques

Depth information is extremely important in scene analysis since it allows one to easily disambiguate most recognition problems encountered in scene analysis when using a single point of view, which are mainly determined by perspective distortions, poor contrast, shadows, etc.

The distance of visible surfaces from the camera can be obtained by active control of the scene illumination, using lasers or projected light patterns. In this approach, which is quite common in many industrial applications, the object elevation from a reference plane is estimated by using triangulation techniques from the known projection and observation points.

Otherwise, 3-D range information is commonly obtained by passive stereovision, according to the general scheme of Figure 11.9, which is closely related to human stereopsis. To obtain stereo measures it is necessary to solve two different problems, that is, finding the pixel correspondence from the two views and computing the distance by triangulation, which is quite simple when the two cameras are accurately calibrated with respect to each other [30]. Figure 11.9 shows the epipolar geometry that makes it possible to reduce the correspondence problem to a 1-D search on the image plane. In fact, the epipolar plane is defined as the plane passing through the two lens centers $C_1$ and $C_2$ and the observed sample point $P$ in 3-D space. Henceforth, given a pixel $P_1$ in the first view, the corresponding point $P_2$ in the second image must be found along the epipolar line 2, which can be precomputed from known spatial position of $C_1$, $C_2$, $P_1$.

Most of the available stereo implementation in machine vision systems make use of a simpler geometry, using a camera system with parallel axes; in this case the epipolar plane intersects the two image planes along parallel lines, which can be either horizontal or vertical. This approach greatly simplifies the correspondence problem, which can be performed along rows and columns of the recorded digital

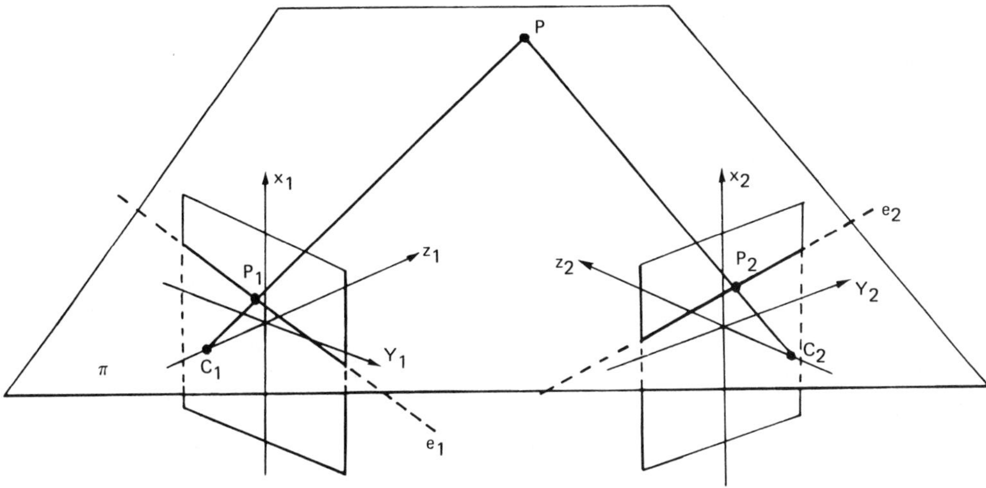


**Figure 11.9** 2-D stereo configuration. The epipolar plane passing through the two lens centers $C_1$ and $C_2$ and the observed 3-D point $P$ intersects the image planes along epipolar lines $e_1$, $e_2$; the correspondence problem is reduced to a 1-D search.

images. Calibration can be achieved either mechanically (for fixed arrangements) or optically (for moving configurations), in which case it is necessary to estimate rotation and translation parameters looking at a reference pattern in the scene.

Available techniques for finding the correct correspondence between the two views can be roughly divided into two classes, depending on whether they are based on feature correspondence or local correlation methods. In the first approach matching is performed on selected features, such as edge points, segments, or corners, by providing some descriptors for similarity checking (edge strength, orientation, first- and second-order derivatives, local contrast, etc.). The other matching scheme is based on correlation of local neighborhoods (1-D or 2-D) using either the image intensity function or some preprocessing result (such as the zero crossing or Laplacian of the Gaussian) [31]. In both cases, errors or ambiguities may arise for many different reasons, from noisy raw data to partial occlusions. To remove these errors it is often necessary to check for consistency of the disparity values within local regions using continuity constraints.

An example of using the correlation approach is shown in Figure 11.10, where a variable-resolution scheme has been applied to local nonoverlapping windows to obtain the necessary information for volume description of a simple block scene. Two pairs of stereo images have been taken, at 90°, to recover 3-D range estimates from two different points of view. This information is then combined to recover the 3-D volume description shown in Figure 11.10c [32].

As a rule, the local correlation approach tends to provide a dense set of averaged range values in the 3-D space and is appropriate for estimating the presence of obstacles and volume occupancy in the scene for path planning. On the other hand, feature-based techniques allow more accurate detection of 3-D positions,

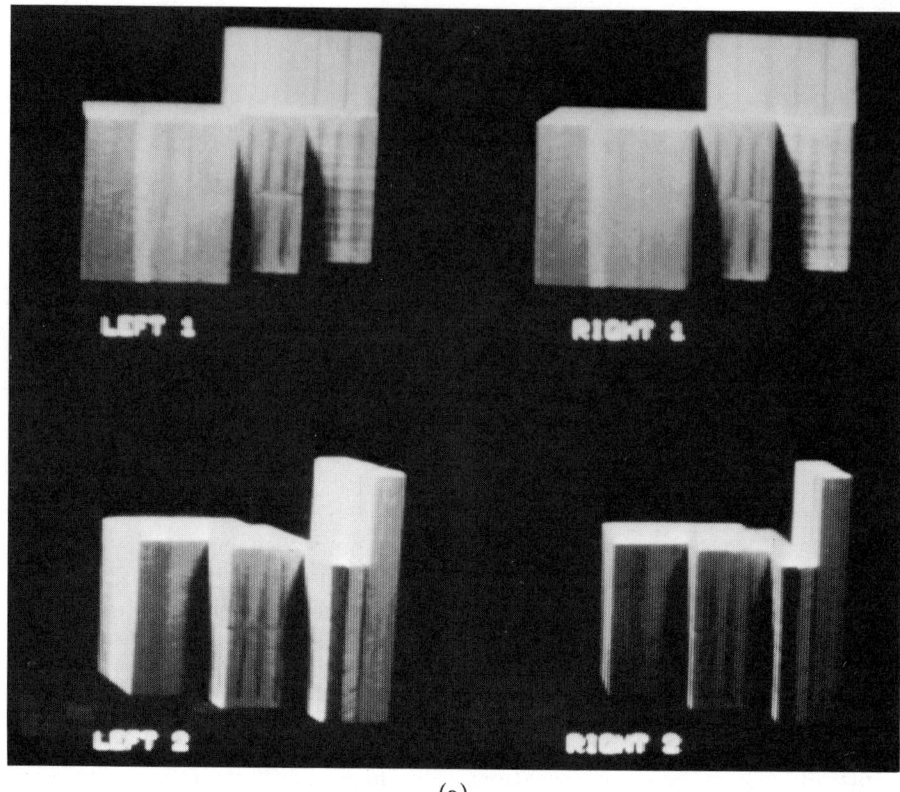


(a)

**Figure 11.10** Example of 3-D reconstruction of a block scene by combining range data from two stereo views: (a) the two pairs of views taken almost at 90°; (b) range data obtained from the two views (intensity is proportional to the distance from the cameras) using correlation measures at different scales; (c) 3-D reconstruction by back-projection of the two sets of range data. (Courtesy of Prof. G. Sandini, Department of Electronic Engineering, DIST, University of Genova.)

at the expense of a sparser set of points or segments in the space (wire-frame description), and are more suitable for 3-D object recognition and handling.

Using more than two cameras makes it possible to simplify the correspondence problem and to deal with arbitrarily oriented axes (convergent to a fixation point of view), to increase the distance from the cameras and eventually improve the accuracy of the estimates [33].

## 4.5 Optical Flow and Motion Analysis

3-D information is perceived by the human visual system not only by stereopsis but also through other sensing capabilities. For instance, the adaptive ability of the human eye to focus automatically on the details of interest allows it to easily detect the relative depth position of objects from the background. Motion is another

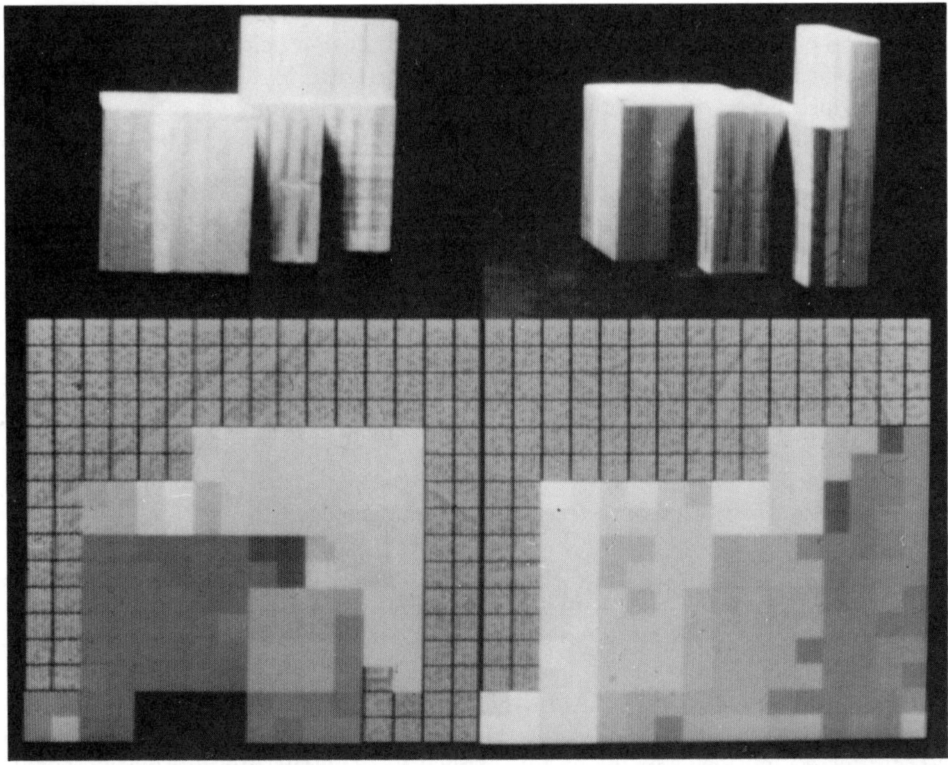

(b)

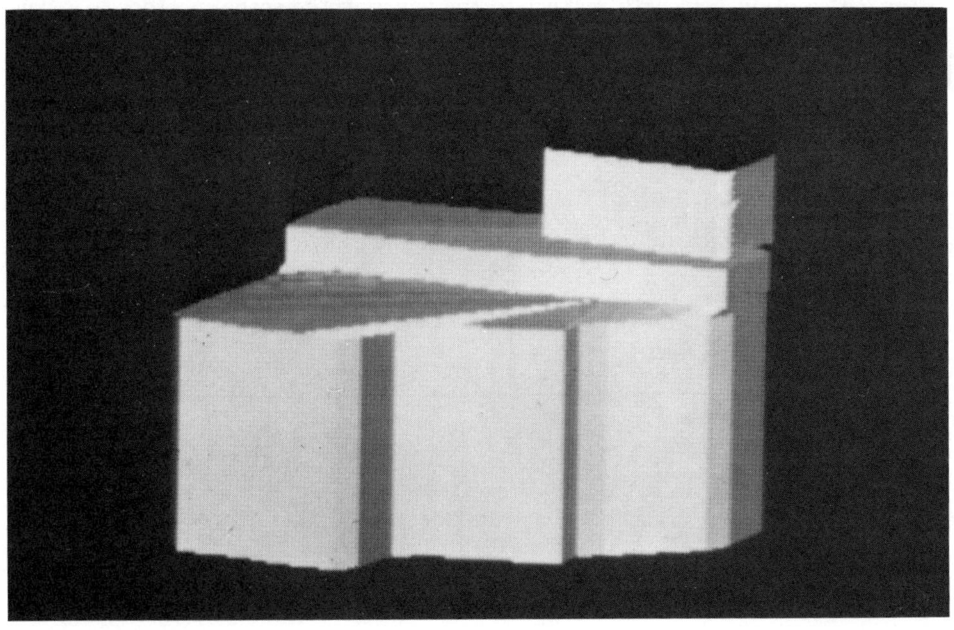

(c)

essential cue for 3-D perception. In this case the available information consists of a 2-D projection into the image plane of the 3-D motion field in the space.

Using the common pinhole camera model of Figure 11.11, where $f$ represents the focal distance of the imaging system, the recorded sample $\underline{r}$ corresponding to the point $\underline{R}$ in the 3-D space, at distance $Z$, is given by

$$\underline{r} = \frac{-f\underline{R}}{Z} \tag{11.19}$$

If we assume that there is a 3-D motion described by a translational component $\underline{V}$ and a rotational velocity $\underline{\Omega}$,

$$\dot{\underline{R}} = \underline{V} + \underline{\Omega} \wedge \underline{R} \tag{11.20}$$

the 2-D motion projected onto the image plane becomes

$$\dot{\underline{r}} = -\frac{f}{Z}\underline{V} - \frac{\underline{r}(\underline{z} \cdot \underline{V})}{Z} + \underline{\Omega} \wedge \underline{r} + \frac{\underline{r}[\underline{z} \cdot (\underline{\Omega} \wedge \underline{r})]}{f} \tag{11.21}$$

where $\underline{z}$ is the unit vector of the optical axis $Z$.

The first two terms in Eq. (11.21) represent the translational component of the 2-D projected motion, which contains explicitly 3-D structure information (depth $Z$). Moreover, in [34] it is shown that it is sufficient to know the normal component of the 2-D motion field with respect to the edges in the scene in order to recover the 3-D structure and rototranslation parameters of rigid motion.

On the other hand, the only term which can be computed from a recorded image sequence is the optical flow, that is, the variation in time of the image

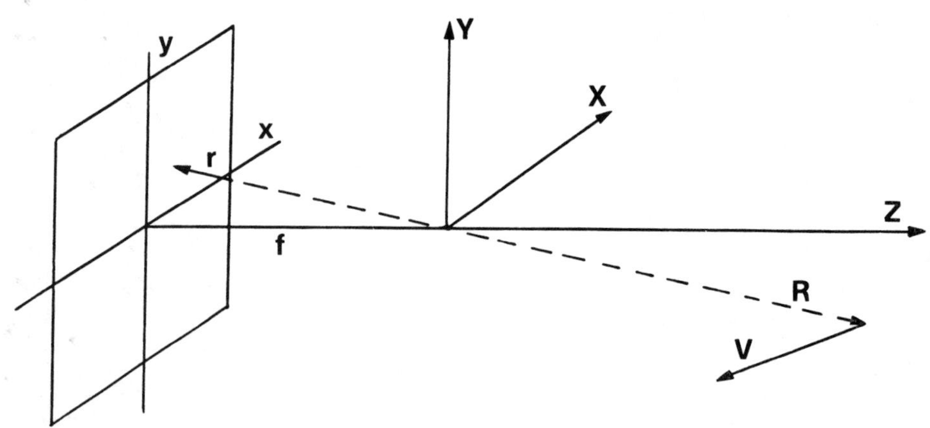


**Figure 11.11** Pinhole camera model of the imaging system; a sample point (vector $R$) in 3-D space is recorded by vector $r$ in the image plane.

intensity $I(x, y, t)$. This function corresponds to the 2-D motion field only under the assumption of Lambertian surfaces and at sharp discontinuities, where the spatial gradient is sufficiently high [35].

The computation of the optical flow $\underline{\sigma} = (u, v)$ is based on the hypothesis that the only changes in the image irradiance function are due to motion, so that

$$\frac{dI}{dt} = 0 \Longrightarrow -I_t = \underline{\nabla}_{x,y} I \cdot \underline{\sigma} \tag{11.22}$$

where $I_t$ represents the partial derivative with respect to time, $\underline{\nabla}_{x,y} I = (I_x, I_y)$ is the gradient vector of the image function, and the dependence on the spatial variables $(x, y)$ has been dropped for notational convenience. From this equation it is possible to estimate the normal component $\underline{o}_{\perp}$ of the optical flow in the direction of the local gradient function $\underline{\nabla}_{x,y} I$.

To evaluate the global flow field $\underline{o}$, a smoothness constraint has been proposed [17] according to the following iterative scheme (3-D recursive filtering):

$$\begin{aligned} u^{n+1} &= \bar{u}^n - \frac{I_x(I_x \bar{u}^n + I_y \bar{v}^n + I_t)}{(\epsilon^2 + I_x^2 + I_y^2)} \\ v^{n+1} &= \bar{v}^n - \frac{I_y(I_x \bar{u}^n + I_y \bar{v}^n + I_t)}{(\epsilon^2 + I_x^2 + I_y^2)} \end{aligned} \tag{11.23}$$

where the current estimates are averaged on a local neighborhood to obtain $(\bar{u}^n, \bar{v}^n)$ and corrected iteratively in the direction of the gradient $(I_x, I_y)$.

Unfortunately, straightforward implementation of this algorithm presents some difficulties. Convergence is achieved after a large number of iterations, proportional to the size of the image. The isotropic smoothness constraint is not appropriate along the edges and discontinuities in the scene, which are the most relevant features in industrial applications. Moreover, irregularities in the surface and recording noise prevent any practical application of this approach.

A very similar problem is also found in the telecommunication field, where it is necessary to estimate the displacement vector to predict the consecutive frames in the image sequence [36]. Other solutions have been proposed [37] using oriented smoothness constraints. For instance, an efficient solution for moving-edge analysis consists of performing an iterative updating of the normal component $\underline{o}_{\perp}$ moving along the 1-D connected contours in the scene. In this way noise sensitivity is strongly reduced because the gradient component is maximum along the edges. Moreover, the number of samples retained is much smaller than the original image size. However, this approach does not allow estimation of the conventional optical flow on low-contrast object surfaces, and suitable edge detection and chaining modules are required.

Up to now we have discussed the problem of range estimation, orientation, and motion of visible surfaces in a scene. A closely related problem is found in the interpretation of the 3-D structure of a collection of moving points, in the rigidity assumption, from their visual motion field (correspondence between consecutive frames) [38]. In fact, any rigid-body motion can be described by an orthonormal $3 \times 3$ rotation matrix and a three-component translation vector.

A similar solution has been proposed in [39], where a linear system of equations in eight unknowns must be solved. This algorithm fails when the selected sample points lie on certain smooth cubic curves in the 3-D space. Anyway, in general, there is some noise in the correspondence and perturbations in the available data. Henceforth, the solutions obtained in this linear scheme are always an approximation to the desired parameters, which can be improved only by increasing the number of matched samples in the scene.

Nonlinear rigidity constraints have been added in a recent paper [40] to improve the motion estimation and to achieve extremely accurate results with a minimum number of data (see Figure 11.12).

## 5 CONCLUSIONS AND FUTURE RESEARCH TRENDS

This chapter has been devoted to examining some image processing techniques which are commonly found in industrial applications. No claim of completeness is made because of the extremely wide field involved. In all practical applications of industrial automation, computer vision is just a part (even though a very important one) of the full system, and the global efficiency is heavily conditioned by

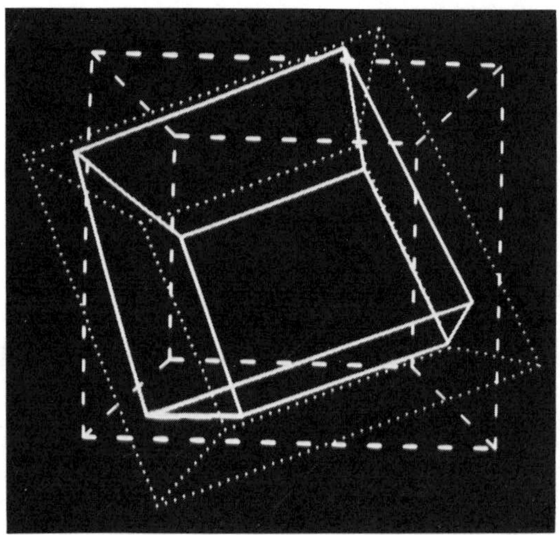

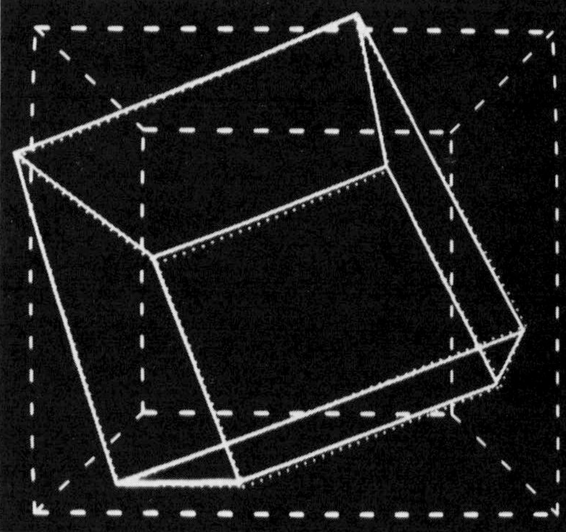

**Figure 11.12** Example of the improvement provided by application of the rigidity constraint to the linear approach in motion estimation from corresponding points. Dashed lines represent the original position of a cube, solid lines show the expected final position, and dotted lines represent the computed position from the estimated motion parameters. The number of corresponding points is $N = 8$, the simulated resolution is 512 pixels, and the applied rotation is 30°. (Courtesy of Prof. C. Braccini, Department of Electronic Engineering, DIST, University Genova).

many other components (cameras, lenses, lighting conditions, processing hardware, human-machine interface, etc.).

Looking at a typical configuration of a robot manipulating arm, as shown in Figure 11.13, it is easy to understand the importance of integration between the machine vision system and the mechanical tools (conveyor line, robot movement, grippers) and eventually other sensing facilities (proximity and force sensors etc.). Most of the available vision systems are confined to solving simple tasks with quite a number of operative constraints (backlighting, 2-D binary models) and strong a priori knowledge of the application.

Key factors in effective improvements in the next few years will be flexibility to function under different working conditions and adequate computing power to accomplish the most demanding tasks, such as universal bin-picking, high-precision inspection, and object identification in an uncontrolled environment.

The large amount of computational power required for such advanced applications represents an unaffordable constraint for conventional technology. On the other hand, image resolution requirements to allow accurate detection of small details in the image are always more severe: this means an increased amount of data to be recorded and processed, possibly at video rate, that is, 25 images per second (European standard). On the other hand, the complexity of many advanced algorithms prevents their practical implementation even on large powerful computers working at speeds over 1 million operations per second. To satisfy such image processing requirements two different solutions are usually found. Highly demand-

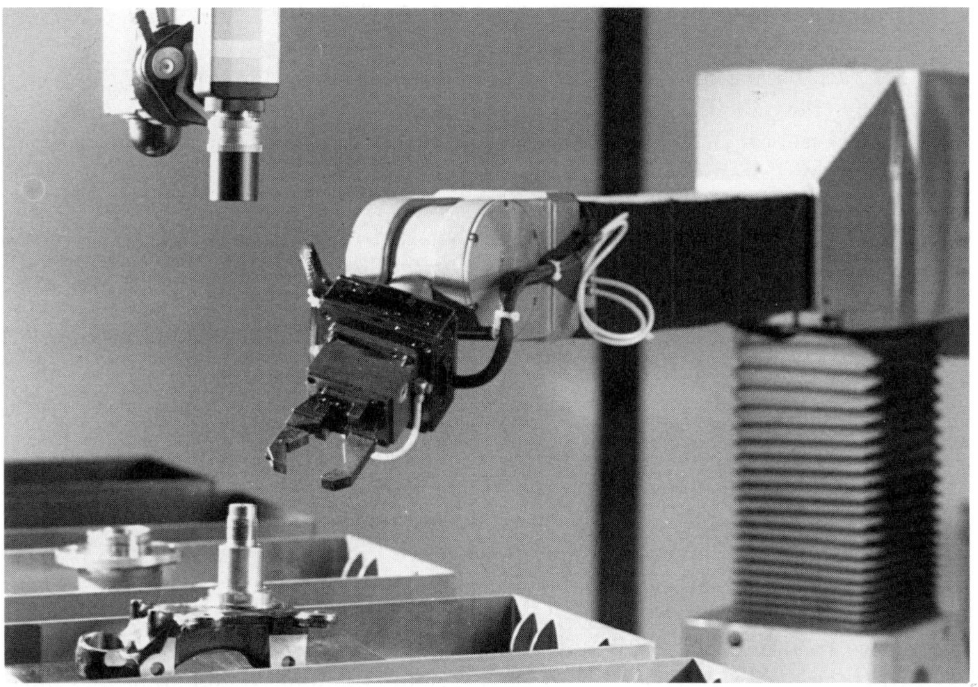

**Figure 11.13** Machine vision system for a robot manipulating arm.

ing repetitive processing techniques, more frequently used at a low level for image feature extraction and segmentation, are often implemented on special hardware modules or through highly parallel machines. The most promising solutions come from pyramidal architectures and systolic units, which are well suited for matrix operation. This approach significantly reduces the overall data rate and generates a highly compressed set of symbols (edges, regions, etc.).

In addition, greater flexibility and increased processing power are obtained with parallel architectures in a multiple instruction, multiple data (MIMD) scheme. High-level languages and processing structures will also be available on such parallel computers to accomplish in a very efficient way, at TV rate, complex tasks of scene recognition and interpretation.

The other major development in the next few years will be the building of more intelligence into the system, which will be made possible by advances in the technology of artificial intelligence and expert systems. This approach will lead to the two complementary goals of improving performance of machine vision in complex dedicated tasks and increasing flexibility and adaptive learning facilities to meet different working conditions.

## REFERENCES

1. M. L. Baird, SIGHT-1: A computer vision system for automated IC chip manufacture, *IEEE Trans. Syst. Man Cybern.*, *8*: 133–139 (February 1978).
2. N. Ayache and O. D. Faugeras, Hyper: A new approach for the recognition and positioning of two-dimensional objects, *IEEE Trans. Pattern Anal. Machine Intell.*, *PAMI-8*(1): 44–54 (1986).
3. M. Yachida and S. Tsui, Industrial computer vision in Japan, *IEEE Comput.*, 50–64 (May 1980).
4. S. M. Pizer, J. B. Zimmermann, and V. E. Staab, Adaptive grey level assignment in CT scan display, *J. Comput. Assist. Tomogr.*, *8*(2): 300–305 (1984).
5. G. Garibotto, "Digital X-Ray Enhancement Through Local Difference Equalization," Proc. SPIE, Cannes, (December 1985).
6. A. V. Oppenheim and R. W. Schafer, *Digital Signal Processing*, Prentice-Hall, Englewood Cliffs, New Jersey (1975).
7. E. I. Jury, Stability of multidimensional scalar and matrix polynomials, *Proc. IEEE*, *66*(9): 1018–1047 (1978).
8. G. Garibotto, "Two-Dimensional Stability Analysis Using the Complex Cepstrum," Proc. ISCAS '82, pp. 868–871, Rome (May 1982).
9. M. M. Fahmy and J. H. Lodge, An optimization technique for the design of half-plane 2-D recursive digital filters, *IEEE Trans. Circuits Syst.*, *CAS-27*(8): 721–724 (1980).
10. G. Garibotto, "Three-Dimensional Recursive Filtering," Proc. of ICASSP '82, pp. 2059–2062, Paris (1982).
11. J. Canny, Finding Edges and Lines in Digital Images, MIT Artificial Intelligence Lab, Report AI-TR-720, Cambridge, Massachusetts (June 1983).
12. O. D. Faugeras, N. Ayache, and B. Faverjon, "Building Visual Maps by Combining Noisy Stereo Measurements," Proc. IEEE Int. Conf. on Robotic and Automation, vol. 3, pp. 1433–1438, San Francisco (1986).

13. N. Mohanty, *Random Signals Estimation and Identification*, Van Nostrand Reinhold, (1986).
14. T. S. Huang, G. J. Yang, and G. Y. Tang, A fast two-dimensional median filtering algorithm, *IEEE Trans. Acoust. Speech Signal Process.*, *ASSP-27*(1): 13–18 (1979).
15. S. Fu, *Syntactic Methods in Pattern Recognition*, Academic Press, New York (1974).
16. D. H. Ballard and C. M. Brown, *Computer Vision*, Prentice-Hall, Englewood Cliffs, New Jersey (1982).
17. B. K. P. Horn, *Robot Vision*, MIT Electrical Engineering and Computer Science Series, Cambridge, Massachusetts (1986).
18. T. Pavlidis, *Structural Pattern Recognition*, Springer-Verlag, Berlin (1977).
19. V. Cantoni, M. Caviglione, G. Musso, and G. Pannunzio, Location and orientation detection of mechanical parts using the Hough technique, *Proc. SPIE*, pp. 229–233, Geneva (1983).
20. C. T. Zahn and R. Z. Roskies, Fourier descriptions for plane closed curves, *IEEE Trans. Comput. 21*: 269–281 (March 1972).
21. E. L. Hall, *Computer Image Processing and Recognition*, Academic Press, New York (1970).
22. T. O. Binford, Inferring surfaces from images, *Artif. Intell. 17*: 205–244 (August 1981).
23. J. D. Boissonnat, Geometric structures for three-dimensional shape representation, *ACM Trans. Graphics*, *3*(4): 266–286 (1984).
24. W. K. Pratt, *Digital Image Processing*, Wiley, New York (1978).
25. V. Torre and T. Poggio, On edge detection, *IEEE Trans. Pattern Anal. Machine Intell. PAMI-8*(2): 147–163 (1986).
26. R. Deriche, Using Canny's criteria to derive an optimal edge detector recursively implemented, *Int. J. Comput. Vision* (in press).
27. D. Marr and E. C. Hildredth, Theory of edge detection, *Proc. R. Soc. London Ser. B 207*: 187–217 (1980).
28. R. Ohlander, K. Price, and D. R. Reddy, Picture segmentation using a recursive splitting method, *Comput. Graphics Image Process.*, pp. 313–333 (1978).
29. L. Borghesi, F. Caccia, M. Caviglione, and R. Toscano, "A Modular Flexible System for Image Processing," Proc. Int. Conf. on Digital Signal Processing, Florence (September 1987).
30. O. D. Faugeras and G. Toscani, "Camera Calibration for 3D Computer Vision," Proc. Int. Workshop on Machine Vision and Machine Intelligence, Tokyo (February 2–5, 1987).
31. H. K. Nishihara, PRISM: A Practical Real-Time Imaging Stereo Matcher, AI Memo 780, Massachusetts Institute of Technology (1984).
32. P. Morasso and G. Sandini, 3-D reconstruction from multiple stereo views, *Proc. Int. Conf. on Image Analysis and Processing*, Plenum, New York, pp. 121–127 (1985).
33. N. Ayache and F. Lustman, "Fast and Reliable Passive Stereovision Using Three Cameras," Int. Workshop on Industrial Applications of Machine Vision and Machine Intelligence, Tokyo, (February 1987).
34. B. F. Buxton et. al., Machine perception of visual motion, *GEC J. Res.*, *3*(3): 145–161 (1985).

35. T. Poggio and A. Verri, "Against Quantitative Optical Flow," MIT Report, Artificial Intelligence Laboratory, Massachusetts Institute of Technology (1986).
36. T. S. Huang, ed., *Image Sequence Processing and Dynamic Scene Analysis*, Springer-Verlag, New York (1983).
37. H. H. Nagel, "Image Sequences—Ten (octal) Years—From Phenomenology towards a Theoretical Foundation," Proc. Int. Conf. Pattern Recognition, Paris, pp. 1174–1185 (October 1986).
38. R. Y. Tsai and T. S. Huang, Estimating three-dimensional motion parameters of a rigid planar patch, II: Singular value decomposition, *IEEE Trans. Acoust. Speech Signal Process.*, *ASSP-30*(4): (1982).
39. H. C. Longuet-Higgins, A computer algorithm for reconstructing a scene from two projections, *Nature (London)*, *293*: 133–135 (1981).
40. C. Braccini, G. Gambardella, A. Grattarola, and S. Zappatore, "Motion Estimation of Rigid Bodies: Effects of the Rigidity Constraints," Proc. the EUSIPCO '86, pp. 645–648, The Hague (September 1986).

# 12

# Advances in Applying Nonlinear Optical Crystals to Optical Signal Processing

YESHAIAHU FAINMAN and SING H. LEE Department of Electrical and Computer Engineering, University of California at San Diego, La Jolla, California

## 1 INTRODUCTION

The photorefractive and Kerr effects in nonlinear crystals can be utilized to perform such elementary optical computing operations as image amplification, two-dimensional optical logic, three-dimensional memory, phase conjugation, and spatial correlation/convolutions. Potential applications of these elementary optical computing operations to analog optical computing (e.g., matrix inversion, solving matrix eigenvector/eigenvalue problems), digital optical processing, pattern recognition, and image processing are described in this chapter.

The major advantage of optical signal processors is their ability to perform parallel computations and global communications. In order to extend the capability of optical processors from the passive domain there is a need for active (linear and

The authors are grateful for research support from the National Science Foundation under Grant ECS-8303107, the Air Force Office of Scientific Research under Grant AFOSR-85-0371, the Rome Air Development Center, Hanscom Field, under grant F-19628-85-K-0039, and the Office of Naval Research under grant ONR N00014-86-K-0697

Portions of this chapter have been presented at the Advanced Institute on Hybrid and Optical Computers, sponsored by SPIE, in Washington D.C., March 24–27, 1986, and the NSF Workshop on Optical Nonlinearities, Fast Phenomena and Signal Processing, at the Optical Science Center, Tucson, Arizona, May 22-23, 1986.

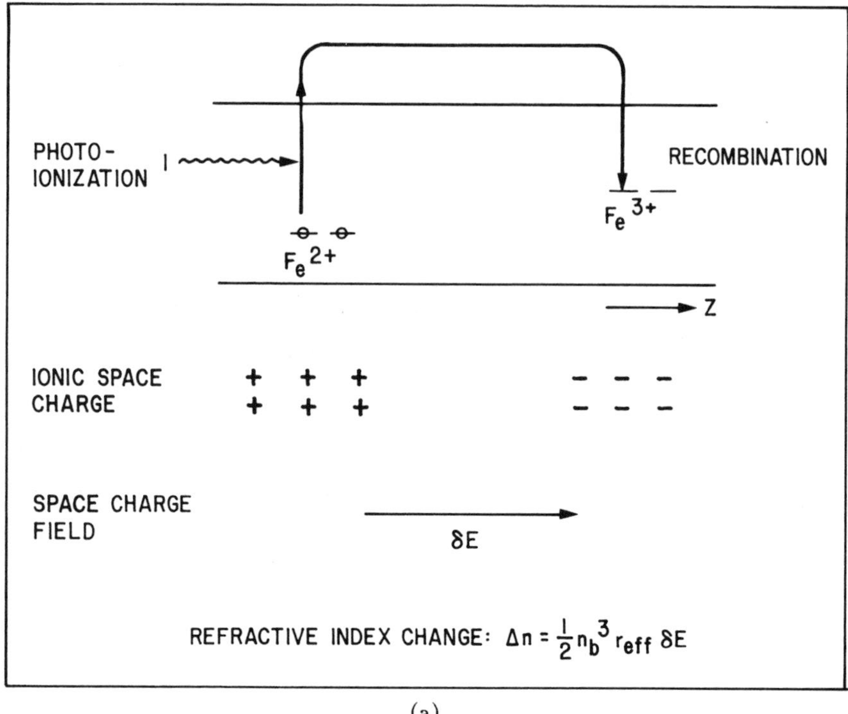


(a)

**Figure 12.1** (a) Schematic of the charge transport model of the photorefractive effect. For example, in lithium niobate, the photoionization excites an electron from $Fe^{2+}$, which drifts or diffuses to a new location and is trapped at $Fe^{3+}$. The resultant ionic space charge induces an internal electric field that modulates the index of refraction [6]. (b) Formation of a photorefractive index grating: light with spatially periodic intensity $I(x)$ rearranges the charge density $\rho(x)$ in the material. The mobile charges, here with positive charge distribution $\rho(x)$, cause a periodic electrostatic field $E(x)$ according to Poisson's equation. This electric field then causes a change in the refractive index $\Delta n(x)$ of the crystal by the linear electro-optic (Pockel's) effect. The photorefractive effect is nonlocal; the maximum refractive index change does not occur at the intensity peak. In this figure the spatial shift between $\Delta n(x)$ and $I(x)$ is one-fourth of the grating period [7].

nonlinear) optical elements, e.g., optical elements with a photorefractive effect [1–7] or a strong Kerr effect [8].

The principle of the photorefractive effect is discussed in [6, 7]; exposure of a photorefractive crystal to a light interference pattern excites electrons from traps to the conduction band (see Figure 12.1a). The electrons diffuse from high to low concentration regions or drift opposite to the electric fields, are retrapped, and set up local charge patterns, which cause changes in the index of refraction through the first-order electro-optic effect. For a sinusoidal interference pattern $I(x)$ we obtain a net space-charge pattern $\rho(x)$ that is positive in regions of low intensity,

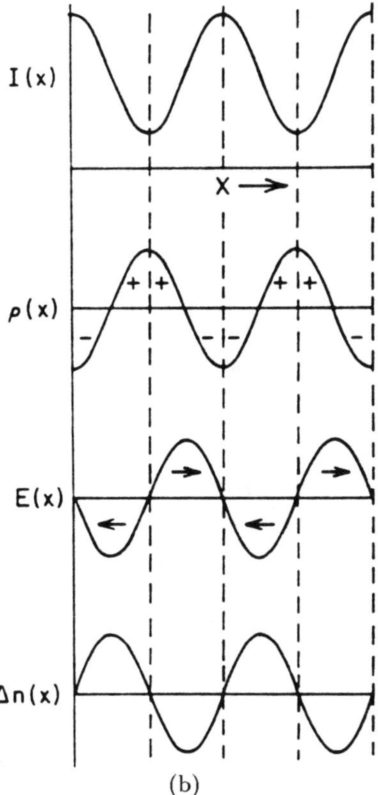

(b)

as shown in Figure 12.1b. The space-charge buildup continues until the electric field $E(x)$ cancels completely the effect of diffusion and drift and makes the current zero throughout. The equilibrium space charge generates a field that modulates the index of refraction $\Delta n(x)$ and gives rise to a phase hologram, which in turn gives rise to two-wave coupling in photorefractive crystals.

The Kerr effect is described by the dependence of refractive index on the electric field according to

$$n = n_0 + n_2 \langle E^2 \rangle \tag{12.1}$$

where $\langle E^2 \rangle$ is the time average of the electric field, $n_0$ is a constant, and $n_2$ is the nonlinear index or the Kerr coefficient. An example of physical mechanisms which can give rise to the Kerr effect is third-order nonlinear polarizability. When two waves interact in a Kerr medium [8], they form an interference pattern that corresponds to spatially periodic variation of the time-averaged field $\langle E^2 \rangle$. Such a periodic intensity produces a phase grating in the Kerr medium. Thus, the problem that we address is closely related to the phenomenon of self-diffraction from an induced grating. The formulation of such a problem [8] is similar to that of the two-wave coupling in photorefractive crystals. However, in photorefractive media the refractive index modulation is proportional to the contrast of the interference fringes, whereas in Kerr media the refractive index modulation is proportional to the field strength. Thus, the coupling strength in Kerr media is proportional to the beam intensities, whereas the coupling strength in photorefractive media is determined by the ratio of beam intensities [1–7].

The photorefractive and Kerr effects can be applied to perform many elementary optical computing operations such as image amplification [4–11], two-dimensional (2-D) optical logic operations [12], three-dimensional (3-D) memory [1, 13, 14], phase conjugation [15], and spatial convolution and correlation [16–18]. Furthermore, these elementary optical computing operations can be employed in optical signal processing applications such as matrix inversion, matrix eigenvector/eigenvalue problems [19, 20], pattern recognition/image processing [11, 18, 21–24], and digital optical processing [12, 25].

## 2 ELEMENTARY OPTICAL COMPUTING OPERATIONS BY NONLINEAR OPTICAL CRYSTALS

A nonlinear optical material can be used to implement different elementary optical computing operations in real time as discussed in the following subsections.

### 2.1 Image Amplification

A photorefractive crystal can be operated as a coherent image amplifier of high gain ($\sim 3600$) for images of large space-bandwidth product ($\sim 10^6$) and wide dynamic range ($> 100$). Its speed varies inversely with the intensity of the inputs. (For $BaTiO_3$ crystals and milliwatts of input power, the speed is typically in the hundred millisecond range.) Proper operation of this amplifier requires optimization of the two-beam interaction in the photorefractive material [11]. The operational geometry is shown in Figure 12.2. Because of the self-induced grating formation and autodiffraction, the process of two-beam interaction allows the weak signal beam to extract energy from the strong pump beam and be amplified according to

$$G_1 = \frac{(1+r)\exp(\Gamma L_{\text{eff}})}{1 + r\exp(\Gamma L_{\text{eff}})} \tag{12.2}$$

The gain in Eq. (12.2) is defined as the intensity ratio of the output signal beam in the presence of a pump beam to that in the absence of a pump beam, $r$ is the intensity ratio of the input beams

$$r = \frac{I_1(0)}{I_2(0)} \tag{12.3}$$

and $\Gamma$ is the exponential gain coefficient related to the coupling strength $\gamma(\theta, \beta)$

$$\Gamma \triangleq 2\,\text{Re}[\gamma(\theta, \beta)] \tag{12.4}$$

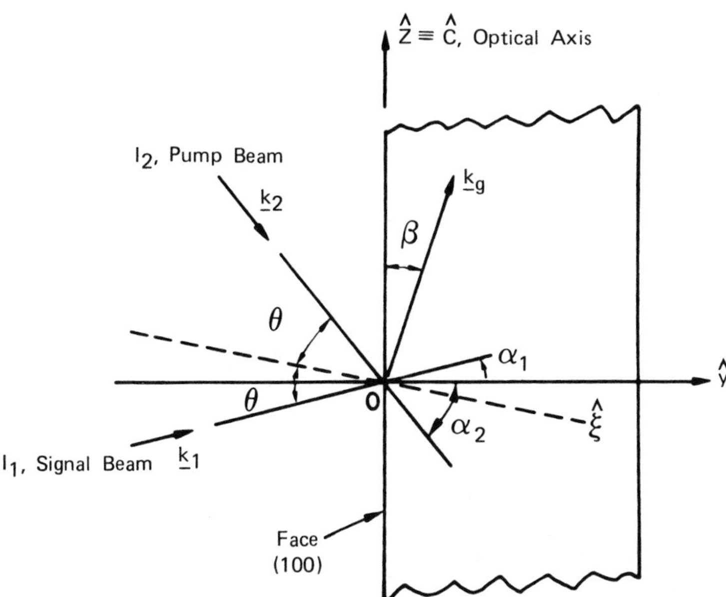


**Figure 12.2** Geometric configuration of two-wave mixing in nonlinear photorefractive material: $\mathbf{k}_1$ and $\mathbf{k}_2$ are wave vectors of the interacting signal beam and pump beam; $\mathbf{k}_g$ is the wave vector of the grating induced by the intensity distribution from the interference of the two intersecting beams; $\alpha_1$ and $\alpha_2$ are the angles of the beams with respect to the surface normal; $\hat{c}$ is the optical axis of the crystal, $2\theta$ is the angle between the two beams; $\beta$ is the angle of $\mathbf{k}_g$ from the $c$-axis; $L$ is the thickness of the crystal; and $\hat{\xi}$ is a unit vector in the direction of the bisector of the two beams [11].

where $\theta$ is half of the angle between the interacting beams, $\beta$ is the angle between the grating wave vector and the crystal optic axis $\hat{c}$, and the effective interaction length is approximately equal to

$$L_{\text{eff}} = \frac{L}{\cos\beta} \tag{12.5}$$

where $L$ is the thickness of the crystal.

To amplify an image, which consists of plane waves (or beams) of various amplitudes propagating at various angles within the range $2\theta_B \pm \Delta/2$ and $\beta_B \pm \Delta/2$, we need to determine the conditions on $r$, $\theta_B$, and $\beta_B$ under which constant amplification can be obtained. Figure 12.3 shows that to maintain uniform amplification of an image the varying intensities of the plane wave components must satisfy the saturation condition

$$\frac{1}{r} \gg \exp(\Gamma L_{\text{eff}}) > 1 \tag{12.6}$$

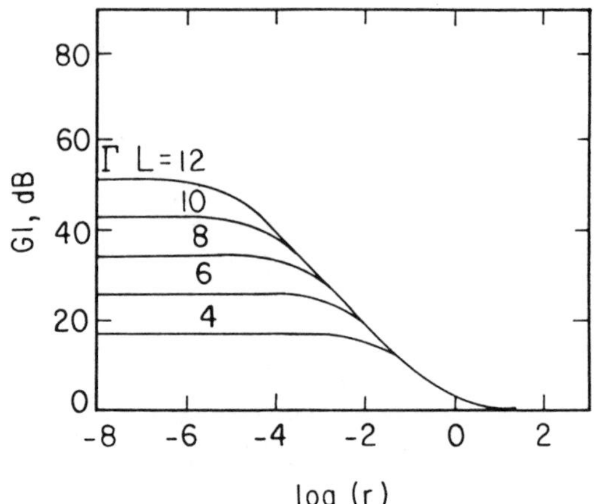


**Figure 12.3** Plot of signal beam gain in dB versus input beam intensity ratio for several values of $\Gamma L$. When the varying intensities of the plane wave components of an image satisfy $\log(r) < -2$ and the crystal is properly oriented at angles $\theta_B$ and $\beta_B$ such that $\Gamma L = 7$, the image will be amplified uniformly with 30 dB or 1000:1 gain.

which gives rise to the saturation gain

$$G_{1\text{sat}} = \exp(\Gamma L_{\text{eff}}) \tag{12.7}$$

Figure 12.4 shows the dynamic range characteristics of an actual coherent image amplifier.

To optimize $\Gamma$ over the angular bandwidth of the input signal (see Figure 12.5), we biased the signal of angular bandwidth $\pm\Delta/2$, at $2\theta_B$ and $\beta_B$. Optimal image amplification can be achieved if the gain remains constant, within 3 dB of $G_{1\text{sat}}$, over the desired signal bandwidth $\Delta$. The photographs of Figures 12.6a and 12.6b illustrate the quality of the amplified U.S. Air Force (USAF) resolution target for an optimized and nonoptimized amplifier, respectively. The optimization procedures yield experimentally a gain of 4000, dynamic range of greater than 100, and resolution of 250 lp/mm. For a 4 mm × 4 mm aperture $BaTiO_3$ crystal we have a space-bandwidth product (SBP) of $10^6$ [11].

## 2.2 2-D Optical Logic

Three nonlinear phenomena of the two-beam interaction in photorefractive materials can be employed to perform different optical logic operations [12].

### Saturation Gain

The saturation gain region shown in Figure 12.3 can be applied to implement optically the logic OR operation by employing the system schematically shown in Figure 12.7a. According to Eq. (12.7), the weak signal beam is amplified by the

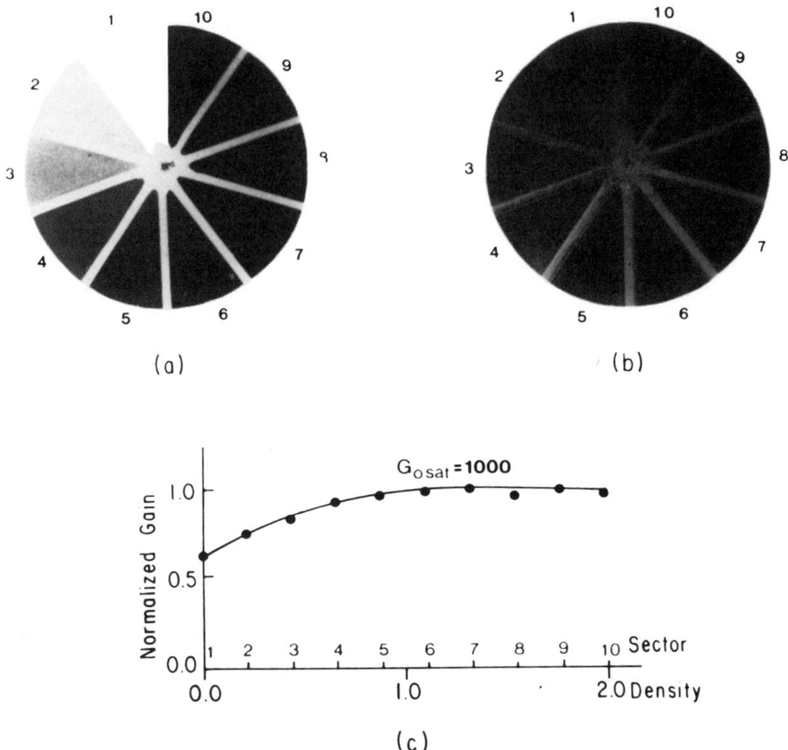


**Figure 12.4** Dynamic range characteristics of the coherent image amplifier. (a) Photograph of the original image used in the experiments, a 10-step gray-level object of maximum density 2.0 (dynamic range 100). (b) Photograph of the amplified gray-level object; the saturation condition was satisfied for the lowest-density segment. (c) Experimentally measured, normalized gain versus density of the gray level object. The saturation gain, $G_{1\text{sat}} = 1000$, was normalized to 1.0.

same constant factor of the saturation gain when one or both of the pump beams are at high intensity levels (logic 1). Given the input functions of Figures 12.8a and 12.8b, the experimental result for the optical logic OR operation is shown in Figure 12.9a. Furthermore, using the schematics of Figure 12.7b, we performed an optical logic AND operation; the experimental result is shown in Figure 12.9b.

## Pump Depletion

When a signal beam is stronger than the pump beam ($r > 1$) and

$$\exp(\Gamma L) \gg 1 \tag{12.8}$$

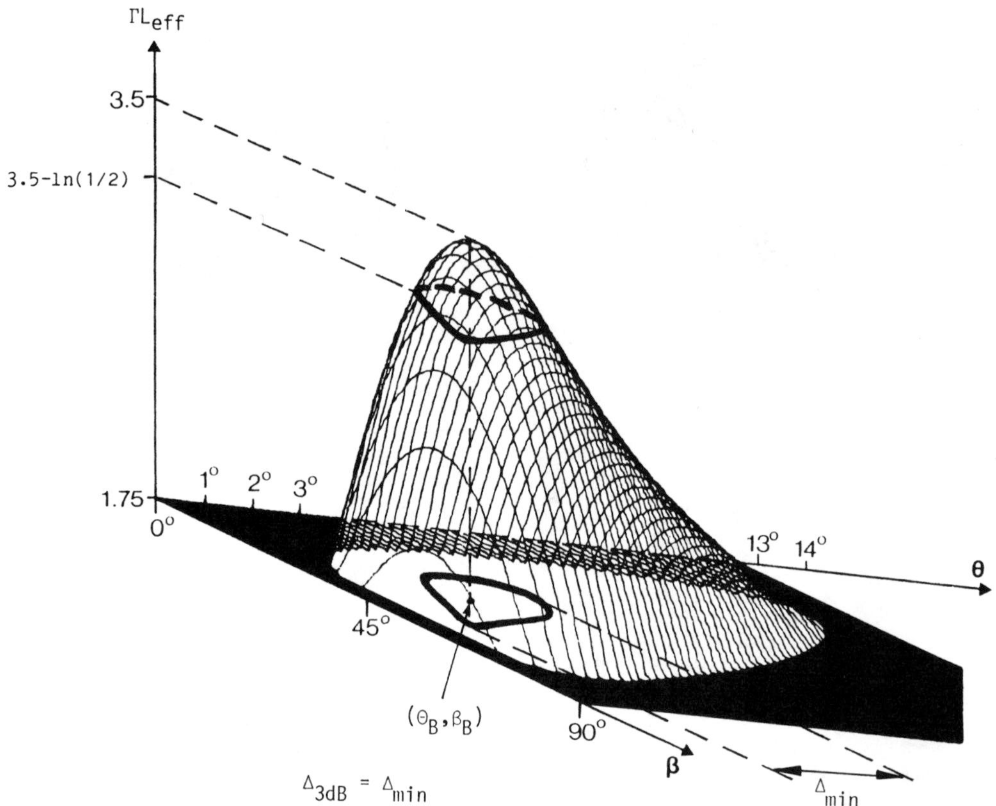


**Figure 12.5** 3-D plot of $\Gamma L_{\text{eff}}$ versus angles $\theta$ and $\beta$. To ensure the 3 dB gain variation, $\frac{1}{2}G_{1\text{sat}}^{\max} \leq G(\theta,\beta) \leq G_{1\text{sat}}^{\max}$, it is necessary to choose the bias angles $(\theta_B, \beta_B)$ such that all spatial frequency components of the image within bandwidth $\pm\Delta/2$ will experience values of $\Gamma L_{\text{eff}}$ within the interval $(\Gamma L_{\text{eff}})_{\max} + \ln(1/2) \leq \Gamma(\theta,\beta)L_{\text{eff}} \leq (\Gamma L_{\text{eff}})_{\max}$.

The pump depletion phenomenon is obtained, as shown in Figure 12.10, with

$$G_{2\text{dep}} = \left(2 + \frac{1}{r}\right)\exp(-\Gamma L) \tag{12.9}$$

where $G_{2\text{dep}}$ is the value of the pump beam attenuation factor. Using the system schematically shown in Figure 12.7c and the crystal operated in the pump beam depletion mode, an optical logic NOR operation was performed to yield the results shown in Figure 12.9c.

## Optically Controlled Two-Beam Coupling

By introducing an additional beam (the third, or control beam) which is incoherent with the previously introduced signal and pump beams (by choice of polarization,

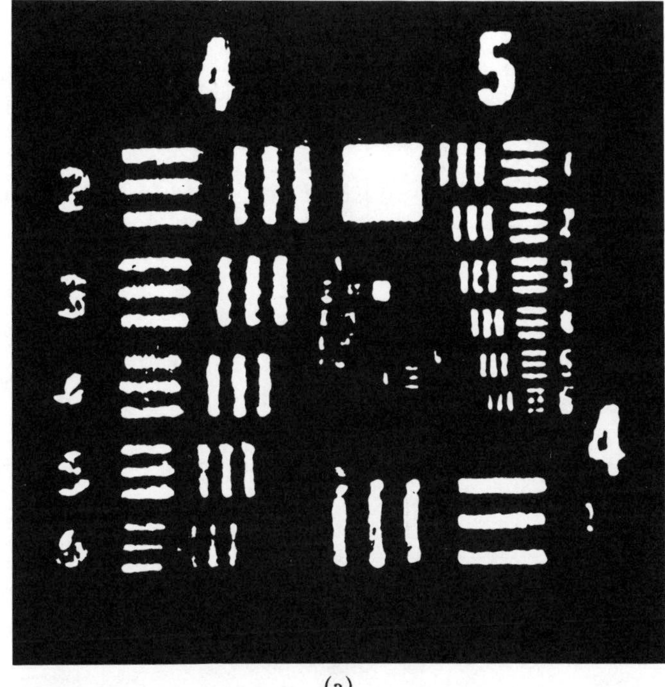


(a)

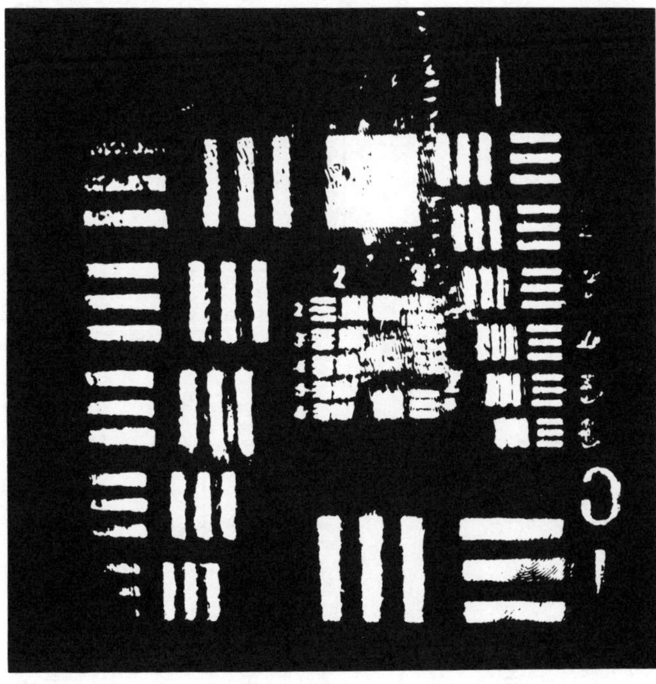


(b)

**Figure 12.6** Coherent image amplification of a USAF resolution target using a poled crystal of $BaTiO_3$ with no externally applied electric field. (a) An optimally biased amplifier can resolve spatial frequencies as high as 250 lp/mm. (b) A nonoptimized amplifier can resolve spatial frequencies no higher than 40 lp/mm [11].

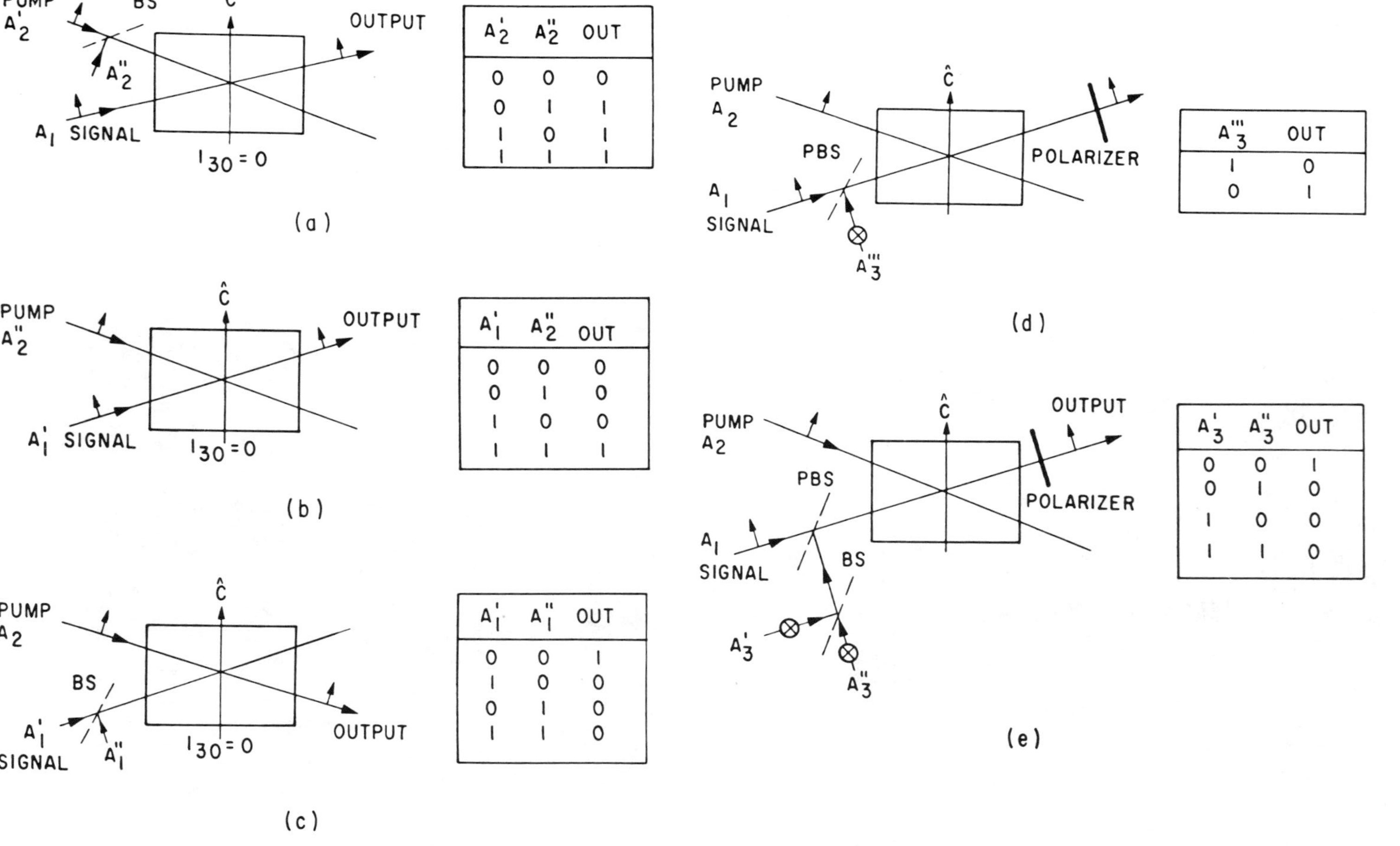


(a)

| $A_2'$ | $A_2''$ | OUT |
|---|---|---|
| 0 | 0 | 0 |
| 0 | 1 | 1 |
| 1 | 0 | 1 |
| 1 | 1 | 1 |

(b)

| $A_1'$ | $A_2''$ | OUT |
|---|---|---|
| 0 | 0 | 0 |
| 0 | 1 | 0 |
| 1 | 0 | 0 |
| 1 | 1 | 1 |

(c)

| $A_1'$ | $A_1''$ | OUT |
|---|---|---|
| 0 | 0 | 1 |
| 1 | 0 | 0 |
| 0 | 1 | 0 |
| 1 | 1 | 0 |

(d)

| $A_3'''$ | OUT |
|---|---|
| 1 | 0 |
| 0 | 1 |

(e)

| $A_3'$ | $A_3''$ | OUT |
|---|---|---|
| 0 | 0 | 1 |
| 0 | 1 | 0 |
| 1 | 0 | 0 |
| 1 | 1 | 0 |

**Figure 12.7** Schematics of (a) OR gate using signal beam saturation, (b) AND gate using signal beam saturation, (c) NOR gate using pump deletion mode, (d) NEGATION gate using controlled two-beam coupling, and (e) NOR gate using controlled two-beam coupling. BS, Beamsplitter; PBS, polarization-selective beamsplitter [12].

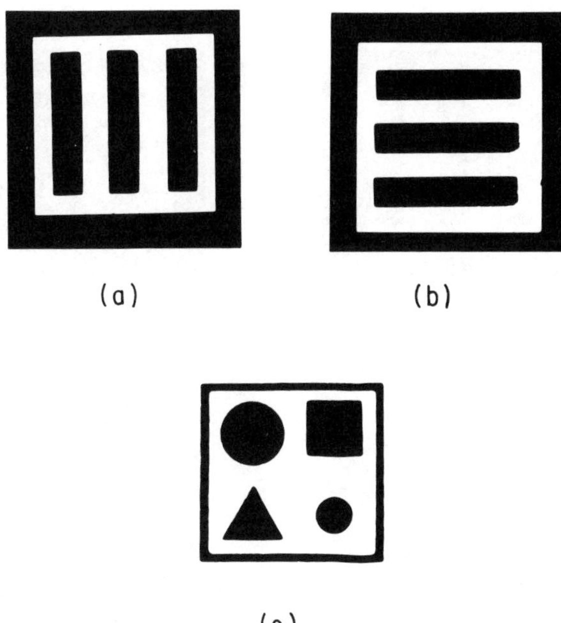


**Figure 12.8** Photograph of the input images (a) $A'$, (b) $A''$, and (c) $A'''$ used in the experiments [12].

coherence length, wavelength, etc.), we can control the coupling phenomenon by erasing the self-induced gratings formed by the signal and pump beam, which gives rise to an exponential gain coefficient

$$\Gamma' = \Gamma \frac{1}{1+p} \tag{12.10}$$

where the parameter $p$ is defined by

$$p \triangleq \frac{I_{30}}{I_0} \tag{12.11}$$

Here $I_{30}$ is the intensity of the third beam at the input, $I_0 = I_{10} + I_{20}$ is the total intensity of the signal and pump beams at the input, and $\Gamma$ is the gain coefficient when $I_{30} = 0$. The dependence of the signal beam gain on the control parameter $p$ is plotted in Figure 12.11. This mode of operation is used to perform a NEGATION operation by employing the schematics of Figure 12.7d and operating the crystal at curve $\Gamma L = 6$. The experimental illustration for the NEGATION function of Figure 12.8c is shown in Figure 12.9d. In another experiment, employing the schematics of Figure 12.7e, we implement an optical logic NOR operation as illustrated in Figure 12.9e.

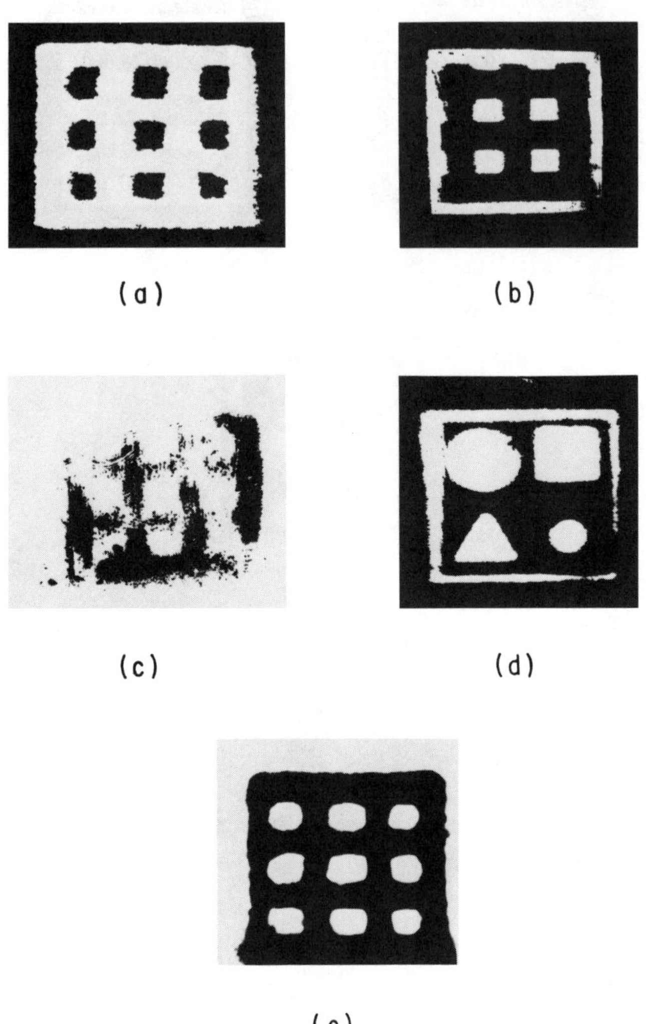


**Figure 12.9** Photographs demonstrating different optical logic operations using the system shown in Figure 12.7 and the inputs shown in Figure 12.8 (a) OR, (b) AND, (c) NOR, (d) NEGATION, and (e) NOR [12].

The 2-D optical logic operations are cascadable. This provides a basis for constructing a flip-flop memory or the arithmetic logic unit of a central processing unit (CPU) in a digital optical processor.

## 2.3 Photorefractive Memory

A photorefractive crystal, being a thick holographic material, can be operated as a memory unit for storing many large SBP images [14]. An example of such a

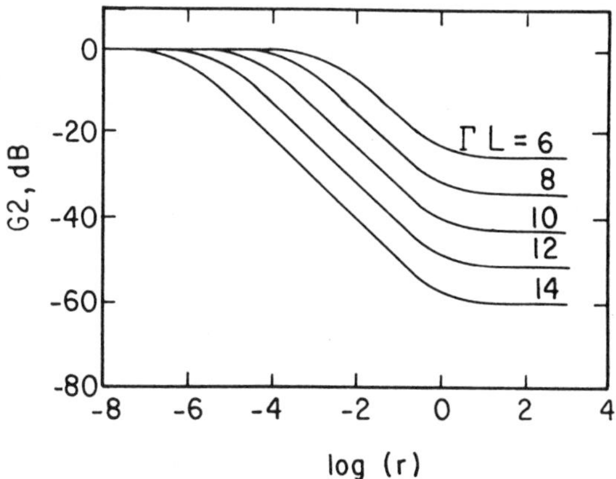


**Figure 12.10** Plot of pump beam gain in dB versus input beam intensity ratio for several values of $\Gamma L$. When the intensity ratio is switched between $\log(r) < -4$ and $\log(r) > 0$, 30 dB or 1000:1 contrast can be obtained for a crystal operated at $\Gamma L = 7$.

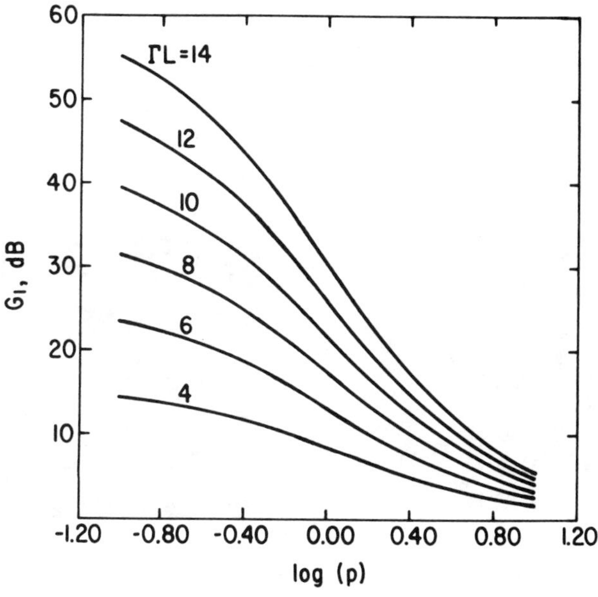


**Figure 12.11** Plot of signal beam gain versus intensity ratio $p = I_3/I_0$ for several values of $\Gamma L$. When the intensity ratio is switched between $\log(p) > 0.4$ and $\log(p) < -0.8$, 30 dB or 1000:1 contrast can be obtained for a crystal operated at $\Gamma L = 8$.

memory unit is shown schematically in Figure 12.12a. When the photorefractive crystal is rotated to an angular position $\omega_i$, the $i$th image is stored via the process of grating formation in photorefractive materials. The angular increment of the photorefractive crystal can be obtained by mounting it on a galvanometric scanner under the control of a microcomputer. To read the $i$th image from such a memory unit, the input is turned off and the pump beam turned on when the photorefractive crystal is rotated to the $\omega_i$ angular position. Figure 12.12b shows the experimental results on recalling the first and tenth stored images.

## 2.4 Phase Conjugation and Convolution/Correlation

Increasing the number of interacting waves from two to four, we have the four-wave mixing geometry shown in Figure 12.13. The four waves are made of two pairs of counterpropagating beams, all of the same frequency, intersecting inside a medium possessing a third-order optical susceptibility. Three of the four optical waves, $U_1$,

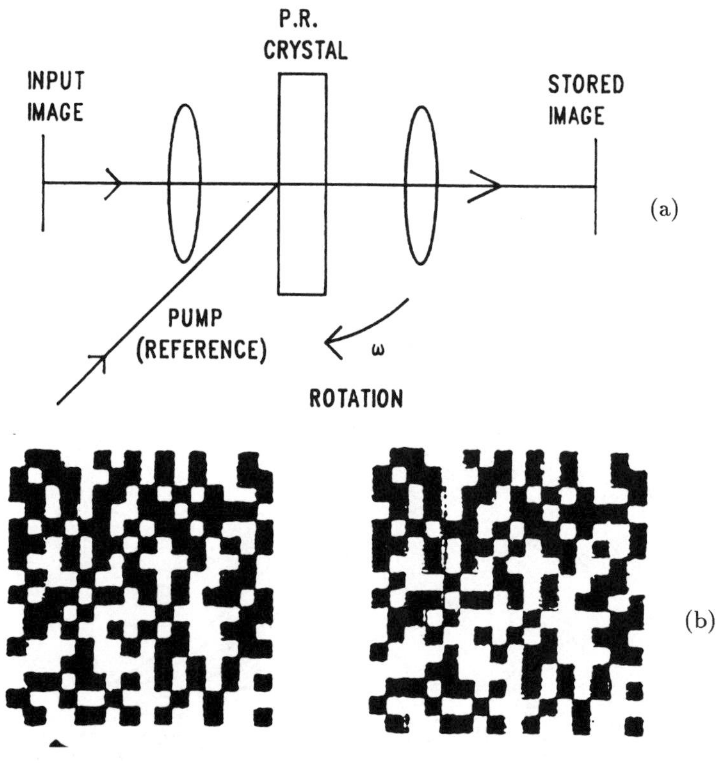


**Figure 12.12** Photorefractive memory unit. (a) When the photorefractive crystal is rotated to the $\omega_i$ angular position, the $i$th image is stored. To read the $i$th image from the memory unit, the input is turned off and the pump beam turned on when the crystal is rotated to the $\omega_i$ angular position. (b) Photograph of the first and tenth images read from the photorefractive memory.

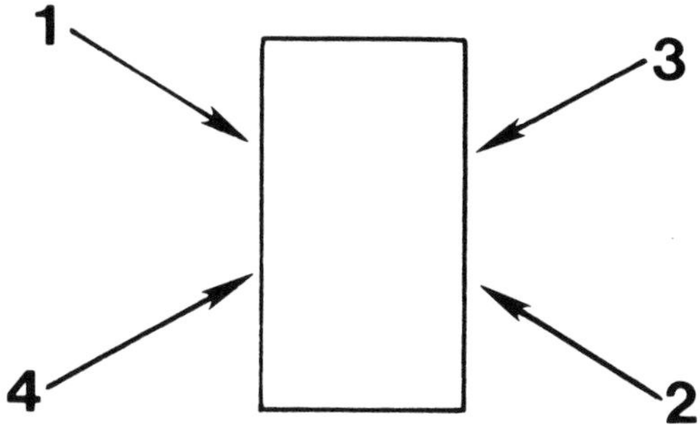


**Figure 12.13** Basic configuration of the four-wave mixing experiment.

$U_2$, and $U_4$, are the inputs to the nonlinear medium to stimulate the radiation of the output beam $U_3$ in accord with [15–17]

$$U_3 \propto U_1 U_2 U_4^* \tag{12.12}$$

When the signals $U_1$ and $U_2$ are set to be plane waves, the photorefractive crystal is operated as a phase conjugate mirror ($U_3 \propto U_4^*$).

Performing a Fourier transformation on both sides of Eq. (12.12), we obtain

$$u_3 \propto u_1 * u_2 \circledast u_4 \tag{12.13}$$

For the special case of $U_1$ being a plane wave, the last equation can be recognized as the correlation of $u_2$ and $u_4$. Hence we can perform real-time spatial filtering.

## 3 APPLICATIONS IN OPTICAL SIGNAL PROCESSING

The elementary optical computing operations discussed in Section 2 can be applied to solve problems of matrix algebra, pattern recognition and image processing, and digital optical processing.

### 3.1 Matrix Inversion/Eigenvalue Problem

An optical feedback system consisting of a confocal Fabry–Perot (CFP) processor has been applied to solve partial differential and integral equations [26]. Recently, it has also been applied to matrix inversion by introducing a photorefractive image amplifier into the feedback loop [19]. The system is shown schematically in Figure 12.14. Under the steady-state condition of the system, the output signal can be related to the input signal by the expression

$$u_{\text{out}}(x, y) = \alpha_1 S_1 u_{\text{in}}(x, y) + \alpha_1 \alpha_2 g S_1 S_2 u_{\text{out}}(x, y) \tag{12.14}$$

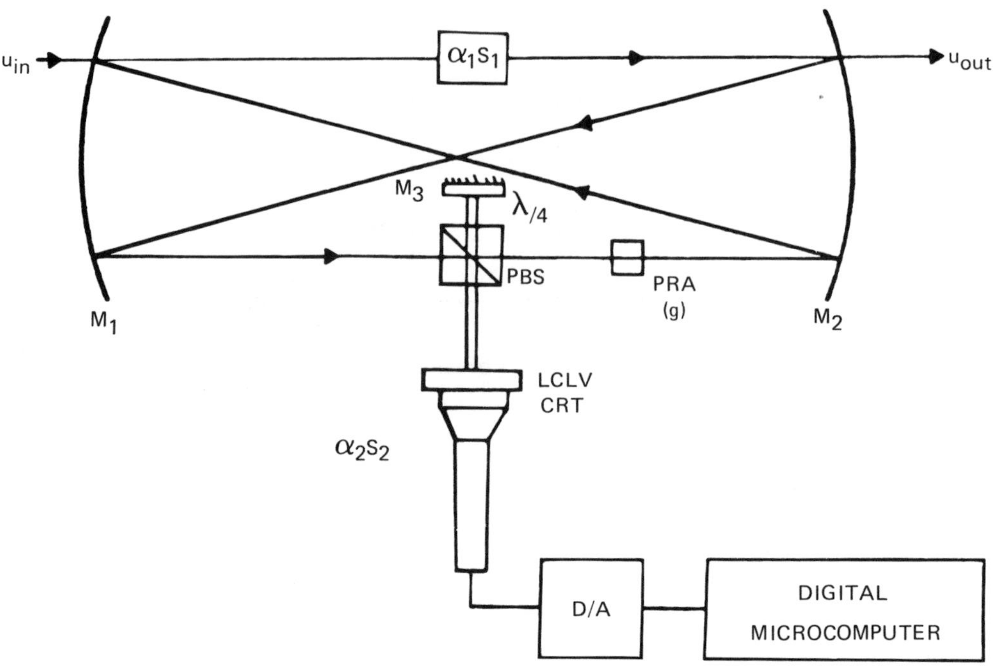


**Figure 12.14** A confocal Fabry–Perot processor: $M_1$ and $M_2$ are specially designed mangin mirrors; $\alpha_1 S_1$ is the open loop operator; $\alpha_2 S_2$ is the feedback loop operator, which is introduced into the CFP by a CRT–LCLV interface to a digital computer; mirror $M_3$, a $\lambda/4$ retardation plate, and a polarization beamsplitter PBS are combined to read the CRT–LCLV; PRA is a photorefractive amplifier of gain $g$.

where $\alpha_1$ and $\alpha_2$ are constants associated with the losses in the open loop and the feedback loop, respectively, $g$ is the amplitude gain of the photorefractive amplifier, and $S_1$ and $S_2$ are the open loop and feedback loop operators, respectively. The transfer function of the system is

$$L = \frac{u_{\text{out}}}{u_{\text{in}}} = \frac{\alpha_1 S_1}{I - \alpha_1 \alpha_2 g S_1 S_2} \tag{12.15}$$

where $I$ is the identity operator. (Typical values of $\alpha_1$ and $\alpha_2$ are 0.98 and 0.05, respectively; $g \leq 60$).

## Matrix Inversion

Let the operator $S_1$ be the identity operator ($S_1 = \mathbf{I}$) and $S_2$ represent a matrix-vector multiplier ($S_2 u = \mathbf{Au}$). Under these assumptions the output of the CFP is given by

$$\mathbf{u}_{\text{out}} = \frac{\alpha_1 \mathbf{I}}{\mathbf{I} - \alpha_1 \alpha_2 g \mathbf{A}} \mathbf{u}_{\text{in}} \tag{12.16}$$

Defining the matrix **B**

$$\mathbf{B} \triangleq \mathbf{I} - \mathbf{A} \tag{12.17}$$

and controlling the gain such that $\alpha_1 \alpha_2 g = 1$, we obtain

$$\mathbf{u}_{\text{out}} = \alpha_1 \mathbf{B}^{-1} \mathbf{u}_{\text{in}} \tag{12.18}$$

Therefore, when the input vector $\mathbf{u}_{\text{in}} = [1\ 0\ 0 \cdots]^T$, it can be shown that the output of the CFP will be the first column of $\mathbf{B}^{-1}$. By indexing the position of the unity element in the input vector $\mathbf{u}_{\text{in}}$ we can generate all the columns of the $\mathbf{B}^{-1}$ matrix [19].

The convergence of this algorithm is governed by the properties of the matrix **A**, requiring all the eigenvalues $|\lambda_i| < 1$. According to Eq. (12.17), the convergence requirements will be fulfilled by any positive definite matrix **B**. The experimental results for a matrix of size $8 \times 8$ are shown in Figure 12.15.

## Matrix Eigenvector/Eigenvalue Problem

When the photorefractive amplifier is pumped harder so that the CFP system oscillates even without any input, the denominator of Eq. (12.15) must vanish:

$$I - \alpha_1 \alpha_2 g S_1 S_2 = 0 \tag{12.19}$$

The resultant output can be determined from

$$(I - \alpha_1 \alpha_2 g S_1 S_2) u_{\text{out}}(x, y) = 0$$

or

$$S_1 S_2 u_{\text{out}}(x, y) = \frac{1}{\alpha_1 \alpha_2 g} u_{\text{out}}(x, y) \tag{12.20}$$

Equation (12.20) has the general form of an eigenvalue problem with $S_1 S_2$ as the operator, $u_{\text{out}}(x, y)$ as the eigenfunction, and $1/\alpha_1 \alpha_2 g$ as the eigenvalue. For $S_1 = \mathbf{I}$

$\underline{\underline{B}} = 10^{-2} \times$

| | | | | | | | |
|---|---|---|---|---|---|---|---|
| 81 | -6 | -8 | -3 | -3 | -2 | -3 | -4 |
| -6 | 84 | -8 | -8 | -6 | -5 | -4 | -2 |
| -8 | -8 | 84 | -3 | -3 | -1 | -2 | -2 |
| -3 | -8 | -3 | 84 | -5 | -6 | -8 | -6 |
| -3 | -6 | -3 | -5 | 82 | -4 | -5 | -3 |
| -2 | -5 | -1 | -6 | -4 | 80 | -4 | -2 |
| -3 | -4 | -2 | -8 | -5 | -4 | 80 | -3 |
| -4 | -2 | -2 | -6 | -3 | -2 | -3 | -81 |

(a)

$B^{-1} = 10^{-2} \times$

| | | | | | | | |
|---|---|---|---|---|---|---|---|
| 127 | 12 | 14 | 9 | 8 | 6 | 8 | 9 |
| 12 | 124 | 14 | 15 | 12 | 11 | 9 | 6 |
| 14 | 14 | 123 | 8 | 6 | 4 | 3 | 5 |
| 9 | 15 | 8 | 124 | 11 | 12 | 15 | 12 |
| 8 | 12 | 6 | 11 | 126 | 9 | 11 | 8 |
| 6 | 11 | 4 | 12 | 9 | 127 | 9 | 6 |
| 8 | 9 | 3 | 15 | 11 | 9 | 129 | 8 |
| 9 | 6 | 4 | 12 | 8 | 6 | 8 | 126 |

(b)

**Figure 12.15** (a) Matrix to be inverted; (b) the inverted matrix; (c) the inverted matrix obtained column by column sequentially as the unity element in the input vector is shifted from the uppermost to the lowermost position. [Note that in (b) and (c) the dynamic range is $\leq 43$.]

and $S_2$ a matrix-vector multiplier, Eq. (12.20) can be rewritten as

$$\mathbf{A}\mathbf{u}_{\text{out}} = \frac{1}{\alpha_1 \alpha_2 g} \mathbf{u}_{\text{out}} \tag{12.21a}$$

which can be compared with

$$\mathbf{A}\mathbf{X}_i = \lambda_i \mathbf{X}_i \tag{12.21b}$$

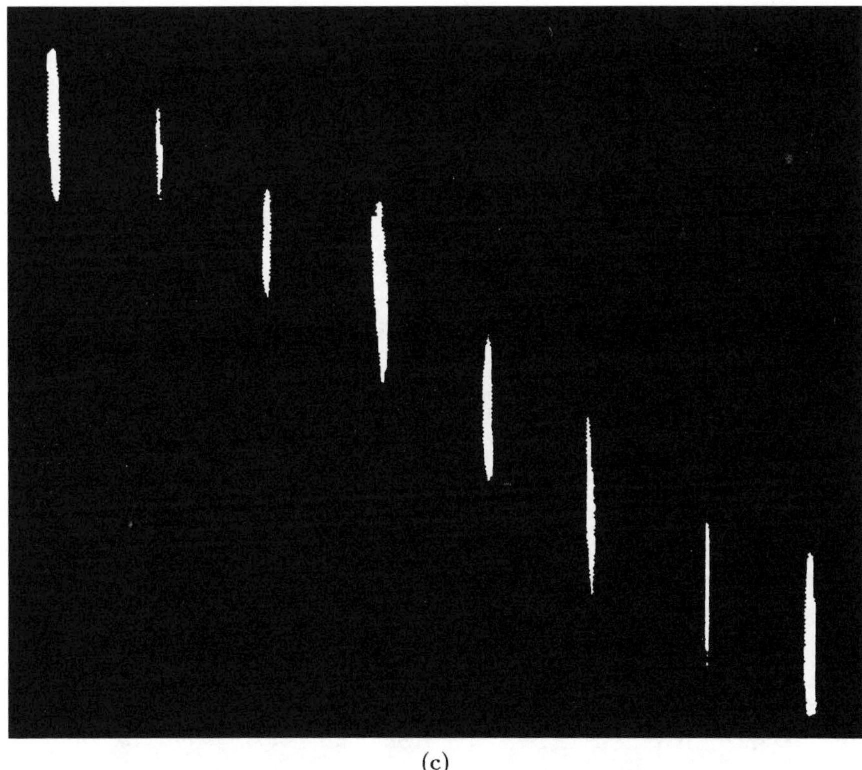

(c)

The solution of the oscillating system $\mathbf{u}_{out}$ is one of the eigenvectors of matrix $\mathbf{A}$, which corresponds to the eigenvalue $1/\alpha_1\alpha_2 g$ [20].

Generally, the analog solutions obtained for matrix inversion or eigenvector/eigenvalue problems are of limited accuracy. Solutions of limited accuracy are often adequate for a variety of applications. However, when solutions of higher accuracy are needed, the approximate solutions from these analog systems can be used as the initial solutions of digital systems and can be helpful in reducing significantly the number of iterations and computing time required on digital systems.

## 3.2 Pattern Recognition and Image Processing

Real-time phase conjugation and correlation operations are attractive for real-time image processing and pattern recognition. The accuracy limitations of analog systems are found not to be important issues if proper recognition or processing algorithms are chosen, as shown in the following examples.

### Defect Detection

Defects in a periodic mask have been detected using four-wave interaction in photorefractive $Bi_{12}GeO_{20}$ (BGO) and $BiSiO_{20}$ (BSO) crystals [21, 22] (see Fig-

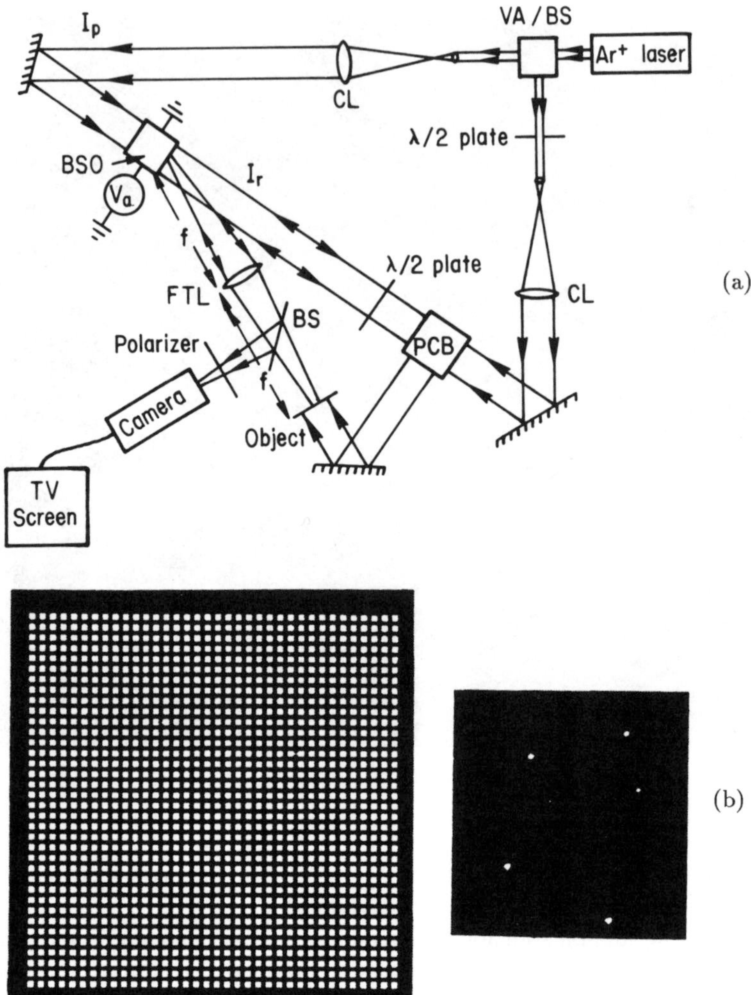


**Figure 12.16** (a) Experimental setup for defect detection. VA/BS, Variable attenuator/beamsplitter; BS, Beamsplitter; CL, collimating lens; PCB, polarizing-cube beamsplitter; FTL, Fourier transform lens. (b) Input mask and output-defect-enhanced image [22].

ure 12.16a). The principle behind this experiment involves controlling the diffraction efficiencies of volume holograms (for the reading beam $I_p$). The diffraction efficiency is maximized when the intensities of writing beams ($I_r$ and $I_0$) are equal and is decreased when the difference in the intensities is increased. Hence, when $I_r$ is chosen such that high-contrast interference fringes are formed with the light from the defects in the mask and low-contrast fringes are formed with that from the periodic structure of the mask, one can achieve real-time enhancement of defects as shown in Figure 12.16b, demonstrating suppression of the periodic structure, leaving the defects clearly visible.

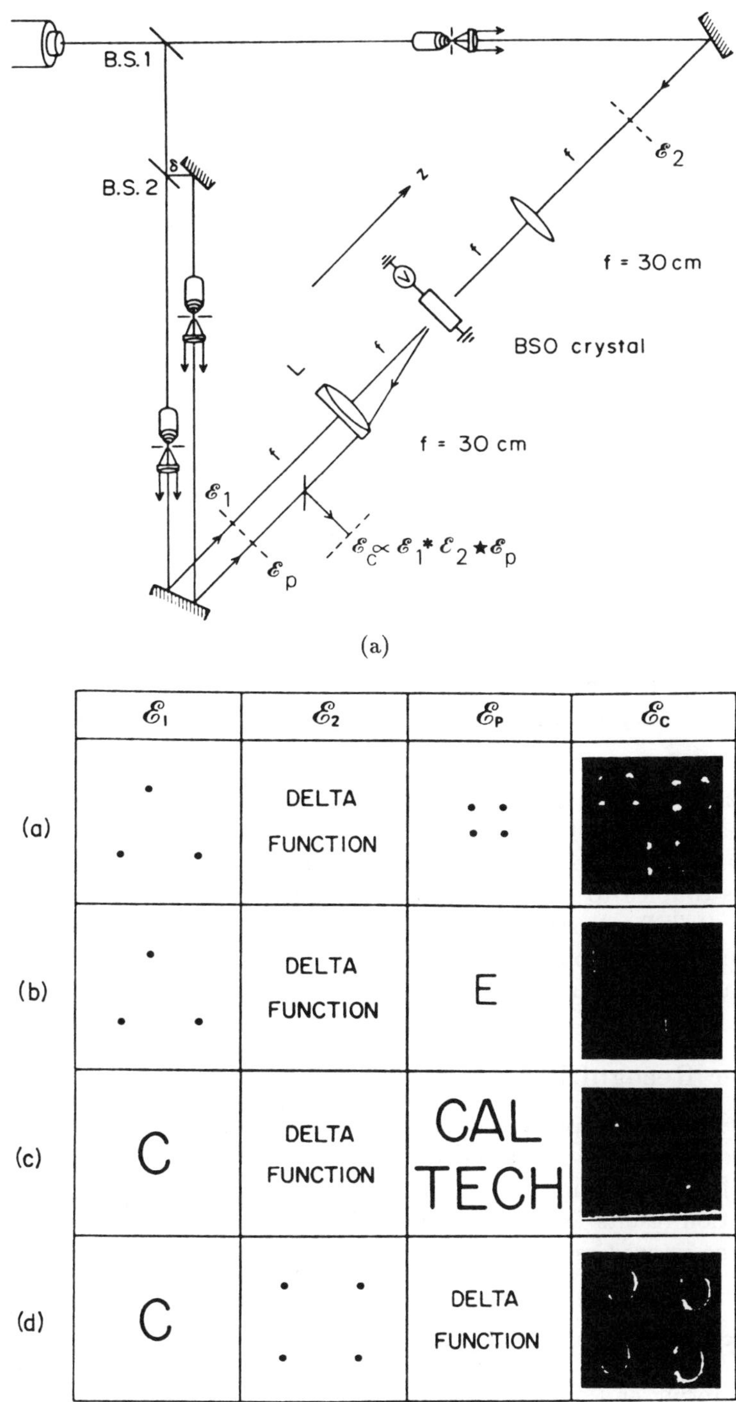


**Figure 12.17** (a) Experimental apparatus for performing real-time correlation via four-wave mixing. Input and output planes are shown by dashed lines. (b) The first three columns are the input objects $E_1$, $E_2$, and $E_p$. The last column shows photographs of the output $E_c$ [17].

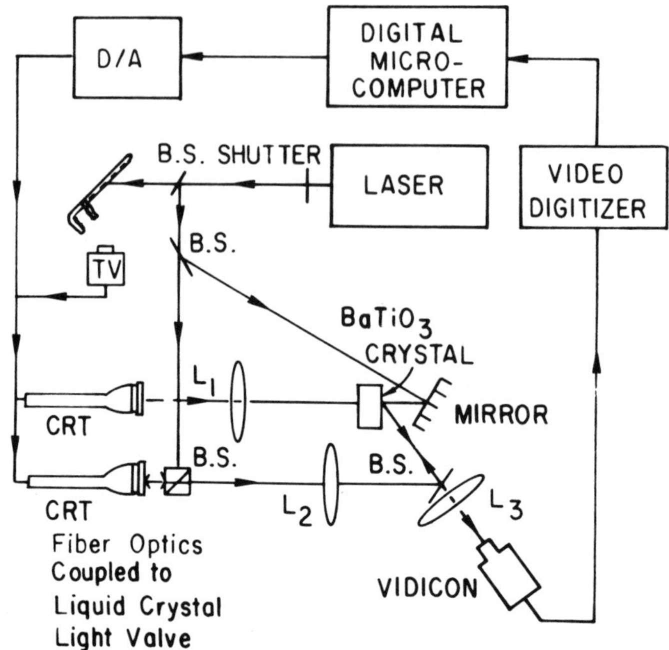


**Figure 12.18** Hybrid optical-electronic system for iterative, adaptive, or recursive processing. The optical processor uses a CRT and is an incoherent processor. The processing functions are programmable using microcomputer-addressable CRT fiber optics coupled to a liquid crystal light valve for writing real-time hologram filters on a $BaTiO_3$ crystal. Interactive processing is possible because the output from the incoherent processor can be input to the microcomputer and the output from the microcomputer can be input to the incoherent processor.

## Character Recognition

A real-time coherent optical correlator has been constructed [17, 18] employing a BSO photorefractive crystal operated in the phase-conjugated mode (see Figure 12.17a). The experiments use various functions for $E_1$, $E_2$, and $E_p$ to obtain the correlation results for $E_c$ as shown in Figure 12.17b. The real-time speed depends on the intensities of the three input beams and the frame rate of the input devices.

## Optical Statistical Pattern Recognition

Matched filtering and optical correlation are optimal for recognition of deterministic patterns. When the pattern varies statistically, new algorithms for statistical pattern recognition must be optically implemented; e.g., the Fukunaga–Koontz transform [27] and the Foley–Sammon transform [28] are good algorithms for two-class problems; and the Hotelling trace criterion transform [29] and the least-squares linear mapping technique are good algorithms for multiple-class problems [30]. Beginning with a reasonable number of training images, one can apply these statis-

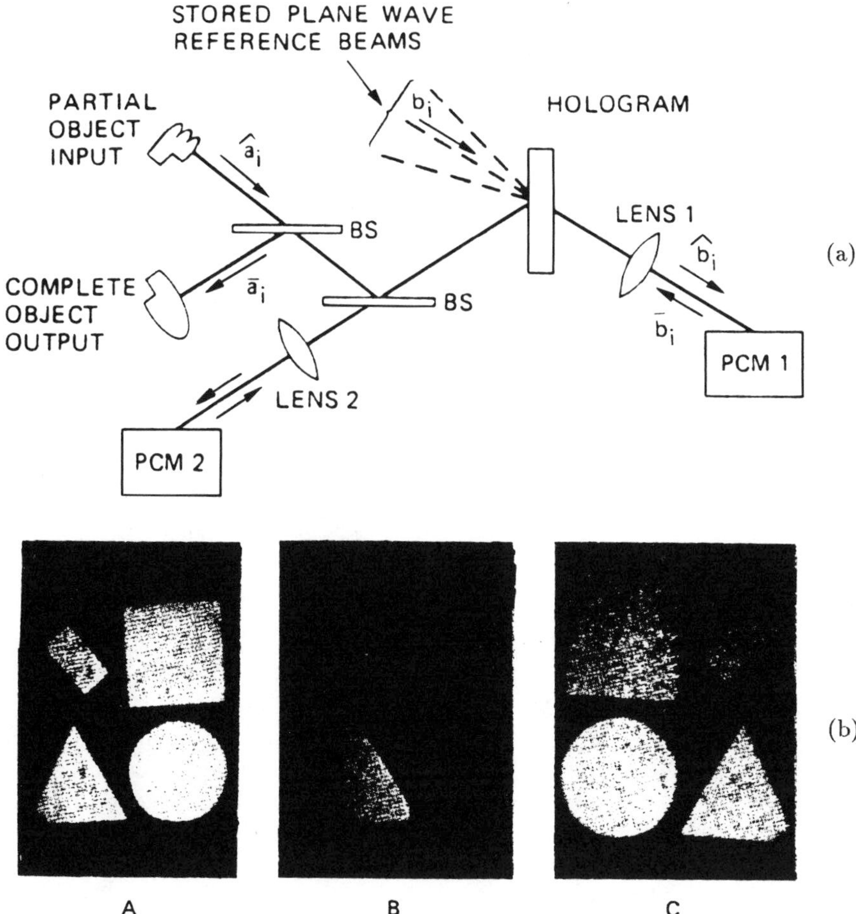


**Figure 12.19** (a) Implementation of an associative holographic memory using PCMs. (b) Experimental results from A, image stored in the memory, and B, incomplete input image. C is the associated output image [24].

tical algorithms to extract off-line the synthetic discriminant functions, which are to be used in the place of matched filters. With a $BaTiO_3$ crystal and a computer-addressable cathode ray tube (CRT) coupled through fiber optics to a liquid crystal light valve as shown in the hybrid system of Figure 12.18, one can perform real-time processing [23].

The optical processing portion of the hybrid system in Figure 12.18 consists of a CRT; lenses $L_1$, $L_2$, and $L_3$; the $BaTiO_3$ crystal; and the vidicon. It works on the principle of an incoherent processor [31]. The filtering information in $BaTiO_3$ can be updated in real time; therefore, it is a real-time processing system. Since the CRTs and the vidicon are connected to the microcomputer, the hybrid system can implement iterative pattern recognition algorithms. For example, we start with the synthetic discriminant function (SDF) $M_{K-1}$, obtained from a set of training

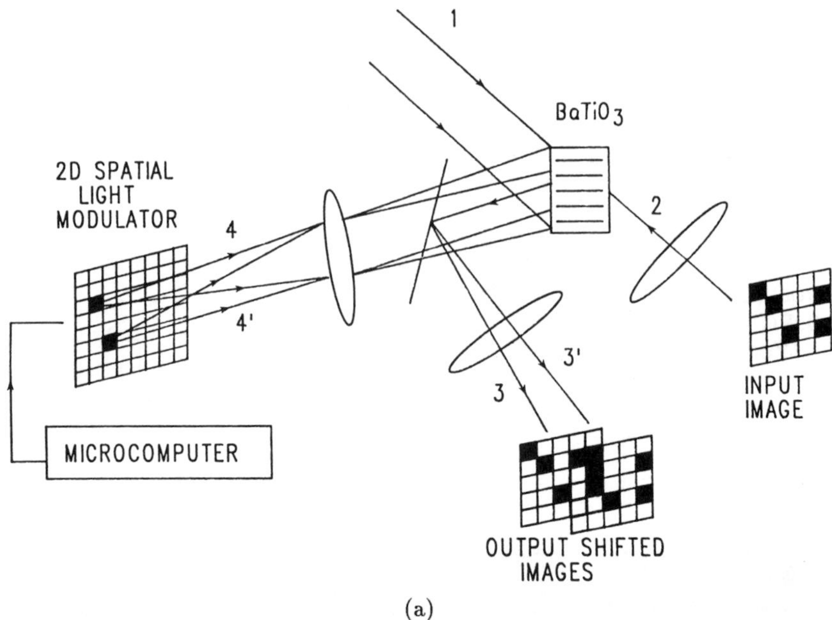


**Figure 12.20** Image shifter using a photorefractive crystal PR. (a) The image shifter is operated in the four-wave mixing configuration. Beam 1 is a plane wave; beam 4 is introduced from a computer-addressable spatial light modulator (SLM). The image to be shifted is introduced as beam 2, resulting in beam 3, which is a shifted version of the image in beam 2. The amount and direction of the shift are determined by the location of the pixel on the SLM under computer control. (b) Experimental results on shifting an image by controlling the location of a pixel displayed on the SLM with a microcomputer.

images and one of the statistical pattern recognition algorithms. Analyzing the results from the test image, we determine the error and change the SDF according to [32]

$$M_k = M_{k-1} + (y_k - M_{k-1}X_k)C_k^T \tag{12.22}$$

where $C_k$ is the gain vector that defines the correction, $X_k$ is the vector representing the $k$th test image, $M_{k-1}X_k$ is the prediction of the signal in the decision plane, and $y_k - M_{k-1}X_k$ can be seen to be proportional to the prediction error. The optimal $M$ is reached if the same $M$ is obtained from successive pairs of $(X_k, y_k)$ in the iterative process.

## Associative Holographic Memory

The principle of information retrieval by association has been suggested as a basis for parallel computing and as the process by which the human memory functions [32]. Various associative memory schemes that use electronic or optical means

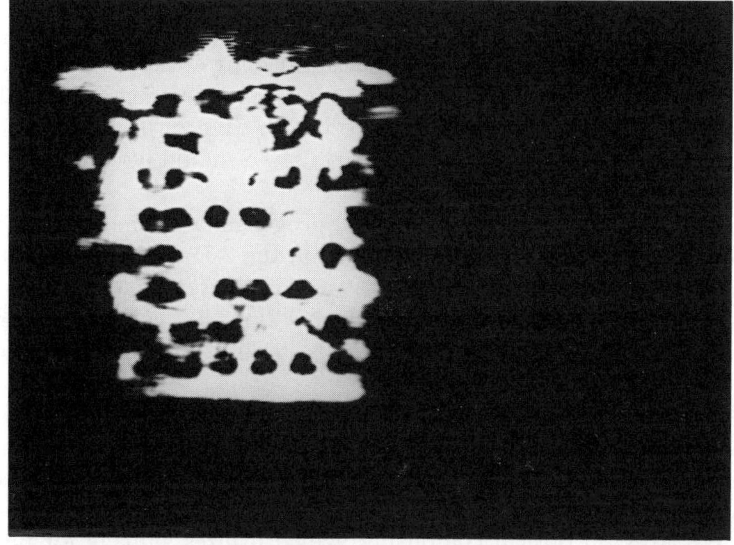

(b)

have been proposed. Recently, an associative holographic memory (thermoplastic) with feedback using phase-conjugated mirrors (PCMs) has been reported [24]; the phase-conjugated mirrors are used for beam retroreflection, gain, and thresholding. When part of the stored object is introduced into the processor, the output reconstructs the entire object associated with the partial input (Figure 12.19). One of the phase-conjugated mirrors is operated in the threshold mode so that the combined effects of optical feedback and thresholding operation will minimize crosstalk in associative recalls.

## 3.3 Digital Optical Processing

The capability of photorefractive crystals in performing optical logic operations [12] was summarized in Section 2.2. The application of photorefractive crystals to holographic memory [14] was discussed in Section 2.3. Hence, if we can also perform shifting operations with photorefractive crystals, we should be able to use them to construct a digital optical computer.

Figure 12.20 shows one possible scheme for programmable image shifting by using a photorefractive crystal. When the ON pixel in the 2-D spatial light modulator is switched from one location to another, gratings of different spatial frequencies will be formed in the photorefractive crystal, which will in turn cause the image to shift according to the principle of four-wave mixing. Figure 12.20b shows photographs of experimental results.

Shifting images is the simplest form of global interconnection. Global interconnections and 2-D input/output format are among the most important features of digital optical computing. Other important features are programmability and solution accuracy. Figure 12.21 shows a single-instruction, multiple-data (SIMD) architecture for digital optical computing, which utilizes shifting for interconnection. Beginning with binary images stored in memory, it can implement parallel algorithms for fast transform and solve partial differential equations [25] and matrix algebra problems [33].

We can construct a digital optical computer, initially with photorefractive crystals, in order to study parallel algorithms and system issues. When multiple quantum well devices [34], silicon/PLZT spatial light modulators [35], or organic polymers [36] are further developed and proved to be better devices, photorefractive crystals can then be replaced.

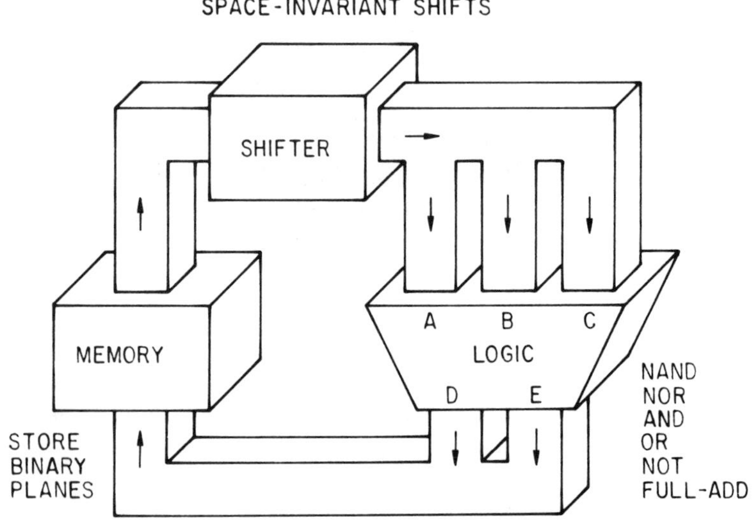


**Figure 12.21** Shift-connected SIMD machine block diagram [25].

## 4 CONCLUSION

The availability of nonlinear optical crystals has added new capabilities to the field of optical signal processing. Their applications to problems of analog optical computing, digital optical processing, pattern recognition, and image processing appear to be very promising. Further improvements in the speed and sensitivity of photorefractive crystals or new developments in organic polymers will stimulate further growth in the areas of nonlinear optics and optical information processing.

## REFERENCES


1. D. L. Staebler and J. J. Amodei, Coupled-wave analysis of holographic storage in $LiNbO_3$, *J. Appl. Phys.*, *43*: 1042–1049 (1972).
2. R. L. Townsend and J. T. La Macchia, Optically induced refractive index changes in $BaTiO_3$, *J. Appl. Phys.*, *41*: 5188–5192 (1970).
3. A. M. Glass, The photorefractive effect, *Opt. Eng.*, *17*: 740–479 (1978).
4. N. V. Kukhtarev, V. B. Markov, S. G. Odulov, M. S. Soskin, and V. L. Vinetskii, Holographic storage in electrooptic crystals, *Ferroelectrics*, *22*: 949–964 (1979).
5. J. Feinberg, D. Heiman, A. R. Tanguay, Jr., and R. W. Hellwarth, Photorefractive effect and light-induced charge migration in barium titanate, *J. Appl. Phys.*, *51*: 1297–1305 (1980).
6. G. C. Valley and M. B. Klein, Optimal properties of photorefractive materials for optical data processing, *Opt. Eng.*, *22*: 704–711 (1983).
7. J. Feinberg, Optical phase conjugation in photorefractive materials, *Optical Phase Conjugation*, (R. A. Fisher, ed.), Academic Press, New York, Chapter 11, pp. 417–443 (1983).
8. P. Yeh, Exact solution of a nonlinear model of two-wave mixing in Kerr media, *J. Opt. Soc. Am. B*, *3*: 747–750 (1986).
9. F. Laeri, T. Tschudi, and J. Albers, Coherent CW image amplifier and oscillator using two-wave interaction in a $BaTiO_3$-crystal, *Opt. Commun.*, *47*: 387–390 (1983).
10. J. P. Huignard, H. Rajbenbach, Ph. Refreigier, and L. Solymar, Wave mixing in photorefractive bismuth silicon oxide crystals and its applications, *Opt. Eng.*, *24*: 586–592 (1985).
11. Y. Fainman, E. Klancnik, and S. H. Lee, Optimal coherent image amplification by two-wave coupling in photorefractive $BaTiO_3$, *Opt. Eng.*, *25*: 228–234 (1986).
12. Y. Fainman, C. C. Guest, and S. H. Lee, Optical digital logic operations by two-wave coupling in photorefractive material, *Appl. Opt.*, *25*: 1598–1603 (1986).
13. J. B. Thaxter and M. Kestigian, Unique properties of SBN and their use in a layered optical memory, *Appl. Opt.*, *13*: 913–924 (1974).
14. Y. Fainman, H. Rajbenbach, and S. H. Lee, Applications of photorefractive crystals as basic computational modules for digital optical processing, *J. Opt. Soc. Am. A.*, *3*: P16 (1986).

15. M. Gonin-Golomb, B. Fisher, J. O. White, and A. Yariv, Theory and applications of four-wave mixing in photorefractive media, *IEEE J. Quant. Electr.*, *QE-20*: 12–30 (1984).
16. D. M. Pepper, J. Au Yeung, D. Fekete, and A. Yariv, Spatial convolution and correlation of optical fields via degenerate four-wave mixing, *Opt. Lett. 3*: 7–9 (1978).
17. J. O. White and A. Yariv, Spatial information processing and distortion correction via four-wave mixing, *Opt. Eng.*, *21*: 224–230 (1982).
18. B. Loiseaux, G. Illiaguer, and J. P. Huignard, Dynamic optical cross-correlation using a liquid crystal light valve and a bismuth silicon oxide crystal in the Fourier plane, *Opt. Eng.*, *24*: 144–149 (1985).
19. H. Rajbenbach, Y. Fainman, and S. H. Lee, Optical implementation of iterative algorithm for matrix inversion, *Appl. Opt.*, *26*: 1024–1031 (1987).
20. H. Rajbenbach, Y. Fainman, and S. H. Lee, Real time eigenvector determination by self-induced coherent optical oscillations, *J. Opt. Soc. Am. A. 3*: P24 (1986).
21. E. Ochoa, L. Hesselink, and J. W. Goodman, Real-time intensity inversion using two-wave and four-wave mixing in photorefractive $Bi_{12}SiO_{20}$, *Appl. Opt. 24*: 1826–1832 (1985).
22. E. Ochoa, J. W. Goodman, and L. Hesselink, Real-time enhancement of defects in a periodic mask using photorefractive $Bi_{12}SiO_{20}$, *Opt. Lett.*, *10*: 430–432 (1985).
23. S. H. Lee, Optical implementation of digital algorithms for pattern recognition, *Opt. Eng.*, *25*: 69–75 (1986).
24. B. H. Soffer, G. J. Dunning, Y. Owechko, and E. Marom, Associative holographic memory with feedback using phase-conjugate mirrors, *Opt. Lett.*, *11*: 118–120 (1986).
25. T. J. Drabik and S. H. Lee, Shift connected SIMD array architecture for digital optical computing systems, with algorithms for numerical transforms and partial differential equations, *Appl. Opt.*, *25*: 4053–4064 (1986).
26. S. H. Lee, Optical analog solutions of partial differential and integral equations, *Opt. Eng. 24*: 41–47 (1985).
27. J. R. Leger and S. H. Lee, Image classification by an optical implementation of the Fukunaga–Koontz transform, *J. Opt. Soc. Am.*, *72*: 556–564 (1982).
28. Q. Tian, M. Barbero, Z. H. Gu, and S. H. Lee, Image classification by Foley–Sammon transform, *Opt. Eng.*, *25*: 834–840 (1986).
29. Z. H. Gu and S. H. Lee, Optical implementation of the Hotelling trace criterion for image classification, *Opt. Eng.*, *23*: 727–731 (1984).
30. Z. H. Gu, J. R. Leger, and S. H. Lee, Optical implementation of the least-square linear mapping technique for image classification, *J. Opt. Soc. Am.*, *72*: 787–793 (1982).
31. A. W. Lohmann, Matched filtering with self-luminous objects, *Appl. Opt.*, *71*: 561–563 (1968).
32. T. Kohonen, *Self-Organization and Associative Memory*, Springer-Verlag, New York, pp. 181–188 (1984).
33. T. J. Drabik and S. H. Lee, Parallel algorithms for matrix algebra problems on shift-connected digital optical single-instruction multiple-data arrays, *J. Opt. Soc. Am. A.*, *3*: P15 (1986).

34. H. M. Gibbs, S. S. Tarng, J. L. Jewell, D. A. Weinberger, K. Tai, A. C. Gossard, S. L. McCall, A. Passner, and W. Wiegmann, Room-temperature excitonic optical bistability in GaAs–GaAlAs superlattice etalon, *Appl. Phys. Lett.*, *41*: 221–222 (1982).
35. S. H. Lee, S. C. Esener, M. A. Title, and T. J. Drabik, Two-dimensional silicon/PLZT spatial light modulators: design considerations and technology, *Opt. Eng.*, *25*: 250–260 (1986).
36. G. M. Carter, Y. J. Chen, and S. K. Tripathy, Intensity dependent index of refraction in organic materials, *Opt. Eng.*, *24*: 609–612 (1985).

# 13

# Adaptive Signal Processing

THOMAS S. FONG Radar Systems Group, Hughes Aircraft Company, El Segundo, California

## 1 INTRODUCTION

The design of a conventional signal processing system for extracting information from an incoming signal that is corrupted by noise requires a priori information about the characteristics of the input data to be processed. The system performance is optimum, in accordance with a preselected criterion, only when the input data match the a priori information. In many practical situations, a system is required to operate in an environment where the input conditions are uncertain, perhaps not even statistical, or the input conditions change from time to time. Under such circumstances, the system can achieve good performance only if it has the flexibility to alter the values of a set of parameters which govern the system performance, making the adjustments based on the measured input data or the estimated statistical properties of the relevant signals. A system which searches for improved performance guided by a computational algorithm for adjustment of the parameters or weights is called an adaptive system. By its very nature, an adaptive system is time-varying.

Much of the early work in the 1960s on adaptive signal processing was associated with adaptive filtering [1–3] and adaptive interference nulling in antennas [4–6]. Recent progress in microcircuit technology has had a great impact on speed and cost in signal processing, and many extremely difficult computational problems of the past have become readily solvable. Over the past two decades, application of adaptive signal processing has expanded to various fields—communication, control, radar, sonar, seismology, and biomedical electronics—and many excellent texts and articles are available [7–14].

An adaptive process may be either open-loop or closed-loop. The open-loop adaptive process first makes measurements on the input data, learns the statistics of the data, and then applies this knowledge to solve for the adjustable parameters that optimize the performance. The closed-loop adaptive process, on the other hand, has a performance feedback feature. This algorithm operates in an iterative manner and updates the adjustable parameters with the arrival of new data and current signal processor performance feedback. In each iteration the system learns more about the characteristics of the input signal and the signal processor and makes adjustments of the current set of parameters based on the latest system performance. The optimum set of values of the adjustable parameters is then approached sequentially. The principle of open-loop and closed-loop adaptive signal processing systems is illustrated in Figures 13.1 and 13.2.

## 2 EXAMPLES OF ADAPTIVE SYSTEMS

The purpose of the following three examples is to illustrate the common features in the technique of formulating the problem for solution in adaptive signal processing. The various methods of obtaining the solution are presented in a later portion of this chapter. Throughout this chapter, only discrete systems are considered. In each case, the input signal sequence $\{x(k)\}$ is operated on by the adaptive processor to produce an output sequence $\{y(k)\}$ whose objective is to match the desired response $\{d(k)\}$, or drive the error sequence $\{e(k)\}$ toward zero.

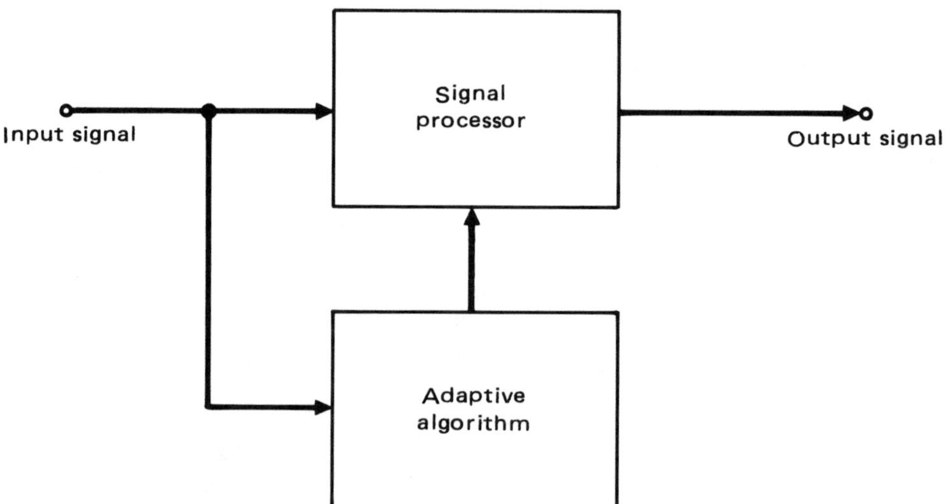


**Figure 13.1** Open-loop adaptive signal processing system.

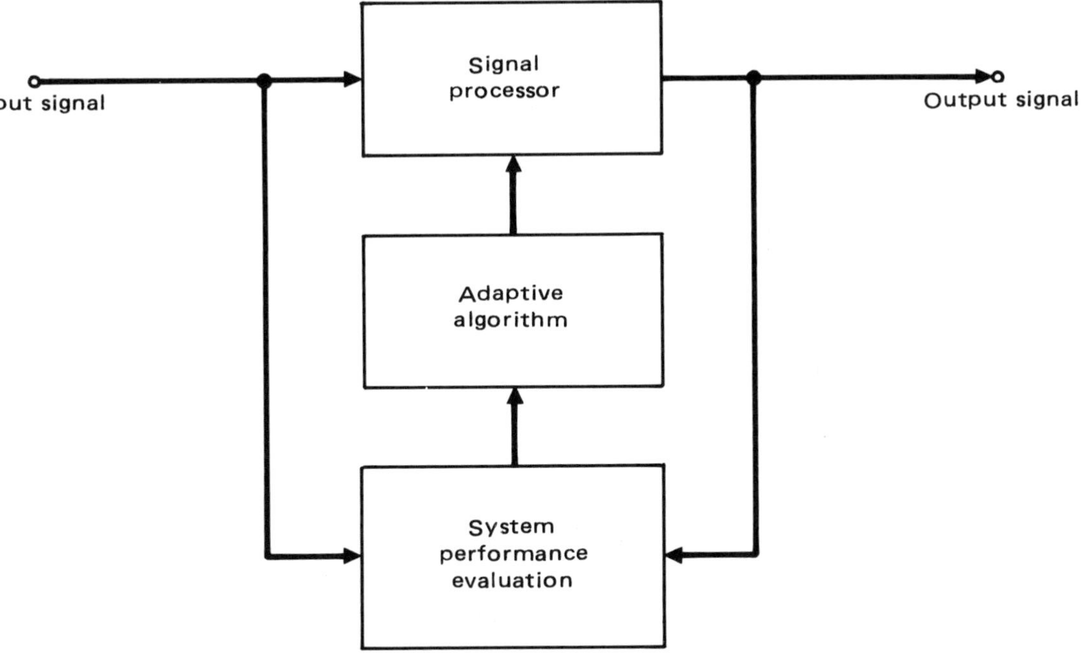


**Figure 13.2** Closed-loop adaptive signal processing system.

**Example 1: Adaptive Prediction Filter.** Suppose that $\{x(k)\}$, $k = 0$, 1, 2, ..., a stationary sequence representing the input signal, is applied to the adaptive system. The requirement of this system is to predict the value of the signal sequence $N$ sampling time intervals into the future. The signal at time $k$, $x(k)$, can be regarded as the desired response, or $d(k)$ in this case. Consider the situation where the delayed input $x(j)$, $0 \leq j \leq k - N$, is applied to the adaptive processor, an adaptive filter here, which has a set of adjustable parameters, and cause its output $y(k)$ to cancel the input $x(k)$ or $d(k)$. The objective is to vary the filter coefficients so that the error signal sequence $\{e(k)\}$ is minimized in accordance with a preselected criterion. The adjustment of the filter coefficients depends on the error signal, and the adjustment procedure is guided by a computational algorithm to approach the optimum performance. Figure 13.3 illustrates the processing concept. The adaptive predictor is useful in many interference cancellation applications.

If the input is nonstationary and varies slowly in comparison with the sampling period, then the algorithm has the added task of adjusting the filter coefficients continually, seeking the optimum system performance. In this situation, the convergence rate of the algorithm becomes extremely important.

**Example 2: System Modeling.** Adaptive filtering is often used in system modeling, or system identification. The objective is to develop a model for the unknown system, or to determine a set of filter coefficients so that the input-output relationships of the unknown system and the adaptive filter will match. Suppose that a broadband (broad enough to cover the bandwidth of the unknown system)

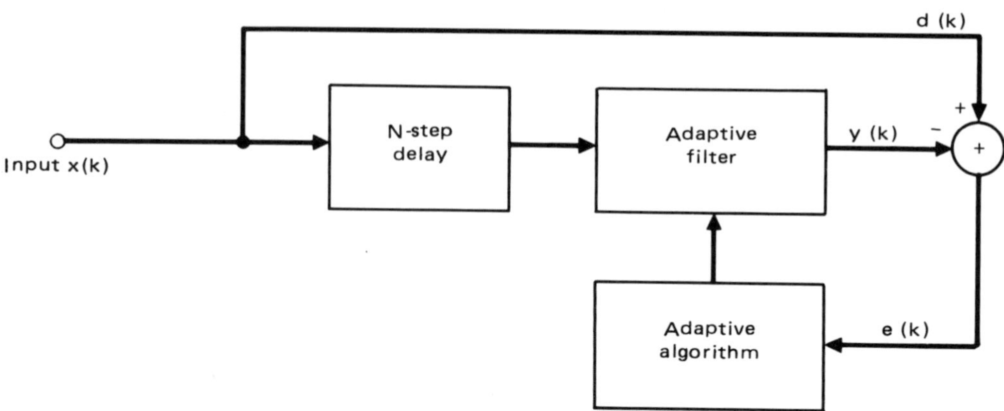


**Figure 13.3** Adaptive predictor.

signal sequence $\{x(k)\}$ is applied to the unknown system and the filter with adjustable coefficients simultaneously as shown in Figure 13.4. Let $\{y(k)\}$ be the output sequence of the filter, and regard the output sequence of the unknown system, $\{d(k)\}$, as the desired response. By comparing $d(k)$ with $y(k)$ for each $k$, an error signal $e(k)$ is produced. Based on the error signal, the filter coefficients are then adjusted following an adaptive algorithm.

**Example 3: Antenna Sidelobe Canceller.** Consider an antenna system, as shown in Figure 13.5, having a high-gain main antenna with low side-

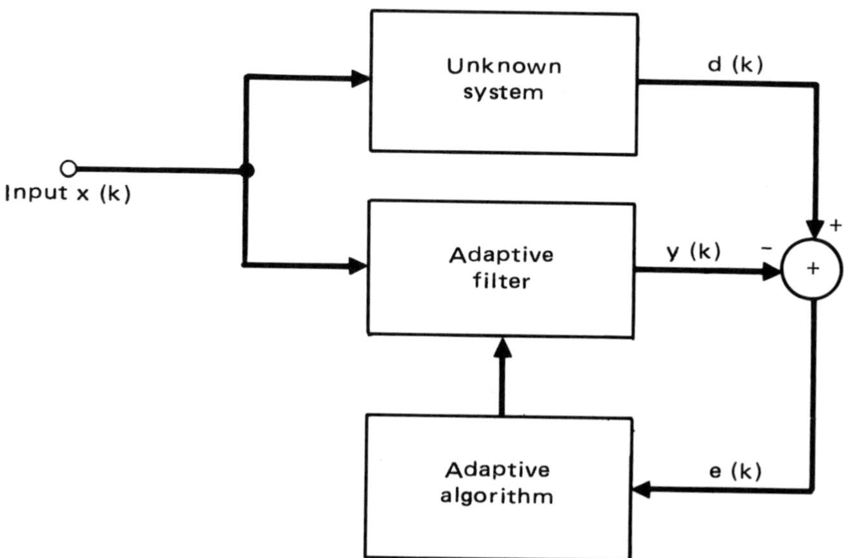


**Figure 13.4** Modeling of an unknown system.

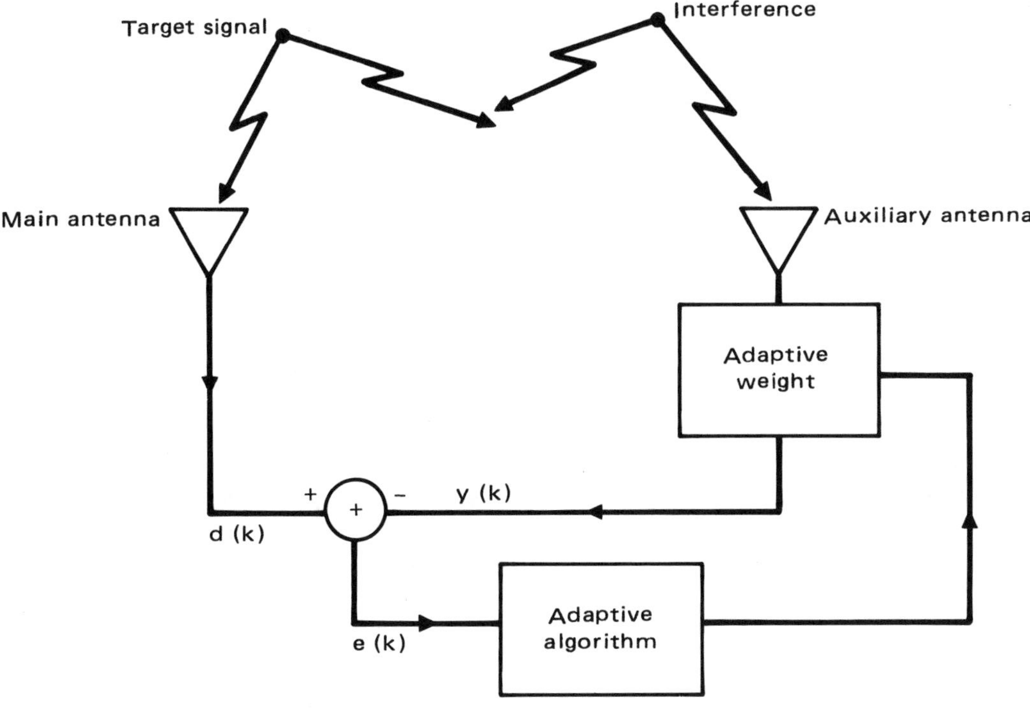


**Figure 13.5** Adaptive antenna sidelobe canceller.

lobes on the order of 30 dB below its maximum, and an auxiliary antenna whose gain approximately equals the gain at the peak of the sidelobe of the main antenna. Suppose the target signal enters the main antenna at its mainlobe peak while the interference comes from a sidelobe direction. Further suppose that the target signal and the interference are uncorrelated. The objective of this adaptive antenna system is to suppress the interference at the combined antenna output.

Suppose the interference is considerably stronger than the target signal in the auxiliary channel. That is, $\{y(k)\}$ as shown in Figure 13.5 is predominantly interference signal. (If the target signal was anywhere near the interference level in the auxiliary channel, then the target signal in the main channel would be dominant over the interference and there would be no need for the sidelobe canceller.) Treating the signal in the main channel as the desired response $d(k)$, an error signal $e(k)$, the difference between $d(k)$ and $y(k)$, is produced. The error signal is, in turn, used to adjust the adaptive weight in the auxiliary channel by an adaptive algorithm.

Since the perturbation of the adaptive process on the mainlobe of the combined antenna system by the low-gain auxiliary antenna is negligible, the target signal detectability is essentially unaffected. After the adaptive processing is successfully completed, the antenna pattern of the combined system will have a very low response in the direction of the interference, or the sidelobe of the main antenna in that direction has been cancelled by the auxiliary antenna.

In each of the three examples, very little is assumed to be known concerning the input signal. A common feature is that the adaptive process is to cause the output sequence $\{y(k)\}$ to agree as closely as possible with the desired response. Because of the presence of the performance feedback, all three examples belong to the closed-loop category. Depending on the application, the choice of the desired response, which leads to the creation of the error signal, often requires some ingenuity on the part of the designer.

## 3 DERIVATION OF THE OPTIMUM WEIGHT VECTOR

Consider the adaptive signal processing systems of the following two configurations:

1. *Single-input system*: The adjustable weights are applied to a sequence of $M$ samples generated from a single source as in a transversal filter or tapped-delay line filter as shown in Figure 13.6.

Let $k$ be the time index. Let the $M$-by-1 complex-valued tap-input vector at time $k$ be

$$\mathbf{x}(k) = \begin{bmatrix} x(k) \\ x(k-1) \\ x(k-2) \\ \vdots \\ x(k-M+1) \end{bmatrix} \tag{13.1}$$

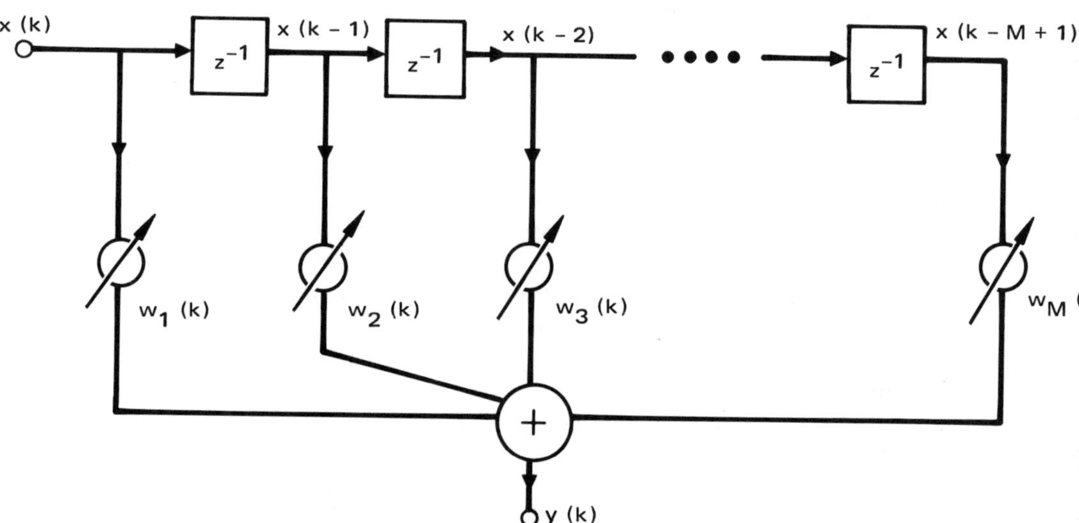


**Figure 13.6** Single-input system, the adaptive transversal filter.

and let the complex-valued weight vector at time $k$ be

$$\mathbf{w}(k) = \begin{bmatrix} w_1(k) \\ w_2(k) \\ w_3(k) \\ \vdots \\ w_M(k) \end{bmatrix} \tag{13.2}$$

The filter output is then given by

$$y(k) = \mathbf{w}^{\mathrm{T}}(k)\mathbf{x}(k) \tag{13.3}$$

where T denotes the transpose operation.

2. *Multiple-input system*: The adjustable weights are applied to $M$ separate channels as in the narrowband signal adaptive antenna illustrated in Figure 13.7. Narrowband here means that the ratio of information bandwidth to carrier frequency is small, say 0.001 or less, and under this condition the propagation time delay is effectively equivalent to phase shift. If we let the $M$-by-1 complex-valued signal vector at time $k$ be $\mathbf{x}(k)$, whose components are the signals appearing in the

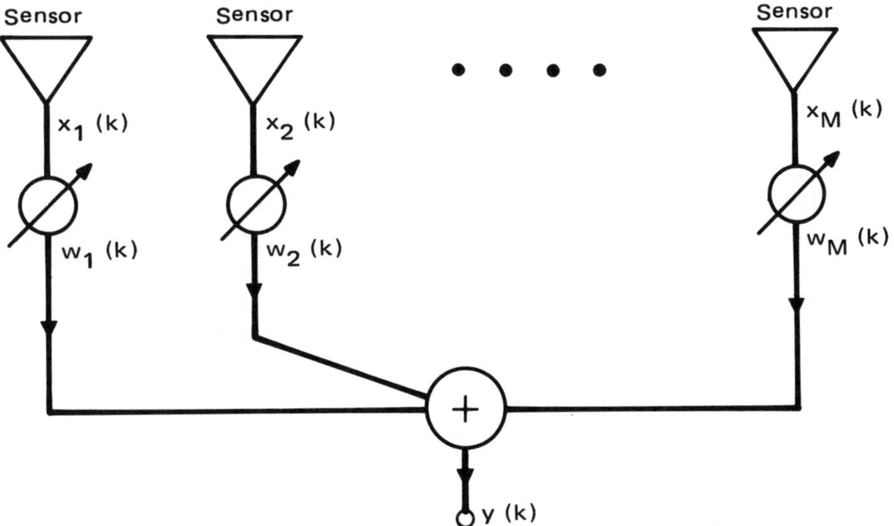


**Figure 13.7** Multiple-input adaptive antenna system.

$M$ channels, that is,

$$\mathbf{x}(k) = \begin{bmatrix} x_1(k) \\ x_2(k) \\ x_3(k) \\ \vdots \\ x_M(k) \end{bmatrix} \tag{13.4}$$

then the system output $y(k)$ is given by

$$y(k) = \mathbf{w}^{\mathrm{T}}(k)\mathbf{x}(k) \tag{13.5}$$

identical to Eq. (13.3). Since the outputs are identical in form, the derivation of the optimum weight vector is therefore applicable to both configurations.

Generally, the system output $y(k)$ is different from the desired response $d(k)$, with the result that the error signal

$$\begin{aligned} e(k) &= d(k) - y(k) \\ &= d(k) - \mathbf{w}^{\mathrm{T}}(k)\mathbf{x}(k) \end{aligned} \tag{13.6}$$

is nonzero. In the following, we assume that $d(k)$ and $\mathbf{x}(k)$ are statistically stationary, and our objective is to determine the weight vector so that the mean-square value of the error signal is minimized.

The orthogonality principle which to be used in the sequel is stated here without proof. For the proof of this subject, we suggest the books by Papoulis [15] and Strang [16].

**Orthogonality Principle.** The error under the best approximation condition is orthogonal to the set of approximants.

In other words, we require the error signal to be orthogonal to the subspace generated by the data. Mathematically, we set the inner product of the error signal and the data to zero. For stochastic quantities, the inner product of $u$ and $v$ is taken as the expected value of $uv$. Figure 13.8 illustrates geometrically the concept of the orthogonality principle associated with our present problem.

In view of the orthogonality principle, to minimize the mean-square value of the error signal as given by Eq. (13.6), we require

$$E\left\{[d(k) - \mathbf{w}^{\mathrm{T}}\mathbf{x}(k)]\bar{\mathbf{x}}^{\mathrm{T}}(k)\right\} = \mathbf{0}^{\mathrm{T}} \tag{13.7}$$

where the overbar denotes the complex conjugate operation and $\mathbf{0}$ is an $M$-by-1 zero vector. For complex signals, complex conjugation must be applied to the second factor on the left-hand side of Eq. (13.7). After taking the transpose, making use of the fact that the transpose of the product of matrices is equal to the product of the

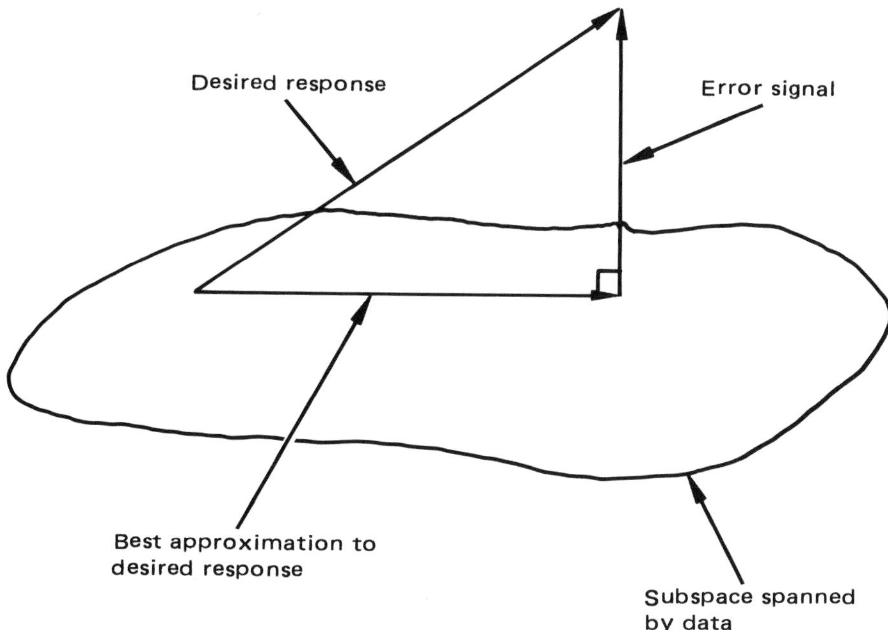


**Figure 13.8** Illustration of the orthogonality principle.

transposes in reverse order, and recalling that $d(k)$ is a scalar quantity, Eq. (13.7) becomes

$$E\{d(k)\bar{\mathbf{x}}(k)\} - E\{\bar{\mathbf{x}}(k)\mathbf{x}^{\mathrm{T}}(k)\}\,\mathbf{w} = \mathbf{0} \tag{13.8}$$

We have deleted the time index $k$ in the weight vector, since we are not interested in adjusting the weights in the present situation.

Let the $M$-by-$M$ correlation matrix be $\mathbf{R}$, that is,

$$\mathbf{R} = E\{\bar{\mathbf{x}}(k)\mathbf{x}^{\mathrm{T}}(k)\} \tag{13.9}$$

Since the transposed complex conjugate of $\mathbf{R}$ is equal to itself, $\mathbf{R}$ is a Hermitian matrix. Next, let us examine the invertibility of the matrix $\mathbf{R}$, which is crucial in the determination of the optimum weight vector. Consider an arbitrary $M$-by-1 vector $\mathbf{v}$, and form the scalar random variable as

$$c = \mathbf{v}^{\mathrm{T}}\bar{\mathbf{x}}(k) \tag{13.10}$$

Equation (13.10) yields

$$\bar{c} = \mathbf{x}^{\mathrm{T}}(k)\bar{\mathbf{v}} \tag{13.11}$$

The mean-square value of the random variable $c$ is

$$\begin{aligned} E\left\{|c|^2\right\} &= E\left\{\mathbf{v}^{\mathrm{T}}\bar{\mathbf{x}}(k)\mathbf{x}^{\mathrm{T}}(k)\bar{\mathbf{v}}\right\} \\ &= \mathbf{v}^{\mathrm{T}}\mathbf{R}\bar{\mathbf{v}} \end{aligned} \tag{13.12}$$

Since

$$E\left\{|c|^2\right\} \geq 0 \tag{13.13}$$

this implies that the quadratic form $\mathbf{v}^{\mathrm{T}}\mathbf{R}\bar{\mathbf{v}}$ is positive semidefinite, and $\mathbf{R}$ is a positive semidefinite matrix. If the quadratic form satisfies the condition

$$\mathbf{v}^{\mathrm{T}}\mathbf{R}\bar{\mathbf{v}} > 0 \tag{13.14}$$

for an arbitrary vector $\mathbf{v}$, then $\mathbf{R}$ is positive definite and consequently invertible. Since in practice only estimates of the correlation matrix are available and such estimates are often based on time samples, for a non-zero-bandwidth process with a sufficient number of samples taken, the matrix $\mathbf{R}$ is almost always positive definite and, therefore, invertible. For further discussion of the subject of positive definiteness of the correlation matrix, see Feller [17] and Jenkins and Watts [18].

Let the $M$-by-1 cross-correlation vector be $\mathbf{p}$, that is,

$$\mathbf{p} = E\left\{d(k)\bar{\mathbf{x}}(k)\right\} \tag{13.15}$$

It follows from Eqs. (13.8), (13.9), and (13.15) that the optimum choice of the weight vector is given by

$$\mathbf{w}^* = \mathbf{R}^{-1}\mathbf{p} \tag{13.16}$$

and this is often referred to as the optimum Wiener solution. An asterisk is used to emphasize this optimum choice.

The inverse of $\mathbf{R}$ can be calculated by many efficient computational algorithms, taking advantage of its Hermitian positive definite properties. The algorithm based on Cholesky's decomposition of the matrix $\mathbf{R}$ is quite attractive, and it is presented below. Only the computational aspect of the algorithm is given here. For a detailed theoretical development of the Cholesky algorithm, see Jain et al. [19] and Strang [16].

**Cholesky Algorithm.** If $\mathbf{R}$ is a Hermitian positive definite matrix, there exists a decomposition $\mathbf{R}$ as

$$\mathbf{R} = \mathbf{C}\mathbf{D}\bar{\mathbf{C}}^{\mathrm{T}} \tag{13.17}$$

where $\mathbf{D}$ is a diagonal matrix and $\mathbf{C}$ is a lower triangular matrix having elements $d_{ij}$ and $c_{ij}$, $i, j = 1, 2, \ldots, M$, respectively.

The solution for $\mathbf{w}$ in Eq. (13.16) may be obtained via the following procedure. The inverse of $\mathbf{R}$ is not explicitly shown in the process. First, determine the elements of $\mathbf{C}$ and $\mathbf{D}$ in terms of the elements of $\mathbf{R}$ or $r_{ij}$, $i, j = 1, 2, \ldots, M$. Let

$$d_{11} = r_{11} \tag{13.18}$$

and

$$c_{jj} = 1 \qquad \text{for } j = 1, 2, \ldots, M \tag{13.19}$$

The remaining elements are then sequentially calculated from the following:

$$b_{ij} = r_{ij} - \sum_{k=1}^{j-1} b_{ik}\bar{c}_{jk} \qquad j = 1, 2, \ldots, i-1 \tag{13.20}$$

$$c_{ij} = \frac{b_{ij}}{d_{jj}} \tag{13.21}$$

$$d_{ii} = r_{ii} - \sum_{k=1}^{i-1} b_{ik}\bar{c}_{ik} \tag{13.22}$$

If the upper limit in the above summations is less than 1, then the sum is defined to be zero. Finally, the elements in the $\mathbf{w}$ vector, $w_i$, $i = 1, 2, \ldots, M$, can be obtained from

$$w_i = \frac{g_i}{d_{ii}} - \sum_{k=i+1}^{M} \bar{c}_{ki} w_k \qquad i = M, M-1, \ldots, 2, 1 \tag{13.23}$$

where

$$g_i = p_i - \sum_{k=1}^{i-1} c_{ik} g_k \qquad i = 1, 2, \ldots, M \tag{13.24}$$

and $p_i$ is the $i$th component of the vector $\mathbf{p}$.

## 4 ERROR PERFORMANCE SURFACE

For the weight vector $\mathbf{w}$ and the assumptions that $d(k)$ and $\mathbf{x}(k)$ are stationary processes, the mean squared error, denoted by $J$ in the following, is given by

$$J = E\left\{(d - \mathbf{w}^{\mathrm{T}}\mathbf{x})(\bar{d} - \bar{\mathbf{w}}^{\mathrm{T}}\bar{\mathbf{x}})\right\} \tag{13.25}$$

On substituting Eqs. (13.9) and (13.15) in (13.25) and making use of the Hermitian property of the matrix $\mathbf{R}$ that $\bar{\mathbf{R}}^{\mathrm{T}} = \mathbf{R}$, we have

$$J = E\left\{|d|^2\right\} - \bar{\mathbf{w}}^{\mathrm{T}}\mathbf{p} - \mathbf{w}^{\mathrm{T}}\bar{\mathbf{p}} + \bar{\mathbf{w}}^{\mathrm{T}}\mathbf{R}\mathbf{w} \tag{13.26}$$

Using $\mathbf{p}$ as given by Eq. (13.16), Eq. (13.26) becomes

$$J = E\left\{|d|^2\right\} - \bar{\mathbf{w}}^{\mathrm{T}}\mathbf{R}\mathbf{w}^* - \bar{\mathbf{w}}^{*\mathrm{T}}\mathbf{R}\mathbf{w} + \bar{\mathbf{w}}^{\mathrm{T}}\mathbf{R}\mathbf{w} \tag{13.27}$$

To obtain the expression for minimum mean squared error, we use the optimum weight vector $\mathbf{w}^*$ for $\mathbf{w}$ in Eq. (13.27). Thus,

$$J_{\min} = E\left\{|d|^2\right\} - \bar{\mathbf{w}}^{*\mathrm{T}}\mathbf{R}\mathbf{w}^* \tag{13.28}$$

Subtracting Eq. (13.28) from (13.27) and rearranging, the mean squared error may be written as

$$J = J_{\min} + \bar{\mathbf{w}}^{*\mathrm{T}}\mathbf{R}\mathbf{w}^* - \bar{\mathbf{w}}^{\mathrm{T}}\mathbf{R}\mathbf{w}^* - \bar{\mathbf{w}}^{*\mathrm{T}}\mathbf{R}\mathbf{w} + \bar{\mathbf{w}}^{\mathrm{T}}\mathbf{R}\mathbf{w} \tag{13.29}$$

or equivalently,

$$J = J_{\min} + (\bar{\mathbf{w}} - \bar{\mathbf{w}}^*)^{\mathrm{T}}\mathbf{R}(\mathbf{w} - \mathbf{w}^*) \tag{13.30}$$

Since $\mathbf{R}$ is positive semidefinite, the quadratic form on the right-hand side of Eq. (13.30) indicates that any departure of the weight vector $\mathbf{w}$ from the optimum $\mathbf{w}^*$ would result in additional positive error. The error performance surface is convex or bowl-shaped and possesses a unique minimum, and this feature is very useful when we utilize search techniques in seeking the optimum weight vector.

## 5 ADAPTIVE ALGORITHMS

When the dimension of the correlation matrix $\mathbf{R}$ is large, the calculation of its inverse to obtain the optimum vector may present a computational burden. We showed previously that finding the optimum weight vector is equivalent to locating the minimum of the performance surface. Therefore, it is desirable to develop more efficient systematic algorithms capable of descending toward the minimum when only measured or estimated data are available. Various gradient-based algorithms such as Newton–Raphson, conjugate gradients, and steepest descent are suitable [20–24]. The selection of an algorithm is normally dictated by the speed of convergence and the computational complexity.

The Newton–Raphson method, in general, involves greater computations and possesses faster convergence. The conjugate gradients technique involves generation of a sequence of expanding subspaces, and the search direction toward the minimum is orthogonal to the previously generated subspaces. The steepest descent method reaches the minimum following the direction in which the performance surface has the greatest rate of decrease. Both the conjugate gradients and steepest descent techniques are simpler computationally and have approximately the same rate of convergence. Because of the similarity of the gradient-based algorithms, only the method of steepest descent is discussed here. In subsequent sections the least-mean-square, sequential matrix inversion, and sequential least-squares fit algorithms for obtaining the optimum weight vector are presented.

### 5.1 Method of Steepest Descent

The steepest descent technique is a powerful iterative computational algorithm for determining extremals of a function. The concept was originated by Cauchy in 1845 (see Forsythe et al. [25]). Geometrically, it is easy to see that with successive corrections of the weight vector in the direction of steepest descent on the convex performance surface, we should arrive at its minimum $J_{\min}$, at which point the weight vector takes on its optimum values. Let $\nabla J$ be the gradient of the performance measure $J$ with respect to weight vector $\mathbf{w}$. $\nabla J$ is a row vector, or a 1-by-$M$ vector. The algorithm proceeds according to the following steps:

1. Select an initial guess of the weight vector where the minimum of the performance measure may be located.
2. Based on the present weight vector, compute the gradient of the performance measure $\nabla J(k)$.
3. Make a correction to the weight vector by

$$\mathbf{w}(k+1) = \mathbf{w}(k) + \mu[-\nabla J(k)]^{\mathrm{T}} \tag{13.31}$$

   where $\mu$ is a positive scalar which controls the size of the correction.
4. Go to step 2 and repeat the process until a predefined stopping rule is satisfied.

For adaptive systems which have a performance measure as given by Eq. (13.26), then

$$(\nabla J)^{\mathrm{T}} = -2\mathbf{p} + 2\mathbf{R}\mathbf{w} \tag{13.32}$$

and the weight vector adjustment in step 3 becomes

$$\mathbf{w}(k+1) = \mathbf{w}(k) + 2\mu\mathbf{p} - 2\mu\mathbf{R}\mathbf{w}(k) \tag{13.33}$$

The signal flow diagram associated with Eq. (13.33) is shown in Figure 13.9, and the feedback nature of the steepest descent algorithm can be seen there. The symbol $z^{-1}$ is the time delay of one iteration period.

For a given correlation matrix **R**, the stability or convergence of the algorithm is governed by the step size parameter $\mu$. Let us examine the restriction on $\mu$ by expressing Eq. (13.33) in its diagonal form, where the various modes are uncoupled [26, 27]. Some facts concerning the Hermitian and positive definite properties of **R** are stated below without proof [16, 28–30].

1. The eigenvalues of **R** are all real and positive.
2. There exists an invertible matrix **C** such that

$$\mathbf{R} = \mathbf{C}^{-1}\mathbf{D}\mathbf{C} \tag{13.34}$$

where **D** is a diagonal matrix consisting of the eigenvalues of **R**.

Consider the transformation of the weight vector **w** by the matrix **C** as

$$\mathbf{w}' = \mathbf{C}\mathbf{w} \tag{13.35}$$

and Eq. (13.33) becomes

$$\mathbf{C}^{-1}\mathbf{w}'(k+1) = \mathbf{C}^{-1}\mathbf{w}'(k) - 2\mu\mathbf{R}\mathbf{C}^{-1}\mathbf{w}'(k) + 2\mu\mathbf{p} \tag{13.36}$$

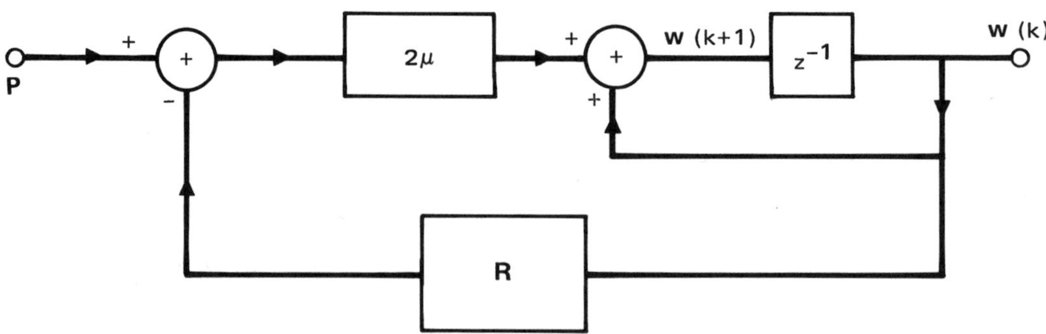


**Figure 13.9** Signal flow diagram of the steepest descent algorithm.

After premultiplication by $\mathbf{C}$, we have the matrix difference equation of (13.36) in its diagonal form as

$$\mathbf{w}'(k+1) = \mathbf{w}'(k) - 2\mu\mathbf{D}\mathbf{w}'(k) + 2\mu\mathbf{C}\mathbf{p} \tag{13.37}$$

or equivalently,

$$\mathbf{w}'(k+1) = [\mathbf{I} - 2\mu\mathbf{D}]\mathbf{w}'(k) + 2\mu\mathbf{C}\mathbf{p} \tag{13.38}$$

where $\mathbf{I}$ is the identity matrix. To determine stability, we examine the homogeneous solution, or the transient response. Therefore, we require all the roots of the following polynomial in $z$

$$\det[z\mathbf{I} - (\mathbf{I} - 2\mu\mathbf{D})] = 0 \tag{13.39}$$

to be less than unity in magnitude, where the symbol det denotes "determinant of" [26, 31]. Since the matrix within the brackets is a diagonal matrix whose diagonal elements are of the form $(z - 1 + 2\mu\lambda)$, where $\lambda$ is an eigenvalue of $\mathbf{R}$, and in view of the positive feature of $\mu$ and $\lambda$, the above stability requirement translates to

$$|-1 + 2\mu\lambda_{\max}| < 1 \tag{13.40}$$

where $\lambda_{\max}$ is the largest eigenvalue of $\mathbf{R}$. Finally, we have the restriction on the step size parameter as

$$\mu < \frac{1}{\lambda_{\max}} \tag{13.41}$$

Next, let us examine the steady-state behavior of the algorithm under the step size restriction of (13.41). The weight vector in the $k$th iteration from solving Eq. (13.38) is

$$\mathbf{w}'(k) = [\mathbf{I} - 2\mu\mathbf{D}]^k\mathbf{w}'(0) + 2\mu\sum_{n=0}^{k-1}[\mathbf{I} - 2\mu\mathbf{D}]^n\mathbf{C}\mathbf{p} \tag{13.42}$$

For the steady-state condition, with $\mu < 1/\lambda_{\max}$, we have

$$\lim_{k\to\infty}[\mathbf{I} - 2\mu\mathbf{D}]^k = \mathbf{0} \tag{13.43}$$

where $\mathbf{0}$ is the square zero matrix, and

$$\lim_{k\to\infty}\sum_{n=0}^{k-1}[\mathbf{I}-2\mu\mathbf{D}]^n = \frac{1}{2\mu}\mathbf{D}^{-1} \tag{13.44}$$

where the convergence of the series is ensured by the condition imposed on $\mu$. In arriving at (13.44), we have used the relationship that

$$(\mathbf{I}-\mathbf{A})^{-1} = \sum_{n=0}^{\infty}\mathbf{A}^n \tag{13.45}$$

under the condition that the sum converges.

It follows from (13.42), (13.43), and (13.44), after a large number of iterations, that the weight vector approaches a fixed value,

$$\lim \mathbf{w}'(k) = 2\mu\left[\frac{1}{2\mu}\mathbf{D}^{-1}\mathbf{Cp}\right] \tag{13.46}$$

and consequently,

$$\begin{aligned}\lim \mathbf{w}(k) &= \mathbf{C}^{-1}\mathbf{D}^{-1}\mathbf{Cp}\\ &= \mathbf{R}^{-1}\mathbf{p}\end{aligned} \tag{13.47}$$

where we have used the fact that the inverse of the product of invertible matrices is equal to the product of the inverses in reverse order. Expression (13.47) shows that the weight vector under the steepest descent algorithm converges to the optimum Wiener solution of Eq. (13.16). From the diagonal form, the primed coordinate system where the various modes comprising the solution are uncoupled, we can see that the mode with the smallest eigenvalue has the slowest decay. The length of the transient interval of the algorithm, for a selected value of $\mu$ satisfying the condition imposed by $\lambda_{max}$ in (13.41), is governed by the smallest eigenvalue. Therefore, the convergence rate of the algorithm, in general, is dictated by the step size parameter $\mu$ and the spread of the eigenvalues of correlation matrix $\mathbf{R}$.

When the step size is too small, the updated vector in step 3 of the algorithm makes little change in each iteration, and consequently a large number of iterations would be needed to reach the minimum of the performance surface, as illustrated in Figure 13.10 for the case of two components in the weight vector. Too large a step size would yield a convergence feature as displayed in Figure 13.11. In each individual step, the direction of descent is normal to the constant contour.

## 5.2 The Least-Mean-Square Algorithm

In many practical applications the statistics on $d(k)$ and $\mathbf{x}(k)$ are stationary but unknown. Therefore, the method of steepest descent cannot be used directly, since

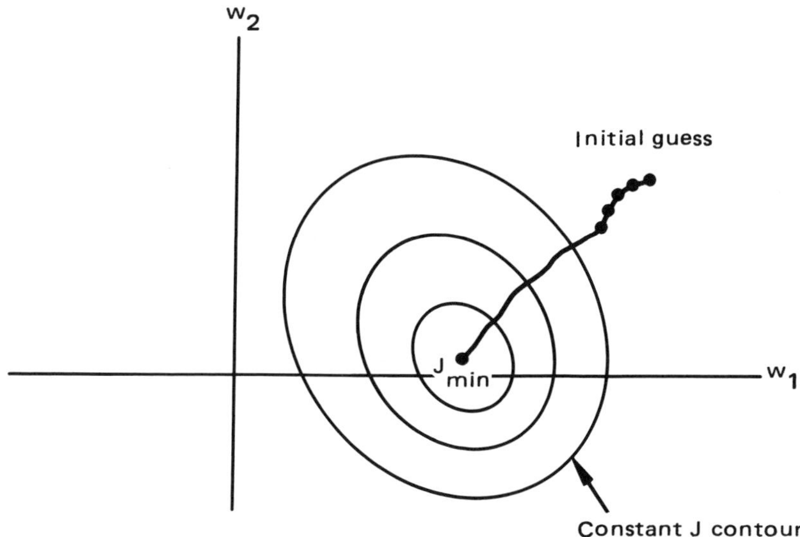


**Figure 13.10** Illustration of convergence of the steepest descent algorithm, small step size.

it assumes exact knowledge of the gradient vector in each iteration. The least-mean-square algorithm makes use of the steepest descent approach, but it derives the estimates of the gradient vector based on a limited number of data samples. Consider the gradient vector as given in Eq. (13.32), and express it in terms of $d(k)$,

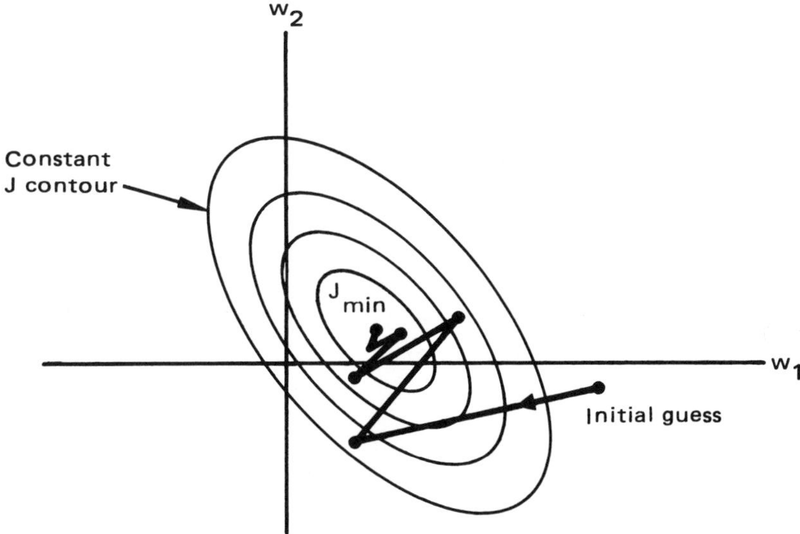


**Figure 13.11** Illustration of convergence of the steepest descent algorithm, large step size.

$\mathbf{x}(k)$, and $e(k)$. We have

$$(\nabla J)^{\mathrm{T}} = -2E\left\{d(k)\bar{\mathbf{x}}(k) - \bar{\mathbf{x}}(k)\mathbf{x}^{\mathrm{T}}(k)\mathbf{w}\right\} \tag{13.48}$$

or alternatively,

$$(\nabla J)^{\mathrm{T}} = -2E\left\{e(k)\bar{\mathbf{x}}(k)\right\} \tag{13.49}$$

Therefore, a reasonable choice for the estimate of the gradient vector based on one time sample is

$$(\widehat{\nabla} J)^{\mathrm{T}} = -2e(k)\bar{\mathbf{x}}(k) \tag{13.50}$$

This estimated gradient is unbiased since its expected value is the actual gradient vector.

The computational process of the least-mean-square algorithm is as follows:

1. Select an initial guess of the weight vector where the minimum of the performance measure may be located.
2. Using the present sample of $e(k)$ and $\mathbf{x}(k)$, obtain the estimate of the gradient vector, $\widehat{\nabla} J$, by Equation (13.50).
3. Update the weight vector making the correction according to

$$\mathbf{w}(k+1) = \mathbf{w}(k) - \mu(\widehat{\nabla} J)^{\mathrm{T}} \tag{13.51}$$

4. Go to step 2 and repeat the process until a predefined stopping rule is satisfied.

At first sight, the least-mean-square algorithm may seem incapable of having satisfactory performance because of the large variance in the instantaneous estimate of the gradient, but the iterative nature of the algorithm tends to smooth out the noisy gradient estimates.

To simplify the statistical analysis on the convergence of the least-mean-square algorithm, let us assume that the time between successive iterations is long enough so that the sampled vectors $\mathbf{x}(k+1)$ and $\mathbf{x}(k)$ are uncorrelated. The weight vector $\mathbf{w}(k)$ depends only on $\mathbf{x}(k-1)$, $\mathbf{x}(k-2)$, ..., $\mathbf{x}(0)$ and the initial guess $\mathbf{w}(0)$. On substituting (13.50) in (13.51),

$$\mathbf{w}(k+1) = \mathbf{w}(k) + 2\mu e(k)\bar{\mathbf{x}}(k) \tag{13.52}$$

Using $d(k) - \mathbf{w}^{\mathrm{T}}(k)\mathbf{x}(k)$ for $e(k)$, Equation (13.52) becomes

$$\mathbf{w}(k+1) = \mathbf{w}(k) + 2\mu d(k)\bar{\mathbf{x}}(k) - 2\mu\bar{\mathbf{x}}(k)\mathbf{x}^{\mathrm{T}}(k)\mathbf{w}(k) \tag{13.53}$$

Taking the expected value on both sides of (13.53) and making use of the above assumption,

$$E\{\mathbf{w}(k+1)\} = E\{\mathbf{w}(k)\} + 2\mu\mathbf{p} - 2\mu\mathbf{R}E\{\mathbf{w}(k)\} \tag{13.54}$$

Equation (13.54) is exactly the same mathematical form as Eq. (13.33) of the steepest descent algorithm. Therefore, we can conclude that the restriction on the step size parameter for convergence is again given by

$$\mu < \frac{1}{\lambda_{\max}} \tag{13.55}$$

where $\lambda_{\max}$ is the largest eigenvalue of the correlation matrix $\mathbf{R}$, and the least-mean-square algorithm converges in the mean to the Wiener solution, that is,

$$\lim_{k\to\infty} E\{\mathbf{w}(k)\} = \mathbf{R}^{-1}\mathbf{p} \tag{13.56}$$

Also, as with the steepest descent algorithm, the rate of convergence is governed by the step size parameter $\mu$ and the spread of the eigenvalues of $\mathbf{R}$. For further discussion of the convergence of the least-mean-square algorithm, the interested reader is referred to the work of Fisher and Bershad [32], Bershad and Qu [33], and Iltis and Milstein [34].

The assumption made above on the uncorrelated successive sampled vectors may be overly restrictive [35]. Griffiths [36] has presented experimental results based on highly correlated successive samples and demonstrated that the least-mean-square algorithm still converges to the Wiener solution except that the mean-squared error is slightly higher than in the case of uncorrelated samples.

As an illustration of the convergence characteristic of the least-mean-square algorithm, consider the following system identification problem as shown in Figure 13.12. This is a closed-loop adaptive process because of its feedback property. Let the transfer function of the unknown plant be $1/(1-0.25z^{-1})$, an infinite impulse response filter, and suppose we model it by a fourth-order finite impulse response filter. Let the unknown plant and the adaptive filter be driven simultaneously by a zero-mean white noise sequence $\{x(k)\}$ with unity variance. Consider two cases, with plant noise represented as additive noise as shown and without plant noise. Let the noise sequence be $\{n(k)\}$, whose mean is zero and variance equal to 1/4, and let the sequences $\{x(k)\}$ and $\{n(k)\}$ be uncorrelated.

The convergence features of the present problem are shown in Figures 13.13 and 13.14a–d. The step size parameter $\mu = 0.1$ was utilized. After the transient interval, or learning period as it is sometimes called, the filter coefficients $w_1$, $w_2$, $w_3$, and $w_4$ approach the first four coefficients of the Maclaurin series for the transfer function of the unknown plant, namely, 1, 1/4, 1/16, and 1/64. In the presence of noise, the filter coefficients in the steady state oscillate about the corresponding values of the noiseless case. The amplitude of these oscillations would be smaller if a smaller step size had been used. The nonzero residue error $e(k)$ that occurred in

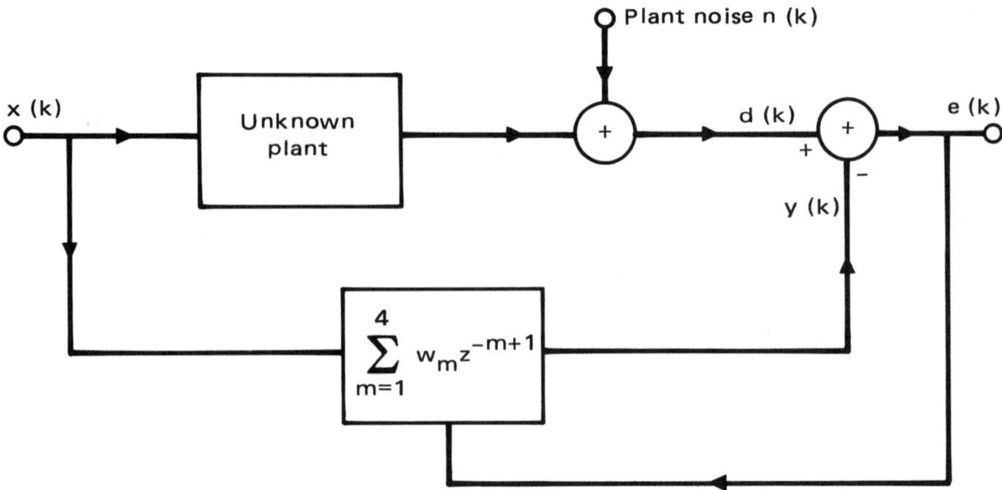


**Figure 13.12** Identifying an unknown plant by an adaptive filter.

the steady state in the noiseless case indicates that the fourth-order adaptive filter is not capable of identifying the plant completely, revealing the fact that a finite impulse response filter cannot have an output identical to that of a filter that has an infinite impulse response. Each curve shown is an average of 100 independent computer runs. The length of the learning period is in the neighborhood of $10M$ to $20M$ samples, where $M$ is the dimension of the weight vector $\mathbf{w}$.

## 5.3 Sequential Inversion of the Correlation Matrix

We have seen that the optimum Wiener solution as given by Equation (13.16) requires evaluation of the inverse of the correlation matrix, a difficult task, especially in a situation with data obtained continuously. The method utilizing the matrix inversion lemma stated below circumvents the computational problem by obtaining the inverse sequentially as the data arrive. This technique has been used in airborne radar [37] and in sonar [38].

**Matrix Inversion Lemma.** Let $\mathbf{A}$ be an $N$-by-$N$ invertible matrix and $\mathbf{v}$ be an arbitrary $N$-by-1 vector. Assume that $\mathbf{A} + \bar{\mathbf{v}}\mathbf{v}^T$ is invertible. Then

$$(\mathbf{A} + \bar{\mathbf{v}}\mathbf{v}^T)^{-1} = \mathbf{A}^{-1} - \mathbf{A}^{-1}\bar{\mathbf{v}}(\mathbf{v}^{\mathrm{T}}\mathbf{A}^{-1}\bar{\mathbf{v}} + 1)^{-1}\mathbf{v}^{\mathrm{T}}\mathbf{A}^{-1} \tag{13.57}$$

The important feature of this formula is that if $\mathbf{A}^{-1}$ is known, then the inverse of $\mathbf{A}$ augmented by a rank-1 matrix can be obtained by a simple modification of the known $\mathbf{A}^{-1}$, the key in sequential computation. The validity of this lemma may be verified in a straightforward manner by either premultiplying or postmultiplying the right-hand side of Eq. (13.57) by $\mathbf{A} + \bar{\mathbf{v}}\mathbf{v}^T$ and demonstrating that the resultant is an identity matrix.

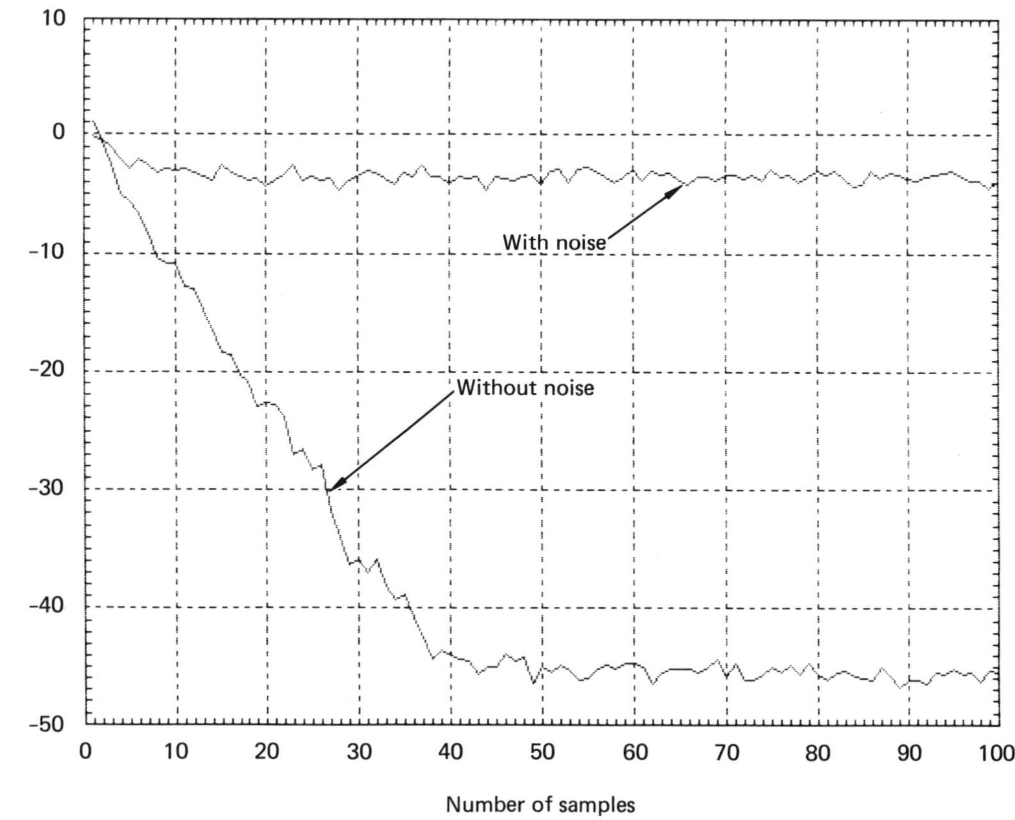


**Figure 13.13** Learning curve using the least-mean-square algorithm.

Suppose that at time $k$, the estimated correlation matrix is $\mathbf{R}(k)$ and its inverse is $\mathbf{R}^{-1}(k)$. On receiving the new data $\mathbf{x}(k+1)$, the updated estimated correlation matrix becomes

$$\mathbf{R}(k+1) = \alpha\mathbf{R}(k) + (1-\alpha)\bar{\mathbf{x}}(k+1)\mathbf{x}^{\mathrm{T}}(k+1) \tag{13.58}$$

where $\alpha$ is selected to express the relative emphasis on the present and past data, and $0 < \alpha < 1$. For a stationary environment, in which each sample vector is weighted equally, then $\alpha = k/(k+1)$. Applying the matrix inversion lemma to the right-hand side of Eq. (13.58), we have

$$\mathbf{R}^{-1}(k+1) = \frac{1}{\alpha}\mathbf{R}^{-1}(k) - \frac{1-\alpha}{\alpha^2}\frac{\mathbf{R}^{-1}(k)\bar{\mathbf{x}}(k+1)\mathbf{x}^{\mathrm{T}}(k+1)\mathbf{R}^{-1}(k)}{1+[(1-\alpha)/\alpha]\mathbf{x}^{\mathrm{T}}(k+1)\mathbf{R}^{-1}(k)\bar{\mathbf{x}}(k+1)} \tag{13.59}$$

Equation (13.59) shows that the updated inverse of $\mathbf{R}$ is given by the previous $\mathbf{R}$ inverse modified by an added revision term that involves only matrix multiplications

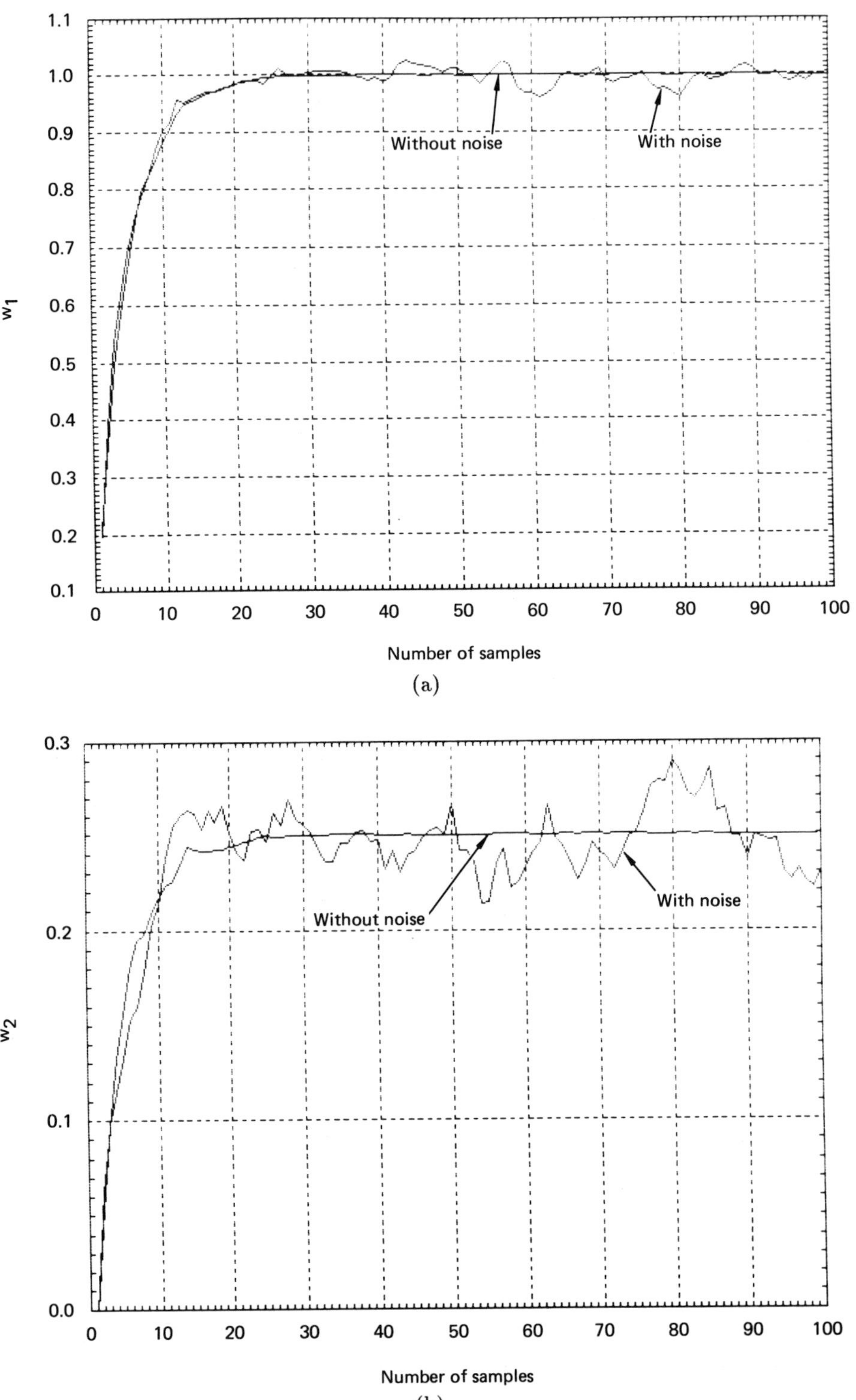


**Figure 13.14** Convergence of the filter coefficients (a) $w_1$, (b) $w_2$, (c) $w_3$, and (d) $w_4$.

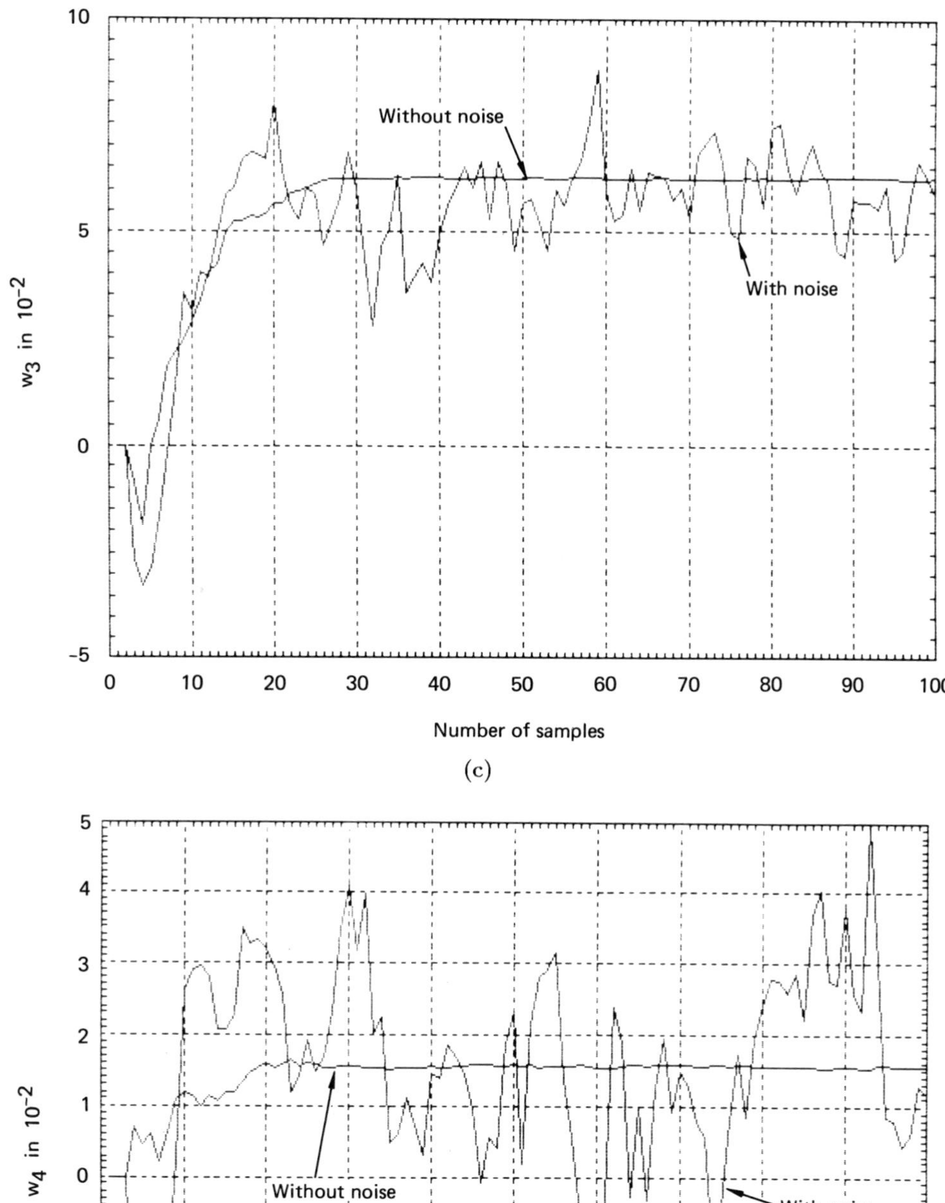


(c)

5
4
3
2
1
0
-1
-2
-3
-4
$w_4$ in $10^{-2}$
Without noise
With noise
0 10 20 30 40 50 60 70 80 90 100
Number of samples

(d)

and no matrix inversion. Initially, we may select $\mathbf{R}(0)$ to be a convenient matrix such as a constant times an identity matrix whose inverse can be evaluated easily. After a sufficient amount of data $\bar{\mathbf{x}}(n)\mathbf{x}^{\mathrm{T}}(n)$, $n = 1, 2, \ldots, N$, has been observed, $\mathbf{R}^{-1}(k)$ approaches the inverse of the weighted time average of $\bar{\mathbf{x}}(n)\mathbf{x}^{\mathrm{T}}(n)$ for $k > N$. This method is very attractive from the computational viewpoint, and there is no stability problem associated with it. With the arrival of the current data $\mathbf{x}(k+1)$, we may obtain the updated estimated weight vector by postmultiplying both sides of Eq. (13.59) by $\mathbf{p}$ and the result is

$$\mathbf{w}(k+1) = \frac{1}{\alpha}\mathbf{w}(k) - \frac{(1-\alpha)}{\alpha^2}\frac{\mathbf{R}^{-1}(k)\bar{\mathbf{x}}(k+1)\mathbf{x}^{\mathrm{T}}(k+1)\mathbf{w}(k)}{1+[(1-\alpha)/\alpha]\mathbf{x}^{\mathrm{T}}(k+1)\mathbf{R}^{-1}(k)\bar{\mathbf{x}}(k+1)} \tag{13.60}$$

The computational process for the sequential matrix inversion algorithm is as follows:

1. Select the constant $\alpha$, $0 < \alpha < 1$, according to the relative importance of the present and past data.
2. Select the initial inverse of the correlation matrix $\mathbf{R}^{-1}(0)$ and the initial weight vector $\mathbf{w}(0)$.
3. Using the current data, obtain the updated estimate of $\mathbf{R}^{-1}$ by Eq. (13.59).
4. Update the estimated weight vector by Eq. (13.60).
5. Go to step 3 and repeat the process until a predefined stopping rule is satisfied.

As an example of the convergence of $\mathbf{R}^{-1}$ using the sequential matrix inversion algorithm, we simulated the following situation on the computer. The components of the signal vector $\mathbf{x}(k)$ are produced from the random number generator with such correlation as to yield the correlation matrix

$$\mathbf{R} = \begin{bmatrix} 1 & 0.25 & 0 \\ 0.25 & 2 & 0 \\ 0 & 0 & 3 \end{bmatrix} \tag{13.61}$$

The ideal inverse of $\mathbf{R}$ is

$$\mathbf{R}^{-1} = \begin{bmatrix} 1.032 & -0.129 & 0 \\ -0.129 & 0.516 & 0 \\ 0 & 0 & 0.333 \end{bmatrix} \tag{13.62}$$

We assume that the random process $\mathbf{x}(k)$, $k = 0, 1, \ldots$, is stationary, and thus select $\alpha = k/(k+1)$ so that each sample vector is equally weighted. Consider the error in the estimate as

$$J = \frac{\|\mathbf{R}^{-1} - \hat{\mathbf{R}}^{-1}\|}{\|\mathbf{R}^{-1}\|} \tag{13.63}$$

where $\hat{\mathbf{R}}^{-1}$ is the estimate of $\mathbf{R}^{-1}$ and the norm of the matrix is taken to be the square root of the sum of the squares of all its elements. Figure 13.15 shows the convergence characteristic of a single realization or trial. The amplitude of the oscillation in the transient interval, or the learning period, depends strongly on the initial guess $\mathbf{R}^{-1}(0)$. However, the length of the transient interval is essentially independent of that choice.

## 5.4 Sequential Least-Squares Fit Algorithm

Consider a set of $N$ complex-valued data $\{d(1), d(2), \ldots, d(N)\}$ and a set of $N$ complex-valued $M$-by-1 data vectors $\{\mathbf{x}(1), \mathbf{x}(2), \ldots, \mathbf{x}(N)\}$. Let $\mathbf{w}$ be an $M$-by-1 complex-valued vector whose values are to be determined. Suppose that $N \geq M$ and we wish to approximate $d(k)$ by $\mathbf{w}^{\mathrm{T}}\mathbf{x}(k)$, $k = 1, 2, \ldots, N$. In general, there may not be such a vector which yields a perfect match at all $N$ points. Consider the compromise, the least-squares fit, and take the performance measure to be the sum of the squares of the magnitudes of the differences between $d(k)$ and $\mathbf{w}^{\mathrm{T}}\mathbf{x}(k)$, with less emphasis on the approximation on the older data [39]. Let the performance

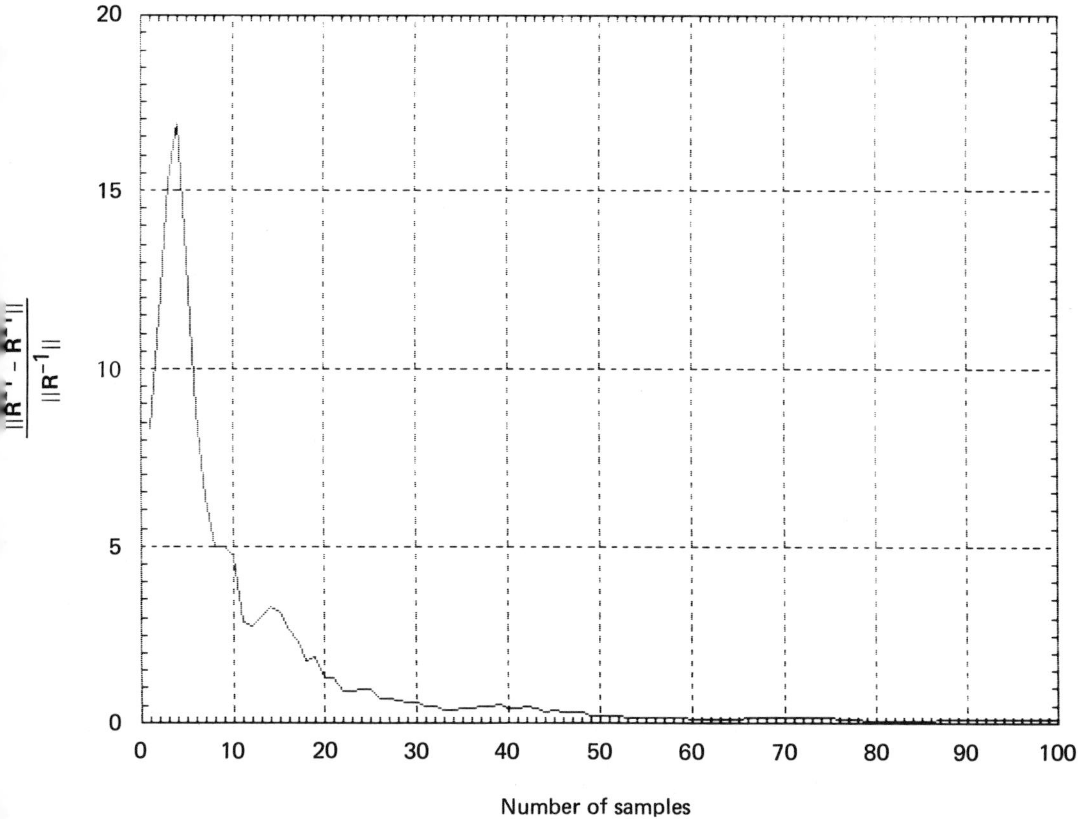


**Figure 13.15** Convergence of the inverse of **R**.

measure based on $k$ observed data be

$$J(\mathbf{w}) = \sum_{n=1}^{k} q^{k-n} |\mathbf{w}^{\mathrm{T}}\mathbf{x}(n) - d(n)|^2 \tag{13.64}$$

where $0 < q \leq 1$. Let $\mathbf{d}(k)$ be a $k$-by-1 vector whose components are the desired signal at 1, 2, ..., $k$, that is,

$$\mathbf{d}(k) = \begin{bmatrix} d(1) \\ d(2) \\ d(3) \\ \vdots \\ d(k) \end{bmatrix} \tag{13.65}$$

and let $\mathbf{X}(k)$ be a $k$-by-$M$ signal matrix whose row vectors are $\mathbf{x}^{\mathrm{T}}(1)$, $\mathbf{x}^{\mathrm{T}}(2)$, ..., $\mathbf{x}^{\mathrm{T}}(k)$, the transposes of the $M$-by-1 signal vectors. Specifically,

$$\mathbf{X}(k) = \begin{bmatrix} x_1(1) & x_2(1) & x_3(1) & \cdots & x_M(1) \\ x_1(2) & x_2(2) & x_3(2) & \cdots & x_M(2) \\ & \vdots & & & \\ x_1(k) & x_2(k) & x_3(k) & \cdots & x_M(k) \end{bmatrix} \tag{13.66}$$

Equation (13.64) can now be written as

$$J(\mathbf{w}) = [\mathbf{X}(k)\mathbf{w} - \mathbf{d}(k)]^{\mathrm{T}} \mathbf{Q} [\bar{\mathbf{X}}(k)\bar{\mathbf{w}} - \bar{\mathbf{d}}(k)] \tag{13.67}$$

where $\mathbf{Q}$ is the $k$-by-$k$ diagonal penalty matrix given by

$$\mathbf{Q} = \begin{bmatrix} q^{k-1} & 0 & 0 & \cdots & 0 \\ 0 & q^{k-2} & 0 & \cdots & 0 \\ 0 & 0 & q^{k-3} & \cdots & 0 \\ & \vdots & & & \\ 0 & 0 & 0 & \cdots & 1 \end{bmatrix} \tag{13.68}$$

Treating $\mathbf{X}(k)\mathbf{w} - \mathbf{d}(k)$ as the error vector and $\mathbf{Q}\mathbf{X}(k)$ as the data or the approximant, in accordance with the orthogonality principle, the vector which minimizes the weighted least-squares performance measure of Eq. (13.64), or equivalently the expression of (13.67), must satisfy

$$[\mathbf{X}(k)\mathbf{w} - \mathbf{d}(k)]^{\mathrm{T}} \mathbf{Q} \bar{\mathbf{X}}(k) = \mathbf{0}^{\mathrm{T}} \tag{13.69}$$

where $\mathbf{0}$ is the $M$-by-1 zero vector. Taking the transpose of (13.69), we have

$$\bar{\mathbf{X}}^{\mathrm{T}}(k)\mathbf{Q}[\mathbf{X}(k)\mathbf{w} - \mathbf{d}(k)] = \mathbf{0} \tag{13.70}$$

and consequently the optimum vector based on $k$ data samples is given by

$$\mathbf{w}(k) = [\bar{\mathbf{X}}^{\mathrm{T}}(k)\mathbf{Q}\mathbf{X}(k)]^{-1}\bar{\mathbf{X}}^{\mathrm{T}}(k)\mathbf{Q}\mathbf{d}(k) \tag{13.71}$$

If $\mathbf{X}(k)$ is of rank $M$, that is, maximal rank, then the inverse of $\bar{\mathbf{X}}^{\mathrm{T}}(k)\mathbf{Q}\mathbf{X}(k)$ is assured.

When additional data samples $\mathbf{x}(k+1)$ and $d(k+1)$ are taken, we must update the foregoing weight vector to meet the demand of the new situation. We will make use of the matrix inversion lemma of (13.57) again to expedite the computation. Let the new $(k+1)$-by-$(k+1)$ signal matrix be partitioned as follows:

$$\mathbf{X}(k+1) = \begin{bmatrix} \mathbf{X}(k) \\ \hline \mathbf{x}^{\mathrm{T}}(k+1) \end{bmatrix} \tag{13.72}$$

Similarly, partition the $(k+1)$-by-1 desired signal vector and the $(k+1)$-by-$(k+1)$ penalty matrix as follows:

$$d(k+1) = \begin{bmatrix} \mathbf{d}(k) \\ \hline d(k+1) \end{bmatrix} \tag{13.73}$$

$$\mathbf{Q}_{\text{new}} = \left[\begin{array}{c|c} & 0 \\ q\mathbf{Q} & 0 \\ & 0 \\ & \vdots \\ & 0 \\ \hline 0 \quad 0 \quad \cdots \quad 0 & 1 \end{array}\right] \tag{13.74}$$

Based on $(k+1)$ observed data, using the orthogonality principle, we have the optimum weight vector of dimension $M$-by-1 given by

$$\mathbf{w}(k+1) = [\bar{\mathbf{X}}^{\mathrm{T}}(k+1)\mathbf{Q}_{\text{new}}\mathbf{X}(k+1)]^{-1}\bar{\mathbf{X}}^{\mathrm{T}}(k+1)\mathbf{Q}_{\text{new}}\mathbf{d}(k+1) \tag{13.75}$$

In view of the above partitioning of the matrices, we have

$$\bar{\mathbf{X}}^T(k+1)\mathbf{Q}_{\text{new}}\mathbf{X}(k+1) = [\bar{\mathbf{X}}^T(k) \,\vdots\, \bar{\mathbf{x}}(k+1)] \left[\begin{array}{ccccc|c} & & q\mathbf{Q} & & & 0 \\ & & & & & 0 \\ & & & & & 0 \\ & & & & & \vdots \\ & & & & & 0 \\ \hline 0 & 0 & 0 & \cdots & 0 & 1 \end{array}\right] \left[\begin{array}{c} \mathbf{X}(k) \\ \\ \hline \mathbf{x}^T(k+1) \end{array}\right]$$

$$= q\bar{\mathbf{X}}^T(k)\mathbf{Q}\mathbf{X}(k) + \bar{\mathbf{x}}(k+1)\mathbf{x}^T(k+1) \tag{13.76}$$

and

$$\bar{\mathbf{X}}^T(k+1)\mathbf{Q}_{\text{new}}\mathbf{d}(k+1) = [\bar{\mathbf{X}}^T(k) \,\vdots\, \bar{\mathbf{x}}(k+1)] \left[\begin{array}{ccccc|c} & & q\mathbf{Q} & & & 0 \\ & & & & & 0 \\ & & & & & 0 \\ & & & & & \vdots \\ & & & & & 0 \\ \hline 0 & 0 & 0 & \cdots & 0 & 1 \end{array}\right] \left[\begin{array}{c} \mathbf{d}(k) \\ \\ \hline d(k+1) \end{array}\right]$$

$$= q\bar{\mathbf{X}}^T(k)\mathbf{Q}\mathbf{d}(k) + \bar{\mathbf{x}}(k+1)d(k+1) \tag{13.77}$$

For convenience, define

$$\mathbf{P}^{-1}(k) \triangleq \bar{\mathbf{X}}^T(k)\mathbf{Q}\mathbf{X}(k) \tag{13.78}$$

and also define

$$\mathbf{P}^{-1}(k+1) \triangleq \bar{\mathbf{X}}^T(k+1)\mathbf{Q}_{\text{new}}\mathbf{X}(k+1) \tag{13.79}$$

The matrix $\mathbf{P}^{-1}$ is Hermitian, and consequently $\mathbf{P}$ is also Hermitian. It follows from (13.76) and (13.78) that (13.79) becomes

$$\begin{aligned} \mathbf{P}^{-1}(k+1) &= q\bar{\mathbf{X}}^{\mathrm{T}}(k)\mathbf{Q}\mathbf{X}(k) + \bar{\mathbf{x}}(k+1)\mathbf{x}^{\mathrm{T}}(k+1) \\ &= q\left[\mathbf{P}^{-1}(k) + \frac{1}{q}\bar{\mathbf{x}}(k+1)\mathbf{x}^{\mathrm{T}}(k+1)\right] \end{aligned} \tag{13.80}$$

Taking the inverse of both sides of Eq. (13.80), according to the matrix inversion lemma of Eq. (13.57) we have

$$\mathbf{P}(k+1) = \frac{1}{q}\left[\mathbf{P}(k) - \frac{\mathbf{P}(k)\bar{\mathbf{x}}(k+1)\mathbf{x}^{\mathrm{T}}(k+1)\mathbf{P}(k)}{q+\mathbf{x}^{\mathrm{T}}(k+1)\mathbf{P}(k)\bar{\mathbf{x}}(k+1)}\right] \tag{13.81}$$

As a consequence of Eqs. (13.77), (13.79), and (13.81), Eq. (13.75) becomes

$$\mathbf{w}(k+1) = \frac{1}{q}\left[\mathbf{P}(k) - \frac{\mathbf{P}(k)\bar{\mathbf{x}}(k+1)\mathbf{x}^{\mathrm{T}}(k+1)\mathbf{P}(k)}{q+\mathbf{x}^{\mathrm{T}}(k+1)\mathbf{P}(k)\bar{\mathbf{x}}(k+1)}\right] \cdot [q\bar{\mathbf{X}}^{\mathrm{T}}(k)\mathbf{Q}\mathbf{d}(k) + \bar{\mathbf{x}}(k+1)d(k+1)] \tag{13.82}$$

Equations (13.71) and (13.78) imply that

$$\mathbf{w}(k) = \mathbf{P}(k)\bar{\mathbf{X}}^{\mathrm{T}}(k)\mathbf{Q}\mathbf{d}(k) \tag{13.83}$$

Substituting Eq. (13.83) into Eq. (13.82), we have the updated expression for the weight vector as follows:

$$\mathbf{w}(k+1) = \mathbf{w}(k) + \mathbf{K}(k+1)[d(k+1) - \mathbf{x}^{\mathrm{T}}(k+1)\mathbf{w}(k)] \tag{13.84}$$

where

$$\mathbf{K}(k+1) = \frac{\mathbf{P}(k)\bar{\mathbf{x}}(k+1)}{q+\mathbf{x}^{\mathrm{T}}(k+1)\mathbf{P}(k)\bar{\mathbf{x}}(k+1)} \tag{13.85}$$

Equation (13.84) indicates that the updated weight vector is given by the previous weight vector plus a correction term based on the current data. The bracketed term in Eq. (13.84) may be viewed as the estimation error and $\mathbf{K}$ as the gain vector. The $\mathbf{P}$ matrix in Eq. (13.85) is evaluated through the recursive relationship of Eq. (13.81), and no matrix inversion is involved in the computational process. The development of this algorithm is purely deterministic, since no consideration has been given to the nature of the observed data $\mathbf{x}(k)$ and $d(k)$. This same algorithm can be used in a stochastic environment without any modification, as in the case where the observed data are corrupted by noise.

Suppose we treat the error vector $\mathbf{e}(k)$, where $\mathbf{e}(k) = \mathbf{d}(k) - \mathbf{X}(k)\mathbf{w}(k)$, and the weight vector $\mathbf{w}(k)$ as stochastic quantities and consider $\mathbf{P}(k)$ and $\mathbf{X}(k)$ as known matrices, since they are based on the observed signal vector $\mathbf{x}(k)$. Let us examine the significance of $\mathbf{P}$ in this situation. Let the error vector corresponding to the optimum weight vector $\mathbf{w}^*$ be

$$\mathbf{e}^*(k) = \mathbf{d}(k) - \mathbf{X}(k)\mathbf{w}^* \tag{13.86}$$

Assume that the components of the error vector are independent and each component has zero mean. Suppose that the error in a given component is inversely proportional to the penalty associated with that component. That is,

$$E\left\{\mathbf{e}^*(k)\bar{\mathbf{e}}^{*\mathrm{T}}(k)\right\} = h\mathbf{Q}^{-1} \tag{13.87}$$

where $h$ is a constant. As a consequence of Eq. (13.86), Eq. (13.71) becomes

$$\mathbf{w}(k) = [\bar{\mathbf{X}}^{\mathrm{T}}(k)\mathbf{Q}\mathbf{X}(k)]^{-1}\bar{\mathbf{X}}^{\mathrm{T}}(k)\mathbf{Q}[\mathbf{X}(k)\mathbf{w}^* + \mathbf{e}^*(k)] \tag{13.88}$$

and it follows from (13.78) that

$$\mathbf{w}(k) = \mathbf{w}^* + \mathbf{P}(k)\bar{\mathbf{X}}^{\mathrm{T}}(k)\mathbf{Q}\mathbf{e}^*(k) \tag{13.89}$$

Hence,

$$\begin{aligned} E\left\{[\mathbf{w}(k) - \mathbf{w}^*][\bar{\mathbf{w}}(k) - \bar{\mathbf{w}}^*]^{\mathrm{T}}\right\} &= E\left\{[\mathbf{P}(k)\bar{\mathbf{X}}^{\mathrm{T}}(k)\mathbf{Q}\mathbf{e}^*][\bar{\mathbf{P}}(k)\mathbf{X}^{\mathrm{T}}(k)\bar{\mathbf{Q}}\bar{\mathbf{e}}^*]^{\mathrm{T}}\right\} \\ &= \mathbf{P}(k)\bar{\mathbf{X}}^{\mathrm{T}}(k)\mathbf{Q}E\left\{\mathbf{e}^*\bar{\mathbf{e}}^{*\mathrm{T}}\right\}\mathbf{Q}\mathbf{X}(k)\mathbf{P}(k) \end{aligned} \tag{13.90}$$

In view of Eqs. (13.78) and (13.87), Eq. (13.90) simplifies to

$$E\left[[\mathbf{w}(k) - \mathbf{w}^*][\bar{\mathbf{w}}(k) - \bar{\mathbf{w}}^*]^{\mathrm{T}}\right\} = h\mathbf{P}(k) \tag{13.91}$$

which indicates that, except for the proportionality constant, $\mathbf{P}$ is the correlation matrix of the weight-error vector.

The signal flow diagram associated with the sequential least-squares fit algorithm is shown in Figure 13.16. The computational process for the sequential least-squares fit algorithm is summarized as follows:

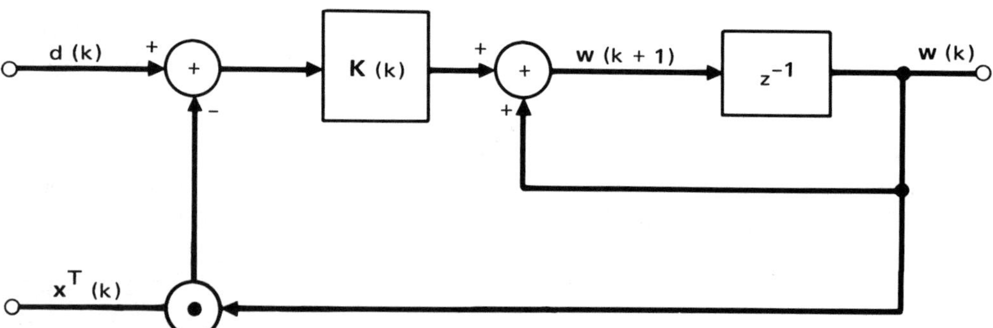


**Figure 13.16** Signal flow diagram of the sequential least-squares fit algorithm.

1. Select a suitable constant $q$, $0 < q \leq 1$, the emphasis to be given to the old data and the recent data.
2. Select the initial values for the correlation matrix of the weight-error vector $\mathbf{P}(0)$ and the initial weight vector $\mathbf{w}(0)$.
3. Compute the gain vector by

$$\mathbf{K}(n) = \frac{\mathbf{P}(n-1)\bar{\mathbf{x}}(n)}{q + \mathbf{x}^{\mathrm{T}}(n)\mathbf{P}(n-1)\bar{\mathbf{x}}(n)} \tag{13.92}$$

4. Using the current data $\mathbf{x}(n)$ and $d(n)$, evaluate the estimation error $e(n) = d(n) - \mathbf{x}^{\mathrm{T}}(n)\mathbf{w}(n-1)$.
5. Update the weight vector according to

$$\mathbf{w}(n) = \mathbf{w}(n-1) + \mathbf{K}(n)e(n) \tag{13.93}$$

6. Update the matrix $\mathbf{P}$ by

$$\mathbf{P}(n) = \frac{1}{q}\{\mathbf{P}(n-1) - \mathbf{K}(n)\mathbf{x}^{\mathrm{T}}(n)\mathbf{P}(n-1)\} \tag{13.94}$$

7. Go to step 3 and repeat the process until a predefined stopping rule is satisfied.

To get an insight into the convergence characteristic of the sequential least-squares fit algorithm, consider the following simple problem. Let

$$x_1(k) = \exp\left(\frac{i2\pi k}{5.2}\right) + 2\exp\left(\frac{i2\pi k}{3.7} + \frac{i\pi}{3}\right) \tag{13.95}$$

$$x_2(k) = 3\exp\left(\frac{i2\pi k}{3.7} + \frac{i\pi}{4}\right) \tag{13.96}$$

$$d(k) = 2\exp\left(\frac{i2\pi k}{5.2} - \frac{i\pi}{3}\right) + \exp\left(\frac{i2\pi k}{3.7} + \frac{i\pi}{2}\right) \tag{13.97}$$

and let the penalty matrix be

$$\mathbf{Q} = \begin{bmatrix} 0.9 & 0 \\ 0 & 1 \end{bmatrix} \tag{13.98}$$

Recall that the objective is to find the complex-valued vector $\mathbf{w}^{\mathrm{T}} = (w_1, w_2)$ so that $\mathbf{w}^{\mathrm{T}}\mathbf{x}(k)$ matches $d(k)$. As the initial guess for the weight-error correlation matrix and the weight vector, we select

$$\mathbf{P} = \begin{bmatrix} 1 & 0 \\ 0 & 1 \end{bmatrix} \tag{13.99}$$

and

$$\mathbf{w}(0) = \begin{bmatrix} 1 \\ 1 \end{bmatrix} \tag{13.100}$$

As the data arrive sequentially, the weight vector $\mathbf{w}$ is updated according to Eq. (13.93). Figure 13.17 shows the learning curve of this algorithm on the current problem. The choice of $\mathbf{P}(0)$ and $\mathbf{w}(0)$ is not a critical factor in the convergence time in general. The learning period is longer when the dimension of the weight vector $M$ increases; it varies approximately as $4M$ for complex coefficients and $2M$ for real coefficients. Figure 13.18 presents the convergence of the weight vector toward $(1 - i\sqrt{3}, -\sqrt{2}/2 + i5\sqrt{2}/6)$, the solution of this problem.

## 6 COMPARISON OF THE ALGORITHMS

The sequential correlation matrix inversion algorithm and the sequential least-squares fit algorithm are similar in structure, and hence both have the same conver-

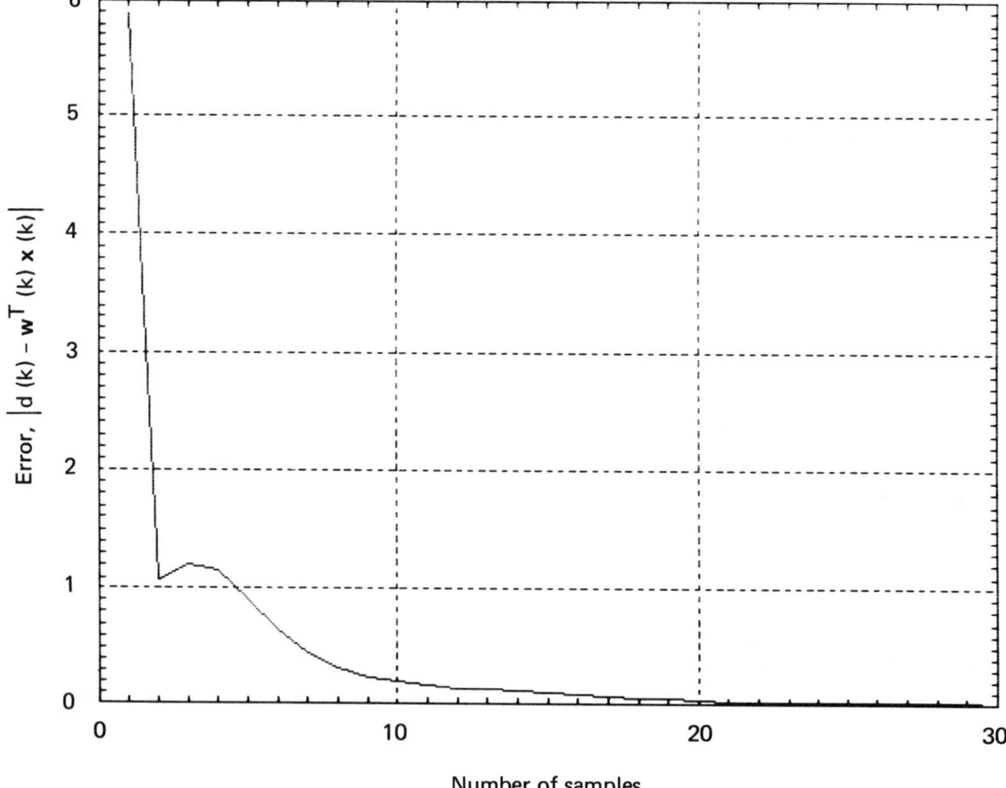


**Figure 13.17** Learning curve using the sequential least-squares fit algorithm.

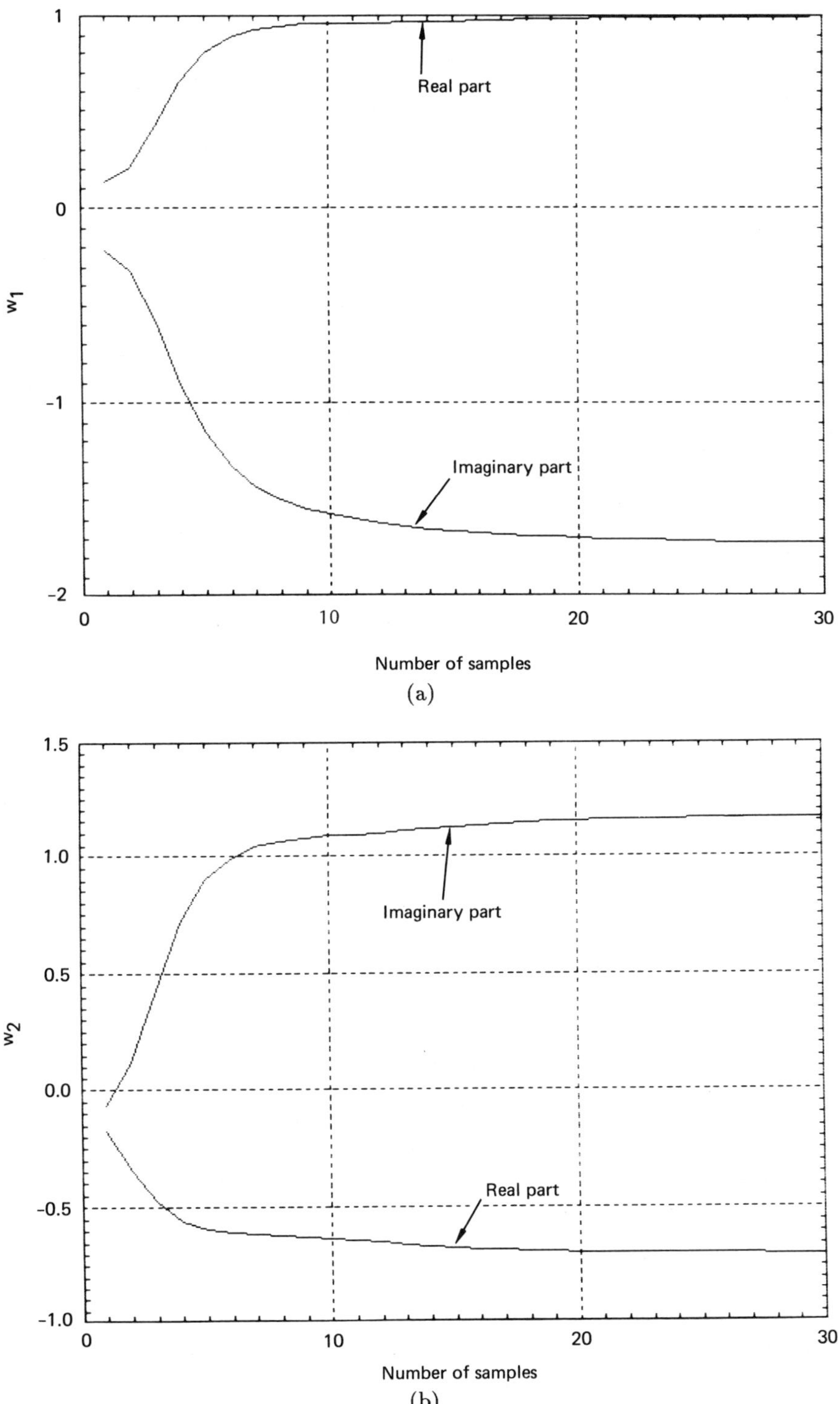


**Figure 13.18** Convergence of (a) the first component and (b) the second component of the weight vector.

gence rate and computational complexity. Both rely heavily on the matrix inversion lemma to circumvent the direct computation of the matrix inverse. To determine the weight vector, the sequential correlation matrix inversion algorithm requires knowledge of the cross-correlation vector $\mathbf{p}$, while the equivalent information in $\mathbf{p}$ enters the sequential least-squares fit algorithm as input data during the computation. Depending on what input data are available, both algorithms can easily be modified to suit the particular situation.

In the recursive expression for the weight vector in the least-mean-square algorithm, the correction in each iteration depends on the product of the step size parameter $\mu$ and the instantaneous gradient estimate based on the current data $e(k)$ and $\mathbf{x}(k)$. The convergence in the mean to the Wiener solution tends to be noisy, as exhibited in Figures 13.14a–d. On the other hand, the sequential improvement of the weight vector in the least-squares fit algorithm (also in the sequential matrix inversion algorithm) utilizes not only the current data but all the past data as well. Also, there is the flexibility of assigning different amount of emphasis to old and recent data. The correction depends on the product of the error vector and the feedback gain vector. The convergence toward the optimum vector is less oscillatory, as demonstrated in Figures 13.18a and 13.18b. The convergence rate is case-dependent, but, in general, the learning period for the least-mean-square algorithm is in the neighborhood of $10M$ to $20M$ samples, where $M$ is the dimension of the weight vector $\mathbf{w}$ (real coefficients). For the sequential least-squares fit algorithm, the learning period is approximately $2M$ (real coefficients), almost an order of magnitude faster. The superiority in convergence rate of the sequential least-squares fit and the sequential correlation matrix inversion algorithms is at the expense of much greater computational complexity.

## 7 APPLICATION—AN ANTENNA SIDELOBE CANCELLER

The antenna sidelobe canceller mentioned earlier can be operated as a closed-loop adaptive system [5]. An open-loop adaptive process can conveniently be used also because only a small dimension is needed for the weight vector. Consider the open-loop adaptive process in the pulse Doppler radar signal processing system for moving-target detection as shown in Figure 13.19.

Let the high-gain main antenna have sidelobes of 30 dB below its maximum, and use an auxiliary antenna whose gain is 3 dB above the main antenna sidelobes. Let the target signal enter the main antenna at its mainlobe peak and the interference come from a sidelobe direction as depicted in Figure 13.20.

To illustrate this sidelobe canceller's performance, a computer simulation is done for the system parameters and assumptions stated below:

1. Pulse repetition frequency 160 kHz.
2. Noiselike interference power 0 dB.
3. Interference bandwidth 95 kHz.
4. Interference spectrum uniform over the interval between 0 and 95 kHz.
5. Target signal power −50 dB.
6. Ground clutter power −5 dB.
7. Receiver noise power −45 dB.

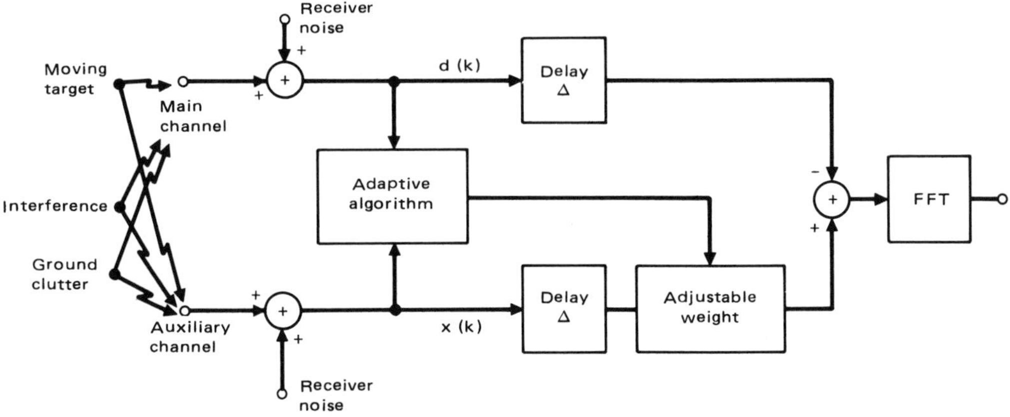


**Figure 13.19** The open-loop adaptive sidelobe canceller.

8. Antenna scan rate 80° per second.
9. Clutter rejection filter, 35 dB attenuation over stopband.
10. Target's Doppler frequency in the passband of the clutter rejection filter.
11. Wavelength 0.12 foot.
12. Separation between antenna centers 1.4 feet.
13. 2048 Doppler filters over the pulse repetition frequency interval.
14. Interference signal, target signal, ground clutter and receiver noise mutually uncorrelated.
15. Components in both channels identical. The issue of channel mismatch is not considered here.

It is easy to see that to null out the interference in this two-channel system (assuming for now a nonscanning antenna), the magnitude of the weighting co-

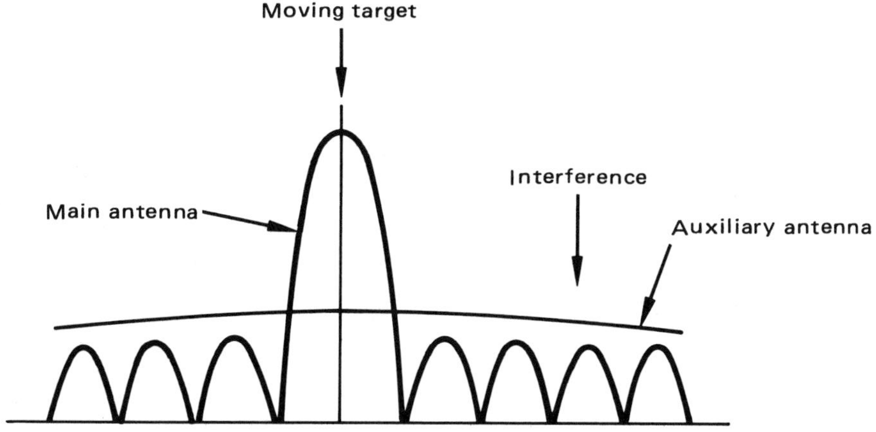


**Figure 13.20** Illustration of main antenna pattern, auxiliary antenna pattern, and signal direction.

efficient $w$ must be the ratio of the main antenna sidelobe gain in the direction of interference to the gain of the auxiliary antenna in that direction, and the phase of $w$ is the propagation delay between the auxiliary antenna and the center of the main antenna. Since there is no a priori knowledge of the interference direction, the weighting coefficient must be determined through the observed data.

For the pulse repetition frequency of 160 kHz, the filter formation time is 0.0128 second. Over this period the antennas scan through roughly 1°, and a significant gain variation in the main antenna on the interference may occur (a time-varying environment). To compensate for this variation, evaluation of many Wiener coefficients becomes necessary. The filter formation period is divided into 20 equal-length subperiods, and a single weighting coefficient is determined for each subperiod. Thus, the delay $\Delta$ as shown in Figure 13.19 may be taken equal to the length of one subperiod, or 0.64 ms.

Since the signal in both the main channel and the auxiliary channel is predominantly interference, we can treat the signal in the main channel as the desired signal $d(k)$ and estimate (or cancel) it by the weighted signal in the auxiliary channel, namely $wx(k)$. Consequently, most of the interference component is canceled. However, the target signal component in $d(k) - wx(k)$, the input to the Doppler filter bank, is essentially the same as that in $d(k)$. Recalling that the target signal in the main channel enters through the mainlobe, it is much stronger than the target signal in the auxiliary channel. Moreover, the magnitude of $w$ is less than unity, since the auxiliary antenna has higher gain than the main antenna in the direction of the interference.

The required quantities for the determination of the Wiener coefficient that appeared in Equation (13.16), namely $\mathbf{R}^{-1}$ and $\mathbf{p}$, can be estimated from $d(k)$ and $x(k)$ for each subperiod. $\mathbf{R}$ and $\mathbf{p}$ are reduced to 1-by-1 matrices or scalars in the present problem. For the first subperiod, we take the Wiener coefficient as

$$w(1) = \frac{\sum_{n=1}^{102} d(n)\bar{x}(n)}{\sum_{n=1}^{102} |x(n)|^2} \tag{13.101}$$

and we can similarly determine the Wiener coefficients for the subsequent subperiods.

Figure 13.21 shows the Doppler filter response, or the spectrum of the sum of the interference, ground clutter, receiver noise, and moving target, before the adaptive process is used. The target signal, being at such a low power level, is not observable there. The low response over the interval between 0 and 25 kHz is due to the clutter rejection filter. Figure 13.22 exhibits the Doppler filter response after the adaptive process is employed. The result indicates that virtually full detection sensitivity is achieved.

In an environment where two or more interference sources are present, more independent auxiliary channels would be required in order to achieve significant reduction in the interference, and the amount of computation needed for the determination of the Wiener coefficients would likewise increase.

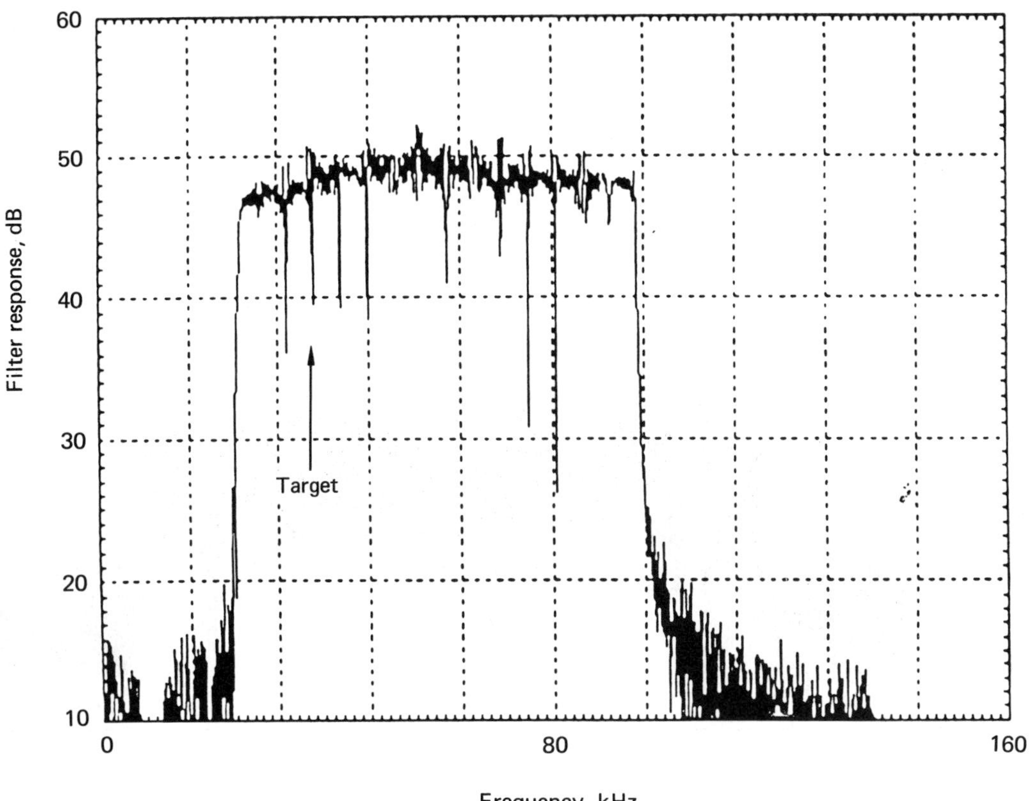


**Figure 13.21** Spectrum without adaptive processing.

## 8 SOME COMMON GRADIENT FORMULAS

The gradient of the performance measure $J$, a real-valued function of complex-valued quantities such as weight vector and correlation matrix, often occurs in adaptive algorithms. The gradients of some fundamental expressions from which the gradients of more complicated performance measures can be obtained are given below. Let the complex-valued vector be represented as

$$\mathbf{w} = \mathbf{u} + i\mathbf{v} = \begin{bmatrix} u_1 \\ u_2 \\ u_3 \\ \vdots \\ u_M \end{bmatrix} + i \begin{bmatrix} v_1 \\ v_2 \\ v_3 \\ \vdots \\ v_M \end{bmatrix} \tag{13.102}$$

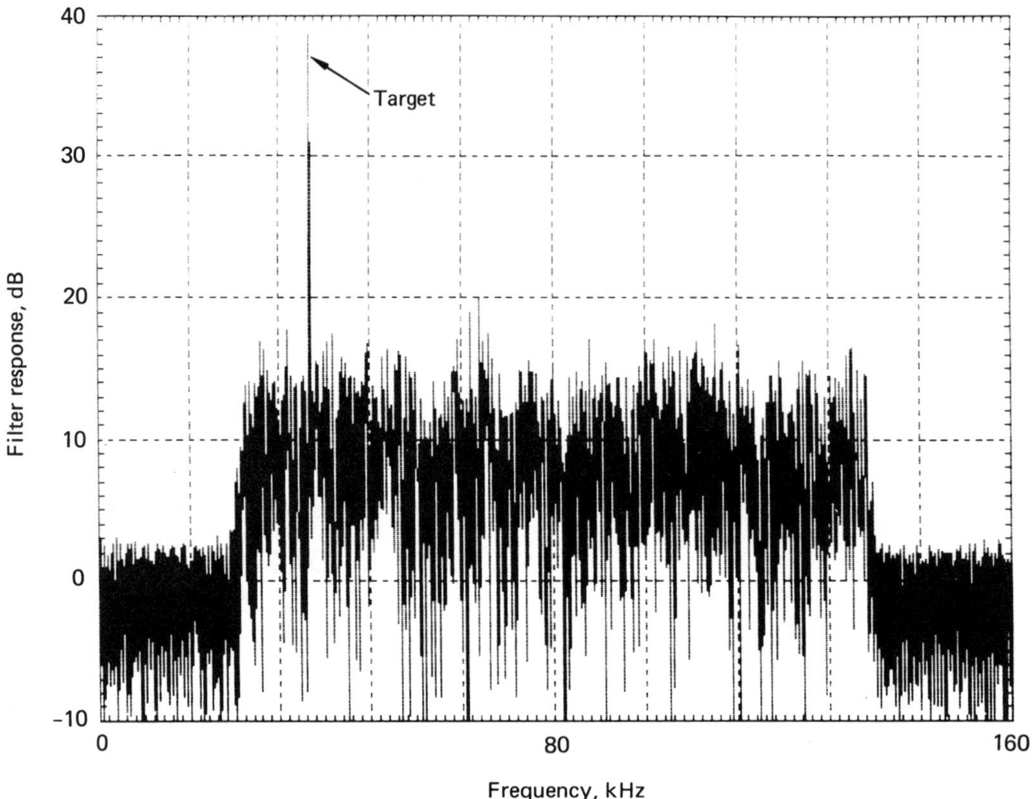


**Figure 13.22** Spectrum after adaptive processing.

where $\mathbf{u}$ and $\mathbf{v}$ are real-valued vectors. We are interested in the rate of change of $J$ with respect to $\mathbf{u}$ and $\mathbf{v}$, respectively, in a recursive expression of the form

$$\mathbf{w}(k+1) = \mathbf{w}(k) - \mu(\nabla J)^{\mathrm{T}} \tag{13.103}$$

where $\nabla J$ is the gradient of $J$ with respect to $\mathbf{w}$. Express the real and imaginary parts of the complex-valued vector in the recursive form as follows:

$$\mathbf{u}(k+1) = \mathbf{u}(k) - \mu \left( \frac{\partial J}{\partial \mathbf{u}} \right)^{\mathrm{T}} \tag{13.104}$$

$$\mathbf{v}(k+1) = \mathbf{v}(k) - \mu \left( \frac{\partial J}{\partial \mathbf{v}} \right)^{\mathrm{T}} \tag{13.105}$$

Thus,

$$\nabla J = \frac{\partial J}{\partial \mathbf{u}} + i\frac{\partial J}{\partial \mathbf{v}} \tag{13.106}$$

1. For $J = c\mathbf{w}^{\mathrm{T}}\bar{\mathbf{w}}$, where $c$ is a scalar,

$$\nabla J = 2c\mathbf{w}^{\mathrm{T}} \tag{13.107}$$

2. For $J = \mathbf{w}^{\mathrm{T}}\bar{\mathbf{A}}\bar{\mathbf{p}} + \bar{\mathbf{w}}^{\mathrm{T}}\mathbf{A}\mathbf{p}$, where $\mathbf{A}$ is a constant $M$-by-$M$ matrix and $\mathbf{p}$ is a constant $M$-by-1 vector,

$$\nabla J = 2\mathbf{p}^{\mathrm{T}}\mathbf{A}^{\mathrm{T}} \tag{13.108}$$

3. For $J = \mathbf{w}^{\mathrm{T}}\bar{\mathbf{p}} + \bar{\mathbf{w}}^{\mathrm{T}}\mathbf{p}$, a special case of item 2,

$$\nabla J = 2\mathbf{p}^{\mathrm{T}} \tag{13.109}$$

4. For $J = \mathbf{w}^{\mathrm{T}}\mathbf{Q}\bar{\mathbf{w}}$, where $\mathbf{Q}$ is a constant $M$-by-$M$ Hermitian matrix,

$$\nabla J = 2\mathbf{w}^{\mathrm{T}}\mathbf{Q} \tag{13.110}$$

## REFERENCES


1. B. Widrow and M. Hoff, Jr., Adaptive switching circuits, IRE WESCON Convention Record, pp. 96–104 (1960).
2. K. Steinbuch and B. Widrow, A critical comparison of two kinds of adaptive classification networks, *IEEE Trans. Electronic Computers, EC-14*(5): 737–740 (1965).
3. R. W. Lucky, Automatic equalization for digital communication, *Bell Syst. Tech. J., 44*: 547–588 (1965).
4. P. W. Howells, Intermediate frequency sidelobe canceller, U.S. Patent 3202990 (1965).
5. S. P. Applebaum, Adaptive arrays, Syracuse University Research Corp., Report SPL TR66-1 (1966).
6. B. Widrow, P. E. Mantey, L. J. Griffiths, and B. B. Goode, Adaptive antenna systems, *Proc. IEEE, 55*: 2143–2159 (1967).
7. B. Widrow and S. D. Stearns, *Adaptive Signal Processing*, Prentice-Hall, Englewood Cliffs, New Jersey (1985).
8. S. Haykin, *Introduction to Adaptive Filters*, Macmillan, New York (1984).

9. C. F. N. Cowan and P. M. Grant, eds., *Adaptive Filters*, Prentice-Hall, Englewood Cliffs, New Jersey (1985).
10. R. A. Monzingo and T. W. Miller, *Introduction to Adaptive Arrays*, Wiley, New York (1980).
11. G. C. Goodwin and K. S. Sin, *Adaptive Filtering, Prediction, and Control*, Prentice-Hall, Englewood Cliffs, New Jersey (1980).
12. S. T. Alexander, *Adaptive Signal Processing Theory and Applications*, Springer-Verlag, New York (1986).
13. W. F. Gabriel, Adaptive arrays—an introduction, *IEEE Proc.*, *64*(2): 239–271 (1976).
14. B. Widrow, J. R. Glover, Jr., J. M. McCool, J. Kaunitz, C. S. Williams, R. H. Hearn, J. R. Zeidler, E. Dong, Jr., and R. C. Goodlin, Adaptive noise cancelling: principles and applications, *Proc. IEEE*, *63*(12): 1692–1712 (1975).
15. A. Papoulis, *Probability, Random Variables, and Stochastic Processes*, McGraw-Hill, New York (1965).
16. G. Strang, *Linear Algebra and Its Applications*, 2nd ed., Academic Press, New York (1980).
17. W. Feller, *An Introduction to Probability Theory and Its Applications*, vol. II, Wiley, New York (1966).
18. G. M. Jenkins and D. G. Watts, *Spectral Analysis and Its Applications*, Holden-Day, San Francisco (1980).
19. M. K. Jain, S. R. K. Iyengar, and R. K. Jain, *Numerical Methods for Scientific and Engineering Computation*, Wiley, New York (1985).
20. D. K. Faddeev and V. N. Faddeeva, *Computational Methods of Linear Algebra* (translated by R. D. Williams), Freeman, San Francisco (1963).
21. G. Dahlquist and A. Bjorck, *Numerical Methods* (translated by N. Anderson), Prentice-Hall, Englewood Cliffs, New Jersey (1974).
22. J. M. Ortega and W. C. Rheinboldt, *Iterative Solution of Nonlinear Equations in Several Variables*, Academic Press, New York (1970).
23. J. A. Cadzow, Recursive digital filter synthesis via gradient based algorithms, *IEEE Trans. Acoust. Speech Signal Process.*, *ASSP-24*: 349–355 (1976).
24. Y. Bard, Comparison of gradient methods for solution of nonlinear parameter estimation problems, *SIAM J. Numer. Anal.* *7*: 157–186 (1970).
25. G. E. Forsythe, M. A. Malcolm, and C. B. Moler, *Computer Methods for Mathematical Computations*, Prentice-Hall, Englewood Cliffs, New Jersey (1977).
26. T. Kailath, *Linear Systems*, Prentice-Hall, Englewood Cliffs, New Jersey (1980).
27. L. A. Zadeh and C. Desoer, *Linear Systems Theory*, McGraw-Hill, New York (1963).
28. R. Bellman, *Introduction to Matrix Analysis*, McGraw-Hill, New York (1960).
29. G. H. Golub and C. F. Van Loan, *Matrix Computations*, John Hopkins University Press, Baltimore, Maryland (1983).
30. B. Noble and J. W. Daniel, *Applied Linear Algebra*, Prentice-Hall, Englewood Cliffs, New Jersey (1977).
31. J. A. Cadzow and H. R. Martens, *Discrete-Time and Computer Control Systems*, Prentice-Hall, Englewood Cliffs, New Jersey (1970).

32. B. Fisher and N. J. Bershad, The complex LMS adaptive algorithm—transient weight mean and covariance with applications to the ALE, *IEEE Trans. Acoust. Speech Signal Process., ASSP-31*: 34–45 (1983).
33. N. J. Bershad and L. Z. Qu, On the joint characteristic function of the complex scalar LMS weight, *IEEE Trans. Acoust. Speech Signal Process., ASSP-32*: 1166–1175 (1984).
34. R. A. Iltis and L. B. Milstein, An approximate statistical analysis of the Widrow LMS algorithm with application to narrow-band interference rejection, *IEEE Trans. Commun. COM-33*: 121–130 (1985).
35. N. J. Bershad and L. Z. Qu, LMS adaptation with correlated data—a scalar example, *IEEE Trans. Acoust. Speech Signal Process., ASSP-32*: 695–700 (1984).
36. L. J. Griffiths, Signal extraction using real-time adaptation of a linear multichannel filter, Ph.D. Dissertation, Stanford University, Stanford, California (1967).
37. L. E. Brennan, J. D. Mallett, and I. Reed, Adaptive arrays in airborne MTI radar, *IEEE Trans. Antennas Propag., AP-24*: 607–615 (1976).
38. T. S. Fong, Two Self-Adaptive Techniques for Sonar Array Processor, Hughes Aircraft Co., IDC 2764.20/40 (1972).
39. R. C. K. Lee, *Optimal Estimation, Identification and Control*, MIT Press, Cambridge, Massachusetts (1964).

# 14

# Digital Time Delay Beamforming

ROGER PRIDHAM Submarine Signal Division, Raytheon Company, Portsmouth, Rhode Island

## 1 INTRODUCTION

Advances in sonar transducer technology have made it feasible to build large arrays consisting of many hydrophones. This has been accompanied by a progressive increase in the complexity of sonar array topology. There has been a transition from simple linear and planar array geometries to cylindrical, spherical, and complex conformal arrays with compound curvatures. Along with these advances, new developments have been required in time delay beamformer technology to exploit the potential of large arrays. This area has seen an evolution from single-beam scanning systems to preformed beam systems where hundreds of sonar beams are simultaneously formed to provide spatial coverage over a wide angular region. During this transition, digital hardware has largely replaced analog hardware in the implementation of beamforming and other sonar functions [1, 2]. In addition, new beamforming techniques have been developed to reduce the cost and complexity of the beamformer electronics required for large aperture arrays. These include interpolation beamforming [3, 4], frequency domain beamforming [5, 6], phase shift beamforming [7], and hybrid techniques which combine time and frequency domain structures [8, 9]. Such techniques are essential because the beamformer throughput is usually far greater than any other computational requirement in the sonar and may even be prohibitive in some applications. This chapter outlines important developments in the design and implementation of time delay beamforming, which is the most widely used sonar array processing technique. Reference [10] gives a comprehensive review of the general sonar array processing problem which addresses such topics as optimum and adaptive beamforming.

We begin by reviewing some of the fundamentals of the sonar beamforming problem. First, it is recalled that the function of an individual hydrophone in an array is to convert a pressure signal into an electrical voltage [11]. If it is desired to combine voltage signals at the output of the various hydrophones in the array, it is necessary to account for the fact that the arrival time of an acoustic wave front to each hydrophone is slightly different. This difference in travel time is illustrated in Figure 14.1, which shows an acoustic wavefront being sampled by four sensors labeled 1, 2, 3, and 4. This figure shows a snapshot of the wave front at a time when it has arrived at hydrophone 3. At this instant, the perpendicular distances from the wave front to hydrophones 1, 2, and 4 are given as 1, 0.5, and 0.75 m, respectively. The wave front, which propagates at a speed of about 1500 m/s in water, arrives at hydrophone 1 about 0.67 ms (= 1/1500) after it arrives at 3. Likewise, it arrives at 2 and 4 about 0.33 and 0.5 ms after its arrival at 3. The beamformer operates on these voltages by first delaying them to compensate for the relative propagation delays and then summing the resulting time-aligned signals.

Figure 14.2 shows the structure of a time delay beamformer for an array of five hydrophones. The beamformer time delays are denoted as $T_1$, $T_2$, $T_3$, $T_4$, and $T_5$. These are also called steering delays because they are used to maximize the beamformer response to a wave front arriving from a specified direction. This

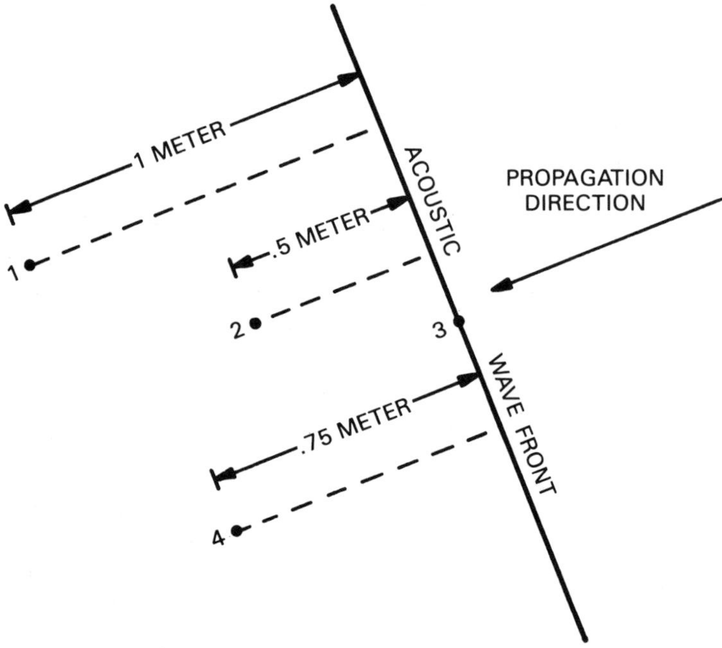


**Figure 14.1** Illustration of relative propagation delays for an array of four hydrophones.

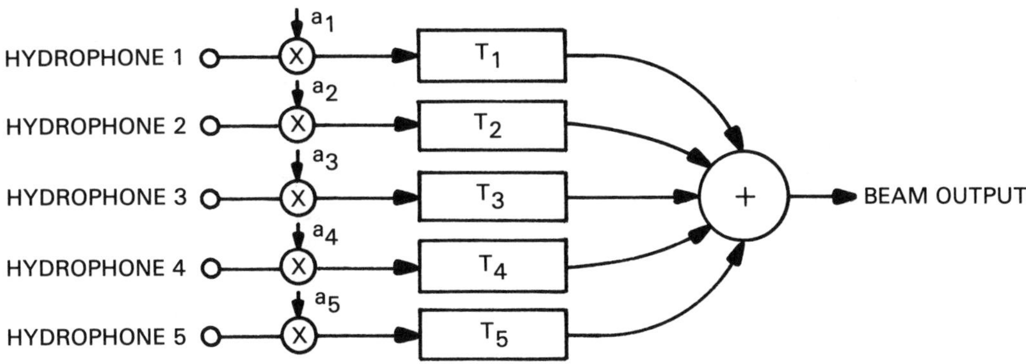


**Figure 14.2** Functional structure of time delay beamformer.

steered direction is called the maximum response axis (MRA) of the beam. The beam output response for wave fronts arriving from directions other than the MRA are reduced in amplitude since they will not be properly time-aligned prior to summation. Figure 14.2 also shows that the hydrophone outputs are multiplied by coefficients denoted by $a_1$, $a_2$, $a_3$, $a_4$, and $a_5$. These coefficients are used to control the beam's sidelobe structure, which is the beam response to wave front arriving well away from the MRA direction.

It is useful at this point to discuss how the response of a beamformer can be visualized. Figure 14.3 gives an illustration of a typical beam output response versus wave front arrival angle. This response curve is called a beampattern. Once again, the response is maximum for the MRA direction. The region in the vicinity of the MRA is called the mainlobe region. The angular span for which the

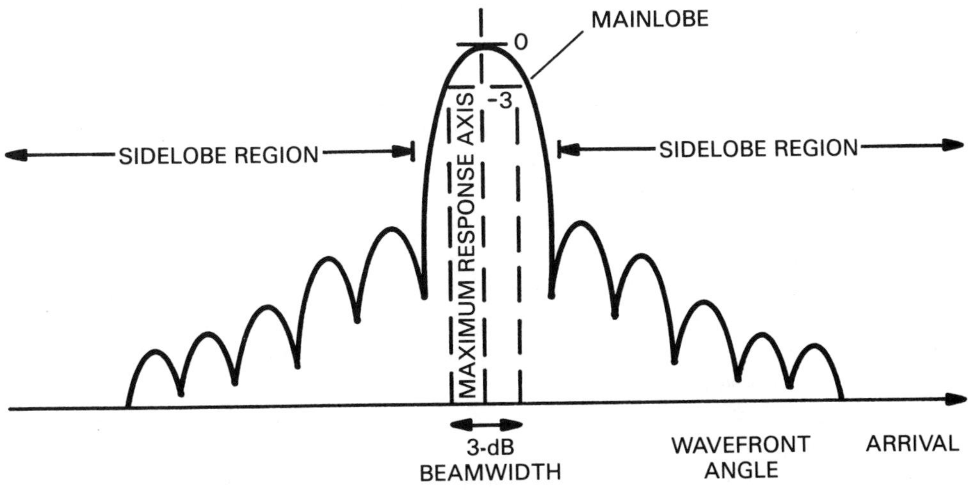


**Figure 14.3** Illustration of array beampattern.

mainlobe response is within 3 dB of the MRA response if the 3 dB beamwidth of the beampattern. This beamwidth defines the angular resolution of the beam. The sidelobe structure is typically well below the MRA in level. For a simple line array of hydrophones, which are uniformly distant in spacing, the peak sidelobe will be down from the mainlobe peak by about −13 dB if the beamformer weighting coefficients (i.e., the weights $a_1$ through $a_5$ in Figure 14.2) are chose as unity [1]. Other sets of weighting coefficients can be chosen which will reduce the height of the peak sidelobe to −30 dB or better at the expense of a wider mainlobe region. Thus, it is possible to trade off beam angular resolution with sidelobe level. For readers who are familiar with the field of spectrum analysis, it can be noted that this technique is identical to the procedure of using temporal weighting to trade off frequency resolution and sidelobe leakage in the frequency analysis of time series [12]. Low sidelobe structure is very important in beamforming so that a weak target signal will not be obscured by a strong interfering target arriving off the MRA.

As a summary of the above discussion, it is noted from Figure 14.2 that the basic operations of a time delay beamformer are 1) to weight each hydrophone voltage with a coefficient $a$, 2) to introduce a time delay $T$ in each channel to align the voltages for a wave front arriving along the beam pointing direction, and 3) to sum the weighted and delayed hydrophone signals to form a signal beam output signal. This procedure forms a single beam pointing in one direction. It is possible to form a set of different beams simultaneously by replicating the structure given in Figure 14.2. This beam set, which can simultaneously provide coverage over a wide angular sector, is referred to as a preformed beam system. The alternative is to form a single scanning beam which provides wide spatial coverage by continuously steering the beam throughout an angular sector.

The sonar beamforming problem outlined above is similar to the radar beamforming problem. However, there are some key differences which lead in general to significantly different beamformer implementations. For example, in most radar applications, the ratio of the signal bandwidth to the center frequency is very small. In this case, the time delays in Figure 14.2 can be implemented with simple phase shifts [13]. In some sonar applications, it is also possible to use a phase shift beamformer [7]. However, in many situations of interest in sonar, the signal spectrum is relatively wide in frequency extent and centered at a low frequency in the audio band. Thus, the phase shift approximation is often not valid [14] and true time delay beamforming is required.

Time delay beamforming can be implemented in either the time domain or the frequency domain. In the time domain implementation, the structure in Figure 14.2 is directly implemented in either analog or digital hardware. For example, the time delays can be realized using analog delay lines, digital shift registers, or random access memory (RAM). In the frequency domain implementation of a time delay beamformer, the hydrophone signals are first converted from a time domain representation using fast Fourier transform (FFT) processing. The time delays are then realized by applying a frequency-dependent phase shift to the Fourier coefficients of each signal. The advent of high-speed digital circuitry and analog-to-digital (A/D) converters has permitted complete digital implementations of time delay beamformers [15]. In such an implementation, the hydrophone signals are first converted to digital sequences. Each hydrophone channel has signal conditioning electronics to prepare the analog signal for conversion. This circuitry includes a preamplifier to boost the signal above the system noise and a filter to suppress out-of-band frequency components which can alias into the processing band due to the time sampling. The analog-to-digital conversion process consists of the time sampling and amplitude quantization of each hydrophone signal. Time sampling is achieved by providing each channel with either a sample and hold or a track and hold circuit. These circuits permit simultaneous or nearly simultaneous sampling of all hydrophone signals in the array. If the sensors are not all sampled at the same time, then a time skew exists between some of the sensor sequences. This time skew can be compensated for later in beam formation by adjusting the beamformer time delays once the sensor voltages have been sampled by the hold circuits. The A/D converter then sequentially quantizes these samples, which are then multiplexed and transmitted to the digital beamformer for subsequent processing. The number of channels which an A/D converter can digitize depends on its throughput capability and the required sampling rate per channel. For example, with current technology, a 12-bit A/D converter may have a 300-kHz throughput, which means that it can multiplex data from 10 channels at a maximum sampling rate of 30 kHz each.

An important feature of a digital beamformer is that the beam steering time delays must be quantized to a time interval which is equal to the input sampling interval as shown in Figure 14.4. For example, if the digital samples of the hydrophone output voltage are available at a 30-kHz rate, then the accuracy of each time delay is about 33.33 $\mu$s ($= 1/30,000$). A required beam steering delay more precise than this cannot be achieved because the corresponding digital sample is not available. When the exact time delay required for beam steering cannot be achieved, a frequency-dependent phase error results which can significantly increase the sidelobe level and, hence, increase the susceptibility interference from target wave fronts which arrive off of the beam's MRA. This phase error, which is equal to $2\pi f\epsilon$, where $f$ is frequency and $\epsilon$ is time delay error, is greatest for the highest-frequency component in the band. Typically, in order to achieve acceptable sidelobe levels the input sampling rate, which is the inverse of the input sampling interval, must be at least four to eight times the highest signal frequency. This is in sharp contrast to the Nyquist rate requirement that a signal must only be sampled at just over twice the highest signal frequency so that sufficient information is retained in the digital sequence to ensure that the signal can be reconstructed from the samples. In fact, in a digital beamformer, the input sample rate is usually much higher than the rate at which the beam output samples are computed. The high input rate is required only to achieve accuracy in the beam steering time delays. Once the required delays are implemented and the hydrophone channel digital sequences are properly aligned, the beam output sequence is computed at a much lower rate which is dictated by Nyquist sampling rate considerations.

In subsequent sections, we review beamformer implementation concepts in more detail. Section 2 examines time domain implementation approaches which involve application of digital interpolation techniques. Section 3 discusses frequency domain implementation concepts. Finally, Section 4 gives a summary of the chapter.

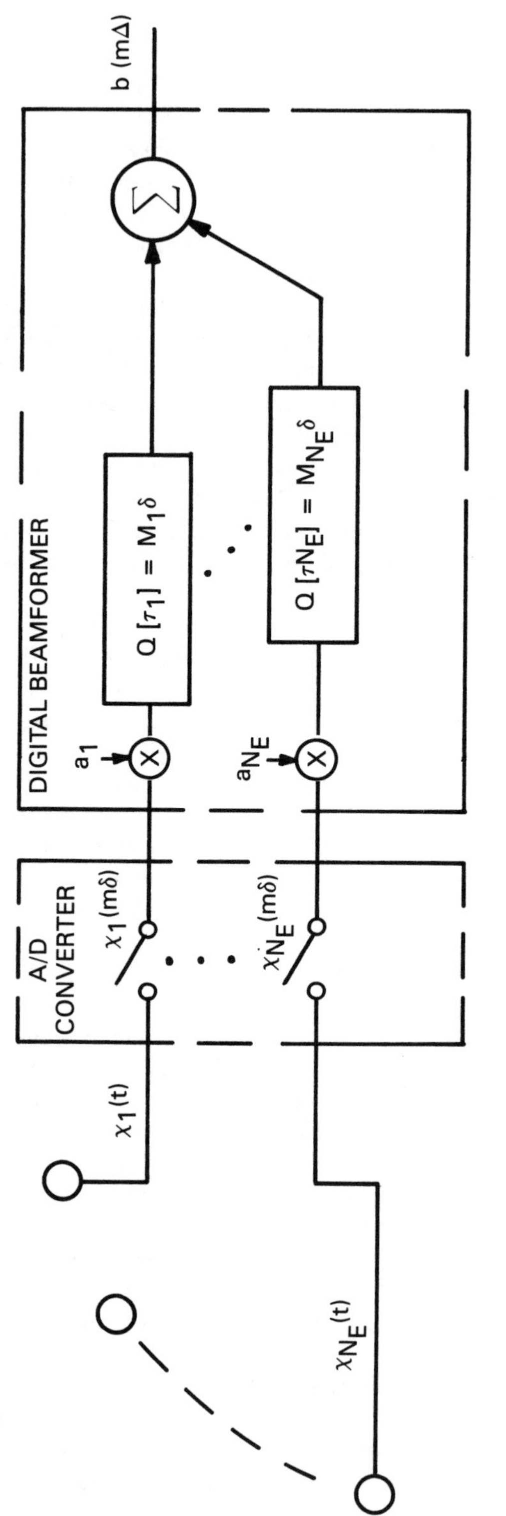


**Figure 14.4** Digital beamforming operation.

## 2 TIME DOMAIN IMPLEMENTATION

The beamforming operation illustrated in figure 2 can be represented by the following equation:

$$b(t) = \sum_{n=1}^{N_E} a_n X_n(t - \tau_n) \tag{14.1}$$

where $b(t)$ is the beam output, $a_n$ is the shading weight for the $n$th sensor, $X_n$ is the signal from the $n$th sensor, $\tau_n$ is the required steering delay for the $n$th sensor, and $N_E$ is the number of sensors. Equation (14.1) represents the response of a continuous time analog delay-sum beamformer.

In a discrete time implementation of Eq. (14.1), the sensor signals $X_n$ are sampled in time and quantized by an A/D converter as shown in Figure 14.4. The sampling interval is denoted as $\delta$ (seconds) and the corresponding input sampling rate is $f_i = \delta^{-1}$ Hz. In a straightforward implementation these sensor samples are stored in RAM prior to beam formation. Since the sensor signals are available only at the rate $f_i$, the beam former time delays are effectively quantized to the interval $\delta$. Therefore, the sensor sequences can be time-aligned only to the accuracy of the quantized time delay:

$$Q[\tau_n] = M_n \delta \tag{14.2}$$

where $M_n$ is an integer. The beam output sequences are obtained by fetching the required sensor samples, multiplying then by the appropriate shading coefficient, and performing the summation operation. As mentioned previously, the rate at which the beam output must be computed is usually substantially lower than $f_i$. If the required beam output rate is denoted as $f_0$, then the operations shown in Figure 14.4 can be represented as

$$\underset{\sim}{b}(m\Delta) = \sum_{n=l}^{N_E} \underset{\sim}{a}_n \underset{\sim}{X}_n(m\Delta - M_n\delta) \tag{14.3}$$

where $\Delta = f_0^{-1}$ is the beam output sampling interval and the tilde $(\underset{\sim}{\ })$ below a symbol denotes amplitude quantization. Figure 14.5 illustrates the time relationship between the input and output sequences for a conventional time domain beamformer.

As described above, all of the required hydrophone samples are first stored in delay memory and subsequently accessed to form the beams. For this implementation, which is called an input storage beamformer, the size of the delay memory is proportional to the high input sampling rate. An alternative approach is called a partial sum beamformer, where the beam output is computed in a progressive fashion as new input sensor samples become available. As in the case of the input storage beamformer, the high channel input rate is required for time delay accuracy.

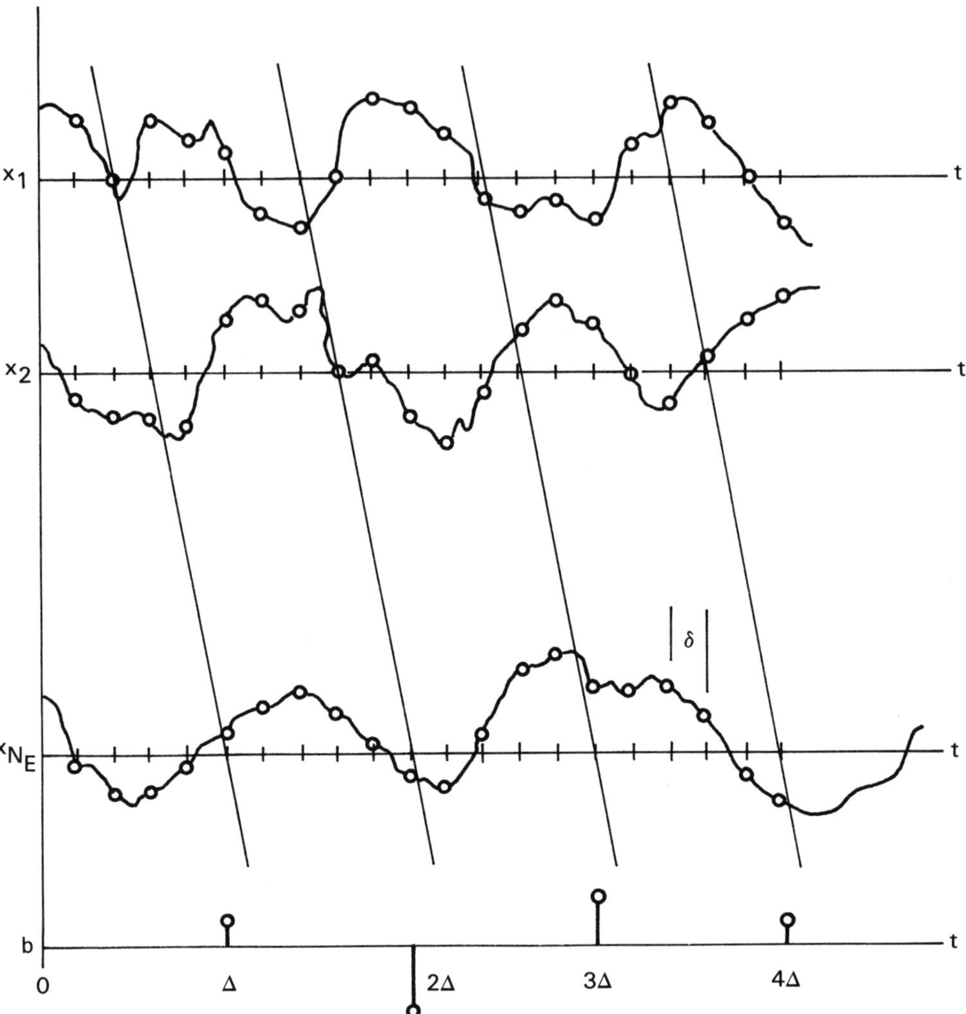


**Figure 14.5** Input/output timing relationship for conventional timing domain digital beamformer.

However, since the input samples are only summed and stored as they are needed to compute a beam sample at the relatively low output rate, the size of storage memory needed to form a single beam is greatly reduced. The reduction factor approximately equal to the ratio of the output to the input sampling rate. Because of this, the partial sum architecture is generally preferred for many beamforming applications.

The computational rate for both of the above beamformer configurations can be made the same by proper design. It is evident from Eq. (14.3) that $N_E$ multiplications and $N_E - 1$ additions are required to compute one beam output sample. If $N_B$ beams must be computed, then the approximate multiply-add rate for large

$N_E$ is

$$R = N_E N_B f_0 \tag{14.4}$$

operations (multiply-adds) per second. It is easy to see that $R$ can become very large as the array size increases. For example, consider an application where $N_E = 600$ hydrophones, $N_B = 40$ beams, and $f_0 = 10$ kHz. In this case $R = 240$ MHz, which can be realized only with high-speed digital circuitry.

Since the weighting coefficients, steering delays, and number of array elements vary with application, it is generally desirable to utilize programmable hardware to implement the beamforming function. The Microprogrammable Beamformer (MPBF) concept [15] is a particularly efficient implementation of the partial sum beamformer algorithm. Figure 14.6 shows a functional diagram of the basic MPBF element. The operation of the MPBF can be described with respect to this figure. First, hydrophone samples from the A/D converter are loaded into input buffer memory. Under control of control memory, a sample (a) is taken from the input buffer memory, multiplied by the appropriate shading coefficient (b), and added to the appropriate partial sum (c) from partial sum memory, which is accumulating the beam. The result (d) is stored in this memory. As a beam computation (e) is completed, it is stored in the output buffer memory to be routed for postbeamformer processing.

The addresses (f) for the various memories are obtained from control memory and associated addressing circuitry. The associated addressing circuitry includes a counter which is incremented as each control word is executed. This counter provides an offset address for the partial sum memory address provided by the control memory. The two addresses are added modulo the partial sum memory size to

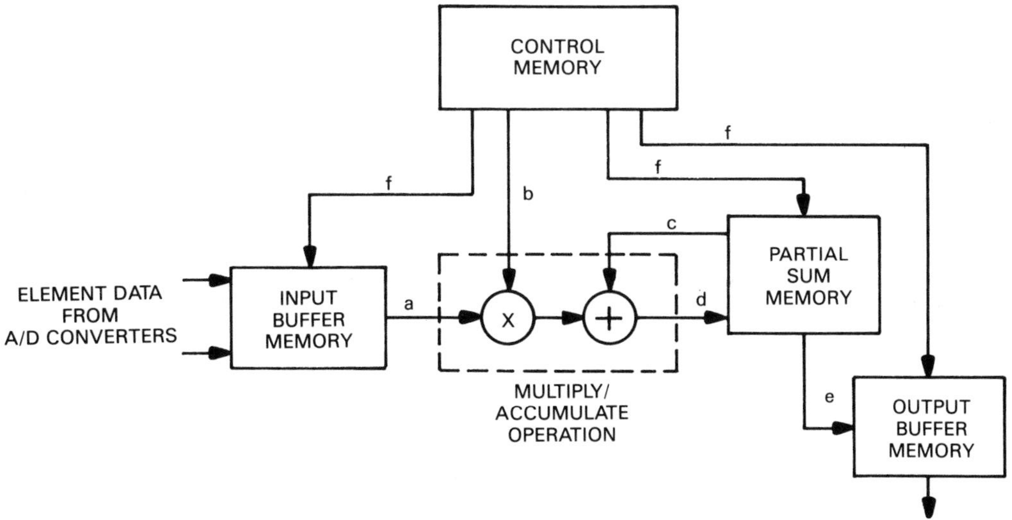


**Figure 14.6** Functional diagram of MPBF element.

compute the actual partial sum memory address. This rotating address technique makes efficient use of the partial sum memory by writing newly computed partial sums into memory locations containing obsolete data. Hydrophone weighting coefficients (b) are provided in the control instructions. Input, output, and control memories are usually double-buffered to permit asynchronous operation with other system elements.

The throughput capability of an MPBF element depends on both the speed of the arithmetic unit and the memory access time. Current state-of-the-art very large scale integration (VLSI) technology is available which can perform the floating-point multiply accumulation operation of Figure 14.6 at a 25-MHz rate. Figure 14.7 is a picture of the Floating-Point Multiply-Accumulate Kernel (FPMAK) chip developed by Raytheon Company. This component is fabricated using 1.25 $\mu$m CMOS technology and can process 24-bit (16-bit mantissa, 8-bit exponent) input samples and accumulate with 33 bits of mantissa precision. A given beamformer application is implemented by partitioning the task among a number of MPBF elements. In order to implement the previously mentioned rate of 240 MHz, a minimum of 10 MPBF elements is required, where each utilizes a single FPMAK chip as the processor.

A useful concept which can reduce total computational loading for some array geometries is multistage beamforming. For example, in a rectangular planar array the columns of elements can be grouped into vertical staves. Each vertical stave can be considered as a line array for which the required depression/elevation (D/E) beams are formed in a first stage of beamforming. The required throughput for the first stage is

$$R_{1S} = (N_E)_v \times N_{D/E} \times N_{\text{STAVE}} \times f_0 \tag{14.5}$$

where $(N_E)_v$ is the number of hydrophones in a vertical stave, $N_{\text{STAVE}}$ is the number of vertical staves, and $N_{D/E}$ is the number of D/E beam directions. In the second stage, the preformed stave outputs are treated as elements and combined to form the azimuth beams at each D/E angle. Required throughput for the second stage is

$$R_{2S} = N_{D/E} \times N_{\text{STAVE}} \times N_{\text{AZ}} \times f_0 \tag{14.6}$$

where $N_{\text{AZ}}$ is the number of azimuth beams. The total throughput for the two stages is

$$R = R_{1S} + R_{2S} = N_B N_E f_0 F \tag{14.7}$$

where $N_B = N_{D/E} \times N_{\text{AZ}}$ is the total number of beams, $N_E = (N_E)_v \times N_{\text{STAVE}}$ is the total number of elements, and $F = [N_{\text{AZ}} + (N_E)_v]/[N_{\text{AZ}} \times (N_E)_v]$ is the multiple stave reduction factor. In order for multiple-stage beamforming to reduce

**Figure 14.7** (Top) FPMAK chip; (bottom) FPMAK packaged as a leadless chip carrier.

computation over the direct implementation rate of $N_B N_E f_0$ the reduction factor $F$ must be less than unity.

Another technique for reducing hardware complexity is interpolation beamforming. This concept addresses the case where the required input sample rate $f_i$ and the number of sensors $N_E$ are so large that the telemetry bandwidth or the number of A/D converters is prohibitive. The basic ideas behind interpolation beamforming are:

1. Sample the hydrophone channels at $f_0$ and use digital interpolation to achieve the high input rate $f_i$ which is needed for beamforming.

2. Use MPBF elements to implement the filtering operation required for interpolation.
3. Commute interpolation filtering and beamforming as required to minimize the extra digital filtering computations.

The use of interpolation beamforming reduces the A/D converter requirements by the ratio $f_i/f_0$.

To discuss this idea further, it is necessary to describe digital interpolation. The concept of digital interpolation can be explained as a two-stage process [16, 17] with reference to Figure 14.8. First, a signal $x(t)$ is sampled at a coarse time interval $\Delta$ by the A/D converter. The zero pad operation involves artificially increasing the rate of the sequence by inserting zeros between the data samples, $x(m\Delta)$. In Figure 14.8, the zero-padded sequence is denoted by $v(m\delta)$, where $\delta$ is the smaller time delay interval which is needed to reduce the beamformer phase errors. This zero-padded sequence is smoothed using a linear digital interpolation filter to compute the samples at the higher rate $f_i = \delta^{-1}$. The key point is that interpolation can be implemented with a digital filter. The structure of such a filter is very similar to beamforming in that input data are weighted by the filter coefficients, delayed, and summed. In fact, the MPBF elements can implement both the beamforming and the interpolation filtering operations. Furthermore, the MPBF can be programmed to take advantage of the periodically occurring zeros. This is done by simply eliminating any microcode instructions that involve multiplications of zero. Thus, the zero-padding operation does not affect either the size of the control memory or the computational throughput requirement of the MPBF in the implementation of the interpolation filter.

Figure 14.9 shows the frequency domain representation of the interpolation process. In Figure 14.9a the zero-padding operation on $X(m\Delta)$ increases the effective sampling rate from $f_0$ to $Kf_0$, where $K$ is the upsampling factor. Figure 14.9b shows how an ideal low-pass filter $H(f)$ rejects the replica spectra at $f_0$, $2f_0$, ..., $(K-1)f_0$ to yield the interpolated sequence spectra $X(f)$ in Figure 14.9c. The digital filter $H(f)$ is ideal in the sense that it has a perfectly flat rectangular passband and a stopband with infinite attenuation.

One can only approximate this ideal filter characteristic with a practical filter design that has ripple in both the passband and stopband as well as a nonzero transition band. Figure 14.10 shows these parameters, which can be specified

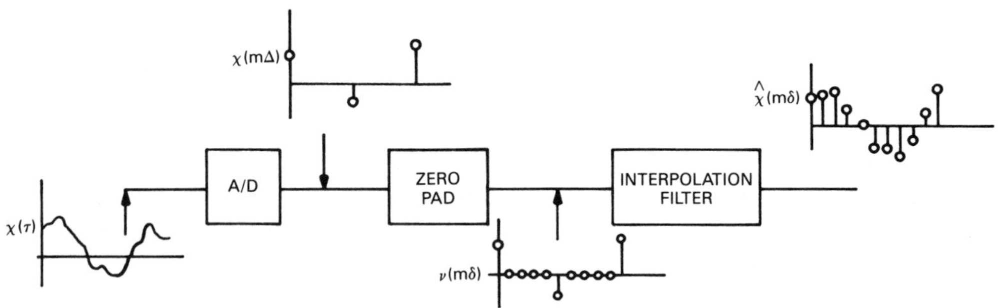


**Figure 14.8** Digital interpolation process for a single data channel.

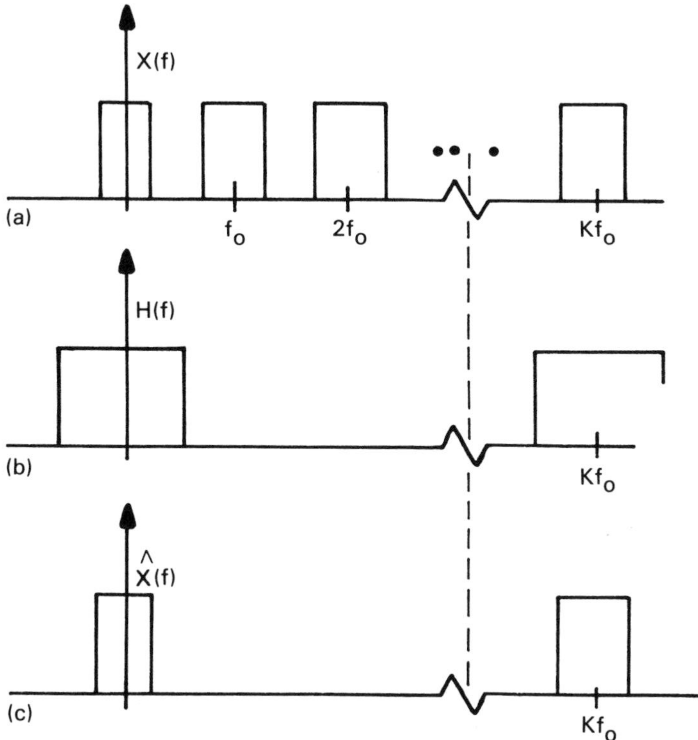


**Figure 14.9** Frequency domain representation of interpolation process for an ideal interpolation filter: (a) zero padding; (b) interpolation filter; (c) interpolated sequence.

to yield both nonrecursive and recursive digital filter designs. In general, the number of filter coefficients increases as the stopband rejection increases and the transition bandwidth decreases. Infinite impulse response (IIR) recursive filters have the general advantage of requiring fewer multiplies per output point for a given filter specification. This is a desirable feature for the beamforming application [18]. However, finite impulse response (FIR) nonrecursive filters are often preferred since FIR filters can be efficiently downsampled and are not as sensitive to finite-precision arithmetic effects such as round-off noise and limit cycles.

Figure 14.11 illustrates the interpolation process for the case of a real filter. The passband ripple of $H(f)$ has distorted the in-band portion of $X(f)$ and attenuated replicas are now present at $f_0$, $2f_0$, $3f_0$, ..., $(K-1)f_0$. For beamforming the interpolation is actually used to resample the signal at $f_0$ but with a different time delay translation. This involves a downsampling operation on the interpolated sequences that folds the attenuated spectra in Figure 14.11c back into the low-pass band of interest. These aliased attenuated spectra produce the most significant interpolation error. This error can be reduced by increasing the stopband rejection, which in turn increases the size of the interpolation filter. It is most important to

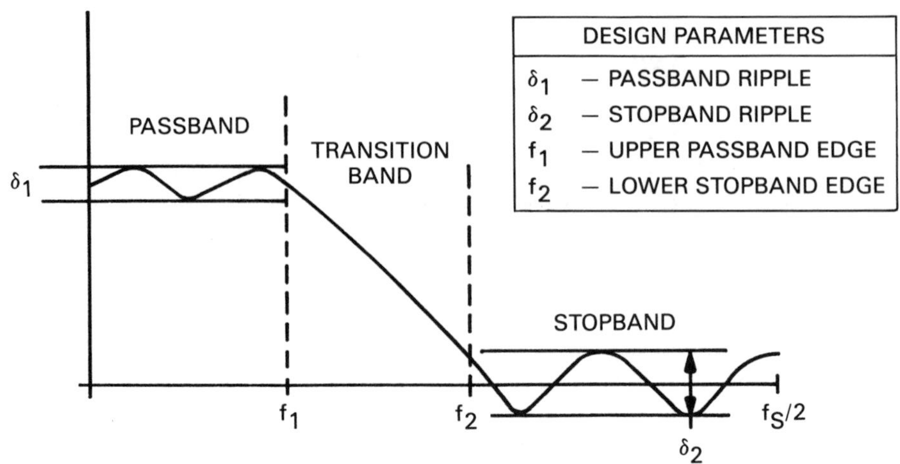


**Figure 14.10** Design parameters for interpolation filter; $f_s$ denotes sampling frequency.

design the filter to minimize the interpolation error at high frequencies [19], where the beampattern sidelobe levels are most sensitive.

Figure 14.12 shows a function diagram of how the interpolation filtering can be performed prior to beam formation. The A/D converters sample the sensors at $f_0$ and the zero-pad and interpolate process achieves the high sampling rate $f_i$ required for the beamformer. Since the filtering process involves simple delay, weight, and sum operations, the interpolation filter can be implemented with the MPBF elements in the same way as the beamformer itself. If the number of filter coefficients is $N_C$, then the throughput required of the filtering shown in Figure 14.12 is $R_{IF} = N_E N_C f_i$. However, since the filter input sequences are zero-padded, the MPBF element is programmed so that multiplications by zero are not performed, as mentioned previously. Therefore, the required computation rate is reduced by the factor $K^{-1}$ so that

$$R_{IF} = N_E N_C f_i K^{-1} = N_E N_c f_0 \tag{14.8}$$

The beamformer rate is still $N_E N_B f_0$ so that the total throughput is

$$R_T = (N_E N_c + N_E N_B) f_0 \tag{14.9}$$

Figure 14.13 shows the alternative implementation where the interpolation filtering is performed at the beamformer output. In this case the beamformer rate is $R_{BF} = N_E N_B f_i$. However, once again the MPBF can take advantage of the zero padding to reduce this by the factor $K^{-1}$ so that

$$R_{BF} = N_E N_B f_i K^{-1} = N_E N_B f_0 \tag{14.10}$$

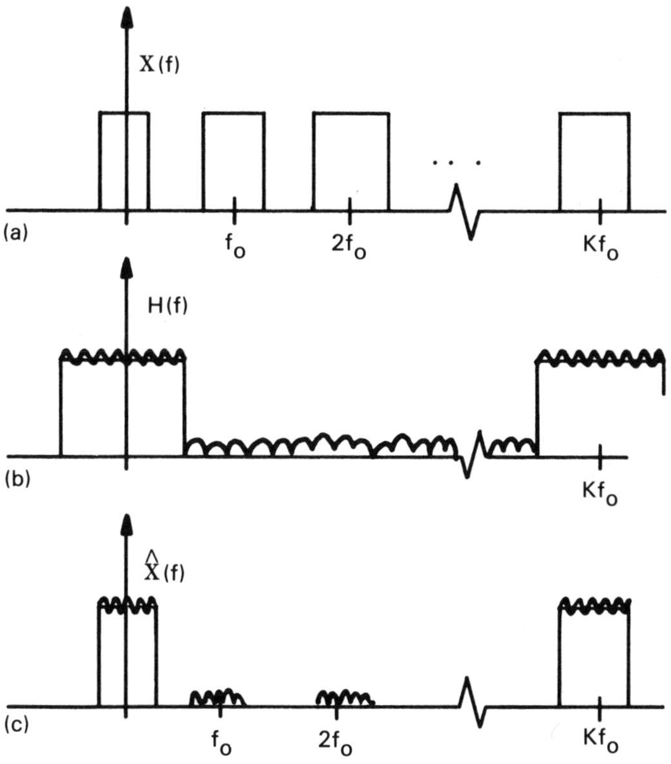


**Figure 14.11** Frequency domain representation of interpolation process for a real interpolation filter: (a) zero padding; (b) interpolation filter; (c) interpolated sequence.

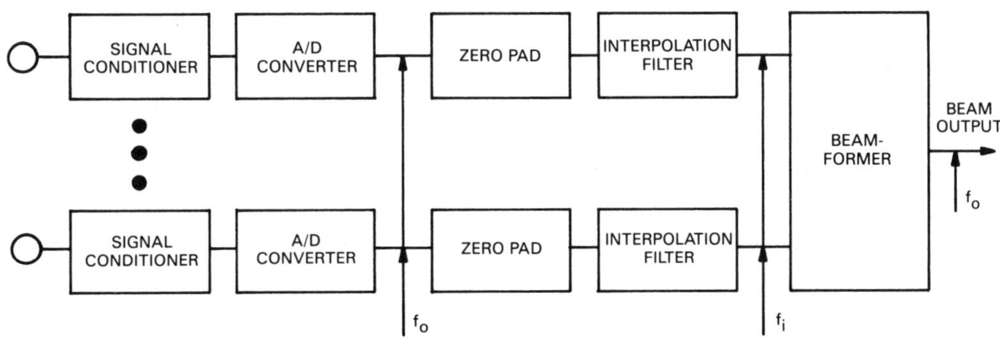


**Figure 14.12** Interpolation beamformer where interpolation is implemented at beamformer input.

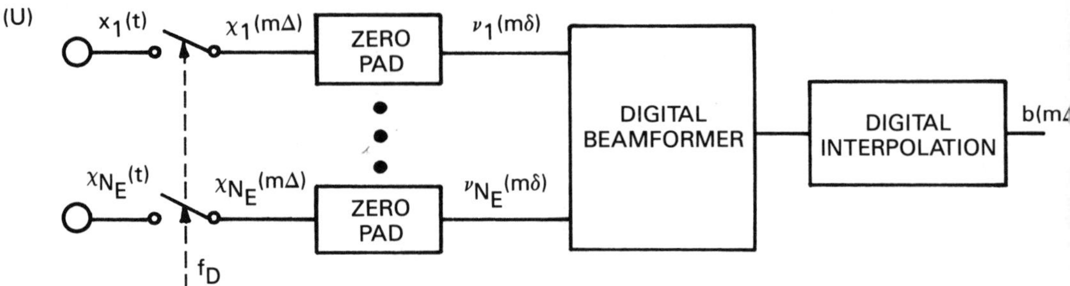


**Figure 14.13** Interpolation beamformer where interpolation is implemented at beam output.

The beam output interpolation filter can be directly downsampled to $f_0$ so that its rate is $N_B N_c f_0$, which yields a total throughput of

$$R_T = (N_B N_c + N_E N_B) f_0 \tag{14.11}$$

Comparison of Eqs. (14.9) and (14.11) demonstrates the advantages of beam output interpolation when $N_B < N_E$.

It should be noted that the operation of the beamformer in Figure 14.12 differs from that in Figure 14.13 in a fundamental way. The structure of Figure 14.12 is referred to as a downsampling beamformer because the output rate is lower than the input rate. In Figure 14.13 the reverse is true; the input rate is $f_0$ (neglecting the padded zeros) while the output rate is $f_i$. This structure is referred to as an upsampling beamformer by analogy to the upsampling filter used in the interpolation process. Figure 14.14 shows a comparison of the beam formation process in a downsampling and upsampling beamformer for a seven-element array with $K = 3$.

The foregoing discussion has concentrated on the simplest example where real sampling is performed on the sensor signals. However, the results can be extended to the case where complex sampling is performed on complex band-shifted sensor data [4, 8] to directly compute the complex band-shifted beam output sequence.

## 3 FREQUENCY DOMAIN IMPLEMENTATION

Delay-sum beamforming can be represented in the frequency domain by taking the Fourier transform of Equation (14.1):

$$B(\omega) = \sum_{n=1}^{N_E} a_n X_n(\omega) e^{-j\omega\tau_n} \tag{14.12}$$

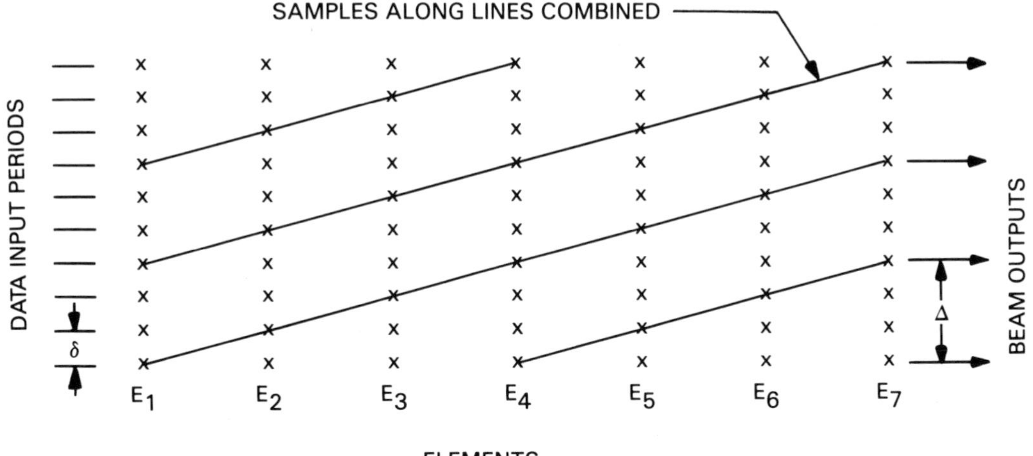


(a) DOWNSAMPLING BEAMFORMER (3 TO 1)

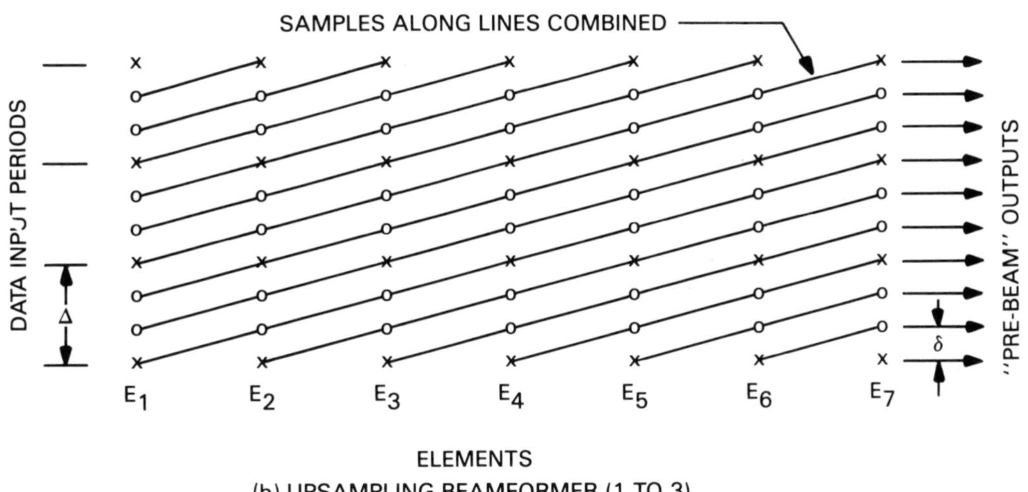


(b) UPSAMPLING BEAMFORMER (1 TO 3)

**Figure 14.14** Examples of (a) downsampling and (b) upsampling partial sum beamformer. (×) data sample; (∘) padded zero.

where $B(\omega)$ and $X_N(\omega)$ are the Fourier transforms of $b(\tau)$ and $x_n(\tau)$, respectively. This expression is efficiently represented in vector notation as

$$B(\omega) = \mathbf{V}^*(\omega)\mathbf{X}(\omega) \tag{14.13}$$

where

$$\mathbf{V}(\omega) = \begin{vmatrix} a_i \exp(-j\omega\tau) \\ \vdots \\ a_{N_E} \exp(-j\omega\tau_{N_E}) \end{vmatrix} \tag{14.13a}$$

$$\mathbf{X}(\omega) = \begin{vmatrix} X_1(\omega) \\ \vdots \\ X_{N_E}(\omega) \end{vmatrix} \tag{14.13b}$$

and the asterisk represents conjugate transpose.

Figure 14.15 shows a diagram of the discrete time implementation of Eq. (14.13) where the sensor Fourier transforms are computed using an FFT algorithm. The sensor signals need only be sampled at the low rate $f_0$ required for sequence reconstruction. The FFT size $M$ is normally an integer power of 2 and the time record length $T = M\Delta$ used in the FFT determines the frequency resolution $W_f$ of the FFT bin:

$$W_f = \frac{k}{M\Delta} \tag{14.14}$$

where $k$ depends on the type of temporal shading used in the FFT. For example, $k = 0.88$ yields $W_f$ as the double-sided 3-dB bandwidth for a rectangular window. The frequency resolution must be sufficiently small that the phase shift approximation to the time delay operation represented by $\mathbf{V}$ is accurate.

The number of operations required to implement beamforming in the frequency domain can be estimated by noting that a complex $M$-point FFT requires approximately $M \log_2 M/2$ complex operations (complex multiply-additions) [20]. Therefore, the total FFT requirement shown in Figure 14.15 is $N_E M \log_2 M/2$ complex operations if there is no overlap of time records prior to the FFT. If it is required that the beam output $B(\omega)$ be transformed back into the time domain, then the total FFT rate requirement is approximately

$$R_{\text{FFT}} = N_E M \log_2 M/(2T) = N_E f_0 \log_2 M/2 \tag{14.15}$$

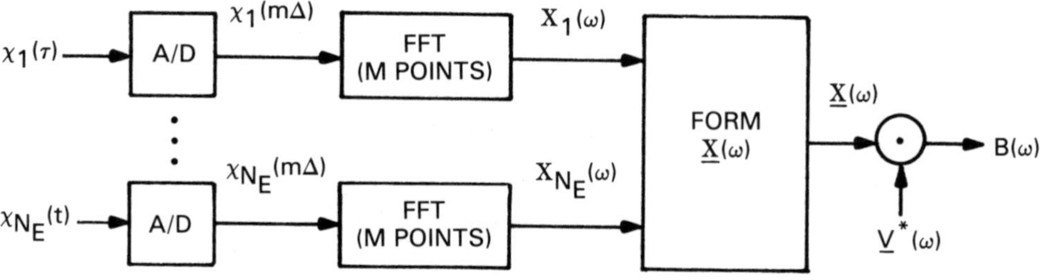


**Figure 14.15** Frequency domain beamforming operation.

complex operations per second since the required FFTs must be performed every $T = M\Delta = M/f_0$ seconds to keep up with the input data rate. The actual beamforming operation is the inner product $\mathbf{V}^*(\omega)\mathbf{X}(\omega)$ which requires $N_E$ complex multiplys and $N_E - 1$ complex adds for each of the $M$ frequency bins and $N_B$ beams. Therefore the rate associated with the actual beam formation is

$$R_{BF} = \frac{MN_BN_E}{T} = N_BN_Ef_0 \tag{14.16}$$

complex operations per second.

The inner production operation $\mathbf{V}^*(\omega)\mathbf{X}(\omega)$ can be performed efficiently using an FFT for very special geometries such as a line array or planar array of uniformly spaced elements [6, 7, 21, 22]. This can readily be shown by considering the form of the discrete Fourier transform (DFT) of the complex sensor samples:

$$\underset{\sim}{B}_m(\omega) = \sum_{n=l}^{N_E} a_n X_n(\omega) \exp\left(-j\frac{2\pi}{N}mn\right) \tag{14.17a}$$

where $N$ is an integer power of 2 such that $N \geq N_E$. For a line array of $N_E$ sensors uniformly spaced at interval $d$, the beam output is

$$B_m(\omega) = \sum_{n=1}^{N_E} a_n X_n(\omega) \exp\left(-j\frac{2\pi fnd}{c}\sin\theta_m\right) \tag{14.17b}$$

where $\theta_m$ is the MRA for the $m$th beam. Comparison of Eqs. (14.17a) and (14.17b) shows that the $\underset{\sim}{B}_m$ correspond to the line-array beam output Fourier coefficients at the steering angles

$$\sin\theta_m = \frac{mc}{Nfd} \tag{14.18}$$

Therefore the FFT algorithm can be used to compute $B_m$; however, the MRA directions vary as a function of frequency. Since the number of beams formed is $N = N_B$, the required number of complex operations is $N_B \log_2 N_B/2$. The problem of MRA variation with frequency can be eliminated by interpolating the complex beam coefficients at each frequency to obtain beam sets with a fixed set of MRAs. This requires approximately five additional operations per beam output [21] for a total requirement of $N_B(\log_2 N_B/2 + 5)$. Use of the FFT results in computational savings if $N_B(\log_2 N_B/1 + 5) < N_EN_B$. As an example, consider a line array where $N_E = N_B - 128$, for which FFT beamforming reduces the inner product computation by the factor

$$\frac{N_E}{\log + 2N_B/2 + 5} \doteq 15$$

## 4 SUMMARY

This chapter has outlined some of the primary implementation considerations associated with digital time delay beamforming. The computational intensity of this sonar function has required the formulation of efficient techniques in both the time and frequency domains. Time domain approaches are generally preferred because of the extra FFT computations of the frequency domain approach. An exception to this occurs for special array geometries where the beamforming inner products can be implemented with the FFT algorithm.

Programmable computing elements such as the MPBF are capable of addressing the most general beamforming and filtering applications. The total beamformer problem can be efficiently partitioned among multiple stages of MPBF kernels implementing various upsampling and downsampling operations. Current VLSI technology provides arithmetic throughput on the order of 25 million floating-point operations per second. At this rate, high-speed memory is required so that memory access time does not limit beamformer throughput.

## REFERENCES


1. W. C. Knight, R. G. Pridham, and S. M. Kay, Digital signal processing for sonar, *Proc. IEEE, 69*(11): 1451–1506 (1981).
2. D. E. Dudgeon, Fundamentals of digital array processing, *Proc. IEEE, 65*(6): 898–904 (1977).
3. R. G. Pridham and R. A. Mucci, A novel approach to digital beamforming, *J. Acoust. Soc. Am., 63*(2): 425–434 (1978).
4. R. G. Pridham and R. A. Mucci, Digital interpolation beamforming for low-pass and bandpass signals, *Proc. IEEE, 67*(6): 904–919 (1979).
5. P. Rudnick, Digital beamforming in the frequency domain, *J. Acoust. Soc. Am., 46*(5): 1089–1090 (1969).
6. J. R. Williams, Fast beamforming algorithm, *J. Acoust. Soc. Am., 44*: 1454–1455 (1968).
7. S. P. Pitt, W. T. Adams, and J. K. Vaughan, Design and implementation of a digital phase shift beamformer, *J. Acoust. Soc. Am., 64*(3): 808–814 (1978).
8. R. G. Pridham and R. A. Mucci, Shifted sideband beamformer, *IEEE Trans. Acoust. Speech Signal Process., ASSP-27*(6): 808–814 (1979).
9. R. A. Gabel and R. R. Kurth, Hybrid time-delay/phase-shift digital beamforming for uniform collinear arrays, *J. Acoust. Soc. Am. 75*(6): 1837–1847 (1984).
10. N. L. Owsley, Sonar array processing, *Array Signal Processing*, S. Haykin, ed.), Prentice-Hall, Englewood Cliffs, New Jersey pp. 115–192 (1985).
11. R. J. Urick, *Principles of Underwater Sound*, McGraw-Hill, New York (1975).
12. F. J. Harris, On the use of windows for harmonic analysis with the discrete Fourier transform, *Proc. IEEE, 66*(1): 51–83 (1978).
13. M. L. Skolnik, *Introduction to Radar Systems*, McGraw-Hill, New York (1962).
14. R. A. Mucci and R. G. Pridham, Impact of beam steering errors on shifted sideband and phase shift beamforming techniques, *J. Acoust. Soc. Am.* (1981).

15. W. J. Martin, *The Microprogrammable Beamformer*, Raytheon Company, Portsmouth, Rhode Island (1974).
16. R. W. Schafer and L. R. Rabiner, A digital signal processing approach to interpolation, *Proc. IEEE*, *61*: 692–720 (June 1973).
17. R. E. Crochiere and L. R. Rabiner, Optimum FIR digital filter implementation for decimation, interpolation and narrowband filtering, *IEEE Trans. Acoust. Speech Signal Process.*, *ASSP-23*: 444–456 (October 1975).
18. R. A. Gabel and R. R. Kurth, "Digital Beamsteering with Recursive Multichannel Filters," Proc. International Conference on Acoustics, Speech, and Signal Processing, pp. 803–806 (1982).
19. M. L. Cohen, "Designing Filters for Interpolation Beamforming," 1985 IEEE International Conference on Acoustics, Speech, and Signal Processing, Tampa, Florida, pp. 89–92 (1985).
20. R. C. Singleton, An algorithm for computing the mixed radix fast Fourier transform, *IEEE Trans. Audio Electroacoust.*, *AU-17*: 93–103 (June 1969).
21. G. L. DeMuth, "Frequency Beamforming Techniques," 1977 IEEE International Conference on Acoustics, Speech, and Signal Processing, Hartford, Connecticut, pp. 713–715 (1977).
22. M. E. Weber and R. Heisler, A frequency-domain beamforming algorithm for wideband, coherent signal processing, *J. Acoust. Soc. Am.*, *76*(4): 1132–1144 (1984).

# 15

# Coherence and Time Delay Estimation

G. CLIFFORD CARTER Signal and Post Processing Branch, Naval Underwater Systems Center, New London, Connecticut

## 1 INTRODUCTION

This chapter is a summary of work done by the author and several coauthors over more than a decade. Section 2 is a review of coherence. Section 3 is a review of the generalized framework for coherence estimation. Section 4 is a summary of statistics of the MSC estimator and has subsections discussing the probability density function, experimental results, bias, receiver operating characteristics, and confidence bounds. Section 5 discusses time delay estimation. Finally, Section 6 discusses the focused time delay beamformer form of passive ranging.

## 2 COHERENCE

This section discusses application of the coherence function. Much of the material in this section is a summary of the work by Carter and Knapp [1]. The coherence function between two wide-sense stationary random processes $x$ and $y$ is equal to the cross-power spectrum $G_{xy}(f)$ divided by the square root of the product of the

---

Portions of this chapter were presented at the Advanced Study Institute on Signal Processing, Les Hoches, France, August 12–September 6, 1985, and in the February, 1987 issue of *Proceedings of the IEEE.*

two auto-power spectra. Specifically, the complex coherence is defined by

$$\gamma_{xy}(f) = \frac{G_{xy}(f)}{\sqrt{G_{xx}(f)G_{yy}(f)}} \tag{15.1a}$$

where $f$ denotes the frequency of interest and where the complex cross-power spectrum

$$G_{xy}(f) = \int_{-\infty}^{\infty} R_{xy}(\tau)e^{j2\pi f\tau}\, d\tau \tag{15.1b}$$

is the Fourier transform of the cross-correlation function

$$R_{xy}(\tau) = E[x(t)y(t+\tau)] \tag{15.1c}$$

Here $x$ and $y$ are real and $E$ denotes the mathematical expectation (for ergodic random processes the ensemble average can be replaced by a time average). The coherence is a normalized cross-spectral density function; in particular, the normalization constrains Eq. (15.1a) so that the magnitude-squared coherence (MSC) defined by

$$C_{xy}(f) \triangleq |\gamma_{xy}(f)|^2 \tag{15.2a}$$

lies in the range

$$0 \leq C_{xy}(f) \leq 1 \tag{15.2b}$$

for all frequencies. Throughout this chapter we use $C$ and $|\gamma|^2$ interchangeably.

The coherence function has uses in numerous areas, including system identification, measurement of signal-to-noise ratio (SNR), and determination of time delay. The coherence—in particular, the MSC—can be put to use only when its value can be accurately estimated. Indeed, it is highly desirable to understand the statistics of the estimator. Therefore, this section addresses interpretations of the coherence function. Following sections address procedures for properly estimating the MSC and statistics of the estimator.

One interesting interpretation of coherence—particularly MSC—is that it is a measure of the relative linearity of two processes. To illustrate this, consider Figure 15.1, in which a sample function $y(t)$ of an arbitrary stationary random process consists of the response $y_0(t)$ of a linear filter plus an error component $e(t)$. When the linear filter is chosen to minimize the mean-square value of $e(t)$, that is, the area under the error spectrum, then $y_0(t)$ becomes the part of $y(t)$ that is linearly related to $x(t)$. The spectral characteristics of $e(t)$ are given by

$$G_{ee}(f) = G_{yy}(f) + G_{xx}(f)|H(f)|^2 - H(f)G^*_{xy}(f) - H^*(f)G_{xy}(f) \tag{15.3}$$

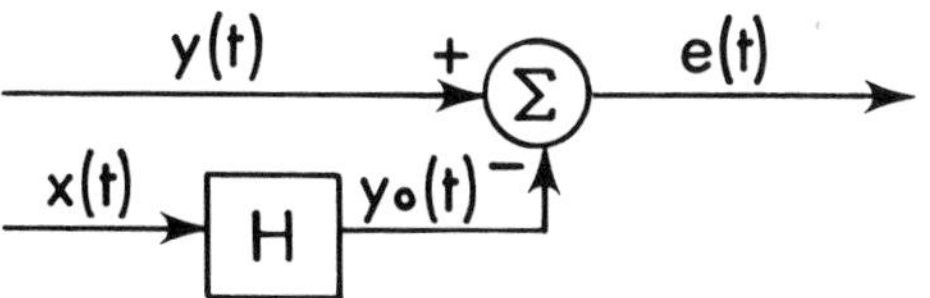


**Figure 15.1** Model of error resulting from linearly filtering $x(t)$ to match any desired signal $y(t)$.

where the $*$ indicates complex conjugation and $H(f)$ is the filter transfer function. The error spectrum

$$G_{ee}(f) = G_{xx}(f)\left|H(f) - \frac{G_{xy}(f)}{G_{xx}(f)}\right|^2 + G_{yy}(f)[1 - C_{xy}(f)] \tag{15.4}$$

Hence, the optimum filter is given by

$$H_0(f) = \frac{G_{xy}(f)}{G_{xx}(f)} \tag{15.5}$$

Note that the coherence is related to the optimum linear filter according to

$$\gamma_{xy}(f) = H_0(f)\sqrt{\frac{G_{xx}(f)}{G_{yy}(f)}} \tag{15.6a}$$

and

$$C_{xy}(f) = |H_0(f)|^2 \frac{G_{xx}(f)}{G_{yy}(f)} \tag{15.6b}$$

These results apply regardless of the source of $y(t)$. When the linear filter is optimum in the mean-square sense, the error is uncorrelated with $x(t)$, that is,

$$G_{xy_0}(f) = H_0(f)G_{xx}(f) = G_{xy}(f) \tag{15.7}$$

Furthermore, the minimum value of $G_{ee}(f)$ is given by

$$G_{ee}(f) = G_{yy}(f)[1 - C_{xy}(f)] \tag{15.8}$$

and

$$G_{y_0y_0}(f) = |H_0(f)|^2 G_{xx}(f) = G_{yy}(f)C_{xy}(f) \tag{15.9}$$

From the identity

$$G_{yy}(f) = C_{xy}(f)G_{yy}(f) + [1 - C_{xy}(f)]G_{yy}(f) \tag{15.10a}$$

we can show that

$$G_{yy}(f) = G_{y_0y_0}(f) + G_{ee}(f) \tag{15.10b}$$

indicating that the MSC is the fraction of $G_{yy}(f)$ contained in the linear component of $y(t)$, and $1 - C_{xy}(f)$ is the proportion of $G_{yy}(f)$ contained in the error, or nonlinear component of $y(t)$.

## 3 GENERALIZED FRAMEWORK FOR COHERENCE ESTIMATION

The purpose of this section is to review a generalized framework for power spectral estimation and to show how three estimation methods fit into this framework. (Beyond the scope of this section are important methods of time-varying spectral estimation and parametric methods of spectral estimation. See, for example, Marple [2] and IEEE Press books by Childers [3] and Kesler [4].)

In the generalized framework of Nuttall and Carter [5] we are concerned with both auto and cross (nonparametric) spectral estimation of stationary random processes; hence, we consider two discrete random processes. Auto spectral estimation is then a special case of cross spectral estimation. In particular, in auto spectral estimation we replace the second time series by a duplicate of the first time series. As is often the case in practice, we are limited to a single time-limited observation of each random process. Within our generalized framework for power spectral estimation, we first partition each time series into $N$ segments, where the segments may be overlapped. Second, each segment is multiplied by a time-weighting function (the weighting function may be unity everywhere within the segment, as in rectangular weighting, or it may be smooth, as in Hanning weighting). Third, fast Fourier transforms (FFTs) are computed for each weighted segment after each segment has been appropriately appended with zeros. Fourth, the FFTs for one segment are multiplied by the complex conjugate of the FFTs for the other time-synchronous segment (or the same segment for auto spectra). As a matter of clarification, both time series are presumed to be aligned, that is, in proper time register, so that the $i$th segment of one segment only interacts with the $i$th segment of the other time series. Fifth, the complex products are averaged at each frequency over the $N$ available segments (one segment if $N = 1$). Sixth, the resultant spectral estimate are fast Fourier–transformed into the correlation or lag domain and multiplied by a lag-weighting function (which may be unity). Finally, the results are transformed back into the frequency domain. (Alternatively, the last two steps can be replaced by a convolution in the frequency domain. Depending on the extent of the frequency domain convolution, the former alternative may be computationally preferable to the latter.) Mathematical details are in the paper by Nuttall and Carter [6].

We now point out how three spectral analysis techniques fit into this generalized framework. All three techniques (to be described) will achieve virtually the same mean and variance of the estimated power at any particular frequency. First, the Blackman and Tukey (BT) method [7] allows for only one segment with rectangular time weighting over the entire record (from each time limited realization (TLR)), and it applies a smooth lag-weighting function in the correlation domain, which goes to zero well before the end of the data record. (We note that the original BT approach estimated the correlation function directly rather than using FFTs.) By adjusting how quickly the lag weighting goes to zero, resolution and stability can be compromised. For example, if the weighting goes to zero quickly the spectral estimates will have coarse resolution and good stability compared with a lag weighting that does not go to zero quickly. The exact shape of the weighting will influence the exact shape of the sidelobes and mainlobe in the frequency domain.

The second method, widely in use today, that falls within the generalized framework is the Welch overlapped-segment averaging (WOSA) technique. The WOSA method is referred to as Welch's method based on the contribution of Welch [8]. In the WOSA method, we apply a smooth multiplicative time weighting to each of a large number of segments and average the FFT products from these overlapped segments to obtain a final spectral estimate, without employing additional lag weighting. The time weighting is typically a smooth weighting such as Hanning, in the WOSA method, because it yields good sidelobe behavior. Overlap is important in the WOSA method in order to realize maximum stability (that is, minimum variance) of the spectral estimate.

A third technique that falls within the generalized framework is the lag reshaping method (see, e.g., Nuttall and Carter [6]). The lag reshaping method recognizes that the number of available data points may be so large as to preclude the normal BT method in practical situations. We segment the data without overlapping and apply unit-gain rectangular time weighting to each segment (this rectangular weighting requires no time-weighting multiplications). Later we will undo the bad sidelobe effects that this rectangular time weighting initially causes, and gain additional stability. Note that the segment-averaged power spectrum will be transformed into the correlation (or lag) domain, where a smooth multiplicative lag-weighting function will be applied before transforming back into the frequency domain. The smooth lag weighting will be the product of two lag weightings, one for the desired window and one for lag reshaping. This lag-domain "reshaping" is an ingenious method for almost completely undoing the bad sidelobe effects of rectangular time weighting.

All three of these techniques that fall within the generalized framework have good statistical properties. The Blackman and Tukey [7] method attains minimum variance spectral estimates and is the benchmark against which other techniques have been measured. Under simplifying assumptions, it can be shown that the WOSA method with overlap can achieve the same stability, or equivalent degrees of freedom (EDF), as the BT spectral estimation method when both methods operate on the same amount of data and are constrained to the same frequency resolution. For many practical time weightings, virtually minimum variance can be attained by a computationally reasonable amount of overlap.

For more complicated spectral measures, such as coherence, the analysis of WOSA becomes unwieldy, and one is driven to simulation. Carter et al. [9] empirically investigated the effect of overlap on the variance of the coherence estimate via the WOSA method. There was a pronounced improvement (about a factor of two in variance reduction) with overlap as opposed to no overlap. In another experiment, Carter and Knapp [1] compared the use of Hanning and rectangular weightings for estimating coherence between a flat broadband input to a second-order digital filter and its output. It was demonstrated that smooth weighting functions are required to obtain good coherence estimates with the WOSA method. (Recall that the WOSA method does not use additional lag shaping.)

The WOSA nonparametric spectral estimation method widely in use today with proper overlap can attain the EDF of the BT method, when both methods operate on the same amount of data and are constrained to have the same frequency resolution. Further, for good time weightings, reasonable amounts of overlap achieve most of the available EDF. Moreover, based on analytic work by Nuttall and Carter [6], the new lag reshaping method can virtually (but not exactly) attain the EDF of the BT method and yield very good sidelobes through the use of unusual lag weighting. It appears highly certain that the lag reshaping method requires fewer computations (perhaps by a factor of two) than the WOSA method in practice and therefore deserves serious consideration as a replacement for (or variation on) the widely used WOSA method.

# 4 STATISTICS OF MSC ESTIMATES OBTAINED VIA THE WELCH OVERLAPPED-SEGMENT AVERAGING (WOSA) METHOD

## 4.1 Introduction

Much of the historical work on the statistics of the MSC estimates centers on the WOSA method; by proper interpretation of variables, these results also apply to the lag reshaping method. Recall that the WOSA method consists of obtaining two finite-time series from the random processes being investigated. Each time series is partitioned into equal-length segments and sampled at equally spaced data points. The segments are overlapped. However, the statistics are analytically developed for nonoverlapped segments. Samples from each segment are multiplied by a weighting function, and the FFT of the weighted sequence is performed. Then the Fourier coefficients for each weighted segment are used to estimate the auto- and cross-power spectral densities. The spectral density estimates thus obtained are used to form the MSC estimate.

Spectral resolution of the estimates varies inversely with the segment length $T$. Proper weighting or "windowing" of the $T$-second segment is also helpful in achieving good sidelobe reduction. On the other hand, for independent segments with ideal windowing, the bias and the variance of the MSC estimate vary inversely with the number of segments $n$. Therefore, to generate a good estimate with limited data, one may be faced with conflicting requirements on $n$ and $T$. Segment overlapping can be used to increase both $n$ and $T$. When the segments are disjoint, that is, nonoverlapping, we call the number of segments $n_d$. As the percentage of overlap increases, however, the computational requirements increase rapidly, while the improvement stabilizes owing to the greater correlation between data segments (see Carter et al. [9]).

## 4.2 Probability Density for the Estimate of the MSC

The first-order probability density and distribution functions for the estimate of MSC, given the true value of MSC, and the number of independent segments $n_d$ are given in Table 15.1. Notationally, recall that $|\gamma|^2 = C$. Equations (15.1b) and (15.1c) are useful because the ${}_2F_1$ hypergeometric function is an $(n_d - 1)$st-order polynomial.

Figures 15.2 and 15.3 illustrate the probability density and distribution functions for several cases, as computed from Table 15.1. The bias and variance of the MSC estimate have been evaluated using a general expression for the $m$th moment of the MSC estimate. (See Carter et al. [9].] Bias and variance expressions obtained are summarized in Table 15.2. Figures 15.4 and 15.5, respectively, show the bias and variance as functions of $N = n_d$ and $|\gamma|^2$. The curves in Figures 15.4 and 15.5 were obtained using the exact formulas. Peaks in the variance curves when the MSC equals one-third are evident. We note, however, there is an additional bias when our FFT size is too small. This second type of bias can be extremely important. It is the subject of texts including those by Koopmans [10] and Brillinger [11] and is discussed in Section 4.4.

**Table 15.1** Probability Density and Distribution Functions

Density Function

$$
\begin{aligned}
p(|\hat{\gamma}|^2 \mid n_d, |\gamma|^2) &= (n_d - 1)(1 - |\gamma|^2)^{n_d}(1 - |\hat{\gamma}|^2)^{n_d - 2} \\
&\quad \cdot {}_2F_1(n_d, n_d; 1; |\gamma|^2|\hat{\gamma}|^2), \qquad 0 \le |\gamma|^2|\hat{\gamma}|^2 < 1 \qquad \text{(Ia)} \\
&= (n_d - 1)(1 - |\gamma|^2)^{n_d}(1 - |\hat{\gamma}|^2)^{n_d - 2} \\
&\quad \cdot (1 - |\gamma|^2|\hat{\gamma}|^2)^{1 - 2n_d} {}_2F_1(1 - n_d, 1 - n_d; 1; |\gamma|^2|\hat{\gamma}|^2) \qquad \text{(Ib)} \\
&= (n_d - 1)\left[\frac{(1 - |\gamma|^2)(1 - |\hat{\gamma}|^2)}{(1 - |\gamma|^2|\hat{\gamma}|^2)^2}\right]^{n_d} \\
&\quad \cdot \frac{(1 - |\gamma|^2|\hat{\gamma}|^2)}{(1 - |\hat{\gamma}|^2)^2} {}_2F_1(1 - n_d, 1 - n_d; 1; |\gamma|^2|\hat{\gamma}|^2) \qquad \text{(Ic)}
\end{aligned}
$$

Distribution Function

$$
\begin{aligned}
P(|\hat{\gamma}|^2 \mid n_d, |\gamma|^2) &= |\hat{\gamma}|^2 \left(\frac{1 - |\gamma|^2}{1 - |\gamma|^2|\hat{\gamma}|^2}\right)^{n_d} \sum_{k=0}^{n_d - 2} \left(\frac{1 - |\hat{\gamma}|^2}{1 - |\gamma|^2|\hat{\gamma}|^2}\right)^k \\
&\quad \cdot {}_2F_1(-k, 1 - n_d; 1; |\gamma|^2|\hat{\gamma}|^2) \qquad \text{(Id)}
\end{aligned}
$$

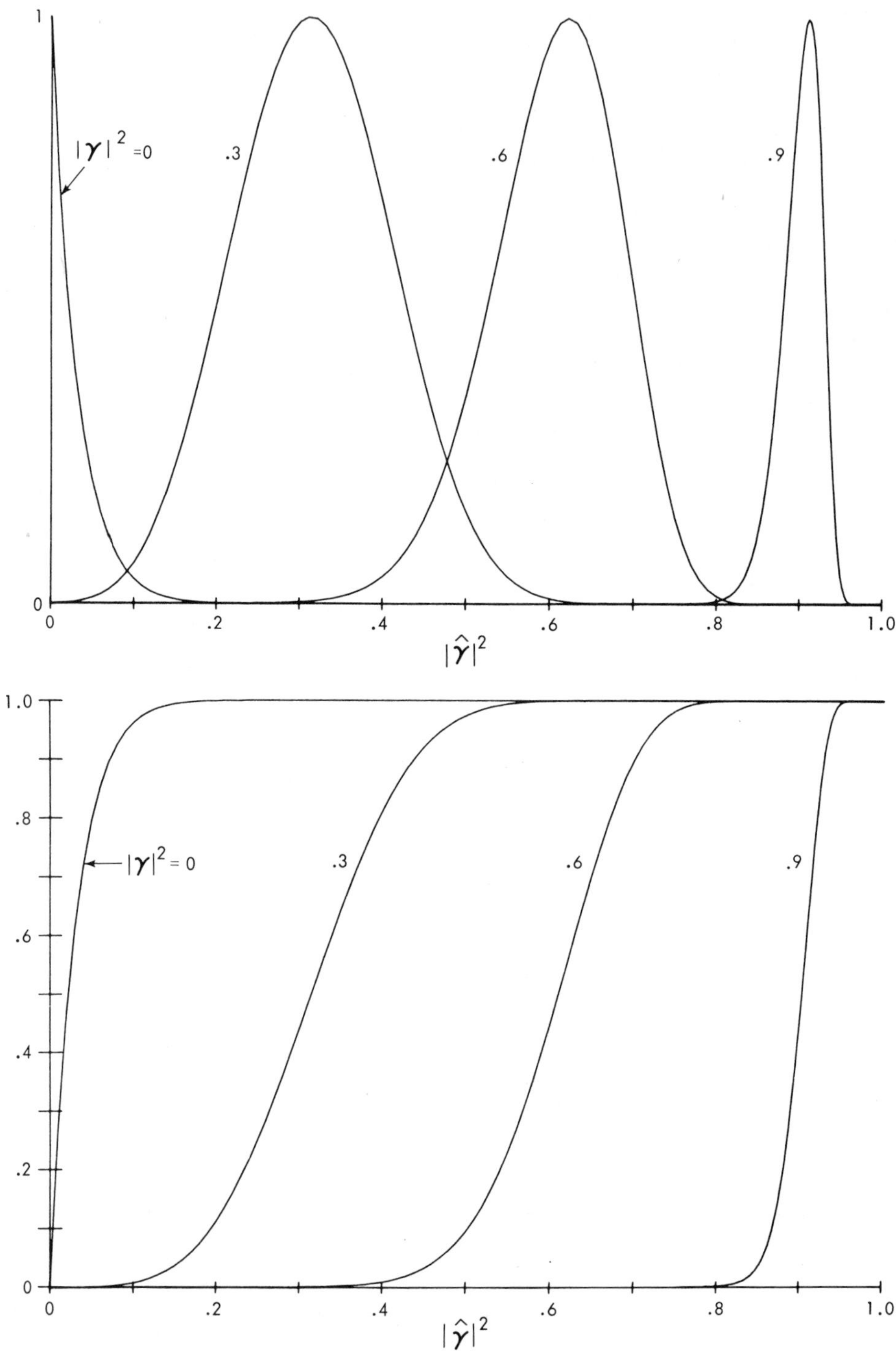


**Figure 15.2** (a) Probability density functions. (Functions have been normalized by maximum values, which are 31.0, 4.13, 5.23, and 17.5.) (b) Distribution functions.

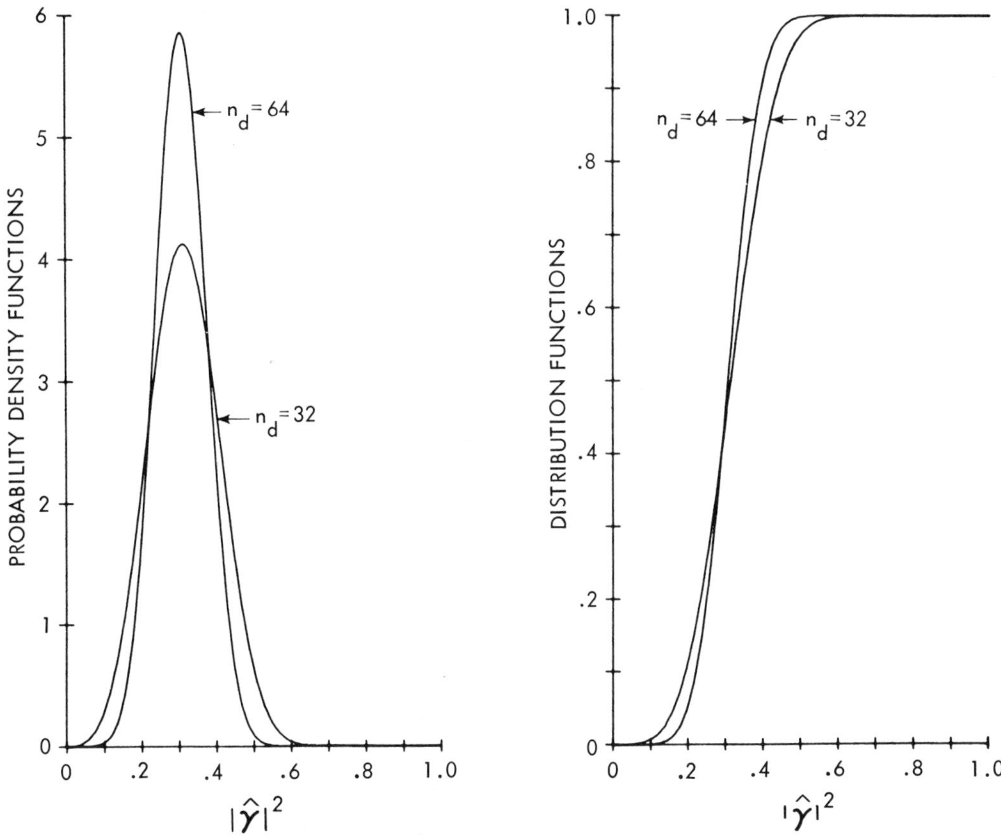


**Figure 15.3** Probability density and distribution functions of $|\hat{\gamma}|^2$ for $|\gamma|^2 = 0.3$.

## 4.3 Experimental Investigation of Overlap Effects

An experimental study has been made of the effect of overlap of data on the MSC estimate. The analytical results presented earlier are only for the case of independent segments, that is, the case of zero overlap. The application of nonoverlapped smooth weighting functions does not make the most efficient use of the data when forming the MSC estimate. An experiment was conducted to examine this inefficiency in terms of bias and variance of the MSC estimate as a function of different amounts of overlap. Details are described in the paper by Carter et al. [9].

Results of the experiment are summarized in Figures 15.6 and 15.7. It is apparent from these results that the bias and variance can be reduced through overlapped processing. In the WOSA method, clearly, as the overlap increases, the computational cost must also increase. It is doubtful that the improvement in the WOSA method derived from using a 62.5% overlap, as opposed to a 50% overlap, will warrant the increased computational costs, except in unusual circumstances. Overlaps of 50% are, however, quite reasonable and widely used with the WOSA method. With the advent of the lag reshaping method other computational efficiencies need to be explored.

**Table 15.2** Bias and Variance Expressions

Exact

Bias

$$B = E[|\hat{\gamma}|^2 \mid n_d, |\hat{\gamma}|^2] - |\gamma|^2 = \frac{(1-|\gamma|^2)^{n_d}}{n_d} {}_3F_2(2, n_d, n_d; n_d+1, 1; |\gamma|^2) - |\gamma|^2$$

$$= \frac{1}{n_d} + \frac{n_d - 1}{n_d + 1}|\gamma|^2 {}_2F_1(1, 1; n_d + 2; |\gamma|^2) - |\gamma|^2 \qquad \text{(IIa)}$$

Variance

$$V = E(|\hat{\gamma}|^4 \mid n_d, |\gamma|^2) - E^2(|\hat{\gamma}|^2 \mid n_d, |\gamma|^2)$$

$$= \frac{2(1-|\gamma|^2)^{n_d}}{n_d(n_d+1)} {}_3F_2(3, n_d, n_d; n_d+2, 1; |\gamma|^2)$$

$$- \left[\frac{(1-|\gamma|^2)^{n_d}}{n_d} {}_3F_2(2, n_d, n_d; n_d+1, 1; |\gamma|^2)\right]^2 \qquad \text{(IIb)}$$

Approximate

$$B_0 \cong \frac{1}{n_d} - \frac{2}{n_d+1}|\gamma|^2 + \frac{1!(n_d-1)}{(n_d+1)(n_d+2)}(|\gamma|^2)^2$$

$$+ \frac{(n_d-1)2!}{(n_d+1)(n_d+2)(n_d+3)}(|\gamma|^2)^3; \quad B \cong \begin{cases} B_0, & B_0 \geq 0 \\ 0, & B_0 < 0 \end{cases} \qquad \text{(IIc)}$$

$$V_0 \cong \frac{(n_d-1)}{n_d(n_d+1)}\left[\frac{1}{n_d} + 2\frac{n_d-2}{n_d+2}|\gamma|^2 - 2\frac{2n_d^3 - n_d^2 - 2n_d + 3}{(n_d+1)(n_d+2)(n_d+3)}(|\gamma|^2)^2\right.$$

$$+ 2\frac{n_d^4 - 6n_d^3 - n_d^2 + 10n_d - 8}{(n_d+1)(n_d+2)(n_d+3)(n_d+4)}(|\gamma|^2)^3$$

$$\left. + \frac{13n_d^5 - 15n_d^4 - 113n_d^3 + 27n_d^2 + 136n_d - 120}{(n_d+1)(n_d+2)^2(n_d+3)(n_d+4)(n_d+5)}(|\gamma|^2)^4\right]; \quad V \cong \begin{cases} V_0, & V_0 \geq 0 \\ 0, & V_0 < 0 \end{cases} \qquad \text{(IId)}$$

Approximation for Large $n_d$

$$B \cong \frac{1}{n_d}[1 - |\gamma|^2]^2 \qquad \text{(IIe)}$$

$$\leq \frac{1}{n_d}[1 - |\gamma|^2] \qquad \text{(IIf)}$$

$$V \cong \begin{cases} \dfrac{1}{n_d^2}, & |\gamma|^2 = 0 \\ \dfrac{2|\gamma|^2}{n_d}(1-|\gamma|^2)^2, & 0 < |\gamma|^2 \leq 1 \end{cases} \qquad \text{(IIg)}$$

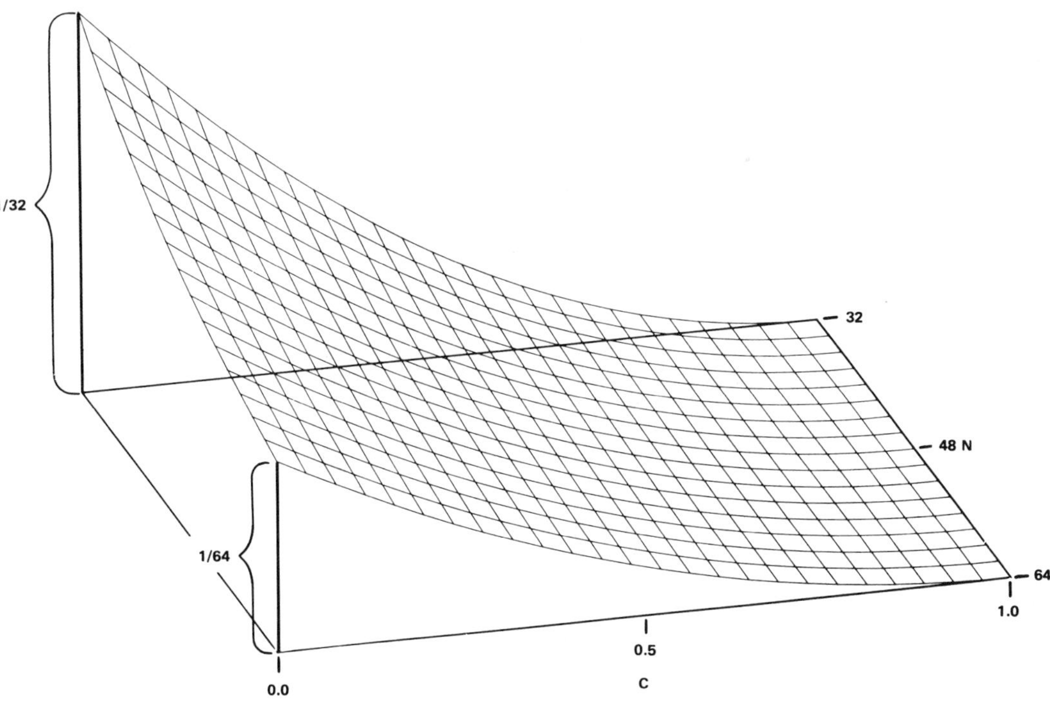


**Figure 15.4** Bias of $|\hat{\gamma}|^2$ versus $\gamma^2$ and $n_d$.

## 4.4 MSC Bias

One type of bias, derived under simplifying assumptions including the assumption that each data segment is sufficiently long to ensure adequate spectral resolution, has been shown by Nuttall and Carter [12] from Table 15.2 to be

$$E[\hat{C}] - C \cong \frac{1}{n_d}(1 - C)^2 \left(1 + \frac{2C}{n_d}\right) \tag{15.11}$$

where $E$ denotes the expected value. Equation 15.11 corroborates the observation by Bendat and Piersol [13] that more than one segment must be used to estimate MSC. It can be shown that for $n_d = 1$ the estimated MSC equals unity regardless of the true value of MSC; however, since Equation (15.11) is an approximation it cannot be used to prove this relationship.

However, there is a second type of bias, described by Koopmans [10], that can be extremely serious. This is the bias due to time delay, misalignment, or rapidly changing phase. In particular, Koopmans [10] notes that if the phase angle of the cross-power spectrum is a rapidly varying function of frequency at the frequency at which the coherence is to be estimated, the estimated coherence (in particular, MSC) can be biased downward to such an extent that a strong coherence is masked. A brief derivation of the effect of misalignment akin to rapidly changing phase will be given later. These results compare favorably with analytical results by Halvorsen

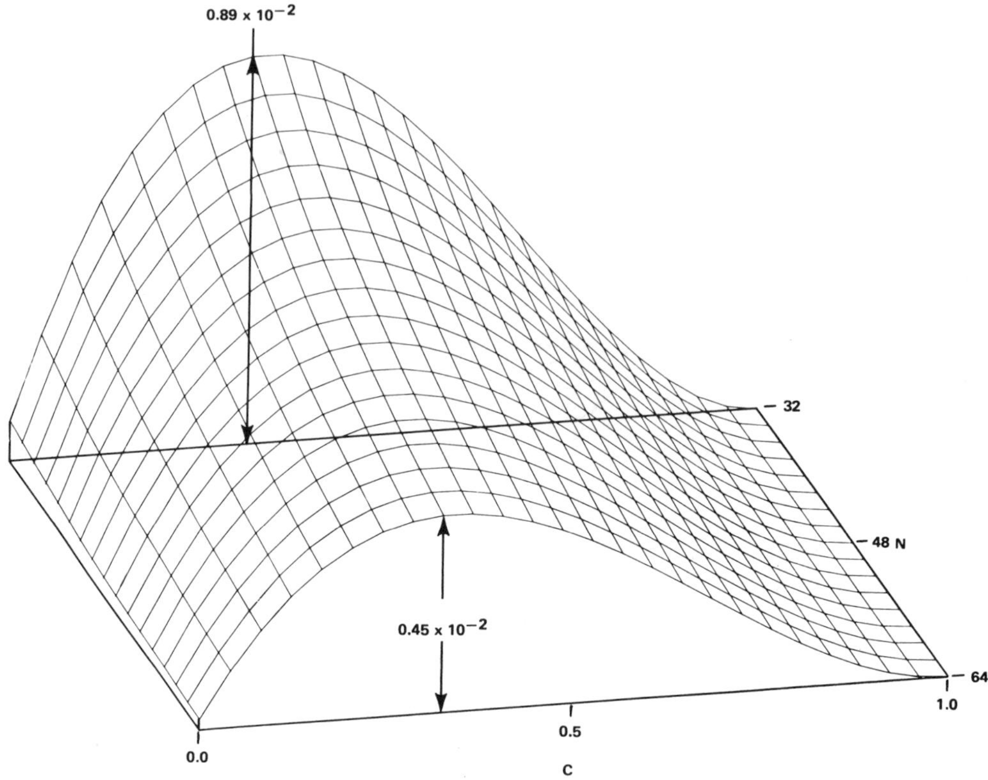


**Figure 15.5** Variance of $|\hat{\gamma}|^2$ versus $\gamma^2$ and $n_d$.

and Bendat [14], empirical results by Carter and Knapp [1], and empirical results to be presented here.

Rapidly varying phase as a function of frequency is caused by a time delay. One way to see this is to consider that the units of the slope of the phase are radians divided by radians per second, or simply seconds. The data can be realigned to compensate for a time delay. As stated by Brillinger [11], the importance of some form of prefiltering cannot be overemphasized, the simplest form being to lag one time series relative to the other (we have noted that the importance of prefiltering is also true for time delay estimation). This procedure for coherence estimation has been suggested by others, including Akaike and Yamanouchi [15], Jenkins and Watts [16], and Koopmans [10]. Important to the concept of prefiltering two time series before estimating the MSC is the fact that (unlike the estimated value of MSC) the (true value of) MSC is invariant under the linear filtering of the two series, as shown, for example, by Carter et al. [9] and Koopmans [10].

The effect of source motion can be seen in the correspondence by Carter and Abraham [17]. The results of that work show that the magnitude of the cross-power spectrum (and cross-correlation) is decreased by a constant factor, depending on the ratio of the delay misalignment to the FFT time duration. Note, though, that the average phase estimate remains unaltered. Further, we note that the constant

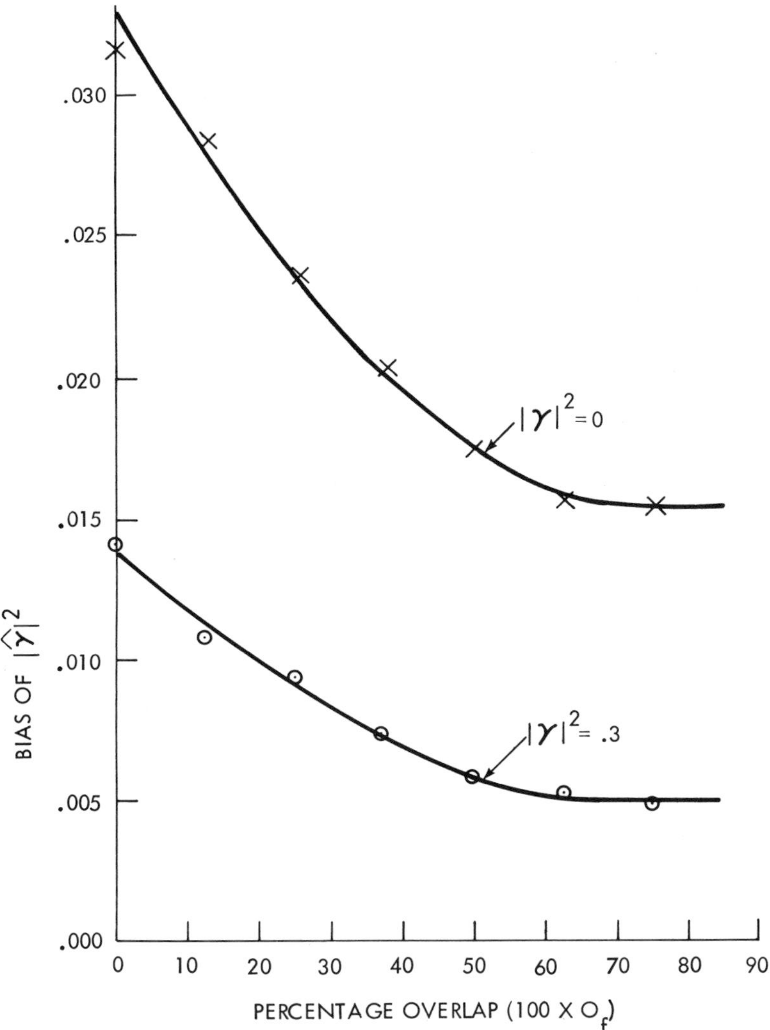


**Figure 15.6** Bias of $|\hat{\gamma}|^2$ when $n_d = 32$.

degradation factor will not appear in either of the auto-power spectral densities. Thus, the complex coherence is degraded by the same factor as the cross spectrum and the MSC is degraded by the square of this factor. That is,

$$E[\hat{C}(f)] \cong \left(1 - \frac{|D|}{T}\right)^2 C(f) \tag{15.12}$$

which agrees with Halvorsen and Bendat [14]. Heuristically, this makes sense because for no delay there is no degradation, and for a delay equal to or greater than the FFT size the estimated MSC is zero. Note that the bias due to misalignment

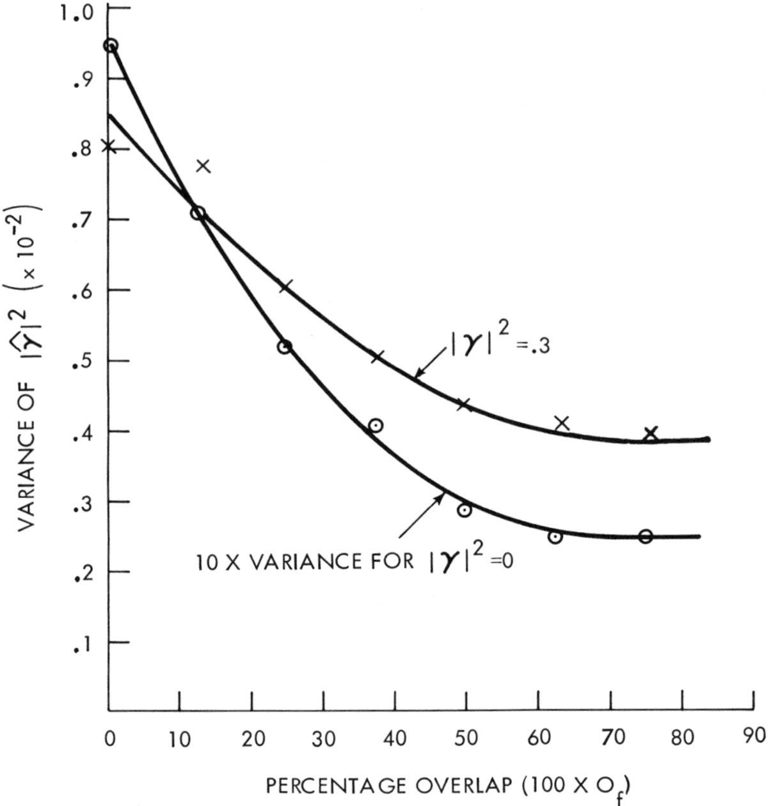


**Figure 15.7** Variance of $|\hat{\gamma}|^2$ when $n_d = 32$.

$D$, with FFT time duration $T$, is

$$E[\hat{C}] = C \cong \frac{-2|D|}{T}C + \left(\frac{|D|}{T}\right)^2 C \tag{15.13a}$$

$$\cong \frac{-2|D|}{T}C, \qquad |D| \ll T \tag{15.13b}$$

For example, if $|D|/T = 0.25$, the expected value of the estimated MSC is about one-half of its true value. Clearly, the effect is important. Indeed, empirical results bear this out. For example, the results of Carter and Knapp [1] demonstrated the need to make the FFT (or equivalent transform) size larger. Empirically, large $T$ was observed to reduce the bias in MSC estimation.

One practical means of reducing the bias due to a single path misalignment is to realign the two time series under investigation before estimating the MSC. The effects of misalignment were evident in an empirical investigation in which a broadband underwater acoustic signal was transmitted through a direct path from a submerged transmitter to a submerged receiver. The recorded signals were processed with a number of different bulk time delays inserted before estimating the MSC. The bulk time delays were quantized to 250 ms; the FFT size was 1.0 s. The effect of degraded MSC estimation is evident in Figures 15.8 and 15.9, computed from 16 independent FFTs. In Figure 15.8, $D = 0.18$ and the estimated MSC appears to be about 0.45 at 250 Hz. In Figure 15.9 introducing another 250-ms bulk delay moves the generalized cross-correlation SCOT peak from $-0.18$ to 0.07. (The SCOT or smoothed coherence transform is the

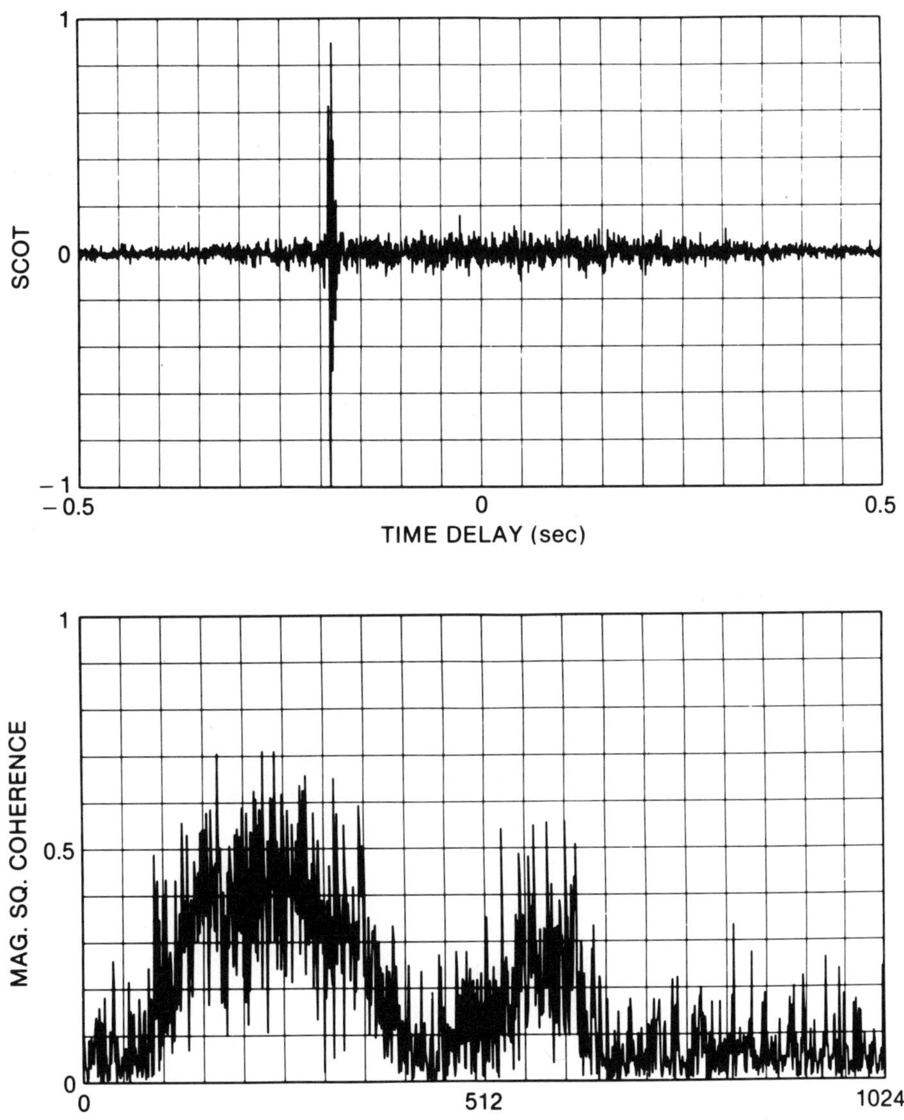


**Figure 15.8** SCOT estimate showing a $-180$-ms delay and corresponding MSC estimate of 0.45 at 250 Hz with a $-180$-ms delay.

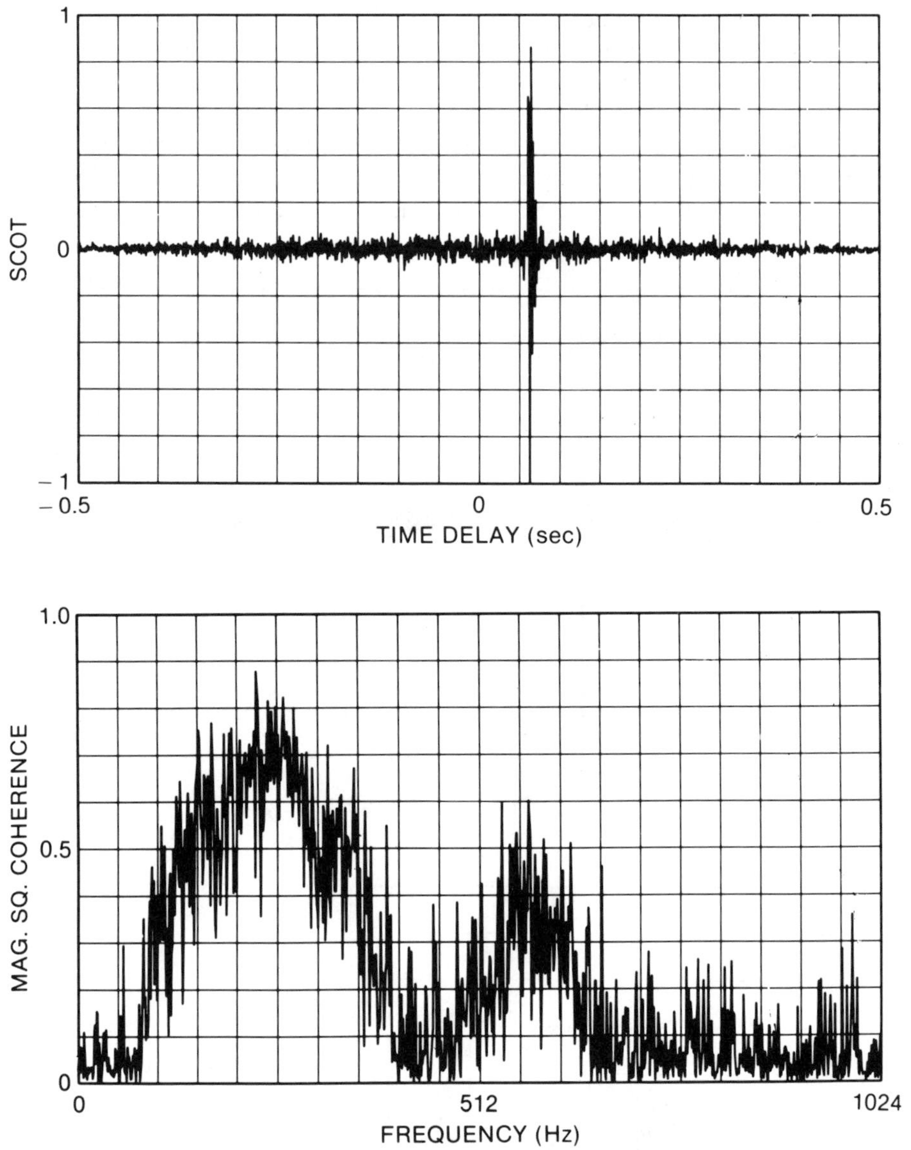


**Figure 15.9** SCOT estimate shown a 70-ms delay and corresponding MSC estimate of 0.7 at 250 Hz with a 70-ms delay.

Fourier transform of the complex coherence; it is therefore the Fourier transform of the cross-power spectrum after weighting with a particular frequency function. See Carter et al. [18], Knapp and Carter [19], and Kuhn [20] for a discussion of the SCOT.) Now the estimated MSC is about 0.7 at 250 Hz in Figure 15.9. There is a notable and predictable increase in the MSC estimate due to realignment.

Thus, we see that even with a large number of FFT segments, estimates of the MSC can be significantly biased downward by unremoved bulk time delays, giving an erroneous indication of the value of the coherence. When the data are realigned and processed, estimates of the coherence are informative descriptors of the extent to which the ocean channel can be modeled by a linear time-invariant filter.

## 4.5 Receiver Operating Characteristics for a Coherence Detector

An algorithm for computing the receiver operating characteristics (ROC) or the probability of detection, $P_D$, versus the probability of false alarm, $P_F$, for a two-channel coherence detector is presented together with an example of an ROC table by Carter [21]. An article by Gevins et al. [22] presents results on using linearly thresholded coherence estimates to detect biomedical phenomena. We present here the results of an algorithm for computing $P_D$ versus $P_F$ for a specified amount of averaging and underlying coherence, using simplifying assumptions and the probability density function obtained from Table 15.1.

The ROC curves for $|\gamma|^2 = C = 0.25$ and $N = n_d = 4$, 8, and 16 independent data segments are given in Figure 15.10. As seen in the figure, performance can be improved by increasing the number of disjoint data segments $n_d$ if a sufficient amount of stationary data exists; if not, $n_d$ can be increased only at the expense of degrading the frequency resolution with its inherent difficulties. If $n_d$ is fixed, performance is determined by the underlying coherence or, equivalently, the SNR (see, e.g., Figure 15.11). For many particular problems, the performance will be desired for different values of $n_d$ and $C$. Because of the large number of possible choices for these parameters, we will not present an exhaustive series of results; a computer program listing is available in Carter [23].

## 4.6 Confidence Bounds for Magnitude-Squared Coherence Estimates

In many applications, two received signals are digitally processed to estimate coherence. Results of computing coherence estimate confidence bounds for stationary Gaussian signals are presented. Computationally difficult examples are given for 80 and 95% confidence with independent averages of 8, 16, 32, 64, and 128. A more complete discussion can be found in Scannell and Carter [24].

The MSC is useful in estimating the amount of coherent power common to two received signals. Therefore, it would be desirable, having estimated a particular value of MSC, to state with certain confidence that the true coherence falls in a specified interval. (A general discussion of confidence intervals is available in Cramer [25].) Early attempts to present 95 percent confidence were reported by Haubrich [26], who apparently used precomputed cumulative distributive function (CDF) curves and used a method of presentation different from the one used here. Related confidence work for the magnitude coherence (MC) is presented by Koopmans [10]. Empirical results for 95% confidence are given by Benignus [27]. The confidence limits given here appear to agree with approximate results in Bendat and Piersol [13] and Enochson and Goodman [28] and with results of Brillinger [11] from tabulated densities. Gosselin [29] compared MSC detectors with other detectors using the notion of ROC curves. A computer program has been written (see

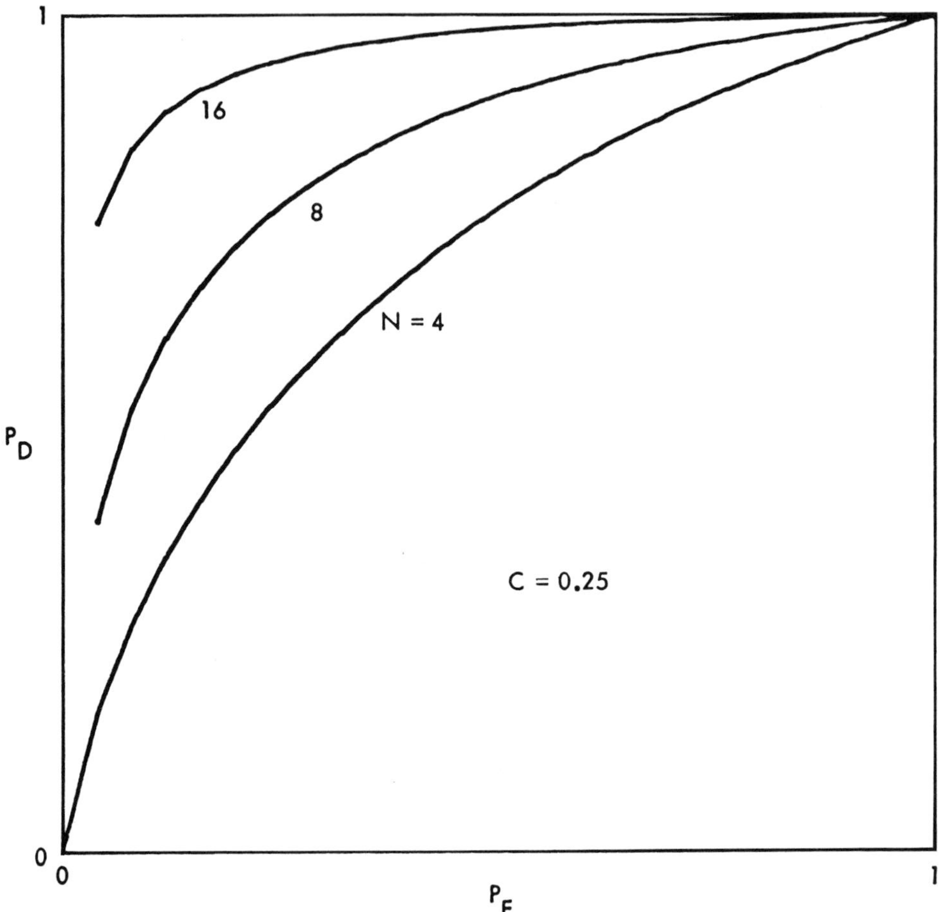


**Figure 15.10** ROC curves for $|\gamma|^2 = C = 0.25$, $n_d = N = 4, 8, 16$.

Scannell and Carter [24]) to evaluate the CDF and confidence limits. Recall that the CDF is a finite sum of ${}_2F_1$ hypergeometric functions, each of which is polynomial, as given in Table 15.1. When $C$ equals zero or unity, CDF values can be computed in closed form.

Figure 15.12 presents computer-generated 80 and 95% confidence limits. (See Scannell and Carter [24].) The five pairs of curves in each part of the figure are for $n_d = 8$, 16, 32, 64, and 128 from outer to inner, respectively. If we make many estimates of MSC and keep applying confidence rules (whether $C$ is random or constant), we will correctly include the true value of $C$ in the determined interval the specified percent of the time. Sometimes the method of applying confidence rules is in doubt; for example, in Figure 15.12, if the estimate comes out to be 0.3 and $n_d = 8$, then a horizontal line does not intersect the upper confidence limit unless we extrapolate it backward. Thus, we could say that with 95% confidence the true MSC is in the region $(-0.1, 0.62)$. Since we know a priori that the true value

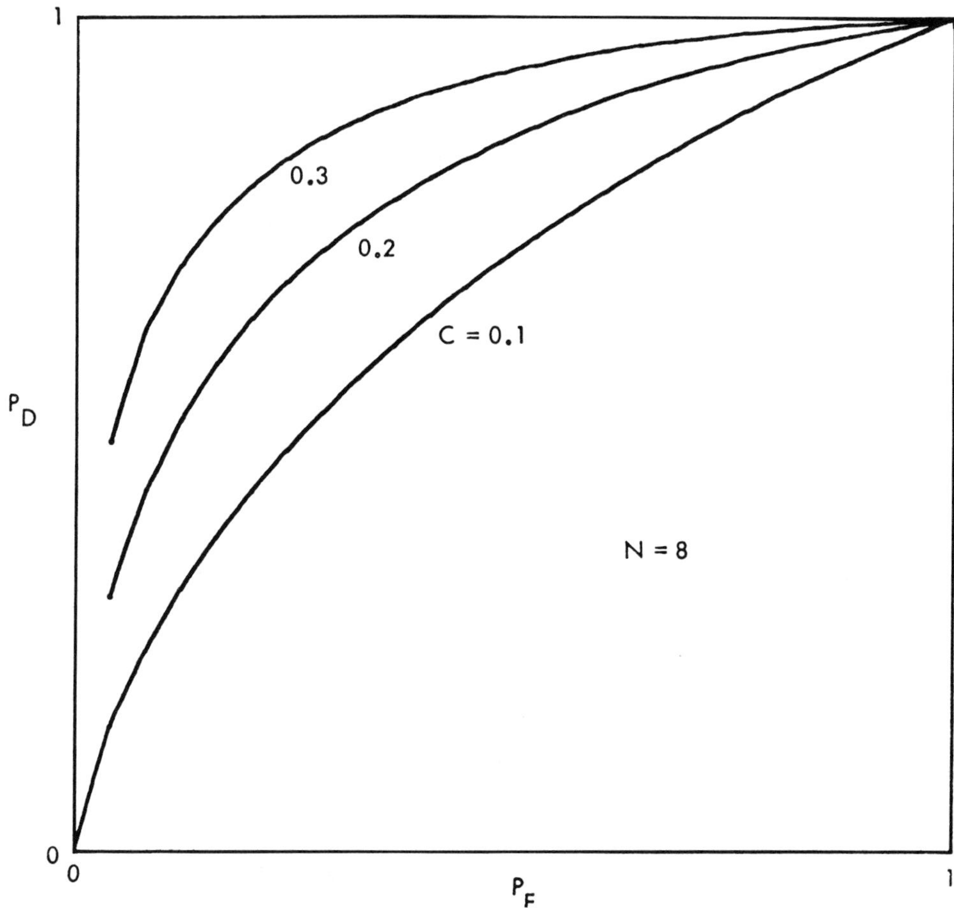


**Figure 15.11** ROC curves for $n_d = N = 8$, $|\gamma|^2 = C = 0.1, 0.2, 0.3$.

of $C$ is nonnegative, we could just as easily say (but with no more confidence) that with 95% confidence (for $n_d = 8$ and $\hat{C} = 0.3$) the true MSC falls in the region (0.0, 0.62). Confidence limits for magnitude coherence are discussed in Koopmans [10].

## 5 TIME DELAY ESTIMATION

### 5.1 Introduction

A coherent signal emanating from an underwater acoustic source and monitored in the presence of noise at two spatially separated sensors can be mathematically modeled (see, e.g., Figure 15.13) in the direct path as

$$x_1(t) = s(t) + n_1(t) \tag{15.14a}$$

$$x_2(t) = s(t - D) + n_2(t) \tag{15.14b}$$

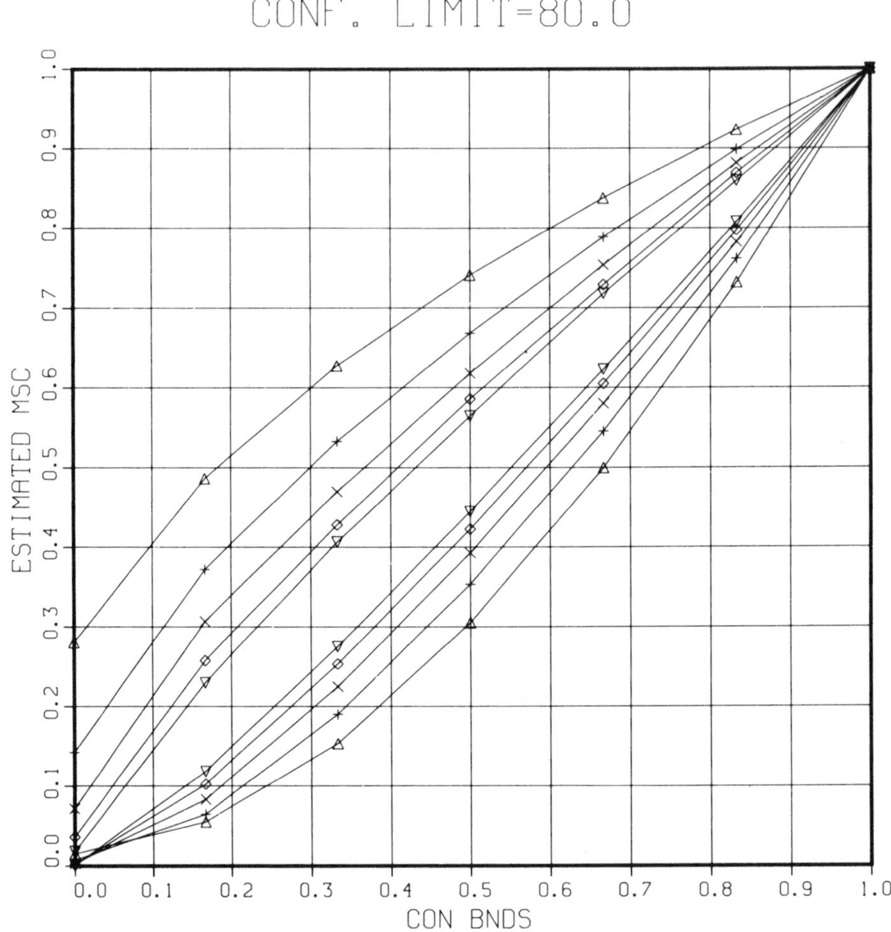


**Figure 15.12** The 80 and 95% MSC estimate confidence bounds for $N = 8$, 16, 32, 64, and 128.

where $s(t)$, $n_1(t)$, and $n_2(t)$ are real, jointly stationary random processes. Signal $s(t)$ is assumed to be uncorrelated with noise $n_1(t)$ and $n_2(t)$ and one wishes to estimate the unknown time delay $D$. We note that the received signals in noise are sometimes denoted by $r_1(t) = x_1(t)$ and $r_2(t) = x_2(t)$.

There are many applications in which it is of interest to estimate the time delay $D$. This section reviews the derivation of a maximum likelihood (ML) estimator given by Knapp and Carter [19]. Although the model of the physical phenomena presumes stationarity, the techniques to be developed here are usually employed in slowly varying environments where the characteristics of the signal and noise remain stationary only for a finite observation time $T$. Studies of more complex effects are given by Bjorno [30], Tacconi [31], Griffiths et al. [32], and Chan [33]. Further, the time delay $D$ may also change slowly, requiring time-varying or adaptive techniques such as those of Griffiths [34], Owsley [35], Picinbono [36], Meyr [37], and Lindsey

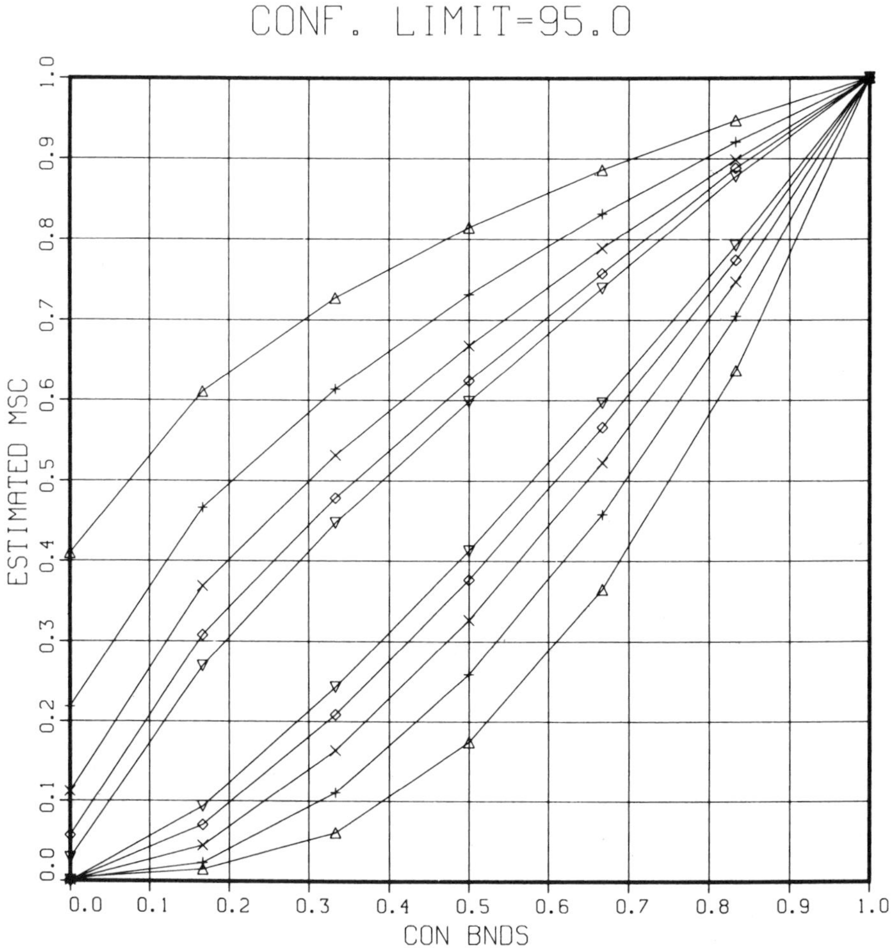


and Meyr [38]. Other investigations of the time-varying case are reported by Knapp and Carter [39], Adams et al. [40], Carter and Abraham [17], and Schultheiss and Weinstein [41].

Another important consideration in estimator design is the available amount of prior knowledge of the signal and noise statistics. In many problems, this information is negligible. For example, in passive detection, unlike the usual communications problems, the source spectrum is unknown or only known approximately.

## 5.2 The ML Estimator

The ML estimator is derived in the paper by Carter (1986). The ML estimator selects as the estimate of delay the value of $\tau$ at which the generalized cross-

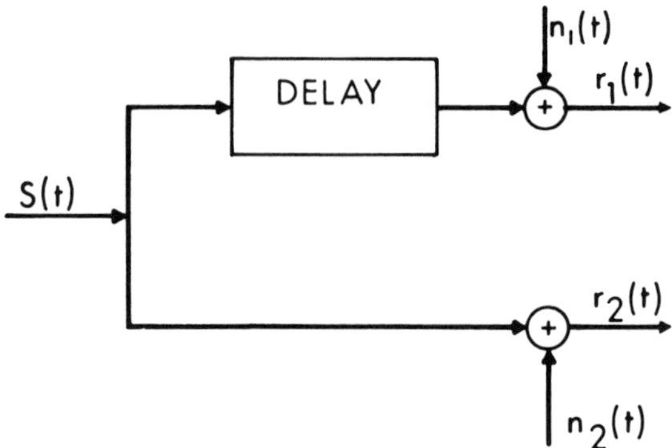


**Figure 15.13** Model of direction signal corrupted with additive noise and processed.

correlator

$$\hat{R}_{y_1 y_2}(\tau) = \int_{-\infty}^{\infty} \hat{G}_{x_1 x_2}(f) \frac{1}{|G_{x_1 x_2}(f)|} \frac{C_{12}(f)}{[1 - C_{12}(f)]} e^{j2\pi f\tau} \, df \tag{15.15}$$

achieves a peak.

These results compare favorably with closely related work of MacDonald and Schultheiss [42], Hannan and Thomson [43–45], Hahn and Tretter [46], and Cleveland and Parzen [47]. A more complete discussion of the derivation and related references can be found in Knapp and Carter [19].

## 5.3 Interpretation of the ML Estimator

One common method of determining the time delay $D$ is to compute the standard cross-correlation function

$$R_{x_1 x_2}(\tau) = E[x_1(t) x_2(t - \tau)] \tag{15.16}$$

where $E$ denotes expectation. The argument $\tau$ that maximizes Eq. (15.16) provides an estimate of delay. Because of the finite observation time, however, $R_{x_1 x_2}(\tau)$ can only be estimated. For example, for ergodic processes an estimate of the cross-correlation is given by

$$\hat{R}_{x_1 x_2}(\tau) = \frac{1}{T - \tau} \int_{\tau}^{T} x_1(t) x_2(t - \tau) \, dt \tag{15.17}$$

where $T$ represents the observation time. To improve the accuracy of the delay estimate $\hat{D}$, it is desirable to prefilter $x_1(t)$ and $x_2(t)$ prior to cross-correlation. We call this simple but very important process generalized cross-correlation. As shown in Figure 15.14, $x_i$ or $r_i$ may be filtered through $H_i$ to yield $y_i$ for $i = 1, 2$. The resultant $y_i$ are cross-correlated, that is, multiplied and integrated, for a range of hypothesized time delays or time shifts until the peak is obtained. We caution the reader that here we use $x_i$ and $r_i$ interchangeably to denote received signal. The time shift causing the peak is an estimate of the true delay. When the filters $H_1(f) = H_2(f) = 1$, for all $f$, the estimate of time delay is simply the abscissa value at which the standard cross-correlation function peaks. Knapp and Carter [19] provided for a generalized cross-correlation through the introduction of the filters $H_1(f)$ and $H_2(f)$ which, when properly selected, can significantly enhance the estimation of time delay.

The cross-correlation between $x_1(t)$ and $x_2(t)$ is related to the cross-power spectral density function by the well-known Fourier transform relationship

$$R_{x_1x_2}(\tau) = \int_{-\infty}^{\infty} G_{x_1x_2}(f)e^{j2\pi f\tau}\,df \tag{15.18}$$

When $x_1(t)$ and $x_2(t)$ have been filtered with filters having transfer functions $H_1$ and $H_2$ respectively, as depicted in Figure 15.14, the cross-power spectrum between the filter outputs is given by

$$G_{y_1y_2}(f) = H_1(f)H_2^*(f)G_{x_1x_2}(f) \tag{15.19}$$

(again we note $x_1 = r_1$ and $x_2 = r_2$).

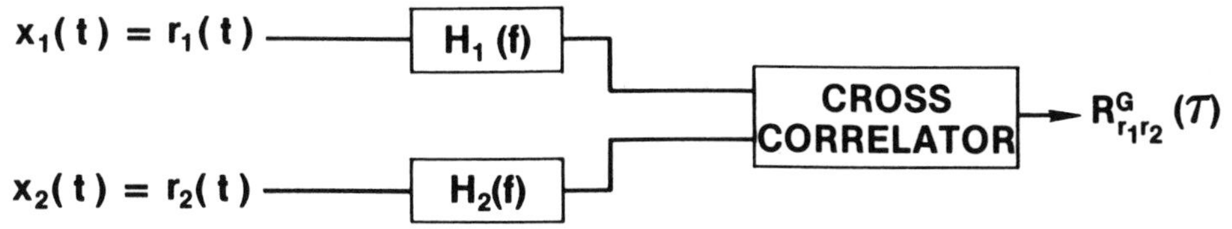


**GCC FUNCTION**

$$R^G_{x_1x_2}(\tau) = R^G_{r_1r_2}(\tau) = \int_{-\infty}^{\infty} W(f)\,G_{r_1r_2}(f)\,e^{j2\pi f\tau}\,df = \int_{-\infty}^{\infty} W_\phi(f)\,e^{j\phi(f)}\,e^{j2\pi f\tau}\,df$$

**WEIGHTING FUNCTION**

$$W(f) = H_1(f)\,H_2^*(f), \quad W_\phi(f) = |G_{r_1r_2}(f)|\,W(f)$$

**Figure 15.14** Generalized cross-correlation (GCC) function.

Therefore, the generalized cross-correlation (GCC) function between $x_1(t)$ and $x_2(t)$ is

$$R_{y_1y_2}(\tau) = \int_{-\infty}^{\infty} W(f)G_{x_1x_2}e^{j2\pi f\tau}\,df \tag{15.20a}$$

where the generalized frequency weighting

$$W(f) = H_1(f)H_2^*(f) \tag{15.20b}$$

In practice, only an estimate of the cross-power spectral density can be obtained from finite observations of the received signals. Consequently, the integral

$$\hat{R}_{y_1y_2}(\tau) = \int_{-\infty}^{\infty} W(f)\hat{G}_{x_1x_2}(f)e^{j2\pi f\tau}\,df \tag{15.21}$$

is evaluated and used for estimating delay. Indeed, depending on the particular form of $W(f)$ and prior information available, it may also be necessary to estimate the generalized weighting. For example, when the role of the prefilters is to accentuate the signal passed to the correlator at frequencies at which the coherence or SNR is highest, then $W(f)$ can be expected to be a function of the coherence or signal and noise spectra, which must either be known or estimated. Besides the ML weighting, there is an entire family of generalized weightings. See Figure 15.15 for

| METHOD | $W(f) = H_1(f)\,H_2^*(f)$ | $W_\phi(f) = W(f)\,\lvert G_{r_1r_2}(f)\rvert$ |
|---|---|---|
| SCC | $1$ | $\lvert G_{r_1r_2}(f)\rvert$ |
| ROTH | $1/G_{r_1r_1}(f)$ | $\lvert G_{r_1r_2}(f)\rvert / G_{r_1r_1}(f)$ |
| WIENER PROCESSOR | $C_{r_1r_2}(f)$ | $C_{r_1r_2}(f)\,\lvert G_{r_1r_2}(f)\rvert$ |
| SCOT | $1/\sqrt{G_{r_1r_1}(f)\,G_{r_2r_2}(f)}$ | $\sqrt{C_{r_1r_2}(f)}$ |
| PHAT | $1/\lvert G_{r_1r_2}(f)\rvert$ | $1$ |
| ML | $\dfrac{C_{r_1r_2}(f)}{[1 - C_{r_1r_2}(f)]\lvert G_{r_1r_2}(f)\rvert}$ | $\dfrac{C_{r_1r_2}(f)}{1 - C_{r_1r_2}(f)}$ |

$$C_{r_1r_2}(f) = \frac{\lvert G_{r_1r_2}(f)\rvert^2}{G_{r_1r_1}(f)\,G_{r_2r_2}(f)}$$

**Figure 15.15** Various GCC functions.

some common GCC weightings. A discussion of when to use these ad hoc weightings is beyond the scope of this chapter; however, the reader should be aware that other weightings have been shown under certain conditions to offer attractive processing capabilities (e.g., tonal rejection via SCOT processing).

The ML weighting that causes minimum-variance time delay estimation (TDE) under Gaussian assumptions is given by

$$W_{\mathrm{ML}}(f) = \frac{1}{|G_{x_1x_2}(f)|}\frac{C_{12}(f)}{[1 - C_{12}(f)]} \tag{15.22}$$

When $G_{x_1x_2}(f)$ and $C_{12}(f)$ are known, this is exactly the proper weighting. When these terms are unknown, they can be estimated via the lag reshaping spectral estimation techniques discussed earlier or using techniques of others, e.g., Marple [2], by the WOSA method of Carter et al. [9], or by classical methods of Jenkins and Watts [16] and Bendat and Piersol [48]. Substituting estimated weighting for true weighting is entirely a heuristic procedure whereby the ML estimator can approximately be achieved in practice. When the noises are uncorrelated but have the same power spectrum we can show that

$$W_{\mathrm{ML}}(f) = \frac{G_{ss}(f)/G_{nn}^2(f)}{[1 + 2G_{ss}(f)/G_{nn}(f)]} \tag{15.23}$$

For small signal-to-noise ratios we have

$$W_{\mathrm{ML}}(f) \cong \frac{G_{ss}(f)}{G_{nn}^2(f)} \tag{15.24}$$

which is the well-known Eckart filter (see Eckart [49]) used in optimum signal detection at low SNR. Thus the prefilters used for optimum signal detection at low SNR are the same as the prefilters used for minimum variance time delay estimation at low SNR.

## 5.4 Fundamental Performance Limits

The Cramer–Rao lower bound (CRLB) is given by

$$\min \operatorname{var} = \left[T\int_{-\infty}^{\infty}(2\pi f)^2\frac{C_{12}(f)}{1 - C_{12}(f)}\,df\right]^{-1} \tag{15.25}$$

This is the minimum variance and that which the ML processor achieves asymptotically for sufficiently large $T$. For constant signal and noise power spectra ratios, denoted here by SNR,

$$\frac{C_{12}(f)}{1 - C_{12}(f)} = \frac{(\mathrm{SNR})^2}{1 + 2\mathrm{SNR}} \tag{15.26}$$

so that

$$\text{min var} = \frac{1}{2T \int_0^B (2\pi)^2 f^2 [(\text{SNR})^2/(1+2\text{SNR})]\, df} \tag{15.27a}$$

or simply (see Quazi [50]).

$$\sigma^2_{\text{CRLB}} = \frac{3}{8\pi^2} \frac{(1+2\text{SNR})}{(\text{SNR})^2} \frac{1}{B^3 T} \tag{15.27b}$$

When the signal and noise have the same flat power spectra, as noted by Scarborough et al. [51], for the time delay estimation problem, this form of the Cramer–Rao lower bound is commonly used as the performance standard. The CRLB yields a lower bound on the variance of any unbiased time delay estimate as a function of several parameters [e.g., the signal and noise power spectra and the integration (observation) time]. Part of the appeal of the CRLB is that, for cases of practical interest, there is a theorem which states that the ML estimate can be made arbitrarily close to the CRLB for sufficiently long integration times; see e.g., Van Trees [52]. However, the theorem does not specify how long the integration time must be. Thus, while the CRLB sets a lower bound on the variance of the time delay estimate, actual performance can be much worse for a given (low) SNR and (short) observation time. This is corroborated by the simulation results of Scarbrough et al. [51, 53] and Hassab and Boucher [54]. Several studies have been conducted to find a bound tighter than the CRLB which would predict performance more accurately. Significant work in this area has been done by Ianniello et al. [55], Weiss and Weinstein [56, 57], Chow and Schultheiss [58], and Ianniello [59]. The following discussion presents some of these results and discusses some implications of them. In particular, the behavior of estimators is considered as a function of the observation time and SNR, and the implications of this behavior are considered in relation to coherent and incoherent signal processing techniques for time delay estimation. In addition, simulation results are presented to support the inferences of the theoretical analysis.

Chow and Schultheiss [58], Scarbrough [60], Betz [61], and Johnson et al. [62] have also studied the low-SNR performance problem. In the work here, consideration will be limited to signal and noise power spectra $G_{ss}(f)$ and $G_{nn}(f)$, respectively, which are flat (constant) over the frequency range $-B$ to $+B$ Hz and zero outside this range. $B$ is then a measure of the source signal bandwidth. In addition, it will be assumed that the bandwidth-observation time product $BT$ is large (say $BT \geq 100$, e.g., for $B = 100$ Hz, $T \geq 1$ s).

In general, proper prefiltering prior to cross-correlation is required to achieve the ML estimate of time delay. Under the special conditions here, the ML estimate of the time delay can be obtained by computing by cross-correlation function $R(\tau)$ between $x_1(t)$ and $x_2(t)$. The ML time delay estimate is the value of $\tau$ which maximizes $R(\tau)$. For other power spectra, a generalized cross-correlator with the proper prefilters is required.

The correlator performance estimator (CPE) was developed by Ianniello [59] to provide more accurate performance prediction for the cross-correlation technique for TDE in the presence of large estimation errors or anomalous estimates. The CPE assumes that the anomalous estimates will be uniformly distributed across the correlation window, say from $-T_0$ to $+T_0$ seconds. The CPE yields the following estimate of the variance of the TDE error:

$$\sigma^2_{\text{CPE}} = P\frac{T_0^2}{3} + (1-P)\sigma^2_{\text{CRLB}} \tag{15.28}$$

where $P$ is the probability of an anomalous estimate. Details of the expression for $P$ can be found in Ianniello [59] and Scarbrough [60].

The CPE and CRLB are compared in Figure 15.16 for the case of $B = 100$ Hz, $T = 8$ s, and $T_0 = 1/8$ s relative to an assumed sampling frequency of 2048 Hz. Both curves in Figure 15.16 are plotted as the base 10 logarithm of $\sigma_D$, the standard deviation of the time delay error, versus the input SNR in dB. The upper curve is the CPE; the lower curve is the CRLB. The CPE is characterized by three regions: 1) at low SNR there is a region where prior information limits the variance (e.g., the maximum observable delay $T_0$ is known); 2) at moderate SNR there is a transition region from the prior information limit to the CRLB; and 3) at high SNR the CPE coincides with the CRLB. The SNR at which the CPE begins to deviate from the CRLB is referred to as the threshold SNR ($\text{SNR}_{\text{th}}$ in Figure 15.16). We note also that while the CPE is not a bound, recent work has been done to derive a bound

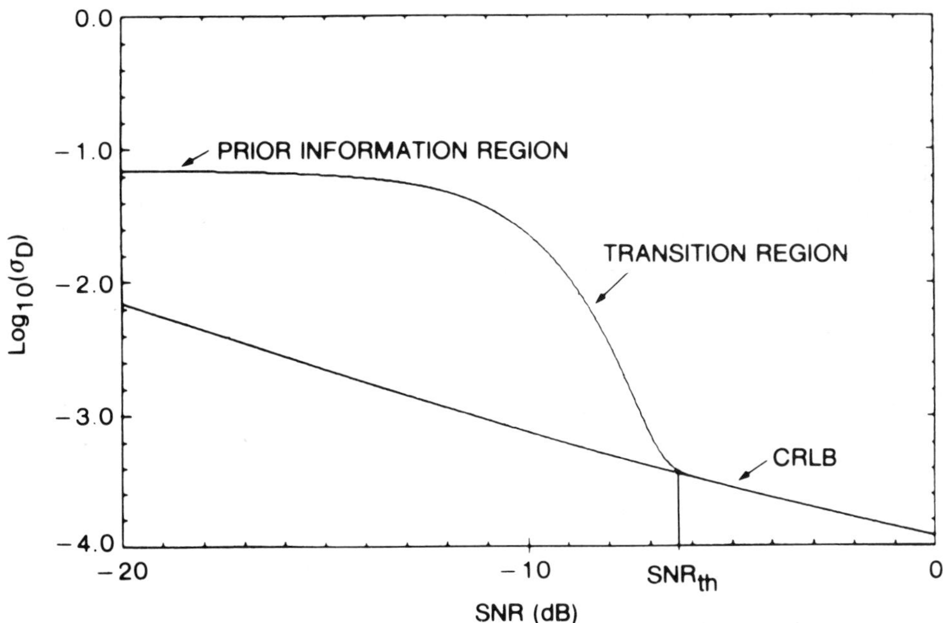


**Figure 15.16** Plot of CPE and CRLB versus SNR ($T = 8$ s, $B = 100$ Hz, $T_0 = 1/8$ s).

tighter than the CRLB. This new bound is close to the CPE, falling just below the CPE on Figure 15.16. This new bound, the Ziv Zakai lower bound (ZZLB), is discussed by Ianniello et al. [55] and Scarbrough [60].

### 5.5 Simulation Description and Results

A computer simulation was conducted to corroborate the theoretical TDE performance predictions. Earlier simulations were conducted by Hassab and Boucher [63]. Good agreement among many earlier simulations was shown in a comparison of a large number of simulation results by Kirlin and Bradley [64]. The cross-correlation of simulated received sequences was computed using the FFT approach described by Oppenheim and Schafer [65]. The particular method is very similar to the lag reshaping method of spectral estimation presented earlier. It is a little more complex in that zero filling on FFT segments is not necessarily equal. The details are not important for the discussion here but may be if attempting to reproduce the results. It is important that the mean and variance of all the noise-only correlation points be the same for all correlation lags. After cross-correlation the estimate of the time delay was obtained by finding the delay value for which the cross-correlation was a maximum. This approach yields the time delay estimate quantized in units of the sampling interval.

A simulation was conducted for two integration times over a range of SNR values. Unlike many earlier simulations, effects of low SNR were studied. Different integration times are obtained by varying the number of cross-power spectra which are averaged before taking the inverse FFT to obtain the estimate of the cross-correlation function. In the simulation 8 and 32 data segments were processed coherently to obtain integration times of 2 and 8 s, respectively, for the case of 512-point data segments and an assumed sampling frequency of 2048 Hz. A total of 2000 trials were conducted at each SNR to obtain the experimental time delay variances. These results are plotted in Figure 15.17 along with the corresponding theoretical curves for the CRLB and the CPE. The symbol size of the experimental points is indicative of the 90% confidence limits. The theory and results are in very close agreement and support the previous analysis. Of profound importance is that the CRLB is a poor predictor of performance at low input SNR. This has significant implications for using techniques to increase coherent processing time as discussed by Scarbrough et al. [51] and Scarbrough [60]. These results are graphically portrayed in Figure 15.18. Note that now we plot the ZZLB as opposed to the CPE. Also, we compare a coherent or tracking approach for $T = 8$ with an incoherent approach. (In the incoherent case we first estimate the delay over short observation intervals and then average the time delay estimates over the available intervals.) The threshold is clearly reduced by coherent processing.

## 6 FOCUSED TIME DELAY ESTIMATES FOR RANGING

### 6.1 Introduction

This section discusses using focused time delay estimates to passively estimate range to a coherent source. Most of the work in this section is a summary of the work in

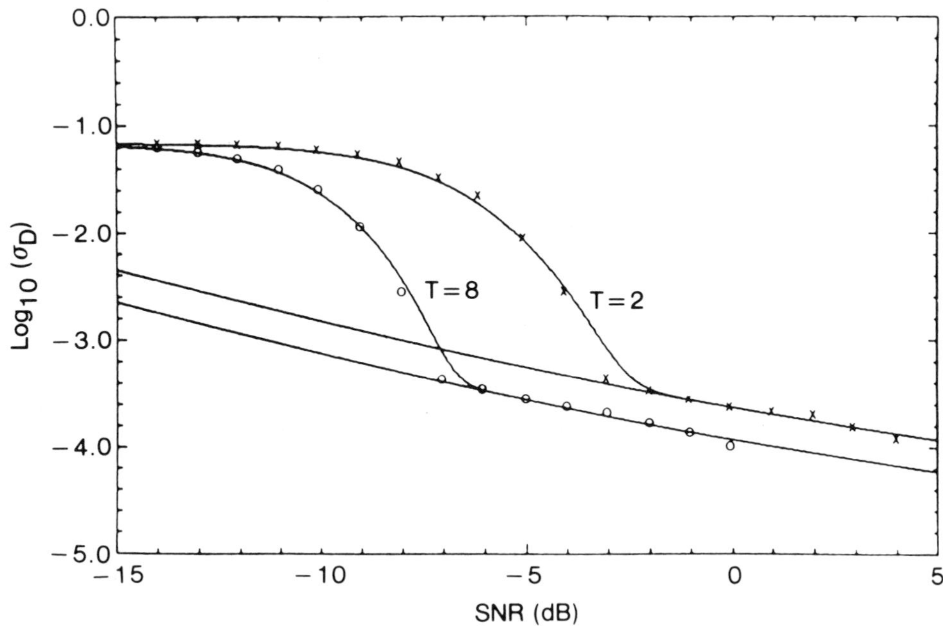


**Figure 15.17** Comparison of CPE, CRLB, and simulation results ($B = 100$ Hz, $T_0 = 1/8$ s, $T = 2$, 8 s).

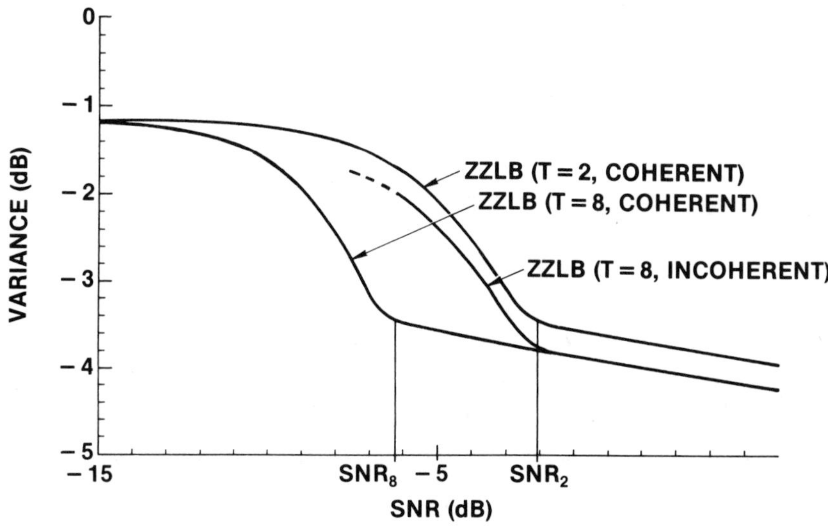


**Figure 15.18** Plot of ZZLB and effects of coherent versus incoherent processing.

Carter [66]. Other related references include Owsley [35], Lynch [67], and Lynch et al. [68].

The ML processor is presented for passively estimating range and bearing to an acoustic source. The source signal is observed for a finite time duration at several sensors in the presence of uncorrelated noise. When the speed of sound in an isovelocity medium and the sensor positions are known, the ML estimator for position constrains the source-to-sensor delays to be focused into a point corresponding to a hypothesized source location. The variances of the range error and bearing error are presented for the optimum processor. It is shown that for bearing and range estimation, different sensor configurations are desirable. However, if the area of uncertainty is to be minimized, then the sensors should be divided into equal groups with one-third of the sensors in each group.

An underwater acoustic point source radiating energy to several collinear receiving sensors is shown in Figure 15.19. The position of the source in two space can be characterized by range $R$ and bearing $B$ from a given frame of reference. The particular geometry of interest is two-dimensional with an acoustic point source whose range and bearing are to be estimated by a fixed number of receivers. For the purposes of this work, we presume that the receiving hydrophones are collinear. However, regardless of the hydrophone positions, a fixed number of sensors have an inherent uncertainty in estimating source location. For our geometries here, this uncertainty region is nominally elliptical, so that by properly defining how range and bearing are measured the estimation errors can be decoupled. For a collinear array of sensors, we measure the bearing as the angle between the line array and the major axis of the uncertainty region.

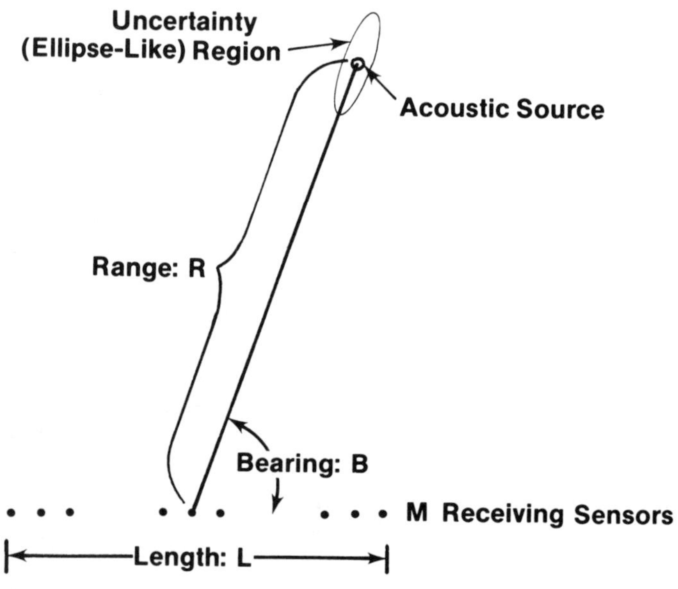


**Figure 15.19** Array geometry used to estimated source position.

For a radiating source distant from an array, the uncertainty in measuring $R$ and $B$ is characterized by an extremely elongated elliptical uncertainty region. The problem addressed here is how to estimate range $R$ and bearing $B$ to a source when $M$ sensors separated by a maximum $L$ (meters) have observed $T$ seconds of received data. We will examine the ML technique for position estimation.

## 6.2 Mathematical Model

For our purposes we assume that each receiving sensor at the $t$th instant in time corresponds to a signal plus noise. Namely, the $i$th (of $M$) sensor outputs is characterized by

$$r_i(t) = s(t + D_i) + n_i(t), \qquad i = 1, M, \ 0 \leq t \leq T \tag{15.29}$$

The signal and noises are uncorrelated and the noises are mutually uncorrelated. Without loss of generality $D_1 = 0$.

For a spatially stationary, that is, nonmoving, source, the signal can be viewed as an attenuated and delayed source signal. However, the problem of estimating the position of a stationary source is considerably easier than that of a moving source. Thus, the results here serve as a bound on performance; still, we will see that it is extremely difficult in the best case to passively estimate source range.

## 6.2 Maximum-Likelihood Performance

The ML estimate for the time delay vector has been derived for stationary Gaussian processes by Hahn [69] and Carter [23]. One can show that an ML estimate for range and bearing when the sensor (element) positions are known is achieved by a variation of the method in Carter [23]. In particular, by focusing all the time-delay elements at many (hypothesized) range and bearing pairs and watching for the peak output of the ML time-delay vector system, the ML position estimate is observed. An ML system realization is shown in Figure 15.20. This figure is sometimes referred to as a focused beamformer. More details are given by Carter [66]. Another way to look at this problem is that we want to maximize a quantity by adjusting a number of delay parameters subject to the constraint that all the delays must intersect in a single hypothesized position. Such a system and its variance have been examined by Bangs and Schultheiss [70].

For equal noise spectra, we should utilize a product of pre- and postfilters with magnitude-squared transfer function, given by Bangs and Schultheiss [70] (see also Knapp and Carter [19]), Hannan and Thomson [44], MacDonald and Schultheiss [42], and Carter [23],

$$|H(f)|^2 = \frac{G_{ss}(f)/G_{nn}^2(f)}{1 + MG_{ss}(f)/G_{nn}(f)} \tag{15.30}$$

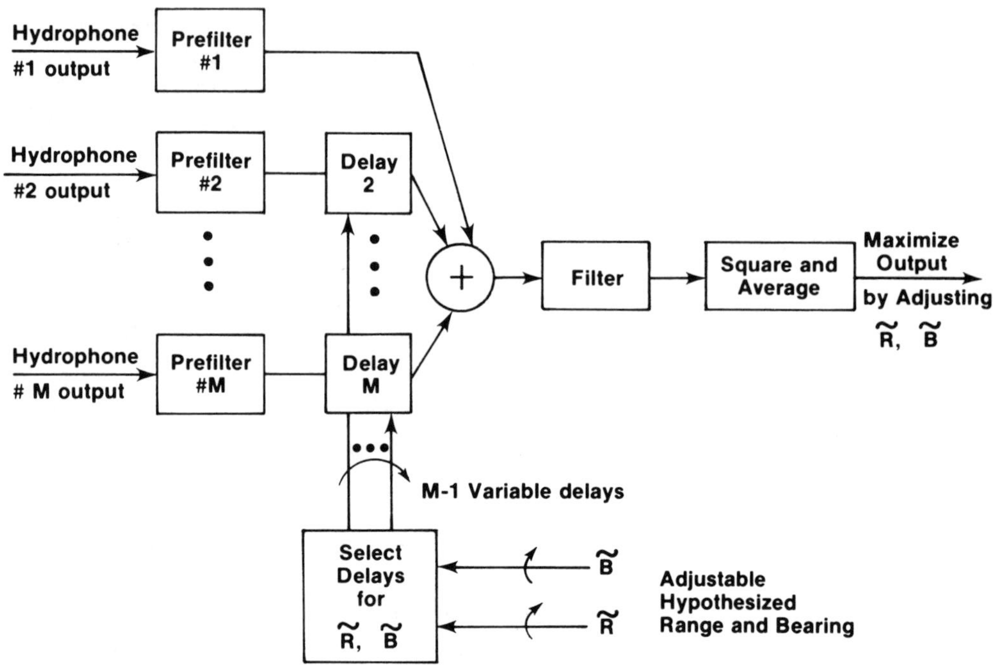


**Figure 15.20** Maximum-likelihood estimator for range and bearing.

in order to minimize the variance of delay estimates. For passive range and bearing estimation the variance of the parameter estimate is more complicated (see Carter [66]) and is summarized in the following subsection.

Having selected proper prefilters for a specified array geometry, the maximum-likelihood estimates of range and bearing (i.e., minimum-variance estimates) are obtained by coherently processing the outputs of the sensing hydrophones. In particular, each hydrophone output is prefiltered to accentuate a high SNR, then delayed and summed. The summed signal is fed to a filter, then squared and averaged for the observation time. The output of this network is maximized through indirect adjustment of the delay parameters. The delay parameters are derived on the basis of two adjustable parameters: hypothesized bearing and hypothesized range. Thus an operator need only adjust the best estimate of bearing and range. From these two inputs, proper delays are inserted in each hydrophone receiving line. The process of delaying and summing is a focused beamformer where the delays used cause the beamformer to presume that the source wave front is curved and not planar. The individual sensor-to-sensor delays inserted are directly related to the hypothesized source and sensor locations.

For a particular array type A, Carter [66] has shown that, at high output SNR, the minimum variances of bearing and range estimates are given by

$$\sigma_A^2(\hat{B}) \cong \frac{K_B^{(A)}}{\mathrm{TMVL}_e^2} \tag{15.31}$$

and

$$\sigma_A^2(\hat{R}) \cong \frac{K_R^{(A)} R^4}{\mathrm{TMVL}_e^4} \tag{15.32}$$

where $\sigma^2(\hat{B})$ is measured in radians squared, the effective array length $L_e = L \sin B$, and the constants $K_R$ and $K_B$ for four array types are given in Table 15.3. $M$ is the minimum of the number of sensors and the array length divided by the design half-wavelength; also, at high output SNR

$$V = \frac{1}{2\pi C^2} \int_0^{\Omega} \frac{G_{ss}(\omega)}{G_{nn}(\omega)} \omega^2 \, d\omega \tag{15.33}$$

where $G_{ss}(\omega)$ is the signal power spectrum, $G_{nn}(\omega)$ is the noise power spectrum, $C$ is the speed of sound in the medium, $\Omega$ is the highest source (or receiver) frequency, and, as earlier, $T$ is the observation time, $R$ the range, and $B$ the bearing.

## 6.4 Discussion

Doubling the number of sensors $M$ or the observation time $T$ will reduce the standard deviation of either the bearing estimate or the range estimate by 1.4. In bearing estimation, we wish to make $L_e$ large and the constant $K_B$ small in order to reduce variance. Note that doubling the array length reduces the variance by four. Thus array length is a more important factor in bearing estimation than either integration time or the number of hydrophones when operating at high output SNR.

The four different array types studies are an equispaced line array and three line arrays; the three line arrays have $M$ elements grouped or clustered at each end and in the middle of the array. MacDonald and Schultheiss [42] have shown that, by placing half of the $M$ elements at each end of a line array in a grouping of $M$ divided by two, zero in the middle, and $M$ divided by two, a bound on bearing variance is obtained. This bound, of course, is for a hypothetical array where the elements are collocated but still sense independent noises. The practical

**Table 15.3** Constants for Four Arrays of Interest

| Array Type | $\sqrt{\frac{K_R}{K_B}}$ | $K_R$ | $K_B$ |
|---|---|---|---|
| Equispaced Line | 7.75 | 360 | 6 |
| $M/2$, 0, $M/2$ | $\infty$ | $\infty$ | 2 |
| $M/3$, $M/3$, $M/3$ | 6.9 | 144 | 3 |
| $M/4$, $M/2$, $M/4$ | 5.7 | 128 | 4 |

implications of MacDonald and Schultheiss's result are that they both provide a bound on how well bearing can be estimated under ideal conditions and suggest how to place a limited number of hydrophones over a large aperture. That is, for bearing estimation half of the hydrophones should be positioned at each end of the array, placed at half-wavelength spacing for the design frequency, and none should be placed in the middle of the available aperture.

It is noteworthy that the variance of the range estimate depends on the fourth power of the range relative to the effective baseline. This fourth-order dependence is a fundamental physical limit and makes the task of passive ranging at long range with short arrays extremely difficult even at high output SNR. The variance of the range estimate is reduced by making the effective array length $L_e$ large. This can be done by making the array length $L$ large or, to a lesser degree, by physically steering the array broadside to the source. The variance can also be reduced by decreasing the range to the source. Of course, reducing the range to the source can also increase SNR, depending on propagation conditions.

The constant $K_R$ depends on the array type. For an equispaced line array, $K_R$ is 360. A bound on passive ranging performance is provided by an array configured with a quarter of the hydrophones at each end and half in the middle. Thus we see that the hydrophone configuration in a line array for bearing estimation and that for range estimation differ. The bearing array should have its elements toward the array ends, while the ranging array should have half of its elements in the central portion. However, a line array with a third of its elements at each end and a third in the middle will minimize the uncertainty region (or the product of cross-range error and ranging error). (See Carter [66].) Thus, in this sense a line array physically segmented into three equal groups of elements will outperform all other arrays for passively locating an acoustic source. This result is summarized in Table 15.4. The optimum processor coherently combines all $M$ hydrophone outputs. However, if only the beamformer output from each subarray is used for coherent processing, a nearly optimum technique is believed to result.

Of considerable concern when attempting to predict the performance of a passive localization technique are values such as SNR, number of sensors, and integration time. It is interesting that these terms, together with a constant such as $2\pi$, can all be attributed to the standard deviation of the bearing estimates (measured in radians). Then the relative range error given by the standard deviation of the range estimate divided by the true range is given by a constant times the standard

**Table 15.4** Optimum Sensor Configurations

| For Estimating | Best Array Configuration | | |
|---|---|---|---|
| Bearing | • • • • • •<br>$M/2$ | | • • • • • •<br>$M/2$ |
| Range | • • •<br>$M/4$ | • • • • • •<br>$M/2$ | • • •<br>$M/4$ |
| Position | • • • •<br>$M/3$ | • • • •<br>$M/3$ | • • • •<br>$M/3$ |

deviation of the bearing estimate times a term that depends linearly on the range to the source relative to the effective array length. In particular,

$$\frac{\sigma_A(\hat{R})}{R} = \sqrt{\frac{K_R}{K_B}\frac{R}{L_e}}\sigma_A(B) \tag{15.34}$$

For example, suppose an equispaced line array had an inherent standard deviation of 1/10 rad (5.7°) and was to estimate the range to a source 10 times as far away as the effective array length. In that case, the relative range error is 7.75, or more than 700%. Hence, we see that it is extremely difficult to passively estimate the range of a distant source even under ideal conditions with high output SNR.

One of the advantages of expressing relative range errors in this form is that the standard deviation of bearing estimates is a term familiar to sonar engineers and signal processors. Moreover, the ocean medium may inherently limit the practical ability to estimate bearing even though theory predicts that, with enough SNR or integration time, the bearing can be measured arbitrarily well. The expression given here clearly points out the need to make the array length large when the source range cannot be reduced. This conclusion is extremely insensitive to the type of array, provided the array has some ranging capability. This can be seen from the similarity of the constants given in Table 15.3.

To summarize, we wish to know how to place a limited number of hydrophones over a baseline of fixed length. The hydrophones should be placed in groups, with the hydrophones in each group placed at half-wavelength spacing for the desired frequency. For passive bearing estimation, half of the $M$ hydrophones should be placed at each end of the array. For passive range estimation, one-quarter of the hydrophones should be placed at each end of the array and half placed in the middle. For simultaneously estimating range and bearing, both passively, the hydrophones are placed in three groups or subarrays, each with $M/3$ hydrophones. If the baseline remained of fixed length and we had more hydrophones to add, we would add the hydrophones at half-wavelength spacing approaching an equispaced line array. On the other hand, if the number of hydrophones was limited but the baseline was not, we would keep the hydrophones at half-wavelengths spacing and increase the distance between subarrays. By keeping the hydrophones at half-wavelength spacing in the subarrays, ambiguities would be minimized.

## 7 SUMMARY

This chapter has presented a review of research on coherence estimation and time delay estimation. References to much of the relevant work in these two fields are included.

## ACKNOWLEDGMENTS

As evident from the text and the list of references, this chapter is a review of research conducted by this author alone and with the following colleagues: Prof. Charles

Knapp, Dr. Kent Scarbrough, Mr. E. Scannell, Dr. Peter Cable, Dr. Albert Nuttall, Prof. N. Ahmed, Mr. Roger Tremblay, and Dr. Philip Abraham. The author has been honored by Prof. G. Tacconi, Prof. L. Bjorno, Prof. Y. T. Chan, Prof. J. Plant, Dr. H. Urban, Prof. T. Durrani, and Prof. J. L. Lacoume to present this work at several meetings over the past decade in Italy, France, the United Kingdom, Denmark, West Germany, and Canada as well as in the United States.

## REFERENCES

1. G. C. Carter and C. H. Knapp, Coherence and its estimation via the partitioned modified chirp-z transform, *IEEE Trans. Acoust. Speech Signal Process.*, *ASSP-23*: 257–264 (June 1975).
2. S. L. Marple, Jr., *Digital Spectral Analysis with Applications*, Prentice-Hall, Englewood Cliffs, New Jersey (1987).
3. D. G. Childers, ed., *Modern Spectrum Analysis*, IEEE Press (1978).
4. S. B. Kesler, ed., *Modern Spectrum Analysis II*, IEEE Press (1986)
5. A. H. Nuttall and G. C. Carter, A Generalized Framework for Power Spectral Estimation, *IEEE Trans. Acoust. Speech Signal Process. ASSP-28*(3): 334–335 (1980).
6. A. H. Nuttall and G. C. Carter, Spectral Estimation using combined time and lag weighting, *Proc. IEEE 70*(9): 1115–1125 (1982).
7. R. B. Blackman and J. W. Tukey, *The Measurement of Power Spectra*, Dover, New York (1958).
8. P. D. Welch, The use of fast Fourier transform for estimation of power spectra: a method based on time averaging over short modified periodograms, *IEEE Trans. Audio Electroacoust.*, *AU-15*: 70–73 (1967).
9. G. C. Carter, C. H. Knapp, and A. H. Nuttall, Estimation of the magnitude-squared coherence function via overlapped fast Fourier transform processing, *IEEE Trans. Audio Electroacoust.*, *AU-21*: 337–344 (1973).
10. L. H. Koopmans, *The Spectral Analysis of Time Series*, Academic Press, New York (1974).
11. D. R. Brillinger, *Time Series Data Analysis and Theory*, Holt, Rinehart, & Winston, New York (1975).
12. A. H. Nuttall and G. C. Carter, Bias of the estimate of magnitude squared coherence, *IEEE Trans. Acoust. Speech Signal Process.*, *ASSP-24*: 582–583 (December 1976).
13. J. S. Bendat and A. G. Piersol, *Random Data: Analysis and Measurement Procedures*, Wiley, New York (1971).
14. W. G. Halvorsen and J. S. Bendat, Noise source identification using coherent output power spectra, *J. Sound Vibr.*, 15–24 (August 1975).
15. H. Akaike and Y. Yamanouchi, On the statistical estimation of frequency response function, *Ann. Inst. Stat. Math.*, *14*: 23–56 (1963).
16. G. M. Jenkins and D. G. Watts, *Spectral Analysis and Its Applications*, Holden-Day, San Francisco (1968).
17. G. C. Carter and P. B. Abraham, Estimation of source motion from time delay and time compression measurement, *J. Acoust. Soc. Am.*, *67*(3): 830–832 (1980).

18. G. C. Carter, A. H. Nuttall, and P. G. Cable, The smoothed coherence transform, *Proc. IEEE, 61*: 1497–1498 (1973).
19. C. H. Knapp and G. C. Carter, The generalized correlation method for estimation of time delay, *IEEE Trans. Acoust. Speech Signal Process., ASSP-24*: 320–327 (August 1976).
20. J. P. Kuhn, "Detection Performance of the Smoothed Coherence Transform (SCOT)," Proc. 1978 International Conference on Acoustics, Speech, and Signal Processing (1978).
21. G. C. Carter, Receiver operating characteristics for a linearly threshold coherence estimation detector, *IEEE Trans. Acoust. Speech Signal Process., ASSP-27*: 90–94 (February 1977).
22. A. S. Gevins, C. L. Yeager, S. L. Diamond, J. P. Spire, G. M. Zeitlin, and A. H. Gevins, Automated analysis of the electrical activity of the human brain (EEG): A progress report, *Proc. IEEE, 63*: 1382–1399 (October 1975).
23. G. C. Carter, Time delay estimation, Ph.D. dissertation, University of Connecticut (1976).
24. E. H. Scannell, Jr., and G. C. Carter, "Confidence Bounds for Magnitude-Squared Coherence Estimates," Proc. IEEE International Conference on Acoustics, Speech, and Signal Processing (1978).
25. H. Cramer, *Mathematical Methods of Statistics*, Princeton University Press, Princeton, New Jersey (1946).
26. R. A. Haubrich, Earth noise 5 to 500 millicycles per second, 1 Spectral stationarity, normality, and nonlinearity, *J. Geophys. Res., 70*: 1415–1427 (1965).
27. V. A. Benignus, Estimation of coherence spectrum of non-Gaussian time series populations, *IEEE Trans. Audio Electroacoust., AU-17*: 198–201 (September 1969) [and *AU-18*: 320 (September 1970)].
28. L. D. Enochson and N. R. Goodman, Gaussian Approximations to the Distribution of Sample Coherence, AFFDL-TR-65-57, (June 1965).
29. J. J. Gosselin, "Comparative Study of Two-Sensor (Magnitude-Squared Coherence) and Single-Sensor (Square-Law) Receiver Operating Characteristics," in Proc. IEEE International Conference on Acoustics, Speech, and Signal Processing, pp. 311–314 (1977).
30. L. Bjorno, ed., *Proceedings of the 1980 NATO Advanced Study Institute on Underwater Acoustics and Signal Processing*, Reidel, Boston (1981).
31. G. Tacconi, ed., *Aspects of Signal Processing*, Reidel, Boston (1977).
32. J. W. R. Griffiths, P. L. Stocklin, and C. Van Schooneveld, eds., *Signal Processing*, Academic Press, New York (1973) (including E. B. Lunde, "Wavefront stability in the ocean," and W. J. Bangs and P. M. Schultheiss, "Space-time processing for optimal parameter estimation").
33. Y. T. Chan, "Time delay estimation in the presence of multipath propagation," Proc. NATO ASI (1984).
34. L. Griffiths, "Time Varying Filtering and Spectrum Estimation," Proc. NATO ASI (1984).
35. N. Owsley, "Overview of Adaptive Array Processing Techniques," Proc. NATO ASI (1984).
36. B. Picinbono, "Adaptive, Robust, and Non-Parametric Methods in Signal Detection," Proc. NATO ASI (1984).

37. H. Meyr, Delay lock tracking of stochastic signals, *IEEE Trans. Commun., COM-24*(3): 331–339 (1976).
38. W. C. Lindsey and H. Meyr, Complete Statistical description of the phase-error process generated by correlative tracking system, *IEEE Trans. Inf. Theory, IT-23*(2): 194–202 (1977).
39. C. H. Knapp and G. C. Carter, Estimation of time delay in the presence of source or receiver motion, *J. Acoust. Soc. Am., 61*: 1545–1549 (1977).
40. W. B. Adams, J. P. Kuhn, and W. P. Whyland, Correlator compensation requirements for passive time delay estimation with moving source or receivers, *IEEE Trans. Acoust. Speech Signal Process., ASSP-28*(2): 158–168 (1980).
41. P. M. Schultheiss and E. Weinstein, Source tracking using passive array data, *IEEE Trans. Acoust. Speech Signal Process.*, 600–607 (June 1981).
42. V. H. MacDonald and P. M. Schultheiss, Optimum passive bearing estimation in a spatially incoherent noise environment, *J. Acoust. Soc. Am., 46*: 37–43 (1969).
43. E. J. Hannan and P. J. Thomson, The estimation of coherence and group delay, *Biometrika, 58*: 469–481 (1971).
44. E. J. Hannan and P. J. Thomson, Estimating group delay, *Biometrika, 60*: 241–253 (1973).
45. E. J. Hannan and P. J. Thomson, Delay estimation, *IEEE Trans. Acoust. Speech Signal Process.*, 485–490 (June 1981).
46. W. R. Hahn and S. A. Tretter, Optimum processing for delay-vector estimation in passive signal arrays, *IEEE Trans. Inf. Theory, IT-19*(5): 608–614 (1973).
47. W. S. Cleveland and E. Parzen, The estimation of coherence, frequency response, and envelope delay, *Technometrics, 17*(2): 167–172 (1975).
48. J. S. Bendat and A. G. Piersol, *Engineering Applications of Correlation and Spectral Analysis*, Wiley, New York (1980).
49. C. Eckart, Optimum rectifier Systems for the Detection of Steady Signals, University of California, Scripps Institute of Oceanography, Marine Physical Laboratory Report S10, Ref. 52-11 (1952).
50. A. H. Quazi, An overview on the time delay estimate in active and passive systems for target localization, *IEEE Trans. Acoust. Speech Signal Process., ASSP-29*: 527–533 (June 1981).
51. K. Scarbrough, R. J. Tremblay, and G. C. Carter, Performance predictions for coherent and incoherent processing techniques of time delay estimation, *IEEE Trans. Acoust. Speech Signal Process., ASSP-31*(5): 1191–1196 (1983).
52. H. L. Van Trees, *Detection, Estimation, and Modulation Theory, Part I*, Wiley, New York (1968).
53. K. Scarbrough, N. Ahmed, and G. C. Carter, On the simulation of a class of time delay estimation algorithms, *IEEE Trans. Acoust. Speech Signal Process., ASSP-29*: 534–540 (June 1981).
54. J. C. Hassab and R. E. Boucher, "A Quantitative Study of Optimum and Suboptimum Filters in the Generalized Correlator," Proc. 1979 IEEE International Conference on Acoustics, Speech, and Signal Processing, pp. 124–127 (1979).
55. J. P. Ianniello, E. Weinstein, and A. Weiss, "Comparison of the Ziv-Zakai Lower Bound on Time Delay Estimation with Correlator Performance," Proc.

1983 International Conference on Acoustics, Speech, and Signal Processing, pp. 875–878 (1983).
56. A. Weiss and E. Weinstein, Composite bound on the attainable mean-square error in passive time-delay estimation from ambiguity prone signals, *IEEE Trans. Inf. Theory, IT-28*: 977–979 (November 1982).
57. A. Weiss and E. Weinstein, Fundamental limitations in passive time delay estimation—Part I: Narrow-band systems, *IEEE Trans. Acoust. Speech Signal Process., ASSP-31*: 472–485 (April 1983).
58. S. K. Chow and P. M. Schultheiss, Delay estimation using narrow-band processes, *IEEE Trans. Acoust. Speech Signal Process., ASSP-29*: 478–484 (June 1981).
59. J. P. Ianniello, Time delay estimation via cross-correlation in the presence of large estimation errors, *IEEE Trans. Acoust. Speech Signal Process., ASSP-30*: 998–1003 (December 1982).
60. K. Scarbrough, Ph.D. thesis, Kansas State University (1984).
61. J. Betz, Ph.D. thesis, Northeastern University (1984).
62. G. W. Johnson, D. E. Ohlms, and M. L. Hampton, "Broadband correlation Processing," Proc. 1983 International Conference on Acoustics, Speech, and Signal Processing, pp. 583–586 (1983).
63. J. C. Hassab and R. E. Boucher, Performance of the generalized cross correlator in the presence of a strong spectral peak in the signal, *IEEE Trans. Acoust. Speech Signal Process.*, 549–555 (June 1981).
64. R. L. Kirlin and J. N. Bradley, Delay estimation simulations and a normalized comparison of published results, *IEEE Trans. Acoust. Speech Signal Process., ASSP-30*: 508–511 (June 1982).
65. A. V. Oppenheim and R. W. Schafer, *Digital Signal Processing*, Prentice-Hall, Englewood Cliffs, New Jersey (1975).
66. G. C. Carter, Variance bounds for passively locating an acoustic source with a symmetric line array, *J. Acoust. Soc. Am.*, 922–926 (1977).
67. J. F. Lynch, On the use of focused horizontal arrays as mode separation and source location devices in ocean acoustics, Part I: Theory, *J. Acoust. Soc. Am. 74*: 1406–1416 (1983)
68. J. F. Lynch, D. K. Schwartz, and K. Sivaprashad, On the use of focused horizontal arrays as mode separation and source locations devices in ocean acoustics. Part II Theoretical and Modeling results, *J. Acoust. Soc. Am. 78*: 575–586 (1985).
69. W. R. Hahn, Optimum signal processing for passive sonar range and bearing estimation, *J. Acoustic. Soc. Am., 58*: 201–207 (1975).
70. W. J. Bangs and P. M. Schultheiss, Space time processing for optimal parameter estimation, *Signal Processing* (J. W. R. Griffiths, P. L. Stocklin, and C. Van Schooneveld, eds.) 577–590 Academic Press, New York. pp. 577–590 (1973).
71. G. C. Carter, A brief description of the fundamental difficulties of passive ranging, *IEEE Trans. OE, OE-3*(3): 65–66 (1978).
72. G. C. Carter, Time delay estimation for passive sonar signal processing, *IEEE Trans. Acoust. Speech Signal Process., ASSP-29*(3): 463–470 (1981).

73. A. H. Nuttall, Theory and application of the separable class of random processes, Ph.D. dissertation, Massachusetts Institute of Technology, Cambridge, RLE Report 343 (1958).
74. A. H. Nuttall, Spectral Estimation by Means of Overlapped FFT Processing of Windowed Data, Report 4169, Naval Underwater Systems Center, New London, Connecticut (1971).

# 16

# Radar Signal Processing

CHARLES EDWARD MUEHE Lincoln Laboratory, Massachusetts Institute of Technology, Lexington, Massachusetts

## 1 INTRODUCTION

The name radar was originally an acronym for "radio detection and ranging." To see how this very limited definition has expanded over the years, consider the list of present-day radar applications in Table 16.1. We cannot hope to describe the signal processing employed in all of these applications but instead will address a short list of generic types of signal processing and refer occasionally to various applications listed in Table 16.1.

This list is conveniently broken into two classes of radars, namely detection-tracking radars and imaging radars. The detection-tracking radars employ filtering or integration to enhance the target, separating it from background clutter, and usually employ some form of automatic thresholding to declare the existence of a target. They often use range, angle, and Doppler measurements of the detected target to form tracks. Imaging radars are employed to form displays of the very background that the detection-tracking radars work to eliminate. The imaging radar is designed to enhance the presentation of images or scenes. Thus synthetic aperture radar (SAR) implements a fine-resolution picture of the earth's surface by employing a wide-bandwidth waveform (for fine range resolution) and a very

This work was sponsored by the Department of the Navy. The views expressed are those of the author and do not reflect the official policy or position of the U.S. government.

**Table 16.1a** Detection-Tracking Radar Functions

| |
|---|
| Air traffic control |
| Airport surveillance radar |
| Enroute radar |
| Air surveillance |
| Airborne |
| Ground-based |
| Ship-based |
| Over-the-Horizon |
| Sea surface surveillance |
| Coastal Defense |
| Ground surveillance |
| Artillery and mortar locators |
| Artillery support |
| Local security |
| Target tracking |
| Air-to-air |
| Air-to-ground |
| Ground-to-air |
| Missile guidance (active, semiactive) |
| Proximity fuzing |
| Range instrumentation |

wide synthetic antenna aperture (for fine angular resolution). Imaging radars often enhance images by integration of multiple looks at the same scene or by various spatial filtering techniques.

One major distinction between the detection-tracking radar and the imaging radar is the size of the target with respect to the radar's resolution. In the former the target is usually smaller than the radar's resolution, whereas the distributed target is much larger than the resolution of the imaging radar.

In the rest of this chapter we will first explore the signal processing employed in detection-tracking radars and then explore some facets of imaging-radar signal processing.

Common to all types of radar are the subjects of radar signal generation; resolution in range, angle, and Doppler; range-Doppler ambiguities; and sampling of radar signals. We discuss these next.

## 1.1 Radar Signals

Radar signals used in radar signal processors are produced by reflecting radar pulses from various objects. A simple radar pulse is generated by modulating a radio

**Table 16.1b** Imaging Radar Functions

| Functions | Resolution | | | |
|---|---|---|---|---|
| | Range | Azimuth | Elevation | Doppler |
| Navigation | | | | |
| Altimetry | × | | | |
| Doppler navigation | | × | × | × |
| Terrain following/avoidance | × | × | | |
| Severe weather avoidance | × | × | × | |
| Airport surface monitoring | × | × | | |
| Marine navigation | × | × | | |
| Hazard warning and scientific functions | | | | |
| Weather monitoring | × | × | × | × |
| Sea surface monitoring | × | × | | × |
| Ornithology and entomology | × | × | × | × |
| Ionospheric studies | × | × | × | × |
| Underground mapping | × | × | × | |
| Range-Doppler imaging | | | | |
| Strip mapping | × | | | × |
| Spotlight imaging | × | | | × |
| Aircraft and satellite imaging | × | | | × |
| Moon and planet surface mapping | × | | | × |
| Topographic mapping | × | | | × |

frequency (RF) carrier so that it has a rectangular envelope. This rectangular pulse is either generated directly using a high-power free-running oscillator (i.e., a magnetron) or generated at a low level and amplified to a high power level. The high-power pulse is then fed into the terminals of an antenna, which radiates the pulse in the form of electromagnetic energy into a narrow beam in space. The pulse is reflected off any object within the antenna beam. The reflected energy is scattered in all directions. Some of the energy falls on and is captured by a receiving antenna (which may be the same antenna used to transmit the pulse). The received pulse is amplified, perhaps heterodyned to a different frequency, and finally enters the signal processor. Signal returns from one or more pulses are utilized by the radar signal processor to detect individual objects (called targets) to measure their ranges and azimuth and elevation angles and, by using the Doppler effect, determine their radial velocities.

The next few sections describe the important properties of radar pulse signals.

## 1.2 Radar Range and Resolution

The range of a target is determined by the round-trip time for the radar pulse to reach the target and return to the receiving antenna. For the usual monostatic radar, employing a single antenna to transmit and receive, the target range is given by $\tau c/2$, where $\tau$ is the round-trip time and $c$ the velocity of light. This range may be measured by observing the location of the target's return with respect to a time base on an analog display or by utilizing a digital counter to measure the round-trip time.

If two targets have nearly the same range, we are interested in the range resolution of the radar. If simple pulses are employed it is clear that the peaks of the pulse returns must be separated in time (or range) to resolve the targets. Thus the resolution is usually defined as the pulse width between 3-dB points on the pulse. When pulse compression is employed the 3-dB width of the compressed pulse is approximately the reciprocal of the bandwidth of the longer uncompressed pulse.

## 1.3 Angular Measurement and Resolution

In an analogous fashion the angular resolution of a radar is simply the 3-dB azimuth and/or elevation beamwidth of the radar. The angular position of a single target is measured either by scanning the antenna past a target and carefully estimating the position of the peak return or by employing a special antenna known as a monopulse antenna [1]. The monopulse antenna employs a conventional beam (called a sum beam) and another beam (called a difference beam) which is antisymmetric about the center of the sum beam, differing in phase by 180° on opposite sides of the centerline. The difference beam is used to measure, on a single pulse, how far the target is from the centerline and on which side. Orthogonal difference beams are utilized to simultaneously measure target azimuth and elevation. Track-while-scan radars measure angular position by utilizing the measured target signal response as the antenna scans by the target. Conically scanned beams can be employed to measure target angular position in both elevation and azimuth.

## 1.4 Target Radial Velocity

A fourth property of a radar target, namely radial velocity, may be measured directly using radar. Indeed, the major application of modern radar signal processing is to resolve targets from each other and from extraneous returns from other objects (called clutter) by filtering in the velocity or Doppler domain. Doppler filtering makes use of the progressive phase shift of an object's return signal on successive pulses. The Doppler effect causes an offset in the carrier frequency of the target's return signal of $2V/\lambda$, where $V$ is the target's radial component of velocity and $\lambda$ is the wavelength of the carrier. Thus the progressive phase shift from pulse to pulse is $4\pi VT/\lambda$, where $T$ is the time between pulses. By measuring the phase shift of the carrier on two successive pulses, the radial velocity can be estimated modulo $\Delta V = \lambda/2T$.

If two targets have nearly the same range, azimuth, and elevation but are moving at slightly different radial velocities, it may be possible to resolve them in velocity. The signal returns from the two objects will add coherently (that is, with due account for their phases) to form the target return for a single pulse. Target returns may be sampled over $N$ pulses and used to form a filter bank (a discrete Fourier transform). The width of each filter in the filter bank is $1/NT$. If the two targets are centered in adjacent filters they are resolved; thus the velocity resolution is $\lambda/2NT$.

## 1.5 Range-Doppler Ambiguities

We observed above that the determination of radar velocity is ambiguous with modulo $\lambda/2T$. Thus if all possible target velocities have a greater spread than $\lambda/2T$ it is necessary somehow to resolve this ambiguity to determine the true target radial velocity. The waveform designer could choose the pulse spacing $T$ small enough so that $\lambda/2T$ spans all possible target velocities. At most microwave frequencies this would imply a very short time (and thus range) between successive pulses and would lead to target range ambiguities. To understand range ambiguities consider the diagram in Figure 16.1. This is a snapshot in time of the range position of various pulses in space whose returns arrive simultaneously at the receiving antenna. We assume the pulse energy is confined to the antenna beamwidth. The return signal is the sum of the returns from all pulses whose return signals reach the receiving antenna at the sample time. In the example shown in Figure 16.1 it cannot be determined whether the target is at $R_1$, $R_2$, or $R_3$, which are spaced by increments of $cT/2$. Thus we say that there is a range ambiguity modulus $\Delta R = cT/2$. Eliminating $T$ between the range and velocity moduli we find:

$$\Delta V \Delta R = \frac{c\lambda}{4} \tag{16.1}$$

In Figure 16.2 is plotted the velocity ambiguity modulus versus the range ambiguity modulus for various wavelengths in the microwave region.

Consideration of the range-velocity ambiguity leads to the definition of three classes of radar that employ Doppler filtering to resolve targets and/or clutter. These are expressed in terms of the pulse repetition frequency (PRF), which is the reciprocal of the interpulse spacing $T$, as follows:

*A low-PRF radar* is unambiguous in target range.
*A high-PRF radar* is unambiguous in target velocity.
*A medium-PRF radar* is ambiguous in both target range and target velocity.

Radars with short enough target range and/or low enough frequency are simultaneously both low- and high-PRF radars. In other words, they are unambiguous in both target range and velocity.

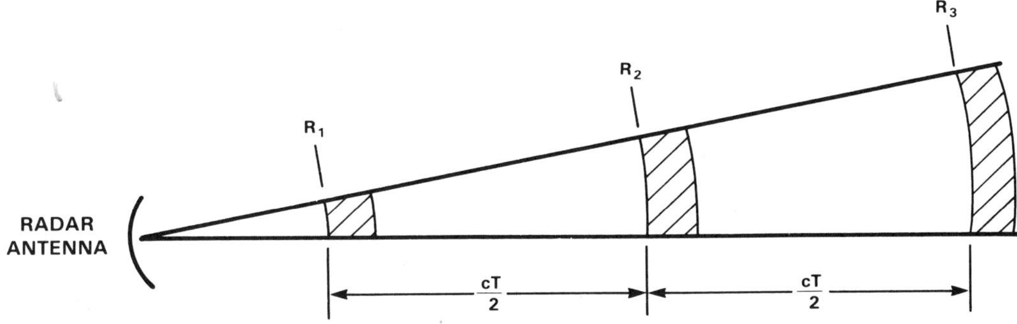


**Figure 16.1** Instantaneous range responses of a radar with interpulse spacing $T$.

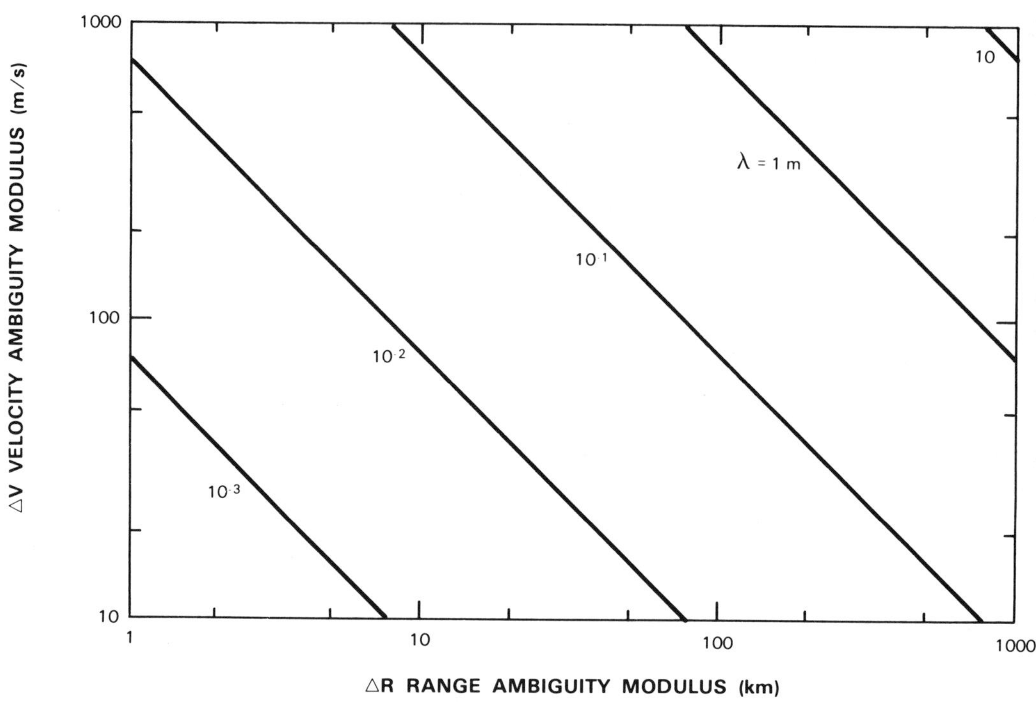


**Figure 16.2** Velocity ambiguity modulus versus range ambiguity modulus for various wavelengths.

The range-velocity ambiguity plays a large part in the choice of radar waveform for any radar employing Doppler filtering. We will consider it again in the design of airborne and synthetic aperture radars.

## 1.6 Sampling the Radar Signal

Most radar signals consist of a carrier on which is imposed some modulation. Its information content lies in the modulation. Analytically, a radar signal can be expressed as

$$s(t) = A(t)\cos[\omega_c t + \varphi(t)] \tag{16.2}$$

where $\omega_c$ is the carrier radian frequency and $A(t)$ and $\varphi(t)$ are the amplitude and phase of a complex modulation. When this signal is reflected from a complex moving target, the return signal is the sum of the returns from the various scattering centers with the amplitude of the incident wave multiplied by that of each scattering center, the phase added, and the time delay adjusted according to the range of each scattering center. For a single point scatterer the reflected signal is

$$s_1(t) = A(t - \tau_1)R_1 \cos[(\omega_c + \omega_{d1})(t - \tau_1) + \varphi(t - \tau_1) + \varphi_1] \tag{16.3}$$

where $R_1$ and $\varphi_1$ are the amplitude and phase of the reflector, $\tau_1$ is its round-trip delay time, and $\omega_{d1}$ is its Doppler radian frequency.

The Doppler modulation frequency is usually small compared to the bandwidth of the modulation, which is also usually a small fraction of the carrier frequency. With this narrowband assumption, by using linear mixers, the signal modulation may be translated to any convenient carrier frequency sufficiently greater than the modulation bandwidth and the signal processing performed there. For instance, acoustic delay lines with delays exactly equal to the interpulse intervals may be used to implement a transversal filter.

Most modern applications (digital and charge-coupled device, CCD) employ filtering of baseband signals which have zero carrier frequency. To preserve the complex nature of the modulation, the signals are mixed in two separate mixers with local oscillator signals $\cos\omega_c t$ and $\sin\omega_c t$. The output of these mixers, filtered to remove higher-frequency mixer products, are equal to the real and imaginary parts of the complex modulation including any Doppler offset. The complex modulation has both positive and negative frequency components and the spectrum is usually not symmetrical about zero. The outputs of these mixers are called in-phase and quadrature videos.

If analog processing is to be employed, the in-phase and quadrature video outputs are sampled at ranges corresponding to multiples of the range resolution and the signals are held constant for the next interpulse period. The output of each range cell is separately filtered to enhance the signal-to-interference ratio.

For digital signal processors the in-phase and quadrature video are sampled at time intervals equal to or shorter than that corresponding to the range resolution and are immediately converted into digital words, which are stored in digital memory for processing.

It has been found that amplitude and phase unbalance in the quadrature video mixers causes a Doppler imaging problem. Small replicas of positive and negative spectral components appear folded about zero Doppler. It is very difficult, if not impossible, to reduce these replicas below about −30 to −40 dB. An alternative method [2] is to translate the signal to a carrier frequency of somewhat more than half the signal bandwidth and sample this single channel at four times the carrier frequency. A very precise digital filter is then employed to simultaneously translate to baseband and generate the in-phase and quadrature video digital samples. Using this method, the spectral folding is reduced to negligible levels.

## 2 DETECTION-TRACKING RADARS

Figure 16.3 is a functional diagram of a typical modern detection-tracking radar employing digital processing. In this example waveform generation takes place at the intermediate frequency (IF) and is translated to RF using a stable local oscillator (STALO). After passing through the circulator to the antenna, the signal is radiated from the antenna. On reflection from the target, the signal enters the receiver and is converted back to IF. Next it is converted to baseband using a coherent oscillator (COHO) at the IF and digitized with the analog-to-digital converters as explained in the preceding section.

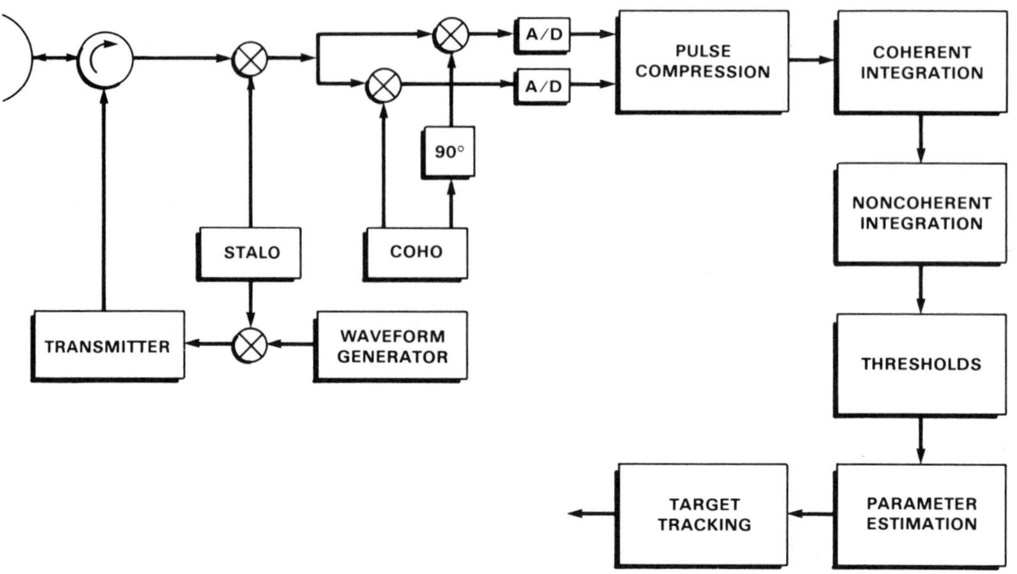


**Figure 16.3** Functional signal flow diagram of a typical modern detection-tracking radar.

Signal processing typically requires some or all of the six steps indicated in the functional blocks after the analog-to-digital converter. Any particular radar may perform all or some of these functions using analog signal processing. In the following sections we will describe how several of these functions are implemented.

The principal role of pulse compression is to allow higher pulse energy without exceedingly high peak power. Pulse compression may be accomplished by either analog or digital methods.

Early radars transmitted a simple rectangular RF pulse and utilized unscanned or mechanically scanned antennas to direct a beam of energy toward the radar target. The signal reflected from the target was collected by the same or a similar receiving antenna, amplified, envelope-detected, and displayed on an oscilloscope. Radar operators interpreted the scope displays detecting and tracking targets. The operators visually filtered targets from any background clutter present. Tens of thousands of these early radars were built during World War II and contributed heavily to its outcome.

The types of signal processing that can be found in modern radars have been developed over the several decades since World War II.

In the following sections we review the basic types of processing utilized in a modern detection-tracking radar. Following this review we present a sample of a complete radar signal processing system and discuss the types of results obtained.

## 2.1 Noncoherent Integration

A radar operator examining the plan-position indicator (PPI) scope of an early air search radar sees an increase in intensity of the scope brightness as the antenna scans past an aircraft. A short bright circular arc appears at the range of the aircraft. The bright arc is made up of a number of bright dots caused by the aircraft's signal return on successive pulses. The radar operator declares a target based on the brightness and shape of this circular arc. The operator's brain is, in effect, noncoherently integrating the total return [3] over the antenna dwell.

The first radar signal processors performed this noncoherent integration using electronic circuits. The video return resulting from envelope detection is sampled by sample-and-hold circuits, one for each time interval (called a range gate) equal to the radar's pulse length. The output of each range gate is passed through an integrator (low-pass filter) whose time constant is chosen to match the antenna dwell time on the target. The output of this noncoherent integrator is then examined by a thresholding device (see next section), alerting the operator to the presence of a target.

The theory of detection utilizing noncoherent integration was studied at an early date and gives rise to the Marcum [4] and Swerling [5] target models so often referred to in the radar literature. Marcum studied detection statistics using noncoherent integration of a steady nonfluctuating target, and Swerling extended this analysis to several types of targets whose radar cross sections vary according to various probability models and at different rates. The results can be found in several books on detection theory [6]. Algorithms for computer calculation of detection statistics are also available [7].

## 2.2 Automatic Detection (Thresholding)

Once the target signal has been integrated to enhance its strength, additional circuitry may be added to alert the operator to the target's presence. If the integrated signal exceeds a threshold a target detection is declared.

To establish the threshold, more signal processing is required. The goal is usually to provide a constant false alarm rate (CFAR). The residual noise is not constant. It consists of receiver noise plus any outside interference such as clutter signals (see next section). The usual CFAR threshold is generated as a multiple of the sum of the noise output of groups of range cells on either side of the cell to be thresholded. This is called a mean-level threshold since it utilizes an estimate of the mean noise. Because of the multiplication by a constant, in many analog CFAR detectors the signals are first converted to their logarithms. Then the sums of the logarithms of the noise output of groups of range cells are added to establish a basis for thresholding. This is called a logarithmic CFAR. It introduces extra CFAR loss and is not required when digital mean-level thresholding is employed.

One of the principal reasons for introducing an automatic detector is the great reduction in information rate this provides. For a typical surveillance radar the data rate before detection is on the order of many megabits per second. It is reduced to the kilobit-per-second range after thresholding. Groups of contiguous threshold crossings can be converted into a range-azimuth target report. The total useful output of a surveillance radar can then be transmitted over a single telephone line to some remote site.

The first application of digital technology to radar signal processing [8] was to noncoherently integrate, threshold, and convert the signals from surveillance radars to low-bandwidth digital target reports. This device, called a common digitizer (because it converts both radar and IFF signals), was originally utilized on the semi-automatic ground environment (SAGE) system for netting together widely spaced surveillance radars; it is still used on similar military systems and by the Federal Aviation Administration (FAA) to net together the system of enroute radars used for air traffic control throughout the United States.

Because it was designed in the 1950s, when digital technology was in its infancy, the common digitizer used every method possible to reduce the number of bits to be handled. The video pulse returns are first converted in each range gate into a zero or one (1-bit digitization) using the first threshold adaptively set to produce a threshold crossing rate on noise alone of about 1%. Next, in each range gate the output of the first threshold is summed (noncoherently integrated) in an azimuth window corresponding to the number of pulses in the antenna dwell time. Finally, a second threshold is applied to the total count in the sliding azimuth window. When the second threshold is passed, a target leading edge is declared. The target azimuth is declared near the center of the leading and trailing edges.

The performance of this type of detector has been the subject of a great deal of analysis. It has received many names: sliding-window detector, $M$-out-of-$N$ detector, coincidence detector, binary integrator, and double-threshold detector [9].

## 2.3 Coherent Integration

As described earlier, coherent integration or Doppler filtering is employed to separate targets from one another or to separate targets from clutter. Two forms of coherent integration have evolved and are commonly called moving target indicator (MTI) and "filter-bank" processing. MTI employs a clutter elimination filter usually followed by noncoherent integration and thresholding. Filter-bank processors utilize a Doppler filter bank in each range cell followed directly by thresholding. We will describe these two approaches in the next two sections.

First, we will describe briefly the nature of clutter which must be overcome by the Doppler filtering and the improvement factor, $I$, used as a measure of success.

There are six major sources of radar clutter. Their properties are listed in Table 16.2. Most will be familiar to the reader.

**Table 16.2** Major Sources of Clutter Signals

| Type | Mean velocity (m/s) | Velocity spread (m/s) | Source of information (references) |
|---|---|---|---|
| Land | 0 | 0.7 | Nathanson [49] |
| Sea | 0–3 | 1.5 | Nathanson [49], Watts [50] |
| Rain | 0–20 | 1–4[a] | Nathanson [49] |
| Aurora | 200–700 | 400 | Blevis [51], Chesnut et al. [52] |
| Chaff | 0–30 | 2–6[a] | Nathanson [49], Butters [53] |
| Birds | 0–30 | 4[b] | Eastwood [33], Vaughn [34] |

[a] Assumes a pencil beam antenna with 5° vertical bandwidth. Will be larger for a vertical fan beam.
[b] For migrating birds.

Aurora refers to reflection from conducting layers in the ionosphere. The strongest reflections are from the E layer, at approximately 110 km in altitude. Strong reflections only occur from the auroral ovals surrounding the north or south magnetic poles and only when the radar beam is pointing nearly perpendicular to the magnetic lines. These conditions occur only in far northern or southern latitudes in radar sectors facing the magnetic poles. Strong reflections occur in the frequency range up to 1000 MHz, dropping off strongly at the higher end of this frequency range.

Chaff is a human-made form of clutter consisting of very lightweight metal or metallized strips cut in length to resonate at frequencies of interest. Chaff and land usually exhibit the strongest backscatter returns.

The measure of effectiveness of a clutter filter is the improvement factor $I$, which in general is a function of target Doppler. It is given by the expression

$$I = \frac{S_o/(C_o + N_o)}{S_i/(C_i + N_i)} \tag{16.4}$$

where $S$ is the desired signal, $C$ the clutter signal, and $N$ the noise signal. The subscripts $i$ and $o$ refer to signals at the input and output of the clutter filter.

The mean improvement factor $I$, averaged over the PRF interval, is employed to describe the effectiveness of an MTI filter which is designed to minimize the clutter signal without regard to the Doppler of the desired signal.

## 2.4 Moving Target Indicator (MTI)

Even before the end of World War II work was in process to provide a moving target indicator, a PPI scope that presents signals from moving aircraft but not from fixed ground clutter. An area MTI approach was first tried in which the amplitude of the ground clutter video was first mapped on one scan and the returns from the next scan subtracted in the hope of eliminating the ground clutter. This did not provide enough suppression of ground clutter because the ground clutter return signals fluctuate from scan to scan of the radar. This amplitude fluctuates because of the motion of individual reflectors within a range gate. The signals from many individual reflectors add coherently (with due account for phase) at the radar input. Even a slight motion of the reflector by a fraction of a wavelength causes large difference in the amplitude of the coherent sum.

This led to investigating methods for subtracting the coherent returns from successive pulses in each range gate [10]. This was achieved using an acoustic delay line of length exactly equal to the interpulse period and subtracting the output IF signal from the input IF signal coherently. The ground clutter signal is largely canceled, while a moving target signal changes in phase during the interpulse period and does not cancel from pulse to pulse. An example of the performance of a two-pulse canceller is shown in Figure 16.4. Note the blind speeds at zero and multiples of the PRF. Also note the uneven response. The repetition of the same response at intervals of the PRF is due to sampling at too low a rate compared to the maximum target Doppler (see Section 1.5). The response may be smoothed somewhat and the blind speed notches reduced by summing more delayed pulses and staggering the pulse repetition intervals (PRIs).

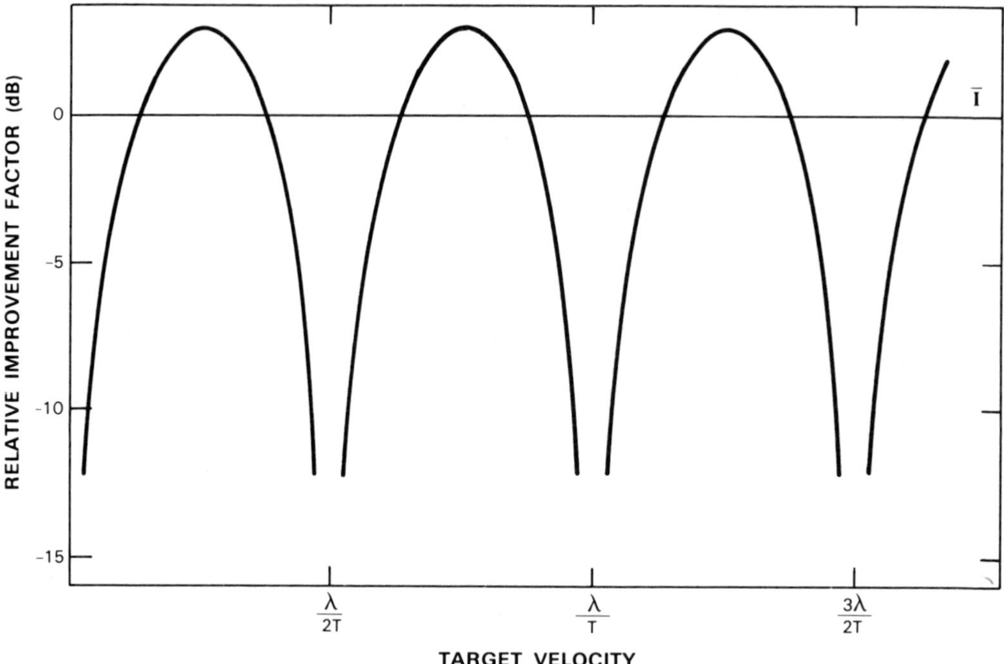


**Figure 16.4** Improvement factor of a two-pulse canceller as a function of target velocity, normalized to the mean improvement factor.

The early MTI processors utilized mercury or quartz analog delay lines, and because of thermal affects it was difficult to keep their lengths adjusted correctly for more than about two delay lines. Recent MTI processors employ very stable digital sampling and digital delay lines and do not experience this difficulty.

Use of staggered PRIs does, however, cause difficulty with so-called second-time-around clutter. This is clutter which is simultaneously illuminated by the last and second-to-last pulses transmitted (see Figure 16.1). The staggering causes the second-to-last-pulse returns to change relative position on successive pulses so that two successive range gate returns, with ranges based on the last pulse transmitted, will have added to them the returns from the second-to-last pulses at different ranges. These will not be the same ranges and thus their clutter returns will not cancel.

MTI processors also experience difficulty rejecting clutter other than zero-velocity land clutter, for which they are usually designed. A second adaptive canceller may be cascaded with the first one to reject rain clutter whose Doppler frequency is determined by the wind field within the antenna beam. Virtually no Doppler discrimination is available to aid in bird rejection, so the target detection across all velocity regions suffers when the gain is reduced or the thresholds raised in the presence of bird clutter.

Finally, MTI processors lose the gain achievable using coherent processing. They form clutter rejection filters utilizing fewer pulses than exist in a beamwidth, so it is common practice to noncoherently integrate the output of the MTI filter over the antenna dwell to buy back some of the loss in gain. If, for automatic detection, the MTI filters are followed by mean-level thresholds and if all of the clutter has not been reduced to receiver noise level, the thresholding will experience a large increase in false alarms due to the so-called azimuth correlation problem. Any residual clutter at the output of the MTI filter will be highly correlated from pulse to pulse [11]. Complete azimuth decorrelation on successive pulses is assumed when the mean-level threshold calculation is made. The existence of azimuth correlation causes a low threshold to be applied with a large increase in false alarms. The pulse-to-pulse correlation properties of the MTI filter output can be measured and the threshold raised appropriately [12]. Unfortunately, this procedure further desensitizes the radar. The azimuth correlation problem is experienced even in L-band long-range surveillance radars in heavy rain. Provision is made in the common digitizer described earlier for operator desensitization of the radar coverage sectors where it occurs. If not desensitized, the false alarms completely overload the phone line digital links to the FAA enroute centers.

The theory of MTI design for any number of pulses shows that to maximize the average improvement factor

$$\bar{I} = \frac{W^T W^*}{W^T C W^*} \tag{16.5}$$

(where $T$ denotes transpose and the asterisk denotes conjugate), one chooses the weight vector $W$ equal to the eigenvector corresponding to the smallest eigenvalue of the interference (clutter plus noise) covariance matrix $C$. These weights whiten, as nearly as possible, the interference spectrum. Performance results as a function of the interference spectrum shown in Hsiao [13].

## 2.5 Filter-Bank Processing

Unlike MTI filtering, which minimizes the clutter return by whitening, as nearly as possible, the clutter plus noise spectrum, filter-bank processing employs a filter bank in each range gate. Each filter is optimized to maximize the target-to-interference (clutter plus noise) ratio for a target within the filter bandwidth [14, 15].

For a deterministic signal whose vector over $n$ samples is $S$, and when the $n \times n$ interference covariance matrix is $C$, the optimum filter weight vector is

$$W = kC^{-1}S^* \tag{16.6}$$

where $k$ is an arbitrary constant. If the interference is white noise, $C$ is a diagonal matrix and the weights correspond to a so-called matched filter. In general, the optimum processor utilizes a set of filters optimized for signals with Doppler frequencies spaced across the PRF interval. Usually the number of filters employed equals the number of complex signal samples available. Implementation of the filter bank may compromise slightly the optimality of the filter bank by employing a clutter reject filter followed by a fast Fourier transform. The shape of the clutter filter is designed so that the cascade of the two provides a nearly optimum set of clutter filters.

Besides providing near-optimum filters, the typical filter-bank radar must deal with range and/or Doppler ambiguous targets as well as nearly blind speeds in the Doppler region of the heavy clutter and/or blind range intervals due to eclipsing of the receiver when the transmitter is turned on. These difficulties are typically overcome in a filter-bank radar by employing multiple PRFs [16]. The antenna dwell time is divided into coherent processing intervals (CPIs) during which pulse trains of different PRFs are transmitted. Detection is required on $M$ out of the $N$ PRFs per antenna dwell. This $M$-out-of-$N$ detection is a form of noncoherent integration which somewhat makes up for the energy loss in dividing the total dwell-time energy among $N$ CPIs. The PRFs are chosen in the ratio of relatively prime numbers such that adequate unambiguous range or Doppler is achieved on the $M$ detections. Care must also be taken that ghost production (caused by incorrect association of threshold crossings on successive CPIs when two or more targets are present) is very small. This usually calls for choice of the relatively prime numbers as low as possible while still meeting the adequate unambiguous range or Doppler requirement.

Airborne or spaceborne air surveillance radars present perhaps the most stressing detection-in-clutter problem [17, 18]. Solutions have been found in the form of high-PRF S-band radars or low-PRF ultrahigh frequency (UHF) radars. Many fighter aircraft employ medium- or high-PRF X-band radars. This demonstrates the diversity of solutions to this difficult problem.

## 3 MOVING TARGET DETECTOR

As an example of a modern radar digital signal processor we discuss the moving target detector (MTD). This processor brings together many of the optimum filtering and thresholding techniques discussed in the previous sections.

The moving target detector is a relatively new class of digital radar signal processor which would have been virtually impossible to construct before the availability of modern digital integrated circuits [19]. The first model was designed and built in the mid-1970s, when medium-scale integrated circuits were available.

The immediate impetus for the MTD's development was the FAA's computer automation in the early 1970s of its terminal air traffic control systems at many airports around the country with the Automated Radar Terminal System III (ARTS-III). The Air Traffic Control Radar Beacon System (ATCRBS) portion of the system, which employs interrogator-responder beacons, was easily automated, but there was no automated output from the primary Airport Surveillance Radar (ASR). See Table 16.3 for characteristics of the ASR. As a result, it was necessary to display primary radar video directly on the controllers' display. It was hoped to generate digital target position reports which could be used to automatically initiate and upgrade radar tracks, particularly of small aircraft not equipped with ATCRBS transponders. Without automatic tracking using primary radar the controllers must manually track these aircraft, and their tracks cannot be employed in automatic conflict detection algorithms.

**Table 16.3** Airport Surveillance Radar, Typical Parameters

| | |
|---|---|
| Transmitter | |
| Type | Klystron |
| Frequency | 2700–2900 MHz |
| Peak power | 500 kW |
| Pulse width | 1 $\mu$s |
| PRF | 1000–1200 Hz |
| Antenna | |
| Type | Parabolic reflector |
| Azimuth beamwidth | 1.4° |
| Elevation pattern | cosec squared to 60 nmi and 25,000 ft |
| Peak gain | 34 dB |
| Rotation period | 4–5 seconds |

The primary reason for this failure to automate the system was the heavy load of false reports generated by clutter signals in the target digitizer and the poor detection when the threshold parameters were adjusted to correct the clutter false alarms. The digitizer was a modern version of the sliding-window common digitizer described earlier.

After some study [20] it was determined that the objectionable false alarms were generated by the processor's handling of signals reflected from ground, rain, birds, surface vehicles, and second-time-around echoes from ground and rain.

To overcome all of these types of clutter simultaneously, the following features are included in the MTD signal waveform and signal processor [21].

1. *Multiple PRF waveform.* The waveform consists of groups of nine pulses occupying a coherent processing interval. Each CPI utilizes a constant PRF chosen from two alternating PRFs approximately 20% apart. Two CPIs fit within an antenna dwell.

2. *Precursor pulse.* On each CPI eight filters are digitally constructed using the signal samples received on the last eight pulses. The first pulse is utilized as a fill pulse so that second-time-around clutter echoes are present on all pulses used to construct filters and are thus eliminated just as well as last-pulse clutter.

3. *Optimal coherent integration.* The eight filters are designed to optimize the signal-to-interference ration (Equation 16.6) at eight Doppler frequencies spread from zero to the PRF. Interference consists of receiver noise plus ground and rain clutter signals.

4. *Mean-level thresholds.* For filters other than the zero-velocity filter, mean-level thresholds are calculated by averaging the interference over approximately a mile in range centered on the cell to be thresholded, excluding that cell and contiguous cells on either side.

5. *Clutter Map.* To combat ground clutter, which has a very uneven spatial distribution, a threshold for the zero-velocity filter and a secondary threshold for

cells contiguous to the zero-velocity filter are generated using a clutter map. This map is generated utilizing the zero-velocity filter output in each range-azimuth cell. A fading memory integrator (1/8 new data added to 7/8 existing data) is employed to update the clutter map on each radar scan.

6. *Report association and interpolation.* A valid target report is generated when at least one threshold crossing exists. When multiple threshold crossings appear contiguous in range, Doppler, or CPI they are combined and interpolated to derive a target report with more accurate range, azimuth, and amplitude.

7. *Bird thresholding.* An area thresholding scheme is employed to limit the number of target reports (principally from birds) entering the tracker.

8. *Tracker.* The tracker is a simple $\alpha - \beta$ tracker with update gains and association areas chosen to match those predicted by a Kalman filter for a target with random accelerations of about 1 $G$. Three detections are required to initiate a track and three misses to drop a track.

Besides the above features, the MTD provides censoring, on a range cell basis, of CPIs that contain signals which overload the analog-to-digital converters or in which one sample is about 15 dB larger than the average during the CPI (this eliminates interfering pulses). The MTD also censors fixed points with extremely high clutter and roads carrying ground vehicles that cause false alarms. This censoring never deletes more than about 1% of the entire coverage.

Two versions of the MTD were built and tested. In the first, all of the logic was hardwired [22]. The second MTD [23, 24], which was microprogrammable [25], was designed for easier maintenance and logistics. Both were tested extensively by the FAA [26].

Figure 16.5 shows the performance of MTD-1 in rain. A small controlled aircraft (Piper Arrow) located in the region from 200 to 220° has made half an elliptical pattern in heavy rain. The probability of detection was 94%. Notice that no dropout occurred when the aircraft flew tangential to the radar (with zero radial velocity).

Figure 16.6 shows nine scans on a digital display of targets in track on MTD-II, which was located at Lexington, Massachusetts. Boston's Logan Airport is about 13 miles east-southeast of the radar. Aircraft 52 has just taken off and aircraft 32 is on its landing approach. As far as we know, every numbered track is a true aircraft.

The simple $\alpha - \beta$ tracker was never overloaded, nor did it produce more than about one short false track per minute when tracking 80 targets within a 60-mile radius. Analysis of several tracks showed the MTD to have an average range error of 0.022 nautical miles and an average angular error of 0.14°. This was compared with a sliding-window detector which produced errors of 0.03 nmi and 0.22°.

Because of the success of these extensive tests, the most recent airport surveillance radar (ASR-9) specifications call for a processor modeled after the MTD [27].

## 4 IMAGING RADARS

The functions performed by imaging radars are quite varied. As emphasized in Table 16.1(b), the various functions require resolution in one or more of the four possible types of radar measurements. The target for imaging radar often consists

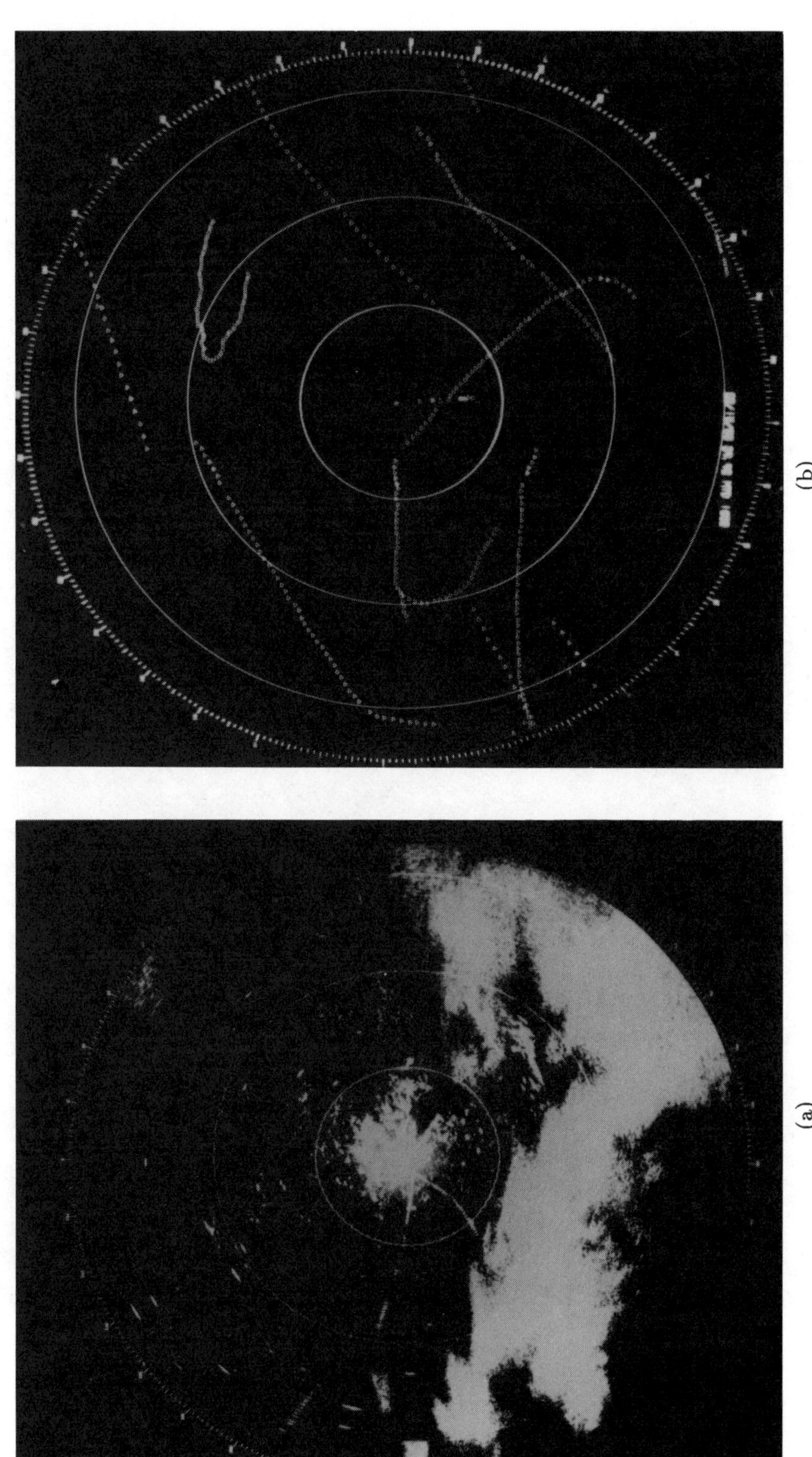


**Figure 16.5** MTD-I performance in rain. (a) Normal video showing rain structure and ground clutter. (b) MTD output tracks for 51 scans. Piper Arrow is flying at about 10 miles and 220° in heavy rain. Five-mile range rings are employed. (a) and (b) were taken about 20 minutes apart. (From Rabinowitz et al. [19].)

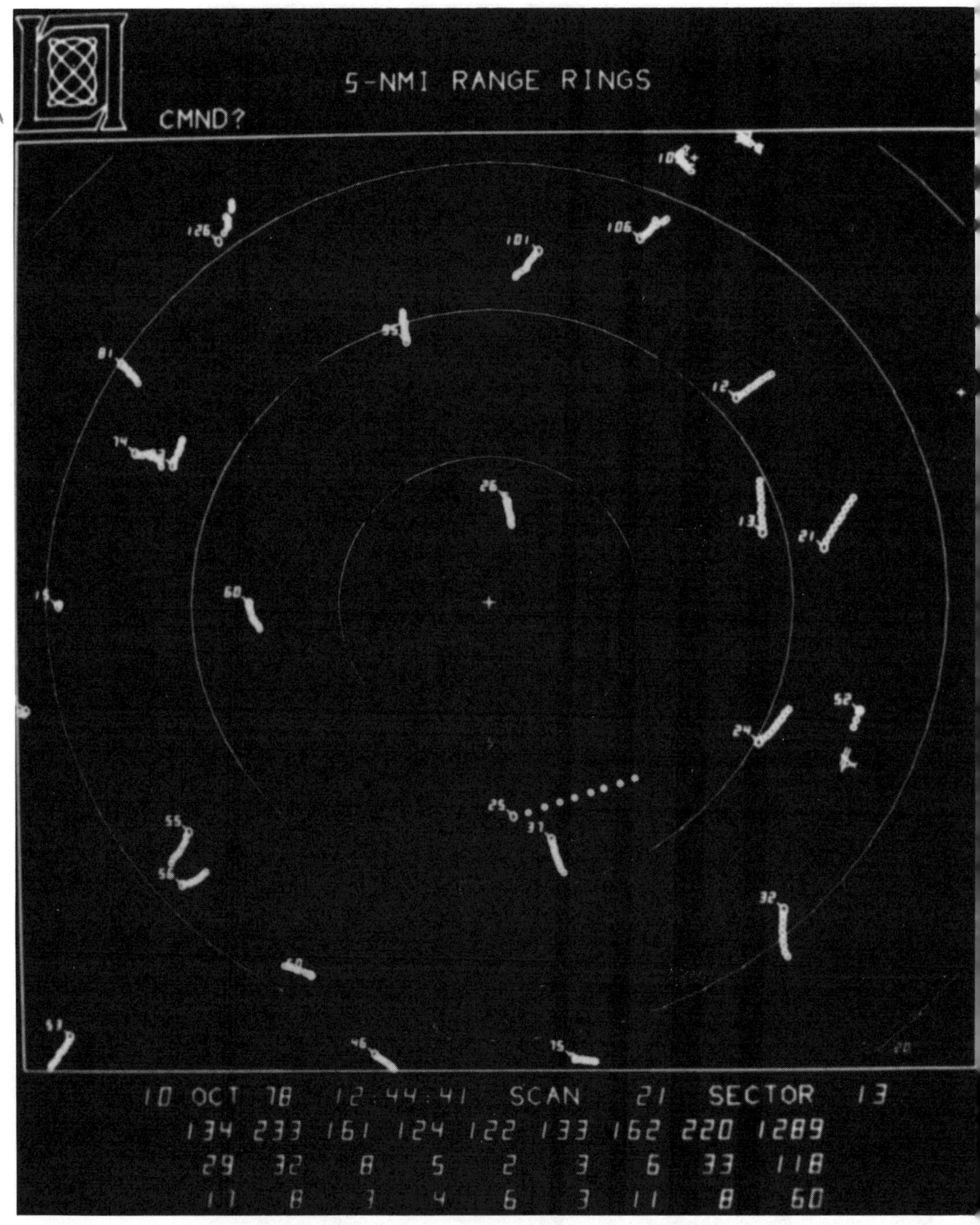


**Figure 16.6** Photograph of MTD-II maintenance display. The numbers at the bottom give the threshold crossings in each filter on the last scan and the number of target reports sent to the tracker for close-in and outer range intervals along with their overall totals. (From Rabinowitz et al. [19].)

of the clutter which is suppressed in detection-tracking radars. Resolution is important in imaging radars because the object to be imaged forms a continuum in radar measurement space. The finer the resolution the more detailed is the information in the image. For each function listed in Table 16.1(b) there exists a certain required resolution to provide an image sharp enough to adequately perform the function.

## 4.1 Navigation Radars

The first five applications concern the navigation and control of aircraft.

Altimetry [28] is perhaps the simplest function listed. Broadbeam transmitting and receiving antennas are employed to form a one-dimensional image (range) of the surface of the earth below an aircraft. Both pulse and continuous-wave (CW) radars have been used. The CW radar is modulated with a linear FM sawtooth modulation which is beat with the received signal to produce a tone whose frequency is proportional to altitude.

The airborne Doppler navigator [28] utilizes four narrowbeam antennas pointing downward in the four quadrants at an angle of about 45° away from the vertical. These antennas receive a reflected CW signal which has been Doppler-modulated differently in each quadrant, depending on the forward and sideways motion of the aircraft. The output Doppler measures the true ground speed of the aircraft and can be used to perform dead reckoning or as an input to an inertial navigator.

Terrain following and avoidance radars employ scanning beam antennas pointing in the forward direction and at low depression angles. The surface of the ground ahead of the aircraft is imaged and the information employed to control the low-altitude aircraft in pitch so as to avoid collision with the ground.

Almost all commercial aircraft carry a nose-mounted radar which is employed to generate a range-azimuth image of rain intensity in the forward quadrant. This presentation is employed by the pilot to steer around areas of severe turbulence.

Finally, each large airport employs a millimeter-wave radar to image the surface of the airport [29]. The resolution is sufficient to show the size and orientation of each aircraft as it taxies about the airport. It also images any other vehicles present.

Another imaging radar, this time ship-mounted, is applied to marine navigation [30]. It produces a range-azimuth map of the shoreline, other ships, and navigation buoys.

## 4.2 Hazard Warning and Scientific Functions

The next group of radars in Table 16.1(b) are employed to provide information useful (and sometimes essential) to carrying out commerce, to give warning of hazardous conditions, or to provide data for scientific studies.

Weather radars, both airborne and ground-based, may use all four forms of radar resolution to locate and map out the severity of storms [31]. Doppler resolution allows mapping of the wind fields within the storms, warning of areas of high turbulence [32], and locating tornadoes.

Over-the-horizon radars reflect high-frequency (3 to 30 MHz) electromagnetic waves from the ionosphere to great distances. By analysis of the Doppler of the reflections from the ocean it is possible to estimate the wave lengths and directions of surface waves and thus infer wind conditions over vast areas.

Ornithologists regularly employ pencil beam radars to follow bird movements [33, 34]. At shorter ranges single birds can be tracked. At longer ranges bird flocks can be counted and estimates made of their number and velocity. Most of the detailed information on worldwide bird migrations has been produced with radar. Besides mapping weather and bird movements, very sensitive pencil beam radars can actually track insects and measure turbulence in the atmosphere.

Lower-frequency radars (UHF and below) are employed to probe the ionosphere and have helped to provide a better understanding of the physical mechanisms which account for its structure.

In another direction, radar has been employed to examine underground structures [35], to locate buried pipes and other objects, to find tunnels, and to investigate natural layered structures, particularly in ice. These radars often utilize a short wide-bandwidth video pulse transmitted into the surface. The antenna is moved along the surface and the return intensity versus depth on each pulse is plotted on a strip chart and used to interpret the underground structure.

Airborne earth-mapping radars are built in two forms, SLAR and SAR. The side-looking airborne radar (SLAR) utilizes a high microwave frequency (perhaps 1.5 cm in wavelength) and a long antenna looking out the side of the aircraft. It is not difficult to achieve 0.25° azimuth beamwidth. The pulse returns are noncoherently integrated over the time it takes for the aircraft to fly one beamwidth and are recorded as strips of intensity modulation across a film strip. The processed film then contains an image of a strip of ground along the flight path. The resolution is sufficient to discern major features such as roads. If a second channel is provided with a simple MTI filter and recorded to the same scale, the background is suppressed and one can then observe moving vehicles on the roads.

To improve the resolution provided by airborne mapping a new kind of radar was invented. It is called synthetic aperture radar (SAR) and is described in the next section.

## 4.3 Range-Doppler Imaging Radar [36]

Performance of all the imaging radars discussed in the preceding section depends heavily on the attainable resolution. Range resolution of a particular range is independent of range and depends only on the bandwidth of the signal. Cross-range resolution depends on the angular width of the radar's antenna pattern and degrades with increasing distance between radar and target. Range-Doppler imaging overcomes the cross-range limits to resolution set by the antenna beamwidth and can provide very fine two-dimensional resolution (in range and cross-range) at very long ranges.

Doppler resolution implies that various points on the imaged object have relative motion with respect to the radar. Following Walker [37], consider an object rotating about a point at a distance $r_a$ from the radar (see Figure 16.7). The range to the point $(x_0, y_0, z_0)$ on the object is given by

$$r = [r_0^2 + r_a^2 + 2r_a r_0 \sin(\theta_0 + \omega t) z_0^2]^{1/2}$$

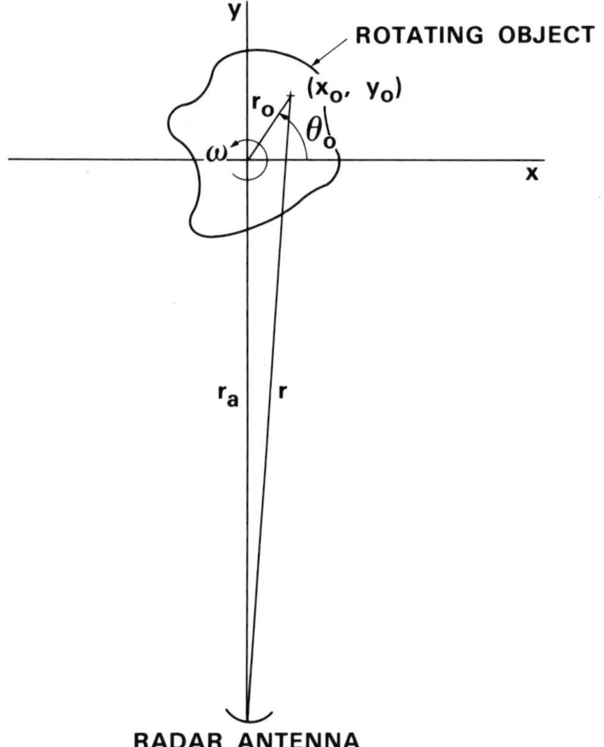


**Figure 16.7** Geometry of rotating object with respect to radar antenna.

where $z$ is the coordinate perpendicular to the paper in Figure 16.7. Making the assumption that the size of the object is very small compared to the distance $(r_a \gg r_0, z_0)$, the above equation may be written

$$r = r_a + x_0 \sin \omega t + y_0 \cos \omega t \tag{16.7}$$

The Doppler frequency from this point is

$$f_d = \frac{2}{\lambda}\frac{dr}{dt} = \frac{2x_0\omega}{\lambda} \cos \omega t - \frac{2y_0\omega}{\lambda} \sin \omega t \tag{16.8}$$

If the processing is performed over a small time interval starting arbitrarily at $t = 0$,

$$r = r_a + y_0 \tag{16.9}$$

$$f_d = \frac{2x_0\omega}{\lambda} \tag{16.10}$$

It is easy to see that at long range the surfaces of constant range and constant Doppler are orthogonal. They are perpendicular and parallel to the line of sight, respectively. The range resolution is

$$\rho_r = \frac{c}{2B} \tag{16.11}$$

where $c$ is the velocity of light and $B$ is the signal bandwidth. The cross-range resolution is

$$\rho_a = \Delta x_0 = \frac{\lambda \Delta f_d}{2\omega} = \frac{\lambda}{2\omega T_c} = \frac{\lambda}{2\Delta\theta} \tag{16.12}$$

where we have employed Eq. 16.10 and the fact that the Doppler resolution is equal to the reciprocal of the coherent integration time $T_c$. $\Delta\theta$ is the angle through which the object must rotate about an axis perpendicular to the line of sight during the coherent integration period $T_c$.

Range and cross-range resolution will be limited by migration of a point on the periphery of the object being imaged through range or Doppler resolution cells during a coherent integration time. This limit is

$$\rho_a^2 > \frac{\lambda D_r}{4} \tag{16.13}$$

$$\rho_a \rho_r > \frac{\lambda D_a}{4} \tag{16.14}$$

where $D_r$ and $D_a$ are, respectively, the maximum range and cross-range dimensions of the imaged object. Several techniques of compensating for motion through resolution cells, allowing finer resolution, have been developed.

Returning again to Table 16.1(b), stripmap SAR [38, 39] utilizes the linear motion of an aircraft with respect to the earth to image a strip parallel to its flight path. An antenna with beamwidth $\Delta\theta$ pointing perpendicular to the flight path will provide a cross-range resolution

$$\rho_a = \frac{\lambda}{2\Delta\theta} = \frac{L}{2} \tag{16.15}$$

according to Equation (16.12). Here, $L$ is the length of the antenna utilized and the beamwidth is assumed to be $\theta = \lambda/L$. Stripmap resolution may be limited by available motion compensation techniques which are required because of the somewhat irregular flight path caused by air turbulence. Atmospheric turbulence may also cause phase fluctuations in the returned signal and thus limit resolution.

The conventional side-looking SAR has been extended to satellite-borne SAR [40]. In satellite-borne SAR the ranges are great and the satellite has a much larger velocity with respect to the earth than does aircraft-borne SAR. The result is severe migration of the imaged points through range and Doppler cells. Correction for both types of cell migration can be accomplished by a two-dimensional correlation of the received signals with a replica of the expected return from a fixed point in each resolution cell.

In spotlight SAR [41, 42] the airborne radar images a spot on the ground. The spot need not be broadside to the flight path. The antenna is continuously pointed at the spot. The situation is the same as that depicted in Figure 16.7, so the same analysis applies. The only difference is that the range to the center of the spot is changing slowly with time and the relative angular rotation of the spot about the line of sight is not perfectly uniform. These restrictions may be compensated so that the best resolution possible is the same as that given by Eqs. (16.11) and (16.12).

The spotlight mode of SAR may be turned around [43, 44] with the relative motion caused by the changing line-of-sight angle of an aircraft or satellite as it passes a fixed ground-based radar. This mode is sometimes called inverse SAR (ISAR).

Equations (16.11) and (16.12) indicate that very fine resolution may be achieved. For instance, by utilizing an X-band radar (10 GHz) with 10% bandwidth, a range resolution of 15 cm is indicated. The object need only rotate 6° about the radar line of sight to produce a comparable cross-range resolution. According to Eqs. (16.13) and (16.14), if the object is larger than 3 m in extent the processing must compensate for range and Doppler cell migration. Otherwise the processing is fairly simple, consisting of estimating the rotation rate of the target to scale the cross-range dimension, accurately removing the effect of translational velocity in range, and using pulse compression in range and fast Fourier transform (FFT) processing in cross-range.

There need be no translational motion at all between the radar and target to image the target [45]. For instance, the roll of a ship could produce sufficient line-of-sight motion for imaging by a fixed radar.

Another interesting example of range-Doppler imaging is the surface mapping of the moon and planets [46, 47] utilizing high-power earth-bound radars. The necessary rotation about the line of sight is caused by a planet's natural rotation. The moon's rotation about its axis is synchronized with its rotation in an orbit about the earth so that it always presents the same face to an observer on the earth. It does, however, rock back and forth slightly about this alignment and this motion is sufficient to allow mapping. As a result, maps with 1 to 2 km resolution have been made of the moon's surface using earth-based radars. This is about the resolution provided by earth-based optical instruments as limited by the earth's atmosphere.

Of particular interest is the mapping of the planet Venus, which is continuously hidden under a blanket of clouds. Mapping of the surface of a planet utilizing range-Doppler imaging presents a particular problem. Range resolution will discriminate signals from an annulus centered on the line of sight, and Doppler resolution will discriminate a strip parallel to the apparent axis of rotation. Unfortunately, the Doppler strip crosses the range annulus at two spots, one in the northern and one in the southern hemisphere. In the case of Venus this dual-spot resolution has been solved by forming an interferometer on receive utilizing a second ground-based antenna about 2 km distant from the transmitting antenna. This interferometer places about eight fringes across the planet. The fringes are not aligned with the rotational axis, so they may be phased to null one of the two spots with the same range-Doppler response. Using this method, radar maps of a large portion of the surface of Venus have been prepared.

The introduction of interferometric signal processing utilizing the coherent signals from two antennas has added a third dimension to range-Doppler processing [47, 48]. It has allowed the production of topographic maps of the surface of the moon as well as an SAR which produces topographic maps of the earth.

## 5 CONCLUSION

The field of radar signal processing encompasses a wide variety of applications (see Table 16.1). The applications may be neatly divided into two groups, target detection-tracking radars and imaging radars.

The detection-tracking radars are operating against essentially point targets and the signal processing attempts to discriminate these points target from the background (clutter) by optimum filtering, taking advantage of all possible discriminants. The detection-tracking radar is faced with the problem of thresholding to decide the presence of the target while avoiding false alarms which would degrade the tracking performance.

The imaging radars come in a great variety of forms. Since their images form a continuum in measurement space, they invariably utilize filters that, through correlation techniques, optimize point reflections at each resolvable point in the image.

In this short chapter we have only been able to review in general terms the signal processing aspects of the many types of radars. We have tried to provide adequate references which will lead the reader into the subtle details which must be considered to design a useful radar signal processor.

Among the important subjects connected with detection-tracking radars and not even discussed are system stability and dynamic range, parameter estimation, theory of clutter rejection performance, adaptive filtering, target tracking, and electronic counter-counter measures against jamming. Subjects connected with imaging radars but not discussed include motion compensation, tropospheric phase errors, speckle reduction, and details of optical and digital processors.

## REFERENCES

1. S. M. Sherman, *Monopulse Principles and Techniques*, Artech House, Dedham, Massachusetts (1984).
2. C. E. Rader, A simple method for sampling in-phase and quadrature components *IEEE Trans. Aerosp. Electron. Syst., AES-20*: 821–823 (1984).
3. J. L. Lawson and G. E. Uhlenbeck, *Threshold Signals*, McGraw-Hill, New York (1950).
4. J. I. Marcum, A statistical theory of detection by pulsed radar, and mathematical appendix, *IRE Trans. Inf. Theory, IT-6*: 59–267 (1960).
5. P. Swerling, Probability of detection of fluctuating targets, *IRE Trans. Inf. Theory, IT-6*: 269–308 (1960).

6. J. V. DiFranco and W. L. Rubin, *Radar Detection*, Artech House, Dedham, Massachusetts (1980).
7. R. L. Mitchell and J. F. Walker, Recursive methods of computing detection probabilities, *IEEE Trans. Aerosp. Electron. Syst.*, *AES-7*: 671–676 (1971).
8. G. P. Dinneen and I. S. Reed, An analysis of signal detection and location by digital methods, *IRE Trans. Inf. Theory*, *IT-2*: 29–38 (1956).
9. J. W. Caspers, Automatic detection theory, *Radar Handbook*, (M. I. Skolnik, ed., chapter 15), McGraw-Hill, New York (1970).
10. W. W. Shrader, MTI radar, *Radar Handbook*, (M. I. Skolnik, ed.), chapter 17, McGraw-Hill, New York (1970).
11. A. J. Bogush, Jr., Correlated clutter and resultant properties of binary signals, *IEEE Trans. Aerosp. Electron. Syst.*, *AES-9*: 208–213 (1973).
12. W. S. Reid, M. R. Saltsman, and L. E. Vogel, IEEE 1975 International Radar Conference, pp. 294–299 (1975).
13. J. K. Hsiao, On the optimization of MTI clutter rejection, *IEEE Trans. Aerosp. Electron. Syst.*, *AES-10*: 622–629 (1974).
14. L. E. Brennan and I. S. Reed, Theory of adaptive radar, *IEEE Trans. Aerosp. Electron. Syst.*, *AES-9*: 237–252 (1973).
15. E. J. Kelly, An adaptive detection algorithm, *IEEE Trans. Aerosp. Electron. Syst.*, *AES-22*: 115–127 (1986).
16. W. H. Long and K. A. Harriger, Medium PRF for the AN/APG-66 radar, *IEEE Proc.*, *73*: 301–311 (1985).
17. J. Clarke, Airborne early warning radar, *IEEE Proc.*, *73*: 312–324 (1985).
18. D. H. Mooney and W. Skillman, Pulse-Doppler radar *Radar Handbook*, (M. I. Skolnik, ed.), Chapter 19, McGraw-Hill, New York (1970).
19. S. J. Rabinowitz, C. H. Gager, E. Brookner, C. E. Muehe, and C. M. Johnson, Applications of digital technology to radar, *IEEE Proc.*, *73*: 325–339 (1985).
20. C. E. Muehe, L. Cartledge, W. H. Drury, E. M. Hofstetter, M. Labitt, McCorison, and V. J. Sferrino, *IEEE Proc.*, *62*: 716–723 (1974).
21. C. E. Muehe, "Moving Target Detector, an Improved Signal Processor," *AGARD Conference Preprint No. 197 on New Devices, Techniques and Systems in Radar*, Paper No. 14 (1976).
22. W. H. Drury, Improved MTI Radar Signal Processor, Project Report ATC-39, MIT Lincoln Laboratory; FAA Report FAA-RD-74-185 (1975).
23. J. R. Anderson and D. Karp, "Evaluation of the MTD in a High Clutter Environment," *IEEE 1980 International Radar Conference*, 219–224 (1980).
24. D. Karp and J. R. Anderson, Moving Target Detector (Mod II) Summary Report, Project Report ATC-95, MIT Lincoln Laboratory; NTIS No. AD-A114-709 (1981).
25. C. E. Muehe, P. G. McHugh, W. H. Drury, and B. G. Laird, The parallel microprogrammed processor (PMP), *IEE Radar*, *77*: (1977).
26. R. S. Bassford, W. Goodchild, and A. De La Marche, *Test and Evaluation of the Moving Target Detector*, Final Report FAA-RD-77-118 (1977).
27. T. W. Taylor, Jr., and G. Brunins, Design of a new airport surveillance radar (ASR-9), *IEEE Proc. 73*: 284–289 (1985).
28. W. K. Saunders, CW and FM radar, *Radar Handbook*, (M. I. Skolnik, ed.), chapter 16, McGraw-Hill, New York, (1970).

29. C. E. Schwab and D. P. Post, Airport surface detection equipment, *IEEE Proc.*, *73*: 290–300 (1985).
30. J. Croney, Civil marine radar, *Radar Handbook*, (M. I. Skolnik, ed.), Chapter 31, McGraw-Hill, New York (1970).
31. R. J. Doviak, D. S. Zrnic, and D. S. Sirmans, Doppler weather radar, *IEEE Proc.*, *67*: 1522–1553 (1979).
32. M. Labitt, Coordinated Radar and Aircraft Observations of Turbulence, Project Report ATC-108, MIT Lincoln Laboratory; FAA-RD-81-44 (1981).
33. Eastwood, *Radar Ornithology*, Methuen, London (1967).
34. C. R. Vaughn, Birds and insects as radar targets: a review, *IEEE Proc.*, *73*: 205–227 (1985).
35. L. J. Porcello, R. L. Jordan, J. S. Zelenka, G. F. Adams, R. J. Phillips, W. E. Brown, Jr., S. H. Ward, and P. L. Jackson, The Apollo lunar sounder radar system, *IEEE Proc.*, *62*: 769–783 (1974).
36. D. A. Ausherman, A. Kozma, J. L. Walker, H. M. Jones, and E. C. Poggio, Developments in radar imaging, *IEEE Trans. Aerosp. Electron. Syst.*, *AES-20*: 363–398 (1984).
37. J. L. Walker, Range-Doppler imaging of rotating objects, *IEEE Trans. Aerosp. Electron. Syst.*, *AES-16*: 23–52 (1980).
38. L. J. Cutrona, Synthetic aperture radar, *Radar Handbook*, (M. I. Skolnik, ed.), McGraw-Hill, New York (1970).
39. R. O. Harger, *Synthetic Aperture Radar Systems: Theory and Design*, Academic Press, New York (1970).
40. C. Elachi, T. Bicknell, R. L. Jordan, and C. Wu, Spaceborne synthetic-aperture imaging radars: Applications, techniques, and technology, *IEEE Proc.*, *70*: 1174–1209 (1982).
41. W. M. Brown and R. J. Fredericks, Range-Doppler imaging with motion through resolution cells, *IEEE Trans. Aerosp. Electron. Syst.*, *AES-5*: 98– (1969).
42. W. M. Brown, Walker model for radar sensing of rigid target fields, *IEEE Trans. Aerosp. Electron. Syst.*, *AES-16*: 104–107 (1980).
43. D. R. Bromaghim and J. P. Perry, A wideband linear FM ramp generator for the long-range imaging radar, *IEEE Trans. Microwave Theory Tech.*, *MTT-26*: 322–325 (1978).
44. C. C. Chen and H. C. Andrews, Target-motion-induced radar imaging, *IEEE Trans. Aerosp. Electron. Syst.*, *AES-16*: 2–14 (1980).
45. M. J. Prickett and C. C. Chen, "Principles of Inverse Synthetic Aperture Radar (ISAR) Imaging," *IEEE 1980 EASCON Record*, pp. 340–345 (1980).
46. A. E. E. Rogers and R. P. Ingalls, Radar mapping of Venus with interferometric resolution of the range-Doppler ambiguity, *Radio Sci.*, *5*: 425–433 (1970).
47. T. Hagfors and D. B. Campbell, Mapping of planetary surfaces by radar, *IEEE Proc.*, *61*: 1219–1225 (1973).
48. L. C. Graham, Synthetic interferometer radar for topographic mapping, *IEEE Proc.*, *62*: 763–768 (1974).
49. F. E. Nathanson, *Radar Design Principles*, McGraw-Hill, New York (1969).
50. S. Watts, Radar detection in sea clutter using the compound K-distribution model, *IEEE Proc.*, *132* (Pt. F): 613–620 (1985).

51. B. C. Blevis, A Statistical Study of the Occurrence and Characteristics of Radar Auroral Echoes, DRTE Report No. 1100, Defence Research Telecommunications Laboratory, Ottawa, Canada (1962).
52. W. G. Chesnut, J. C. Hodges, and R. L. Leadabrand, Auroral Backscatter wavelength Studies, Stanford Research Institute Technical Report No. RADC-TR-68-286 (1968).
53. B. C. F. Butters, Chaff, *IEEE Proc.*, *129* (Pt. F): 197–201 (1982).
54. J. R. Riley, Radar cross section of insects, *IEEE Proc.*, *73*: 228–232 (1985).

# 17

# Speech Signal Analysis

DALE E. VEENEMAN Telecommunications Research Laboratory, GTE Laboratories, Inc., Waltham, Massachusetts

## 1 INTRODUCTION

The purpose of speech is communication. Traditionally, speech has been used to communicate concepts acoustically from one human to another, although in the past hundred years or so it has been possible to transform the acoustic speech signal to other media such as electric current or magnetic recording. In addition, more recently, it has been possible to use speech to communicate concepts between humans and machines.

Although speech is an analog signal, the advent of digital computing and the techniques of digital signal processing have allowed the analysis and processing of speech signals by digital means. This is known as digital speech processing and is the major thrust of this chapter. The chapter is divided into three main areas. First, the physical nature of speech production is introduced, including speech production anatomy, the acoustic theory of speech production, and a survey of phonetics (Section 2). Second, a digital model of the speech production process is constructed that incorporates elements of the physical process and allows an understanding of some of the techniques used in the digital analysis of speech (Section 3). Third, a number of digital analysis techniques are introduced, including time domain analysis, short-time Fourier analysis, linear predictive analysis, and homomorphic analysis (Sections 4–7).

It is the goal of this chapter to introduce only the major techniques used in digital speech processing. Detailed descriptions of applications of these techniques such as speech coding or speech recognition are not covered, but references to the literature are made when appropriate. Also, the background theory of these techniques (such as the Fourier transform) is not covered in detail; see other chapters of this book, or other references [1, 2]. It is assumed that the reader has some basic knowledge of digital signal processing, including a familiarity with the frequency domain, basic digital filtering, and the concept of the $Z$ transform [1].

### 1.1 The Speech Signal

Speech is the result of a very complex and not completely understood process. A "concept" that is formed in the brain is somehow converted to neural signals that travel to the muscles of the speech production components (lungs, vocal folds, jaw, etc.) that make up the speech production mechanism. These components then produce an acoustic waveform that is radiated out from the head as the speech signal.

Speech can be thought of as being composed of elements of discrete information. These basic elements of speech are known as phonemes (this will be discussed more in Section 2). The English language has about 42 phonemes, so the set can be represented with 6 bits. Since the maximum physical rate of production of the articulators is about 10 phonemes/sec, this gives an information rate of speech of around 60 bits/sec. However, this representation of speech is as if it were written text; it has no information about the identification of the speaker or the rate, pitch, stress, or amplitude of the speech. This type of information requires a much greater rate, from below 300 bits/sec to beyond 60,000 bits/sec, depending on the naturalness and quality.

### 1.2 Applications of Digital Speech Processing

Digital techniques allow much more power and control in speech processing than was ever available with analog processing. Following is a list of typical application areas of digital speech processing:

- Speech transmission. This covers areas such as digital network switching, including packet or burst switching; bandwidth compression techniques for reduced bit rates [3, 4]; and sophisticated encryption techniques.
- Speech storage. Similar to speech transmission, speech compression may be used to reduce storage requirements; in addition, digital storage allows enhanced routing techniques used in electronic voice mail or "voice store and forward."
- Speech recognition. This and speech synthesis are known as man-machine communication [5]. Speech recognition can be divided into speaker-dependent versus speaker-independent techniques, or isolated word versus continuous speech techniques [6], and has application in areas such as process control or speech-to-text systems.
- Speech synthesis. Already in wide use, speech synthesis allows efficient storage of speech for use in applications of information systems and text-to-speech systems [7].

- Speaker recognition or verification. Whereas speech recognition seeks to ignore speaker differences, speaker recognition seeks to identify a speaker from a number of possible speakers, based on his/her speech [8]. The more limited technique of speaker verification seeks to authenticate a single speaker and is used in access control.
- Speech enhancement. This covers a number of applications to improve the quality or characteristics of speech, such as reduction of noise, removal of echoes, replacement of lost packets of speech due to transmission errors, or speech time-scale modifications [9].
- Aids to the handicapped. A large number of applications are in use including visual or tactile displays of speech, speech-to-text or text-to-speech ("talking book") systems, or speech processing systems for use with cochlear implants ("artificial ears").
- Clinical speech diagnosis. Speech processing can aid in disease or disorder detection and in speech theory.

## 2 ACOUSTIC THEORY OF SPEECH PRODUCTION

It will be useful at this point to introduce the fundamentals of the speech production process and the modeling of this process to enable a better understanding of the digital processing techniques used for speech signals. A number of more complete references are available in this area, including excellent texts by Fant [10] and Flanagan [11].

A complete acoustic description of speech sound production is not available. It is much too complex a problem to account for all the variables and degrees of freedom. However, it is possible to make some simplifying assumptions that allow the acoustic problem to be separated into more tractable basic elements. This type of analysis still does a good job of describing the physical phenomenon of speech, although there is some evidence that the traditional acoustic theory of speech may be substantially deficient [12]. This section starts with some human anatomy, next covers descriptions of speech sounds, and then develops acoustic models that try to fit both.

### 2.1 Speech Production Anatomy

The speech production mechanism consists of subglottal components, the larynx, the vocal tract, and the articulators (see Figure 17.1). The subglottal components are the lungs, which supply the energy for sound production in the form of air pressure, and the trachea, which channels the air to the larynx. Contained in the larynx are the vocal folds, soft tissues composed of muscle and mucous membranes, that extend into the air passageway. The opening between the vocal folds is called the glottis. The vocal folds (or equivalently, glottis) serve to modulate the air flow for voiced sounds (for example, vowels). The vocal tract, composed of the pharynx and oral cavity, acts as a resonant tube to filter the sound. In addition, when the velum is lowered, the nasal cavity is acoustically coupled to the vocal tract to produce nasal speech sounds. The articulators, which include the velum (or soft palate), tongue, jaw, and lips, configure the vocal tract to determine which fre-

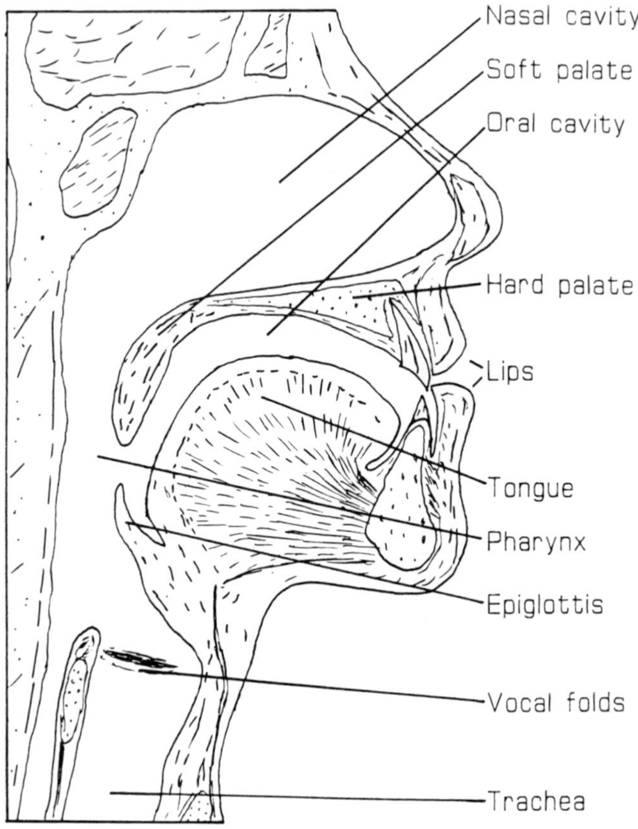


**Figure 17.1** A sagittal section of the vocal tract.

quencies of the sound source are passed to produce various speech sounds, which are then radiated by the lips or nostrils. The average length of the vocal tract, from the glottis to the lips, is about 17 cm in males, and the cross-sectional area of the vocal tract can vary from zero to 20 cm$^2$, depending on the positions of the articulators.

## 2.2 Phonetics

As mentioned in the introduction, the elements of information in speech are transmitted as a sequence of speech sounds. Linguistics is the study of the rules of language in the arrangement of these speech sounds for human communication. The study and classification of these speech sounds themselves is known as phonetics. Speech sounds may be broadly classified by their mode of excitation or, in more detail, by their individual characteristics as phonemes. The text by Chomsky and Halle [13] is an excellent reference in this area.

## Speech Sounds Classified by Mode of Excitation

In the processing of speech signals, it is generally useful to classify the various sounds by their mode of excitation because of the different characteristics of the sound source independent of vocal tract filtering. There are three classes of sounds: voiced, unvoiced, and plosive.

- Voiced sounds are produced by glottal excitation. This excitation is caused by vocal fold vibration that releases quasi-periodic pulses of air through the larynx.
- Unvoiced or fricative sounds are produced by turbulent (noiselike) excitation. This is caused by a constriction somewhere in the vocal tract.
- Plosive sounds are produced by a burst of acoustic energy. This is caused by a closure of the vocal tract, a buildup of pressure behind the closure, and then sudden release of the pressure.

As an example, the three classes of excitation are evident in the word "SIT." The "S" is unvoiced, the "I" is voiced, and the "T" is plosive. As a further example, an illustration of a speech signal waveform is given in Figure 17.2. This represents about 1 second of speech (a male saying the phase "Oak is strong"). Note the

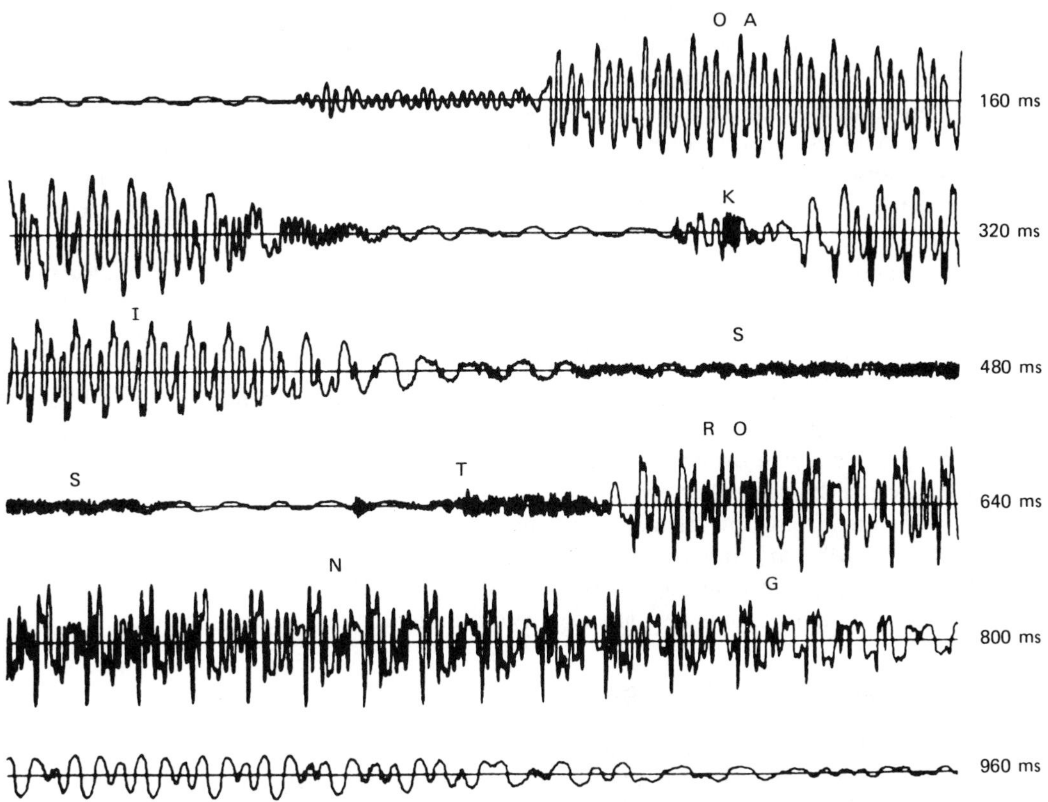


**Figure 17.2** Speech signal waveform: "Oak is strong."

periodic nature of the voiced speech in the "OA" or "I," the noiselike nature of unvoiced speech in the "S," and the attenuation and then burst of energy in the "K" and "T."

### Phonemes

A phoneme is defined as the smallest unit of sound that, when substituted, will alter the linguistic content of speech. For example, the words "BIT" and "PIT" each have three phonemes but differ in the first. American English has 42 phonemes, including the four broad classes vowels, diphthongs, semivowels, and consonants (which include fricatives, stops, nasals, affricates, and aspirants). Phonemes are also broadly classed as continuants and noncontinuants. Continuants (which include vowels, fricatives, and nasals) are produced by a fixed vocal tract. Noncontinuants (diphthongs, semivowels, stops, and affricates) are produced by a time-varying vocal tract. Following is a description of English phonemes and their number:

- Vowels (12)—produced by glottal excitation in a fixed vocal tract with no constrictions. They differ by resonant characteristics that depend on the articulators. Examples are /E/ as in "pet" and /A/ as in "pat."
- Diphthongs (6)—start with vocal tract in position for one vowel and proceed with the vocal tract smoothly changing toward another vowel. Examples are /eI/ as in "pay" and /aU/ as in "pow."
- Semivowels (4)—include glides /R/, /Y/, and liquids /W/, /L/. They are like voiced vowels, but the vocal tract is somewhat more constricted. They are also transitional, in that they precede a vowel and exhibit movement toward it.
- Fricatives (8)—produced by a constriction in the vocal tract that causes turbulent noise. They are characterized the location of the constriction (for example, teeth-lip /V/ or teeth-teeth /S/. They may be voiced /V/, /Z/ or unvoiced /S/, /F/.
- Stops (6)—produced by a complete closure of the vocal tract, which allows pressure to build, then a sudden release of pressure, which causes a turbulent noiselike excitation. They may be voiced, as in /B/, or unvoiced, as in /P/.
- Nasals (3)—produced by glottal excitation with a closure of the oral cavity and a lowering of the velum, which opens the channel to the nasal cavity. The sound radiation occurs mostly at the nostrils, while the closed oral cavity is coupled as a side branch resonator. Examples are /M/ and /N/.
- Affricates (2)—like the vowel combinations that form diphthongs, the affricates are formed from stop-fricative combinations. There are two affricates, unvoiced /tS/ as in "char" and voiced /dZ/ as in "jar."
- Aspirant (1)–the aspirant /H/ is produced by glottal turbulence, without vocal fold vibration. This is also the excitation mode during whispered speech.

## 2.3 Source, Filter, and Radiation

An acoustic model of speech production can be formed by an analysis of the physical elements of the production system. There are three parts to the system: the sound source, the filter, and the radiation load. Although certain assumptions are made (for example, the system is linear and the source and filter are separable), the traditional acoustic model does a good job of describing speech production.

## The Sound Source

The sound source is what generates the acoustic energy for speech and can be thought of as the excitation applied to the system. The excitation sound can take two forms: voiced and fricative.

Voiced sound is produced by the vibration of the vocal folds that are contained in the larynx (see Figure 17.3). With the glottis closed, air pressure builds behind the folds until they are forced apart. The Bernoulli effect (due to the increased air velocity between the folds) decreases the pressure between the folds, drawing them together again. This natural oscillation repeats at a fundamental frequency that depends on the air pressure behind the folds, the mass of the folds, and the tension applied to them. The result is that the air stream is modulated by the glottis and is released as a series of pulses (see Figure 17.4a). The fundamental frequency ranges from 80 to 200 Hz in males, 150 to 300 Hz in females, and 200 to 500 Hz in children. The spectrum of the glottal air volume velocity shows harmonics of the fundamental frequency, with a magnitude rolloff of about $-12$ dB/octave (see Figure 17.4b). The fundamental frequency is commonly called the pitch and the harmonics are called pitch harmonics, although, technically speaking, pitch is actually the perceived fundamental frequency.

Fricative or unvoiced sound (which includes plosives) is produced by turbulent air flow. This is caused by a constriction somewhere in the vocal tract that causes the air velocity to exceed the value (related to the Reynolds number) where it becomes turbulent. It gives a noise sound source with an approximately flat power spectrum and is often modeled as white noise. Both types of sound source can be present at the same time, as in /Z/.

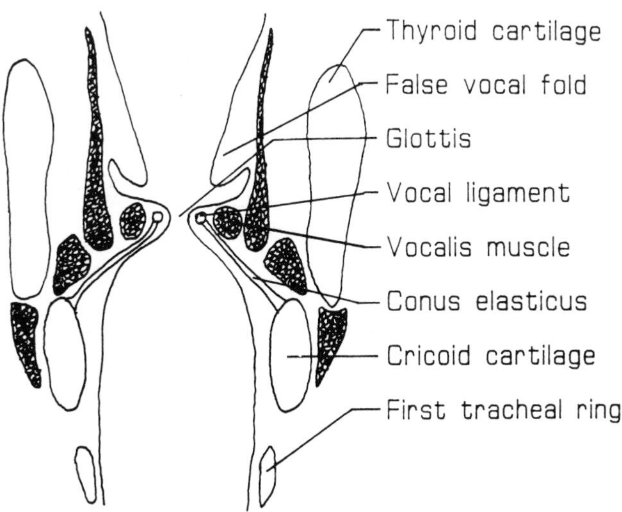


**Figure 17.3** Coronal section of the larynx.

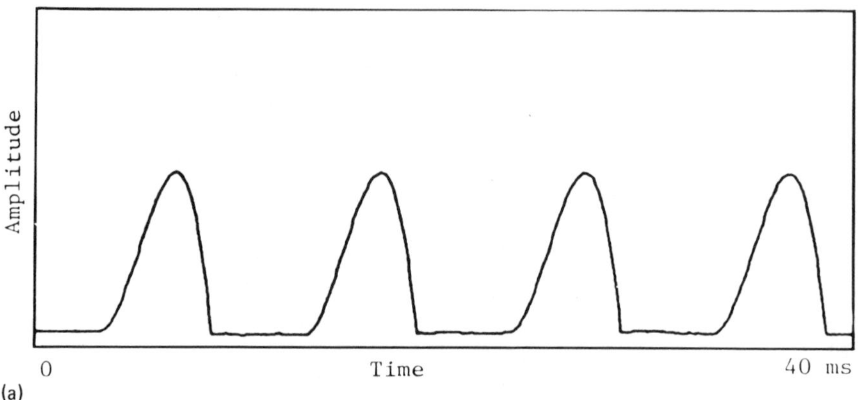


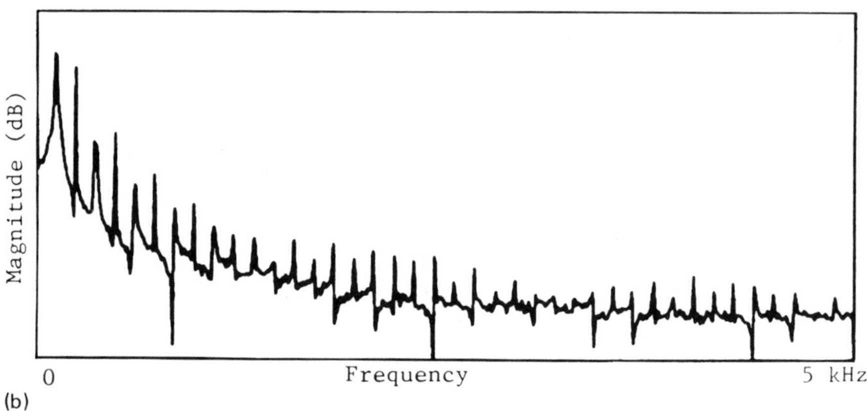


**Figure 17.4** Glottal volume velocity: (a) time waveform; (b) magnitude spectrum.

## Vocal Tract Filtering

The vocal tract acoustically filters the source sound and allows some frequencies to pass, while attenuating others. Perhaps the simplest example of a vocal tract model is the uniform lossless tube (see Figure 17.5). This is a hard-walled cylindrical tube of constant area $A$ and length $l$ that is driven by a piston at one end and is open at the other. If it is assumed that the piston provides an ideal source of volume velocity flow, that there are only variations in volume velocity, not pressure, at the open end, that there are no losses due to viscosity or thermal conduction, and that there are only plane waves in the tube, the sound waves in the tube satisfy the following pair of partial differential equations:

$$-\frac{\partial p}{\partial x} = \frac{\rho}{A}\frac{\partial u}{\partial t} \qquad \text{and} \qquad -\frac{\partial u}{\partial x} = \frac{A}{\rho c^2}\frac{\partial p}{\partial t} \tag{17.1}$$

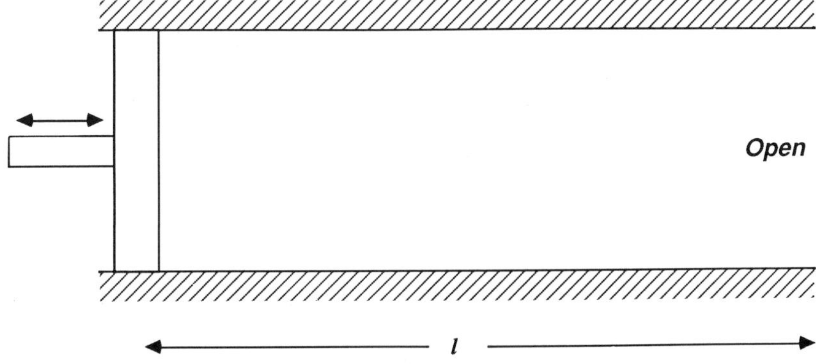


**Figure 17.5** Uniform lossless tube of length $l$ and area $A$.

where $p$ is the pressure and $u$ is the volume velocity, both a function of distance $x$ and time $t$, $\rho$ is the air density, and $c$ is the velocity of sound. This system is analogous to a lossless uniform electric transmission line with an ideal current source at one end and a short-circuit termination at the other end, where the acoustic pressure is represented by voltage and the volume velocity by current. The acoustic tube is best characterized in the frequency domain, by solving Eq. (17.1) for the transfer function that relates the volume velocity at the open end $U_o$ to the volume velocity at the source $U_s$:

$$\frac{U_o}{U_s} = V(j\Omega) = \frac{1}{\cos(\Omega l/c)} \tag{17.2}$$

This frequency response is illustrated in Figure 17.6 for $l = 17.5$ cm and $c = 35000$ cm/sec. The poles of $V(j\Omega)$ (where the denominator goes to zero) are the resonant frequencies of the acoustic tube and in this case lie on the $j\Omega$ axis. In speech, the resonant frequencies of the vocal tract are called formants.

A more realistic model may be obtained by concatenating a number of acoustic tubes of different areas together to represent the vocal tract (see Figure 17.7). It may be assumed here, without loss of generality, that the tubes are equal in length, which will help to lead to the discrete time modeling of speech. This model may be entirely characterized by the traveling waves that are partially propagated and partially reflected at each junction [2, 11]. The reflection coefficient for the $k$th junction is given by

$$r_k = \frac{A_{k+1} - A_k}{A_{k+1} + A_k} \tag{17.3}$$

where each $r_k$ represents the amount of traveling wave reflected at that junction. The transfer function of the model is determined by the set of $\{r_k\}$ and will contain only poles. Thus, the frequency response of the vocal tract model is determined by either the set of reflection coefficients, the area function (the set of areas of the

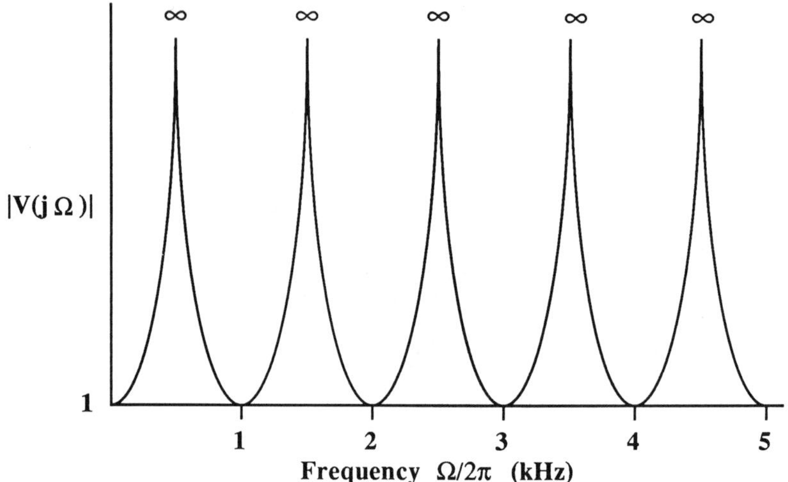


**Figure 17.6** Frequency response of uniform lossless tube of length 17.5 cm and velocity of sound 35,000 cm/sec.

concatenated tubes), or the poles of the transfer function. Any of these can be calculated from each other and, as will be seen later, the set of filter coefficients for a digital filter describing the vocal tract can also be calculated from any of these. A vocal tract frequency response measured from actual speech (a sustained vowel /A/) is illustrated in Figure 17.8.

When the velum is lowered for nasalized vowels, the nasal branch is included in the vocal tract. The oral branch is also closed (for example, at the lips for /M/), making a resonant cavity that traps acoustic energy at "antiresonant" frequencies. This introduces zeros into the transfer function (where the numerator goes to zero), in addition to the poles. In the transmission line analogy, this is like an open-circuited branch on the line. In addition, the format bandwidths of a nasalized vowel are somewhat broadened.

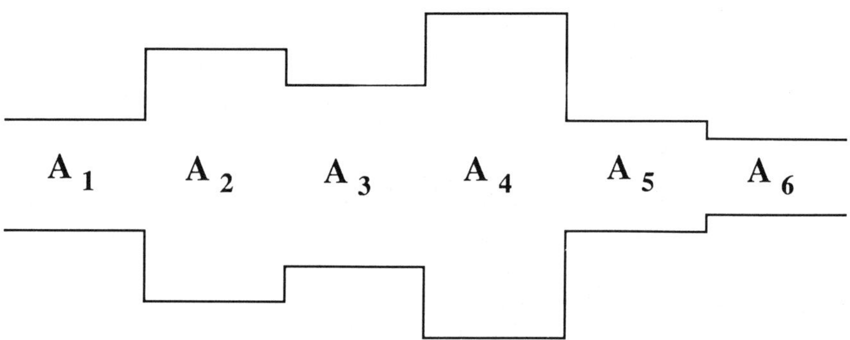


**Figure 17.7** Concatenated acoustic tubes (six sections).

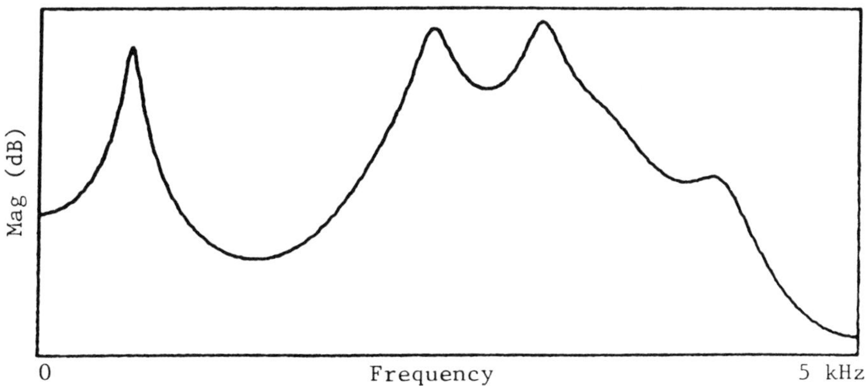


**Figure 17.8** Vocal tract frequency response for sustained vowel /A/.

## Radiation at the Lips

The relationship between the volume velocity $U(j\Omega)$ and the pressure $P(j\Omega)$ at the lips can be modeled by a plane baffle with opening area $A$ (Figure 17.9):

$$P(j\Omega) = R(j\Omega)U(j\Omega) \tag{17.4}$$

where the radiation impedance is given by

$$R(j\Omega) = \frac{j\Omega r L}{r + j\Omega L} \tag{17.5}$$

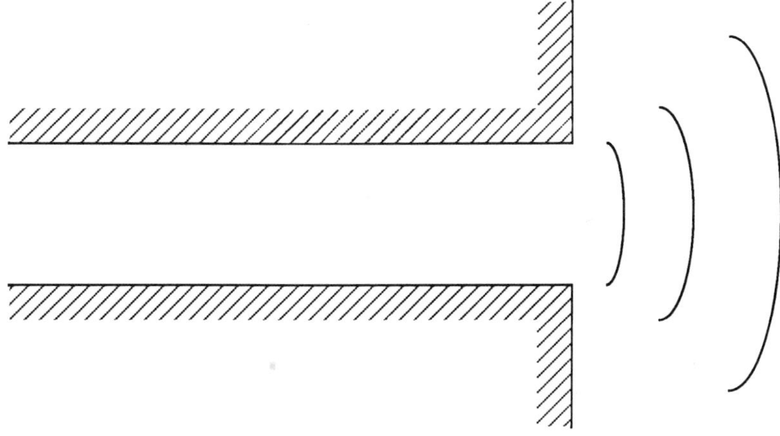

**Figure 17.9** Infinite plane baffle.

with radiation resistance $r = 128/9\pi^2$ and radiation induction $L = 8A/3\pi c$. The net effect of the lip radiation load in going from volume velocity to pressure is usually simply modeled as a high-pass filter with a single zero at DC (see Figure 17.10).

### Complete System Example

The three elements can now be combined into a complete system wherein the sound source is filtered by the vocal tract and radiation transfer functions. Figure 17.11 illustrates this for a sample of steady voiced speech. Here the glottal excitation spectrum is multiplied by the vocal tract and radiation frequency responses to produce the speech spectrum. Also shown are the time waveforms of the source excitation and the resulting speech. Note that the $-12$ dB/octave rolloff of the glottal excitation, when combined with the $+6$ dB/octave response of the radiation, gives the characteristic $-6$ dB/octave slope of voiced speech.

## 2.4 The Speech Spectrogram

Since speech has both spectral and temporal aspects (in other words, the speech spectrum varies with time), it is desirable to somehow visualize both simultaneously. An analog device, the speech spectrograph (sometimes called a sonograph), was developed in the 1940s to do this [14]. A magnetic recording of about 2 seconds of speech is made on the edge of a rotating disk. Then the recording is played back repetitively through a sliding bandpass filter with the output applied to a stylus on heat-sensitive paper that is fixed to the same drum as the recording. The stylus is incremented vertically with frequency, while the paper moves under it horizontally with time.

There are normally two bandwidth settings for the filter. The wideband setting (300 Hz bandwidth) gives good temporal resolution but poor frequency resolution (see Figure 17.12b). Note that the pitch periods are visible with time and the formants are broad. The narrowband setting (45 Hz bandwidth) gives good frequency resolution but poor temporal resolution (Figure 17.12c). In this case, note that the

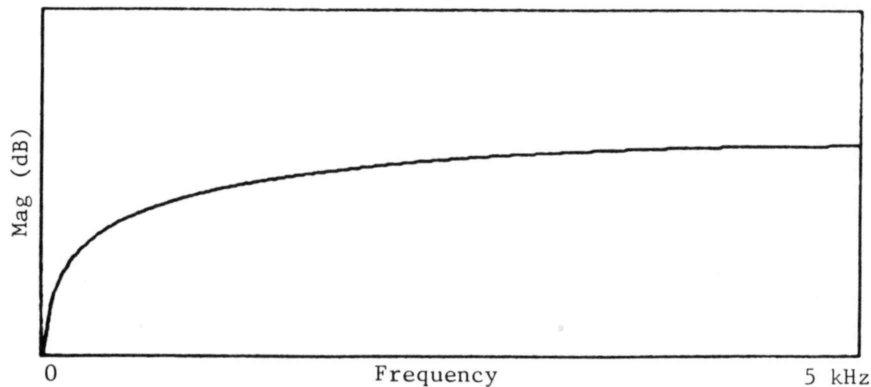


**Figure 17.10** Radiation load frequency response.

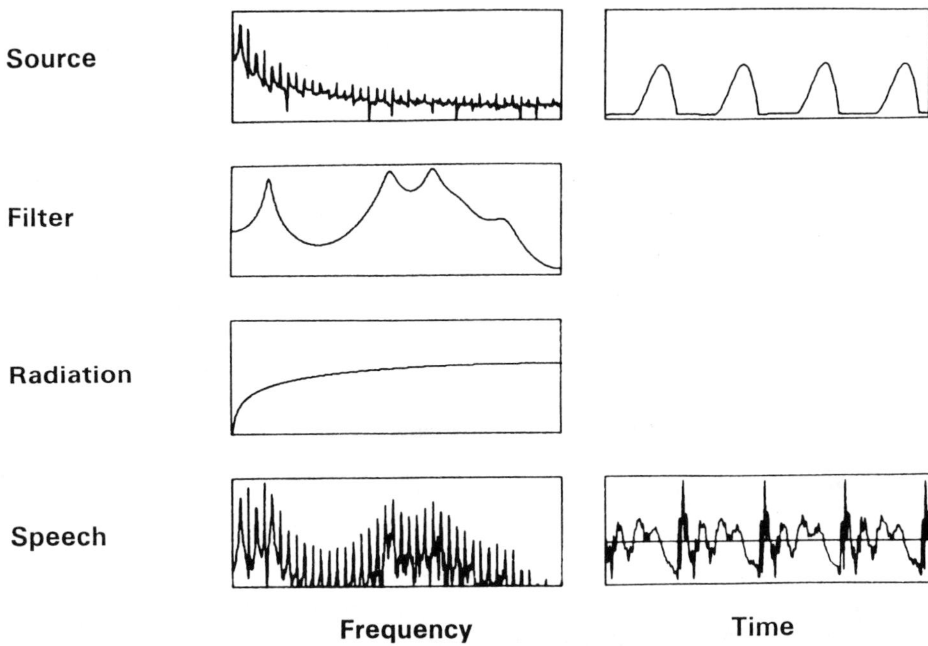


**Figure 17.11** Speech production system example for steady voiced speech (using elements from Figures 17.4, 17.8, and 17.10).

pitch harmonics are now visible with frequency, but the pitch periods are no longer visible with time.

Both spectrograms were made from the same recording (Figure 17.12a), a male speaking the phrase "Oak is strong and also gives shade." Note the vowels with their characteristic formants (e.g., the /O/ is "oak" and /A/ in "shade"), the fricatives /S/ in "strong" and /SH/ in "shade," the stops /T/ in "strong" and /G/ in "gives,' the glide /R/ in "strong," and the liquid /L/ in "also." Quite remarkably, it is possible for some people to become skilled enough to be able to "read" spectrograms [14]. In addition, it is currently possible to produce speech spectrograms by digital means, using sampled speech [15].

## 3 DIGITAL MODEL OF SPEECH PRODUCTION

This section is concerned with the digital representation of the acoustic elements of speech production. The model is very general to allow insight into digital speech analysis and synthesis but at the same time gives an adequate representation of speech signals. Because the analysis is in sampled time (and sampled frequency), the elements are necessarily band-limited to the Nyquist frequency (one-half the sampling frequency). As before, the system is assumed linear (source/filter separable).

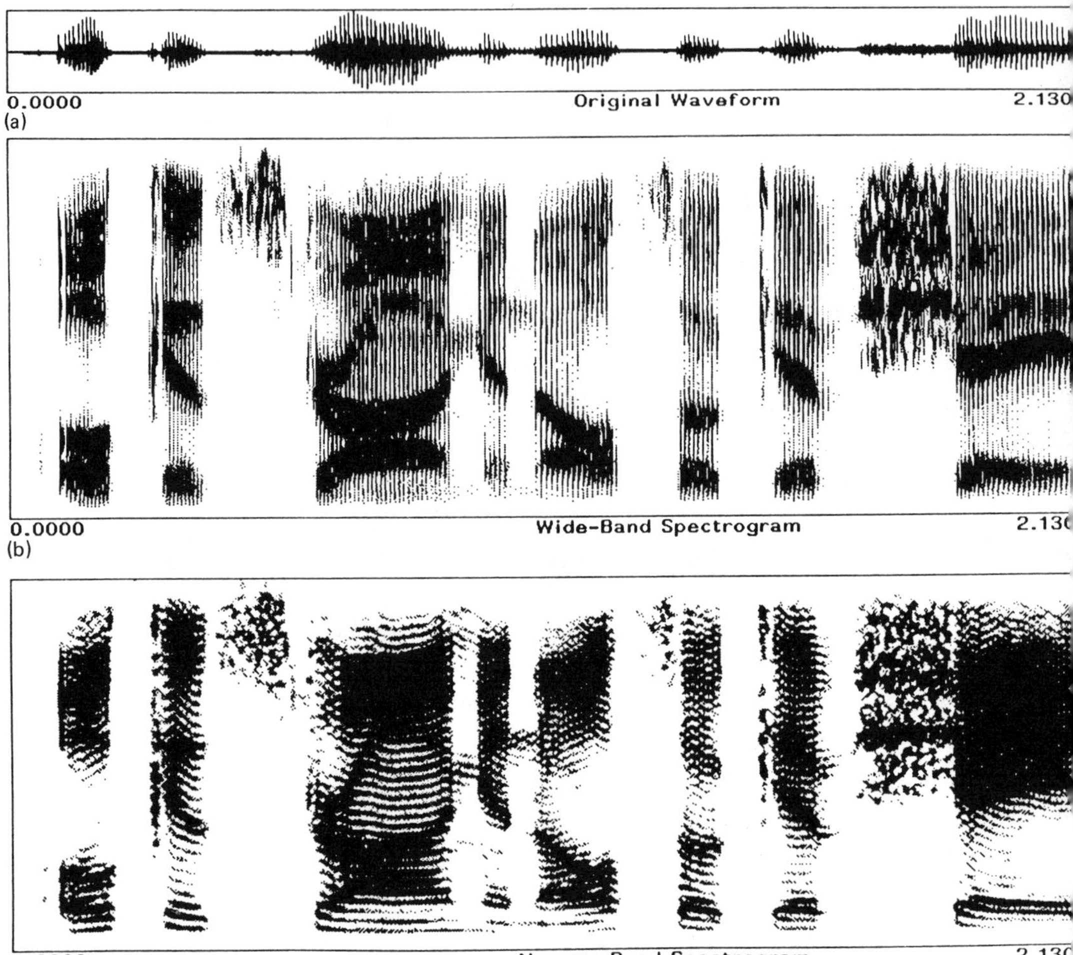


**Figure 17.12** Speech spectrograms for utterance "Oak is strong and also gives shade." (a) Time waveform; (b) wideband spectrogram; (c) narrowband spectrogram.

### 3.1 Excitation

The excitation is applied to the speech system to produce speech sounds. The speech sounds may generally be modeled with two types of excitation: voiced and unvoiced.

For voiced speech sounds, glottal pulse excitation is needed. This is modeled by starting with a train of impulses separated by the pitch period (see Figure 17.13). The impulse train is then filtered by a glottal pulse model $G(z)$ and multiplied by an amplitude control $A$. Ideally, $G(z)$ should be a finite impulse response (FIR) filter that contains only zeros, because the glottal pulses are known to be of finite duration between periods of glottal closure. However, since the glottal excitation

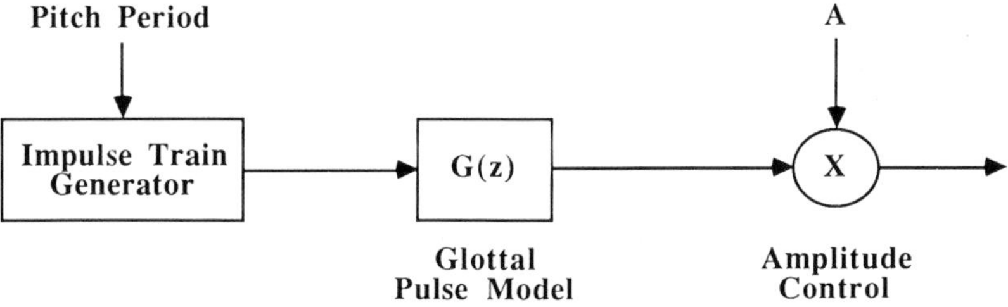


**Figure 17.13** Voiced excitation model for digital speech.

has a −12 dB/octave rolloff, $G(z)$ is often modeled as an infinite impulse response (IIR) filter with two poles:

$$G(z) = \frac{1}{(1 - g_1 z^{-1})(1 - g_2 s^{-1})} \tag{17.6}$$

where $g_1$ and $g_2$ are either both real or complex conjugate.

For unvoiced speech sounds, turbulent noise excitation is needed. This is simply provided by a random (Gaussian) noise generator with an amplitude control. In addition, it may be desired to allow a mix of voiced and unvoiced excitation in some cases.

## 3.2 Vocal Tract Filtering

The excitation signal is filtered by the vocal tract model. It was shown previously that the transfer function of the concatenated acoustic tube (nonnasalized) is determined by the reflection coefficients $r_k$ and has only poles. The digital form of the vocal tract filter is thus given by

$$V(z) = \frac{1}{1 - \sum_{k=1}^{p} a_k z^{-k}} = \frac{1}{A(z)} \tag{17.7}$$

where the poles are the roots of the denominator $A(z)$ and $p$ is the number of poles. Since the filtered waveform is real, the complex poles will appear in symmetric pairs representing the $p/2$ formants. $A(z)$ can be found from the set of $\{r_z\}$ by the recursive procedure:

$$\begin{aligned} A_0(z) &= 1 \\ A_k(z) &= A_{k-1}(z) + r_k z^{-k} A_{k-1}(z^{-1}), \qquad k = 1, 2, \ldots, p \\ A(z) &= A_p(z) \end{aligned} \tag{17.8}$$

and, conversely, the reflection coefficients can be found from $A(z)$ by a similar procedure [2].

If the vocal tract is nasalized, it may be necessary to add zeros to the transfer function:

$$V(z) = \frac{\sum_{k=0}^{q} b_k z^{-k}}{1 - \sum_{k=1}^{p} a_k z^{-k}} = \frac{B(z)}{A(z)} \tag{17.9}$$

Alternatively, the nasalized vocal tract may be modeled by increasing the number of poles in Eq. (17.7). This is because a pole-zero or autoregressive moving average (ARMA) transfer function can be approximated by an all-pole (autoregressive) function by increasing the number of poles (the match is exact as the number of poles approaches infinity). However, the inclusion of many more poles to account for the presence of zeros will require more parameters to characterize the filter.

### 3.3 Radiation

The radiation load function is commonly modeled by a digital approximation of $R(j\Omega)$:

$$R(z) = 1 - z^{-1} \tag{17.10}$$

This is simply a digital differentiator with a single zero at DC.

### 3.4 Complete Model

The complete model is illustrated in Figure 17.14, where a switch determines the mode of excitation. Since the individual elements are assumed to be linear, a single "lumped" transfer function may be formed from the elements:

$$H(z) = G(z)V(z)R(z) \tag{17.11}$$

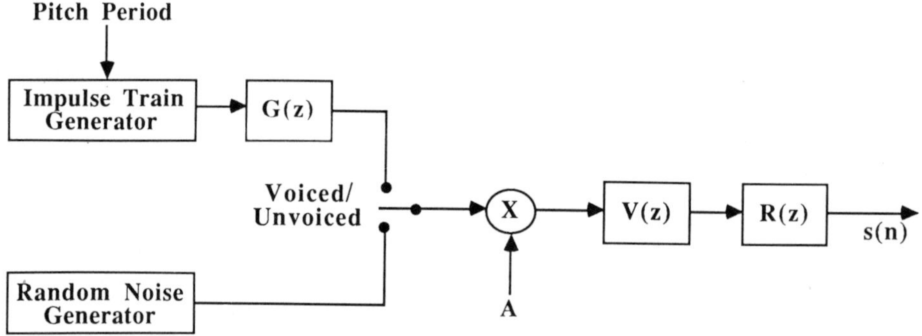


**Figure 17.14** Complete digital speech production model.

In this case, either an amplitude scaled pitch train or random noise sequence is applied to the filter $H(z)$ to produce the speech signal $s(n)$. To produce different speech sounds, the model parameters (pitch period, voiced/unvoiced decision, amplitude, and filter characteristics of $H$) may be updated to vary with time. Indeed, this is how commercial speech synthesis systems work.

## 4 TIME DOMAIN ANALYSIS

This section introduces actual analysis methods for speech signal processing. These are time domain methods using the time waveform, as opposed to transform domain (such as frequency domain) methods or linear prediction methods. First, means for acquiring the digital speech signal are discussed, then a number of "short-time" measures are introduced, and finally an example of speech classification using some of these measures is illustrated.

### 4.1 Sampling and Quantizing Continuous Speech

The acoustic speech signal exists as pressure variations in the air. A microphone converts these pressure variations into an electric current that is related to the pressure (similarly, the ear converts the pressure variations into a series of nerve impulses that are transmitted to the brain). To process the speech signal digitally, it is necessary to make the analog waveform discrete in both time (sample) and amplitude (quantize).

#### Sampling

The sampling theorem states that a signal must be sampled at at least twice the rate of the highest frequency contained in the signal to avoid aliasing. This is usually accomplished by low-pass (or "antialias") filtering the signal at a frequency less than one-half that of the sample rate. One-half the sampling frequency is known as the Nyquist frequency and is the maximum frequency contained in the discrete-time signal. Some common sampling rates for speech are as follows:

- Digital telephone. The current standard has the signal sampled at 8 Hz, giving a Nyquist frequency of 4 kHz.
- General digital processing. A common sampling rate for speech processing (such as speech recognition) is 10 kHz, giving a Nyquist frequency of 5 kHz (which filters out some of the higher-frequency energy of fricatives).
- Full fricative or pathological. For complete analysis of fricative and some pathological speech signals that have frequencies up to 10 kHz, a sampling rate of 20 kHz is used.
- Compact disk. As an example of "studio" quality recording, the sampling rate for compact disk is approximately 44 kHz, giving a Nyquist frequency of around 22 kHz, completely covering the range of human hearing.

It is important to understand, that to prevent aliasing distortion, the signal at the Nyquist frequency ideally should be attenuated 30–40 dB lower than the mean signal energy (in other words, the antialias filter's −3-dB low-pass cutoff frequency should not be set at the Nyquist frequency, but below). For example, in sampling signals for digital telephone (8 kHz sampling rate), the attenuation begins at 3.4 kHz, with a slope of about −60 dB/octave.

### Quantizing

It is necessary to represent the continuum of sampled signal amplitude values by the finite number of values provided in the digital format, given by $B$ bits. For example, if the signal voltage (from an amplified microphone) that ranges from −5 to 5 V is quantized to 10 bits, 1024 discrete values are used to represent that range: −5.12, −5.11, ..., −0.01, 0, +0.01, ..., +5.11 V. (Normally it is not necessary to maintain units such as voltage; integer representation is usually sufficient.) The error that occurs in "rounding" the analog value to the nearest discrete value is called quantization error or noise. For uniform (or linear) quantization that uses equal step sizes between discrete values, the signal-to-noise ratio (SNR) for the quantization noise (in dB) may be approximated [2] by

$$\text{SNR (dB)} = 6B - 7.2 \tag{17.12}$$

This approximation assumes that the quantization noise is stationary and white (meaning the input signal should fluctuate continuously over the range), that the noise is uncorrelated with the signal (meaning that the step size is sufficiently small), and that the quantizer range matches the peak signal range. Because of the difference between the energies of voiced and unvoiced speech, it is considered necessary to use 11- to 12-bit uniform quantization for high-quality speech representation (compact disk uses 16-bit quantization for very wide dynamic range in both voice and music).

In uniform quantization, since the step size remains constant over the signal range, the SNR is actually better for high-amplitude values than for low-amplitude values. However, it is not necessary to use uniform quantization. Logarithmic quantizing (sometimes called instantaneous companding) uses a larger step size for larger values and maintains SNR over the range of signal amplitudes. In digital telephone quantization, $\mu$-law coding (a segmental log approximation) uses 8 bits per sample to give the perceptual quality of uniform quantizing at 12 bits. Many more complex (and more optimal) quantization schemes exist, including adaptive and block methods, but these usually start with uniformly quantized digital waveforms (see [3]).

## 4.2 Concept of "Short-Time" Analysis

Short-time analysis is a central concept in digital speech processing. It is generally required for most analysis methods that a signal's properties be time-invariant. For example, it is difficult to define the pitch of a conversational speech segment that is 5 minutes long.

Although the speech signal is *not* time-invariant, properties of the signal change "relatively" slowly with time. Thus, short segments of the speech signal can be isolated and processed as though they *are* time-invariant (segment analysis frames may overlap). The desire is to have enough data in each frame to characterize the property of interest, balanced by the desire not to "smear" the property. An illustration of this is pitch extraction. By definition, it is necessary to have at least two pitch periods to define the pitch (more periods give better reliability), but if the pitch is rapidly changing, too many periods considered will not give an accurate pitch track. Very generally, representative values for speech signal analysis are a frame length of about 30 msec and a frame shift on the order of 10 msec, although these values can range greatly.

This is the same concept as discussed in the spectrograph section (Section 2.4). The wideband setting (short time-frame) smeared the frequency resolution but not the time resolution. The narrowband setting (long time-frame) had the opposite effect of smearing the time resolution but giving sharpened frequency resolution.

The short-time concept also applies to speech synthesis using the model of Figure 17.14. The model parameters would be fixed for some period of time (a portion of a phoneme, perhaps) and then changed for another period of time to produce a different sound. If the periods of time are short enough, natural-sounding speech results, even though the properties of true speech are continually changing.

## 4.3 Short-Time Measures

The following measures or functions are termed short-time because they are performed on a finite-length signal segment. These are standard measures that are commonly used in speech analysis [16].

### Peak Amplitude

The peak amplitude is simply the largest absolute amplitude value occurring in the analysis frame. It is generally useful as an aid to distinguishing between voiced and unvoiced speech or as a method for gain normalization in a fixed point processing frame.

### Energy

Short-time energy is defined as the sum of the squares of the signal values over the analysis frame ($N$ samples, ending at sample $n$):

$$E_n = \sum_{m=n-N+1}^{n} s^2(m) \tag{17.13}$$

An energy function is illustrated in Figure 17.15b for the speech utterance of Figure 17.15a.

It is convenient at this point to introduce the use of windowing, so that Eq. (17.13) can be equivalently written:

$$E_n = \sum_{m=-\infty}^{\infty} [s(m)w(n-m)]^2, \qquad \begin{aligned} w(n) &= 1, \quad 0 \le n \le N-1 \\ &= 0, \quad \text{otherwise} \end{aligned} \tag{17.14}$$

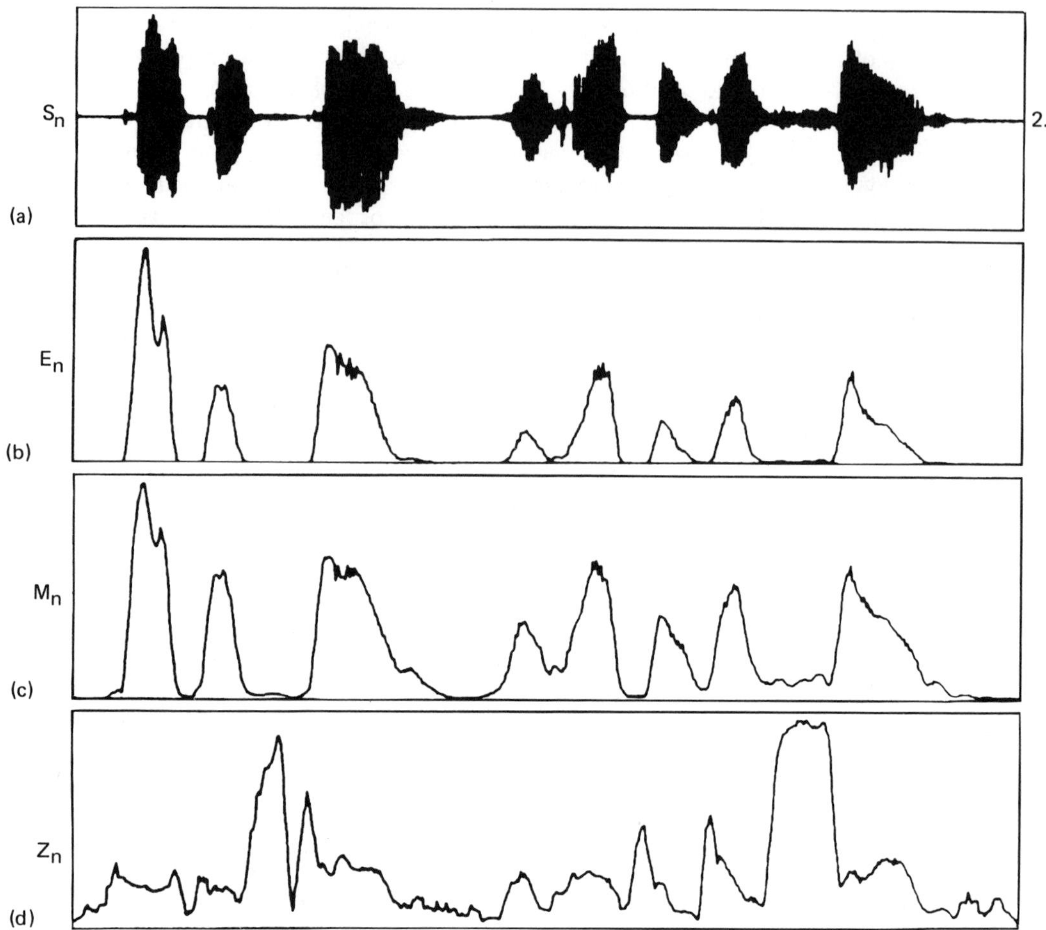


**Figure 17.15** Short-time measure for utterance "Oak is strong and also gives shade" (30-msec window and 5-msec shift). (a) Time waveform; (b) energy; (c) average magnitude; (d) zero-crossing rate.

where $w(n)$ is in this case a rectangular window of length $N$ and amplitude 1 (the amplitude may be set to $1/N$ for normalization). Also, if desired, windows of other shapes (for example, triangular) may be used.

If $N$ is very short, the energy function will show rapid fluctuations due to the presence of individual pitch periods, whereas if $N$ is longer, rapid changes in energy will be averaged and a smoother function will result. However, if $N$ is too long, the function will have little variation and will not reflect the changing properties of the speech signal. The functions presented in Figure 17.15 were all calculated with a 30-msec window, shifted 5 msec each time.

## Average Magnitude

The average magnitude function is similar to the energy function but without the squared weighting (which may cause the energy function to be sensitive to large signal amplitudes):

$$M_n = \sum_{m=-\infty}^{\infty} |s(m)w(n-m)| \tag{17.15}$$

Both the energy and average magnitude functions (see Figure 17.15c) are useful for voiced/unvoiced decisions and are straightforward to implement in hardware.

## Zero-Crossing Rate

Another simple measure is the zero-crossing rate, defined as the number of sign changes occurring in the analysis frame:

$$Z_n = \sum_{m=-\infty}^{\infty} [(\text{sign changes})\, w(n-m)] \tag{17.16}$$

The zero-crossing rate gives a general indication of frequency content as higher-frequency signals (such as unvoiced speech) give a higher zero-crossing rate (see Figure 17.15d). Thus the function is useful (with energy) for voiced/unvoiced/silence detection. It is also very easy to implement in hardware.

## Autocorrelation Function

The general autocorrelation function (ACF) is defined as

$$\Phi(k) = \sum_{m=-\infty}^{\infty} s(m)s(m+k) \tag{17.17}$$

The ACF is large when a lagged version of the signal is similar to the signal (i.e., correlated), where $k$ = lag (the ACF is maximum at $k = 0$ and is equivalent to the energy). Thus the ACF is large at lags equal to periodicities contained in the signal. If, for example, the signal $s$ is periodic with period $P$, then $\Phi(0) = \Phi(P) = \Phi(2P) \cdots$.

To calculate the short-term ACF, windowing is required:

$$R_n(k) = \sum_{m=-\infty}^{\infty} [s(m)w(n-m)][s(m+k)w(n-k-m)] \tag{17.18}$$

Note here that $R_n(0) = E_n$ (the short-term energy). The short-time ACF is illustrated in Figures 17.16 and 17.17. Note that in voiced speech (Figure 17.16b) the ACF has a peak at $k$ equal to the pitch (and also at pitch multiples), whereas in unvoiced speech (Figure 17.17b) little correlation is evident. Also note the triangular weighting of the ACF that occurs due to the multiplication of smaller

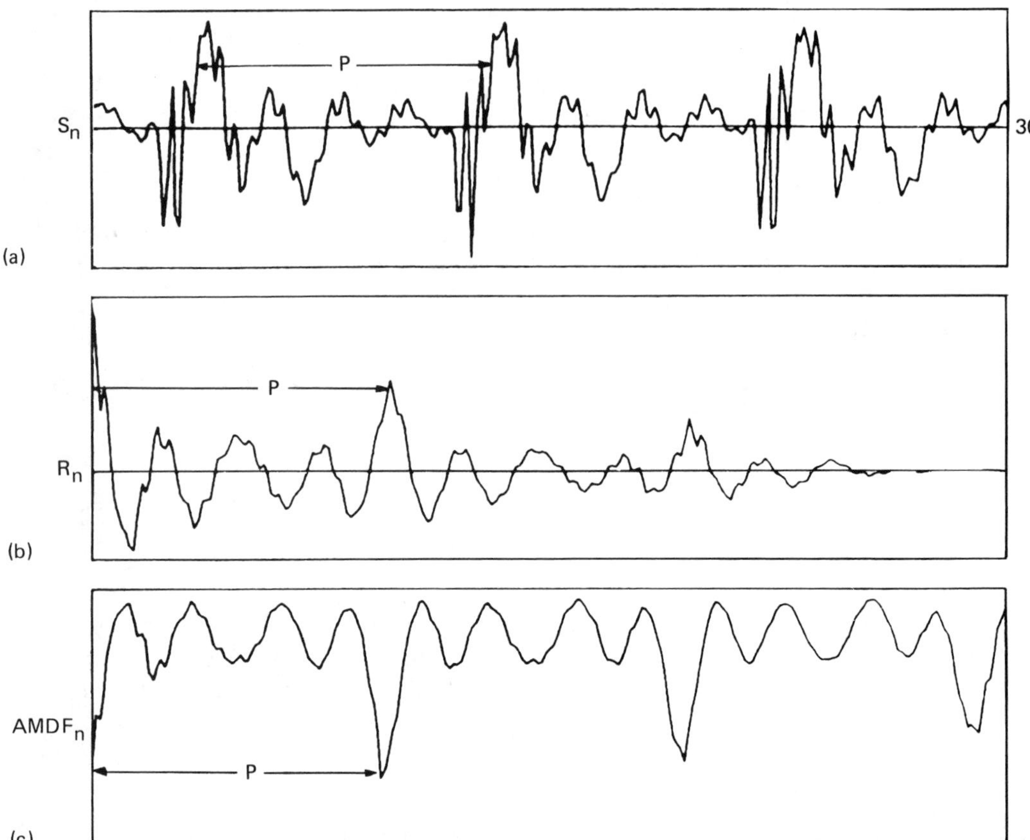


**Figure 17.16** Short-time autocorrelation (ACF) and average magnitude difference (AMDF) functions for voiced speech. (a) Time waveform; (b) ACF (c) AMDF.

portions of the rectangular windows as the lag increases (in effect the convolution of RECT $*$ RECT).

The ACF is more difficult to calculate than the previous functions. It requires on the order of $N^2$ multiplies and adds.

## Average Magnitude Difference Function

Similar to the ACF, the average magnitude difference function (AMDF) is defined as

$$\mathrm{AMDF}_n(k) = \sum_{m=-\infty}^{\infty} |s(n+m)w(m) - s(n+m-k)w(m-k)| \qquad (17.19)$$

where instead of multiplying, the lagged signal is subtracted and the absolute value taken. In this case, periodicities (positive correlations) will appear as small values

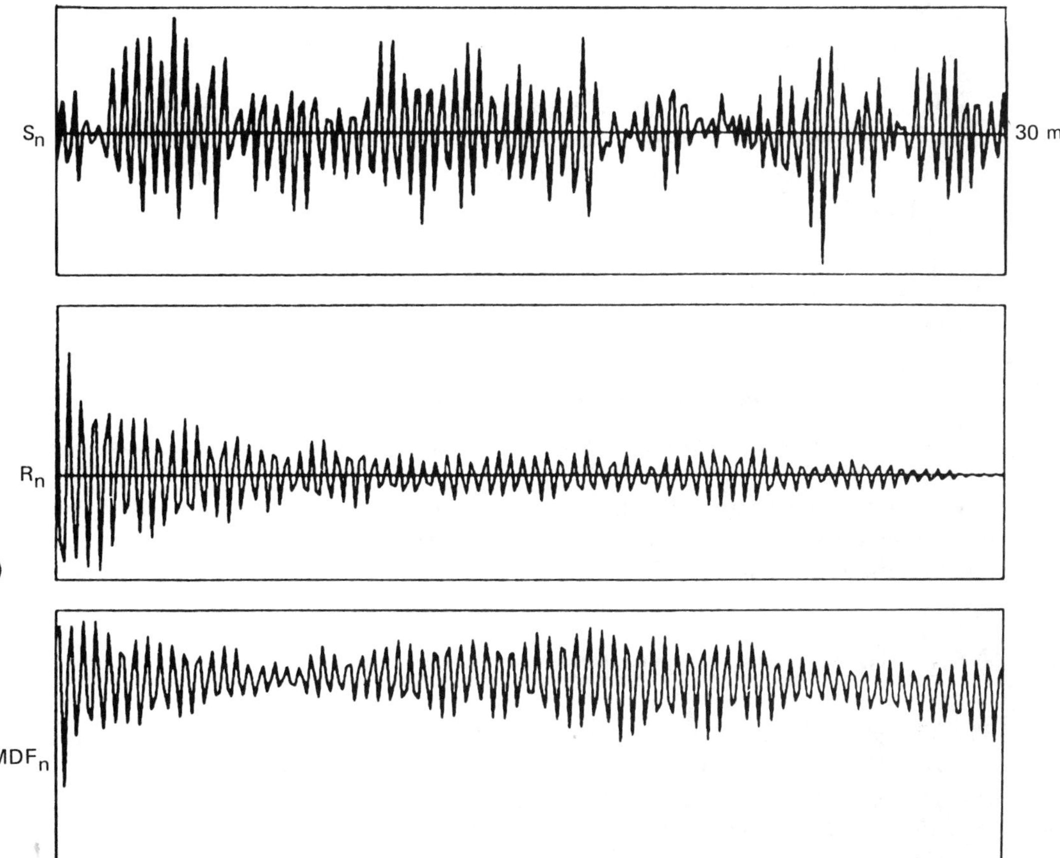


**Figure 17.17** Short-time autocorrelation (ACF) and average magnitude difference (AMDF) functions for unvoiced speech. (a) Time waveform; (b) ACF; (c) AMDF.

(see Figure 17.16c and 17.17c). The AMDF is generally less difficult to calculate, as the multiplications necessary for the ACF are avoided.

## 4.4 Voiced/Unvoiced/Silence and Pitch Detection

This is intended as an illustration of how the above functions can be used to classify speech segments as voiced, unvoiced, or silence and to detect the pitch of voiced speech. Following are some general forms for accomplishing this:

- Silence. The energy (or average magnitude) and the zero-crossing rate are below some threshold. This works for "clean" speech; with signals corrupted by substantial levels of noise, it is very difficult to determine nonspeech from unvoiced portions.
- Unvoiced. The zero-crossing rate is above some threshold, since unvoiced speech has dominant high frequencies. In addition, a segment may be clas-

sified as unvoiced if the value of the maximum ACF (in some range of lags above zero) is below some fraction of the segment's energy (the value of the ACF at lag zero).

- Voiced. The energy is above some threshold (and the zero-crossing rate is below). Also, if the peak of the ACF (in some range of lags above zero) is above some fraction of the energy (indicating strong periodicity), the segment may be classified as voiced.
- Pitch. For voiced speech, the pitch may be estimated by the lag of the maximum value of the ACF in some allowable range (for example, 60–350 Hz) or, similarly, by the lag of the minimum of the AMDF. It is sometimes helpful to first remove some of the sample-to-sample correlation (which gives rise to extra peaks in the ACF) by a center-clipping operation [2]. Methods using peak detection have also had some success [16].

These are just a few basic methods. More sophisticated signal processing techniques may also be used. One such technique uses homomorphic analysis (Section 7) to estimate the pitch and classification of speech from the cepstrum. In addition, linear predictive techniques (Section 6) may be used to estimate the pitch from the inverse filtered residual waveform.

## 5 SHORT-TIME FOURIER ANALYSIS

The Fourier transform is probably the most familiar technique used for spectral analysis [1]. It is a linear transform that takes a signal from the time domain to the frequency domain (and back again) and is useful for indicating the frequencies at which a signal's energy occurs. It can be thought of as equivalent to the analog technique of passing a signal through a bank of bandpass filters (as in the analog spectrograph). The short-time Fourier transform is defined as

$$S_n(e^{jw}) = \sum_{m=-\infty}^{\infty} w(n-m)s(m)e^{-j\omega m} \tag{17.20}$$

### 5.1 Windowing Considerations

As before, the window determines which portion of the signal is processed (in this case, transformed). However, because the window is multiplied with the signal in the time domain, it is equivalent to the transform of the window being convolved with the transform of the signal in the frequency domain. The Fourier transform of a rectangular window is the $\sin(x)/x$ (or sinc) function, and the longer the rectangular window, the narrower the main lobe of the sinc. Thus, for good frequency resolution a long window is desired, but for good temporal resolution a short window is desired. This is illustrated in Figure 17.18 for voiced speech. Figure 17.18b shows the magnitude (in dB) of the Fourier transform of 5 msec of speech (Figure 17.18a). The formant structure is evident, but not any fine frequency details. Figure 17.18d shows the Fourier transform of 37.5 msec of sustained speech (Figure 17.18c) starting at the same point as Figure 17.18a). Here the fine harmonic structure is evident

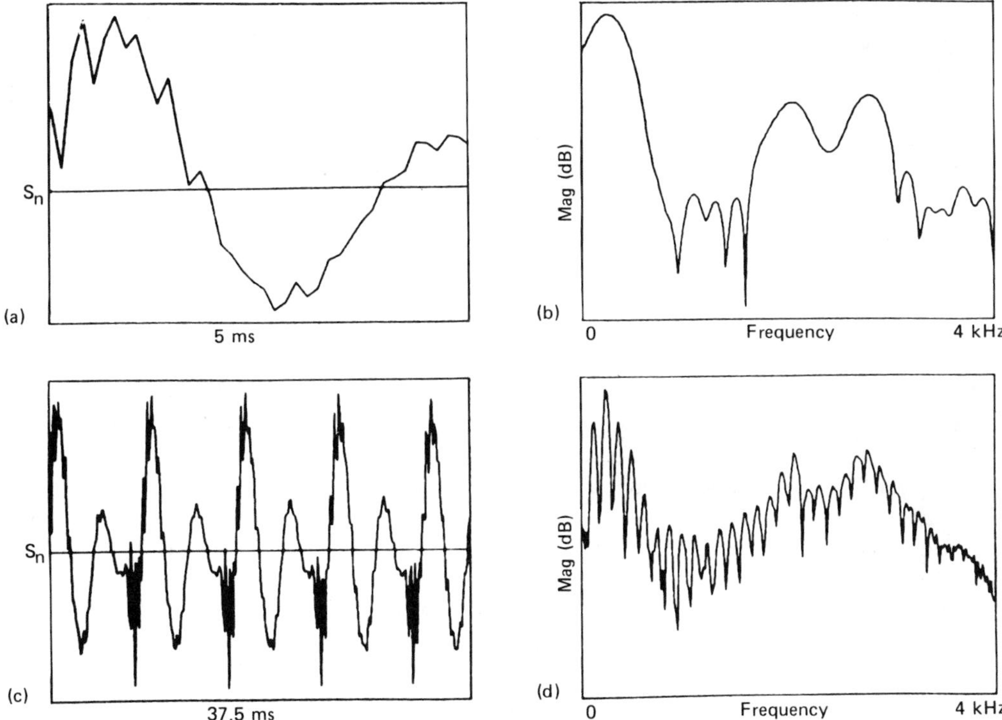


**Figure 17.18** Fourier transforms of voiced speech segments of different lengths. (a) 5-msec segment; (b) Fourier transform (log magnitude) of (a); (c) 37.5-msec segment; (d) Fourier transform of (c).

in addition to the formant structure. However, if, instead of sustained speech, a changing segment had been transformed, the spectrum would not have shown the stationary formants and harmonics, but they would have been broadened (smeared).

The window shape is also important. The convolution with the sinc function (the transform of the rectangular window) in the frequency domain leads to "leakage" of spectral energy to surrounding frequencies. This is because of the high sidelobes of the sinc function. It is usually better to use a window that has lower sidelobes in the frequency domain (for example, a Hamming window [1]) to control leakage. The cost is a wider mainlobe, which decreases frequency resolution.

## 5.2 The Discrete Fourier Transform

The discrete Fourier transform (DFT) is the Fourier transform of an $N$-point time domain sequence, as in Eq. 17.20, resulting in $N$ equally spaced frequency domain samples (distributed from DC to the sampling frequency):

$$S(k) = \sum_{n=0}^{N-1} s(n)e^{-j2\pi nk/N}, \qquad k = 0, 1, \ldots, N-1 \tag{17.21}$$

where here it is assumed that $s(n)$, $n = 0, 1, \ldots, N-1$, is a windowed portion ($N$ samples) of the signal waveform.

The Fourier transform is a complex transform. If the time domain sequence is *real* (i.e., the imaginary part is zero, as in speech signals), the complex frequency domain sequence has an *even* real part and an *odd* imaginary part. Thus it is only necessary to retain frequencies up to the Nyquist (folding) frequency. In addition, it may be more convenient to display the magnitude (the square root of real squared + imaginary squared), and the phase (the arctangent of imaginary/real) of the frequency domain sequence, or to display the power spectrum of the signal, which is the magnitude squared. The magnitude is commonly displayed as 20 log(mag), expressed in (relative) dB.

A relationship of some importance is that between the autocorrelation and the Fourier transform. It can be shown [1] that if a sequence of $N$ samples is padded with an equal number of zeros, the Fourier transform of the ACF of the longer sequence is equivalent to the magnitude squared of the Fourier transform of the sequence (the power spectrum). Thus, using efficient algorithms for the DFT, it is usually less costly to calculate the autocorrelation function by first calculating the power spectrum using the DFT, then obtaining the ACF by an inverse DFT of the power spectrum.

### 5.3 Fast Fourier Transforms

A fast Fourier transform (FFT) is simply an efficient algorithm for calculating the DFT [17]. This is done in the case of the standard Cooley-Tukey FFT by recursively calculating smaller FFTs for sequences that have a length that is a power of two. However, there are other specialized FFTs (such as those with higher or mixed radixes) or FFTs for sequences with special properties (for example, for real sequences such speech signals or symmetrical sequences). FORTRAN programs for a number of FFTs can be found in [18].

The short-time Fourier transform is commonly used for speech spectral analysis, including digital spectrograms that mimic analog spectrograms, but with much more processing versatility [15]. Figure 17.19 illustrates a digital spectrogram, using a three-dimensional display, of a male-spoken utterance "Oak is strong." The short-time Fourier transform is also used for voice coding techniques, such as channel vocoding [2], where frequency domain parameters are coded and transmitted, and for speech signal enhancement and modification [9].

## 6 LINEAR PREDICTIVE ANALYSIS

Linear predictive analysis (commonly called linear prediction coding or LPC) is one of the most powerful speech analysis tools [19, 20]. One reason for this is that it fits so well with the digital speech production model developed in Section 3. Other reasons are the speed of computation and accuracy of LPC methods.

The basic idea behind linear prediction is that a speech sample $s(n)$ is predicted from the sum of $p$ linearly weighted previous samples:

$$\hat{s}(n) = a_1 s(n-1) + a_2 s(n-2) + \cdots + a_p s(n-p) \tag{17.22}$$

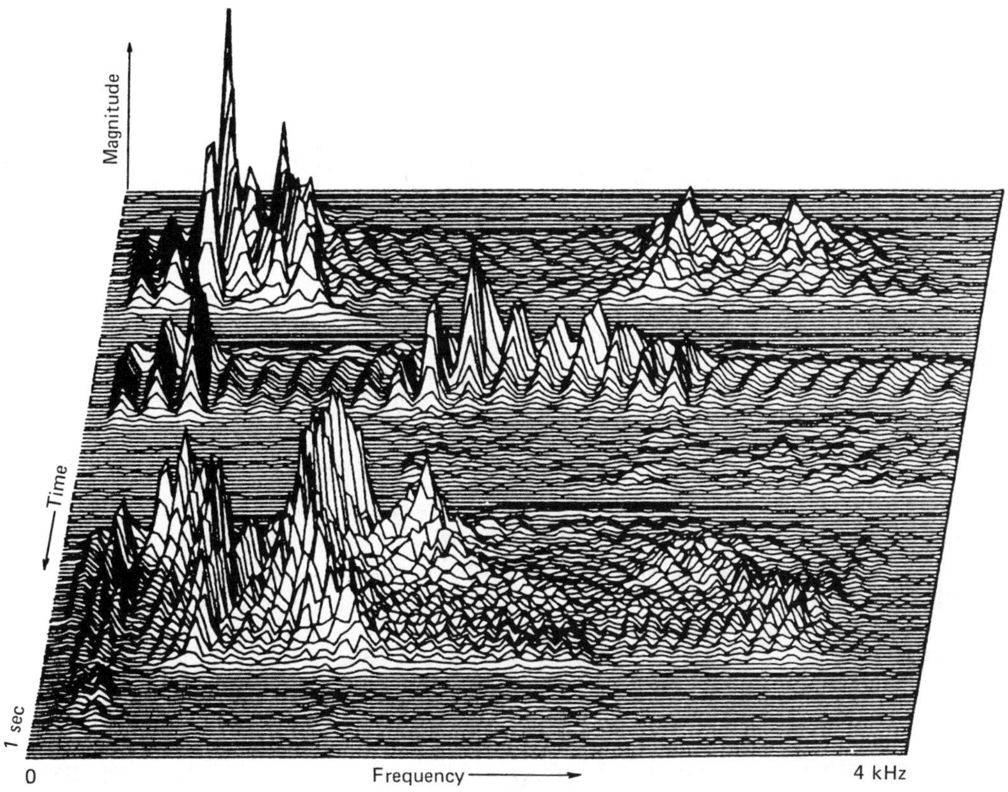


**Figure 17.19** Digital spectrogram (using FFT) of utterance "Oak is strong."

where the set of $\{a_k\}$ are the prediction coefficients, determined by minimizing the mean squared difference between the actual speech samples and the linearly predicted ones. The linear prediction equation (17.22) may be alternatively written:

$$\hat{s}(n) = \sum_{k=1}^{p} a_k s(n-k) \tag{17.23}$$

## 6.1 Basic Formulation

The lumped-parameter model of speech production developed in Section 3.4 (Eq. 17.11) is shown in Figure 17.20. Here the transfer function of the digital filter $H(z)$ that produces the output speech $s(n)$ given the input excitation $u(n)$ (either an impulse train or random noise) is given by

$$H(z) = \frac{S(z)}{U(z)} = \frac{G}{1 - \sum_{k=1}^{p} a_k z^{-k}} = \frac{G}{A(z)} \tag{17.24}$$

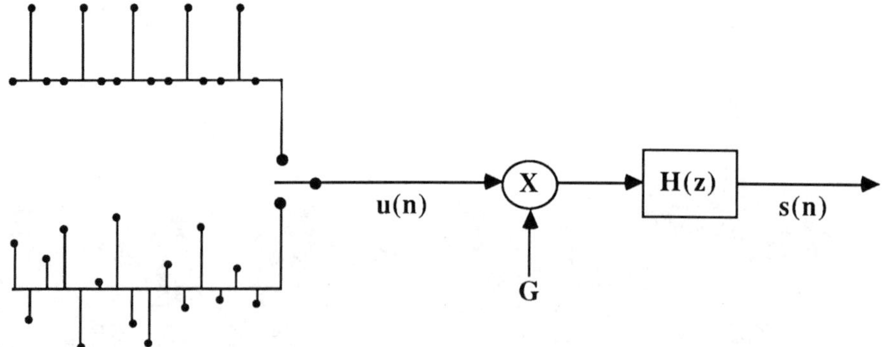


**Figure 17.20** Lumped-parameter model of digital speech production.

where $G$ is the gain applied to the filter. This is the all-pole model of speech production, where the roots of the denominator polynomial $A(z)$ are the poles of the system. $A(z)$ is termed the inverse filter, because it is the inverse system of $H(z)$ [i.e., $H(z)A(z) = G$].

In the time domain, the equation for the speech production model (17.24) is written in difference form:

$$s(n) = \sum_{k=1}^{p} a_k s(n-k) + Gu(n) \tag{17.25}$$

Note the similarity to the linear prediction equation (17.23). Other than the excitation term in Eq. (17.25), they are the same . Thus, if the predictor coefficients in Eq. (17.23) are equal to the filter coefficients in Eq. (17.25), the linear prediction equation will match the lumped-parameter speech production model very well. All that is needed is the excitation form [remember that from Eq. (17.11) the effects of the glottal pulse shape are included in the lumped-parameter model].

$A(z)$, when used to inverse-filter a speech sequence:

$$S(z)A(z) = U(z)H(z)A(z) = U(z)G \tag{17.26}$$

results in a sequence (called a residual sequence or residue) that is an approximation to the excitation applied: either an impulse train or a noise sequence. Since either a single impulse or a noise sequence has a flat spectrum, the inverse filter $A(z)$ is also called a "whitening" filter, because it flattens the speech spectrum (i.e., removes the effects of the formants, glottal rolloff, radiation, etc.). This is the basis of a frequency domain spectral-matching derivation of linear prediction given in [19].

It is necessary to find the set of $\{a_k\}$ so that $A(z)$ is the inverse of $H(z)$. This is done by using Eq. (17.23) to find an estimate $\hat{s}(n)$ for the speech sample $s(n)$.

The error is then given by

$$e(n) = s(n) - \hat{s}(n) = s(n) - \sum_{k=1}^{p} a_k s(n-k) \tag{17.27}$$

The standard approach for obtaining the predictor coefficients is based on minimizing the average squared prediction error over a short segment of the speech waveform. That is, the $a_k$ are found that minimize

$$E_n = \sum_m [e_n(m)]^2 \tag{17.28}$$

over some range of $m$, as yet unspecified. Following the standard least-squares approach of setting $\partial E_n/\partial a_i = 0$, $i = 1, 2, \ldots, p$, the following set of $p$ normal equations is obtained:

$$\sum_{k=1}^{p} a_k \sum_m s_n(m-i)s_n(m-k) = \sum_m s_n(m-i)s_n(m), \qquad i = 1, 2, \ldots, p \tag{17.29}$$

Two different formulations are possible, depending on the range of $m$ over which the minimization is to occur: the autocorrelation method and the covariance method [20].

## 6.2 Autocorrelation Method

In the autocorrelation method it is assumed that $-\infty < m < \infty$ and the speech waveform is windowed so that it is zero outside the range of interest: $s_n(m) = s(m+n)w(m)$, where $w(m)$ is a finite-length ($N$-point) window covering the desired range. Because the autocorrelation method attempts to predict the first $p$ samples from speech samples that are outside the window (which are arbitrarily zero), a large error may result. To reduce this error a window is applied which smoothly tapers the signal to zero at the ends of the window (such as a Hamming window).

The normal equations (17.29) then become

$$\sum_{k=1}^{p} a_k R_n(|i-k|) = R_n(i), \qquad i = 1, 2, \ldots, p \tag{17.30}$$

where

$$R_n(k) = \sum_{m=0}^{N-1-k} s_n(m)s_n(m+k) \tag{17.31}$$

is the short-time autocorrelation function of Section 4.3 (thus the name autocorrelation method).

The matrix that results in the left side of Eq. (17.30) is symmetric, positive definite, and Toeplitz (the elements along any diagonal are equal). A very efficient method called the Levinson recursion [20] can be used to solve this system of equations.

## 6.3 Covariance Method

In the covariance method, $m$ is considered only in the range $0 \leq m \leq N-1$ and the speech segment length is augmented by $p$ samples to enable the first $p$ samples to be predicted from speech samples outside the segment. Thus no windowing is required with this method.

The normal equations (17.29) then become

$$\sum_{k=1}^{p} a_k \varphi_n(i,k) = \varphi_n(i,0), \qquad i = 1, 2, \ldots p \tag{17.32}$$

where

$$\varphi_n(i,k) = \sum_{m=0}^{N-1} s_n(m-i) s_n(m-k), \qquad i = 1, 2, \ldots, p, \ k = 0, 1, \ldots, p \tag{17.33}$$

which is similar to a covariance matrix (thus the name).

The matrix of Eq. (17.33) is symmetric and positive definite (but not Toeplitz). The Cholesky decomposition [20] (sometimes called the square root method) is an efficient method used to solve this system of equations.

In both the autocorrelation and covariance methods, the gain $G$ for the model can be easily calculated, as it is related to the residual energy:

$$G^2 = E_n \tag{17.34}$$

and can be calculated during the solution for the predictor coefficients [20].

## 6.4 Lattice (PARCOR) Method

In both the autocorrelation and covariance methods of LPC, the processing has two stages: the calculation of the correlation matrix, and the solution of the resulting set of linear equations. However, in lattice methods, the two stages have been effectively combined into a recursive procedure for determining the LPC parameters, where not just one but $p$ forward and $p$ backward linear predictors are used and the parameters are calculated one stage at a time (see [21] for a complete development).

The forward prediction at the $i$th stage may be defined by

$$A_i(z) = 1 - \sum_{j=1}^{i} a_{j,i} z^{-j} \tag{17.35}$$

and, using a recursive definition similar to Eq. (17.8), may be written

$$A_i(z) = A_{i-1}(z) - k_i z^{-i} A_{i-1}(z^{-1}) \tag{17.36}$$

where $k_i$ is a parameter similar to the reflection coefficient of Eq. (17.8). The forward prediction error at the $i$th stage is then given by

$$E_i(z) = A_i(z)S(z) = A_{i-1}(z)S(z) - k_i z^{-i} A_{i-1}(z^{-1})S(z) \tag{17.37}$$

If the second term of Eq. (17.37) is defined as

$$B_i(z) = z^{-i} A_i(z^{-1})S(z) \tag{17.38}$$

it can be seen to be a "backward" prediction error sequence [the sample $s(n-i)$ is predicted by the following $i$ samples, the same samples that are used in the forward prediction]. The prediction error sequence (17.37) can then be written in the time domain as

$$e_i(n) = e_{i-1}(n) - k_i b_{i-1}(n-1) \tag{17.39}$$

and by substituting (17.36) into (17.38):

$$B_i(z) = z^{-1} B_{i-1}(z) - k_i E_{i-1}(z) \tag{17.40}$$

or in the time domain:

$$b_i(n) = b_{i-1}(n-1) - k_i e_{i-1}(n) \tag{17.41}$$

Equations (17.39) and (17.41) can be implemented by means of the lattice illustrated in Figure 17.21, where at each stage the forward and backward prediction errors are defined in terms of the errors of the previous stage and with initial errors $e_0(n) = b_0(n) = s(n)$.

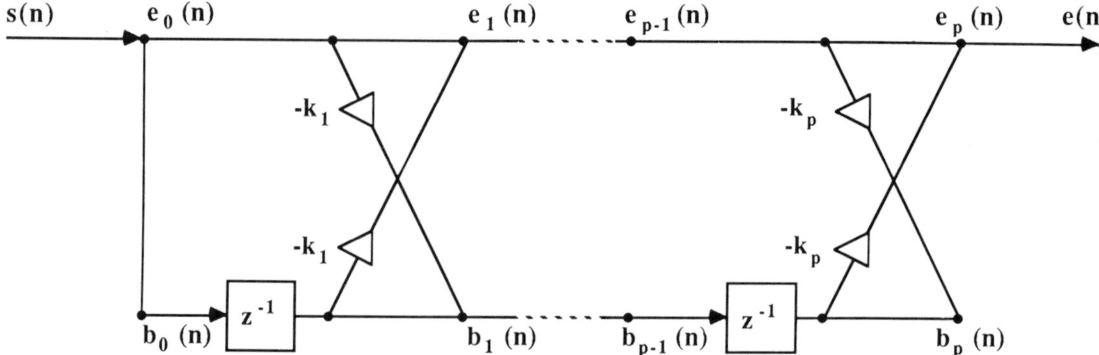


**Figure 17.21** Lattice network for implementing PARCOR LPC analysis.

The $k$ parameters are calculated at each stage by the partial correlation (thus the term PARCOR) between the forward and backward errors:

$$k_i = \frac{\sum_{n=0}^{N-1} e_{i-1}(n) b_{i-1}(n-1)}{\left\{\sum_{n=0}^{N-1} [e_{i-1}(n)]^2 \sum_{n=0}^{N-1} [b_{i-1}(n-1)]^2\right\}^{1/2}} \tag{17.42}$$

The PARCOR parameters are actually the negative of the reflection coefficients discussed in Section 3.2. In addition, a similar lattice structure, which in Figure 17.21 implements $A(z)$, can be used for the speech synthesis filter $H(z)$.

## 6.5 Analysis Considerations

LPC methods determine the inverse or whitening filter $A(z)$, and since the speech spectrum has a $-6$ dB/octave rolloff, $A(z)$ has to account for this. Thus it is usually desirable to equalize the spectrum before LPC analysis, and "prewhitening" is usually applied by a first-order preemphasis filter (i.e., +6 dB/octave) $1 - az^{-1}$, where $a$ is on the order of 0.9 to 1.0.

The linear prediction filter order $p$ determines the number of resonances that will be found, where each real resonance requires two complex conjugate roots of $A(z)$. A rule of thumb to be used for the filter order is to allow for one formant (two poles) per kilohertz of frequency range, plus three or four poles for radiation and glottal effects [20]. Thus, for a sampling rate of 10 kHz (frequency range of 5 kHz), a filter order of 13 to 14 is appropriate (8 to 12 are sometimes used for coding applications). Figure 17.22 illustrates the spectra of $1/A(z)$ for different LPC filter orders of a segment of voiced speech, compared to the FFT spectrum. In addition, Figure 17.23 presents an LPC spectrogram of the same utterance "Oak is strong" used for the FFT spectrogram of Figure 17.19. Note that the pitch harmonics are now absent, while the formants are more visible.

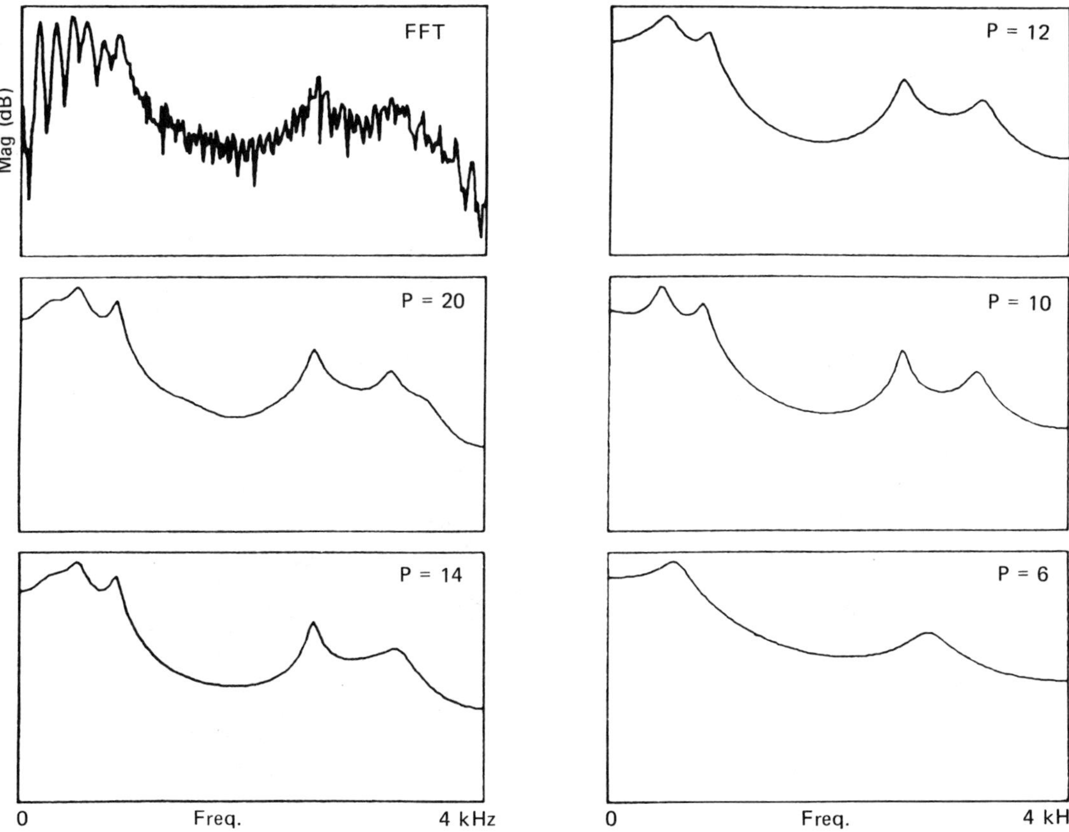


**Figure 17.22** Effect of LPC filter order. Shown are the log magnitude spectra for the FFT and $1/A(z)$ for five different LPC filter orders.

FORTRAN programs for the above LPC methods are available in [18]. In general, the autocorrelation method requires the least computation, and the covariance method requires slightly more. However, because of the windowing, the autocorrelation method may require a larger waveform segment length, so the two methods are generally considered of equivalent complexity. The lattice methods require somewhat more computation. In addition, the autocorrelation and lattice methods guarantee stable synthesis filters [the roots of $A(z)$ will always be contained within the unit circle], while this is not guaranteed for the covariance method.

LPC analysis is very useful for speech coding. Only a few parameters need to be transmitted (the filter coefficients, gain, pitch, and voiced/unvoiced decision determined every 10 msec) at a rate of 1200 to 2400 bits/sec to produce (admittedly synthetic-sounding) speech. More natural-sounding speech may be obtained by coding the error (residual) sequence $e(n)$ in some manner, at rates from 4800 to 16,000 bits/sec [4]. LPC analysis has also found wide use in speech recognition systems [6] and speech synthesis systems [7].

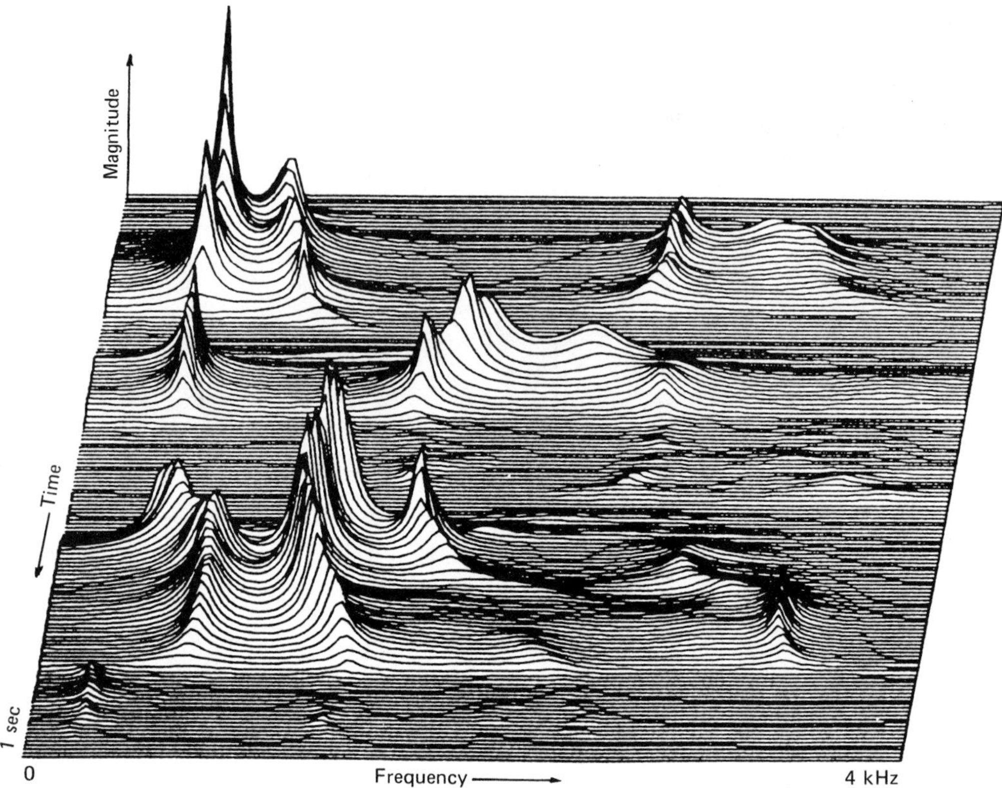


**Figure 17.23** Digital spectrogram (using LPC analysis filter of order 14) of utterance "Oak is strong."

## 7 HOMOMORPHIC ANALYSIS

Homomorphic analysis is a nonlinear extension of the superposition principle of linear systems and has found use in the analysis of signals that are produced by a convolution of separate signals. Since speech may be assumed (in short-time analysis) to be the result of the convolution of the excitation signal with the vocal tract system, homomorphic analysis may be used to separate the component signals (often called "deconvolution") [22].

### 7.1 Description Using Duality

Informally, the concept of duality applied to the time and frequency domains is basically that it does not matter which domain you are in, signal processing techniques work in the same manner. Using this concept, the idea of homomorphic analysis may be described as the following.

When looking at the log magnitude spectrum of a voiced speech segment, the pitch harmonics that are present look somewhat like a periodic component of some frequency, while the formants look like a somewhat lower-frequency component of the spectral waveform. If that spectrum is then thought of as a "time domain" sequence and Fourier-transformed, the frequencies ("quefrencies") of the original spectral components will be apparent in the resulting "time domain" spectrum ("cepstrum," i.e., the spectrum of a spectrum).

This process is illustrated in Figure 17.24. The windowed speech segment of Figure 17.24a is transformed by an FFT and the log magnitude is taken to give the spectrum of Figure 17.24b. The spectrum is then inverse-transformed to produce the cepstrum of Figure 17.24c (since the magnitude spectrum is real and even, the cepstrum will also be real and even, so only the first half of the real part of the complex sequence is shown; the imaginary part is zero).

The cepstrum is obviously useful for detecting the pitch, as it is the first strong quefrency in the cepstrum. In addition, the cepstrum can be divided into a low-time portion (corresponding to the spectral envelope) and a high-time part (corresponding to the excitation). If the high-time portion is set to zero (shown by the dashed line in Figure 17.24c) and the sequence is Fourier-transformed again, the result is the cepstrally smoothed spectrum of Figure 17.24d (in effect, low-pass filtering the spectrum). This then represents the magnitude of the vocal tract response, with the effect of the excitation removed. Since LPC analysis represents only the poles of the system, the cepstral approach can do a better job of representing nasalized speech.

## 7.2 Homomorphic System for Convolution

More formally, the previous discussion was an example of a homomorphic system for convolution (see Figure 17.25). The system $D_*[\ ]$ is the characteristic system for homomorphic deconvolution that obeys a generalized principle of superposition where the input operation is convolution (symbolized by $*$) and the output operation is addition:

$$D_*[x_1(n) * x_2(n)] = D_*[x_1(n)] + D_*[x_2(n)] = \hat{x}_1(n) + \hat{x}_2(n) \tag{17.43}$$

and $D_*^{-1}[\ ]$ is the inverse characteristic system:

$$D_*^{-1}[\hat{y}_1(n) + \hat{y}_2(n)] = D_*^{-1}[\hat{y}_1(n)] * D_*^{-1}[\hat{y}_2(n)] = y_1(n) * y_2(n) \tag{17.44}$$

The system $L$ is a conventional linear system. $D_*[\ ]$ then must be defined so that its input operation is convolution and its output operation is addition.

The $Z$ transform (of which the Fourier transform is a special case) is a system that has convolution as an input operation and multiplication as the corresponding output operation (i.e., convolution in the time domain is multiplication in the frequency domain):

$$Z[x_1(n) * x_2(n)] = Z[x_1(n)] \cdot Z[x_2(n)] = X_1(z) \cdot X_2(z) \tag{17.45}$$

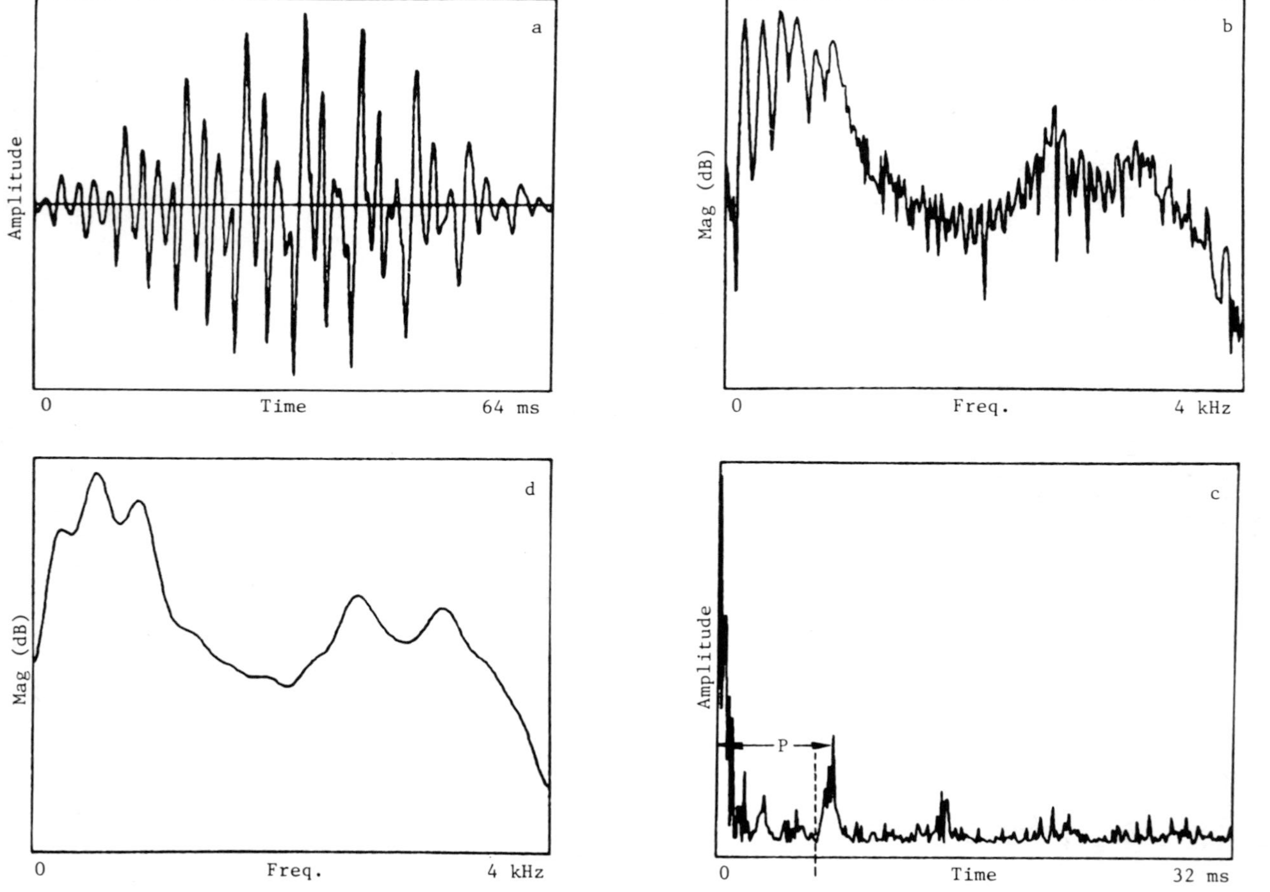


**Figure 17.24** Homomorphic analysis of voiced speech. (a) Windowed time waveforms; (b) log magnitude spectrum (using FFT); (c) cepstrum [inverse FFT of (b)]; (d) cepstrally smoothed spectrum [FFT of (c) with portion above dashed line set to zero].

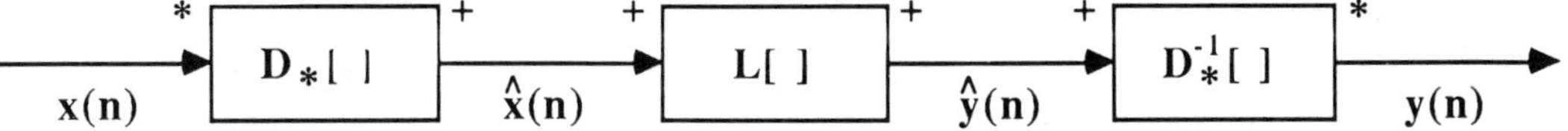


**Figure 17.25** Homomorphic system for convolution.

Likewise, the logarithm of a product can be defined as the sum of the logarithms of the individual terms:

$$\log[X_1(z) \cdot X_2(z)] = \log[X_1(z)] + \log[X_2(z)] \tag{17.46}$$

Thus $D_*[\,]$ (the first system in Figure 17.25) can be defined as in Figure 17.26a and has the desired input operation of convolution and output operation of addition to produce the cepstrum $\hat{x}(n)$ (note that this is essentially what was done in Figures 17.24a–c). The inverse characteristic system (the final system in Figure 17.25) $D_*^{-1}[\,]$ can be defined as in Figure 17.26b.

Since the $Z$-transform is, in general, a complex transform, the logarithm of $X(z)$ must be a complex logarithm [for a positive real $X(z)$, such as the magnitude, the standard log function may be used]. To compute the complex logarithm for $X(z)$ evaluated on the unit circle (i.e., the Fourier transform) the continuous "unwrapped" phase is needed, not the argument of the phase as normally calculated [1]. FORTRAN programs are available in [18] to unwrap phase, as well as perform other homomorphic techniques. The term "complex cepstrum" is used for the inverse transform of the complex logarithm of the complex $X(z)$, whereas the term "cepstrum" is used for the inverse transform of the logarithm of the magnitude of $X(z)$.

The complex cepstrum $\hat{x}(n)$ can be used to deconvolve $x_1(n) * x_2(n)$ by setting $\hat{x}_1(n)$ to zero [or likewise setting $\hat{x}_2(n)$ to zero]. Thus the high and low quefrencies of $\hat{x}(n)$ can be extracted by the simplest linear operation of windowing (the operation

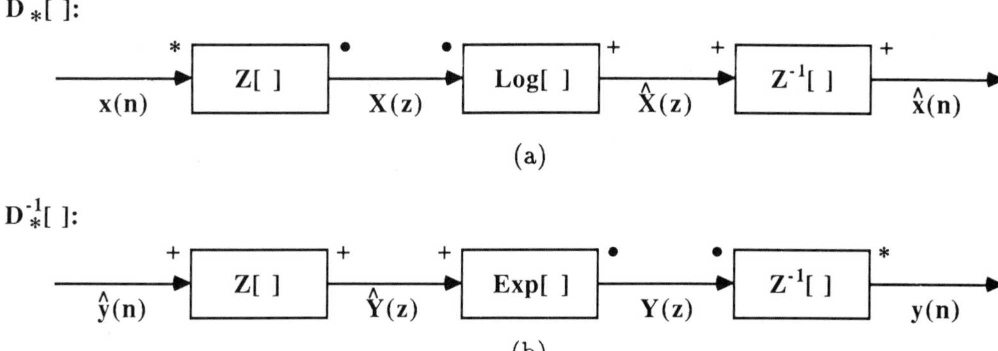


**Figure 17.26** (a) Characteristic system for homomorphic deconvolution; (b) inverse characteristic system.

$L[\,]$ in Figure 17.25) to separate the speech excitation sequence and the vocal tract filter. Homomorphic analysis is used for speech spectrum analysis, pitch period, and formant frequency estimation and has been used in speech coding (the homomorphic vocoder).

## REFERENCES


1. A. V. Oppenheim and R. W. Schafer, *Digital Signal Processing*, Prentice-Hall, Englewood Cliffs, New Jersey (1975).
2. L. R. Rabiner and R. W. Schafer, *Digital Processing of Speech Signals*, Prentice-Hall, Englewood Cliffs, New Jersey (1978).
3. N. S. Jayant and P. Noll, *Digital Coding of Waveforms: Principles and Applications to Speech and Video*, Prentice-Hall, Englewood Cliffs, New Jersey (1984).
4. J L. Flanagan, M. R. Schroeder, B. S. Atal, R. E. Crochiere, N. S. Jayant, and J. M. Tribolet, Speech coding, *IEEE Trans. Commun., COM-27*: 710–736 (April 1979).
5. J. Allen, ed., Man-machine speech communication (Special Issue), *Proc. IEEE, 73* (November 1985).
6. W. A. Lea, ed., *Trends in Speech Recognition*, Prentice-Hall, Englewood Cliffs, New Jersey (1980).
7. D. H. Klatt, Review of the science and technology of speech synthesis, *Overviews of Emerging Research Techniques in Hearing, Bioacoustics, and Biomechanics: Proceedings of the 1981 Meeting*, National Academy Press, Washington, D. C. (1982).
8. N. R. Dixon and T. B. Martin, eds., *Automated Speech and Speaker Recognition*, IEEE Press, New York (1979).
9. J. S. Lim, ed., *Speech Enhancement*, Prentice-Hall, Englewood Cliffs, New Jersey (1983).
10. G. Fant, *Acoustic Theory of Speech Production*, Mouton, The Hague (1970).
11. J. L. Flanagan, *Speech Analysis, Synthesis and Perception*, Springer-Verlag, New York (1972).
12. J. F. Kaiser, "Some Observations on Vocal Tract Operation from a Fluid Flow Point of View," Conference on Physiology and Biophysics of Voice, Iowa City, Iowa (May 7, 1983).
13. N. Chomsky and M. Halle, *The Sound Pattern of English*, Harper & Row, New York (1968).
14. R. K. Potter, G. A. Kopp, and H. G. Kopp, *Visible Speech*, Dover, New York (1966).
15. A. V. Oppenheim, Speech spectrograms using the fast Fourier transform, *IEEE Spectrum, 7*: 57–62 (August 1970).
16. R. W. Schafer and J. D. Markel, eds., *Speech Analysis*, IEEE Press, New York (1978).
17. E. O. Brigham, *The Fast Fourier Transform*, Prentice-Hall, Englewood Cliffs, New Jersey (1974).

18. Digital Signal Processing Committee, eds., *Programs for Digital Signal Processing*, IEEE Press, New York (1979).
19. J. Makhoul, Linear prediction: A tutorial review, *Proc. IEEE, 65*: 561–580 (April 1975).
20. J. D. Markel and A. H. Gray, Jr., *Linear Prediction of Speech*, Springer-Verlag, New York (1976).
21. J. Makhoul, Stable and efficient lattice methods for linear prediction, *IEEE Trans. Acoust. Speech Signal Process., ASSP-25*: 423–428 (October 1977).
22. A. V. Oppenheim and R. W. Schafer, Homomorphic analysis of speech, *IEEE Trans. Audio Electroacoust., AU-16*: 221–226 (June 1968).

# 18

# Vibration Analysis

MOHAMMAD N. NOORI Mechanical Engineering Department, Worcester Polytechnic Institute, Worcester, Massachusetts

HOSSEIN HAKIMMASHHADI Electrical Engineering Department, Worcester Polytechnic Institute, Worcester, Massachusetts

## 1 INTRODUCTION

The subject of vibration deals with the oscillatory motion of dynamic systems and the forces associated with them. A dynamic system is a combination of matter which possesses mass and whose parts are capable of relative motion. Mass is inherent in the bodies comprising the system, and the elasticity is a function of the material properties and shape of the body or bodies. most engineering machines and structures experience vibration to some degree, and their design generally requires consideration of their oscillatory behavior. The oscillatory motion of the system may be objectionable or desirable, depending on the purpose of the machine.

The objective of the designer is to control or minimize the vibration when it is objectionable and to utilize and enhance the vibration when it is desirable. Objectionable vibration in a device may cause fatigue failure, the loosening of parts, malfunctioning, or eventual failure of a mechanical system. On the other hand, shakers in foundries and vibrators in testing machines require vibration. The proper functioning of many instruments depends on the proper control of the vibrational characteristics of the devices.

The primary objective of this chapter is to present some basic concepts of mechanical vibrations and provide an analytical understanding of the oscillatory motion of dynamic systems. For each subject discussed herein, references are provided for further in-depth study of related and more advanced topics.

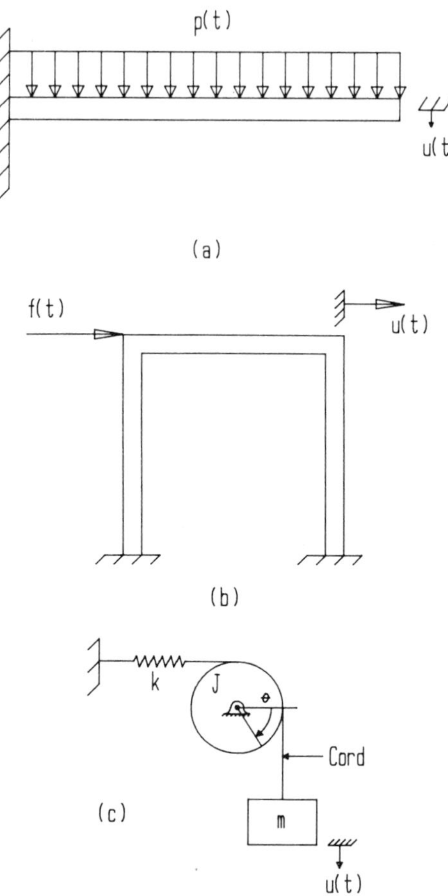


**Figure 18.1** Examples of systems with a single degree of freedom.

## 2 BASIC CONCEPTS AND CLASSIFICATION OF VIBRATION

Vibration can be classified in various ways. This classification depends on the behavior of the physical *system* under study as well as the nature of the *dynamic forces* acting on the system and the response of the system to these loads.

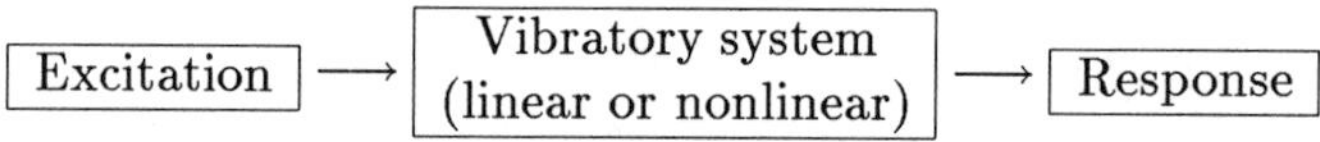


Oscillatory systems can be characterized as *linear* or *nonlinear*. If the basic components of a vibratory system behave linearly, the resulting vibration is known as *linear vibration*. On the other hand, if any of the basic components behave nonlinearly, the vibration is called *nonlinear vibration*. The governing differential equations corresponding to linear and nonlinear vibration are linear and nonlinear,

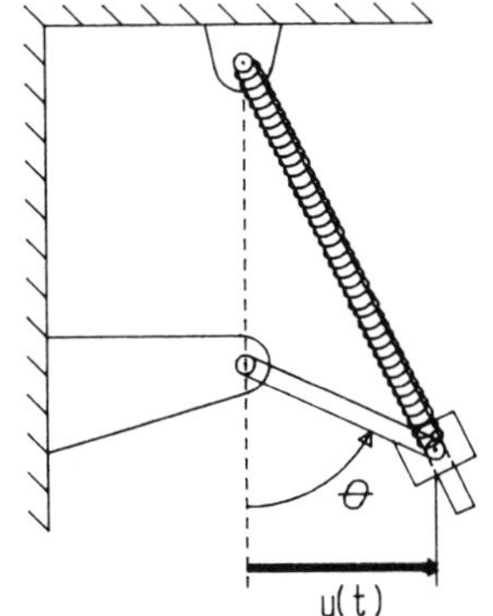


(d)

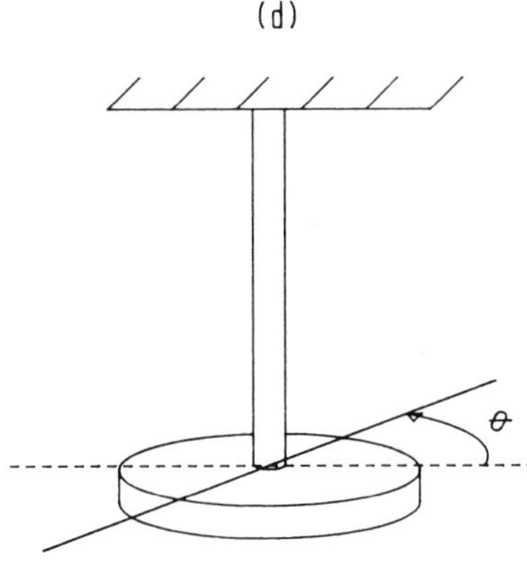


(e)

respectively. For linear vibration, the principle of superposition holds, and the mathematical techniques of analysis are well developed. In contrast, techniques for the analysis of nonlinear systems are less well known and are difficult to apply. Since all vibratory systems tend to behave nonlinearly with increasing amplitude of oscillation, some knowledge of nonlinear vibration is desirable. In this chapter the focus will be on the analysis of linear systems.

In the analysis of dynamic systems, the minimum number of independent parameters required to define the distance of all the masses from their reference positions at any time is called the number of *degrees of freedom.* For example, if there are $N$ masses in a system constrained to move only in the $x$ and $y$ directions, the system has $2N$ degrees of freedom. A continuous structure such as a beam has an infinite number of degrees of freedom. Nevertheless, the process of idealization or selection of an appropriate mathematical model permits the reduction of degrees of freedom (DOF). Systems with a finite number of degrees of freedom are called *discrete* or *lumped-parameter* systems, and those with an infinite number of DOF are called *continuous* or *distributed* systems. Figure 18.1 shows some examples of structures which may be represented for vibration analysis as one-

DOF systems, that is, structures modeled as systems with a single displacement coordinate. These *single-degree-of-freedom* (SDOF) systems can be described conveniently by the mathematical model shown in Figure 18.2, which has the following elements: 1) a mass element $m$ representing the mass and inertial characteristics of the physical system or the kinetic energy-storing element, 2) a spring element $k$ representing the elastic restoring force and potential energy capacity of the structure, 3) a damping element $c$ representing the frictional characteristics and energy losses of the structure, and 4) an excitation force $f(t)$ representing the external forces acting on the physical system. In adopting this *ideal* mathematical model, it is assumed that each element in the system represents a single property; that is, the mass $m$ represents only the property of inertia and not elasticity or energy dissipation, while the spring $k$ represents exclusively elasticity and not inertia or energy dissipation. Finally, the damper $c$ represents the dissipation of energy.

In spite of the limitations, the lumped-parameter model approach to the study of vibration problems is well justified for the following reasons: 1) many physical systems are essentially lumped-parameter systems, 2) the concepts can be extended to analyze the vibration of elastic bodies, 3) many physical systems are too complex to be investigated analytically as elastic bodies and are often studied through the use of their equivalent lumped-parameter systems and 4) the assumption of lumped parameters greatly simplifies the analytical effort required to obtain a solution. The information acquired from the analysis of the mathematical model of Figure 18.2 may very well be sufficient for an adequate understanding of the dynamic behavior of the physical system, including design and safety requirements.

The three elements comprising the SDOF mathematical model spring, mass, and damper require some elaboration. The *spring element*, which is a linear spring, is a type of mechanical link generally assumed to have negligible mass and damping. A force is developed in the spring whenever there is relative motion between the two ends of the spring. The spring force is proportional to the amount of deformation and is given by

$$f_s = kx \tag{18.1}$$

where $f_s$ is the spring force, $x$ is the deformation, and $k$ is called the *spring stiffness.* If a graph of $f_s$ versus $x$ is plotted, the result is a straight line according to Eq. (18.1). The work done in deforming a spring is stored as strain or potential energy in the spring. Elastic elements like beams also behave as springs. For ex-

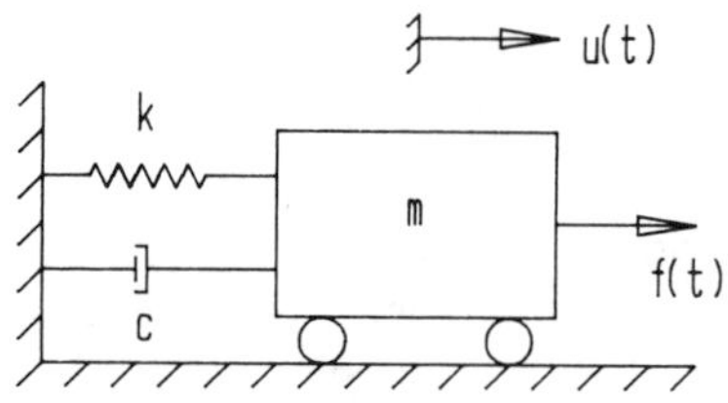


**Figure 18.2** Schematic model of an SDOF system.

ample, consider a cantilever beam with a load $P$ acting at the end, as shown in Figure 18.3. If we assume that the mass of the beam is negligible, from the strength of materials the static deflection of the beam at the free end is given by

$$\delta = \frac{Pl^3}{3EI} \tag{18.2}$$

where $E$ is Young's modulus of elasticity, $I$ is the area moment of inertia of the cross section of the beam, and $l$ is the length of the beam. Hence the stiffness of the beam, defined as the force required to cause a unit deflection, is

$$k = \frac{P}{\delta} = \frac{3EI}{l^3} \tag{18.3}$$

A similar approach, based on the direct definition of stiffness, can be used to obtain the stiffness coefficient for various types of structural or machine elements to be modeled as SDOF systems.

The spring system of a structure of $N$ degrees of freedom can be defined completely by a set of $N^2$ stiffness coefficients. A stiffness coefficient $k_{ij}$ is the change in spring force acting on the $j$th degree of freedom when only the $i$th degree of freedom is displaced a unit amount. This definition is a generalization of the linear elastic spring defined by Eq. (18.1). In most practical applications, several elastic elements (springs) are used in combination. These springs can be combined into a single equivalent spring. If the combination is such that the same force acts through all the springs in the system, the springs are in *series.* The equivalent stiffness in this case is

$$\frac{1}{K_{\text{eq}}} = \sum_i \frac{1}{k_i} \tag{18.4}$$

If the combination is such that each individual spring goes through the same amount of deformation, the springs are in *parallel.* In this case the equivalent stiffness of the system is

$$K_{\text{eq}} = \sum_i k_i \tag{18.5}$$

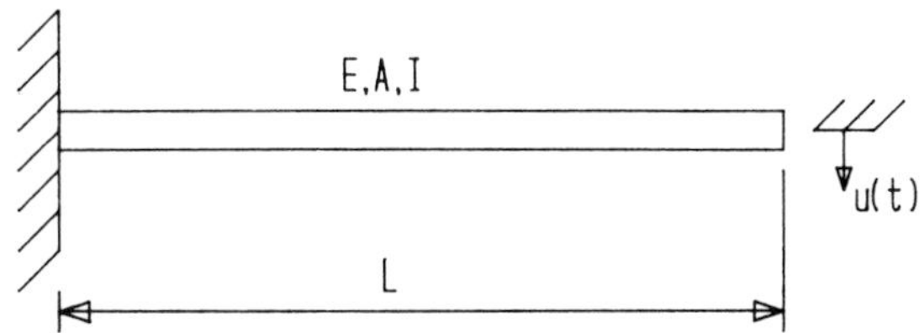


**Figure 18.3** Cantilever beam.

Equation (18.4) indicates that when a very stiff spring is in series with a weak spring, the weak spring will dominate the natural frequency of the system which is directly related to its stiffness. Thus, in practical design, the existence of a weak link should be recognized. Note that beefing up the rest of the system will not alleviate the problem of stiffness. However, as Eq. (18.5) shows, when a stiff spring is in parallel with a weak spring, the stiff spring will dominate. Procedures and tables for the evaluation of the stiffnesses of various types of structural and mechanical elements are given in [1, 2].

The mass or the inertia element is assumed to be a rigid body; it gains or loses kinetic energy whenever the velocity of the body changes. The work done on a mass by the applied force is stored in the form of kinetic energy of the mass.

In most cases, a mathematical model is needed to represent the actual vibrating system, and there are often several possible models. The analysis determines which mathematical model is appropriate. Once the model is chosen, the mass or inertia elements of the system can easily be identified. For example, consider the cantilever beam with a tip mass shown in Figure 18.4a. Suppose the purpose of the analysis is to estimate the natural frequency of this structure. For a quick and reasonably accurate analysis, the mass and damping of the beam can be disregarded; the system can be modeled as a spring-mass system, as shown in Figure 18.4b. The tip mass $m$ represents the mass element, and the elasticity of the beam, as given by Eq. (18.3), denotes the stiffness of the spring. As another example, consider a multistory building subjected to an earthquake. Assuming that the mass of the frame is negligible compared to the masses of the floors, the building can be modeled as a multi-degree-of-freedom system, as shown in Figure 18.5. The masses at the various floors represent the mass elements, and the elasticities of the column members denote the spring elements. As a third example, a vehicle suspension is considered. An automobile has many degrees of freedom. If simplifying the motion of a car, we may assume that the car moves in the plane of the paper and its motion consists of the vertical motion of the

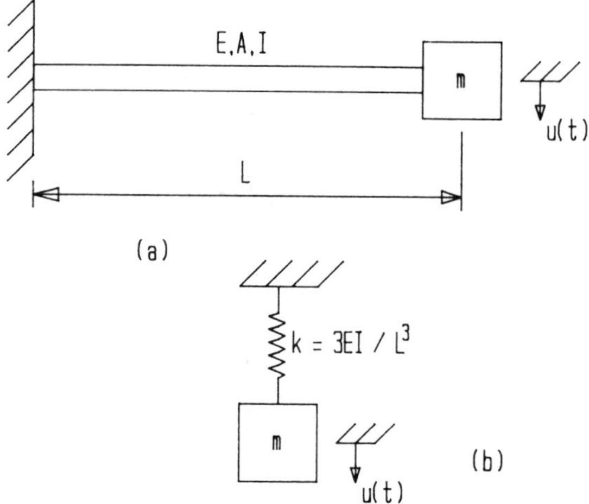


**Figure 18.4** Cantilever beam with an end mass.

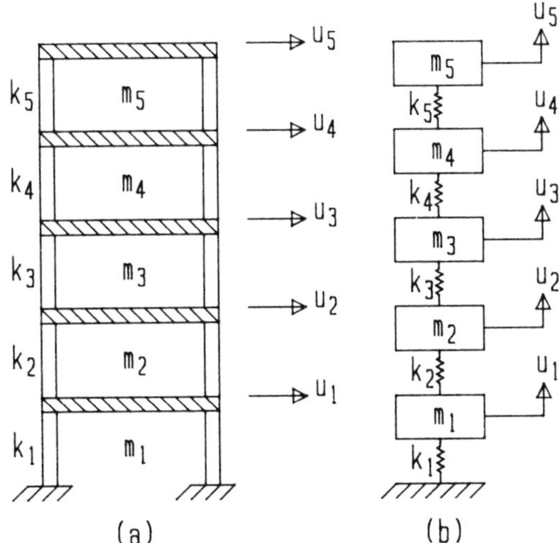


**Figure 18.5** Multistory structure modeled as an MDOF system.

car body, the rotational pitching motion of the body about its center of gravity and the vertical motion of the wheels. Because of the large separation of natural frequencies between the wheels and the car body, the model representing the motions of the car body can be further simplified as shown in Figure 18.6, which is a two-DOF system, where the vertical motion of the body, $x$, and the angular rotation about the center of gravity, $\theta$, describe the two degrees of freedom.

As in spring systems, masses can be combined in practical systems. In general, an energy approach can be used to derive the equivalent mass of a system of lumped masses.

A damper element is assumed to have neither mass nor elasticity, and damping force exists only if there is relative velocity between the two ends of the damper. The energy or work input to a damper is converted into heat; hence the damping element is nonconservative. The damping may be a *viscous damping*, a *Coulomb or dry friction damping*, or a *hysteretic or structural damping.* Viscous damping is the most widely used damping mechanism in vibration analysis. In this linear model, the damping is proportional to the velocity of the vibrating body. This type of damping is present whenever a viscous fluid flows through a slot, around a piston in a cylinder, or around the journal in a bearing. In Coulomb damping, the damping force is constant in magnitude but opposite in direction to the motion of the vibratory system. It is caused by kinetic friction between two sliding surfaces. Hysteretic or structural damping represents the energy dissipation or absorption in materials as they are deformed. The hysteresis action is due to friction between the internal planes, which slip or slide as the deformation takes place. This is very complicated behavior, and various types of hysteresis models have been developed for representing energy dissipation behavior of different types of materials. References on hysteretic behavior and vibration damping are presented at the end of this chapter [3–14].

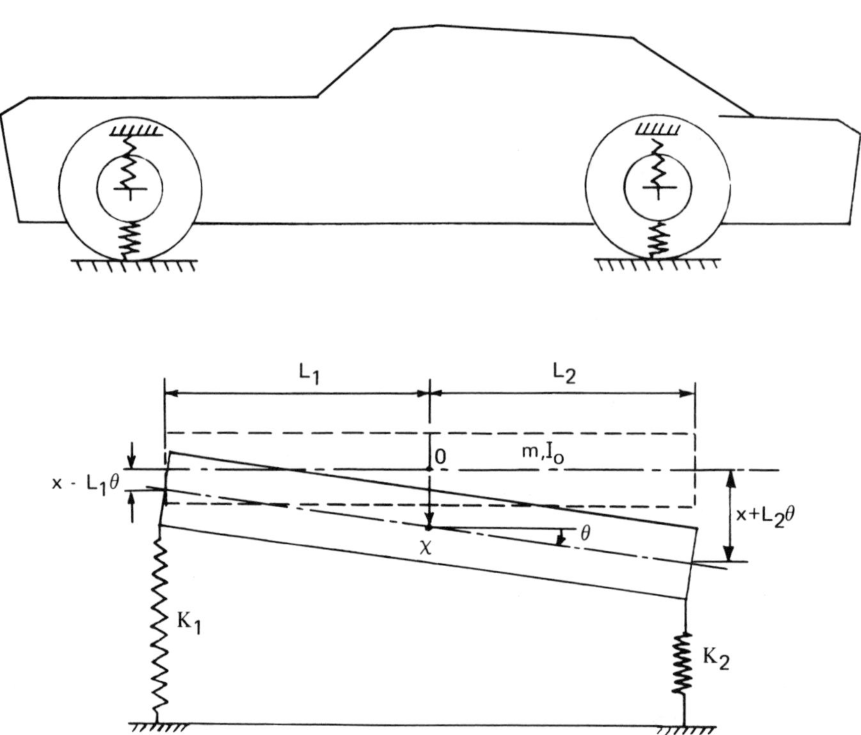


**Figure 18.6** Simplified vibrating system of an automobile body.

So far, only systems with rectilinear motion have been discussed. In the case of systems with rotational motions, such as the one shown in Figure 18.1e, the system parameters are the mass moment of inertia $J$, the torsional spring with spring constant $k_t$, and the torsional damper with damping coefficient $C_t$. An angular displacement $\theta$ is analogous to a rectilinear displacement $u$, and an excitation torque $T(t)$ is analogous to an excitation force $f(t)$.

Classification of vibration can also be based on the nature and behavior of the input to the system and the corresponding response. There are two general classes of vibrations in this regard—free and forced vibrations. *Free vibration* takes place when a system oscillates under the action of forces inherent in the system itself and when, after an initial disturbance, the system is left to vibrate on its own and no external force acts on the system. This represents the behavior of a system as it relaxes from an initial state of constraint to its equilibrium state. The constraints are called the *initial conditions* imposed on the system. For example, the mass of the SDOF system shown in Figure 18.1 may be given an initial displacement and/or an initial velocity to set it into motion. If the system possesses damping, owing to the dissipation of energy in the damper, this vibratory motion will eventually die out. Thus the equilibrium state corresponds to the *static equilibrium* position of the system. Under idealized conditions, if the system does not possess damping, the vibratory motion will not diminish with time. The system will then oscillate about its static equilibrium position. The oscillation of a simple pendulum is an example of free vibration. The system under free vibration will vibrate at one or more of its *natural frequencies*, which are properties of the dynamical system established by its mass and stiffness distribution.

If a system is excited by an external force, the resulting vibration is called *forced vibration.* The oscillation that arises in machines such as rotating shafts due to unbalance is an example of forced vibration. The general classification of forced vibration problems is shown below. This classification is discussed in more detail in the following.

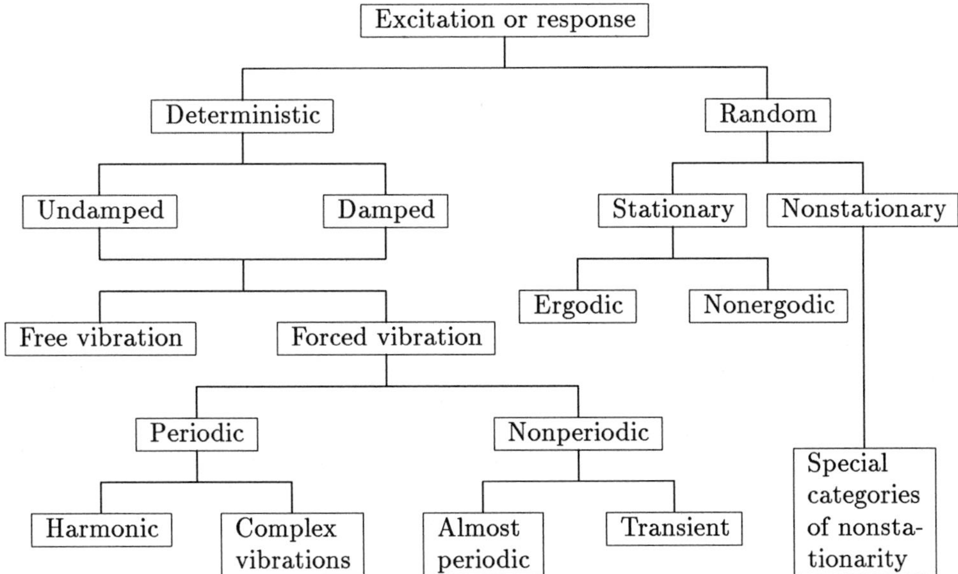


When the excitation is oscillatory, the system is forced to vibrate at the excitation frequency. If the frequency of excitation coincides with one of the natural frequencies of the system, a condition of *resonance* is encountered, and the system undergoes dangerously large oscillations. Failures of major structures, such as bridges, buildings, and airplane wings, have been associated with the occurrence of resonance. Thus, the calculation of the natural frequencies is of major importance in the study of vibration.

Vibration may also be described and classified as *deterministic* or *random.* This classification reflects the deterministic or random nature of the excitation and the response.

If the magnitude of the excitation (force or motion) acting on a vibratory system varies in accordance with a prescribed function at any given time, the excitation is deterministic. The resulting vibration is known as *deterministic vibration.* To be more precise, if the vibration is deterministic, it follows an established pattern so that the value of the vibration at any designated future time is completely predictable from the past history.

In deterministic vibration, the input excitation time history may have a *regular* and *repeating* pattern as shown in Figure 18.7a. In this case the vibratory

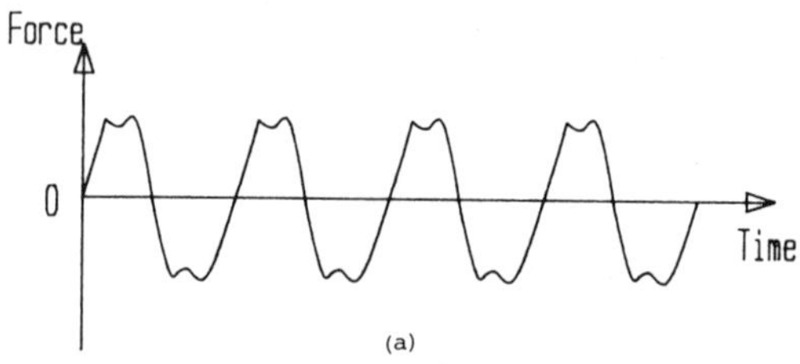


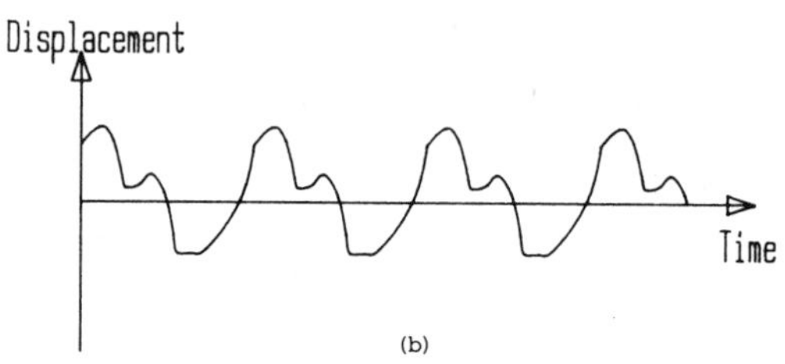


**Figure 18.7** Deterministic periodic excitation.

motion is called *periodic* and the system repeats its motion at equal intervals of time. The time required for the system to repeat its motion is called a *period* $T$, which is the time required to complete one *cycle* of motion. Frequency $f$ is the number of times the motion repeats itself per unit time. Figure 18.7b represents a deterministic periodic vibration (piston acceleration of a combustion engine).

The simplest type of periodic motion is *harmonic motion.* This motion can be demonstrated by a mass suspended from a light spring, as shown in Figure 18.8. If the mass is displaced from its rest (equilibrium) position and released, it will oscillate up and down. If the distance of the mass from equilibrium position is plotted against time, the motion follows a sine curve as shown in Figure 18.8. This

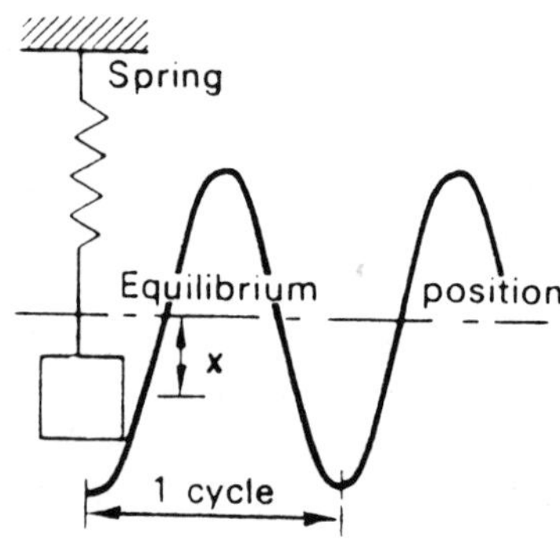


**Figure 18.8** Free oscillation of a mass and spring system.

motion can be expressed by the equation

$$u = A \sin \frac{2\pi t}{T} = A \sin \omega t \tag{18.6}$$

where $A$ is the amplitude, the maximum displacement, or the peak value of oscillation measured from the equilibrium position of the mass, $T$ is the period, and $\omega$ is the frequency of the oscillation.

Harmonic motion can be represented as the projection on the vertical axis of a point moving on a circle at constant angular velocity, as shown in Figure 18.9. With the angular velocity of the vector **OP** designated by $\omega$, the displacement $u$ can be written as

$$u = A \sin \omega t \tag{18.7}$$

The period ($T$) and the circular frequency ($\omega$) of the simple harmonic motion are usually measured in seconds and cycles per second (hertz), respectively.

The velocity and acceleration of harmonic motion can be simply determined by differentiation of Eq. (18.7). This results in

$$\dot{u} = \omega A \cos \omega t = \omega A \sin\left(\omega t + \frac{\pi}{2}\right) = A_v \sin\left(\omega t + \frac{\pi}{2}\right) \tag{18.8a}$$

$$\ddot{u} = -\omega^2 A \sin \omega t = \omega^2 A \sin(\omega t + \pi) = A_a \sin(\omega t + \pi) \tag{18.8b}$$

where $A_v$ and $A_a$ are called the *peak values* of velocity and acceleration, respectively. Thus, velocity and acceleration are also harmonic with the same frequency of oscillation, but they lead the displacement by $\pi/2$ and $\pi$ radians, respectively. Figure 18.10 shows both the time variation and the vector phase relationship between

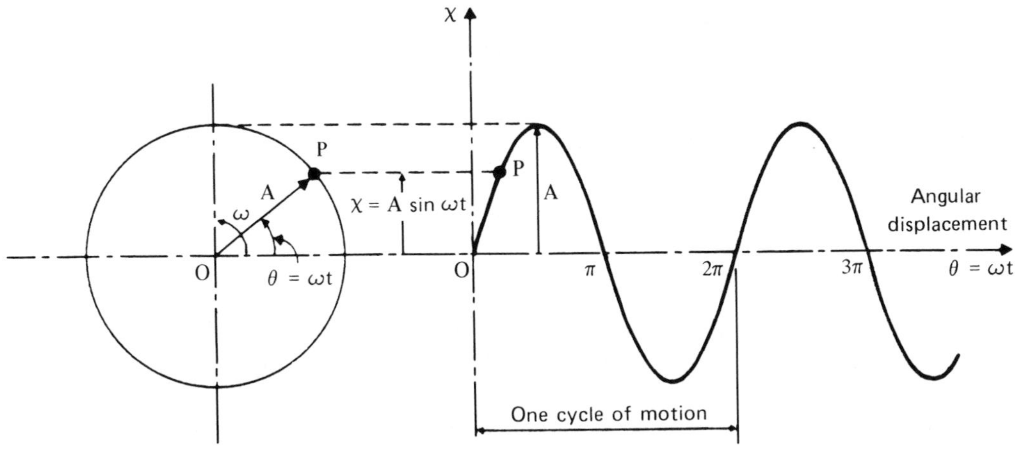


**Figure 18.9** Harmonic motion as the projection of a rotating vector.

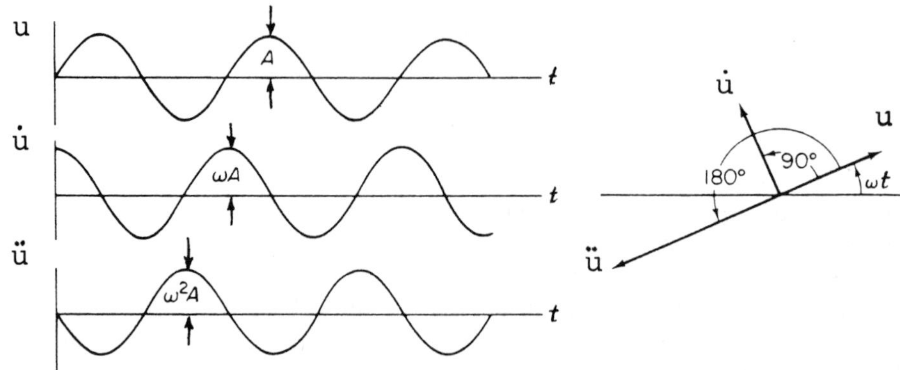


**Figure 18.10** Phase relationships between displacement, velocity, and acceleration for harmonic motion.

the displacement, velocity, and acceleration in harmonic motion. For more complex vibrations, the phase angle can have any value, depending on the characteristics of the system.

Substituting from Eq. (18.7) in Eq. (18.8b) results in

$$\ddot{u} = -\omega^2 u \tag{18.9}$$

so that in harmonic motion the acceleration is proportional to the displacement and is directed opposite to $u$.

Amplitude measurement is particularly important with plain bearings, since it indicates the risk of abrasion. Velocity is related, through the number of stress reversals, to potential fatigue failure. Acceleration is related to force and thus to possible development of wear. Further details on the application of these measurements are provided in [15–25].

A few terms used in the magnitude description are now introduced. The arithmetic *average* value of the amplitude ($u_{\mathbf{avg}}$) is that which produces the same time history as the actual motion. It is defined as

$$u_{\mathbf{avg}} = \frac{2}{T} \int_0^{T/2} |u| \, dt \tag{18.10}$$

Since a full cycle produces a zero integral, time values are halved from $t = 0$ to $T/2$. Average values for velocity and acceleration can be determined in a similar manner.

Energy is proportional to velocity squared, and thus the squared value of displacement etc. is in general more significant than the arithmetical average. Simple vibration measurement instruments produce the *root mean square* (RMS) value.

The root mean square amplitude $U_{\mathrm{RMS}}$ is

$$U_{\mathrm{RMS}} = \sqrt{\frac{1}{T}\int_0^T u^2\,dt} \tag{18.11}$$

A significance of this quantity is it simple relationship to the power content of vibration. For the case of a pure harmonic motion given by Eq. (18.7), it can be shown that

$$U_{\mathbf{avg}} = \frac{2}{\pi A} \tag{18.12}$$

$$U_{\mathrm{RMS}} = \frac{1}{\sqrt{2}A} \tag{18.13}$$

which yields

$$U_{\mathrm{RMS}} = \frac{\pi}{2\sqrt{2}}U_{\mathbf{avg}} \tag{18.14}$$

Figure 18.11 shows an example of a harmonic vibration with indication of the peak, the RMS, and the average absolute value. For complex vibrations the relationships between the peak, average, and RMS have different values, normally expressed as

$$U_{\mathrm{RMS}} = F_f U_{\mathbf{avg}} = C_f A \tag{18.15}$$

where $F_f$ is called the *form factor* and $C_f$ the *crest factor*. Both $F_f$ and $C_f$ give some indication of the waveform of a complex vibration. From Eqs. (8.12)–(18.15) it can be seen that for simple harmonic motion

$$F_f = \frac{\pi}{2\sqrt{2}} = 1.11\ (\approx 1\ \mathrm{dB}) \qquad \text{and} \qquad C_f = \sqrt{2} = 1.414\ (\approx 3\ \mathrm{dB})$$

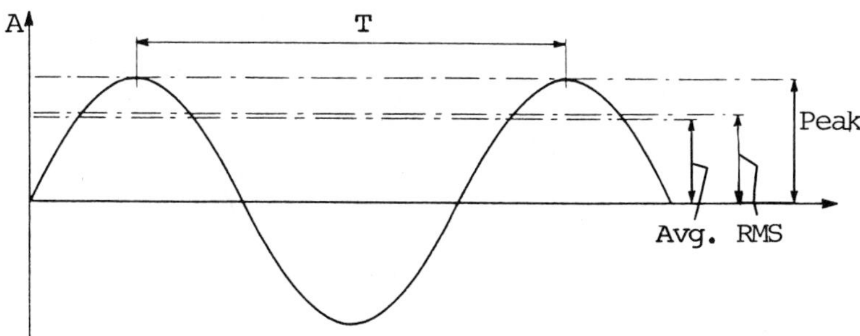


**Figure 18.11** Harmonic vibration with indication of the peak, RMS, and average values.

Since harmonic motion is a common type of vibration encountered in practice, introduction of a few more terms and a discussion of characteristics of this oscillatory motion are presented in the following.

If we consider two harmonic motions given by

$$u_1 = A_1 \sin \omega t \tag{18.16a}$$

$$u_2 = A_2 \sin(\omega t + \varphi) \tag{18.16b}$$

these two motions are called *synchronous* because they have the same frequency. Two synchronous oscillations need not have the same amplitude, and they need not attain their maximum values at the same time. The angle $\varphi$ is the *phase angle.* This means that the maximum of the second motion would occur $\varphi$ radians earlier than that of the first one. In place of maxima, any other corresponding points can be taken for finding the phase angle. The two motions are said to have a *phase difference* of $\varphi$.

When two harmonic vibrations such as $u_1 = A_1 \cos 2\omega t$ and $u_2 = A_2 \cos \omega t$ are combined, the resulting waveform of $u = A_1 \cos 2\omega t + A_2 \cos \omega t$ shows the interrupting influence of one vibration on another, as shown in Figure 18.12. When other vibrations are combined the effect is presented in Figure 18.13.

When out-of-phase vibrations are combined, such as $u_1 = \cos \omega t$ and $u_2 = \cos(3\omega t + \varphi)$, the resulting waveform may be distorted in the manner shown in Figure 18.14, where phase angles of 0°, 40°, 80°, and 120° have been plotted. These illustrations represent waveforms compared to those experienced in practice, where the periodicity may not coincide as in these examples and the phases and peak values may differ. Vibrations which occur in practice may be analyzed to predict their effect on individual components or connected elements of a machine or a structure by using frequency analysis based on Fourier transformation.

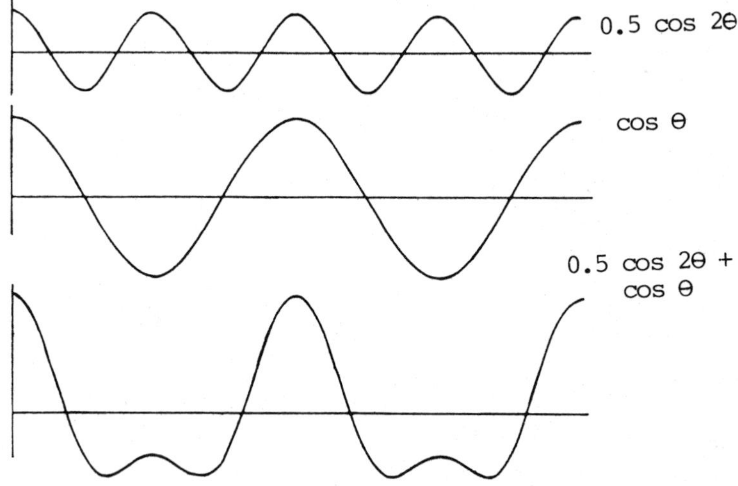


**Figure 18.12** Example of a complex vibration.

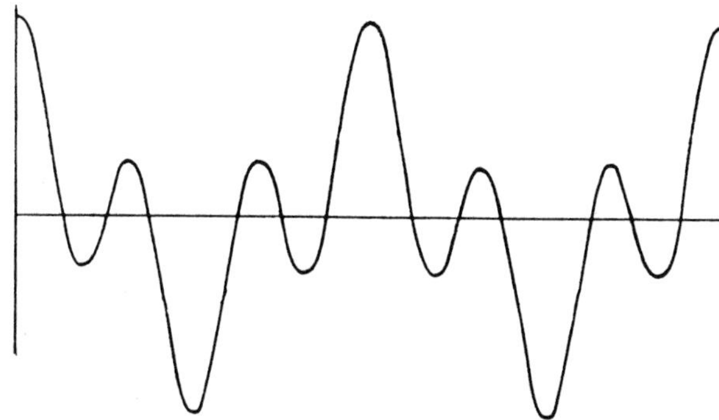

**Figure 18.13** Example of other types of complex vibrations combined.

Any periodic function of time can be represented by Fourier series as an infinite sum of sinusoidal functions. Thus, for a complex vibration all the frequencies can be resolved into harmonically related sine waves. If $x(t)$ is any periodic function with period $T$, it can be represented as an infinite series of sinusoids in the form

(i)
$$x(t) = a_0 + \sum_{n=1}^{\infty}(a_n \cos n\omega t + b_n \sin n\omega t) \tag{18.17a}$$

where $\omega = 2\pi/T$ and

$$a_0 = \frac{1}{T}\int_0^T x(t)\,dt$$

$$a_n = \frac{2}{T}\int_0^T x(t)\cos n\omega t\,dt, \qquad b_n = \frac{2}{T}\int_0^T x(t)\sin n\omega t\,dt \tag{18.17b}$$

Alternatively, $x(t)$ can be written as

(ii)
$$x(t) = C_0 + \sum_{n=1}^{\infty} C_n \cos(n\omega t - \varphi_n) \tag{18.18a}$$

where

$$C_n = (a_n^2 + b_n^2)^{1/2}, \qquad \varphi_n = \tan^{-1}\frac{b_n}{a_n} \tag{18.18b}$$

or

(iii)
$$x(t) = \sum_{-\infty}^{\infty} X_n e^{in\omega t} \tag{18.19a}$$

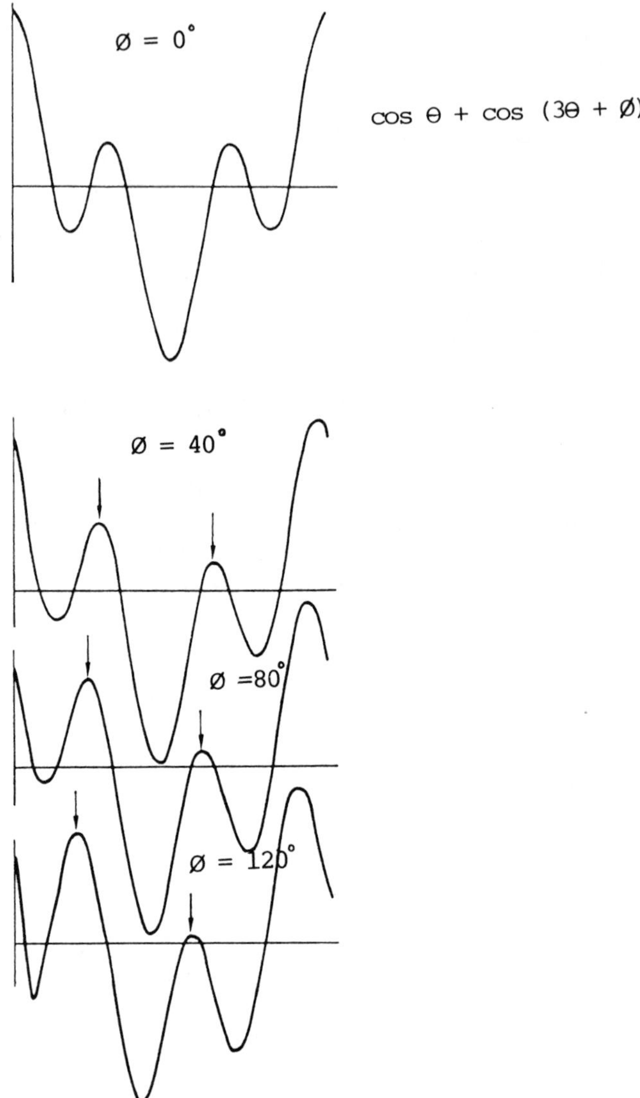


**Figure 18.14** Complex vibrations made of out-of-phase components.

where

$$X_n = \frac{1}{T}\int_0^T x(t)e^{-in\omega t}\,dt \tag{18.19b}$$

and

$$\mathrm{Re}(X_n) = \frac{a_n}{2}, \qquad \mathrm{Im}(X_n) = \frac{-b_n}{2}, \qquad |X_n| = \frac{C_n}{2}$$

The harmonic functions $a_n \cos n\omega t$ or $b_n \sin n\omega t$ are called the *harmonics* of order $n$ of the periodic function $x(t)$. The harmonic of order $n$ has a period $T/n$. These harmonics can be plotted as vertical lines on a diagram of amplitude ($a_n$ and $b_n$ or $C_n$ and $\varphi_n$) versus frequency ($n\omega$) called the *frequency spectrum* or *spectral diagram.* Figure 18.15 shows a typical frequency spectrum.

A typical analysis of a complex vibration would resolve it into a frequency spectrum, separating out the individual components in terms of peak, average, or RMS values plotted against frequency. Thus the displacement frequency spectra of the vibrations plotted in Figure 18.16 may be plotted in the form shown in Figure 18.17.

Vibrations from machines contain a wide range of sinusoidal and random components so that the frequency spectrum may look like that shown in Figure 18.18. Such a frequency spectrum is extensively used in vibration monitoring since the significant frequencies ($f_1$, $f_2$, $f_3$, etc.) correspond to the resonant or discrete frequencies of particular components of the machine. Changes in the vibration level of these discrete frequencies indicate changes in condition (possibly wear) of the relevant components [21, 25–28]. A specific feature of periodic vibrations, which becomes clear by looking at Figure 18.16–18.18 is that their spectra consist of dis-

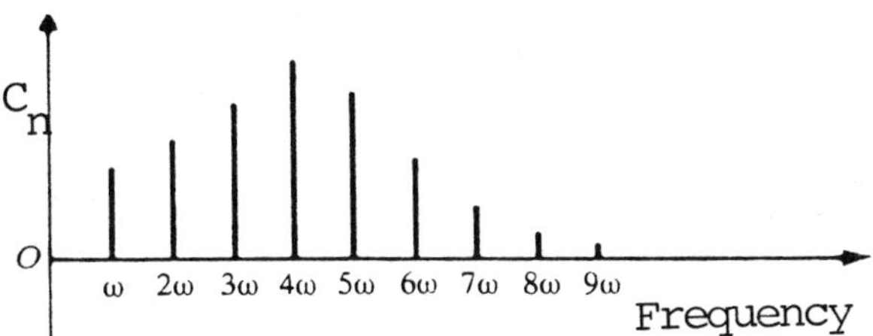


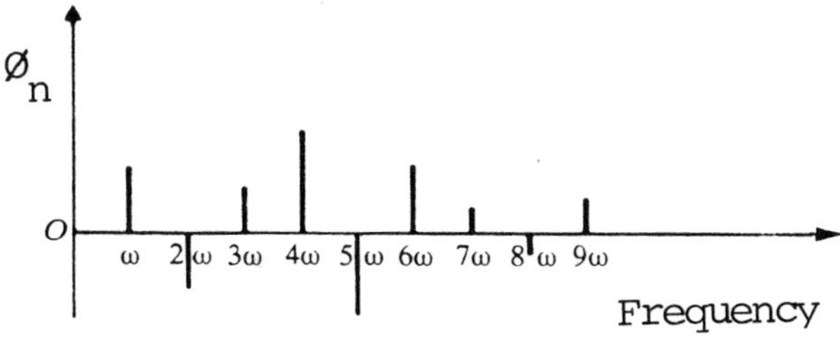


**Figure 18.15** Frequency spectrum of a typical periodic function of time.

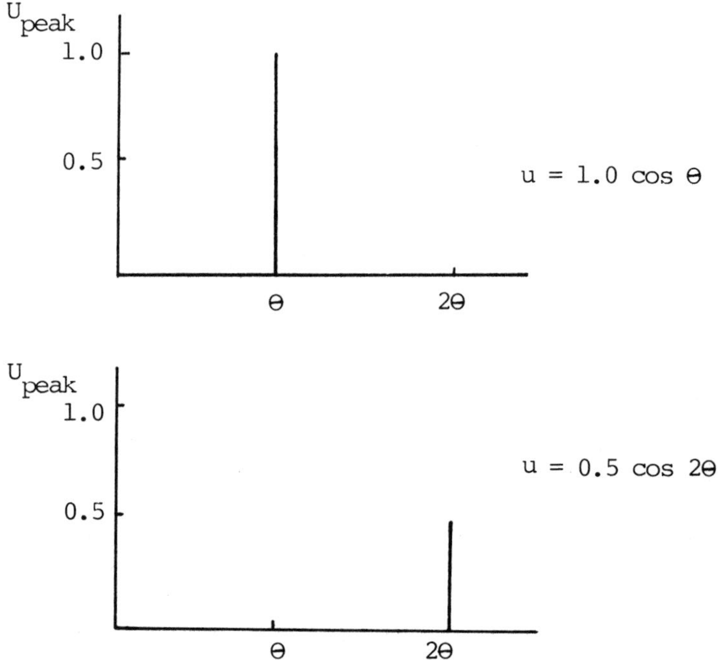


**Figure 18.16** Frequency spectra of simple vibrations.

crete lines when presented in the frequency domain. That is in contrast to random vibrations, which show continuous frequency spectra.

In the classification of deterministic vibration another important category is shock vibration. *Shock* is a somewhat loosely defined type of vibration in which the excitation is nonperiodic, e.g., in the form of a pulse or a step. The word shock implies a degree of suddenness and severity. Form the analytical viewpoint, the important characteristic of shock is that the motion of the system on which the shock acts includes both the frequency of the shock excitation and the natural frequency

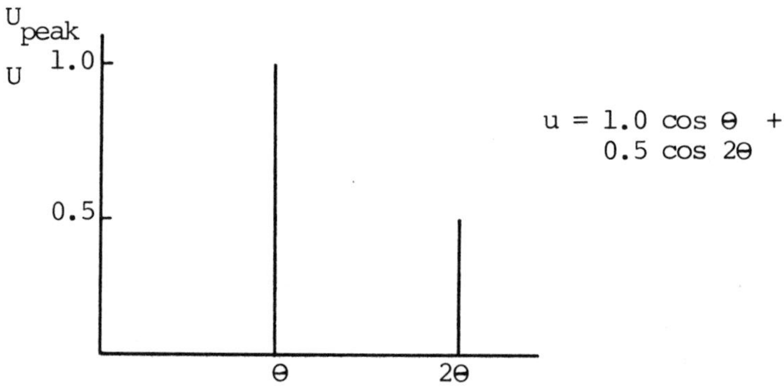


**Figure 18.17** Frequency spectra for the two simple harmonics combined.

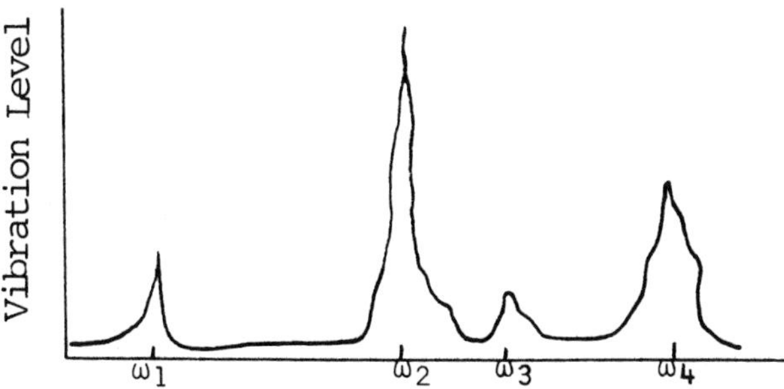


**Figure 18.18** A characteristic spectrum of machine frequency.

of the system. If the excitation is brief, the continuing motion of the system is free vibration at its own natural frequency. Shocks originate from such widely different releases of energy as rough handling of equipment, explosions, and supersonic motions. For more detail on shock and transient phenomena see [9, 13, 22–35].

In some cases, the excitation is nondeterministic or *random*; the value of the excitation at a given time cannot be predicted. This defines a class of vibration problems referred to as *random* or *stochastic vibrations*. Examples of random excitation are wind, velocity, road roughness, and ground motion during earthquakes. In the case of random vibration, the vibratory response of the system is also random; in other words, its future value is unpredictable except on the basis of probability. The instantaneous magnitude of a random vibration cannot be uniquely described at any given instant of time by ordinary mathematics or analysis. The analysis of random vibration involves certain mathematical and physical concepts that are different from those applied to the analysis of deterministic vibration.

Random vibration problems can be classified into two categories, stationary and nonstationary. Stationary random vibrations may, in practice, be defined as those whose statistical characteristics do not change with time. On the contrary, for nonstationary random vibrations these statistical properties vary within time intervals considered essential for their proper description.

One of the statistical measures in the analysis of random vibration response in the time domain is the autocorrelation function $R_x(\tau)$. A practical application of this function is that it describes how a particular instantaneous amplitude value depends on previously occurring instantaneous amplitude values. The autocorrelation function for an ergodic process is defined as

$$R_x(\tau) = \lim_{T\to\infty} \left(\frac{1}{T}\right) \int_{-T/2}^{T/2} x(t)x(t+\tau)\,d\tau \tag{18.20}$$

where $x(t)$ is the magnitude of the vibratory process at an arbitrary instant of time $t$, and $x(t+\tau)$ designates the magnitude of the same process observed at a time $\tau$ later. This basic concept is illustrated in Fig-

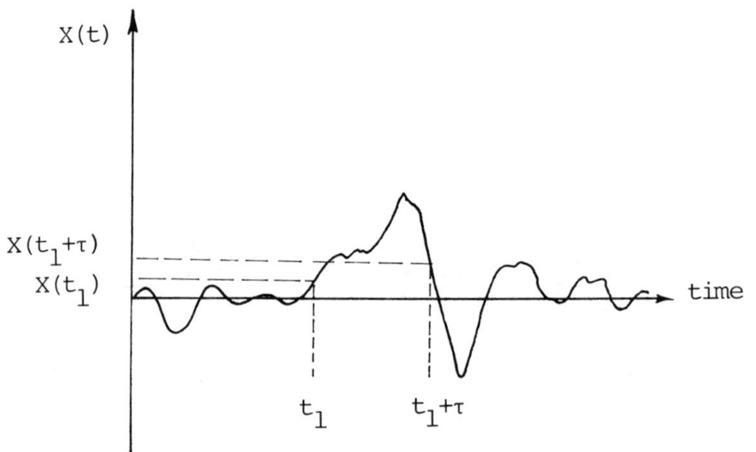


**Figure 18.19** Basic concepts involved in deriving the autocorrelation function.

ure 18.19. When the time interval $\tau$ separating the two measuring points is zero, then

$$R_x(\tau = 0) = \lim_{T \to \infty} \left(\frac{1}{T}\right) \int_{T/2}^{T/2} x^2(t)\, dt \tag{18.21}$$

and is just equal to the mean-square value of the process. In the case of an ideal stationary random process (white noise) the autocorrelation function would consist of an infinitely narrow impulse function around $\tau = 0$, as shown in Figure 18.20a. In such a process each instantaneous amplitude value is independent of all other instantaneous amplitude values. In practice, however, the autocorrelation functions for stationary random vibrations are not ideal delta functions, as can be seen in Figures 18.20b and 18.20c. The reason for this behavior is that all practical random processes are frequency-limited, and the narrower the frequency limits the more spread out are the corresponding autocorrelation functions, because the rate at which a signal can change from its current value is more limited.

Another important function in deciding the random vibrations is the Fourier transform of $R_x(\tau)$, which has a certain resemblance to the Fourier frequency spectra for periodic vibrations described earlier. This function is given by

$$S_x(\omega) = \left(\frac{1}{2\pi}\right) \int_{-\infty}^{\infty} R_x(\tau) e^{-i\omega\tau}\, d\tau \tag{18.22}$$

where $S_x(\omega)$ is called the *spectral density function* or *power spectral density function* of the process $x(t)$ and is a function of angular frequency $\omega$. From the properties

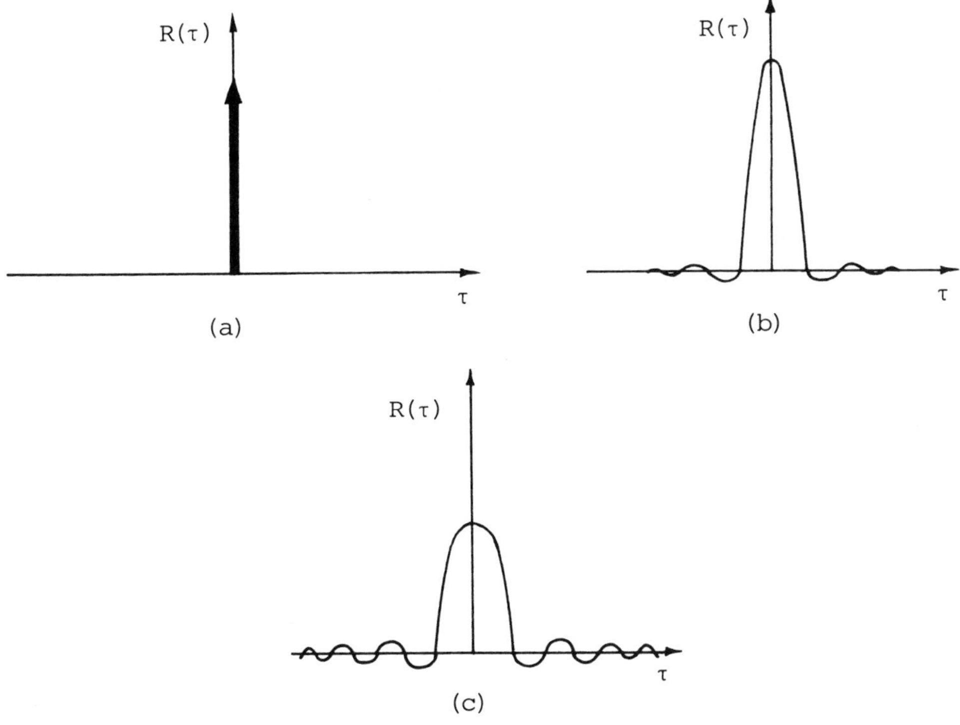


**Figure 18.20** Examples of autocorrelation functions.

of the Fourier transform and its inverse, $R_x(\tau)$ can thus be defined by

$$R_x(\tau) = \int_{-\infty}^{\infty} S_x(\omega) e^{i\omega t}\, d\omega \tag{18.23}$$

The most important property of $S_x(\omega)$ becomes apparent when we put $\tau = 0$. From Eq. (18.23) it can be shown that

$$R_x(\tau = 0) = \int_{-\infty}^{\infty} S_x(\omega)\, d\omega \tag{18.24}$$

Comparison of this equation with Eq. (18.21) shows that the mean-square value of a stationary random process $x$ is given by the area under a graph of spectral density $S_x(\omega)$ against $\omega$, Figure 18.21. The units of $S_x(\omega)$ are accordingly those of mean square per unit of frequency and a more complete name for $S_x(\omega)$ is the *mean-square spectral density.* Since the value of $R_x(\tau = 0)$ in the time domain is given by Eq. (18.21), in which the squaring is involved in the integration of the time function, $S_x(\omega)$ has been designated as the *power spectral density.* It can also be seen that the expression given by

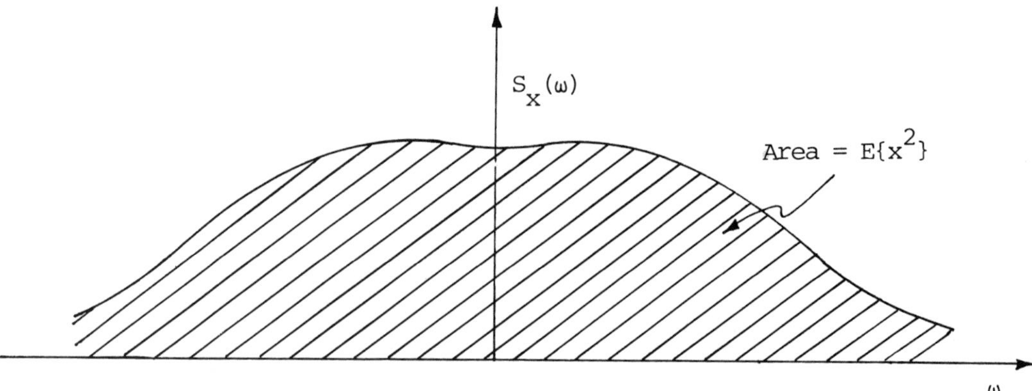


**Figure 18.21** Plot of spectral density versus $\omega$. The area under the curve is equal to the mean-square value of a stationary process $X(t)$.

Eq. (18.21) has a close resemblance to the square of the expression previously used to define an RMS value of a periodic signal, Eq. (18.11). This means that the description of a complex signal in terms of its overall RMS value is equally meaningful whether the signal has a periodic or a random character.

When it comes to spectral description, however, a periodic vibration may well be described in terms of the RMS values of its various components, whereas random vibrations are best described in terms of power spectral density functions. This is due to the fact that random vibrations have continuous frequency spectra and the RMS value measured within a certain frequency band will therefore depend on the width of the band. Detailed measurement evaluation techniques are discussed in the literature [22, 24, 27, 36–37].

Theoretically, all kinds of random vibrations encountered in practice are nonstationary because their statistical properties vary with time. If a process has a beginning or ending, then theoretically it is not stationary. However, if a signal lasts for a long time (long compared with the period of its lowest-frequency spectral component), it may be sensible to assume that the process is approximately stationary over most of its lifetime. Strictly, this is a contradiction in terms. However, it means that sample records of the signal may be thought of as finite lengths cut from records of sample functions of a stationary process which are infinite in length. In other cases it may only be reasonable to assume that a process is stationary over part of its lifetime. For instance, the random pressure fluctuations on the casing of a rocket are initially very violent on launching, but they die down to a much lower level when the rocket leaves the earth's atmosphere and thrust is reduced. It is reasonable to assume that this process was stationary for most of its duration.

To analyze nonstationary random vibrations it is necessary to introduce the concept of *ensemble averaging.* An ensemble average is an average taken over a large number (an ensemble) of spectral experiments. An ensemble average can be taken at any particular instant of time, $t_1$, $t_2$, $t_3$, etc., and when the average values are plotted against time a more or less complete description of the vibration is obtained. This method of description is not, however, very useful in practice.

It is therefore necessary to find other methods of description, and in general some sort of time averaging is used. However, certain limitations are imposed on this kind of time averaging. The response and averaging time of the measurement equipment employed should preferably be small relative to important time trends in the nonstationary data. This may lead to considerable statistical uncertainties in the measurements. In-depth discussion of nonstationary random vibrations can be found in [22, 24, 36–39].

In all the categories discussed above, the vibratory system may be assumed and modeled to be either damped or undamped. If it is assumed that no energy is lost or dissipated in friction or other resistance during oscillation, the vibration is called *undamped vibration*. On the other hand, if any energy is lost in this way, it is called *damped vibration*. In many physical systems, the amount of damping is so small that it can be disregarded for most engineering purposes. However, consideration of damping becomes extremely important in analyzing vibratory systems near resonance. Proper treatment and modeling of damping in vibration analysis is an important and relatively complicated task. A comprehensive discussion of this issue can be found in [6–8, 10–12, 14, 27].

Further discussion of the topics discussed in this section can be found in [1, 2, 7–9, 13, 20–27, 29–31, 33–40].

## 3 VIBRATION-RELATED PROBLEMS AND CAUSES OF VIBRATION

Vibrations result from dynamic forces which produce a series of motions within the system. These motions may be linear or nonlinear or a combination of both types. A wide range of vibratory motions may occur in practice. Typical examples of some of the more characteristic vibratory modes are given in Table 18.1.

Machinery vibrations can be measured and analyzed to provide a cost-effective predictive method of diagnosing the causes for development of failure conditions. Typical machinery faults which may be discovered by vibration analysis are listed in Table 18.2.

Because of the wide range of vibration-related problems and causes of vibration, a brief discussion will be presented here on only two of the most important types of vibration problems. The first problem is the vibration of structural systems (structural dynamics), namely, ground motion or earthquake, and the second is one of the most important classes of mechanical vibrations, namely rotor dynamic problems.

The most important aspect of an earthquake's ground motions is the effect they will have on structures, that is, the stresses and deformations or the amount of damage they will produce. This damage potential is at least partly dependent on the "size" of the earthquake, and a number of measures of size are used for different purposes. The most important measure of size from a seismological point of view is the amount of strain energy released at the source, and this is indicated quantitatively as the *magnitude*. By definition, magnitude is the base 10 logarithm of the maximum amplitude, measured in micrometers, of the earthquake record

**Table 18.1** Typical Vibratatory Modes [22]

- Bending vibration
  - Aircraft wings
  - Belts
  - Bridges
  - Chains
  - Pipes
  - Propellers
  - Rails
  - Shafts
  - Transmission lines
- Extensional and shear vibration
  - Belts
  - Electric motors (hum)
  - Punch presses
  - Transformers (hum), electric
- Flexural and plate-mode vibration
  - Aircraft fuselage
  - Bridges
  - Circular saws
  - Floors
  - Gear blanks
  - Turbine blade
  - Walls
- Intermittent vibration (shocks)
  - Blasting operations
  - Catapults
  - Chipping hammers
  - Drop forges
  - Earthquakes
  - Gunshots
  - Impact wrenches
  - Rachets
  - Shapers
- Random and miscellaneous
  - Combustion
  - Earthquakes
  - Fluid motion
  - Gas motion
  - Ocean waves
  - Tides
  - Turbulence
  - Winds
- Torsional or twisting
  - Compressor shafts
  - Fan blades
  - Gears
  - Propellers
  - Pumps
  - Turbines
- Translateral, axial, or rigid-body vibration
  - Air hammers
  - Compressors
  - Diesel engines
  - Petrol engines
  - Punch presses

obtained by a seismograph, corrected to a distance of 100 km. This magnitude rating has been related empirically to the amount of earthquake energy released, $E$, given by

$$\log E = 11.8 + 1.5M$$

in which $M$ is the magnitude. According to this formula, the energy increases by a factor of 32 for each unit increase of magnitude. More important to engineers, however, is the empirical observation that earthquakes of magnitude less than 5 are not expected to cause structural damage, whereas for magnitudes greater than 5 potentially damaging ground motions will be produced.

**Table 18.2** Typical Machinery Faults

| | |
|---|---|
| Absence of lubrication | Insecure components |
| Bent shafts | Mechanical slackness |
| Broken blades | Onset of cavitation |
| Damaged or misaligned drives | Presence of solid bodies |
| Damaged bearings | (in pump fluids) |
| Eccentricity | Static or dynamic unbalance |
| Fretting corrosion | Worn bearings |
| Incorrect assembly | Worn or damaged gears |

Another important quantity which must be considered in indicating whether structural damage can be expected is *earthquake intensity.* This is a measure of the severity of the ground motion observed at any point.

The three components of a ground motion recorded by a strong-motion accelograph provide a complete description of the earthquake which would act on any structure at that site. However, the most important features of the record obtained for each component, such as Figure 18.22, from the standpoint of its effectiveness in producing structural response, are the amplitude, frequency content, and duration. The amplitude generally is characterized by the peak value of acceleration or sometimes by the number of acceleration peaks exceeding a specified level. Ground velocity may be a more significant measure of intensity than acceleration, but it generally is not available without supplementary calculations. The frequency content can be represented roughly by the number of zero crossings per second in the accelerogram, and the duration by the length of time between the first and last peaks exceeding a threshold level. All these quantitative measures taken together,

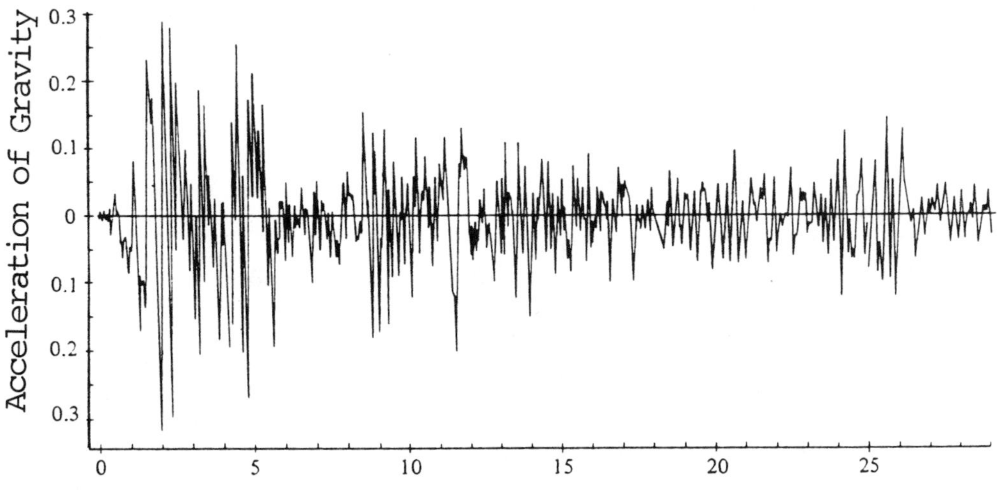


**Figure 18.22** Accelerogram from the El Centro earthquake of May 18, 1940 (NS component).

however, provide a very limited description of the ground motion and certainly do not quantify its damage-producing potential adequately. A detailed discussion of earthquake analysis techniques can be found in [9, 30, 41].

Rotor vibration problems may arise from a number of practical causes. Of these, the most common are residual unbalance and instability. Rotor unbalance causes a rotating force to act in synchronism with running speed (synchronous unbalance). Rotor instability is a self-excited vibration which may arise from bearing fluid film effects, shaft hysteresis, electromagnetic effects, flow effects, or some combination of these factors. Several other mechanisms have also caused rotor vibration problems. For example, different shaft lateral stiffnesses are a known cause of unstable whirling in electrical machinery, a and nonlinear machine foundation properties have led to vibrations in certain speed ranges. Inmost instances, practical "fixes" can be applied to reduce or eliminate these undesirable vibrations once the problem has been correctly diagnosed.

A rotor is said to whirl when the center of gravity of any cross section traces out an orbit in time, instead of remaining at a fixed point. If identical whirl orbits are traced out with successive shaft rotations, the whirl is said to be stable. If the orbit increases in size with successive rotations, the whirl is said to be unstable, and it may grow until the orbit becomes bounded either by a corresponding increase in the internal constraint forces of the system or by some external constraint, e.g., bearing rub, guard ring, or shut-down. Smooth machine operation is characterized by small, stable rotor whirl orbits and by the absence of any instabilities within the machine operating range.

Modern design of smooth-running rotating machinery requires that each of the above potential vibration problem areas be carefully analyzed as part of the design process to ensure that such problems are not built into the design. In most instances, avoidance of dangerous vibrations can be achieved quite readily if adequate allowance is made for dynamic characteristics of the machine during the design process.

The dynamic characteristics of interest in rotating machinery are

1. Rotor critical speeds which may influence operation
2. Unbalance response amplitudes at critical speeds
3. Threshold of resonant-whip instability
4. Bearing transmitted forces
5. System torsional critical speeds
6. Gear dynamic load spectrum
7. Disk natural frequencies (compressor, turbine, gear)
8. Bucket, blade, and impeller natural frequencies and modes
9. Blade flutter frequencies
10. Rotating stall and surge thresholds

Many other important vibration topics such as noise and structural vibration of rotating machinery could be added to this list. Each of the above subjects has an extensive published literature. A list of relevant publications is given at the end of this chapter. Reference [42] in particular provides a useful glossary of the terminology used in rotor dynamics.

The following tables provide a troubleshooting reference guide for rotor vibration problems. Table 18.3 lists a set of general guidelines for judging machine condition.

**Table 18.3** Criteria for Vibration Measurements (10–10,000 Hz)

| Measure overall velocity RMS and allow for the following machine types: | For New Machines | | | | For Worn Machines (full speed & power) | | | |
|---|---|---|---|---|---|---|---|---|
| | Long life (1000-10,000 hrs) | | Short life (100-1000 hrs) | | Check (recondition) level[1] | | Recondition to new (Oct. analysis)[2] | |
| | VdB* | mm/s | VdB* | mm/s | VdB* | mm/s | VdB* | mm/s |
| Gas Turbines | | | | | | | | |
| (over 20,000 HP) | 138 | 7.9 | 145 | 18 | 145 | 18 | 150 | 32 |
| (6 to 20,000 HP) | 128 | 2.5 | 135 | 5.6 | 140 | 10 | 145 | 18 |
| (up to 5,000 HP) | 118 | 0.79 | 130 | 3.2 | 135 | 5.6 | 140 | 10 |
| Steam Turbines | | | | | | | | |
| (over 20,000 HP) | 125 | 1.8 | 145 | 18 | 145 | 18 | 150 | 32 |
| (6 to 20,000 HP) | 120 | 1.0 | 135 | 5.6 | 145 | 18 | 150 | 32 |
| (up to 5,000 HP) | 115 | 0.56 | 130 | 3.2 | 140 | 10 | 145 | 18 |
| Compressors | | | | | | | | |
| (free piston) | 140 | 10 | 150 | 32 | 150 | 32 | 155 | 56 |
| (HP air,air cond.) | 133 | 4.5 | 140 | 10 | 140 | 10 | 145 | 18 |
| (LP air) | 123 | 1.4 | 135 | 5.6 | 140 | 10 | 145 | 18 |
| (refridge) | 115 | 0.56 | 135 | 5.6 | 140 | 10 | 145 | 18 |
| Diesel Generators | 123 | 1.4 | 140 | 10 | 145 | 18 | 150 | 32 |
| Centrifuges, Oil Separators | 123 | 1.4 | 140 | 10 | 145 | 18 | 150 | 32 |
| Gear Boxes | | | | | | | | |
| (over 10,000 HP) | 120 | 1.0 | 140 | 10 | 145 | 18 | 150 | 32 |
| (10 to 10,000 HP) | 115 | 0.32 | 135 | 5.6 | 140 | 10 | 150 | 32 |
| (up to 10 HP) | 110 | 0.32 | 130 | 3.2 | 140 | 10 | 145 | 18 |
| Boilers (Aux.) | 120 | 1.0 | 130 | 3.2 | 135 | 5.6 | 140 | 10 |
| Motor Generator Sets | 120 | 1.0 | 130 | 3.2 | 135 | 5.6 | 140 | 10 |
| Pumps | | | | | | | | |
| (over 5 HP) | 123 | 1.4 | 135 | 5.6 | 140 | 10 | 145 | 18 |
| (up to 5 HP) | 118 | 0.79 | 130 | 3.2 | 135 | 5.6 | 140 | 10 |
| Fans | | | | | | | | |
| (below 1800 rpm) | 120 | 1.0 | 130 | 3.2 | 135 | 5.6 | 140 | 10 |
| (above 1800 rpm) | 115 | 0.56 | 130 | 3.2 | 135 | 5.6 | 140 | 10 |
| Electric Motors | | | | | | | | |
| (over 5 HP or below 1200 rpm) | 108 | 0.25 | 125 | 1.8 | 130 | 3.2 | 135 | 5.6 |
| (up to 5 HP or above 1200 rpm) | 103 | 0.14 | 125 | 1.8 | 130 | 3.2 | 135 | 5.6 |
| Transformers | | | | | | | | |
| (over 1 kVA) | 103 | 0.14 | - | - | 115 | 0.56 | 120 | 1.0 |
| (1 KVA or below) | 100 | 0.10 | - | - | 110 | 0.32 | 115 | 0.56 |

*Ref. $10^{-6}$ mm/s. An older specification for VdB gave values 20 dB smaller than those found here (due to use of a different dB reference level).

[1]When this level is reached, service is called for. Alternatively, perform frequent octave analysis and refer to next column.

[2]When this level is exceeded in any octave band, repair immediately.

*Source*: Extracted from Canadian Government Specification CDA/MS/NVSH 107.

This table is based on the Canadian government specification for "Vibration Limits for Maintenance." This table uses RMS velocity levels and covers the frequency range 10 to 10,000 Hz. Table 18.4 is a guideline on common faults and their characteristic frequencies in terms of rotation speeds in rotor dynamics machines. Further references are provided at the end of the chapter [42–46].

## 4 VIBRATION ANALYSIS TECHNIQUES

The vibratory motion with which a machine oscillates is a complex vibration containing a wide range of forcing harmonics, shocks, and random excitations interacting with the tendency of elastic bodies to vibrate naturally in the form of simple harmonic motion. The frequency of this free and natural vibration is an important dimension in vibration analysis and monitoring. In this section a brief outline of the general concepts used in calculating these frequencies is presented. Most of this section deals with the theoretical foundations of the vibration analysis techniques for general linear systems. These concepts form the basis of experimental analysis methods.

### 4.1 Basic Definition of Frequency Response Functions

Any structure which vibrates under the influence of forces is capable of vibrating freely after it has been released from a displaced position. Without loss of energy, this vibration would continue indefinitely. In practice, vibrations are damped as a consequence of energy loss and therefore die out.

For an undamped SDOF system with mass $m$ and stiffness $k$, the governing equation of motion is

$$m\ddot{u} + ku = 0 \tag{18.25}$$

The trial solution $u(t) = Ue^{i\omega t}$ leads to the requirement that

$$(k - m\omega^2) = 0 \tag{18.26}$$

This results in a natural frequency $\omega_0$ given by $(k/m)^{1/2}$. Now, turning to a frequency response analysis, we can consider an excitation of the form $f(t) = Fe^{I\omega t}$ and assume a solution of the form $u(t) = Ue^{i\omega t}$, where $U$ and $F$ are complex to accommodate both the amplitude and phase information. Now the equation of motion is

$$(k - m\omega^2)e^{i\omega t} = Fe^{i\omega t} \tag{18.27}$$

from which the frequency response function (FRF) in the form

$$\frac{U}{F} = \frac{1}{k - m\omega^2} = H(\omega) \tag{18.28}$$

## Table 18.4 Vibration Troubleshooting Chart

| Nature of Fault | Frequency of Dominant Vibration (Hz=rpm/60) | Direction | Remarks |
|---|---|---|---|
| Rotating Members out of Balance | 1 x rpm | Radial | A common cause of excess vibration in machinery |
| Misalignment & Bent Shaft | Usually 1 x rpm often 2 x rpm Sometimes 3&4 x rpm | Radial & Axial | A common fault |
| Damaged Rolling Element Bearings (Ball, Roller, etc.) | Impact rates for the individual bearing component*<br>Also vibrations at high frequencies (2 to 60 kHz) often related to radial resonances in bearings | Radial & Axial | Uneven vibration levels, often with shocks.<br>* Impact-Rates:<br>Contact Angle $\beta$; Ball Dia (BD); Pitch Dia (PD)<br>**Impact Rates f (Hz)**<br>For Outer Race Defect $f(Hz) = \frac{n}{2} f_r \left(1 - \frac{BD}{PD} \cos\beta\right)$<br>For Inner Race Defect $f(Hz) = \frac{n}{2} f_r \left(1 + \frac{BD}{PD} \cos\beta\right)$<br>For Ball Defect $f(Hz) = \frac{PD}{BD} f_r \left[1 - \left(\frac{BD}{PD} \cos\beta\right)^2\right]$<br>$n$ = number of balls or rollers<br>$f_r$ = relative rev./s between inner & outer races |
| Journal Bearings Loose in Housing | Sub-harmonics of shaft rpm, exactly 1/2 or 1/3 x rpm | Primarily Radial | Looseness may only develop at operating speed and temperature (e.g. turbomachines). |
| Oil Film Whirl or Whip in Journal Bearings | Slightly less than half shaft speed (42% to 48%) | Primarily Radial | Applicable to high-speed (e.g. turbo) machines. |
| Hysteresis Whirl | Shaft critical speed | Primarily Radial | Vibrations excited when passing through critical shaft speed are maintained at higher shaft speeds. Can sometimes be cured by checking tightness of rotor components. |
| Damaged or Worn gears | Tooth meshing frequencies (shaft rpm x number of teeth) and harmonics | Radial & Axial | Sidebands around tooth meshing frequencies indicate modulation (e.g. eccentricity) at frequency corresponding to sideband spacings. Normally only detectable with very narrow-band analysis and cepstrum. |
| Mechanical Looseness | 2 x rpm | | Also sub- and interharmonics, as for loose Journal bearings. |
| Faulty Belt Drive | 1.2.3&4 x rpm of belt | Radial | The precise problem can usually be identified visually with the help of a stroboscope. |
| Unbalanced Reciprocating Forces and Couples | 1 x rpm and/or multiples for higher order unbalance | Primarily Radial | |
| Increased Turbulence | Blade & Vane passing frequencies and harmonics | Radial & Axial | Increasing levels indicate increasing turbulence |
| Electrically Induced Vibrations | 1 x rpm or 1 or 2 synchronous frequency | Radial & Axial | Should disappear when turning off the power |

can be extracted. This FRF is called the *system receptance*. Note that this FRF is independent of the excitation.

If we add a viscous damper $c$, which models the energy-dissipating influences, the vibratory motion if *damped* and the equation of motion for free vibration becomes

$$m\ddot{u} + c\dot{u} + ku = 0 \tag{18.29}$$

Here we sue a more general trial solution of the form $u(t) = Ue^{st}$, where $s$ is complex rather than pure imaginary as before. With this we obtain the condition

$$ms^2 + cs + k = 0 \tag{18.30}$$

This in turn leads to

$$\begin{aligned} s_{1,2} &= -\frac{c}{2m} \pm \frac{1}{2m}(c^2 - 4km)^{1/2} \\ &= -\omega_0\zeta \pm i\omega_0(1-\zeta^2)^{1/2} \end{aligned} \tag{18.31}$$

where $\omega_0^2 = k/m$ and $\zeta = c/c_c = c/2km = 2m\omega_0$. This implies a solution of the form

$$u(t) = Ue^{-\omega_0\zeta t}e^{i(\omega_0[1-\zeta^2]^{1/2})t} = Ue^{-dt}e^{i\omega_d t} \tag{18.32}$$

which is a single mode of vibration with a complex natural frequency having two parts: the imaginary or oscillatory part, a frequency of $\omega_d = \omega_0(1-\zeta^2)^{1/2}$, and the real or decaying part, a damping rate of $a = \omega_0\zeta$. The physical significance of these two parts of this modal behavior is shown in the typical free-response plot in Figure 18.23.

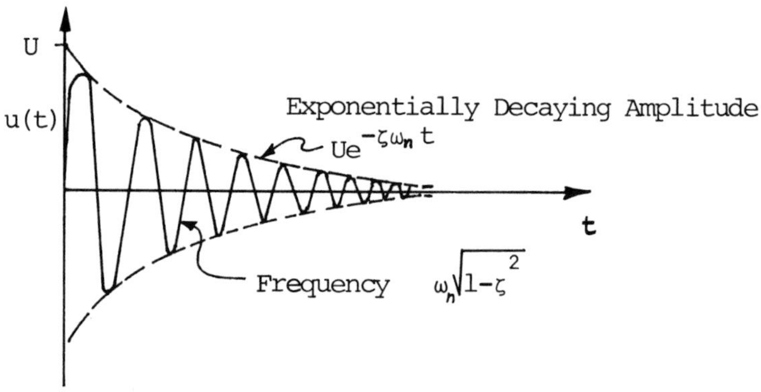


**Figure 18.23** Free vibration behavior of damped SDOF system.

For this system, if we consider the forced response when $f(t) = Fe^{i\omega t}$ and, as before, assume $u(t) = Ue^{i\omega t}$, the equation of motion gives

$$(-\omega^2 m + i\omega c + k)Ue^{i\omega t} = Fe^{i\omega t} \tag{18.33}$$

This results in a receptance FRF of the form

$$H(\omega) = \frac{1}{(k - m\omega^2) + i\omega c} \tag{18.34}$$

which is complex, containing both magnitude and phase information. Note that

$$|H(\omega)| = \frac{1}{(k - m\omega^2)^2 + (\omega c)^2]^{1/2}} \tag{18.35a}$$

and

$$\angle\theta_H(\omega) = \angle u - \angle f = -\tan^{-1}\frac{\omega c}{k - m\omega^2} \tag{18.35b}$$

By selecting different *output* quantities, other forms of FRFs can be defined.

If the response velocity $\dot{u}(t)$ is considered as the output, an alternative FRF—mobility—is defined as

$$M(\omega) = \frac{\dot{U}e^{i\omega t}}{Fe^{i\omega t}} = \frac{\dot{U}}{F} \tag{18.35c}$$

For sinusoidal motion the velocity is proportional to the product of displacement and frequency, i.e.,

$$\dot{u}(t) = \omega i U e^{i\omega t} \tag{18.36}$$

so

$$M(\omega) = i\omega H(\omega) \tag{18.37}$$

Thus

$$|M(\omega)| = \omega |H(\omega)| \tag{18.38}$$

and

$$\theta_M = \theta_H + 90°$$

Therefore, mobility is closely related to receptance. Similarly, by using acceleration $\ddot{u}(t)$ as the response parameter (since it is customary to measure acceleration in tests), a third FRF—inertance or accelerance—is defined as

$$A(\omega) = \frac{\ddot{U}}{F} = -\omega^2 H(\omega) \tag{18.39}$$

These functions can also be defined in an inverse way, namely as force/displacement, called *dynamic stiffness*, force/velocity, called *mechanical impedance*, and force/acceleration, called *apparent mass*. The motional impedance of a mechanical element is similar in characteristics to the impedance of electrical components, for mechanical impedance calculations are based on the variables of force and velocity. These latter forms of FRF are discouraged, except in special cases, as they can lead to confusion and error if improperly used in MDOF systems. Table 18.5 gives details of these six FRF parameters and the names and symbols used for them.

Mechanical impedance and mobility concepts have two important uses in vibration analysis: 1) the overall performance of a complex mechanical system can be predicted from a knowledge of the characteristics of the components, and 2) forces acting within the system can be reliably estimated. In a mechanical system, impedance and mobility are regarded as *point* values if force and velocity are measured at the same point. They are regarded as *transfer* impedance or *transfer* mobility if the force is measured at one point and velocity at another. To evaluate the impedance of a whole system, impedance values are calculated for all the individual elements and applied in a manner identical to that for series-parallel arrangements in electrical circuits. Typical identities are given in Table 18.6, and equivalent applications are shown in Figure 18.24.

Practical measurements of the impedance of a structure make use of an impedance head. This electromechanical device contains two types of electromechanical

**Table 18.5** Frequency Response Functions

| Response parameter | Standard | Inverse |
|---|---|---|
| Displacement | Receptance<br>Admittance<br>Compliance<br>Dynamic flexibility | Dynamic stiffness |
| Velocity | Mobility | Mechanical impedance |
| Acceleration | Interance<br>Accelerance | Apparent mass |

**Table 18.6** Mechanical/electrical identities

| Mechanical | Electrical |
|---|---|
| Mass ($m$) | Inductance ($L$) |
| Spring compliance | Capacitance ($C$) |
| Spring stiffness | 1/Capacitance ($1/C$) |
| Damping force | Resistance ($R$) |
| Velocity | Current ($I$) |
| Force | Voltage |

transducer. One transducer measures the force which is applied to the structure, while the second measures the motion of the point to which the force is applied. The actual construction of an impedance head is shown in Figure 18.25. It consists basically of an accelerometer and a force gauge.

At this point it is appropriate to introduce another parameter, *transmissibility*, understanding of which is important in vibration monitoring. The transmissibility $TR$ is defined as the ratio of the transmitted force to the applied force. The numerical value of $TR$ can be calculated from

$$TR^2 = \frac{1 + (r/Q)^2}{(1 - r^2)^2 + (r/Q)^2} \tag{18.40}$$

where $r = \omega/\omega_0$, the ratio of the forcing frequency to the system natural frequency, and $Q$ is the quality factor. For a lightly damped system with the damping ratio $\zeta$, the quality factor is given by $Q = 1/(2\zeta)$.

A graph relating transmissibility and frequency ratio is given in Figure 18.25 for an SDOF system. Below resonance, $TR$ exceeds 1.0. Above resonance, $TR$ decreases progressively, becoming unity when $r = \sqrt{2}$ and then decreasing to values less than 1.0 according to the amount of damping.

## 4.2 Graphical Representation of Frequency Response Data

In practice, several different graphical representations are utilized for the complex FRF. These representations convey the same information, usually the real or imaginary parts of the FRF or the magnitude and phase angle for a range of values of the excitation frequency. The three most common forms of presentation are 1) magnitude of FRF versus frequency and phase angle versus frequency, Bode plot; 2) real part of FRF versus frequency and imaginary part versus frequency, Co-Quad plot; 3) real part versus imaginary part, which does not explicitly contain frequency information, Nyquist plot.

A typical Bode plot is shown in Figure 18.26 for a damped SDOF system. As can be seen from this figure, at resonance $\omega = \omega_0$, the magnitude is a maximum and is limited only by the amount of damping in the system. The phase ranges from 0 to 180° and the response lags the input by 90° at resonance.

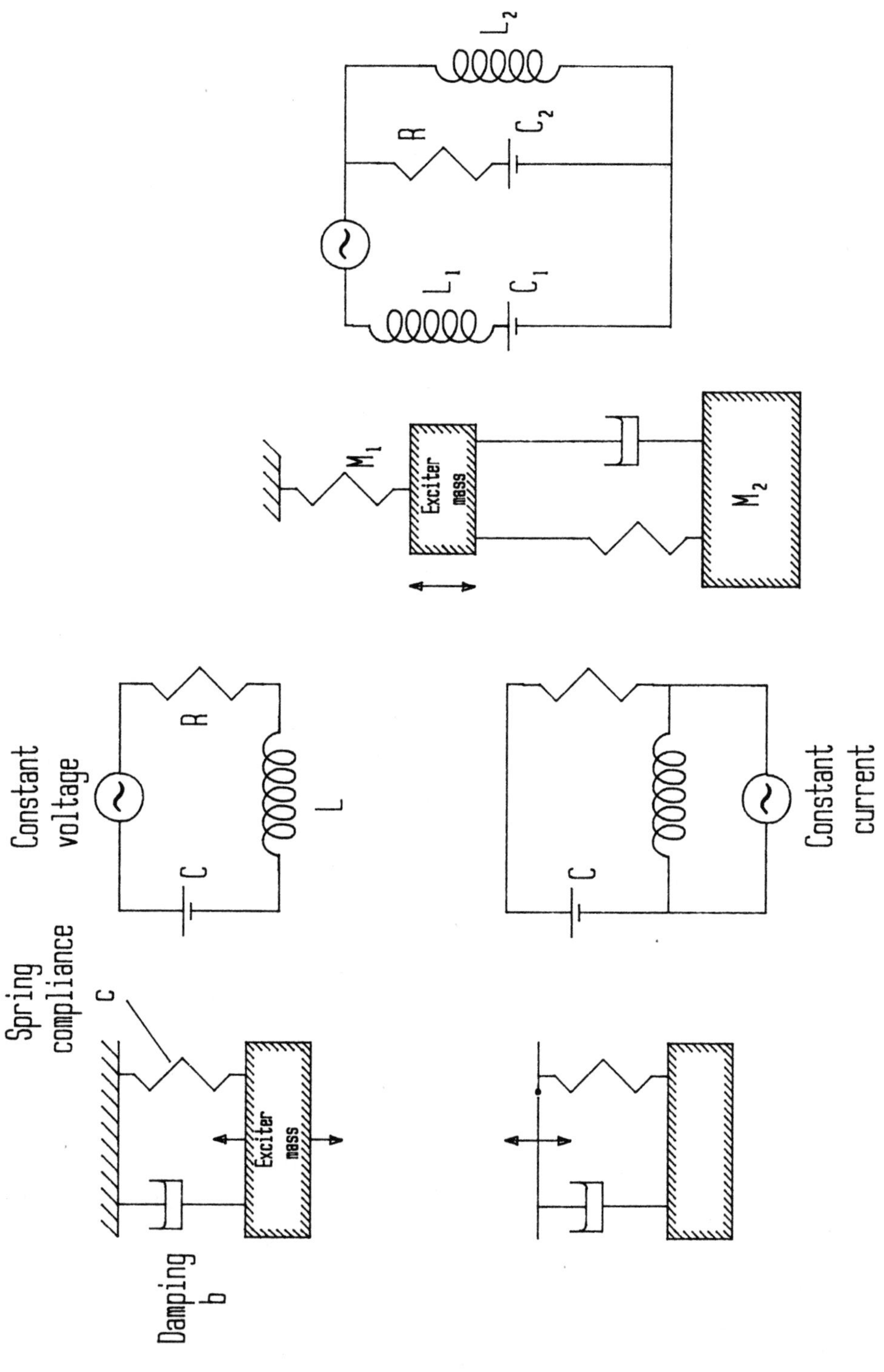

**Figure 18.24** Mechanical and electrical systems analogs.

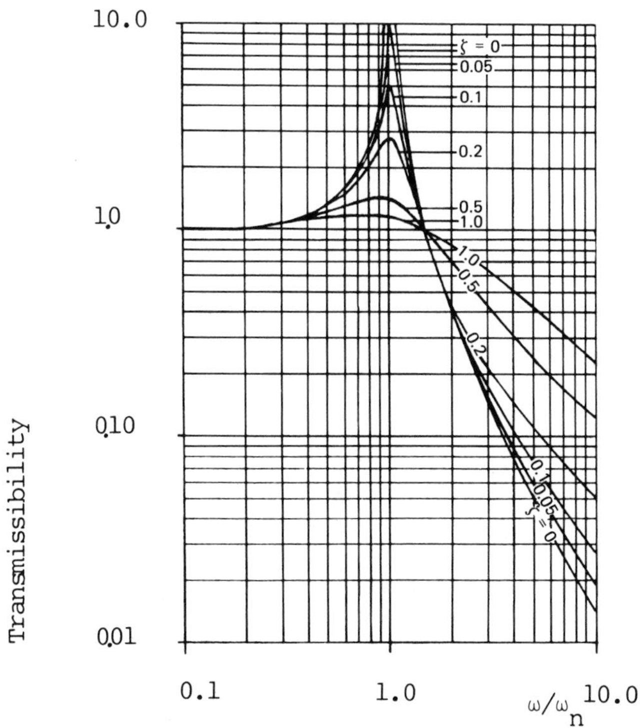


**Figure 18.25** Force transmissibility for an SDOF system.

By plotting the magnitude in decibels versus logarithmic (log) frequency, it is possible to cover a wider frequency range and conveniently display the range of amplitude. This type of plot has some useful parameter characteristics which are described in the following plots.

When $\omega \ll \omega_0$ the frequency response is approximately equal to the asymptote shown in Figure 18.27. This asymptote is called the stiffness line and has a slope of 0, 1, or 2 for displacement, velocity, or acceleration responses, respectively. When $\omega \gg \omega_0$ the frequency response is approximately equal to the asymptote also shown in Figure 18.27. This asymptote is called the mass line and has a slope of −2, −1, or 0 for displacement, velocity, or acceleration responses, respectively.

Another method of representing data is to plot the rectangular coordinates, the real part and the imaginary part, versus frequency. For a proportionally damped system, the imaginary part is maximum at resonance and the real part is 0.

Typical plots of this type called Co-Quad or companion plots for real parts versus frequency and imaginary part versus frequency are shown in Figure 18.28 for the SDOF system with light viscous damping. It can be seen that the phase change through the resonance region is characterized by a sign change in one part accompanied by a maximum or minimum value in the other part. In this case the use of a logarithmic scale is not feasible because it is necessary to accommodate both positive and negative values.

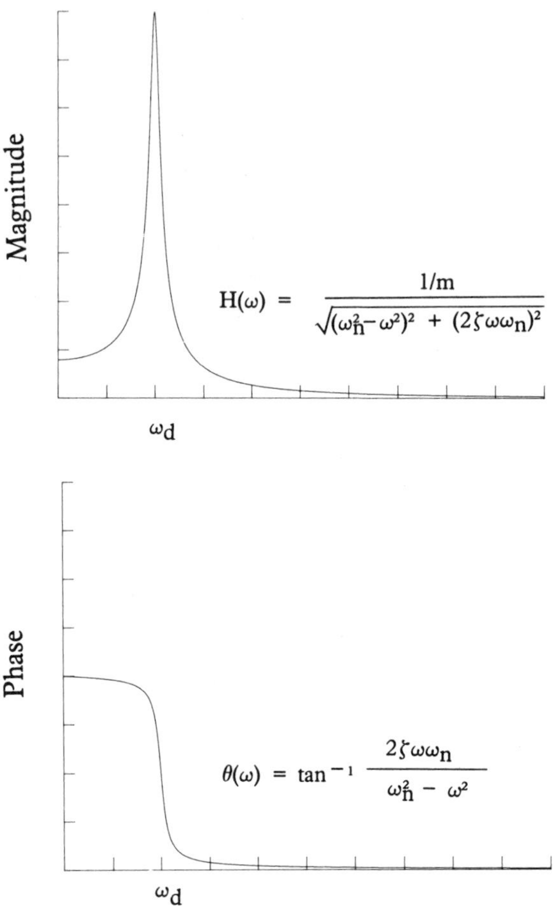


**Figure 18.26** Frequency response—Bode plot.

As discussed before, data from frequency response measurements are normally presented as plotted spectra with either linear or logarithmic presentation of the amplitude and frequency axes. Logarithmic amplitude displays tend to compress the data so that the detailed information is available at both large and small output/input ratios. Linear displays of amplitude tend to accentuate resonant points (antinodes) in compliance, mobility, and inertance presentations and antiresonances (nodes) in apparent stiffness, impedance, and apparent mass presentations. Presentation of the real and imaginary components is especially useful in the precise location of resonant points.

When dealing with systems exhibiting nonlinear damping characteristics, a cross-plot of the imaginary versus real components of the transfer relationship may be used to characterize the nature of the damping relationship. This cross-plot is frequently called a Nyquist diagram.

The Nyquist plot is widely used and is an effective way to display the important resonance region in some detail. Figure 18.29 shows a typical Nyquist plot for a

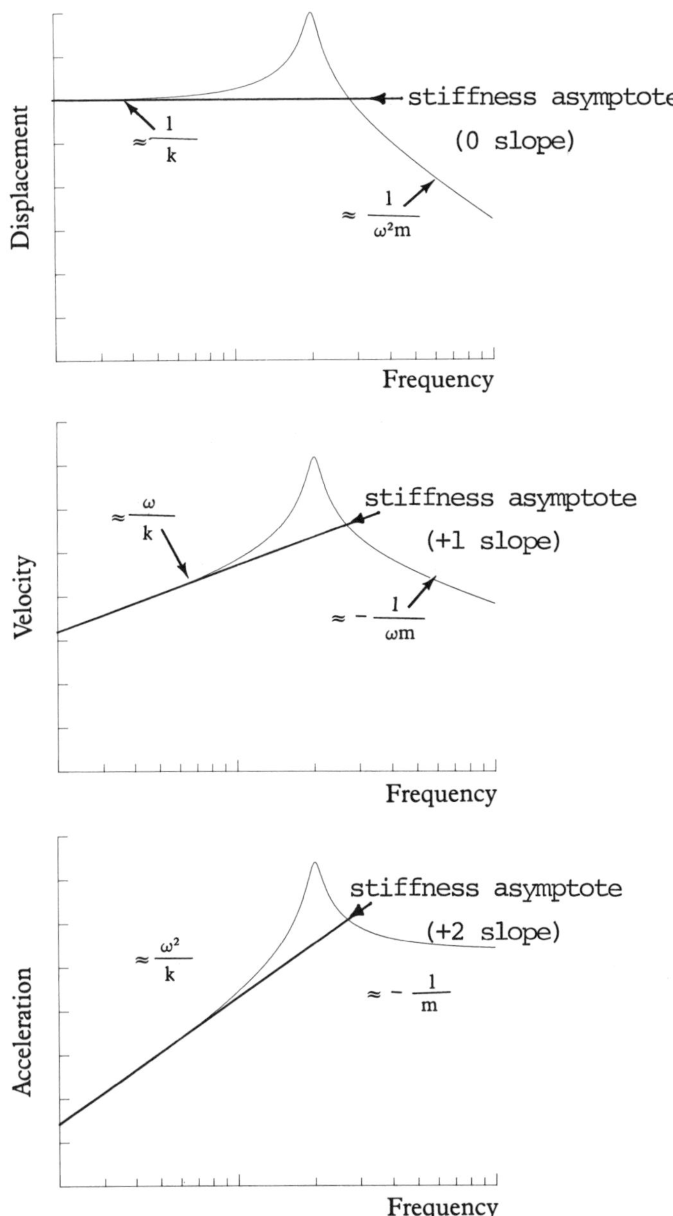


**Figure 18.27** Different forms of frequency response.

viscously damped SDOF system. In these plots usually specific points on the curve at regular increments of frequency are indicated. In the figure shown, only the frequency points closest to resonance are clearly identifiable because those away from this area are very close together. Indeed, it is this feature—of distorting the plot to focus on the resonance area—that makes the Nyquist plot so attractive for

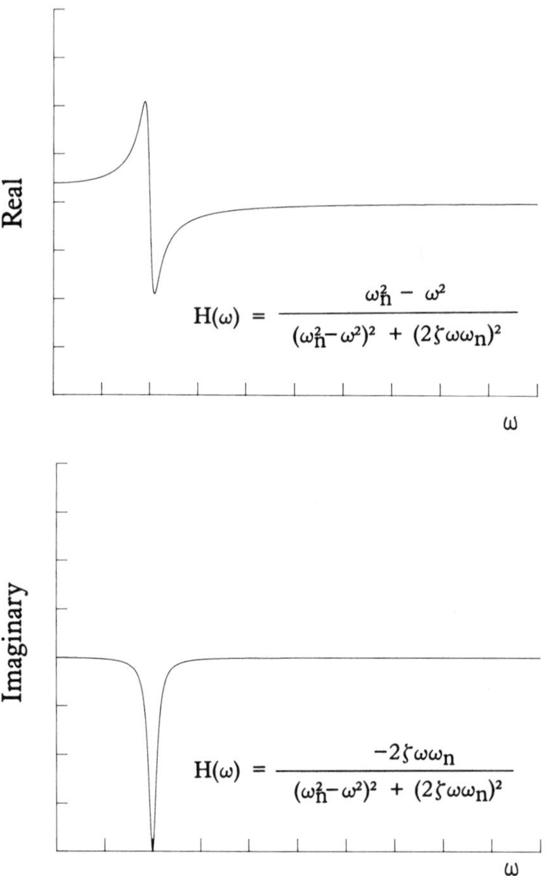


**Figure 18.28** Frequency response—Co-Quad plot.

modal testing applications. Resonance frequencies can be detected in this cross-plot by finding the portion of the display that exhibits the maximum change in arc length with change in frequency.

Plotting any of the frequency response functions described above provides a means of detecting the resonance frequencies. In addition, these plots can be used as a means of directly estimating mass, stiffness, and damping parameters in systems characterized by far coupled modes. The Bode plot specifically has applications in this regard. Further discussion of graphical description of frequency response data can be found in [21, 24, 37, 47–50].

## 4.3 Frequency Response for Arbitrary Excitation

It can be shown that the frequency response method is also effective in computing the response to arbitrary applied forcing functions such as shock and blast loadings. The basis of this use of the frequency response function information is the relationship between the frequency and the time domain responses of a vibrating

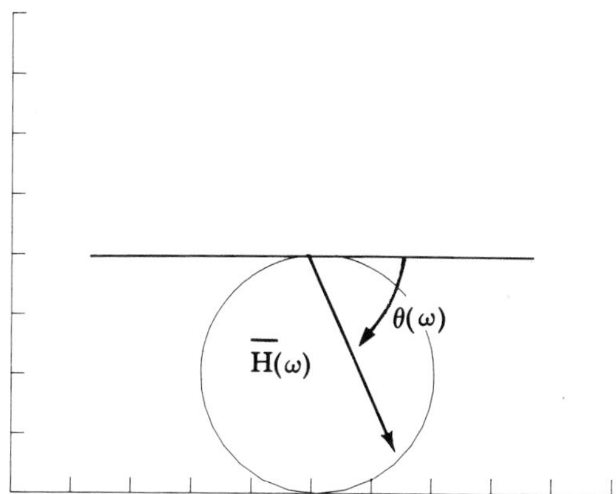


**Figure 18.29** Nyquist plot of frequency response.

system. Figure 18.30a shows a vibration system with its input and output. The system of Figure 18.30b is characterized by the impulse response function $h(t)$ given by

$$h(t) = \frac{1}{\omega_0\sqrt{1-\zeta^2}} e^{-\omega_0\zeta t} \sin\sqrt{1-\zeta^2}\,\omega_0 t \tag{18.41}$$

while the system of Figure 18.30b is described by its FRF, $H(\omega)$:

$$H(\omega) = \frac{1}{1 - (\omega/\omega_0)^2 + i2\zeta\omega/\omega_0} \tag{18.42}$$

The time domain response of Figure 18.30a can be expressed by a Duhamel integral

$$u(t) = \int_0^t f(\tau)h(t-\tau)\,d\tau \tag{18.43}$$

where $f(t)$ is the applied excitation and $h(t-\tau)$ is the response due to a unit impulse at $t = \tau$. Since no impulse is applied for $\tau > t$, it follows that $h(t-\tau) = 0$ for $\tau > t$. Also, $f(\tau)$ is zero for $\tau < 0$. Hence Eq. (18.43) can be written as

$$u(t) = \int_{-\infty}^{\infty} f(\tau)h(t-\tau)\,d\tau \tag{18.44}$$

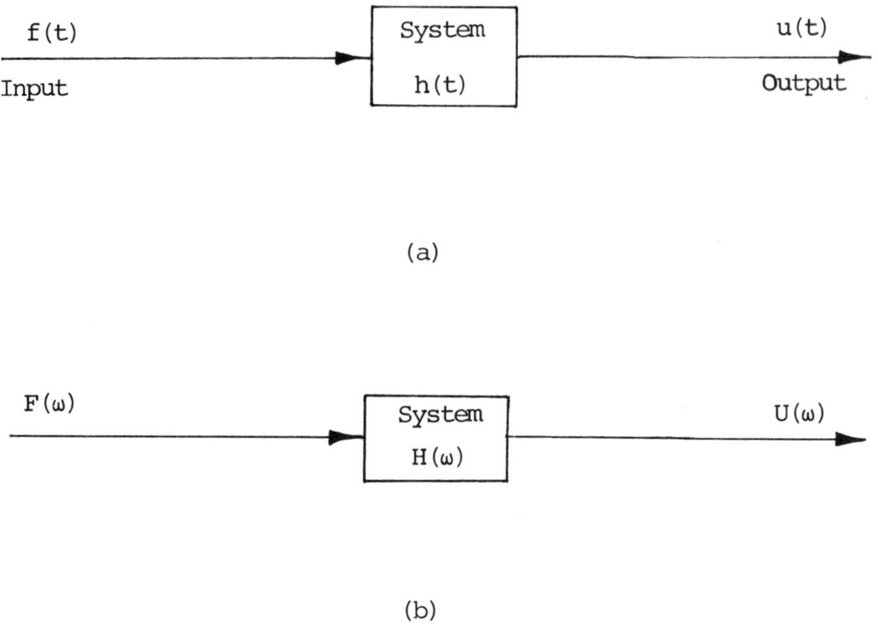


**Figure 18.30** Input-output relationship of a vibration system.

The right-hand side of this equation is the convolution integral of two time functions, $f(t)$ and $h(t)$. From the convolution theorem for Fourier transform

$$u(t) = \int_{-\infty}^{\infty} f(\tau)h(t-\tau)\,d\tau \longleftrightarrow F(\omega)H(\omega) = U(\omega) \tag{18.45}$$

it can be shown that the function $H(\omega)$ is the FRF. Thus

$$h(t) \longleftrightarrow H(\omega) \qquad \text{or} \qquad h(t) = \mathcal{L}^{-1}H(s) \tag{18.46}$$

where, in the Laplace transform, $s = i\omega$. This means that $h(t)$, the impulse response function, and $H(\omega)$, the FRF, form a Fourier transform pair. This result has significant implications in that it forms the basis of a common method of experimental determination of the FRF.

The Fourier transform of the forcing function and the inverse Fourier transform of Eq. (18.45) are obtained by using various computational procedures which are discussed in Chapter 4 of this book. In engineering practice, the FRF is often measured experimentally. Of the many approaches available, *impact testing* appears to be efficient in the sense that the FRF is evaluated rapidly. In this test, an instrument hammer is used to strike the structure. The time history of the input forcing function and the time response of the structure at a measured point are stored and analyzed to compute (using the FFT) their Fourier transforms. The ratio of these Fourier transforms is the system FRF, which is plotted [24, 74, 75].

## 4.4 Vibration Analysis Techniques for MDOF Systems

The extension of SDOF concepts to a more general MDOF system is a straightforward process. For an MDOF system with $n$ degrees of freedom the governing equations of motion can be written in matrix form as

$$[m]\{\ddot{u}\} + [c]\{\dot{u}\} + [k]\{u\} = \{f(t)\} \tag{18.47}$$

where $[m]$ and $[k]$ are $n \times n$ mass and stiffness matrices, respectively, $[C]$ is the damping matrix, and $\{u\}$ and $\{f\}$ are $n \times 1$ vectors of time-varying displacement and forces.

The solution of Eq. (18.47) with no excitation (free vibration) leads to a matrix eigenproblem. Eigenvalues of this equation are the natural or resonance frequencies. The eigenvectors are the mode shapes or characteristic deformation shape of the structure associated with each natural frequency. A typical free vibration response is illustrated in Figure 18.31.

The equations of motion for the forced vibration case also lead to a frequency response of the system which can be written as a weighted summation of the responses of SDOF systems, as shown in Figure 18.32. The weighting, often called the modal participation factor, is a function of excitation and mode shape coefficients at the input and output degrees of freedom. The participation factor identifies the amount each mode contributes to the total response at a particular point. An example with three degrees of freedom showing the individual modal contributions is shown in Figure 18.32.

The frequency response of an MDOF system can be presented in the same form as in the SDOF case. There are other definitional forms and properties of frequency response functions, such as a driving point measurement. These are related to

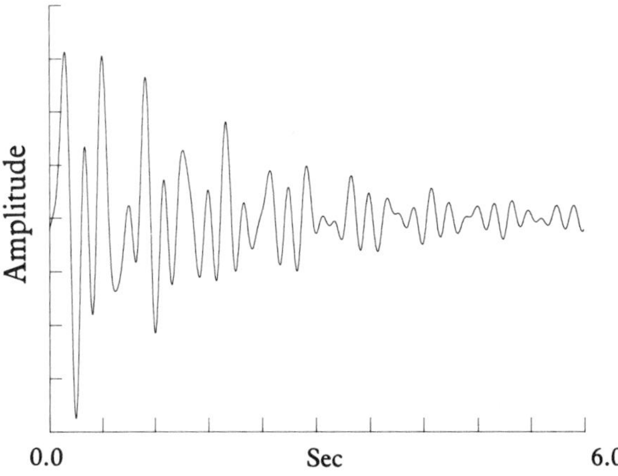


**Figure 18.31** Free response of an MDOF system (time domain).

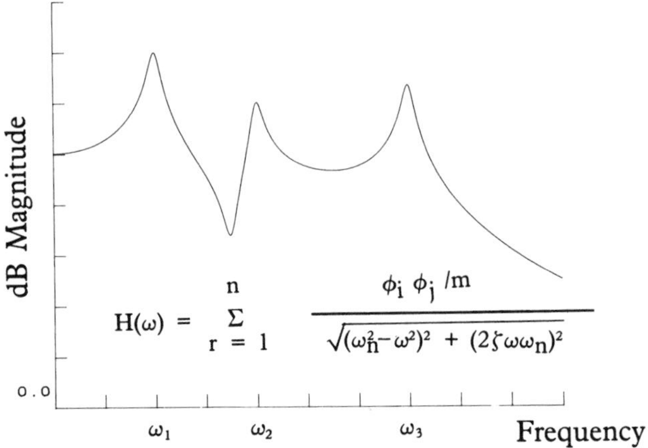


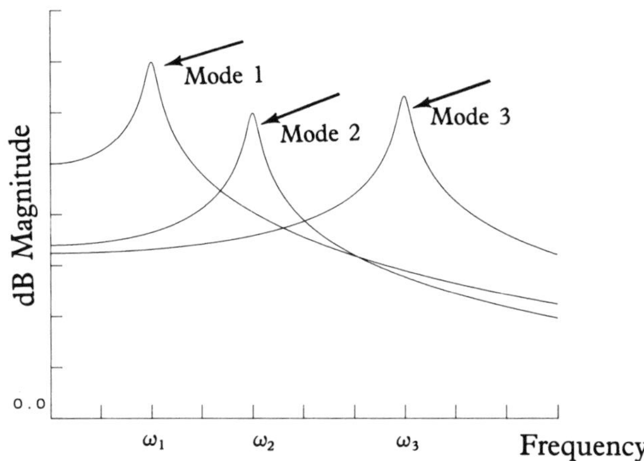


**Figure 18.32** An MDOF frequency response showing the SDOF modal contributions.

specific locations of frequency response measurements and are discussed in [24, 51–54].

The resonance frequencies of a structure are independent of the spatial location from which they are measured. Each modal shape is a global property of the structure (as are the resonance frequencies) and affects motion over the entire structure. It is often desirable to determine the shape of deformation at a particular resonance to better understand how that resonance frequency may be shifted to another frequency location. It is necessary to find the areas of the structure that undergo maximum motion in a particular mode, as these are the areas on which corrective stiffness or mass terms will have maximum effect (and therefore maximum opportunity to correct the problem).

Mode shapes are also of fundamental value in the analytical modeling of structures. Mathematical techniques exist which allow prediction of the dynamic behavior of complicated composite structures based on knowledge of the dynamic behavior of *individual* components. Substructures may be tested separately and mathematically "connected" together to predict the dynamic behavior of the composite structure.

Mode shapes can be estimated directly from frequency response functions. These estimated mode shapes give first-order understanding of the nature of the deformation of a structure and are frequently sufficiently precise to allow the placement of corrective stiffness, damping, or mass components. This is accomplished by exciting the structure at a single point and measuring its response at several points. Alternatively, the structure may be sequentially excited at a series of locations and the response measured at a single point. The two testing techniques are equivalent because of structural reciprocity.

In the following section a technique which is widely used in acquiring frequency response measurements, modal testing, is discussed.

## 4.5 Modal Testing

Modal testing is a method for experimentally determining the dynamic characteristics of a structural system. Even though the dynamic properties of a structure can sometimes be determined with computer techniques, for example, using the finite-element method, experimental verification of the analytical results is still necessary in most cases. The results of modal testing can also be used either directly in design load and response analysis or for the determination of weak points of the structure.

In modal testing, the structure is often assumed to be linear; i.e., the response of the structure is linearly proportional to the input force. Also, the structure is assumed to be time-invariant; i.e., the dynamic characteristics of the structure do not change during the period of the test.

Usually, the dynamic parameters to be identified from the modal test are the natural frequencies, mode shapes, and modal damping of the structure. To accomplish this, one or more dynamic forces and sometimes initial displacements are applied to excite the structure. The dynamic parameters are then determined from the input force and output response measurements.

Modal testing techniques have been developed, mainly since World War II, in the aircraft and space vehicle fields. Prior to 1960, almost all modal testing was carried out using only analog equipment. The multipoint sine dwell test (or multishaker resonance test) was the most popular modal testing method. Several classical papers were published which discuss the calculation of force ratios among shakers [55–58]. This topic was still under discussion in more recent years [59–62].

With the introduction of the fast Fourier transform algorithm in 1965 [63], the invention of low-cost minicomputers, and the development of software programs for modal testing [54, 64], more and more tests were done using digital equipment. The key to conducting modal tests using a digital signal processing technique is measurement of the frequency response function between the input forces and the responses at different locations in the structure. Modal parameters for defining the dynamic characteristics of the structure are then extracted from these measured functions. A substantial literature has dealt with this approach [51–53, 65–67].

The state of art of modal testing has advanced very rapidly over the past decade. Today, modal tests are used not only in the aerospace industry but also in many other fields, e.g., packaging, transportation, automobiles, machinery, and instrumentation design.

Several techniques are available for conducting a modal test. The input force can be random, periodic, transient [54], or impulse [24] and can act at a single point or simultaneously at several points of the structure. The various test methods may be classified according to whether testing is done in the time domain, with only analog equipment, or the data are transformed into the frequency domain with the help of a digital computer.

## Time Domain Modal Testing

In this approach, the recorded time history test data are used directly to extract the dynamic parameters of the test structure. The most commonly used method in this approach is the multipoint sine dwell modal test.

The basic idea in the multipoint sine dwell modal test is to apply multishakers simultaneously to the structure so that the structure responds with one of its natural modes. The mode shape then can be determined directly from the response measurements at different locations of the structure. Usually, the input force is sinusoidal with the frequency of the excitation forces equal to the frequency of the mode to be excited. These frequencies must be determined before applying the multishaker test. This could be accomplished by a single-point sine sweep test, i.e., applying a single shaker to the structure and varying the frequency of the input force so that it sweeps through the range of interest. The natural frequencies of the structure are usually identified from the peaks in the driving point quadrature plot. This plot can be obtained from the imaginary part of the Fourier transform of the response time history data or directly from the Co-Quad analyzer. Usually, several sweep tests are needed with shakers located at different points and directions of the structure in order to completely identify the relevant resonance frequencies of the structure. It should be noted that these peaks in the quadrature plot correspond to the resonance frequencies of an undamped system. For a structure with damping, the actual resonance frequencies will be slightly away from these peaks. Fortunately, for most structures with light damping, this distortion is small and can be neglected for practical purposes.

With the natural frequencies determined, the multishaker can be applied to selected points of the structure. The magnitude of the force at each shaker is adjusted so that it will force the structure to respond with one of its natural modes. Once a natural mode is excited, the mode shape can be determined directly from the response measurements at different locations of the structures. The usual way to achieve excitation of a natural mode is to apply the master shaker to the structure first and adjust the frequencies of the applied forces so that they reinforce the master quadrature peak while the individual quadrature responses peak at the same time [68]. Essentially, this is a trial-and-error procedure. However, methods are available for calculating the force ratio among shakers [56, 57, 61].

Lewis and Wrisley [56] proposed a procedure for exciting natural modes of a structure and devised equipment for its implementation. The equipment consists of 24 independently controllable electromagnetic shakers, each with an attached accelerometer. The forces of the shakers are adjusted by an iteration procedure so that the magnitude of each force is proportional to the product of the mass of the structure on which it acts and the amplitude of response of that mass in the mode being excited.

Modal damping coefficients for each mode of the structure can be determined by suddenly disconnecting the shaker and using the recorded time history decay data. The idea of using a free-decay curve to predict the damping is explained in [24].

The multipoint sine dwell test is till the standard modal test method in the aerospace industry, where high accuracy of mode shapes is desired. The advantages of this test method are the simplicity of its data reduction and the fact that a structural linearity check is readily made by simply changing the amplitude of the shaker force and checking the response measurements. However, this method has the disadvantage that it is time-consuming. Also, there is difficulty in implementing the mode-tuning process, especially for modes having close frequencies. Moreover, complex fixturing and test facilities are required for the test.

Another technique which has been gaining more attention recently is Ibrahim's time domain method. The theory of Ibrahim's method can be found in [69–71].

## Frequency Response Modal Testing

Conceptually, this method is more involved than the time domain approach of the preceding section. It requires transformation of the measured data from the time domain to the frequency domain and hence involves the use of digital processing techniques. The key in this method is the measurement of a response function between various points of the structure. This function is defined as the ratio of the Fourier transform of the response to the Fourier transform of the input force. If a Laplace transform of the test data is used instead of a Fourier transform, this function is called a *transfer function.* Modal parameters of the structure are then determined from these measured frequency response functions using curve-fitting techniques. The major steps involved in this test method can be summarized in general as follows:

1. Input a measurable force. This force can be sinusoidal, random, pseudorandom, transient, or impulse in nature.
2. Measure the response at a selected point, or points, on the structure.
3. Fourier-transform the input force and output response measurement. The fast Fourier transform algorithm may be used to perform the transformation.
4. Compute the frequency response functions.
5. Estimate the modal parameters from these functions.

The concept of a transfer function can be introduced by taking the Laplace transform of the governing equations of motion for a general linear damped $N$-degree-of-freedom system. Thus

$$\mathcal{L}\left[[m]\{\ddot{u}\}+[c]\{\dot{u}\}+[k]\{u\}\right]=\mathcal{L}\{f\} \tag{18.48}$$

Assume the initial displacement and velocity of the system are zero. Then

$$([m]s^2 + [c]s + [k])\{U(s)\} = \{F(s)\} \tag{18.49}$$

where

$$U(s) = \mathcal{L}\{u(t)\}, \qquad F(s) = \mathcal{L}\{f(t)\}, \qquad s = \sigma + i\omega$$

If we define

$$[B(s)] = [m]s^2 + [c]s + [k] \tag{18.50}$$

then

$$\{U(s)\} = [H(s)]\{F(s)\} \tag{18.51}$$

where

$$[H(s)] = [B(s)]^{-1} \tag{18.52}$$

The quantity $H(s)$ is referred to as the system transfer function. Note that the Laplace variable $s$ is a complex quantity and hence the transfer function is also complex-valued; i.e., it has real and imaginary parts. Figure 18.33 shows the real and imaginary parts of a typical transfer function. When the transfer function is evaluated along the frequency axis, that is, $\sigma = 0$, it is called the *frequency response function.*

From Eqs. (18.50)and (18.52) it can be seen that the system transfer function is an $N \times N$ matrix which can be expressed as

$$[H(s)] = \begin{bmatrix} h_{11}(s) & \cdots & h_{1N}(s) \\ \vdots & & \vdots \\ h_{N1}(s) & \cdots & h_{NN}(s) \end{bmatrix} \tag{18.53}$$

The general element $h_{ij}(s)$ in this matrix can be obtained by exciting the system with $F_j(s)$, setting the rest of the forces to zero, and measuring the response $U(s)$. Since the matrix consists of $N^2$ elements, we may have to make $N^2$ measurements, which is a very time-consuming process. Fortunately, it can be shown that only one row or column of the system transfer function $[H(s)]$ has to be measured to completely determine the mode shapes of the system. Thus we need only $N$ measurements to define an $N$-degree-of-freedom system, instead of $N^2$ measurements.

Several points must be considered in obtaining the frequency response function of a system:

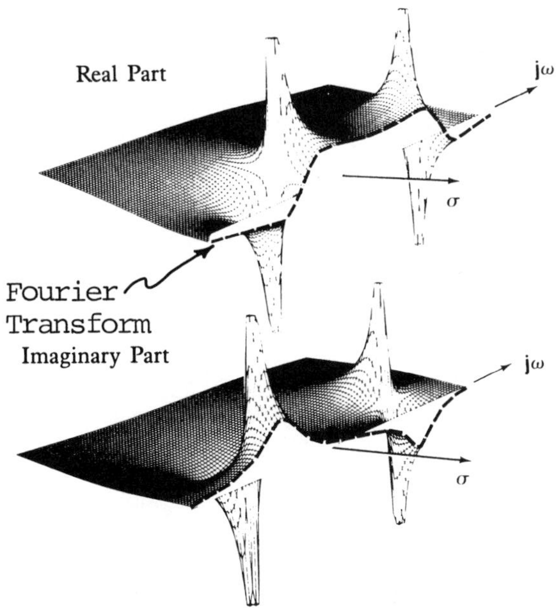


**Figure 18.33** Real and imaginary parts of a transfer function.

1. Although the frequency response function is defined as the Fourier transform of the response divided by the Fourier transform of the input force, it cannot be calculated directly from this definition because of the noise involved in the usual test environment [48]. In general, there is always some source of noise $N(s)$ contained in the measurements and hence, instead of Eq. (18.51), we will have

$$\{U(s)\} = [H(s)]\{F(s)\} + N(s) \tag{18.54}$$

We want to estimate $[H(s)]$ from these measurements. One technique commonly used is least-squares estimation. If $[H(s)]$ is the estimated value of $[H(s)]$ from least-squares estimation, one can obtain

$$H = \frac{\sum_{k=1}^{n}(UF^*)_k}{\sum_{k=1}^{n}(FF^*)_k} \tag{18.55}$$

The cross-power spectrum $G_{UF}$ and the input auto-power spectrum $G_{FF}$ are defined as

$$G_{UF} = \sum_{k=1}^{\infty} UF^* \tag{18.56}$$

$$G_{FF} = \sum_{k=1}^{\infty} FF^* \tag{18.57}$$

In practice, $n$ will be a finite number in Eq. (18.55) and hence

$$\hat{H} = \frac{\hat{G}_{UF}}{\hat{G}_{FF}} \tag{18.58}$$

where $\hat{G}_{UF}$ and $\hat{G}_{FF}$ are the estimations of the cross-power and auto-power spectra.

2. Usually, a coherence function is also calculated with the measurements of the frequency response function to indicate the degree of noise contamination in the measurements [26]. This function is defined as

$$\Gamma^2 = \frac{\text{response power caused by applied load}}{\text{measured response power}} \tag{18.59}$$

The measured response power spectrum is

$$\begin{aligned} G_{UU} &= (\hat{H}F + N)(\hat{H}F + N)^* \\ &= |\hat{H}F|^2 + \hat{H} * NF^* + \hat{H}FN^* + NN^* \\ &= \hat{H}^2 U_{FF} + \hat{H}\hat{G}_{NF} + \hat{H}\hat{G}_{NF} + \hat{G}_{NN} \end{aligned} \tag{18.60}$$

where $\hat{G}_{NN}$ is the estimated noise auto-power spectrum and $\hat{G}_{NF}$ the cross-power spectrum between the input force and noise. This spectrum will be zero if $N$ and $F$ are incoherent. Hence

$$\hat{G}_{UU} = \hat{H}^2 \hat{G}_{FF} + \hat{G}_{NN} \tag{18.61}$$

We can see from this equation that the output power comprises two parts: $|\hat{H}|^2\hat{G}_{FF}$ is directly related to the input and $\hat{G}_{NN}$ is due strictly to the noise. We define the part of $G_{UU}$ that is coherent with the input as

$$\Gamma^2 \hat{G}_{uu} = |\hat{H}|^2 \hat{G}_{FF} \tag{18.62}$$

The coherence function can be expressed as

$$\Gamma^2 = \frac{|\hat{G}_{UF}|^2}{\hat{G}_{UU}\hat{G}_{FF}} \tag{18.63}$$

Since in general $\hat{G}_{NF} = 0$, we cannot measure $\Gamma^2$ but can only obtain an estimate from Eq. (18.63).

The coherence function will be equal to one if the measured response force is totally caused by the input power. If some noise source has contributed to the output power, the coherence value will be less then one. In this case more measurements and more averaging may be needed for calculating the frequency response function or some smoothing techniques may have to be applied to remove the noise.

3. In performing the Fourier transformation to the input force and response time history data, the discrete Fourier transform must be used on a limited time record. Two major errors may result [72]. One is associated with sampling a continuous analog signal at a discrete time, which may introduce a form of amplitude distortion, called aliasing, that converts high-frequency energy to low frequencies. The other error is caused by the finite recording time of the test data. This may introduce a distortion, called leakage, which converts energy at each frequency into energy within a relatively narrow band nearby [73].

Once a set of frequency response functions has been measured for the test structure, the modal parameters, i.e., the natural frequencies, mode shapes, and modal damping coefficients, can be estimated from these functions. Figure 18.34 shows a typical frequency response function plot. Note that a broadband input force may excite many of the structural modes simultaneously. The measured function actually represents the sum of the motions of all the excited modes of the structure, as indicated by the plots in Figure 18.35. The degree of this mode overlap depends on the amount of damping of the modes and their frequency separation. Figure 18.36a shows a frequency response function with light modal overlap and Figure 18.36b shows a function with heavy overlap. Different curve-fitting techniques should be used for these two different functions [53]. In cases where the modal overlap is light, the structure responds as if it were a single-degree-of-freedom system on each of its modes. A single-degree-of-freedom curve-fitting technique can then be used. On

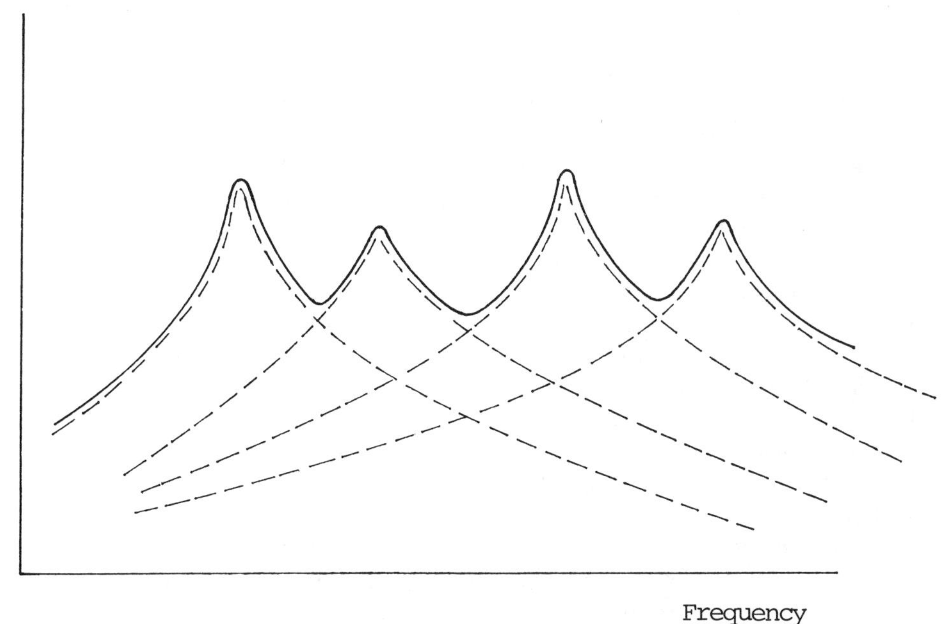


**Figure 18.34** Typical frequency response function.

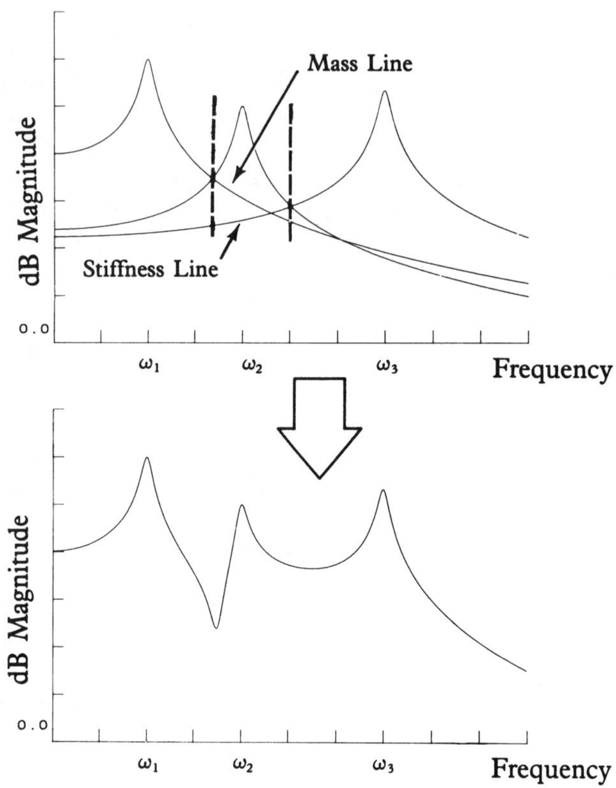


**Figure 18.35** Frequency response function showing single-mode summation.

the other hand, if the frequency response function shows a heavy overlap between the modes, a multi-degree-of-freedom curve-fitting technique should be used.

One of the SDOF approaches for identifying the mode shape is to use a Nyquist, i.e., real versus imaginary, plot of the frequency response function. In general, the measurements are made by a narrowband frequency sweep in the vicinity of the modal resonance frequency. The mode shape of the system is obtained by fitting a circle to these measurements for an entire row or column of the frequency response matrix. This method, which is referred to as the Kennedy–Pacu method, is generally more accurate than simply using the quadrature response.

The modal damping of the structure can be estimated in several ways. The half-power method [24] can be used on the frequency response curve. The other commonly used method is to estimate the damping coefficient from the peaks of the Co-Quad plot [24, 74–75].

The frequency response approach for modal testing has the advantages that the test arrangement is simplified and less time is required than for the multipoint sine dwell test. However, the data reduction and analysis phase of the test can be time-consuming and complex. A linearity check is also more difficult to achieve.

The different modal test methods are compared in the literature, e.g., [48–49, 76–79]. Although the frequency response approach looks promising at this time, no

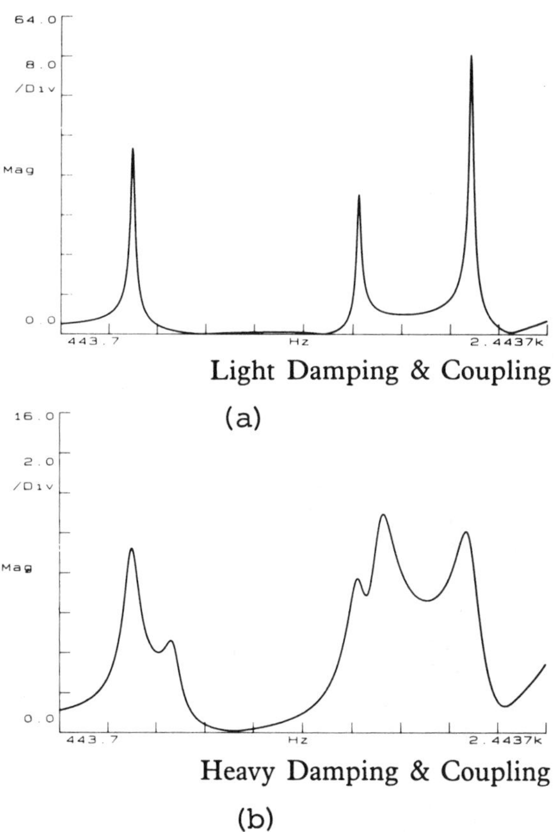


**Figure 18.36** Damping and modal overlaps.

definitive conclusion can be drawn. The selection of a test method depends much on the test goal, test environment, type of test structure, and experience of test personnel.

## 4.6 Brief Review of Vibration Signal Processing

It has been well recognized that the vibration of mechanical systems carries a great deal of information about the internal conditions of these systems. Therefore the vibration signal has been proposed as an indicator of the conditions of mechanical systems which are subject to vibration. However, the vibration signal that is generated at the output of a vibration sensor is raw data and its relation to the conditions of a vibrating mechanical system is not completely visible from its time domain representation. To make the vibration data useful, one must process the data and extract the information which is related to the conditions of the mechanical system. Only then can this information be used to detect and identify specific problems in the mechanical system. In the following sections, significant parameters and techniques of processing the vibration signal are reviewed. Applications of these techniques are also discussed with emphasis on techniques used in condition monitoring and diagnosis of rotating machinery.

## Vibration Signal Amplitude

Since the vibration of mechanical systems reflects the conditions of these systems, the vibration level can be monitored to detect developing faults. This idea has been used successfully for predictive maintenance of rotating machinery. As shown in Figure 18.36b, most machine failures are preceded by an increased level of vibration. Therefore, the vibration level can be measured and monitored as a basis for scheduling machine maintenance so that machine breakdowns are avoided. Figure 18.37 shows a typical vibration acceleration signal for a compressor. As indicated in this figure, the vibration can be measured in terms of peak values or RMS value of the signal. In general, peak values, when measured in terms of displacement, are indicative of maximum stress levels at the point of measurement. In mechanical systems, appreciable displacements occur only at low frequencies; therefore displacement measurements are of limited value in the general study of mechanical vibration and overall machine condition. Where small clearances between machine elements are being considered, vibratory displacement is, of course, an important consideration. Displacement is often used as an indicator of unbalance in rotating machine parts because relatively large displacements usually occur at the shaft rotational frequency, which is also the frequency of greatest interest for balancing purposes. On the other hand, the RMS value of vibration when measured in terms of velocity is the best indicator of severity of vibration for the frequency range 10–

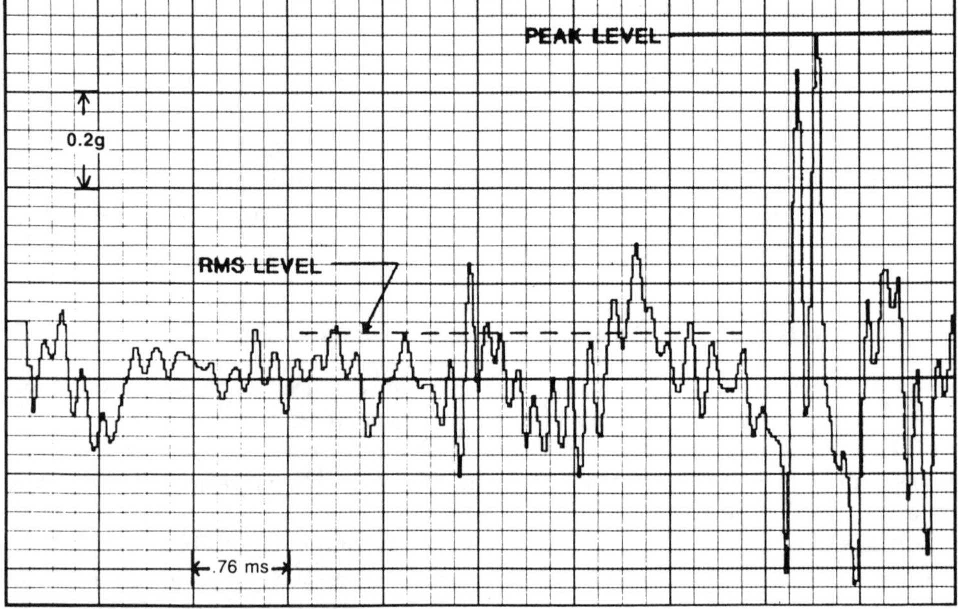


**Figure 18.37** Typical vibration acceleration signal of an air-conditioning compressor.

10,000 Hz. This is due to the fact that the RMS value of the vibration velocity is related to the vibratory energy and is therefore a measure of the destructive effect of the vibration. For the frequency range below 10 Hz, vibration measurement in terms of displacement is recommended, because the vibration velocity or acceleration might not have a satisfactory signal-to-noise ratio in this range. For the same reason vibration measurement for the frequency range above 10,000 Hz in terms of acceleration is recommended [15, 27].

Usually, machine manufacturers provide the range of vibration for different modes of machine operation, such as smooth, good, fair, rough, unacceptable. By comparing the measured level of vibration with these prespecified ranges, a general assessment of machine condition can be obtained. In the absence of manufacturer's recommendations, vibration guidelines such as those in Figure 18.38 can be used.

## Vibration Power Spectrum

In a vibrating mechanical system, the vibrations of different components are transmitted throughout the system. As a result, the vibration at each sensor location contains some information about the vibrations of all components in the system. The various components of a system do not necessarily vibrate at the same frequency. Therefore, the RMS value of the vibration signal is indicative only of the general condition of the system and cannot be used for diagnosis of the entire mechanical system. A more comprehensive approach is to decompose the mean-

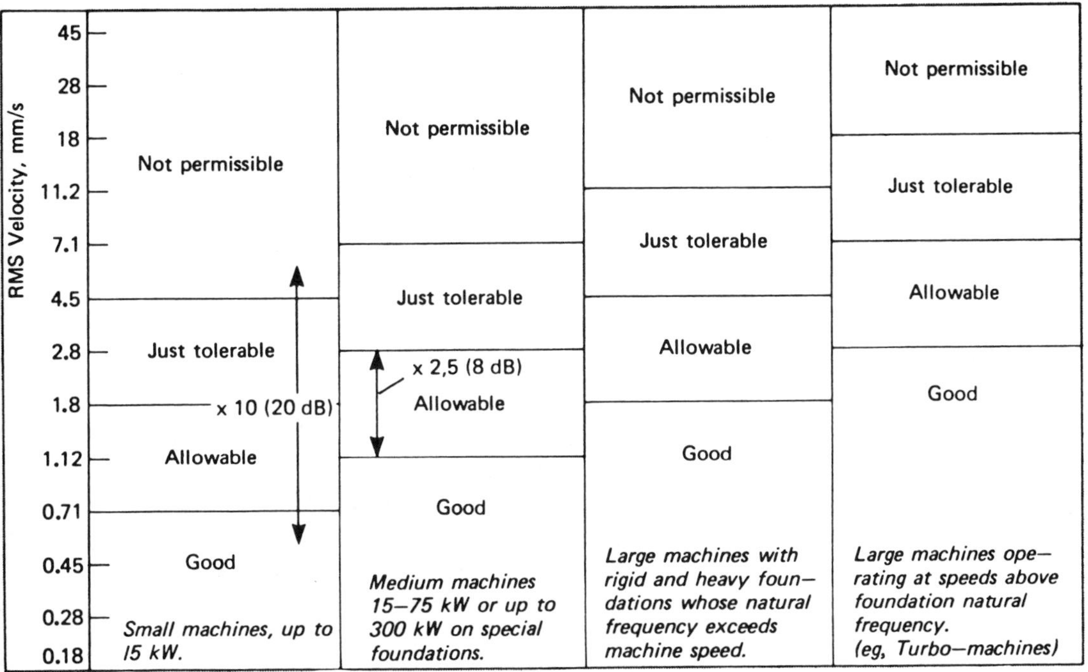


**Figure 18.38** Vibration severity criteria (10 Hz to 1 kHz) in line with VDI 2056, ISO 2372, and BS 4675.

square value of the vibration signal into vibration power within specific frequency bands. This representation of vibration power versus frequency is called the vibration power spectrum. The power spectrum correlates the vibration power at specific frequencies with causes within specific components or subsystems and provides information about the sources of the vibration. Moreover, low-level vibrations, which are produced during the early stages of some defects, cannot be detected by the RMS value of the overall vibration. These defects can be detected by substantial relative increase of the vibration at a specific frequency [16].

As an example, Figure 18.39 shows the vibration power spectrum of a small electric motor. Some of the prominent frequency components and their sources have been identified. This spectrum is characteristic of this particular motor. The level of each vibration frequency component indicates the condition of the motor. As long as the motor is operating properly and there are no faulty parts, its vibration levels will fall around some normal expected value. However, when the motor has a fault such as worn-out bearings, unbalance, or loose assembly, the levels of its characteristic frequency components will increase. Different defects will cause specific frequency components to increase. In an electric motor, for example, unbalance increases the level of the once-per-revolution component. Stator eccentricity will increase the 120-Hz component. Loose rotor blades, worn-out bearings, and other faults show up at other frequencies. Thus the spectrum indicates not only which units are defective but also the nature of the fault. Table 18.4 lists various faults of machinery and shows the vibration frequencies associated with these faults. In applying the vibration spectrum to machine diagnostics, the machine spectrum is compared with a baseline spectrum and changes at different frequencies are recorded. Based on the recorded information and guidelines such as Table 18.4, the condition of the machinery can be evaluated and incipient faults

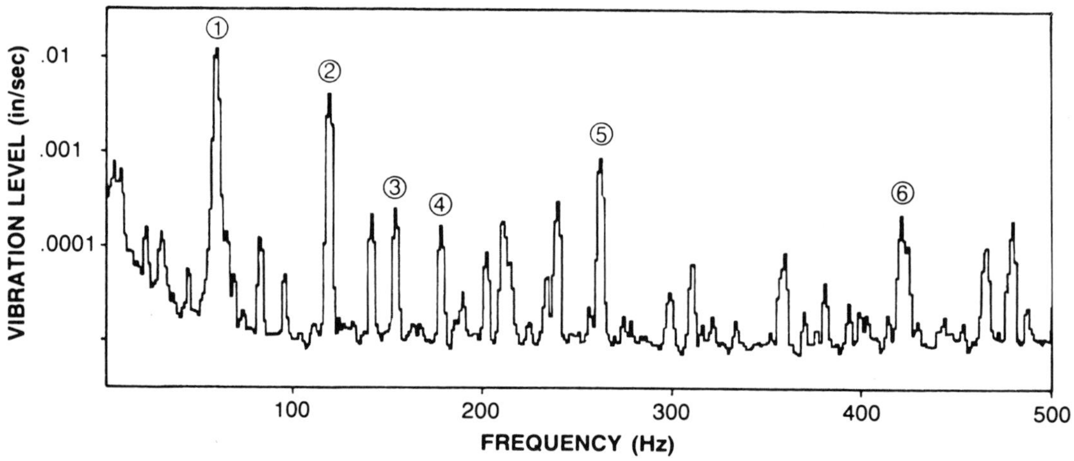


**Figure 18.39** Vibration spectrum of a 1/15 HP electric motor. 1) 60 Hz, rotor unbalance; 2) 120 Hz, electrically induced vibration; 3) 155 Hz, bearing outer race; 4) 180 Hz, misalignment or bent shaft; 5) 262 Hz, bearing inner race; 6) 420 Hz, bearing rolling elements.

can be predicted [16]. In practice, the vibration spectrum is computed from one finite segment of the vibration signal. This means that the resultant spectrum only approximates the true spectrum and is not necessarily accurate enough for the proposed application in vibration analysis. A number of techniques have been proposed to deal with this problem and enhance the signal-to-noise ratio of the vibration spectrum. Ensemble averaging and synchronous averaging are two of the techniques which are more commonly used [15–17, 27, 36, 80]

### Vibration Phase

The vibration power spectrum has only amplitude information and does not contain any information about the phase of vibration. Since the vibration phase contains important information about the relative motion of various parts of mechanical systems such as rotating machinery, it can be used to detect and identify problems in these systems. The main application of phase measurement is in balancing the rotating machinery. It can also be used in detecting misalignment due to bent shaft, determining resonance frequencies, pinpointing mechanical looseness, and identifying types of unbalance [15].

A number of other techniques have been recommended for application in vibration analysis. These include cepstrum analysis [81], the shock pulse technique [82], high-frequency analysis [18, 28], and order tracking [83]. Details of these techniques can be found in the references cited.

## 5 VIBRATION MEASUREMENT INSTRUMENTS

In this section a brief review of vibration measurement instruments is presented. Vibration meters are used for monitoring the vibratory motion of a machine or a structure or for data acquisition.

Three principal stages are involved in the overall monitoring of a machine: 1) *Monitoring* the signals produced by the machine or system. This is done by means of sensors and transducers and is called the data acquisition stage. 2) *Signal/data processing* by means of meters, visual displays, or printouts, all of which may be coupled with alarm systems of lights, sounds, or cutout switches. In principle, this stage manipulates the raw signals to provide significant vibration data. 3) *Condition assessment.* This is the decision stage, at which the system condition can be assessed and operative decisions made.

It is an important requirement of condition monitoring that the reliability of the sensor should be superior to that of the system under surveillance.

The choice of a suitable vibration pickup involves 1) displacement sensed by a proximity transducer with an output signal proportional to displacement, 2) velocity sensed by a seismic-type transducer, or 3) acceleration sensed by an accelerometer. The strongest influence on selection of the appropriate parameter is the frequency at which measurement is important; in any event, a large signal is required so that the signal-to-noise ratio is high. As a rule, displacement measurements record large signals at low frequencies, while acceleration measurements are effective at high frequencies.

A wide range of equipment is commercially available for monitoring purposes. Because of the many problems of use and interpretation, it is advisable to start with relatively simple and inexpensive equipment and gradually progress to more involved systems as experience and knowledge are acquired. The starting point is to decide the immediate objective of monitoring, i.e., whether it is intended for

1. Condition checking or trend monitoring. This is a broad appraisal or confirmation of whether deterioration is occurring without necessarily identifying the source of trouble. This is the usual initial stage and makes use of overall vibration-level meters. For special purposes, shock-pulse meters may be used.
2. Fault diagnosis. Vibration signatures are analyzed and compared. At the simplest level this can involve octave filters or narrow-bandwidth filters; at the most advanced level it may involve advanced data processing techniques aided by computers and used in conjunction with other diagnostic information.

In the following the basic and most common types of vibration measurement equipment are introduced.

**Accelerometers.** Piezoelectric accelerometers operate on the principle that the mass moves and stresses a piezoelectric disk. Acceleration forces cause the amount of applied force on the disk to vary cyclically and this, by the nature of the piezoelectric effect, produces an electrical output signal proportional to the compression. Since this stress is itself directly related to acceleration, such a signal is proportional to the vibratory acceleration of the surface on which the device is mounted.

Stresses and resultant deformations of the piezoelectric element may be 1) thickness or face shear or 2) thickness or transverse compression. Design and construction of the accelerometer vary according to the deformation mode. Figure 18.40 shows the two types of accelerometers in common use. The two piezoelectric disks are arranged so that their positively polarized faces are in contact with a conductor plate, to which one of the output terminals is attached. This assembly sits on a base, with a mass spring clamped on top. The whole mechanism is protected by a housing. The base/housing combination is connected to the other output terminal. When it is wished to measure low or high acceleration levels, the dynamic range of the accelerometer should be considered. The lower limit is determined by electrical noise from connecting cables and amplifier circuitry. This limit is normally as low as 0.01 m/s with general-purpose instruments. The upper limit is determined by the accelerometer's structural strength. A typical general-purpose accelerometer is linear from 50,000 to 100,000 $m/s^2$, that is, well into the range of mechanical shocks. Detailed descriptions of different types of accelerometers and their characteristics and uses can be found in the literature [22, 24, 27, 49].

**Velocity transducers.** Inductive electromotive force (emf) is set up when a permanent magnet surrounding a moving coil is vibrated. This forms the basis of an inductive magnetic or velocity transducer since the emf is an alternating current with a velocity dependence. Machine vibration causes the magnet to move within the coil, thus converting the motions of the machine into an electrical signal. The signal is proportional to the vibrational velocity. Most instruments also indicate displacement by passing the signal through an integrating amplifier. The usual frequency range is 10–10,000 Hz and is considerably influenced by the method adopted to attach the transducer to the machine.

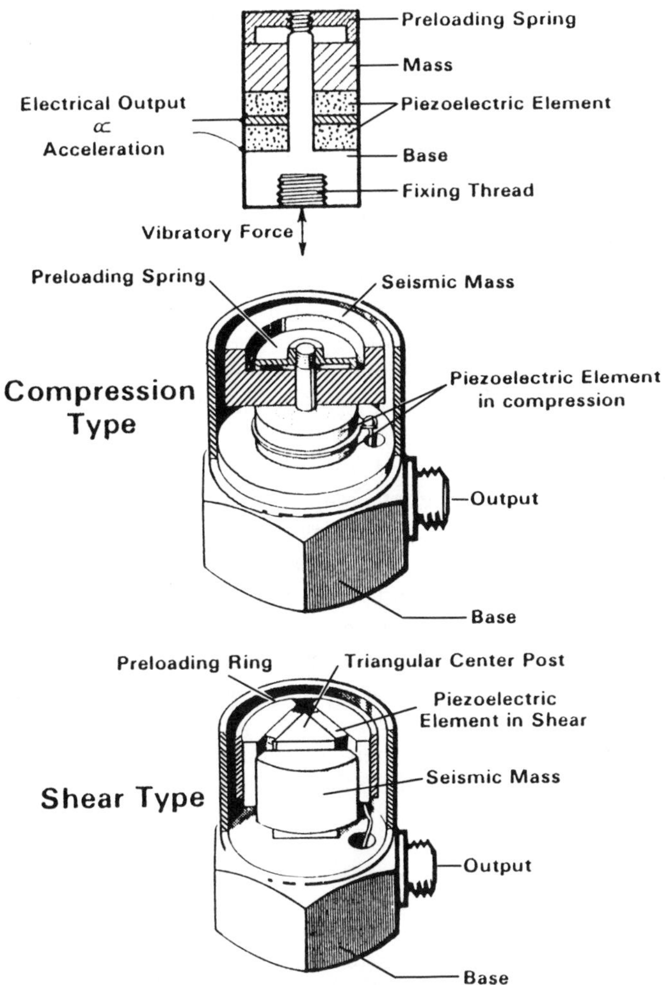


**Figure 18.40** Two types of accelerometers in common use.

**Proximity transducers.** These are sometimes called eddy current transducers, which reflects their operating principles, although the proximity effect may also be sensed by capacitive or inductive effects. An eddy current proximity transducer is an insulated probe with a response coil in the tip. It is fed with a high-frequency carrier signal (> 500 kHz). The signal becomes modulated according to the size of gap it must cross. The output is then modulated to give a voltage proportional to gap size. An output of 100 mV/mil, equivalent to 4 V/mm, is the calibration specified by the Americal Petroleum Institute. The probe is generally about 7.5 mm (0.3 in.) in diameter with a small coil of fine wire at its tip from which the magnetic field is generated by the high-frequency carrier signal. Eddy currents induced by this magnetic field have an opposing influence and reduce the carrier amplitude in proportion to gap distance; the demodulator converts this change in carrier amplitude to a low-impedance, calibrated voltage output. The eddy current sensor and its companion oscillator demodulator therefore constitute a gap-to-voltage measuring system.

**Signal preamplification.** An accelerometer signal is of very low power and is rapidly dissipated; it also has a high electrical impedance. To adjust these conditions, preamplifiers may be fitted. They may also be used to compensate for nonstandard transducer sensitivity by conditioning the signal to a reference sensitivity. Also, by building integration networks into the preamplifiers, either velocity or displacement can be measured from the acceleration signal.

**Magnetic tape recorders.** Signals from vibration sensors or transducers can be recorded on magnetic tape recorders which store these signals for later reproduction and analysis. By speeding up or slowing down the tape it is possible to achieve some expansion or compression of the time scale. Multichannel recording can also preserve the time and phase relationship of a number of signals. Direct recording with high-frequency bias and frequency-modulated (FM) recording are the two most common techniques. With subsequent signal processing using simple spectrum analysis of each signal independently, a direct recording technique is adequate and economical if the frequency range of interest falls between 30 Hz and 10 kHz. If the signals to be processed are of a lower frequency or the signal processing is such that the phase and time relationship of a number of signals are relevant, FM recording is far superior and usually essential.

Other vibration detection instruments are used for various diagnostic purposes. These include, for example, capacitive vibration sensors, vibraswitch malfunction detectors, stroboscopes, electronic stethoscopes, and tuned-reed frequency meters. Descriptions of these instruments, their functions, and their applications can be found in the literature [22, 24, 27, 49, 84].

## 6 CONCLUSION

In this chapter, a brief review of the fundamentals of mechanical vibrations was presented. The topics covered here can serve both as an overview of the subject and as general guidelines for practicing engineers who deal with the vibration of mechanical systems. The aim of this chapter was to present the subject of mechanical vibrations in such a manner that the reader can gain a basic understanding of this field and the methods that are peculiar to its investigation. An extensive list of references is provided at the end of the chapter for those who wish to obtain a more in-depth familiarity with each of the subjects discussed.

## REFERENCES

1. R. D. Blevins, *Formulas for Natural Frequency and Mode Shape*, Van Nostrand Reinhold, New York (1979).
2. W. D. Pilkey and P. Y. Chong, *Modern Formulas for Statics and Dynamics*, McGraw-Hill, New York (1978).
3. M. N. Noori, H. Davoodi, et al., Zero and nonzero mean random vibration analysis of a new general hysteresis model, *Probabil. Eng. Mech.* (in press).

4. T. T. Baber and M. N. Noori, Modelling general hysteresis behavior and random vibration application, *ASME. Vibr. Acoust. Stress Reliabil. Des. 108*(4): 411–420 (1985).
5. T. T. Baber and M. N. Noori, Random vibration of degrading structures with pinching hysteresis, *ASCE J. of Eng. Mech. Div. 3*(8): 1010–1026 (1985).
6. S. H. Crandall, The role of damping in vibration theory, *J. Sound Vibr., 11*1: 3–18 (1970).
7. R. H. Scanlan and A. Mendelson, Structural damping, *AIAA J., 1*: 938–939 (1963).
8. A. D. Nashif, D. I. G. Jones, and J. P. Henderson, *Vibration Damping*, Wiley, New York (1985).
9. R. W. Clough and J. Penzien, *Dynamics of Structures*, McGraw-Hill, New York (1975).
10. A. S. Veletos and C. E. Ventura, Modal analysis of non-classically damped linear systems, *Earthquake Eng. Struct. Dyn., 14*: 217–243 (1986).
11. C. E. Ventura and A. S. Veletos, Steady-state and transient responses of non-classically damped linear systems, *Earthquake Eng. Struct. Dyn., 14*: 595–608 (1986).
12. T. Ikusa, A. D. Kiureghian, and J. L. Sackman, Modal decomposition method for stationary response of non-classically damped systems, *Earthquake Eng. Struct. Dyn., 12*: 121–136 (1984).
13. L. Meirovitch, *Elements of Vibration Analysis*, McGraw-Hill, New York (1986).
14. J. P. Den Hartog, Forced vibration with combined Coulomb and viscous friction, *J. Appl. Mech. (Trans. ASME), 53*: 107–115 (1931).
15. H. P. Bloch and F. K. Geitner, *Practical Machinery Management for Process Plants*, vol. 2, *Machinery Failure Analysis and Trouble Shooting*, Gulf Publishing Co., Houston, Texas (1983).
16. H. Hakimmashhadi, Evaluation of the Application of Vibration Analysis in Machine Condition Monitoring and Diagnostic Systems, IRD Project Report 3.1-1-1, Aircraft Instruments Department, General Electric Co., Wilmington, Massachusetts (1986).
17. Effective Machinery Maintenance Using Vibration Analysis, Hewlett Packard, Application Note 243-1 (1983).
18. L. J. Mertaugh, "Evaluation of Vibration Analysis Techniques for the Detection of Gear and Bearing Faults in Helicopter Gearboxes," 41st Meeting of the Mechanical Failure Prevention Group, Patuxent River, Maryland (October 1986).
19. M. H. Richardson and K. A. Ramsey, Integration of dynamic testing into the product design cycle, *Sound Vibr. 15*(11): 14–27 (1981).
20. N. F. Rieger, The literature of vibration engineering, *Shock Vibr. Dig. 14*(1): 5–13 (1982).
21. G. F. Lang, Understanding vibration measurements, *Sound Vibr. 10*: 26–37 (March 1976).
22. R. A. Collacot, *Vibration Monitoring and Diagnosis*, Wiley, New York (1979).
23. R. J. Harker, *Generalized Methods of Vibration Analysis*, Wiley, New York (1983).
24. D. J. Ewins, *Modal Testing: Theory and Practice*, Wiley, New York (1984).

25. C. E. Crede and C. M. Harris, *Shock and Vibration Handbook*, McGraw-Hill, New York, vol. 3, pp. 44.1–44.51 and 50.1–50.32 (1961).
26. C. E. Crede and C. M. Harris, *Shock and Vibration Handbook*, McGraw-Hill, New York, vol. 1, pp. 1.1–1.25 and 2.1–2.27 (1961).
27. J. T. Brock, *Mechanical Vibration and Shock Measurements*, Bruel & Kjder Instruments, Inc., Copenhagen, Denmark (1984)
28. R. J. Drago and D. B. Board, "High Frequency Vibration Monitoring Techniques for Gear/Bearing Systems Failure Detection," AGMA Paper No. 109.36, American Gear Manufacturers Association Meeting, Montreal, Canada (October 1975).
29. S. P. Timoshenko, D. H. Young, and W. Weaver, *Vibration Problems in Engineering*, 4th ed., Wiley, New York (1974).
30. C. H. Dowding, *Blast Vibration Monitoring and Control*, Prentice-Hall, Englewood Cliffs, New Jersey (1979).
31. S. S. Rao, *Mechanical Vibrations*, Addison-Wesley, Reading, Massachusetts (1986).
32. Y. Matsuzaki and S. Kibe, Shock and seismic response spectra in design problems, *Shock Vibr. Dig. 15*: 3–10 (October 1983).
33. M. Lalanne, P. Berthier, and J. D. Hagopian, *Mechanical Vibrations for Engineers*, Wiley, New York (1983).
34. W. T. Thomson, *Theory of Vibration with Applications*, Prentice-Hall, Englewood Cliffs, New Jersey (1981).
35. F. S. Tse, I. E. Morse, and R. T. Hinkle, *Mechanical Vibrations*, Allyn & Bacon, Boston (1963).
36. J. S. Bendat and A. G. Piersol, *Engineering Applications of Correlation and Spectral Analysis*, Wiley-Interscience, New York (1980).
37. J. S. Bendat and A. G. Piersol, *Random Data. Analysis and Measurement Procedures*, Wiley, New York (1986).
38. V. V. Bolotin, *Random Vibration of Elastic Systems*, Martinus Nijhoff, Amsterdam, Netherlands (1984).
39. D. E. Newland, *Random Vibrations and Spectral Analysis*, Longman, New York (1975).
40. W. Zambrano, A brief note on the determination of the natural frequencies of a spring-mass system, *Int. J. Mech. Eng. Educ.*, *9*: 331–334 (October 1981); *10*: 216 (July 1982).
41. B. O. Skipp, *Vibration in Civil Engineering*, Butterworths, London (1966).
42. N. F. Rieger, H. Poritsky, and J. W. Lund, Rotor-Bearing Dynamics Design Technology, vols. 1–9, Wright-Patterson Air Force Base Aero Propulsion Laboratory Reports (1965–1968).
43. N. F. Rieger and A. W. Kimber, "Dynamic Interaction Between a Vertical Pump and Its Piping System," Proc. International Conference on the Hydraulics of Pumping Stations, BHRA, Fluid Engineering Center, Manchester, England, pp. 253–266 (September 17–19, 1985).
44. J. E. Corley, The effects of pedestal stiffness on rotor dynamics, *Vibrations*, *2*(3): 14–15 (1986).
45. N. F. Rieger, Notes on the development of balancing techniques, *Vibrations*, *2*(1): 3–7 (1986).

46. W. R. Campbell, Diagnosing alternating current electric motor problems. Part 2: Electromagnetic problems, *Vibrations, 1*(3): 12–15 (1985).
47. R. W. Potter, "Measuring Linear System Parameters: Single Input/Output Transfer and Coherence Functions," Short Course in Modal Analysis, University of Cincinnati, Ohio, Summer, 1979.
48. R. W. Mastain, "Survey of Modal Vibration Test/Analysis Techniques," SAE Paper 760870 (1976).
49. G. A. Hamma, S. Smith, and R. C. Stroud, "An Evaluation of Excitation and Analysis Methods for Modal Testing," SAE Paper 760872, 1976.
50. The Fundamentals of Modal Testing, Hewlett Packard Application Note 243-3 (May 1986).
51. K. A. Ramsey, Effective measurements for structural dynamic testing, Part I, *Sound Vibr.*, pp. 26–42 (November 1975).
52. K. A. Ramsey, Effective measurements for structural dynamic testing, Part II, *Sound Vibr.*, pp. 34–48 (April 1976).
53. A. Klosterman and R. Zimmerman, "Modal Survey Activity Via Frequency Response Functions," SAE Paper 751068 (1975).
54. J. D. Favour, M. C. Mitchell, and N. L. Olson, Transient Test Techniques for Mechanical Impedance and Modal Survey Testing, *Shock Vibr. Bull.*, no. 42, part I (January 1972).
55. C. C. Kennedy and C. D. P. Pancu, Use of vectors in vibration measurement and analysis, *J. Aeronaut. Sci.*, *4*11: 605–625 (1947).
56. R. C. Lewis and D. L. Wrisley, A system for the excitation of pure natural modes of complex structures, *J. Aeronaut. Sci.*, *17*(11): 705–723 (1950).
57. G. W. Asher, "A Method of Normal Mode Excitation Utilizing Admittance Measurement, Dynamics and Aeroelasticity," Proc. Institute of the Aeronautical Sciences, pp. 69–76 (1958).
58. R. E. D. Bishop and G. M. L. Gladwell. An investigation into the theory of resonance testing, *Philos. Trans. R. Soc. London Ser. A 225*: 242–280 (1963).
59. R. R. Craig and Y. W. T. Su, On multi-shaker resonance testing, *J. of Am. Inst. of Aeronautics and Astronautics*, *12*(7): 924–931 (1974).
60. P. Ibanez, "Force Appropriation by Extended Asher's Method," SAE Aerospace Engineering and Manufacturing Meeting, Paper No. 760873 (November 1976).
61. G. Moroscow, Exciter force apportioning for modal vibration testing using incomplete excitation, Ph.D. thesis, University of Colorado, Boulder (1977).
62. W. L. Hallauer and J. F. Stafford, On the distribution of shaker forces in multiple-shaker modal testing, *Shock Vibr. Bull.* no. 48, part I, pp. 49–63 (1978).
63. J. W. Cooley and J. W. Tukey, An algorithm for the machine calculation of complex Fourier series, *Math. Comput.*, *19*: 297–301 (April 1965).
64. The Hewlett-Packard Modal Analysis System, Hewlett-Packard (September 1974).
65. M. Richardson and R. Potter, "Identification of the Modal Properties of an Elastic Structure from Measured Transfer Function Data," Proc. 20th International Instrumentation Symposium, Albuquerque, New Mexico, pp. 239–246 (May 1974).

66. M. Richardson, "Modal Analysis Using Digital Test Systems," Seminar on Understanding Digital Control and Analysis in Vibration Test Systems, Shock and Vibration Information Center, Cincinnati, Ohio (May, 1975).
67. D. Brown, R. Allemang, R. Zimmerman, and M. Mergeay, "Parameter Estimation Techniques for Modal Analysis," SAE Paper 790221 (1979).
68. R. W. Budd, "A New Approach to Modal Vibration Testing of Complex Aerospace Structure," *J. Environ. Sci.*, *No. 2*, pp. 14–19 (1969).
69. S. R. Ibrahim and E. C. Mikulick, A time domain modal vibration test technique, *Shock Vibr. Bull.*, no. 43, part 4, pp. 21–37 (June 1973).
70. S. R. Ibrahim and E. C. Mikulick, The experimental determination of vibration parameters from time response, *Shock Vibr. Bull.*, no. 46, part 5, pp. 187–196 (1976).
71. S. R. Ibrahim and E. C. Mikulick, A method for the direct identification of vibration parameters from the free response, *Shock Vibr. Bull.*, No. 42, pp. 183–198 (1972).
72. Hewlett Packard, *Sampling Window Error*, Fourier Analyzer Training Manual, Section III (1972).
73. R. W. Potter, Complication of Time Windows and Line Shapes for Fourier Analysis, Hewlett Packard Co. (1972).
74. Dynamic Signal Analyzer Applications, Hewlett Packard Application Note 243-1 (October 1983).
75. Gen Rad, Time/Data Division, *Modal Analysis and Modelling System—User Manual* (March 21, 1977).
76. E. L. Leppert, S. H. Lee, F. D. Day, P. C. Chapman, and B. K. Wada, "Comparison of Modal Test Results: Multipoint Sine Versus Single-Point Random," SAE Paper 760879 (1976).
77. R. C. Stroud, C. J. Bonner, and G. J. Chambers, "Modal Testing Options for Spacecraft Developments," SAE paper 781043 (1978).
78. C. V. Stahle, Modal test methods and applications, *J. Environ. Sci.*, No. 1, pp. 17–21 (January/February 1978).
79. T. Comstock, J. Niebe, G. Wylie, and F. H. Chu, "Systematic Approach to Experimental Modal Analysis Using Multi-Point Excitation," presented at the ASME Annual Winter Meeting, New York (1979).
80. G. M. Jenkins and D. G. Watts, *Spectral Analysis and Its Applications*, Holden-Day, San Francisco (1968).
81. R. B. Randall, Cepstrum Analysis and Gearbox Fault Diagnosis, Application Notes, Bruel & Kjder Instruments, Marlborough, Massachusetts (1980).
82. T. C. Mayer, F. Covill, J. A. George, and J. T. Harrington, Applications of the Shock Pulse Technique to Helicopter Diagnostics, Park College of St. Louis University, USA AVSCOM TR 77–21, Directorate for Research Development and Engineering, U. S. Army Aviation Systems Command, St. Louis, Missouri (February 1975).
83. H. Herlufsen, Order Analysis Using Zoom FFT, Application Notes, Bruel & Kjder Instruments, Marlborough, Massachusetts (1981).
84. N. L. Baxter and R. L. Eshleman, Portable magnetic tape recorders for vibration analysis and monitoring, *Vibrations*, *1*(3): 4–11 (1985).

# 19

# Geophysical Signal Processing

PETER SCARLETT* and ANASTASIOS N. VENETSANOPOULOS Department of Electrical Engineering, University of Toronto, Toronto, Ontario, Canada

Reflection seismology is the technique of echo sounding the earth's subsurface for the delineation of oil, gas, and other mineral deposits. Oil in particular is usually found pooled under geological discontinuities that may be as deep as 3 km. The detection and location of such subtle features is a considerable problem that has been addressed with increasing success, primarily due to advances in signal processing. While seismic exploration has long been cost-effective, rising costs elsewhere, particularly for land leasing and drilling, have tended to reduce its proportion of the total costs even more. Seismic processing is therefore highly leveraged, increasing the total costs only slightly while greatly reducing the chances of an expensive dry well.

The problem is that raw seismic data are not immediately intelligible; in the same way that the human eye needs a lens to focus reflected light from a scene, the human interpreter needs a technique for seismic *migration* to focus the reflected acoustic energy. Two further complications are the severe linear distortions introduced by the earth and the large amounts of noise often present.

In this chapter, after a brief overview of the various steps involved in a typical seismic survey, we shall discuss in more detail those areas where multidimensional techniques have had the largest impact. Dereverberation and velocity filtering are discussed first and then migration is analyzed. Emphasis is placed on the theoretical background and algorithmic development.

---

Present affiliation:

*Advanced Systems Development, Raytheon Canada Limited, Waterloo, Ontario, Canada

## 1 DATA ACQUISITION

In reflection seismology, a near-surface explosion generates a high-energy acoustic impulse to measure the impulse response of the earth's subsurface. As the wave field spreads through the earth, it is attenuated by absorption and dispersion. Wherever inhomogeneities occur, as at geological interfaces, the wave is reflected and refracted. Multiple reflection, total internal reflection, and wave conversion complicate matters even more. As a result of these processes, a very small amount of the input energy eventually reaches the surface as a long distorted series of echoes that are received by a set of geophones.

The 24, 48, or 96 receivers are spaced 15 to 300 m apart in a linear or grid pattern and displaced from the explosive source for safety. They are sensitive to the earth's surface motion velocity, either directly or via water pressure, and thus sample this aspect of the surface wave field. Data are digitized at rates from 250 to 1000 samples/sec to be consistent with the earth's transmission response, which limits the usable frequencies, above the noise at about −80 dB, to the range 5 to 80 Hz in shallow sections above 2000 m and 5 to 35 Hz below 3000 m [1]. Typically, data are recorded for about 6 sec because later arrivals are due to either extremely deep reflections or excessively complicated interbed reflections, neither of which is of much interest.

Such a set of space-time data from a single source is called a *common shotpoint* (CSP) *gather* and consists of diffraction patterns from the various inhomogeneities in the earth along with additive noise. The dynamic range of the surface wave field is often 80 dB, due to attenuation of the later arrivals, so it is common to apply heuristic time-varying amplification to the data. This results in a nonstationary signal-to-noise ratio even though the noise itself is usually stationary and Gaussian. To get a broader picture, the entire network of source and receivers is laterally displaced (effectively) by one receiver spacing and the process repeated. Thus there is an overlap, with each location being used to record $N$ (the number of receivers) different surface waveforms corresponding to the $N$ different source locations. Data are recorded in this way for many miles to produce a seismic survey line (Figure 19.1).

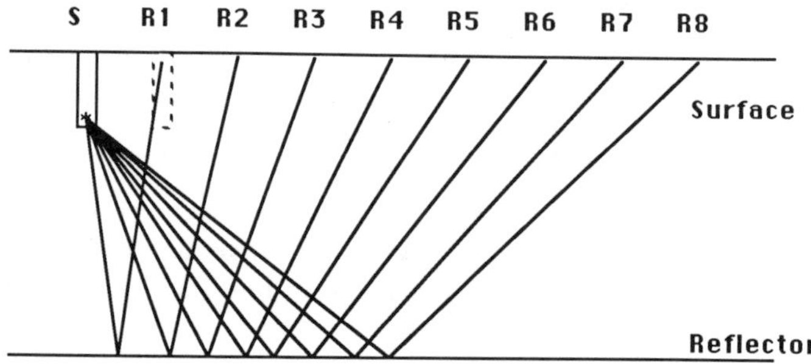


**Figure 19.1** Configuration of the explosive source and the receiver array. Each receiver R# consists of an array of geophones. The location of the next shot is shown dashed; the receivers, being stationary, are allocated from 1 to $N$ when the data are recorded so that the present R2 will become R1 for the next shot.

The volume of data recorded and processed is immense; a single-shot, 24-receiver CSP record can easily comprise 1 Mbyte and there are usually several hundred of these records in a single two-dimensional seismic survey line. Three-dimensional surveys require a grid of seismic sections with an attendant squaring of the data. One of the major problems in practice is that of data management.

## 2 INITIAL CORRECTIONS

The raw recorded data are then statically corrected so that all the records are consistent. In particular, the effects of elevation changes and near-surface variations in the earth must be removed by applying a constant time and space shift to each signal from the array of receivers. At shallow depths, the earth is highly inhomogeneous and sound velocities are very low, leading to a chaotic waveform. As this shallow stratum is usually of no interest, common practice is to ignore it

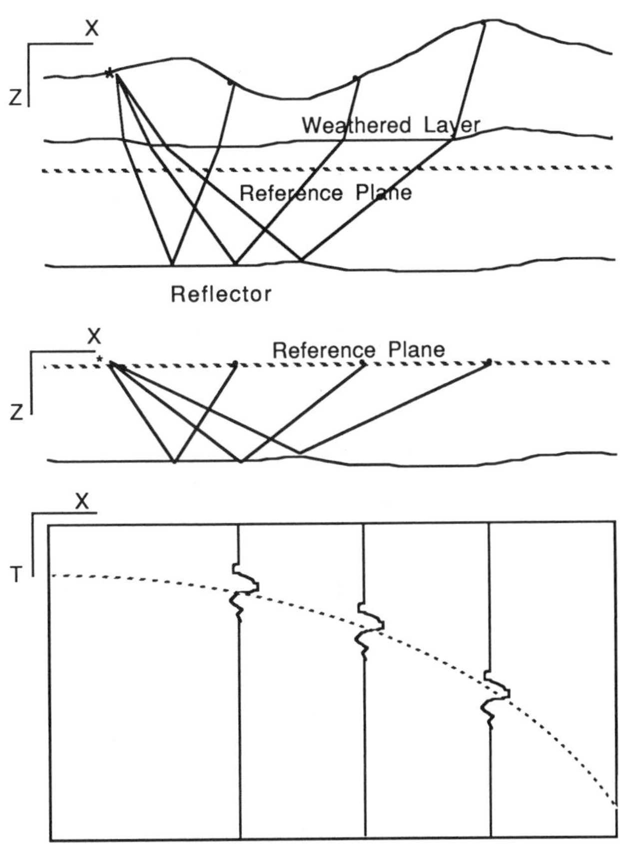


**Figure 19.2** Static corrections for the low-velocity weathered zone. (Derived from [1].)

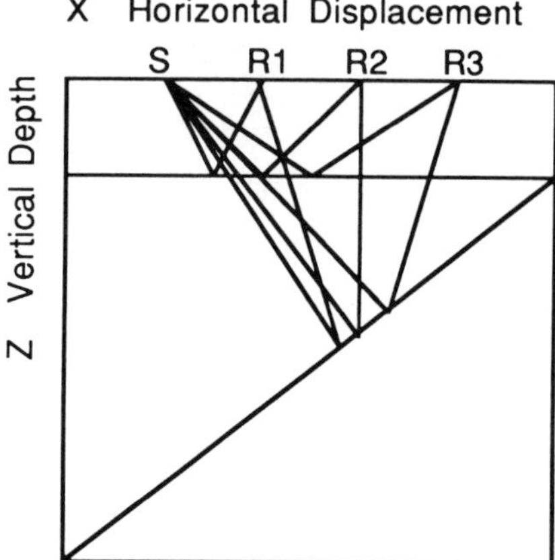


**Figure 19.3** Common source point (CSP) gather, showing ray paths and diffuse focus.

by truncating the signal so that the receivers are effectively on a plane below this "weathered zone" [2] (Figure 19.2).

The trouble with such raw CSP data is that there is little obvious relationship between the received reflections and the underlying geologic structure. For the flat reflector in Figure 19.3, the reflection travel time $T$ is expressed as

$$T^2 = \left(\frac{2Z_s}{V}\right)^2 + \left(\frac{X}{V}\right)^2 \tag{19.1}$$

where $Z_s$ is the reflector depth below the shot point, $X$ the lateral displacement of the receiver from the shot point, and $V$ the velocity of the earth medium. This

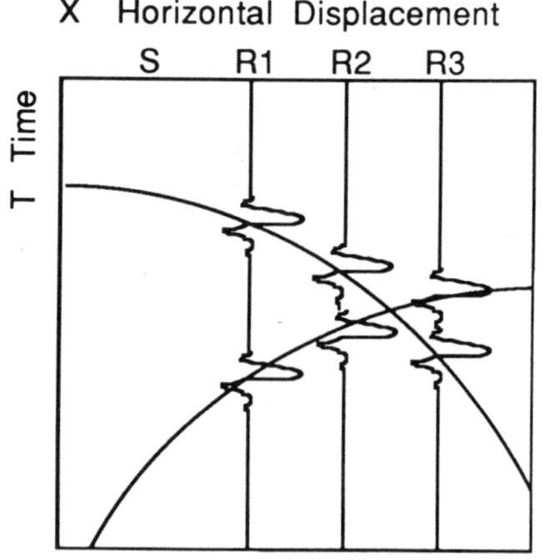


**Figure 19.4** CSP gather showing two-way travel-time section.

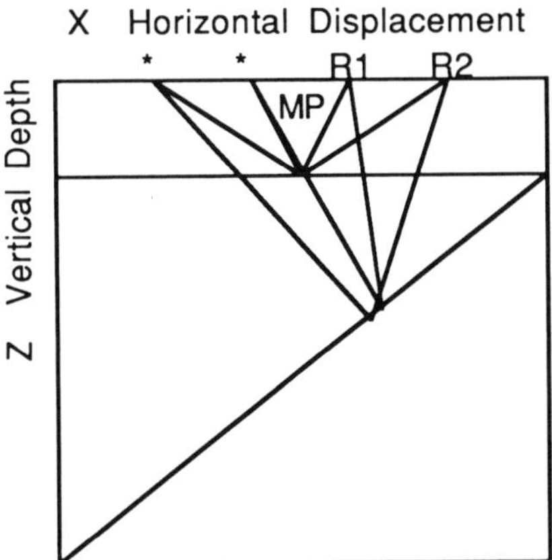


**Figure 19.5** Common midpoint (CMP) gather, showing simplified ray paths and localized focus, particularly for horizontal reflectors.

equation describes a hyperbola in $X - T$ space centered beneath the shot point with an apex at $2Z_s/V$ sec. For the $\theta°$ inclined reflector, the travel-time equation

$$T^2 = \left(\frac{2Z_s \cos^2\theta}{V}\right)^2 + \left(\frac{X + 2Z_s \sin\theta\cos\theta}{V}\right)^2 \tag{19.2}$$

describes a similar but offset hyperbola. Estimating the geological structures from these hyperbolas is clearly a problem even for the simplest case of a uniform-velocity earth (Figure 19.4).

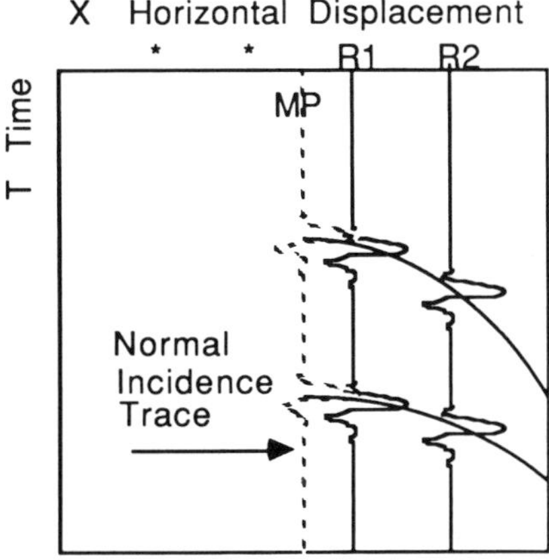


**Figure 19.6** CMP time section showing near-hyperbolic reflections. The hypothetical result of a coincident source-receiver at the midpoint (MP) is the Normal Incidence Trace, shown dashed.

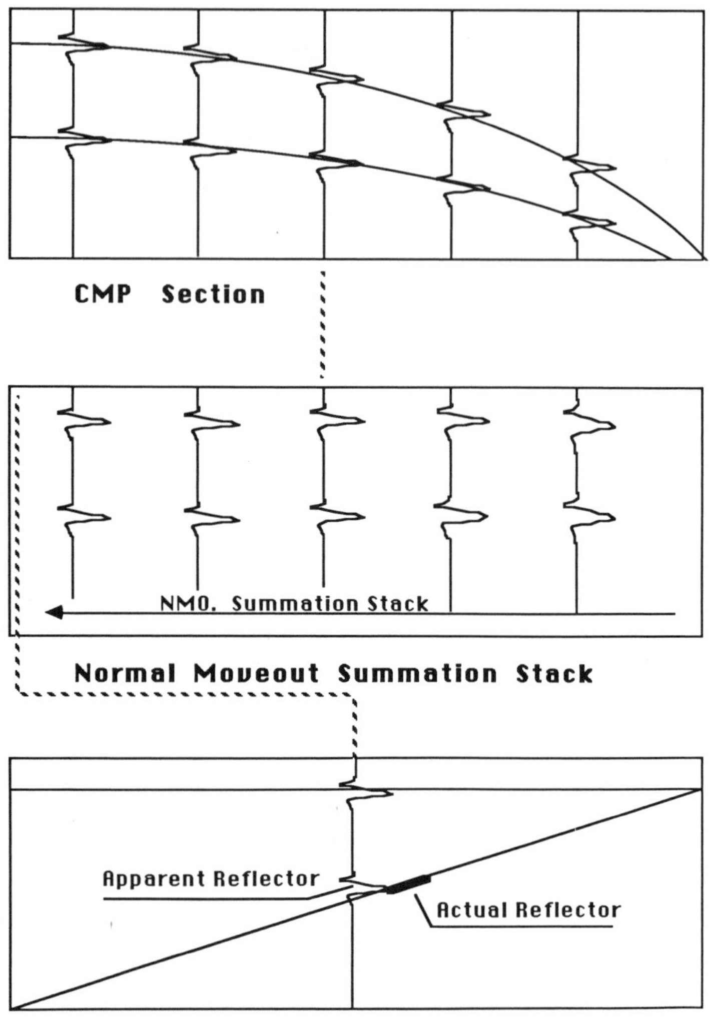


**Figure 19.7** Normal moveout (NMO) stack showing the progression from the CMP section, through removal of the hyperbolic travel times, to summation stacking for the normal incidence trace. Note that the NIT will underestimate the reflector dip.

However, considerable simplification is possible by using symmetries. The many CSP gathers (one per shot) are reshuffled so that only source-receiver pairs with the same midpoint are displayed together in a *common midpoint* (CMP) *gather.* This has the important advantage that only one point of a horizontal reflector is effectively illuminated (Figure 19.5) rather than the confusion of CSP illumination points (Figure 19.3). This arises from Huygens' principle of geometric optics and becomes only approximately true as the reflector dip (i.e., inclination) increases (Figure 19.5). Because a CMP gather is symmetric about the midpoint, all the

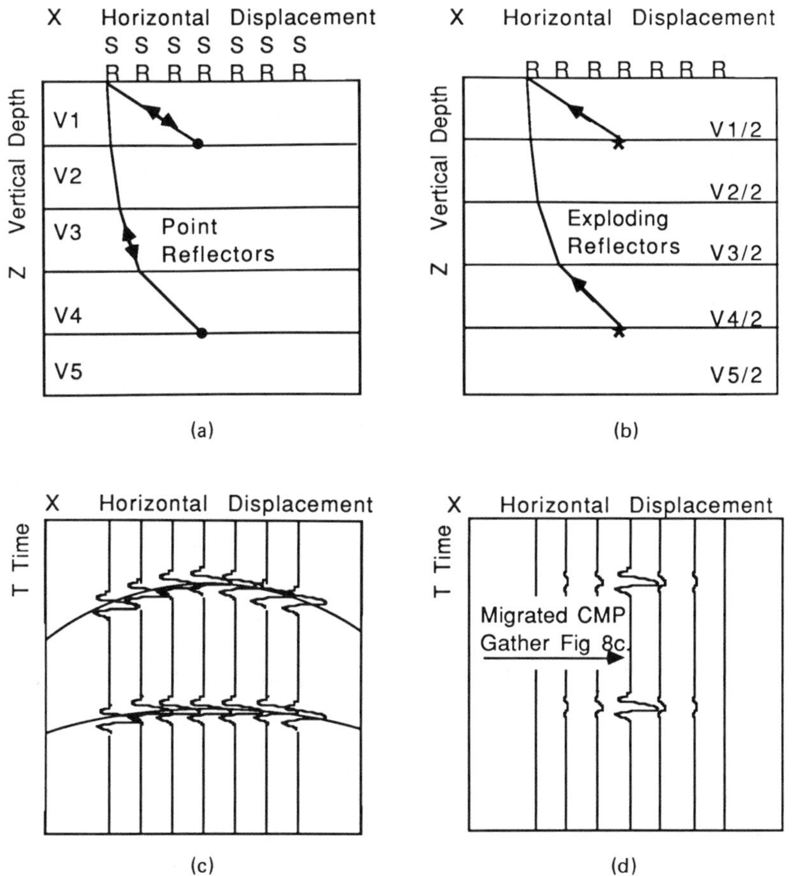


**Figure 19.8** (a) Two point reflectors in layered earth. Ray paths for a single coincident source-receiver giving an NIT. (b) Exploding reflector model, showing identical response of one-way ray with halved velocity. (c) Normal incidence section, showing the hyperbolic response to point reflectors. (d) Migrated NIT section, showing correct placement of imaged point reflectors.

hyperbolas will have the apex at the midpoint position, where the hypothetical signal from a coincident source-receiver pair would be located (Figure 19.6).

Since such a signal cannot be measured in practice, because of the explosive nature of the source, the obvious next step is to try to synthesize this ideal response by removing the hyperbolic travel times imposed by the earth on the offset receivers. This can be done either by a simple shift and stretch of the individual signals followed by a summation, the *normal moveout* (NMO) stack (Figure 19.7), or by inverting the earth's transmission response, the *migration* stack. The resulting response is the *normal incidence trace* (or NIT), so called because with the source and receiver coincident, the ray must arrive normal to every reflector if it is to arrive back at the receiver.

The advantages of using this stacked normal incidence trace are important:

1. By using $N$ independently received traces to create one composite trace, the signal-to-noise ratio (SNR) for incoherent noise increases by $\sqrt{N}$.
2. A vastly simplified solution of the earth's response is possible if the velocity contrasts are approximately flat.
3. If the normal incidence traces are calculated for each midpoint along the survey, they can be displayed side by side to represent a composite image composed of ideal soundings of the earth.

Consider the horizontally layered earth model of Figure 19.8a with coincident sources and receivers arrayed along the surface; clearly, the ray paths will result in normal incidence traces so that the upward and downward rays will coincide. Because these two rays are exact conjugates of each other, there is no need for their duplication. Thus we can consider an equivalent *exploding reflector model* where the single ray starts at the (effectively exploding) reflector at time zero and travels upward at half the speed (since it must travel half the two-way distance) to the receiver (Figure 19.8b). It will be seen in the next section that considering only one-way transmission greatly reduces the complexity of the earth model and simplifies its solution.

## 3 THE WAVE EQUATION AND ITS SOLUTION

In order to make any sense of the wave field received at the surface, it is essential to understand how it arose. While the geometric arguments presented above are approximately valid, they do not account for the diffraction, spherical spreading, and high-frequency attenuation seen in practice. To do this, we must solve the wave equation for a propagating wave field.

The earth is a medium that is well described at the low frequencies of interest by the *acoustic wave equation* for inhomogeneous fluids [3].

$$\nabla^2 p - \frac{1}{c^2}\frac{\partial^2 p}{\partial t^2} = \nabla p \cdot \nabla \ln \rho \tag{19.3}$$

where $p = p(x, y, z, t)$ is the wave field pressure, $c = c(x, y, z)$ the medium velocity, and $\rho = \rho(x, y, z)$ the medium density. While this is an accurate representation, there is no general closed-form solution and so, for analytic reasons, some approximations are necessary.

The first simplification arises from the assumption that the density variation term $\nabla p \cdot \nabla \ln \rho$ is negligible. This is justified by the, typically, slow spatial variation of the density relative to the wavelengths of the frequencies; in particular, if $|\nabla \ln \rho| \ll 2\pi/\lambda$, then the *scalar wave equation* is a valid approximation [3]:

$$\nabla^2 p - \frac{1}{c^2}\frac{\partial^2 p}{\partial t^2} = 0 \tag{19.4}$$

Because the scalar wave equation partially decouples the density from the velocity,* reflectivity and velocity changes are not necessarily coincident. Therefore a layered earth need not imply layered reflectors, though extreme inconsistencies between the two can lead to erroneous results. Moreover, if we assume such a horizontally layered earth, the velocity $c$ varies only with depth, permitting the Fourier transform (FT) in $x$, $y$, $t$ of the scalar wave equation, to result in the *Helmholtz equation*

$$\frac{d^2P}{dz^2} + k_z(z)P = 0 \tag{19.5}$$

where $P = P(k_x, k_y, z, \omega) = FT\{p(x, y, z, t)\}$, $k_z(z) = \sqrt{k^2 - k_x^2 - k_y^2}$ is the wave number function, and $k = \omega/c$ is the angular wave number. For a medium homogeneous between $z_i$ and $z_{i-1}$, this has the exact forward propagation solution

$$P(k_z, k_y, z_{i-1}, \omega) = P(k_z, k_y, z_i, \omega)\exp(-j|z_i - z_{i-1}|k_z) \tag{19.6}$$

To solve for the layered earth case we make the further approximation that the wave number function $k$ varies slowly relative to the depth, implying that reflections due to velocity contrasts can be ignored, leaving only the transmission effects to be considered. This is the Wentzel–Kramers–Brillouin (WKB) or geometric optics assumption, which has the $m$-layer forward propagation solution†

$$P(k_z, k_y, z_{i-m}, \omega) = P(k_z, k_y, z_i, \omega)\exp\left[-j\int_{z_i}^{z_{i-m}} k_z(z)\,dz\right] \tag{19.7}$$

Both of these solutions have the same structure, with the wave field at $z_i$ being multiplied by a phase shift operator that depends only on the medium between $z_i$ and $z_{i-m}$. Since the layered earth model is usually composed of thick layers of near-uniform internal velocity, the final step is to replace the integral with a summation of phase shifts

$$\begin{aligned}P(k_x, k_y, z_{i-m}, \omega) &= P(k_x, k_y, z_i, \omega)\exp\left[-j|z_i - z_{i-m}|\sum_{j=i}^{i-m+1} k_z(z_j)\right]\\ &= P(k_x, k_y, z_i, \omega)\exp[-j|z_i - z_{i-1}|k_z(z_i)]\\ &\quad\cdots\exp[-j|z_{i-m+1} - z_{i-m}|k_z(z_{i-m+1})]\\ &= P(k_x, k_y, z_i, \omega)W_iW_{i-1}\cdots W_{i-m+1}\end{aligned}$$

*Velocity $c = k/\rho$ is derived from compressibility $k$ and density $\rho$ so that density is still important for the large-scale diffraction effects.

†More exactly, an additional amplitude term due to the reflectivity should scale the equation but this term $[k(z_i)/k_z(z_i)]^{1/2}$ is usually neglected in favor of heuristic corrections for spherical divergence and deconvolution [4].

$$\neq P(k_x, k_y, z_i, \omega) \exp\left[-j|z_i - z_{i-m}|\left(\frac{\omega^2}{c_{\mathrm{mig}}^2} - k_x^2\right)^{1/2}\right] \quad (19.8)$$

Each phase shift operator $W_j$ models the propagation effect of its particular layer between $z_j$ and $z_{j-1}$.

Wave propagation across an $m$-layered medium is therefore very conveniently represented either by a cascade of linear operators, each corresponding to a single uniform layer, or by a single consolidated operator. Note, however, that several layers usually cannot be accurately modeled as a single layer with an effective *migration velocity* $c_{\mathrm{mig}}$ because the necessary phase shift is a sum of square roots which is not separable.

Although the wave equation solution is expressed in terms of pressure $P$ throughout an $m$-layered medium, it is most convenient to differentiate between the propagated or surface wave field $Y$ and the source wave field $X$. In order to model generation of the entire surface pressure wave $Y = P(k_z, 0, \omega)$, we must consider the upgoing waves from all the reflectors in the $m$-layered earth. Thus the wave fields $X_i$ at each depth $z_i$, when transformed to the $xt$ domain, will consist at time zero of those reflectors $x_{i,0}$ at that depth, and at later times of the diffracted waves from the deeper reflections. The propagation of all these waves to the surface can be considered either as an additive progression from the deepest reflector up (Figure 19.9) or as a surface summation of independent individual reflector wave fields. In both cases, independent noise $N$ is added at the surface. This can be expressed in the Fourier domain as

$$\begin{aligned} P(k_x, 0, \omega) = Y &= W_1(X_{1,0} + \cdots + X_{m-2,0} W_{m-2}(X_{m-1,0} + W_m X_{m,0})) + N \\ &= \sum_{i=1}^{m} H_i X_{i,0} \end{aligned} \quad (19.9)$$

where

$$H_i = \prod_{j=1}^{i} W_j$$

In this formulation, $X_{i,0}$ is the Fourier transformed wave field at depth $i$ and time zero corresponding to the reflectors at that depth. The complete upgoing wave field $X_i$ is simply the sum at depth $i$ of all the propagated reflections from depth $i$ and below.

Examination of a typical single-layer operator $W_i$ reveals a two-fold fan filter structure

$$W_i = \exp\left(-j|\Delta z_i|\sqrt{k^2 - k_x^2}\right) = \begin{cases} \exp(-j\theta) & \text{if } k^2 = \omega^2/c^2 \geq k_x^2 \\ \exp(-\theta) & \text{if } k^2 = \omega^2/c^2 < k_x^2 \end{cases} \quad (19.10)$$

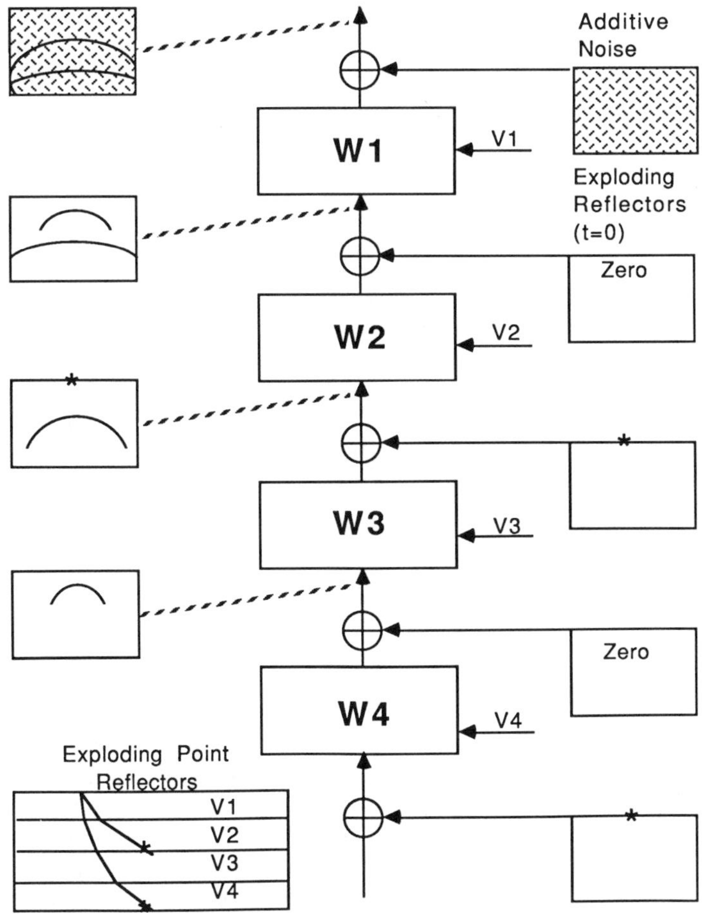


**Figure 19.9** Layered earth model of seismic propagation showing exploding reflectors at $t = 0$ propagating by phase shifts $W_n$ to the surface, where uncorrelated noise is added.

with a unit-gain phase shift corresponding to the *traveling wave* and a real exponential attenuation for the *stationary* (or *evanescent*) *wave*. The evanescent wave is a near-field component that rapidly approaches zero for reasonably large values of $\Delta z$ and is generally neglected because of its susceptibility to noise.

This Fourier domain model can be transformed to the space-time domain. Propagation is represented as a two-dimensional convolution of the earth's impulse response with the reflectors. As expected, this impulse response will have the shape of a hyperbolic diffraction pattern with apex at $x = 0$, $t = \Delta z_i/c$ and asymptotes at $x = \pm ct$ as described by $c^2t^2 - x^2 = \Delta z_i^2$. The dual symmetry of $W_j$ about $\omega$ and $k_x$ ensures that the impulse response is also real. Convolution can be thought of as a summation of the diffraction patterns that would arise from each reflecting point in the subsurface taken separately. This is nothing more than Huygens' principle expressed mathematically.

This formulation is perfectly tailored to the exploding-reflector models; only transmission effects are considered and as a result the wave field propagation from time zero at the reflector up to the surface can be clearly described without the complications of multiple reflections. The reflectivity of a geological horizon is therefore a matter of the amplitude at time zero and need not correspond to the velocity change in either strength or position.

The CMP gather for flat reflectors is an indication of the reflectivity at all points directly beneath the midpoint and can therefore be considered to be the result of an exploding reflector at those points. This model is therefore applicable to both CMP gathers and normal incidence sections, which correspond respectively to the stacked raw data and the ideally echo-sounded composite survey section.

In practice, mainly two-dimensional line surveys are performed for economic and data-handling reasons. Since simple geologies consistent with the layered earth model produce little off-axis energy, the representation of a 3-D earth by a 2-D section remains approximately valid. Despite the many simplifications, it is this model that has been the basis of almost all seismic work to date and for this reason we too shall use it in this chapter. In any case, techniques developed here can be extended to more involved models if desired.

## 4 DEREVERBERATION

Seismic surveys rely on acoustic impedance contrasts within the earth to produce upwardly propagating waves; unfortunately, in the case of very large contrasts, severe reverberations can result. As an example of this, offshore the water column is bounded above by air and below by compacted sediment so that the changes in acoustic impedance (product of wave velocity and density) effectively create a waveguide. On land, the presence of a low-velocity weathered zone can lead to the same phenomenon. This causes large-amplitude, slowly decaying echoes of the initial impulse.

Because these reverberations are contained within a waveguide of fixed dimensions, they are of constant period and can be modeled as a filter. Referring to Figure 19.10, the $z$-transform of the reverberation is [1, 5]

$$R(z) = 1 - cz^n + c^2 z^{2n} - c^3 z^{3n} + \cdots \tag{19.11}$$

Since the magnitude of the reflection coefficient $c$ is less than 1, the series is convergent and can be expressed as a minimum delay filter

$$R(z) = \frac{1}{1 + cz^n} \tag{19.12}$$

Thus to remove this echo we apply the inverse filter $1/R(z)$ [1, 5].

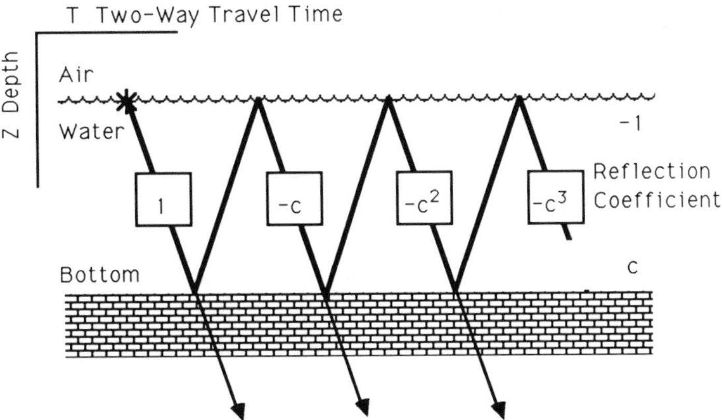


**Figure 19.10** Reverberation in the water-column. (Adapted from [1]).

## 5 VELOCITY FILTERING

Seismic processing is based on the analysis of the reflections due to the propagation of a compressional wave (or, in rare cases, a shear wave) through the formation. Unfortunately, other propagating waves are introduced by the source and the formation which can cause severe coherent additive noise.

The propagation of acoustic waves in a solid medium is quite complex with several modes coexisting, each having its own characteristics [2].

Compressional P waves

- The desired mode, in which all motion is parallel to the direction of wave travel.
- Highest velocity of any mode,

$$V_p = \sqrt{\frac{k + 4u/3}{\rho}}$$

where $k$ is the modulus of elasticity, $u$ the modulus of rigidity, and $\rho$ the density.

- Conversion to the SV mode occurs at boundaries and is proportional to the angle of incidence.

Shear S waves

- Particle motion is perpendicular to the direction of wave travel.
- Lower velocity than P mode,

$$V_s = \sqrt{\frac{u}{\rho}}$$

- At a planar interface an incident S wave has two components: 1) SV waves with particle motion perpendicular to the boundary leading to P wave conversion proportional to the incidence angle, and 2) SH wave with particle motion parallel to the boundary and thus no mode conversion.

Rayleigh R waves

- Surface waves propagating along a free plane boundary (such as air/earth) at a low velocity independent of frequency, $V_r(\text{typical}) = 0.9V_s$.
- Particle has retrograde elliptical motion which decreases exponentially from the boundary.

Love SH waves

- Special case of large-amplitude SH waves propagating in the waveguide formed by the weathered zone.
- Usually low frequency and velocity.

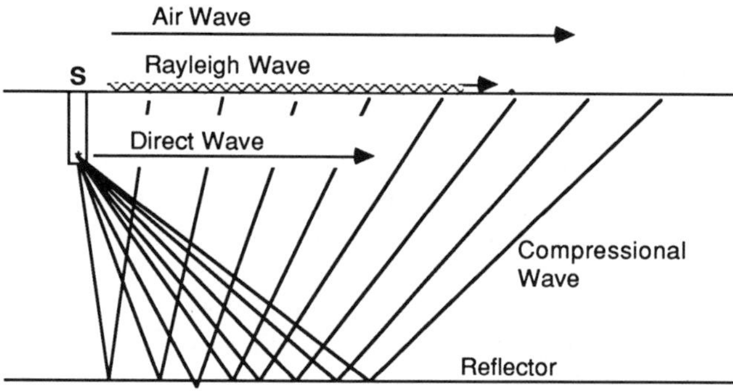


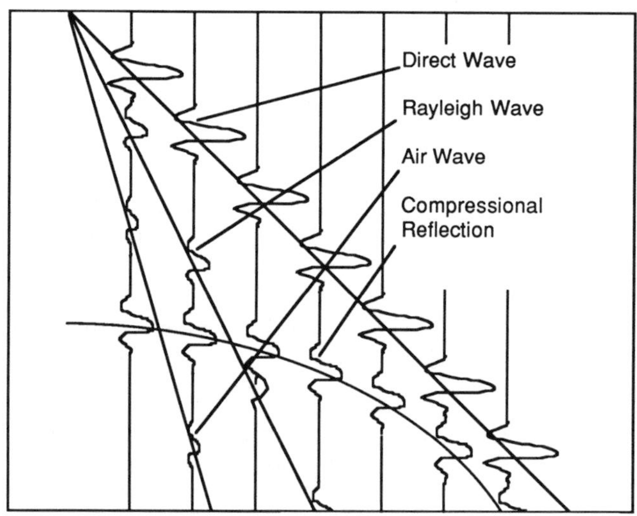


**Figure 19.11** Single shot point and corresponding coherent noise ray paths. Slow surface waves appear as steep linear arrivals on the CSP gather.

Pseudo-Rayleigh waves (Ground Roll)

- Waveguided P and SV waves in the weathered zone with large mode conversion effects.
- Low frequency (8–60 Hz) and low lateral velocity doe to multiple reflections within waveguide.

Although surface waves such as the Rayleigh waves and ground roll often have very high amplitudes, they can be discriminated by their lower velocity and removed. In A CMP gather, all the receivers share, at least approximately, the same common reflection point and thus all compressional arrivals will have a near-horizontal hyperbolic characteristic across the gather. The various surface waves, on the other hand, travel without reflection at such low velocities that they appear as steeply dipping straight lines which can easily be filtered in frequency-wave number ($FK$) space by a fan filter (Figure 19.11). The only major problem occurs at large source-receiver spacings, where the shallower hyperbolic curves may slope steeply enough to be confused with the surface noise as seen at the right of Figure 19.11.

A 2-D Fourier transform maps a space-time line of constant slope to an $FK$ line of reciprocal slope through the origin. If we consider the space-time line to be a delta function $\delta(t - px - T)$ along the line $t = px + T$, then the Fourier transform is (Figure 19.12).

$$\iint_{-\infty}^{\infty} \delta(t - px - T)e^{-j(\omega t - k_x x)}\, dx\, dt = 2\pi e^{-j\omega t}\delta(\rho\omega - k_x) \tag{19.13}$$

Clearly, lines of excessive slope can be discriminated by a linear velocity fan filter and in this way removed.

A more involved technique is to optimize the velocity filter on the basis of the primary's known hyperbolic characteristic by using the classical least-squares method [6–10]. In contrast to the linear velocity filter, the autocorrelation and cross-correlation functions inherent to the normal equations must be computed nu-

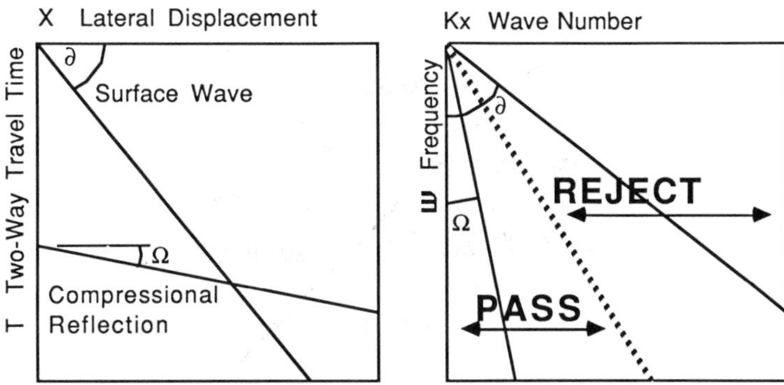


**Figure 19.12** Fourier transform of sloping lines showing conversion to reciprocal slope and rejection of the steepest linear event by a fan filter.

merically. Computation times for the filter coefficients are comparable to those for the optimized fan filters. Because the hyperbolic characteristics will change with depth and root-mean-square velocity, it is necessary to design filters that are applicable over a time gate. Care must be taken so that the overlaps between filters are smooth and do not themselves introduce extraneous noise. Because of their complexity and limited use, these filters are not examined in this survey.

## 5.1 FIR Implementation by Windowing

Designs for such velocity filters have been used since first formulated by Fail and Grau [11]. We shall follow here the computationally efficient approach, taken by Treitel et al. [12] and Peacock [13], of formulating the filter as a sparse 2-D finite impulse response (FIR) followed by a 1-D Hilbert transform operator.

The first step is to determine the desired cutoff velocity on the basis of the data to be used. In practice, the cutoff velocity is normalized to the ratio of the spatial and temporal sample increments ($V_c = \Delta x/\Delta t$) so that both the Nyquist frequencies are the same and neither under- nor oversampling occurs [13]. A lower cutoff velocity will have aliasing problems as well as spatially aliased input data due to the absence of any spatial antialiasing filtering of the data. A higher cutoff velocity leads to an oversampled impulse response that is computationally more burdensome. This is, however, easily remedied by interpolating the data and resampling.

Following Peacock [13], the Fourier domain $(f_x, f_t)$ fan filter of Figure 19.13 is Fourier-transformed from the $xt$ domain to the $f_x$, $f_t$ domain

$$
\begin{aligned}
h(x,t) &= \int_{-f_t(\max)}^{f_t(\max)} \int_{-|f_t/V_c|}^{|f_t/V_c|} \exp(j2\pi(f_t t + f_x x)\, df_x\, df_t \\
&= \int_{-f_t(\max)}^{f_t(\max)} \exp(j2\pi f_t t)\,\mathrm{sgn}\, f_t \int_{-f_t/V_c}^{f_t/V_c} \exp(j2\pi f_x x)\, df_z\, df_t \\
&= \frac{1}{\pi x}\int_{-f_t(\max)}^{f_t(\max)} \exp(j2\pi f_t t)(j\,\mathrm{sgn}\, f_t) j \sin\left(\frac{2\pi f_t x}{V_c}\right) df_t \qquad (19.14)
\end{aligned}
$$

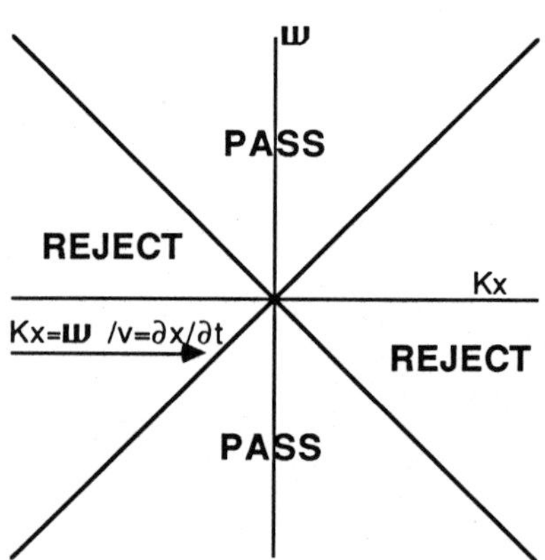


**Figure 19.13** Linear velocity filter in the frequency-wave number domain.

This can be expressed as the product of a time-invariant term with an inverse Fourier-transformed (indicated by $\mathbf{F}^{-1}$) term

$$h(x,t) = g(x)\mathbf{F}^{-1}[A(f_t)B(f_t)] \tag{19.15}$$

This is clearly expressible as a convolution

$$\begin{aligned} h(x,t) &= g(x)[a(t) * b(t)] \\ &= -\frac{1}{\pi x}\left\{\left[\frac{\delta}{2}\left(t+\frac{x}{V_c}\right) - \frac{\delta}{2}\left(t-\frac{x}{V_c}\right)\right] * b(t)\right\} \end{aligned} \tag{19.16}$$

between the delta functions $\delta\,(t \pm x/V_c)$ and the Hilbert operator $b(t)$. It is emphasized that both $b$ and $\delta$ are $f_t(\max)$ band-limited versions of the operators. To implement this impulse response, consider the second section

$$h_2(x,t) = -\frac{1}{\pi x}\left[\frac{\delta}{2}\left(t+\frac{x}{V_c}\right) - \frac{\delta}{2}\left(t-\frac{x}{V_c}\right)\right] \tag{19.17}$$

where

$$\delta\left(t - \frac{x}{V_c}\right) = \begin{cases} \dfrac{1}{2\Delta t} & \text{if } j = k \\ 0 & \text{if } j \neq k \end{cases}$$

Therefore, since both $x$ and $t$ are sampled, $t = k\Delta t$ and $x = j\Delta x$, while $V_c = \Delta x/\Delta t$ so that

$$\delta(k-j) = \begin{cases} \dfrac{1}{2\Delta t} & \text{if } j = k \\ 0 & \text{if } j \neq k \end{cases} \tag{19.18}$$

leading to a sparsely sampled response. Clearly, a nondiagonal $V_c$ would lead to a much more complex impulse response with all the associated computational overhead.

$$h_2(k,j) = \begin{cases} \pm\dfrac{1}{2\pi j\Delta x\Delta t} & \text{if } k = \pm j \\ 0 & \text{else} \end{cases} \tag{19.19}$$

The band-limited Hilbert operator is given by Peacock [13] as

$$\begin{aligned} b(t) &= \frac{1}{\pi t}[\cos 2\pi f_t(\max)t - 1] \\ b(k) &= \frac{-1}{\pi k\Delta t} \end{aligned} \tag{19.20}$$

which can now be convolved with the partial response $h_2(t)$ to implement the filter $h(x,t)$. Gibbs' phenomenon will be apparent because of the steep transition along the $V_c$ line so that a temporal windowing will be necessary. The Kaiser window (or similar) gives good results without overly broadening the transition zone. A sharper cutoff can be achieved by increasing the number of points in the impulse response.

The problem of aliasing is inherent in all digital filter design since the response, delineated by the spatial and temporal Nyquist frequencies, is replicated symmetrically throughout the $FK$ plane. Referring to Figure 19.14, it is clear that, because of the specific shape of a fan filter, velocities below $V_c$ will be aliased due to insufficient spatial sampling; the smaller the velocity, the smaller is the alias-free frequency band. Thus low-velocity but high-frequency energy may be passed without attenuation. The solution is to sample the time axis densely so that any high-frequency aliasing is beyond the actual signal.

Filtering is achieved by convolving the $N$-point FIR filter with $M$ spatially distributed channels of time data. The first $M$ output channels (of the $M + N - 1$ outputs) are the filtered result.

## 5.2 IIR Implementation by Rotation

Mulk et al. [14] proposed an efficient procedure for designing the velocity filter of Figure 19.13 by rotating a simple diagonally symmetric half-plane infinite impulse response (IIR) filter through 45°. Because of this rotation the axes lose their direct correspondence with $k_x$ and $\omega$ and are simply labeled $\omega_1$ and $\omega_2$. Such a recursive filter can have a better magnitude response with fewer terms than an equivalent FIR filter and, moreover, will need less storage and computation.

The prototype 2-D low-pass filter is generated by cascading two identical 1-D low-pass filters along the two axes (Figure 19.15). The cutoff frequency $\omega_c$ of these filters is chosen as $\pi/2T$ so that the resulting separable filter can be shifted from the

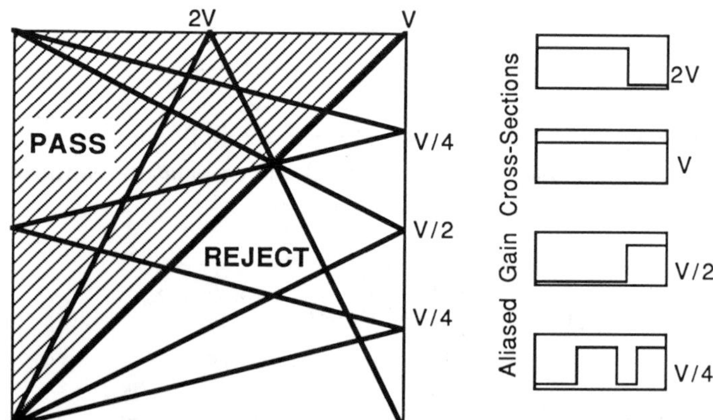


**Figure 19.14** Profiles across the sampled velocity filter at velocities between $2V$ and $V/4$. Note the high frequencies passed for steeply dipping (low-velocity) arrivals due to aliasing.

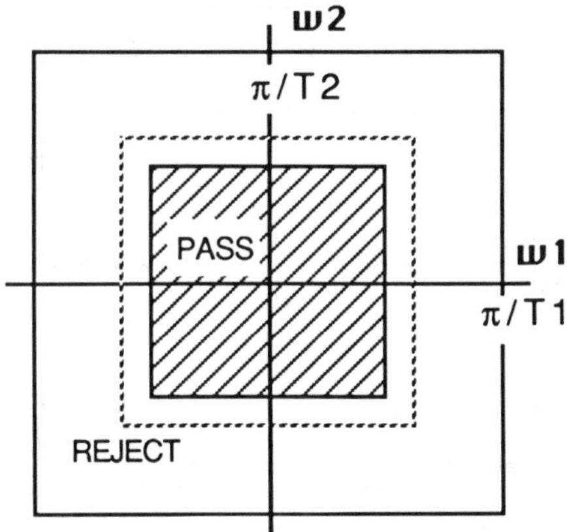


**Figure 19.15** Magnitude response of the prototype 2-D low-pass filter. (Derived from [14].)

center position to the first and third quadrants, as in Figure 19.16 by a coordinate change of $\pm\pi/2T$

$$\begin{aligned} H(\omega_1 T_1, \omega_2 T_2) &= |H(z_1^{-1}, z_2^{-1})|_{z_1 = e^{j\omega_1 T_1}, z_2 = e^{j\omega_2 T_2}} \\ &= |H(\omega_1 T_1)||H(\omega_2 T_2)| e^{j\theta_1} e^{j\theta_2} \end{aligned} \tag{19.21}$$

so that the first and third quadrant filters are

$$\begin{aligned} H_1 &= |H(\omega_1 T_1)| e^{j\theta_1} \\ H_2 &= |H(\omega_2 T_2)| e^{j\theta_2} \end{aligned} \tag{19.22}$$

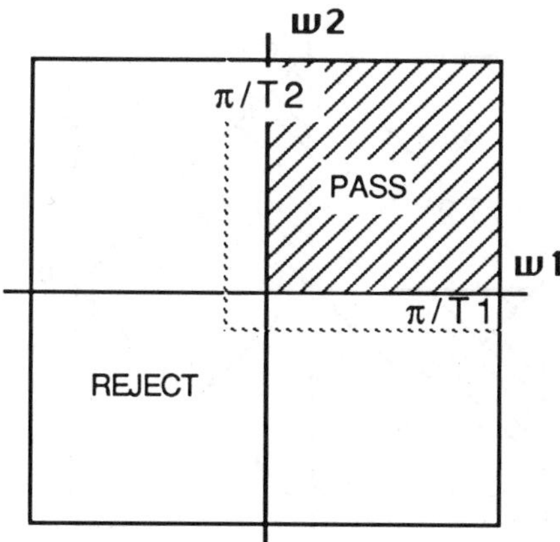


**Figure 19.16** First-quadrant pass filter realized by shifting the low-pass filter of Figure 19.15. (Derived from [14].)

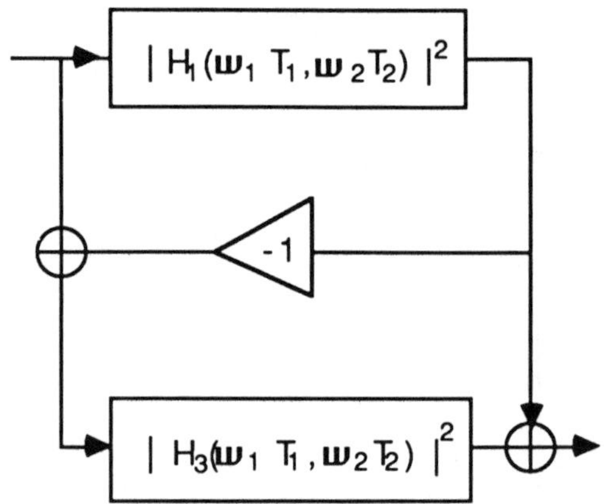


**Figure 19.17** Realization of the diagonally symmetric half-plane filter. (Derived from [14].)

These filters can then be placed in parallel (Figure 19.17) so that a diagonally symmetric half-plane filter results (Figure 19.18). This filter can now be rotated by 45° through the transformation

$$\begin{aligned} \omega_1 T_1 &\longrightarrow \tfrac{1}{2}(\omega_1' T_1' - \omega_2' T_2') \\ \omega_2 T_2 &\longrightarrow \tfrac{1}{2}(\omega_1' T_1' + \omega_2' T_2') \end{aligned} \tag{19.23}$$

The final fan filter is

$$\begin{aligned} H(\omega_1' T_1', \omega_2' T_2') &= H_{13}\left[\tfrac{1}{2}(\omega_1' T_1' - \omega_2' T_2'), \tfrac{1}{2}(\omega_1' T_1' + \omega_2' T_2')\right] \\ &= |H_1'|^2 + |H_3'|^2 - |H_1'|^2 |H_3'|^2 \end{aligned} \tag{19.24}$$

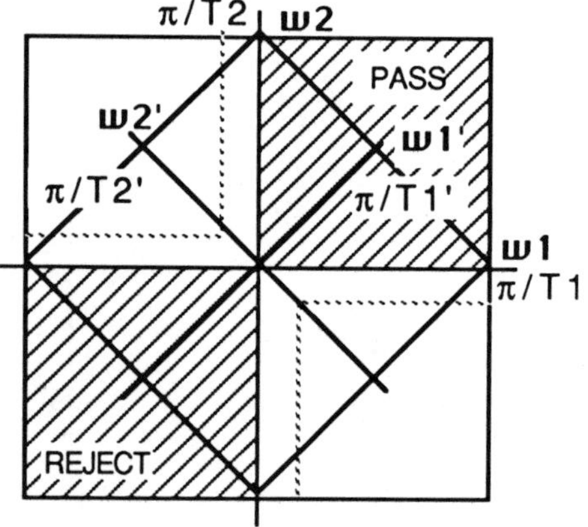


**Figure 19.18** Fan filter realized by rotating the half-plane through 45°. (Derived from [14].)

composed of

$$H_1' = H\left[\frac{1}{2}(\omega_1'T_1' - \omega_2'T_2') - \frac{\pi}{2}, \frac{1}{2}(\omega_1'T_1' + \omega_2'T_2') - \frac{\pi}{2}\right]$$

$$H_3' = H\left[\frac{1}{2}(\omega_1'T_1' - \omega_2'T_2') + \frac{\pi}{2}, \frac{1}{2}(\omega_1'T_1' + \omega_2'T_2') + \frac{\pi}{2}\right]$$

This two-stage transformation affects the variable $z$ as though the delay unit was changed from $z^{-1}$ to $z^{-1/2}$. Thus a stable prototype will result in a stable fan filter, although the direction of recursion and the sampling interval must be changed.

$$z_1^{-1} \longrightarrow j\sqrt{\frac{z_1'^{-1}}{z_2'^{-1}}} \quad \text{and} \quad z_2^{-1} \longrightarrow j\sqrt{z_1'^{-1} z_2'^{-1}} \qquad \text{for } H_1(z_1^{-1} z_2^{-1})$$

$$z_1^{-1} \longrightarrow -j\sqrt{\frac{z_1'^{-1}}{z_2'^{-1}}} \quad \text{and} \quad z_2^{-1} \longrightarrow j\sqrt{z_1'^{-1} z_2'^{-1}} \qquad \text{for } H_3(z_1^{-1}, z_2^{-1}) \tag{19.25}$$

Therefore, the fan filter is related to the prototype by

$$H(z^{-1})\,|_{z^{-1}=e^{-j\omega T}} \Longleftrightarrow H(z^{-1/2})\,|_{z^{-1/2}=e^{-j\omega T/2}} \tag{19.26}$$

This transfer function can be realized by expanding the input data frequency spectrum or by interpolating the input-output data.

Since the transformed filter has a frequency response twice that of the prototype, one implementation is to redefine the input data

$$x'(n_1T_1', n_2T_2') = \begin{cases} x\left(\frac{n_1T_1}{2}, \frac{n_2T_2}{2}\right) & n_1 = 2m_1, n_2 = 2m_2 \\ & m_1, m_2 = 0, 1, \ldots \\ 0 & \text{else} \end{cases} \tag{19.27}$$

so that the transformed and equivalent data are the same

$$X'(e^{j\omega_1}, e^{j\omega_2}) = \tfrac{1}{4}X(e^{j\omega_1}, e^{j\omega_2}) \tag{19.28}$$

Using the corrected data introduces a two-fold frequency expansion, which must be removed by a 2-D low-pass filter of $(\pi/T_1, \pi/T_2)$ cutoff either before or after the fan-pass filter $H(z_1^{-1/2}, z_2^{-1/2})$.

Alternatively, we must redefine the input and output data along the new recursion directions introduced by the 45° rotation. As can be seen from Figure 19.19,

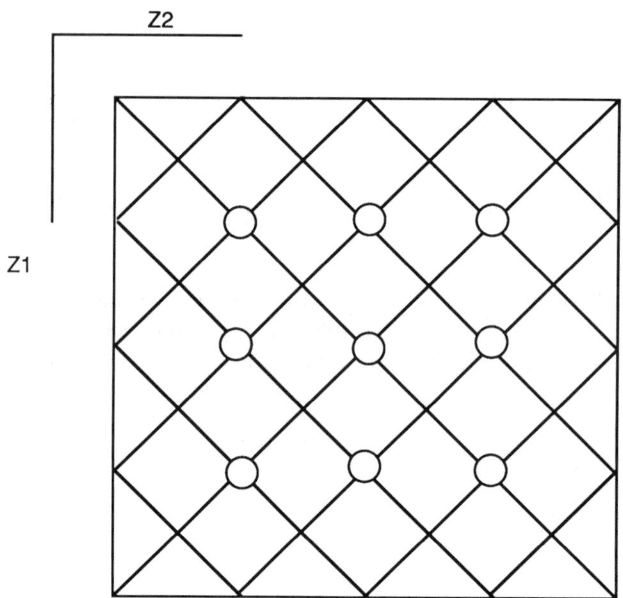


**Figure 19.19** Grid points of the original (circles) and interpolated arrays. (Derived from [14].)

the original data points are still valid, but we must now interpolate the intermediate values. An effective hexagonal sampling matrix is the result.

This implementation is a cascade of the input interpolator, the fan-pass filter $H_{\text{fan}}(z_1^{1/2}, z_2^{-1/2})$, and the output interpolator. The specific form of input-output interpolation depends on the approach chosen for data normalization.

This straightforward design can easily be extended to partial quadrant filters by the use of $z^{-1/3}$ or $z^{-1/4}$. The major problem is the lack of any control over the phase (apart from the choice of the prototype filter); in this respect it is similar to the windowed designs of Peacock [13] and Treitel et al. [12]. Results however, are excellent for such a simple technique.

## 5.3 IIR Implementation by 1-D Frequency Slices

An interesting technique for the design of stable suboptimal IIR filters, with minimal computational complexity, has been developed by Murray [15], whose treatment is followed below.

The symmetric IIR half-plane filter with transfer function

$$H(z_1, z_2) = \frac{A}{1 + \sum_{n=1}^{N} \sum_{m=1}^{M} b_{mn} z_1^m z_2^n} \tag{19.29}$$

is causal, which permits the parallel processing of an entire row of input data at once. Moreover, stability is guaranteed as long as the denominator is nonzero within the stability set $S = \{(z_1, z_2) \in \mathbb{C}^2 \mid |z_1| = 1 \text{ and } |z_1| \leq 1\}$. However, these filters are not general since their magnitude function $H_m(\omega_1, \omega_2)$ must have a constant average gain over the range of $\omega_2$ for a fixed $\omega_1$ [16].

$$\frac{1}{2\pi}\int_{-\pi}^{\pi} \ln H_i(\omega_1, \omega_2) d\omega_2 = \text{constant} \tag{19.30}$$

Therefore, to generalize the magnitude response, we must cascade a 1-D filter $C(z_1)$ to correct for this restriction.

The design of the symmetric half-plane filter starts with the specification of the desired frequency response $H_m(\omega_1, \omega_2)$, for $|\omega_1| < \pi$ and $|\omega_2| < \pi$, as a series of frequency slices in $\omega_2$ for a fixed $\omega_1$. For a 90° fan filter, each slice $H_{\omega_1}(\omega_2)$ is an ideal low-pass filter. These slices are then implemented as cascades of first- and second-order sections of 1-D IIR filter in $\omega_2$.

$$H_{\omega_1}(\omega_2) = \prod_{k=1}^{n} \frac{a_k(\omega_1) + a_{k,1}(\omega_1)z_2 + a_{k,2}(\omega_1)z_2^2}{b_k(\omega_1) + b_{k,1}(\omega_1)z_2 + b_{k,2}(\omega_1)z_2^2} \tag{19.31}$$

The constant terms can be factored out so that the equation is the product of a function of $\omega_1$ (which can be approximated by a filter in $z_1$) with a symmetric half-plane filter in $z_2$.

$$\begin{aligned} H_{\omega_1}(\omega_2) &= \prod_{k=1}^{n} \frac{a_k(\omega_1)}{b_k(\omega_1)} \prod_{k=1}^{n} \frac{1 + \tilde{a}_{k,1}(\omega_1)z_2 + \tilde{a}_{k,2}(\omega_1)z_2^2}{1 + \tilde{b}_{k,1}(\omega_1)z_2 + \tilde{b}_{k,2}(\omega_1)z_2^2} \\ &= C(\omega_1)P(\omega_1, z_2) \end{aligned} \tag{19.32}$$

The half-plane filters are now specified by coefficients $\tilde{a}_{k,i}(\omega_1) = a_{k,i}(\theta_1)/a_k(\theta_1)$ and $\tilde{b}_{k,i}(\omega_1) = b_{k,i}(\theta_1)/b_k(\theta_1)$. These are functions of $\omega_1$ and must therefore be replaced with trigonometric polynomials $\hat{a}_{k,i}(\omega_1)$ and $\hat{b}_{k,i}(\omega_1)$ without destabilizing the filter. The obvious solution of windowing the Fourier-transformed $\tilde{b}_{k,i}(\omega_1)$ leads to large errors near singularities, so Murray used a composite approach. Define

$$\begin{aligned} \alpha_k &= \sqrt{1 + \tilde{b}_{k,1} + \tilde{b}_{k,2}} \\ \beta_k &= \sqrt{1 - \tilde{b}_{k,1} + \tilde{b}_{k,2}} \end{aligned} \tag{19.33}$$

and window (for stability; $W(n)$ is positive and of unit area) the first few terms of the Fourier-transformed composite coefficients to get $\hat{\alpha}_k$, $\hat{\beta}_k$

$$\begin{aligned} \hat{b}_{k,1} &= \frac{1}{2}(\hat{\alpha}_k^2 - \hat{\beta}_k^2) \\ \hat{b}_{k,2} &= \frac{1}{2}(\hat{\alpha}_k^2 + \hat{\beta}_k^2) - 1 \end{aligned} \tag{19.34}$$

This results in a stable filter $P(\omega_1, z_2)$ with approximately double the number of coefficients. Because the coefficient polynomial approximation introduces a significant error, the design of the compensating 1-D filter $C(\omega_1)$ is best done with reference to the actual $P(\omega_1, z_2)$ achieved. This error in the approximation is in fact the main problem with this method, since it forces the correction to be based on the actual as opposed to the ideal frequency response, with an attendant degradation in the accuracy.

## 5.4 IIR Design by Spectral Factorization

The problem of optimizing the design of recursive filters was first approached by Maria and Fahmy [17], followed by Bednar [18] and Twogood and Mitra [19]. Although good results were achieved, difficulties with complex and computationally intensive algorithms were universal. The spectral factorization approach of Ekstrom and Woods [16] permits the design of general half-plane recursive filters without an excessive computational burden.

Ekstrom and Woods considered a half-plane IIR filter of the form

$$y(k,l) = \sum_m \sum_n a(m,n)x(k-m,l-n) - \sum_{\substack{m=0 \\ m+n\neq 0}}^{M} \sum_{n=0}^{N} b(m,n)y(k-m,l-n)$$

$$- \sum_{m=-1}^{-M} \sum_{n=1}^{N} b(m,n)y(k-m,l-n)$$

$$H(z_1,z_2) = \frac{A(z_1,z_2)}{B(z_1,z_2)} = \frac{\sum_m \sum_n a(m,n)z_1^{-m}z_2^{-n}}{\sum_m \sum_n b(m,n)z_1^{-m}z_2^{-n}} \tag{19.35}$$

By defining a "stability error" $J_s$ as a rough measure of the instability of a given design along with the usual amplitude error $J_a$, an overall error $J = J_a + \alpha J_s$ is minimized by nonlinear optimization. The $J_s$ is based on the difference between the designed denominator and a minimum phase polynomial with the same autocorrelation as the denominator. The minimum phase polynomial is determined by a spectral factorization of the autocorrelation (via the cepstrum) followed by its antilog. The filter's numerator and denominator are iteratively derived by successive approximation until convergence, within an allowable error, is achieved.

This method is general to half-plane filters and is therefore an excellent solution to the more complex problems where various interfering waves of different velocities are superimposed on the data. However, for most applications involving simple velocity filters, the extra expense is not necessary.

# 6 MIGRATION

With coherent noise removed from the data, it now remains to derive the true location of the underlying geological features. Examining the normal incidence section of Figure 19.8, it is clear that there is still considerable distortion of the apparent reflector positions. This arises from their display as though they originated directly below the source-receiver midpoint. In fact, simple geometry guarantees that this assumption is true only for flat reflectors. Point reflectors are shown as hyperbolas and planar reflectors are shown as planes of incorrect inclination. The technique for correcting these effects is called *migration* and can be considered as a focusing of the acoustic image.

The problem of migration can be stated as, "given a known velocity profile $c(z)$ and a noise-corrupted surface wave field $P(k_x, z = 0, \omega)$, how can we estimate the true position of the underlying reflectors?" Clearly, since we are given the velocity profile, the earth's transfer function $W$ is determined for each layer and the system can be considered as a known corrupting channel with unknown input.

To estimate the reflector strength and position, the wave field $P(k_x, z = 0, \omega)$ can be propagated backward in time so that the wave field $P(k_x, z_i, \omega)$ is estimated for each depth of the discretized layered earth. Because of the exploding reflector model, each depth's wave field corresponds at time zero to the reflector strength in the horizontal line at $z_i$. Therefore migration is divided into two steps, first the back projection to estimate the wave fields at depth and then the imaging to extract each depth's reflector estimates at time zero.

Migration of the surface wave field therefore entails the inversion of the forward structure discussed in Section 4. Since the propagation of the wave field at $z_i$ upward to $z_{i-1}$ is described by a phase shift in the $k\omega$ domain,

$$P(k_x, z_{i-1}, \omega) = P(k_x, z_i, \omega) \exp\left(-j|z_i - z_{i-1}|\sqrt{k^2 - k_x^2}\right) \tag{19.36}$$

migration reverses this phase shift and effectively back-propagates the wave field by some form of inversion which we will call $\tilde{W}^{-1}$. Such techniques are called *phase shift migration* [20–26]. The estimation of each layer's wave field $P(k_x, z_i, \omega)$ can be considered as part of either a downward recursion

$$\hat{X}_i = \tilde{W}_i^{-1} \hat{X}_{i-1} \tag{19.37}$$

or a series of incremental multilayer inversions

$$\hat{X}_i = \tilde{H}_i^{-1} Y \qquad \text{where } \tilde{H}_i^{-1} = \prod_{j=1}^{i} \tilde{W}_j^{-1} \tag{19.38}$$

The obvious approach to back projection in the absence of any a priori knowledge of noise or signal is to minimize the residual norm $||y - Hx||^2$ by taking the inverse of the known transfer function $W$ over some $\Delta z_i$ layer

$$W^{-1}(k_x, \Delta z_i, \omega) = \begin{cases} \exp(-j\Delta z_i \sqrt{k^2 - k_x^2}) & \text{if } k^2 \geq k_x^2 \\ \exp(\Delta z_i \sqrt{k_x^2 - k^2}) & \text{if } k^2 < k_x^2 \end{cases} \tag{19.39}$$

Clearly, the positive exponential, in attempting to recreate the evanescent wave, results in an unstable solution. Moreover, these high gains occur where the SNR is lowest so the well-known problems of noise amplification would occur in any case.

A more practical solution is to apply the matched filter to maximize the SNR for white noise

$$W^*(k_x, \Delta z_i, \omega) = \begin{cases} \exp(-j\Delta z_k\sqrt{k^2 - k_x^2}) & \text{if } k^2 \geq k_x^2 \\ \exp(-\Delta z_i\sqrt{k_x^2 - k^2}) & \text{if } k^2 < k_x^2 \end{cases} \tag{19.40}$$

which gives a stable transfer function of large bandwidth at small $\Delta z_i$. This bandwidth extends over the evanescent wave field, which typically does nothing to improve the solution because of the locally low evanescent SNR [20]. Even so, this technique is used after transformation to the $x\omega$ domain in the *Kirchhoff summation migration* [3].

In order to best use those wave field components of high SNR, *Wiener filtering* to minimize the mean-square error has been suggested [3]

$$W_w(k_x, \Delta z_i, \omega) = \frac{P_X W^*}{P_X + P_N} \tag{19.41}$$

where $P_X$ and $P_N$ are the transformed autocorrelations (i.e., power spectra) of the wave field and noise, respectively. Unfortunately, even though $P_N$ can often be estimated from the evanescent zone, ignorance of $P_X$ usually makes implementation impractical.

Instead, a form of pseudo-Wiener filtering is often implemented by assuming that all evanescent waves have such low SNR that they can be neglected. This is the *band-limited matched filter* [3]

$$W^*_{BL}(k_x, \Delta z_i, \omega) = \begin{cases} \exp(-j\Delta z_i\sqrt{k^2 = k_x^2}) & \text{if } k^2 \geq k_x^2 \\ 0 & \text{if } k^2 < k_x^2 \end{cases} \tag{19.42}$$

which is the most commonly used simple phase-shift migration operator.

## 6.1 IIR Phase Shift Migration

Because of the difficulty of designing IIR filters in both phase and magnitude, Garibotto [27] implemented phase-shift migration by cascading a fan filter for the amplitude response with an IIR all-pass filter for the phase shift. The required matched filter amplitude response is

$$|W^*(k_x, \Delta z_i, \omega)| = \begin{cases} 1 & \text{if } k^2 \geq k_x^2 \\ \exp(-\Delta z_i\sqrt{k_x^2 - k^2}) & \text{if } k^2 < k_x^2 \end{cases} \tag{19.43}$$

with a linear phase constraint which can be implemented as a two-dimensional filter using any of the FIR or IIR techniques discussed in Section 5. Note that the highly attenuated evanescent field is retained as a form of windowing to avoid the oscillations associated with abrupt truncation.

The phase response

$$\varphi(k_x, \Delta z_i, \omega) = \exp(-j\Delta z_i\sqrt{k^2 - k_x^2}) \tag{19.44}$$

is implemented as an IIR all-pass filter. Because the impulse response is known to be a hyperbola with symmetry about the $t$ axis, a half-plane IIR implementation is necessary. Unfortunately, design of such filters to meet phase specifications is difficult at best. For this reason, Garibotto implements a straightforward single-quadrant IIR filter and then rotates it by 45° to give the required symmetry. The variable transformation

$$\begin{aligned} x &= \frac{\sqrt{2}}{2}(x_2 - x_1) \\ t &= \frac{\sqrt{2}}{2}(x_1 + x_2) \end{aligned} \tag{19.45}$$

is imposed with scaling of the sampling intervals so that $c\Delta t = \Delta x = \Delta$ and the two axes $x_1$, $x_2$ superimpose on the hyperbolic impulse response's asymptotes $x = ct$. The impulse response is therefore defined over the single quadrant defined by $x_1$ and $x_2 \geq 0$ with the transformed hyperbola locus

$$2x_1x_2 = \Delta z_i^2 \tag{19.46}$$

Since rotation in space translates identically to the frequency domain

$$\begin{aligned} k_x &= \frac{\sqrt{2}}{2}(\omega_2 - \omega_1) \\ \omega &= \frac{\sqrt{2}}{2}(\omega_2 + \omega_1) \end{aligned} \tag{19.47}$$

and the resulting phase response is

$$\varphi(\omega_1, \Delta z_1, \omega_2) = \exp(-j\Delta z_i\sqrt{2\omega_1\omega_2}) \qquad \text{if } \omega_1 \text{ and } \omega_2 \geq 0 \tag{19.48}$$

If the transfer function is

$$P(z_1, z_2) = \frac{z_1^N z_2^N D(z_1^{-1}, z_2^{-1})}{D(z_1, z_2)} \tag{19.49}$$

where

$$D(z_1, z_2) = 1 + \sum_{\substack{n_1=0 \\ n_1+n_2\neq 0}}^{N} \sum_{n_2=0}^{N} d(n_1, n_2) z_1^{n_1} z_2^{n_2}$$

the coefficients $d(n_1, n_2)$ can be determined by minimizing a weighted error function $J(d)$ over a fixed set of $k$ frequencies $(\omega_{1k}, \omega_{2k})$. The first partial derivatives of $J(d)$ are set to zero with respect to the $d(n_1, n_2)$ and the resulting set of linear equations are evaluated. The accuracy of the filter is a function of the number $k$ of evaluation frequencies; in the limit, the approximation is exact for a given $\Delta z_i$.

$$\varphi_{\Delta z_i}(e^{-j\omega_i}, e^{-j\omega_2}) = \lim_{\substack{\Delta z \to 0 \\ k\Delta z = \Delta z_i}} [\varphi_{\Delta z}(e^{-j\omega_1}, e^{-j\omega_2})]^k \tag{19.50}$$

Clearly a trade-off is necessary between accuracy and computational expense.

The depropagation operator for an incremental $\Delta z$ can be used to migrate over any depth interval $\Delta z_i$ by cascading $N = \Delta z_i/\Delta z$ filters in series. Although the mean-squared error will increase with the power of $N$, the incremental error per section decreases even faster so that it is desirable to reduce $\Delta z$.

An advantage of this incremental approach is the ease with which spatial inhomogeneities in the velocity distribution can be introduced.

## 7 AFTERWORD

Seismic imaging is presently undergoing a revolution, primarily because the past decade has seen an exponential growth in computing power sufficient to permit large-scale use of multidimensional techniques. Moreover, the associated surge of research activity has revealed a wealth of powerful new techniques well suited to the specific problems of huge spatial data sets and an inhomogeneous medium. We are only now beginning to approach the problem as a unified inversion as opposed to a set of discrete and heuristic steps (such as NMO, CMP stacking, and migration).

The most promising techniques are those using the full wave equation [28–30] to invert the surface data directly without stacking. The problem is that the differential equation is indeterminate because of the incomplete boundary conditions and thus needs some a priori knowledge. In complicated structures, a reliable solution can be derived only if the velocity input and migrated output are defined in depth so that an interpretive approach, with model refinement, is possible. Automated parameter picking from the data [31] and feature detection [32] thus become essential for practical and economic implementation. Expert systems are expected to come to the fore as the actual migration becomes less of a problem.

## REFERENCES


1. E. A. Robinson and S. Treitel, *Geophysical Signal Analysis*, Prentice-Hall, Englewood Cliffs, New Jersey (1983).
2. K. H. Waters, *Reflection Seismology*, Wiley, New York (1980).
3. A. J. Berkhout, *Seismic Migration—Imaging of Acoustic Energy by Wave Field Extrapolation*, Elsevier, Amsterdam (1982).
4. E. A. Robinson and M. T. Silvia, *Digital Foundations of Time Series Analysis*, vol. 2, *Wave Equation Space-Time Processing*, Holden-Day, San Francisco (1981).
5. M. T. Silvia and E. A. Robinson, *Deconvolution of Geophysical Time Series*, Elsevier, Amsterdam (1979).
6. H. Ozdemir, Optimum hyperbolic moveout filters with applications to seismic data, *Geophys. Prospect.*, *29*: 702–714 (1981).
7. K. L. Sengbush and M. R. Foster, Optimum multichannel velocity filters, *Geophysics*, *33*: 11–35 (February 1968).
8. J. N. Galbraith and R. A. Wiggins, Characteristics of optimum multichannel stacking filters, *Geophysics*, *33*: 36–48 (February 1968).
9. P. Hubral, Characteristics of azimuth independent optimum velocity filters designed for 2-D arrays, *J. Geophys.*, *41*: 265–279 (1975).
10. B. Seeman and L. Horowitz, Vertical seismic profiling: Separation of upgoing and downgoing acoustic waves in a stratified medium, *Geophysics*, *48*: 555–568 (May 1983).
11. J. P. Fail and G. Grau, Les filtres en eventail, *Geophys. Prospect.*, *11*: 9–163 (June 1963).
12. S. Treitel, J. L. Shanks, and C. W. Frasier, Some aspects of fan filtering, *Geophysics*, *32*: 789–800 (October, 1967).
13. K. Peacock, On the practical design of discrete velocity filters for seismic data processing, *IEEE Trans. Acoust. Speech Signal Process.*, *ASSP-30*: 52–60 (February 1982).
14. M. Z. Mulk, K. Obata, and K. Hirano, Design of digital fan filters, *IEEE Trans. Acoust. Speech Signal Process.*, *ASSP-31*: 1427–1433 (December 1983).
15. J. J. Murray, A design method for 2-D recursive digital filters, *IEEE Trans. Acoust. Speech Signal Process.*, *ASSP-30*: 45–51 (February 1982).
16. M. P. Ekstrom and J. W. Woods, 2-D spectral factorization with applications in recursive digital filtering, *IEEE Trans. Acoust. Speech Signal Process.*, *ASSP-24*: 115–128 (April 1976).
17. G. Maria and M. M. Fahmy, An lp design technique for 2-D recursive filters, *IEEE Trans. Acoust. Speech Signal Process.*, *ASSP-22*: 15–21 (February 1974).
18. J. B. Bednar, Spatial recursive filter design via rational Chebyshev approximations, *IEEE Trans. Circuits Syst.*, *CAS-22*: 572–574 (1975).
19. R. E. Twogood and S. K. Mitra, Computer-aided design of separable 2-D digital filters, *IEEE Trans. Acoust. Speech Signal Process.*, *ASSP*: 165–169 (April 1977).
20. A. J. Berkhout and D. W. Van Wulffen Pathe, Migration in terms of spatial deconvolution, *Geophys. Prospect.*, *27*, *1*: 261–291 (1979).
21. W. A. Schneider, Integral formulation for migration in two and three dimensions, *Geophysics*, *43*: 49–76 (February 1978).

22. J. Gazdag, Wave equation migration with the phase shift method, *Geophysics, 43*: 1342–1351 (September 1978).
23. J. H. Chun and C. A. Jacewitz, Fundamentals of frequency domain migration, *Geophysics, 46*: 717–733 (May 1981).
24. J. Gazdag, Modelling of the acoustic wave equation with transform methods, *Geophysics, 46*: 854–859 (June 1981).
25. D. D. Kosloff and E. Baysal, Forward modelling by a Fourier method, *Geophysics, 47*: 1402–1412 (October 1982).
26. A. J. Herman, R. M. Anania, J. H. Chun, C. A. Jacewitz, and R. E. F. Pepper, A fast 3-D modelling technique and fundamentals of 3-D frequency-domain migration, *Geophysics, 47*: 1627–1644 (December 1982).
27. G. Garibotto, 2-D recursive phase filters for the solution of 2-D wave equations, *IEEE Trans. Acoust. Speech Signal Process., ASSP-27*: 367–373 (August 1979).
28. D. D. Kosloff and E. Baysal, Migration with the full wave equation, *Geophysics, 48*: 677–687 (June 1983).
29. E. Baysal, D. D. Kosloff, and J. Sherwood, Reverse time migration, *Geophysics, 48*: 1514–1524 (November 1983).
30. R. J. Castle, Wave-equation migration in the presence of lateral velocity variations, *Geophysics, 47*: 1011–1011 (July 1982).
31. I. Pitas and A. N. Venetsanopoulos, "Baysian Estimation of Medium Properties in Wavefield Inversion Techniques," Proc. IEEE International Conference on Acoustics, Speech, and Signal Processing, pp. 830–833 (1985).
32. I. Pitas and A. N. Venetsanopoulos, Towards a knowledge-based system for automated geophysical interpretation of seismic data (AGIS), *Signal Processing, 13*(3): 229–253 (October 1987).

# 20

# Marine Geophysical Signal Processing

FERIAL EL-HAWARY Faculty of Engineering, Technical University of Nova Scotia, Halifax, Nova Scotia, Canada

## 1 INTRODUCTION TO MARINE SEISMIC PROCESSING

Underwater seismic exploration using acoustic arrays poses a number of challenging problems in geophysical signal analysis and processing. The aim of such studies is to establish the structure of the underwater layered media in terms of geometry and material attributes that are significant to a given exploration task. Identifying hydrocarbon formations is one important target of marine seismic analysis. Classical acoustic exploration techniques have come into increasing use in marine sediment classification and layer identification in the recent past. Excellent treatments of the theoretical foundations of the estimation process are provided by a number of research monographs [1–3] and a recent tutorial [4]. A technique for extracting the subsurface features using a system theoretic approach is given in [5].

Layer identification and signal interpretation procedures have been subjective and practiced by capable and seasoned marine geologists who have relied primarily on the interpretation of acoustic gray scale graphic records. Identifying subsurface structures on this basis is a process that involves a great deal of human judgment and knowledge. The interpretation problem is essentially an imaging problem that involves three individual stages:

1. Image (or data) acquisition
2. Image (or signal) processing
3. Pattern recognition (or interpretation)

In this chapter, we discuss a number of important aspects of the marine geophysical signal analysis and processing task.

## 2 MARINE SEISMIC DATA ACQUISITION

The first step in the marine geophysical process is that of acquiring seismic data records, which are then further processed to obtain estimates and enhanced imagery of the underwater media. We presently discuss the basic elements of the seismic data acquisition system.

### 2.1 Marine Energy Sources

For marine subsurface mapping purposes, pulses fired from marine energy sources are used to impart energy to the underwater media. Marine energy sources are either explosive or nonexplosive. The air gun is the most used nonexplosive source. High-pressure air is suddenly introduced into the water, creating a pressure pulse. These pulses usually suffer from air bubbles. The pressure waveform is usually nonperiodic and may be approximated by a damped sinusoidal wave. In practice, the optimum waveform is obtained after the firing pulse has traveled for about 50 m in the water column. The wave hits the seabed and propagates through the ocean floor. The method uses information from reflections collected by acoustic receivers called hydrophones to classify and identify the subsurface layers using a number of interpretation procedures.

### 2.2 Marine Sensors: Source and Sensor Arrays

Hydrophones are pressure sensors that use piezoelectric crystals to generate a voltage proportional to the instantaneous water pressure associated with the seismic signal. Hydrophones are contained in cables up to 3200 m long. A typical cable would contain 48 recording segments (channels). Each channel receives signals from a linear array of many hydrophones. Present technology converts the received signals to digital form where they are received within the cable. Digital signals are transmitted via the cable to the recording instruments on board the ship or a remotely controlled vehicle. The data collected by the array of sensors are essentially a raw image (or map) of the structure. The data used in our studies have been collected by a shallow marine seismic system referred to as the Deep Towed System (DTS). The acoustic signal source is a broadband boomer of the displacement type. A typical waveform of the source is shown in Figure 20.1. The typical boomer has a maximum energy output of 600 joules and a bottom penetration capability of the subsurface to 80 m. The return signals are received by a hydrophone positioned near the source to produce approximately the normal incidence response of the subsurface media.

### 2.3 Frequency Ranges for Operation

Operating frequency ranges in marine acoustic applications are classified as deep seismic in the range of 10 to 100 Hz, whereas in the shallow seismic method the

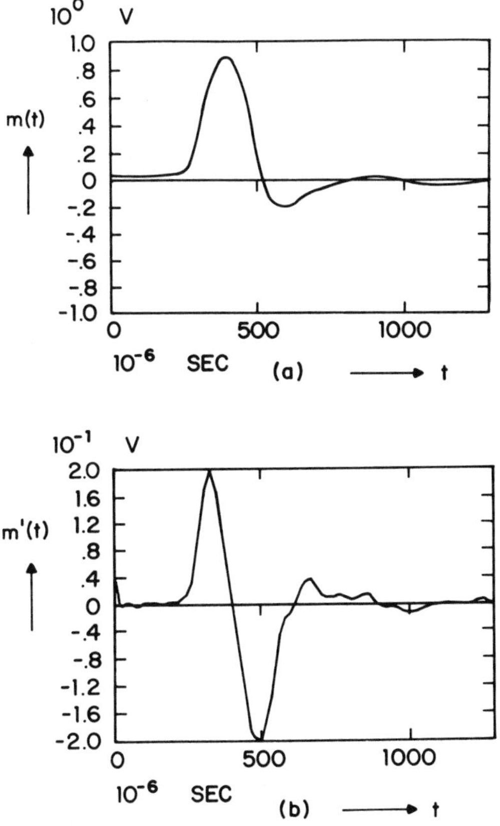


**Figure 20.1** Typical acoustic source signal waveform.

frequency range is from 100 Hz to about 10 kHz. It is worth noting that higher frequencies of up to 100 kHz are used in echo sounding positioning, telemetry, and remote control, scanning sonar, fish finding, and telephony. Doppler logs can operate at frequencies up to 1 MHz. Figure 20.2 shows the frequency ranges encountered in underwater acoustic applications. Our emphasis here is on the shallow marine seismic application.

## 3 PROBLEMS IN MARINE SEISMIC SIGNAL ANALYSIS

As mentioned earlier, in carrying out shallow marine seismic explorations, to determine the structure of the media underwater, a deep towed acoustic signal source and hydrophone receiver are employed. The source imparts energy to the water and underlying media, which then undergoes multiple transmissions and reflections at the layers' boundaries. The signals received by the hydrophone receivers contain valuable information about the layer structures. The received signals contain noise components that must be filtered out prior to parameter extraction and imaging. In

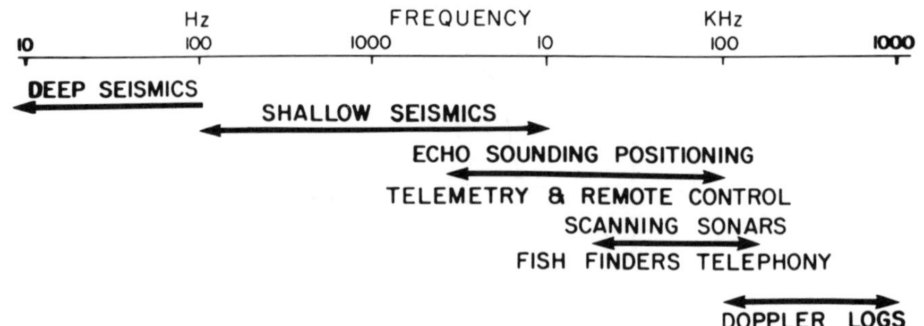


**Figure 20.2** Typical frequency ranges encountered in underwater acoustic applications.

the present section we discuss a major component which is unique to the underwater application.

## 3.1 Dynamic Motion Caused by Towing Sensors

The ship's dynamics, coupled to the towed body (fish) through the towing cable and the hydrodynamics of the towed body cause vertical motions of the source and sensor. These additional motion components have the effect of a varying acoustic wave travel path to the sea floor and to the subbottom reflectors between successive pings of the source. The motion's effects show on the aggregate of reflection records along the ship track as undulations of the sea floor and of the subbottom reflectors. These effects, commonly referred to as heave effects, introduce errors in estimates of delay times.

## 3.2 Preprocessing Requirements

Removal of the heave component is an important preprocessing task for improving gray scale displays of the raw and filtered reflection data in order to extract media parameters such as reflection coefficients and reflector depths. These effects can be partially removed by use of estimates of the vertical source motions from hydrostatic pressure and motion sensors to delay or advance the pulse firing instants relative to a clock pulse reference. This is done such that the source-bottom-sensor pulse travel time corresponds to that of a source and sensor located at a constant depth relative to the mean water surface level. Further details on this aspect are given by Hutchins [6].

The literature in marine hydrodynamics, such as Bhattacharya [7], Price and Bishops [8], and McCormick [9], provides details of modeling the heave hydrodynamics. A second-order transfer function model of the following form can be postulated on the basis of the frequency response of the heave record [10]:

$$T(s) = \frac{cs}{s^2 + as + b} \tag{20.1}$$

The system parameters $c$, $b$, and $a$ are related to the more familiar natural frequency $\omega_0$ and quality factor $Q_0$ by

$$a = \frac{\omega_0}{Q_0}$$
$$b = \omega_0^2$$
$$c = \frac{k\omega_0}{Q_0}$$

It is clear that from an accuracy point of view, higher-order models can be expected to provide less error in modeling the process. We can assume that the heave process is modeled using the following transfer function [11]:

$$H(s) = \frac{KN(s)}{D(s)} \tag{20.2}$$

The system gain is denoted by $K$. The numerator function $N(s)$ has $M$ zeros $z_i$ and is therefore given by

$$N(s) = (s - z_1)(s - z_2) \cdots (s - z_M)$$

The denominator function has $N$ poles $p_i$ and is therefore given by

$$D(s) = (s - p_1)(s - p_2) \cdots (s - p_N)$$

The heave process model estimation task resolves into finding the optimal values of poles and zeros. Ideally, we should attempt to determine a priori the optimal model order, but this is quite an involved process. We process the given records for an assumed model configuration; i.e., a series of $M$ and $N$ values are assumed and the resulting optimal models are compared in terms of accuracy.

The parameters of the transfer function model are obtained on the basis of time domain heave response records extracted from successive reflection records. The parameter estimation task is performed in the frequency domain using the Fourier transform of the heave record and least-squares estimation procedures. The time spacing of samples in the required heave record should be chosen to avoid aliasing. The frequency spectrum of a typical heave record contains a number of dominant peaks, indicating that the process can be represented as the output of a narrowband filter due to a purely random input. Figure 20.3 shows the frequency spectrum of the heave record in a typical case study.

It is important to note that the parameter estimates of the heave response model are defined by nonlinear equations, which makes it necessary to use iterative algorithms to obtain the desired heave model parameters [12]. Least-squares estimation algorithms are extensively used for the modeling process. An alternative

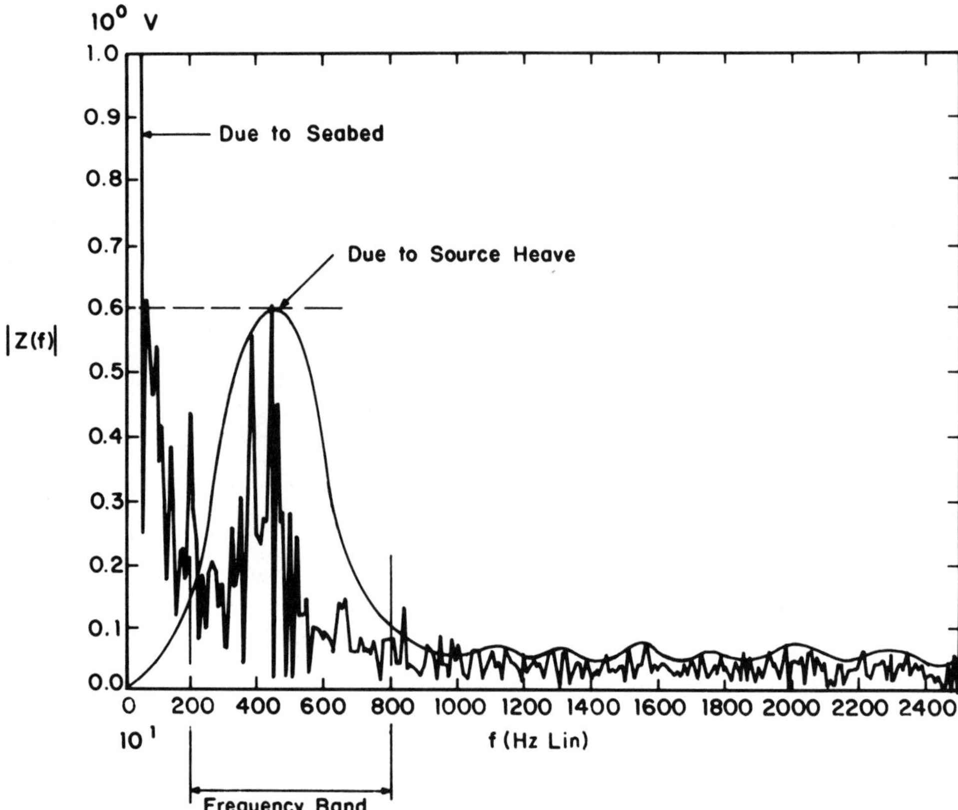


**Figure 20.3** Frequency spectrum of a heave record in a typical case study.

approach [13] uses least $p$th optimization to obtain model parameters by assuming that the estimation criterion is the sum of the $p$th power of the magnitude of the errors.

## 3.3 Heave Removal

Compensation for source heave can best be done using Kalman filtering. The design of the Kalman filter enables reducing the residual heave effects, i.e., delaying and advancing the recording trigger on successive firings to effect a smoothing or removal of such undulations. The filtering can be applied to postexperiment data records, or preferably in a real-time mode during acquisition of the reflection responses. Kalman filter applications are well known for some different physical situations, as reported in Gelb [14] and Meditch [15]. The application to optimum filtering and smoothing of buoy wave data is reported by Severance [16].

Heave component extraction details for the shallow marine seismic application are presented in [17]. This process involves essentially two steps. First, a model of the heave phenomenon is obtained on the basis of available data as outlined in the preceding section. The type of model and its order are important considerations.

In early applications, [17, 18], a linear second-order model of the phenomenon as given by Eq. (20.1) was found to be satisfactory. In this step the estimation of the model parameters is an important aspect. The resulting estimation software uses an iterative procedure based on Newton's method [17]. The model then provides the basis for formulating the heave extraction problem as one of Kalman filtering in the theory of optimal linear estimation. Second, once a model and its parameters have been identified, the actual process of Kalman filtering is carried out. Two versions exist. In the early implementation [18] conventional Kalman filtering is used satisfactorily in a majority of cases. Figure 20.4 shows an example of the subsurface image before and after heave compensation using conventional Kalman filtering for a section in Outer Placentia Bay off the coast of Newfoundland.

The heave motion model of Eq. (20.1), can be written in discrete state space form as

$$x(k+1) = \varphi(k+1,k)x(k) + \Gamma(k+1,k)w(k) \tag{20.3}$$

The state transition matrix $\varphi(k+1,k)$ and the matrix $\Gamma(k+1,k)$ are constants and we therefore write

$$x(k+1) = \varphi x(k) + \Gamma w(k) \tag{20.4}$$

The input sequence $w(k)$ is assumed to be a Gaussian white sequence with zero mean and a covariance matrix $Q(k)$, being positive semidefinite. The initial state is assumed to be a Gaussian random vector with zero mean and known covariance matrix $P(0)$. It is further assumed that $w(k)$ is independent of $x(0)$. The record of the heave component is assumed to be the basis for the measurement model given by

$$z(k+1) = Hx(k+1) + v(k+1) \tag{20.5}$$

The measurement error sequence $v$ is assumed to be Gaussian with zero mean and a covariance matrix $R(k)$.

Assume that measurements $z(1)$, $z(2)$, $\cdots$, $z(j)$ are available, from which we wish to estimate $x(k)$, denoted by $x(k \mid j)$. In filtering we have $j = k$, and we therefore wish to find $x(k \mid k)$. We utilize the standard predictor-corrector form of a Kalman filter given by as follows.

**Predictor.** In the predictor stage we obtain a prediction of the state based on the previous optimal estimate:

$$x_k(-) = \varphi_{k-1} x_{k-1}(+) \tag{20.6}$$

In addition, we obtain for the error covariance matrix

$$P_k(-) = \varphi_{k-1} P_k(+) \varphi_{k-1}^T \tag{20.7}$$

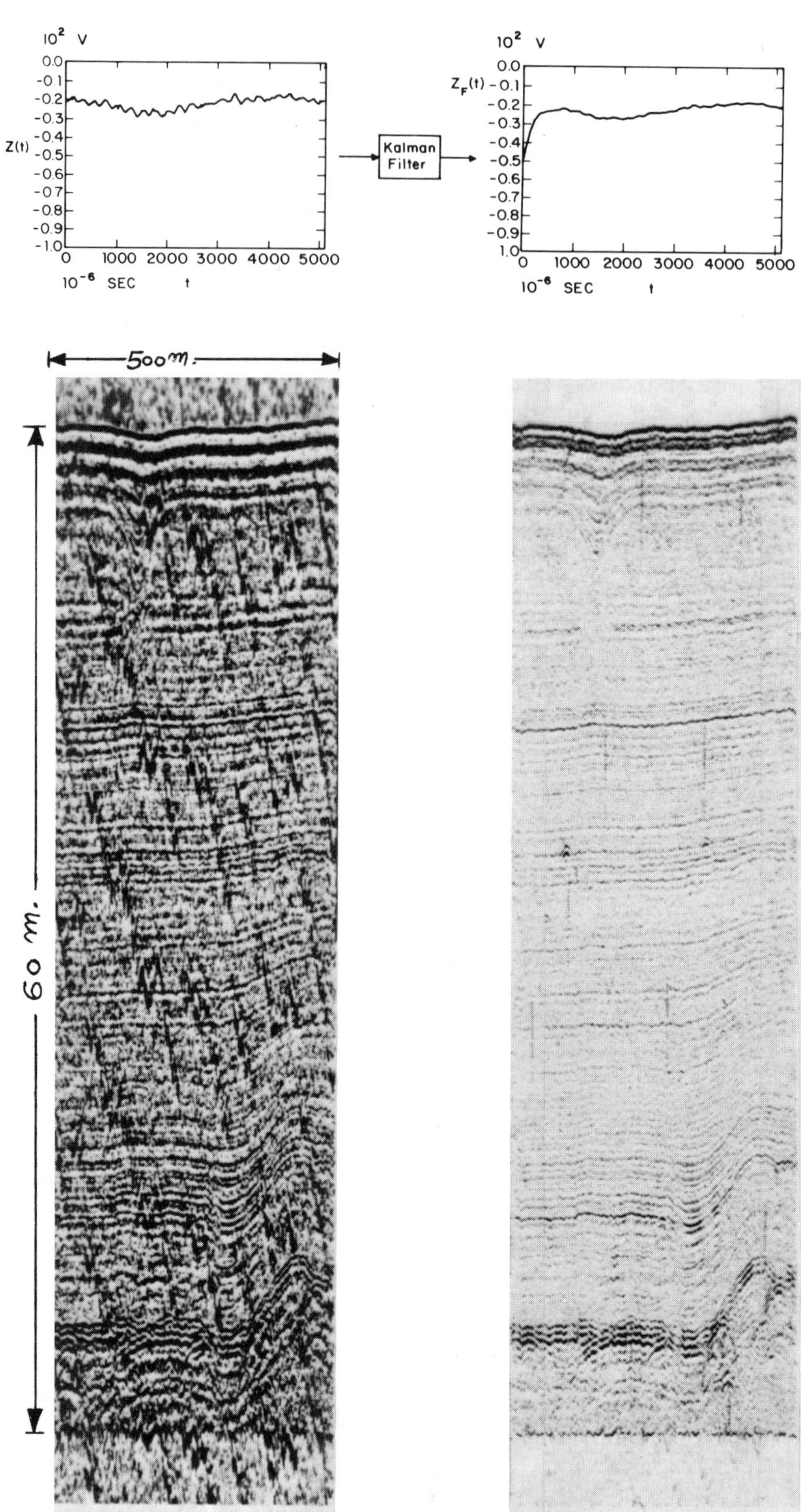

10² V
0.0
-0.1
-0.2
-0.3
-0.4
Z(t)
-0.5
-0.6
-0.7
-0.8
-0.9
-1.0
0 1000 2000 3000 4000 5000
10⁻⁶ SEC t
Kalman Filter
10² V
$Z_F(t)$
0.0
-0.1
-0.2
-0.3
-0.4
-0.5
-0.6
-0.7
-0.8
-0.9
1.0
0 1000 2000 3000 4000 5000
10⁻⁶ SEC t
500 m.
60 m.

**Corrector.** In the corrector stage we obtain an updated state estimate

$$x_k(+) = x_k(-) + K_k[y_k - H_k x_k(-)] \tag{20.8}$$

In addition, we obtain an update of the covariance matrix as

$$P_k(+) = [I - K_k H_k]P_k(-) \tag{20.9}$$

Here $K$ is the Kalman gain matrix given by

$$K_k = P_k(-)H_k^T[H_k P_k(-)H_k^T + R_k]^{-1} \tag{20.10}$$

## Parallel Kalman Filtering

The real-time application of the standard Kalman filter is limited by the filter's computational complexity. It is evident that faster implementations are desirable and parallel versions of Kalman filtering are required. Kalman filtering is a sequential process that evaluates the predictor equations prior to carrying out the computation involving the corrector equations. This coupling introduces computational delays. Delays can be avoided by using a parallel Kalman filter. This is based on decoupling the predictor and corrector stages by forcing the corrector to lag the predictor by one discrete step. The resulting filter is called the parallel or (decoupled) Kalman filter. The filter equations are

**Predictor.** In the predictor stage we have

1. A prediction of the state

$$x_{k+1}(-) = \varphi_k \{\varphi_{k-1} x_{k-1}(+)\} \tag{20.11}$$

This is a modified version of the standard Kalman filtering process

$$x_{k+1}(-) = \varphi_k \{x_k(+)\} \tag{20.12}$$

2. In addition, we have for the error covariance matrix

$$P_{k+1}(-) = \psi_k P_{k-1}(+)\psi_k^T \tag{20.13}$$

where

$$\psi_k = \varphi_k \varphi_{k-1} \tag{20.14}$$

---

**Figure 20.4** Example of subsurface image before and after Kalman filtering for heave compensation.

This is a modified version of the standard Kalman filter.

**Corrector.** In the corrector stage we have

1. An updated state estimate

$$x_k(+) = x_k(-) + K_k[y_k - H_k x_k(-)] \tag{20.15}$$

2. In addition, we have an update of the covariance matrix as

$$P_k(+) = [I - K_k H_k] P_k(-) \tag{20.16}$$

The Kalman gain matrix is given by

$$K_k = P_k(-) H_k^T [H_k P_k(-) H_k^T + R_k]^{-1} \tag{20.17}$$

The predictor is now decoupled from the corrector. For a computational point of view, it is convenient to define

$$\begin{aligned} F_{11} &= I - K_k H_k \\ F_{12} &= K_k \end{aligned}$$

As a result we define

$$\begin{aligned} F_k^T &= [F_{11}^T \mid F_{12}^T] \\ S_k^T &= [x_k^T(-) \mid Z_k^T] \end{aligned}$$

We thus have a compact form for the optimal estimate given by

$$x_k(+) = F_k S_k$$

The covariance matrix is now given by

$$P_k(+) = F_{11} P_k(-)$$

The computations can be carried out concurrently using two processors.

Several recent contributions to the literature on signal processing, computer architecture, and very large scale integration (VLSI) design show that systolic array processing techniques are extremely useful for designing special-purpose devices to solve problems in linear system analysis [19, 20]. Physiologists use the term *systole* to describe the contracting operation of the heart and arteries which pulses blood through the body in a rhythmically recurring pattern. A systolic system is a network of processors which rhythmically compute and pass data through the system. Every processor pumps data in and out regularly, each time performing some short computation, so that a regular flow of data is kept up in the network. The motivation of systolic architecture is that for special-purpose hardware a systematic means of design is required and therefore a methodology for mapping high-level computation into hardware structures. Systolic array architectures can be designed to take advantage of the parallel Kalman filter. In a recent paper, we apply Kalman filtering theory and parallel Kalman filters to the exploration problem [21]. Particular emphasis is given to the multireceiver case as an important application of array processing methodology. Parallel Kalman filtering is designed to take advantage of systolic array implementation. In the initial work the heave model is assumed to be of second order.

### 3.4 Higher-Order Models

Investigations concerning higher-order model implementations using Eq. (20.2) are of current interest because of the potential for improved accuracy in modeling the heave phenomenon. Preliminary results reported in [11] indicate that more accuracy is obtained if we use a very high order model or a fifth-order model. The results for a second-order model are very reasonable and, from an end-use point of view, this model is preferable in Kalman filtering applications.

## 4 INFORMATION EXTRACTION FROM MARINE SEISMIC DATA

The goal of the geophysical signal processing task is to extract physically related information about the subsurface media from the reflected image records received by the sensors subsequent to heave compensation. The acoustic pressure wave travels through the subsurface layered media and undergoes multiple reflections as it impinges on boundaries between successive layers. The received signal at the sensors (hydrophones) contains replicas of the original source signal which are corrupted by noise components from various sources. Assuming a linear lossless model of the wave propagation, one can express the received signal as the sum of $N$ terms, each consisting of a delayed version of the source signal with an amplitude scale factor as well as an additive noise term

$$y(t) = \sum_{i=1}^{\infty} a_i m(t - \tau_i) + v(t) \tag{20.18}$$

The amplitude scale factors $a_i$ are related to the layers' reflection coefficients $r_i$, and the delay times $\tau_i$ are related to the one-way delays $D_i$ obtained as the ratio of the travel distance $X_i$ and the velocity of sound propagation in the given medium, $C_i$. The amplitude scale factors $a_i$ are related by the following recursive relation assuming primary reflections only:

$$a_i = \frac{a_{i-1} r_i (1 - r_{i-1}^2)}{r_{i-1}} \tag{20.19}$$

Here $a_0 = r_0$. The delay times are given by

$$\tau_i = \tau_{i-1} + 2D_i \tag{20.20}$$

The case of higher-order reflections has been treated in [22].

### 4.1 Correlation with Physical Parameters

In essence, the required information about the media is extracted using peak and delay parameter detection techniques.

The reflection coefficient at a boundary is related to the acoustic impedances of the layers by the fundamental relation

$$r = \frac{Z_2 - Z_1}{Z_2 + Z_1} \tag{20.21}$$

The acoustic impedance $Z$ is the product of the layer density $\rho$ and the compressional wave velocity $C$. An approximation used in [23] is

$$r = (1/2)\Delta \ln Z \tag{20.22}$$

As a result, the amplitude scale factors are related to the acoustic impedance by

$$a_i = r_i = (1/2) \ln \left( \frac{Z_{i+1}}{Z_i} \right) \tag{20.23}$$

Alternatively, we can find the acoustic impedances from the amplitude scale factors from the inverse relation

$$Z_{i+1} = Z_i \exp \{2a_i\} \tag{20.24}$$

Usually $Z_1$ corresponds to propagation through the water with known $\rho$ and $C$. It is clear that knowledge of the acoustic properties of the media can be used to find geotechnical properties. For further details, see [24] and the references cited therein.

Performing the required information extraction task in terms of amplitude and delay parameters appears to be straightforward. Many detection procedures are available, and we next review some practical schemes for performing this task. It is our experience, however, that the outcome of an algorithm may not provide satisfactory results that conform with other verification results, depending on the source of the given data. It is therefore important early in the analysis phase to determine which scheme to use. An expert system [25–28] recognizing the selection criteria and limitations of each scheme is therefore of extreme importance.

### 4.2 Peak and Delay Estimate via Sequential Correlation-Based Detection

The estimation of delay and amplitude parameters can be carried out effectively using cross-correlation processing combined with minimum variance analysis [5, 29]. Two steps are involved:

1. Estimate peak and delay times using the cross-correlation between source and received signals. The result is a cross-correlation function with a peak at the unknown delay times. A direct search for the extrema results in an unrefined estimate of the delay parameters. In this step it is important that a threshold be established on the amplitude of the extrema considered. In practical implementations one relies on past experience in a trial-and-error procedure to set a threshold that is appropriate for the area surveyed. A data base of prior knowledge is an important component of an automated implementation of this fast procedure.
2. Refine estimates using minimum variance information. For this purpose, the problem is one of finding an optimal parameter estimate of the amplitude scale factors with optimality defined in the sense of minimum error variance. This can also be achieved by use of a matched filter implementation, where the impulse response of the filter is a reverse time replica of the source waveform.

Without a priori knowledge, an estimate of peak amplitude is given by the ratio of the cross-correlation of source and receiver signals to the autocorrelation of the source signal. Figure 20.5 shows a sequence of the steps involved, from the raw data record shown in part (A) to the estimate of the reflection sequence shown in part (G).

### 4.3 Linearized Delay Model Detection

In this procedure [29], it is assumed that each delayed replica of the source signal in Eq. (20.18) is approximated by a first-order Taylor expansion about the time delay. It is further assumed that a rough estimate of the delays involved is available. The delayed replica is then represented by the values of the source signal and its derivative, both evaluated at the estimate of the delay. The problem therefore is one of linear parameter estimation, which is then solved efficiently using recursive algorithms.

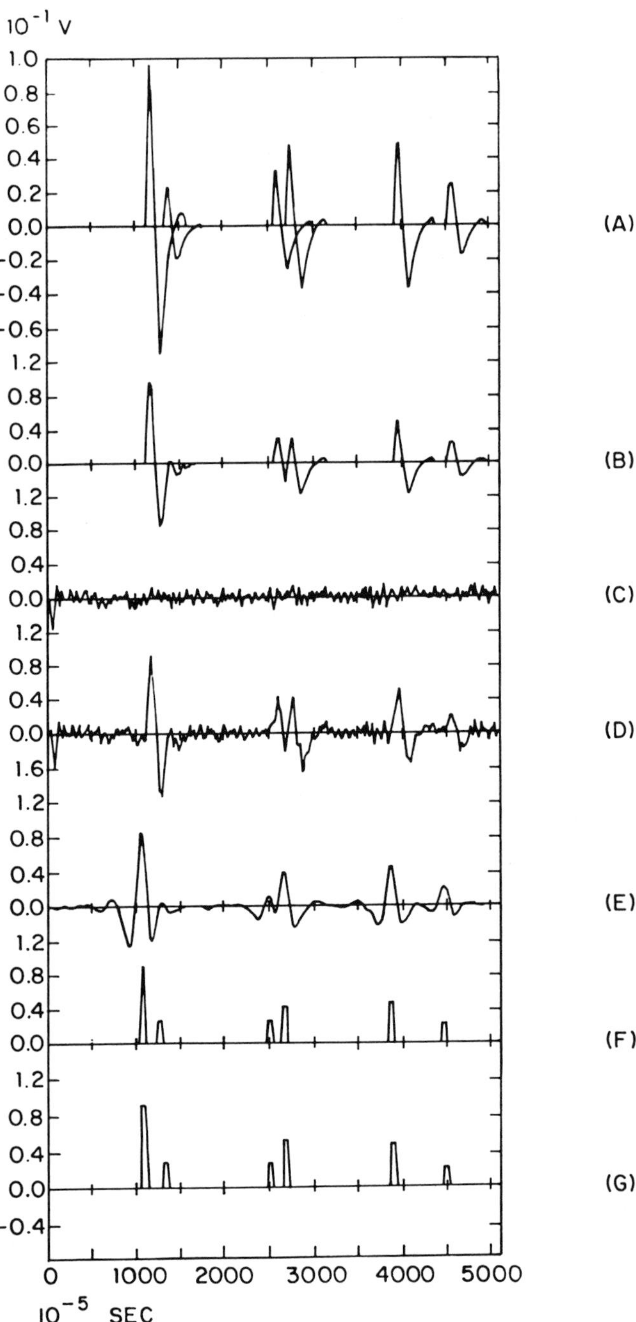


**Figure 20.5** Steps showing amplitude and delay parameter estimation.

It is important to realize that the choice of the estimate of the delays plays a central role in determining the success of this procedure [30]. Aids in choosing these values include consistency between successive records, and in some instances (especially in the initial processing phase) it may be necessary to use results of a sequential correlation detector. Figure 20.6 shows typical results of applying this procedure.

## 5 CONCLUSIONS

We discussed a number of problems in marine geophysical signal processing using underwater acoustic methods for seismic signal analysis and interpretation on the basis of reflections and multiple reflections from the seabed and the underlying media. We reviewed some basic issues involving acquisition and processing of the seismic signals. We emphasized the main software modules incorporated in

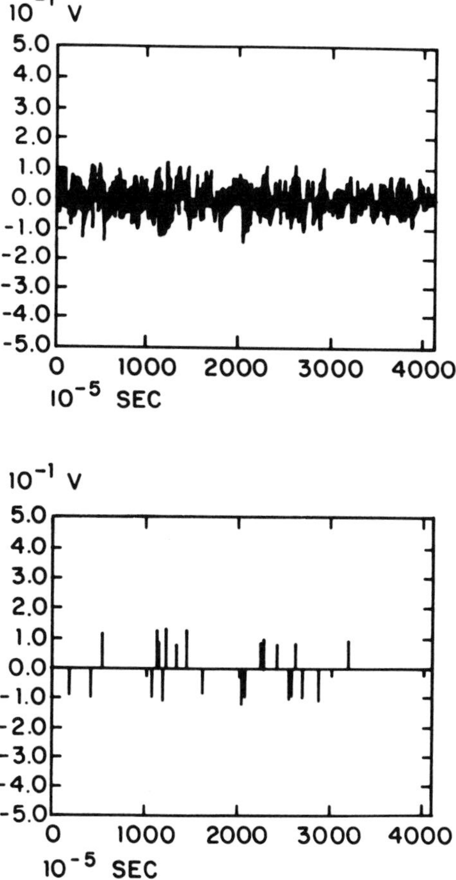


**Figure 20.6** Effect of linear estimation for amplitude and delay parameter estimation.

the system, including compensation for the dynamic heave component present in the signals. Conventional and parallel Kalman filtering proved to be useful tools. Considerations for modeling the heave process were highlighted. A number of information extraction algorithms that are central to the interpretation process were discussed. In particular, we treated delay and amplitude parameter estimation using a combined cross-correlation–minimum variance filter and linearized recursive estimation. This work shows the importance of signal processing methods in marine geophysical applications.

## REFERENCES

1. E. A. Robinson, *Multichannel Time Series Analysis with Digital Computer Programs*, Holden-Day, San Francisco (1967).
2. J. M. Mendel, *Optimal Seismic Deconvolution: An Estimation Based Approach*, Academic Press, New York (1983).
3. E. A. Robinson and T. S. Durrani, *Geophysical Signal Processing*, Prentice-Hall, Englewood Cliffs, New Jersey (1986).
4. J. M. Mendel, Some modeling problems in reflection seismology, *IEEE Trans. Acoust. Speech Signal Process.*, *3*(2): 4–17 (1986).
5. F. El-Hawary, An approach to seismic information extraction, *Time Series Analysis: Theory and Practice 6* (O. D. Anderson, J. K. Ord, and E. A. Robinson, eds.), Elsevier Science Publishers, Amsterdam, pp. 223–238 (1985).
6. R. W. Hutchins, "Removal of Tow Fish Motion Noise from High Resolution Seismic Profiles," presented at SEG-U.S. Navy Symposium on Acoustic Imaging Technology and on Broad Data Recording and Processing Equipment, National Space Technology Laboratory, Bay St. Louis, Mississippi (August 1978).
7. R. Bhattacharya, *Dynamics of Marine Vehicles*, Wiley-Interscience, New York (1978).
8. W. G. Price and R. E. D. Bishops, *Probabilistic Theory of Ship Dynamics*, Wiley-Interscience, New York (1974).
9. M. E. McCormick, *Ocean Engineering Wave Mechanics*, Wiley-Interscience, New York (1973).
10. F. El-Hawary, An approach to extract the parameters of source heave dynamics, Can. Elect. Eng. J. (January 1987).
11. F. El-Hawary and T. Richards, "Heave Response Modelling Using Higher Order Models," Proc. IASTED Conference on Applied System Simulation, Santa Barbara, California (1987).
12. J. M. Ortega and W. C. Rheinboldt, *Iterative Solution of Nonlinear Equations in Several Variables*, Academic Press, New York (1970).
13. X. Lu and J. K. Vandiver, "Damping and Natural Frequency Estimation Using the Least *p*th Optimization Technique," Paper OTC 4283, Proc. 14th Offshore Technology Conference, Houston, Texas (May 1982).
14. A. Gelb, Applied Optimal Estimation, MIT Press, Cambridge, Massachusetts (1974).
15. J. S. Meditch, *Stochastic Optimal Linear Estimation and Control*, McGraw-Hill, New York (1969).

16. R. W. Severance, Optimum filtering and smoothing of buoy wave data, *J. Hydronaut.*, *9*: 69–74 (April 1975).
17. F. El-Hawary and W. J. Vetter "Heave Compensation of Shallow Marine Seismic Reflection Records by Kalman Filtering," IEEE Oceans' 81, Boston (September 1981).
18. F. El-Hawary, Compensation for source heave by use of Kalman filter, *IEEE J. Ocean. Eng.*, *OE-7*(2): 89–96 (1982).
19. S. Y. Kung, VLSI signal processing: From transversal filtering to concurrent array processing, *VLSI and Modern Signal Processing*, (S. Y. Kung, H. J. Whitehouse, and T. Kailath, eds.), Prentice-Hall, Englewood Cliffs, New Jersey, pp. 127–152 (1985).
20. R. H. Travassos, Application of systolic array technology to recursive filtering, *VLSI and Modern Signal Processing*, (S. Y. Kung, H. J. Whitehouse, and T. Kailath, eds.), Prentice-Hall, Englewood Cliffs, New Jersey, pp. 375–388 (1985).
21. F. El-Hawary and K. M. Ravindranath, "Application of Array Processing for Parallel Linear Recursive Kalman Filtering in Underwater Acoustic Exploration," Proc. IEEE Oceans '86, Washington, D.C., vol. 1, pp. 336–340 (September 1986).
22. W. J. Vetter and F. El-Hawary, "A Layer-Indexed Reflection Model for Multilayered Media," IEEE Conference Record, Oceans '79, San Diego, California, pp. 21–29 (1979).
23. R. A. Peterson, W. R. Fillipone, and F. B. Crocker, The synthesis of seismograms from well log data, *Geophysics*, *20*: 516–538 (1955).
24. F. El-Hawary, "Seismic Signal Processing for Geotechnical Properties Applied to Marine Sediments," Proc. Workshop on Atlantic Coastal Erosion and Sedimentation, National Research Council of Canada, Halifax (November 1982).
25. F. El-Hawary, Image analysis methods from sea-bed reflections and multiple reflections, *Int. J. Pattern Recogn. Artif. Intell.*, *1*(2): 85–96 (1987).
26. J. H. Justice, D. J. Hawkins, and G. Wong, Multidimensional attribute analysis and pattern recognition for seismic interpretation, *Pattern Recogn.*, *18*(6): 391–408 (1985).
27. C. H. Chen, Recognition of underwater transient patterns, *Pattern Recogn.*, *18*(6): 485–490 (1985).
28. J. C. Hassab and C. H. Chen, On constructing an expert system for contact localization and tracking, *Pattern Recogn.*, *18*(6): 465–474 (1985).
29. F. El-Hawary and W. J. Vetter, Spatial parameter estimation for ocean subsurface layered media, *Can. Electr. Eng. J.*, *5*(1): 28–31 (1980).
30. F. El-Hawary and W. J. Vetter, Event enhancement on reflections from subsurface layered media, *IEEE J. Ocean. Eng.*, *OE-7*(1): 51–58 (1982).

# 21

# Signal Processing in Nondestructive Evaluation of Materials

C. H. CHEN Electrical and Computer Engineering Department, Southeastern Massachusetts University, North Dartmouth, Massachusetts

## 1 INTRODUCTION

Nondestructive evaluation (NDE) of materials is playing an increasingly important role in a number of industries. Signal processing techniques have significantly advanced the state of the art of nondestructive evaluation and testing. This chapter presents the major techniques of signal processing in NDE.

The development of modern technologies has increased the demand for product testing using NDE methods. NDE is used in automated defect detection and characterization, materials characterization, and process control. In improving the capability of NDE instrumentation, signal processing is the key to automating the inspection process and improving inspection reliability as well as providing information for making objective decisions.

Until about a decade ago, the major effort to improve NDE was on increasing the resolution of the instrumentation. Emphasis was on improved electronics for analog signal processing and transducers to detect the smallest possible defects. Much of the progress in the past 10 years has centered on the application of digital signal processing, already successfully applied in other areas such as radar, sonar, and geophysics as described in other chapters of this book, to enhance the capability of conventional NDE. The specific goals of digital signal processing in NDE are as follows 1) to improve inspection reliability, 2) to improve defect detection, 3) to improve defect characterization, and 4) to generate information about the material properties to assess the remaining life of a structure. Advanced signal processing techniques are needed to achieve these objectives. So far, only partial success can be reported. The scope of this chapter therefore is not the full range of NDE activities using signal processing. The chapter examines signal acquisition and enhancement, time and frequency domain analysis, deconvolution and Wiener filtering, and pattern recognition. Emphasis is placed on the ultrasonic spectroscopy that relies on the frequency analysis method of defect characterization from ultrasonic pulse echoes.

## 2 SIGNAL ACQUISITION

Besides the popular ultrasonic method, eddy current, acoustic emission, X-ray radiography, microwave, magnetic, and optical methods have also been used in NDE. A typical ultrasonic signal acquisition system is shown in Figure 21.1a. The transducer center frequencies typically are 5 and 15 MHz, with a sampling rate as high as 100 MHz. The specimen evaluated typically may be an aluminum plate. The 17 defect geometries considered in our experiment are listed in Figure 21.1b. The pulse echo is supposed to provide information about the defect characteristics. It is difficult, however, to go through the mathematical "inverse" problems to estimate the defect parameters from the echo. The signal processing and pattern recognition considered in this chapter provide solutions in defect characterization, classification, and prediction without going through the inverse problems.

Analog signal enhancement is performed by use of proper filters which remove certain frequency components of the data and reduce some system noise. The analog-to-digital conversion and signal conditioning required are discussed in Chapter 2. Complete systems including signal acquisition, computer, supporting data base, and pattern recognition signal processing software have been developed and are available commercially. One such product is the TestPro developed by Informatrics. However, with the continued interest in more sophisticated NDE, improved systems are always in demand. For detailed information on other signal acquisition systems, see, e.g., [1, 2].

## 3 TIME DOMAIN ANALYSIS

The time domain information is the reflected pulse echo itself, which has many digitized data. The duration of the event portion of the waveform is an example of time domain knowledge. A pulse duration feature can be defined as the time difference between the intercepts of the pulse envelope with a line at 10% of the peak amplitude in each waveform. Another measure of duration is determined by using the segmentation algorithm [3] that determines the starting and ending points of the event portion of the waveform. The pulse duration is somewhat proportional to the angle in angle-cut defects.

Another important time domain feature is the amplitude ratio, defined as the ratio of the cumulative sum of the absolute amplitudes of the test specimen to

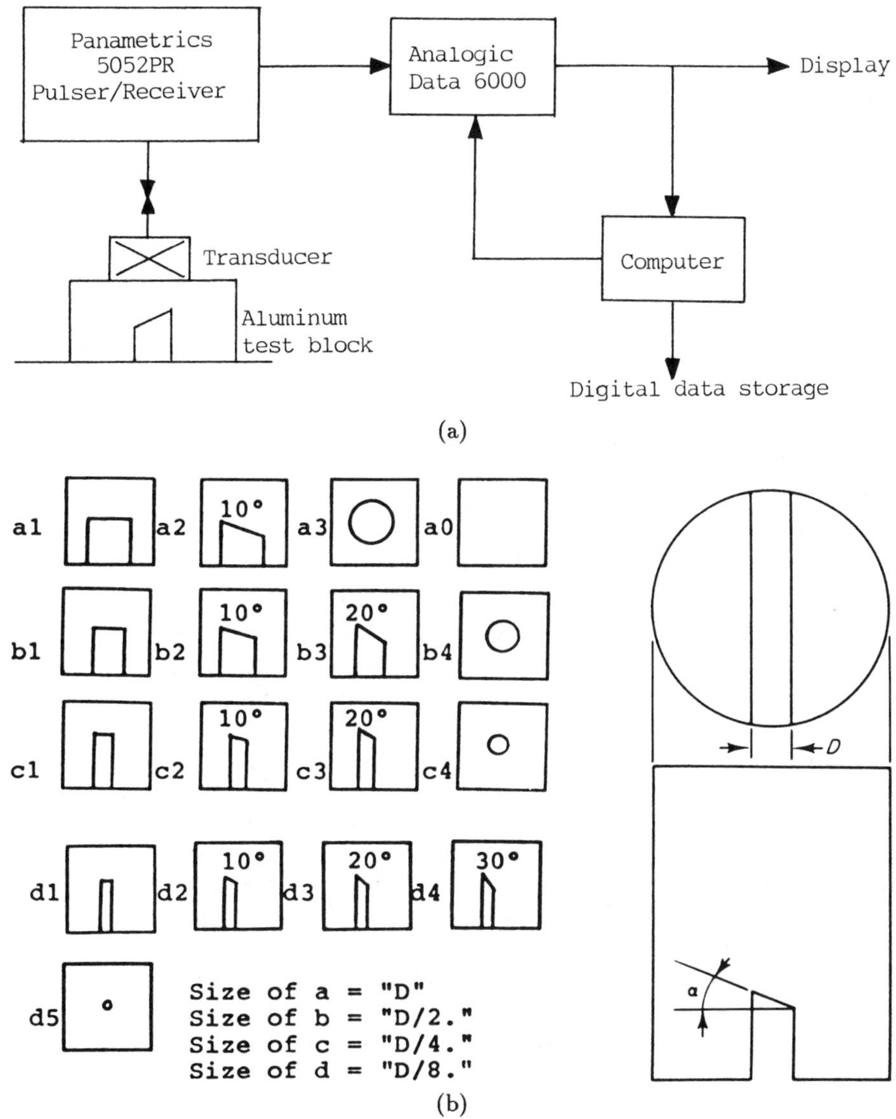


**Figure 21.1** (a) Ultrasonic signal acquisition system. (b) Defect type considered (left) and configuration of aluminum test block (right) with angle cut of angle $\alpha$ and width $D$.

that of the reference specimen. The amplitude ratio tends to be larger for circular defects than for flat defects.

Ray tracing is possible to identify various portions of the waveform with different reflections and refraction paths. This is difficult in most cases as the noise and interferences greatly obscure the association of waveform segments with right paths.

## 4 FREQUENCY DOMAIN ANALYSIS

In his pioneering work on ultrasonic spectroscopy, Gericke [4, 5] experimented with a contact mode of spectral analysis and demonstrated that the spectrum of the pulse echo varies with defect geometry. An early version of Figure 21.1a showed essentially the instrumentation Gericke employed in an ultrasonic spectroscopy study which has been continued at the U.S. Army Materials Technology Laboratory for over 20 years [6]. Typical examples of Gericke's early work are shown in Figures 21.2 and 21.3; these examples demonstrate the diagnostic possibilities of his technique in materials evaluation and in characterization of simulated flaws in metal components. Figures 21.2 and 21.3 show the effect on the spectrum of introducing flaws into an aluminum plate with various widths $D$ and angles $\alpha$ as shown in Figure 21.1b. In Figure 21.2, where the base of the plate is perpendicular to the axis of the beam, a greater proportion of higher-frequency than of low-frequency energy is specularly reflected from the base back to the transducer; the high-frequency humps in the spectrum thus increase in height relative to the low-frequency ones as the width $D$ decreases. Conversely, for angular cuts, as in Figure 21.3, the high-frequency components are preferentially reflected in directions which miss the transducer. As Gericke recognized, the spectral response of a known defect geometry could be explained on this basis, but there was no claim to be able to characterize fully an unknown defect from its spectral response.

Gericke's work was verified by us [7] by using the high-resolution maximum entropy spectral analysis software package supplied by Information Research Laboratory, Inc. Figure 21.4a shows a typical pulse echo from a transducer of center frequency 15 MHz. The vertical lines indicate the segmentation boundaries. Figure 21.4b is the corresponding maximum entropy power spectrum. An example of a spectral domain feature is the spectral ratio based on the powers in two frequency bands as a and b. The mean and second moments of the amplitude spectrum are also useful features. The kurtosis and skewness which are normally used as parameters of probability density can be used as measures of wave shape and spectrum shape. Phase spectrum and other transform have also been examined for possible feature extraction. The main problem with the high-resolution spectral analysis is that the all-pole modeling may not provide an adequate representation of the signal.

## 5 THE CORRELATION DOMAIN

The cross-correlation between the NDE waveform for a defect specimen and a reference waveform for zero-defect geometry provides several useful features. The largest correlation peak value and the root-mean-square (RMS) value of the correlation function are both features in defect classification. The maximum peak of correlation is always negative for angular cut defects and positive for other defects [8]. It is also possible to identify the ray paths from the autocorrelation structure, but again serious difficulty is caused by noise and interference between paths. Figure 21.4c is the cross-correlation structure corresponding to Figure 21.4a.

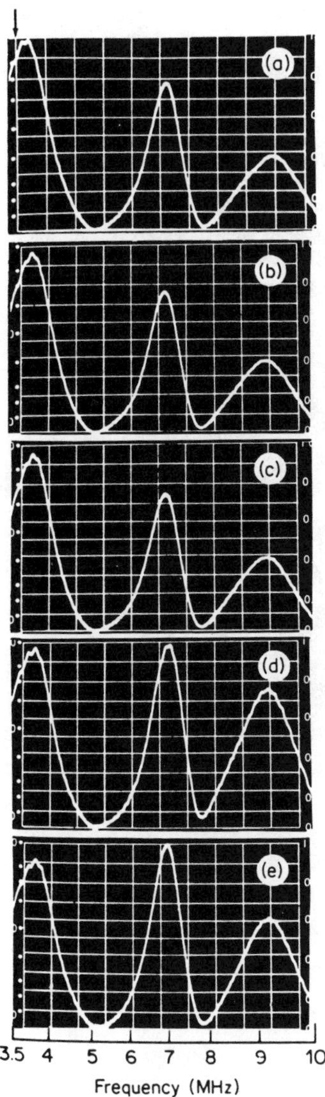


**Figure 21.2** Ultrasonic spectra obtained from slots in the aluminum test block of Figure 21.1 with flat top; i.e., the reflecting surface is normal to the beam axis ($\alpha = 0$). (a) Basic pulse spectrum; (b) $D = 2$ mm; (c) $D = 4$ mm; (d) $D = 8$ mm; (e) $D = 16$ mm. (Courtesy of Dr. Otto Gericke.)

## 6 THE IMPULSE RESPONSE VIA DECONVOLUTION

In the pulse-echo method of ultrasonic testing, if the input signal is denoted by $x(t)$, then the reflected signal, under the time-invariant system assumption, can be

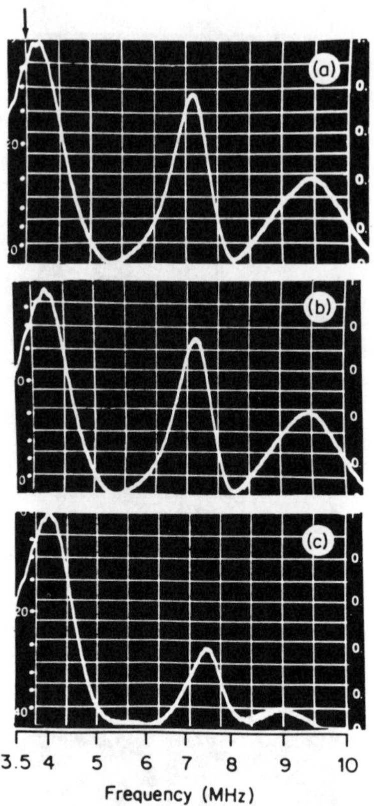


**Figure 21.3** Ultrasonic spectra obtained from slots in the aluminum test block of Figure 21.1. (a) Basic pulse spectrum; (b) $D = 8$ mm, $\alpha = 0°$; (c) $D = 16$ mm, $\alpha = 10°$. (Courtesy of Dr. Otto Gericke.)

expressed as a convolution integral,

$$y(t) = \int_{-\infty}^{\infty} h(\alpha)x(t-\alpha)\,d\alpha = h(t) * x(t) \tag{21.1}$$

where the $*$ denotes convolution and $h(t)$, called the impulse response, is a function of time related to the reflections that occur as the sound propagates through the medium in the direction along the beam. The problem of finding $h(t)$ or $x(t)$ given $y(t)$ is called deconvolution. The problem is complicated because both $x(t)$ and $h(t)$ are unknown. A more complete model of the received signal $y(t)$ is

$$y(t) = u(t) * T_i(t) * P_1(t) * h_1(t) * P_2(t) * T_2(t) \tag{21.2}$$

again based on the assumption of a time-invariant system. Here $u(t)$ is the electrical impulse driving the transducer, $T_1(t)$ the forward transducer impulse response,

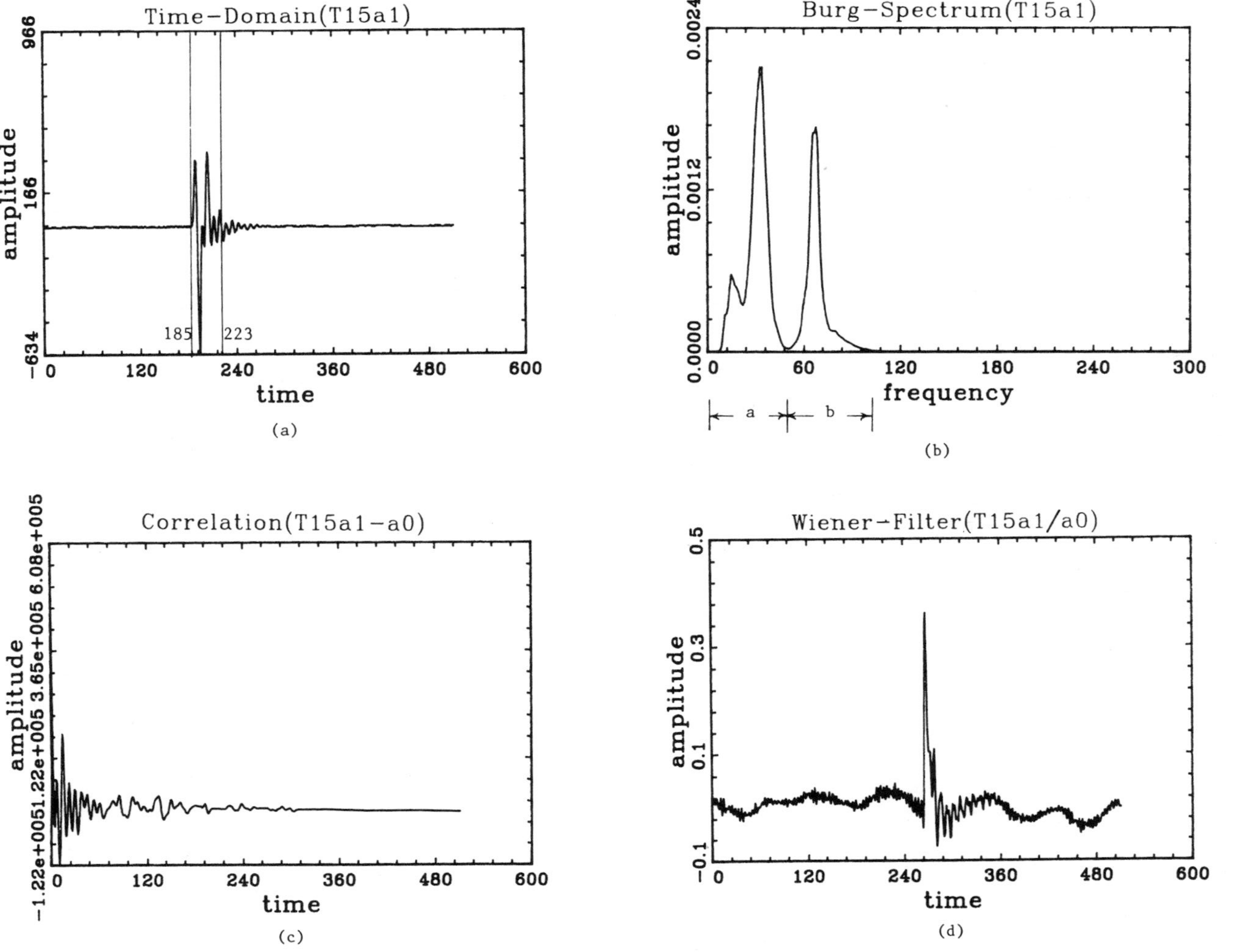


**Figure 21.4** The four domains: (a) time domain signal; (b) Burg's maximum entropy spectrum (a, low-frequency part; b, high-frequency part); (c) correlation structure; (d) impulse response from Wiener filter.

$P_1(t)$ the forward propagation path impulse response, $h_1(t)$ the impulse response of the scatterer of interest (e.g., a defect), $P_2(t)$ the return propagation path impulse response, and $T_2(t)$ the backward transducer impulse response. By letting $x(t) = u(t) * T_1(t) * T_2(t)$ and $h(t) = h_1(t) * P_1(t) * P_2(t)$, Eq. (21.2) is reduced to Eq. (21.1). $x(t)$ is considered as a source signal. The impulse response $h(t)$ obtained by deconvolution that removes the source signal from the received pulse echo presumably provides a complete characterization of the material tested. Thus defects should show up in the impulse response and different defects should have different impulse responses. However, the true impulse response, just like the true spectrum, is really unknown. The accuracy of the impulse response obviously depends on the deconvolution method and the background noise (Figure 21.4d).

Three versions of Wiener filter deconvolution have been examined with similar results. Taking the Fourier transform of Eq. (21.1) and adding the noise term, we have a measurement model in the frequency domain given by

$$Y(f) = X(f)H(f) + N(f) \tag{21.3}$$

where $Y(f)$, $X(f)$, $H(f)$, and $N(f)$ are, respectively, the spectra of the measured specimen, reference specimen, desired impulse response, and noise. The three Wiener filter solutions for $H(f)$ are described as follows.

*Version 1*

$$H(f) = \frac{S_{xy}(f)}{S_{xx}(f) + Q} \tag{21.4}$$

where $S_{xy}(f)$ is the cross (power) spectrum between $x(t)$ and $y(t)$, both digitized, $S_{xx}(f)$ is the auto spectrum of $x(t)$, and $Q$ is a constant set at average noise power.

*Version 2*

$$H(f) = \frac{Y(f)X^*(f)}{X(f)X^*(f) + Q} \tag{21.5}$$

where $X^*(f)$ is the complex conjugate of $X(f)$.

*Version 3*

$$H(f) = \left| \frac{\sum_i a_i Z^i}{\sum_i b_i Z^i + Q} \right| \qquad Z = \exp(-j2\pi i f T_s) \tag{21.6}$$

where $T_s$ is the sampling period and $a_i$ represents the Burg coefficients when $x(t)$ and $y(t)$ are represented in $Z$ transform by an autoregressive model with coefficients computed by Burg's maximum entropy spectral analysis. $x(t)$ has $a_i$ coefficients while $y(t)$ has $b_i$ coefficients.

The second version takes the least amount of time, and the results of all three versions are quite similar.

## 7 OTHER REPRESENTATIONS

Besides the four domains of knowledge described above and illustrated in Figure 21.4, the spectrogram, a plot of spectrum amplitude of a given time segment versus frequency versus time segments, is a useful signal representation. Figure 21.5 shows the spectrograms for T15A0 and T15A1.

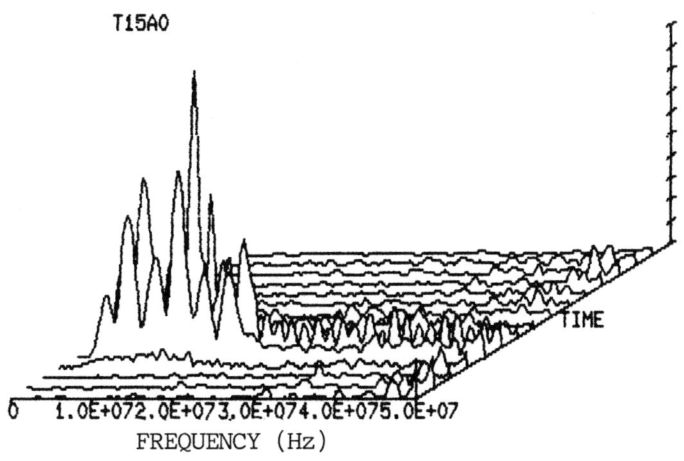


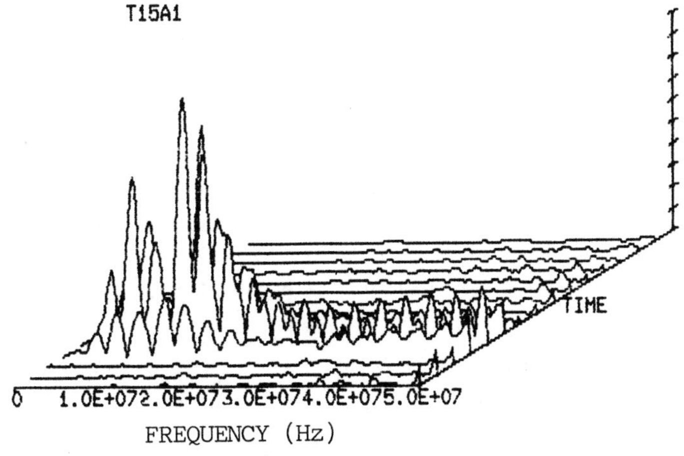


**Figure 21.5** Spectrograms for T15A0 and T15A1.

Among other representations, the wavelet transform, the Hartley transform, and the Hilbert transform have all been examined [9]. The wavelet transform, developed by Morlet and co-workers [10, 11], provides an effective multiple-scale representation of a waveform. A simple case of the wavelet transform for a signal $s(t)$ is to correlate it with a Gaussian pulse,

$$g(t) = \exp\left(-\frac{t^2}{a}\right)\exp(jw_0 t) \tag{21.7}$$

with the resulting transform given by

$$S(b,a) = \frac{1}{\sqrt{a}}\int g\left(\frac{u-b}{a}\right) S(u)\, du \tag{21.8}$$

for $a > 0$ and $b$ arbitrary. The inverse transform for reconstruction is given by

$$s(t) = k\iint g\left(\frac{t-b}{a}\right) S(b,a)\, da\, db \tag{21.9}$$

where $k$ is a constant scale factor for proper normalization. The multiple-scale representation is obtained by computing Eq. (21.7) for several values of $a$. The peak value of the wavelet for certain $a$ has been found to be a useful feature for differentiating different geometries. Figure 21.6a shows the wavelet transform of T5B3 for $w_0 = 1$ and $a = 2$, 4, 6, 8, 10, 12, 14, 16. Figure 21.6b shows the results of reconstruction using up to $a = 2$, 4, 6, 8. The reconstruction is very good even with the use of $a = 2$.

The Hartley transform for a real signal $f(t)$ is given by

$$H(f) = \int f(t)(\cos 2\pi ft + \sin 2\pi ft)\, dt \tag{21.10}$$

which differs from the Fourier transform only by the absence of $(-j)$ in front of the $\sin 2\pi ft$ term. $H(f)$ is real for a real signal and thus contains all the information of the original signal. $H(f)$ shows more difference among defect geometries which are very similar in the time domain (Figure 21.7). It is computationally as efficient as the fast Fourier transform, and in fact it can easily be obtained simply by adding the real and imaginary parts of the Fourier transform of real signal.

## 8 OTHER DECONVOLUTION METHODS AND COMPARISONS

For multipath data through a layered specimen, an important deconvolution objective is to identify the layer interfaces and to determine the thickness of certain

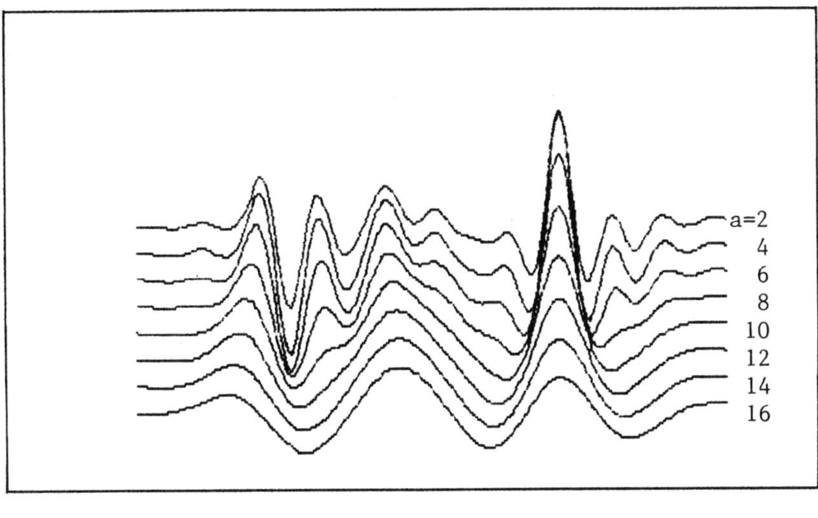


(a)

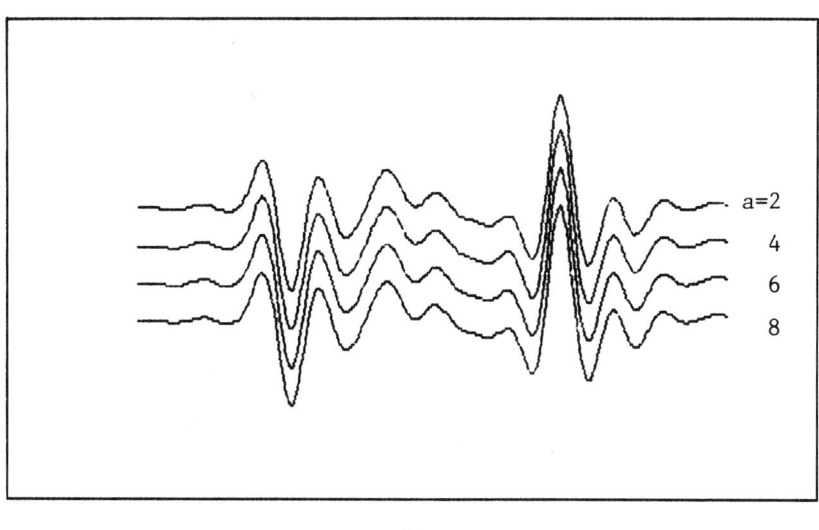


(b)

**Figure 21.6** (a) Forward wavelet transformation and (b) reconstruction results.

intermediate layers. Two deconvolution methods based on pulse shaping are presented in this section.

The first method employs a spiking filter $h$ which is determined by using linear prediction modeling of the signal and the least-squares criterion [12]. Define $Y$, $h$, and $x$ as the input signal matrix, desired spiking filter response, and reference

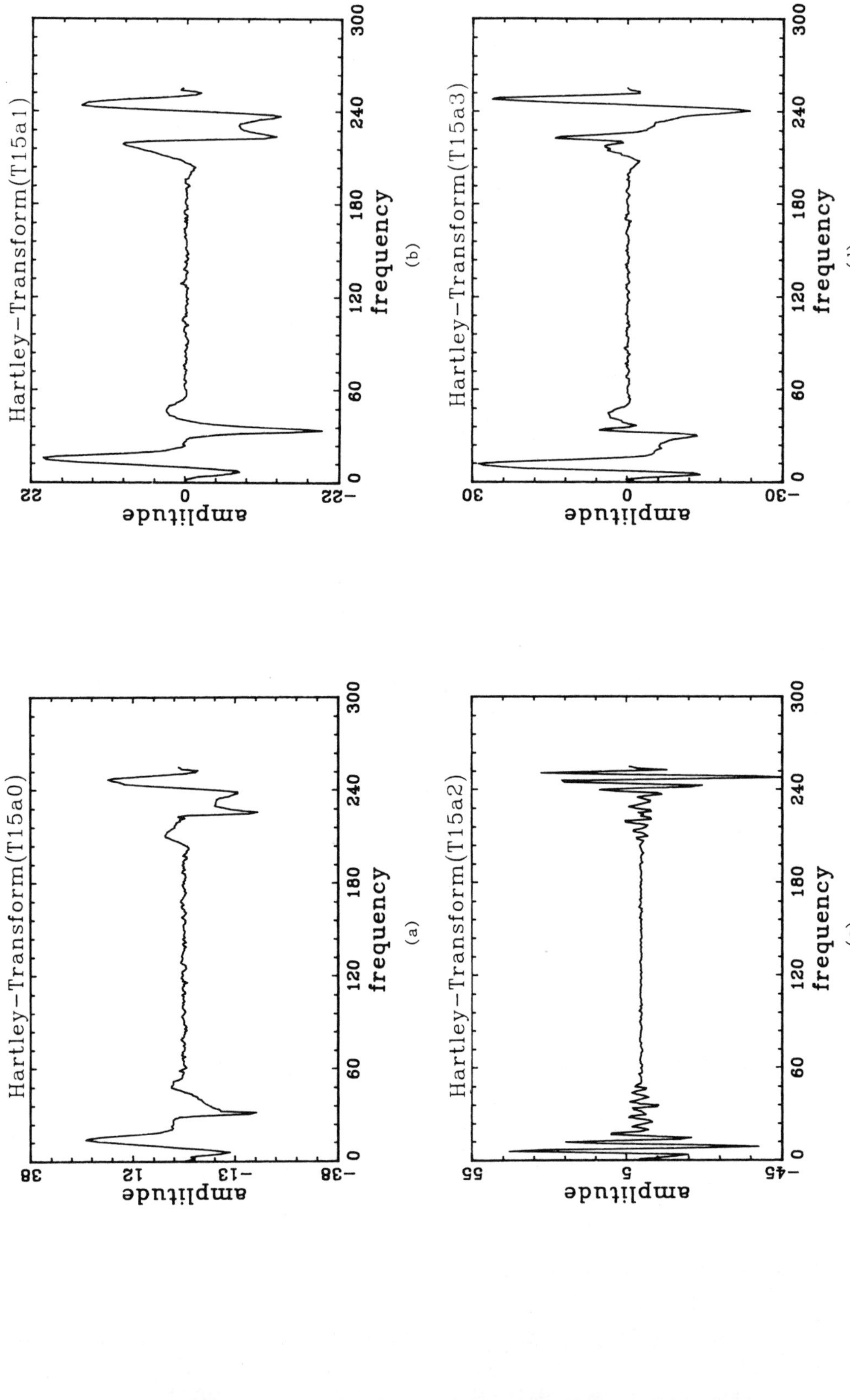

**Figure 21.7** Hartley transforms for (a) T15A0, (b) T15A1, (c) T15A2, and (d) T14A2

waveform, respectively.

$$Y = \begin{bmatrix} y_0 & 0 & \cdots & 0 \\ y_1 & y_0 & \cdots & 0 \\ \vdots & & & \\ y_n & & \cdots & y_{N-M} \\ 0 & & & \\ \vdots & & & \\ 0 & & \cdots & y_N \end{bmatrix}, \qquad h = \begin{bmatrix} h_0 \\ \vdots \\ h_M \end{bmatrix}, \qquad x = \begin{bmatrix} x_0 \\ x_1 \\ \vdots \\ x_{N+M} \end{bmatrix}$$

Then the solution is

$$h = (Y^T Y)^{-1} Y^T x = R^{-1} \gamma$$

where $R = Y^T Y$ is the autocorrelation matrix and $\gamma = Y^T x$. Figure 21.8 shows an example using simulated signals,

$$f_1(t) = f_r(t) + 0.6 f_r(t - 100)$$
$$f_2(t) = f_r + 0.3 f_r(t - 10) + 0.6 f_r(t - 100)$$

where $f_r(t)$ is a single pulse. The deconvolved results $x_i = f_i * h$, $x_2 = f_2 * h$ are shown also in Figure 21.8. A program listing of the use of the spiking filter is given in Appendix C.

The second method makes use of the recursive pulse-shaping procedure due to Simpson [13]. The algorithm outline is presented in Appendix D. Figure 21.9 shows a simulated multiple pulse being shaped to three lines with no background variations. This second method is called time domain deconvolution (TDD) method. The following is a comparison of deconvolution methods discussed in this section and Section 6.

For the Wiener filter deconvolution, and for the same 512-point waveform, version 1 takes 6 minutes, version 2 takes 1.5 minutes, and version 3 takes 15 minutes on the same IBM personal computer (PC). Version 1 requires two correlations to be processed while version 3 requires two Burg's coefficient extractions, and thus both versions take much more time than version 2. The shortcoming of the Wiener filter is that it deconvolves signals in the frequency domain, and indirectly that causes reduction of resolution for deconvolved signals in the time domain.

The time domain deconvolution shows good resolution in the time domain. But its takes as much as 20 minutes to do 200 recursions on a VAX 11/750 for 512 points. In practical use the TDD is unduly sensitive to the selection of the starting function. Furthermore, the approach is probably too conservative in the application of the convergence criterion (Appendix D). These shortcomings limit the usefulness of TDD.

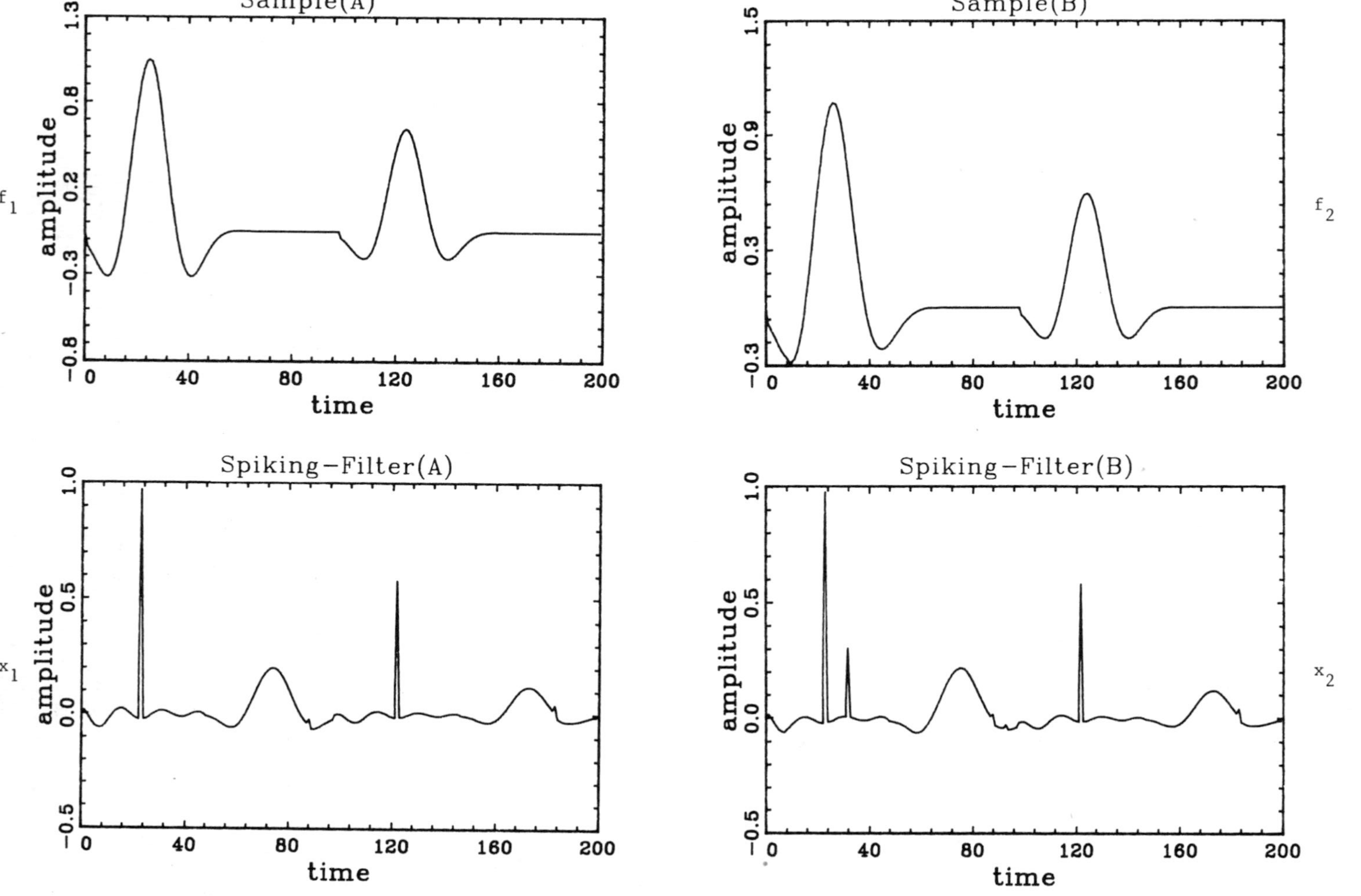


**Figure 21.8** Spiking filter deconvolution. $f_1(t) = f_r(t) + 0.6f_r(t-100)$; $f_2(t) = f_r(t) + 0.3f_r(t-10) + 0.6f_r(t-100)$; $x_1 = f_1 * h$; $x_2 = f_2 * h$

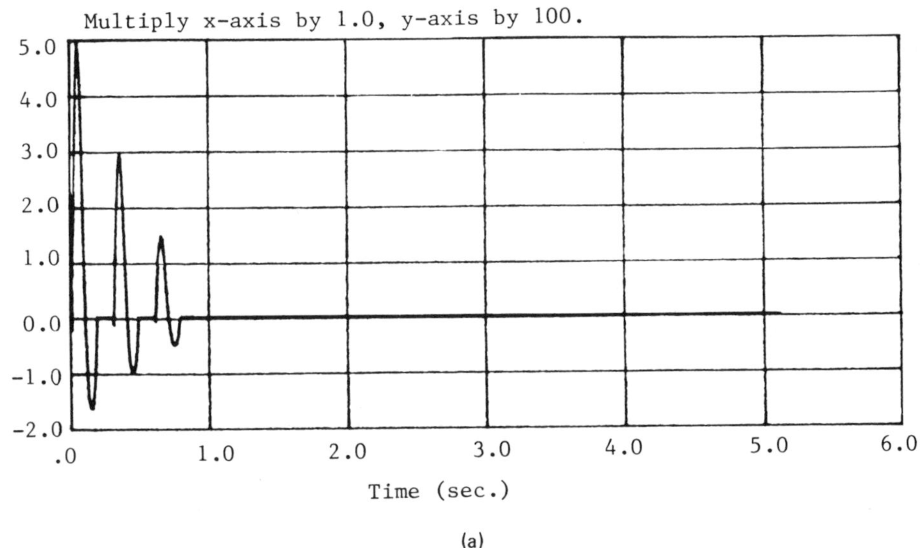


(a)

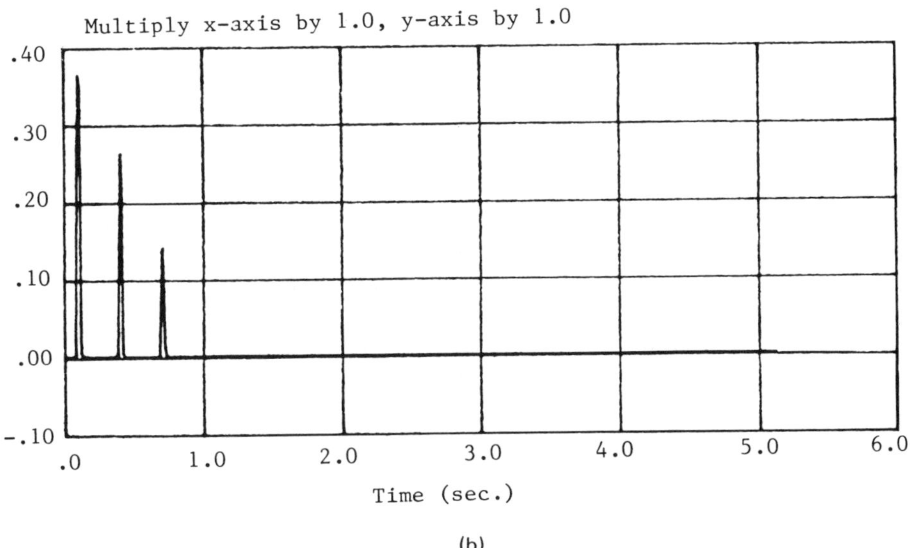


(b)

**Figure 21.9** (a) Simulated multiple pulse; (b) result of time domain deconvolution.

The spiking filter, on the other hand, is a linear prediction technique using the least-squares criterion. It shows good resolution in the time domain. It takes 8 minutes running time on the VAX to extract the inverse filter from the reference signal. Using this inverse filter, we can find the deconvolved waveform for any waveform under consideration, simply by convolving it with the inverse filter. The convolution procedure itself takes only a few seconds on a PC.

The computation times cited in the above paragraphs obviously need improvement for real-time processing and interpretation purposes, but they do give us an idea of the relative efficiency of several techniques.

## 9 PATTERN RECOGNITION WITH ULTRASONIC PULSE-ECHO SPECTROSCOPY

For the ultrasonic pulse-echo data obtained with the acquisition system of Figure 21.1, a complete statistical pattern recognition system has been developed [8]. The major components of a recognition system are feature extraction and classification. Features must be extracted that provide effective discrimination among pattern classes considered, in this case various defect geometries. A classification procedure must be selected that will minimize the number of incorrect decisions or interpretations.

The effective features identified are listed in the following. Features can be evaluated according to the Fisher ratio, which is the ratio of the squared difference between the means to the sum of the variances, with the assumption of a Gaussian distribution.

| Feature | Fisher or scatter ratio |
|---|---|
| Kurtosis (time domain) | 4.5 |
| Skewness (time domain) | 4.18 |
| Amplitude ratio (ratio of cumulative sum of a pulse echo to the cumulative sum of a reference, with sum taken over the amplitudes) | 2.3 |
| T15/T5 amplitude ratio (i.e., ratio of the two amplitude ratios measured by transducers T15 and T5) | 3.129 |
| Frequency ratio (ratio of the area of the power spectrum at the upper frequency band to that at the lower frequency band, with the bands adjacent) | 2.67 |
| Maximum or peak correlation value | |
| Maximum value in wavelet transform | 1.45 |
| Duration through segmentation algorithm | |
| Duration measured from the major pulse in the impulse response | |
| Second moment of the Hartley transform | |
| Bandwidth | |
| Second moment of Burg's spectrum | |
| RMS value of the correlation function | |
| etc. | |

*Note*: The scatter ratio roughly shows the feature effectiveness, but it is not reliable as the individual feature may not be Gaussian-distributed. Normally, a larger scatter ratio shows that the feature is more useful.

It should be pointed out that the duration in terms of the number of data points obtained through segmentation provides a good indication of the angular variation in angle-cut flaws. The following are some numerical results:

| | |
|---|---|
| T5B1 (10°) | Duration = 47 |
| T5B2 (20°) | 92 |
| T5C1 (10°) | 69 |
| T5C2 (20°) | 86 |
| T5D1 (10°) | 54 |
| T5D2 (20°) | 131 |

The angle cut flaws cause the reflected signals to spread along the time domain. Thus a flaw with a larger angle will have a larger event duration, as shown above.

The dimension of feature sets chosen is usually greater than two. In pattern recognition, it is often necessary to have two-dimensional display of all the data considered. Data belonging to the same class must be grouped together, whereas those from different classes must be separated as much as possible. These are the typical problems in cluster analysis. One method, due to Sammon [14], provides a nonlinear mapping of multidimensional data onto a two-dimensional space with the data structure essentially preserved [15]. Figure 21.10 shows the result of nonlinear mapping of T15 data in which data belonging to the three defect geometries are completely separated.

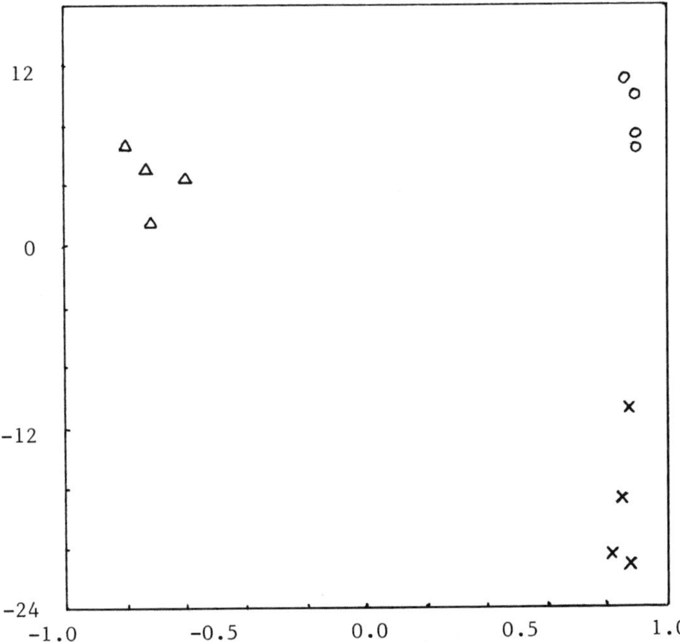


**Figure 21.10** Nonlinear mapping of T15 data onto a two-dimensional space, showing that samples from the three are completely separated. (△) Angle cut; (○) circles; (×) flat top.

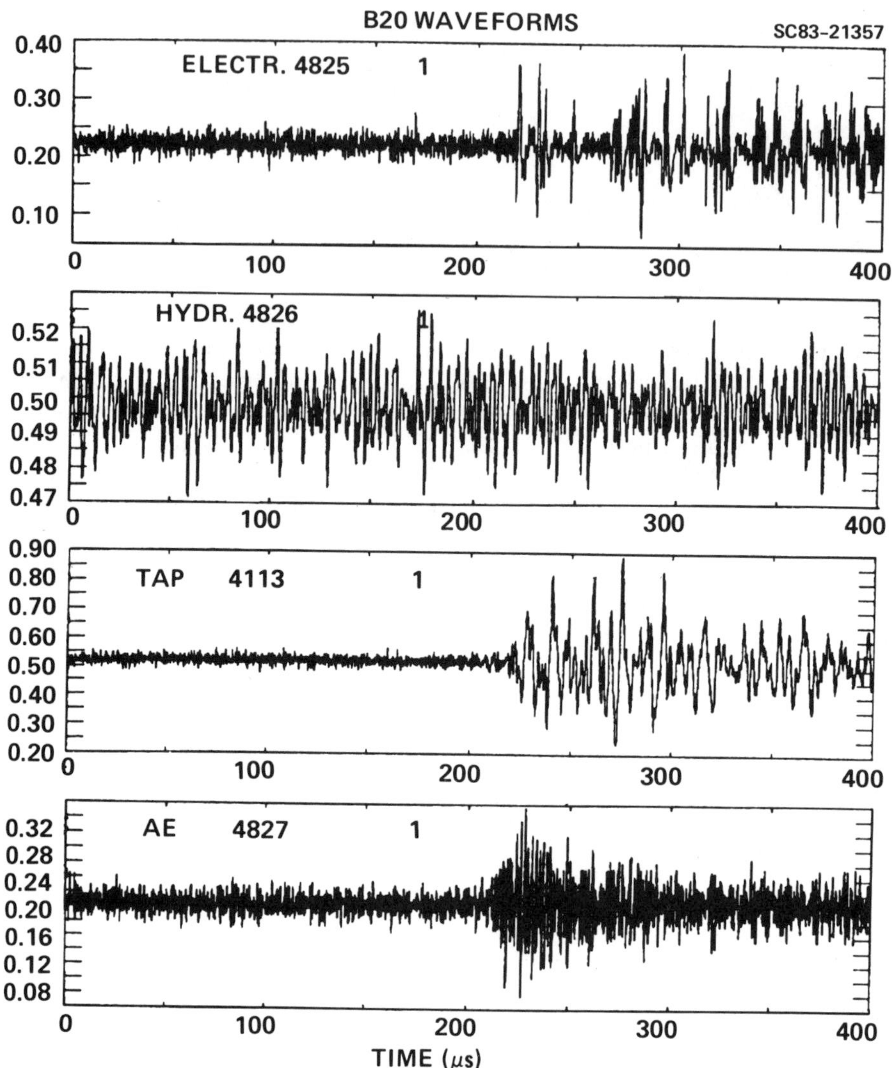


**Figure 21.11** Acoustic emission waveform B20. (Courtesy of Dr. Lloyd Graham.)

For the features selected, the best classification procedure is found to be one that employs multistage classification. In the first stage the peak of the correlation between a specimen and the reference (one without defect) is examined. If it is positive, the decision is that the geometry has a flat top or a circular hole. Otherwise, it has a declining top for which the angle must be determined among a finite number of choices. The discrimination between flat and circle is based on the kurtosis and frequency ratio features. For angular cuts, the pulse duration can be related to the angle empirically. The size or the width of the artificial flaw can be estimated from the impulse response resulting from Wiener filter deconvolution [7].

The multistage classification identifies correctly the defect geometry of T15 data and estimates accurately the angle and size.

## 10 PATTERN RECOGNITION IN ACOUSTIC EMISSION EXPERIMENTS

Pattern recognition methods have been used for classifying acoustic emission (AE) signals according to their source types [16, 17]. Simple time and frequency domain features of the AE waveforms are used in the classification to distinguish one type from another. Sources of AE in the monitoring application considered are crack growth, crack face rubbing, fastener fretting, mechanical impacts, electrical transients, and hydraulic noise, as shown in Figures 21.11 and 21.12 for B20 and B17 waveforms, respectively. For each AE event, two simultaneous waveforms were recorded, one from each transducer. For the waveform of an individual transducer, the major features are arrival time and amplitude of the peak, amplitude in six time intervals around and after the trigger, energy in several frequency bands, and total energy in the spectrum. The two-transducer features include ratio of peak amplitudes, ratio of energy in signals, ratio of energies in frequency bands, etc.

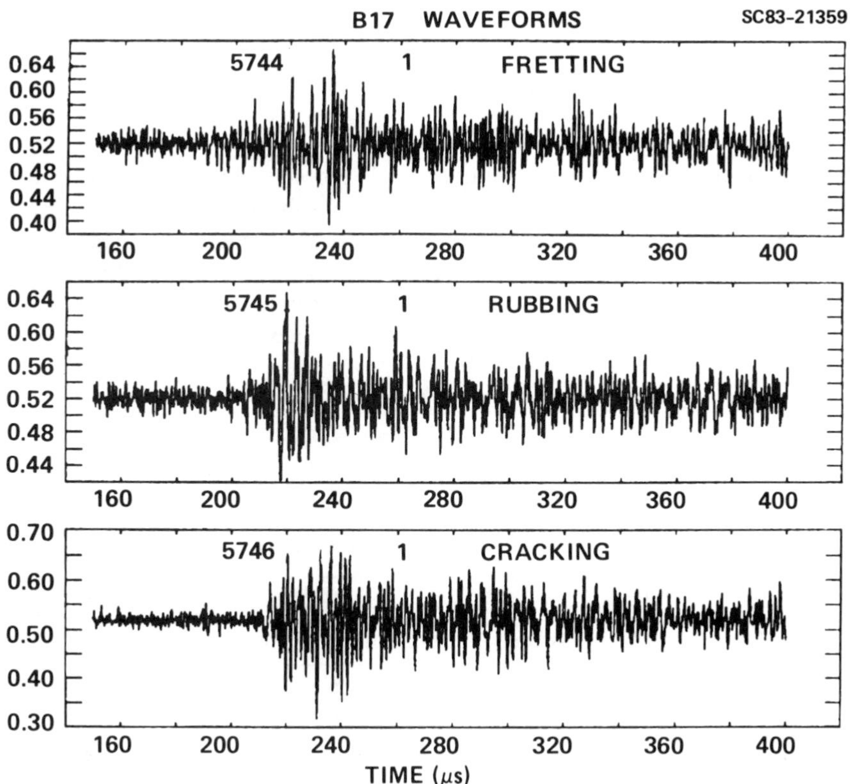


**Figure 21.12** Acoustic emission waveform B17. (Courtesy of Dr. Lloyd Graham.)

The mean and variance of each feature are computed, from which the distance of each event from a given class is computed. The decision is based on the minimum distance. The distance squared of event $j$ from class $i$ is given by

$$D_{ij}^2 = \sum_k \frac{(x_{jk} - \mu_{ik})^2}{\sigma_{ik}^2}$$

where $\mu_{ik}$ is the mean of the $k$th feature of the $i$th class, $\sigma_{ij}^2$ is the corresponding variance, and $x_{jk}$ is the value of the $k$th feature of event $j$. The distance from a given AE event to each class is computed and the event is assigned to the nearest class.

Cluster analysis was also performed, based on a function which measures the density of events in the neighborhood of a given event in the vector space of the features. The idea is somewhat similar to the $k$-mean algorithm in cluster analysis. Two data sets, B20 and B21, were used in the clustering experiment, which employed two different feature sets with 10 features in each set. The results, shown in Table 21.1, indicate little difference in accuracy between the two feature sets selected.

## 11 CONCLUDING REMARKS

Application of signal/image processing to NDE problems has recently received considerable attention. It provides many useful NDE solutions without going through the complex wave equations. The advances in signal processing in other areas such as seismic, speech, sonar, and radar signals described in other chapters of this book also greatly affect the progress of signal processing in NDE. A forthcoming book [18] on this subject provides an up-to-date review of research progress in this area.

**Table 21.1** Separation of AE Classes by Cluster Analysis

| Test description | No. of events in data set | | | | Percent correct | |
|---|---|---|---|---|---|---|
| | Total crack | Rej. crack | Total noise | Rej. noise | Crack | Noise |
| B20, algorithm no. 1 | 270 | 6 | 485 | 5 | 81.8 | 87.1 |
| B20, algorithm no. 2 | 270 | 6 | 485 | 5 | 88.3 | 91.0 |
| B21, algorithm no. 1 | 239 | 6 | 514 | 11 | 93.6 | 86.1 |
| B21, algorithm no. 2 | 239 | 6 | 514 | 11 | 91.4 | 86.5 |

After Ref. [17]. Courtesy of Dr. Lloyd Graham, Rockwell International.

## ACKNOWLEDGMENTS

I am indebted to Dr. Otto R. Gericke of the U.S. Army Materials Technology Laboratory for support, advice, and encouragement in the work reported in this chapter. Thanks are also due to Dr. Lloyd Graham of Rockwell International for kindly supplying the photos for Figures 21.11 and 21.12 and for his permission to use these figures as well as Table 21.1.

## REFERENCES

1. J. Szilard, *Ultrasonic Testing*, Wiley, New York (1982).
2. D. O. Thompson and D. E. Chimenti, eds., *Review of Progress in Quantitative Nondestructive Evaluation*, vols. 5A and 5B, Plenum, New York (1986).
3. C. H. Chen, On a segmentation algorithm for seismic signal analysis, *Geoexploration, 23*: 35–40 (1984).
4. O. R. Gericke, Determination of the geometry of hidden defects by ultrasonic pulse analysis testing, *J. Acoustic Soc. Am., 35*: 364–368 (1963).
5. O. R. Gericke, "Theory and NDT Applications of Ultrasonic Pulse-Echo Spectroscopy," Proc. Symposium on the Future of Ultrasonic Spectroscopy (P. M. Reynolds ed.), paper no. 1 (October 1970).
6. O. R. Gericke, "Ultrasonic Spectroscopy," invited lecture given at Southeastern Massachusetts University, North Dartmouth, Massachusetts (June 1984).
7. C. H. Chen, A Signal Processing Study of Ultrasonic Nondestructive Evaluation of Materials, Report MTL TR 87-11, prepared for U.S. Army Materials Technology Laboratory under Contract DAAL04-86-k-002 (February 1987).
8. C. H. Chen, Pattern analysis for the ultrasonic nondestructive evaluation of materials, *Int. J. Pattern Recogn. Artif. Intell., 1*(2): 251–260 (July 1987).
9. C. H. Chen, High Resolution Spectral Analysis NDE Techniques for Flaw Characterization, Prediction and Discrimination, prepared for U.S. Army Materials Technology Laboratory under Contract DAAL02-86-C-0125 with Information Research Laboratory (April 1987).
10. J. Morlet, Sampling theory and wave propagation, *Issues on Signal/Image Processing for Underwater Acoustics*, (C. H. Chen, ed.), Springer-Verlag, New York (1983).
11. R. Kronland-Martinet, J. Morlet, and A. Grossman, Analysis of sound patterns through wavelet transforms, *Int. J. Pattern Recogn. Artif. Intell., 1*(2): 273-302 (July 1987).
12. S. J. Orfanidis, *Optimum Signal Processing, an Introduction*, Macmillan, New York (1985).
13. W. A. Simpson, Jr., Time domain deconvolution: A new technique to improve resolution for ultrasonic flaw characterization in stainless steel welds, *Mater. Eval., 44*: 998–1003 (July 1986).
14. J. W. Sammon, Jr., A nonlinear mapping for data structure analysis, *IEEE Trans. Comput. C-18*: 401–409 (May 1969).
15. C. H. Chen, Computerized pattern analysis, *Encyclopedia of Microcomputers* (A. Kent, ed.), Marcel Dekker, New York (1987).

16. L. J. Graham and R. K. Elsley, AE source identification by frequency spectral analysis for an aircraft monitoring application, *J. Acoust. Emission*, *2*(112): 47–55 (1983).
17. R. K. Elsley and L. J. Graham, "Pattern Recognition in Acoustic Emission Experiments," Proc. SPIE Symposium on Pattern Recognition and Acoustic Imaging, Newport Beach, California (February 1987).
18. C. H. Chen, ed., *Proceedings of NATO Advanced Research Workshop on Signal Processing and Pattern Recognition in Nondestructive Evaluation of Materials*, held August 1987 in Quebec City, Canada, Springer-Verlag, New York (1988).

# 22

# Signal Processing in Telecommunications

KAVEH PAHLAVAN Department of Electrical Engineering, Worcester Polytechnic Institute, Worcester, Massachusetts

## 1 INTRODUCTION

There are numerous applications of signal processing in telecommunications. Many areas of signal processing such as adaptive filtering, digital filtering, voice and image compression, and design of VLSI signal processors have been motivated with application in telecommunications. As a result, several topics discussed in this book are also treated in the telecommunication literature. Considering the variety of interesting signal processing applications in telecommunications, it is difficult to cover all topics in one chapter. The emphasis in this chapter is on the applications of signal processing algorithms in modem design technology.

The purpose of telecommunications is to transfer information—usually data, voice, or video—between two locations via electrical signals. Problems associated with telecommunications are divided into the following three subareas: pretransmission processing or source coding, transmission, and networks.

The purpose of source encoding is to minimize the volume of transferred information to save transmission time and expenses. Speech and image coding techniques are usually treated in signal processing and communication literature; the data compression is a pure communication problem. The signal processing techniques, such as linear prediction coding, vector quantization, pulse code modulation, differential pulse code modulation, and delta modulation, are parts of this class of problems which are partly covered in the other chapters of this book. However, for further information on speech coding the reader can refer to [1, 2], for voiceband encoding techniques to [3, 4], for image coding to [5, 6], and for data compression to [7].

Basic queuing theory and various multiple accessing techniques and protocols are the main body of the problems in networks and make up a separate part of communication literature which is normally referred to as computer communication or communication networks. The only signal processing issue regarding networks is related to echo suppression and echo cancellation in telephone networks, which are treated in [8–10].

A typical transmission link consists of the following three basic elements: channel, transmitter, and receiver. For two-way communication, normally a transmitter and a receiver are combined in a device called a modem (modulator-demodulator). The channel could be a telephone line, a coaxial cable, or a fiber-optic line, or it could be a radio link over troposphere, ionosphere, microwave line-of-sight, or satellite. Regardless of the difference in physical characteristics between these channels, the basic principles of digital communications are the same for all channels. The binary input data after source coding are mapped to pulse waveforms at a given frequency and transmitted through the channel. The channel delivers a distorted form of the transmitted signal to the receiver, and the receiver recovers the transmitted digits from the received signal.

## 2 SIGNAL PROCESSING IN MODEMS

Today, digital communication is the dominant means of transferring data and most of the nontrivial signal processing in telecommunications has been developed in this area. Digital communication is more immune to noise and provides flexibility in combining various information sources such as video, voice, and data in a unified network. Throughout this chapter only modems for digital communications are considered.

Figure 22.1 shows a block diagram of a transmitter and a receiver for a typical voiceband modem. The transmitter consists of a coder, a pulse-shaping filter, and a modulator. The receiver consists of a decoder, the second part of the pulse-shaping filter, a timing recovery circuit, a phase recovery circuit, a demodulator, and a box called an adaptive processor.

Groups of transmitted data bits are mapped into a complex digit $a_k$. Associated with each complex digit, a complex waveform of shape $h_1(t)$ is formed which is modulated over a carrier frequency $\omega_0$. The mathematical representation of the transmitted signal is given by

$$y_t(t) = \mathrm{Real}[y(t)e^{j\omega_0 t}]$$

where

$$y(t) = \sum_k a_k h_1(t - kT), \tag{22.1}$$

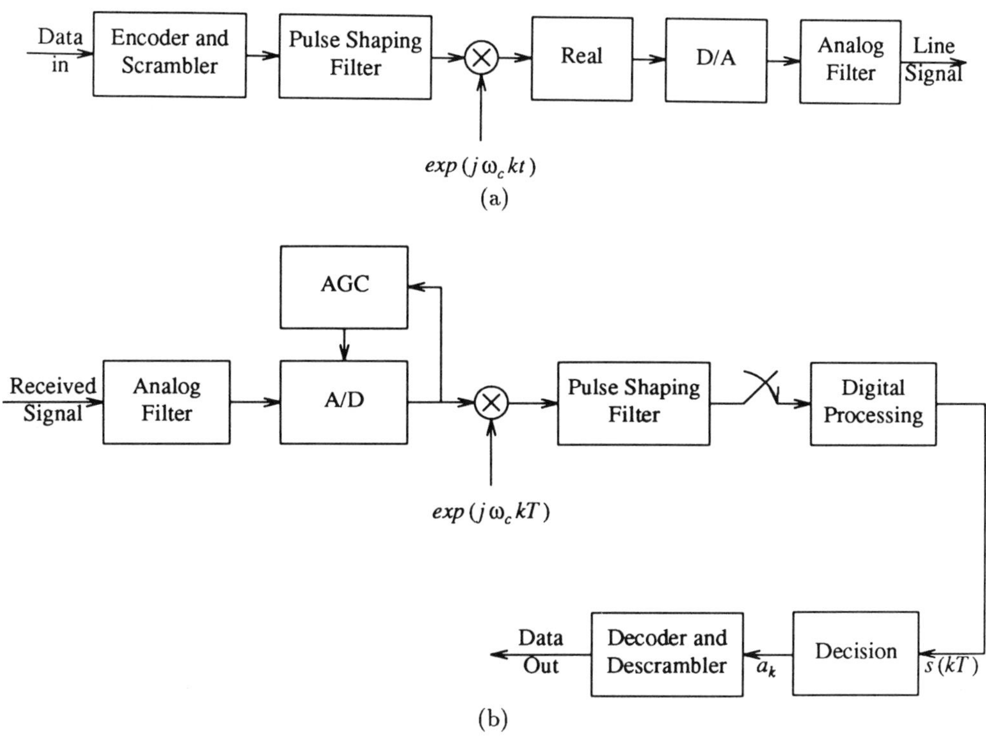


**Figure 22.1** Block diagram of a typical Voiceband modem: (a) transmitter; (b) receiver.

The representation of all possible complex digits $a_k$ in the complex plane is referred to as the signal constellation.

At the receiver, an automatic gain control (AGC) circuit normalizes the received input power so that the fluctuations of the received signal power do not result in erroneous decisions on the detected symbols. Then the signal is demodulated and passed through the receiver pulse-shaping filter, $h_2(t)$. The signal after pulse shaping is processed with an adaptive processor, which counteracts the instantaneous effects of channel amplitude and phase distortions. This part of the modem is used either for intersymbol interference (ISI) cancellation or echo cancellation. The processed signal is then sampled according to the timing provided by the timing recovery circuit to detect the transmitted symbol. The phase recovery circuit provides the reference phase for tracking the transmitted carrier phase.

In the above description the scrambler and the coder were not mentioned. Scrambling provides a source of pseudodata for adjustment of the timing and AGC during the silence time between bursts of transmitted data. The error-correcting coding works by adding parity bits to the scrambled data for possible gain in the performance.

## 2.1 Implemental Considerations

Data communications over voiceband telephone channels involves lower data rates, and the channel does not fade. As a result, most of the advanced signal processing algorithms are first applied to voiceband modems and then extended to radio communications and other applications. This section concentrates on the digital implementations of voiceband modems in which most complex signal processing algorithms are used [11]. In practice, the sampling rate for digital signal processing in these modems is at least four times the symbol rate.

Because of the rapid growth of semiconductor technology, the past two decades witnessed simultaneous reductions in the size and price of the voiceband modems. Digital technology and large-scale integrated (LSI) devices resulted in more cost-efficient implementation with lower internal processing noise. Microprocessors facilitated the design and promoted the inclusion of ever-growing network control and management features inside the modem box. Finally, superior number-crunching capabilities of digital signal processors allow low-cost implementation of sophisticated trellis code modulation techniques, which require enormous computational power for decoding.

In the past decade or more, implementation of voiceband modems has rapidly changed. In the mid-1970s, custom MOS/LSI circuits were used for 9600 bits per second (bps) modems over four-wire telephone lines [12, 13]. In the late 1970s, microprocessor-based designs were becoming more interesting [14–16]; however, off-the-shelf microprocessors were too slow for the implementation of a 9600 bps modem and special-purpose microprocessors were too expensive for the modem design [17]. Hardware implementation of a typical 9600 bps modem in this period included a bit slice processor, a fast multiplier (number cruncher), and an ordinary microprocessor for additional features in the networks.

Today, for low-speed modems, simple general-purpose microprocessors such as the Intel 8085 or Motorola 6800 are used. For higher speeds, inclusion of other digital signal processor chips such as the TI TMS320 or NEC 7720 [18] or custom VLSI chips [19, 20] is customary. Surveys of modem integrated circuits (ICs) and digital signal processing chips are available in [21, 22].

The general-purpose VLSI digital signal processors are designed to provide a low-cost solution to signal processing problems associated with the design of adaptive filters, particularly for applications in modems. The architecture of these processors is suited for fast multiply and add operations, which are normally required for tap gain adjustment of the equalizers. The size of the internal random access memories (RAMs) is suited for storing tap gains of an equalizer and other variables used for AGC, timing, and phase recovery. Two pointer registers are used efficiently to keep the track of both tap gains and the tapped value in an equalizer.

A typical architecture for a modem using two microprocessors is shown in Figure 22.2. The general microprocessor can be used for network control, scrambling, mapping, detection, and modulation, and the digital signal processor for pulse shaping, demodulation, equalization, timing recovery, phase recovery, and AGC.

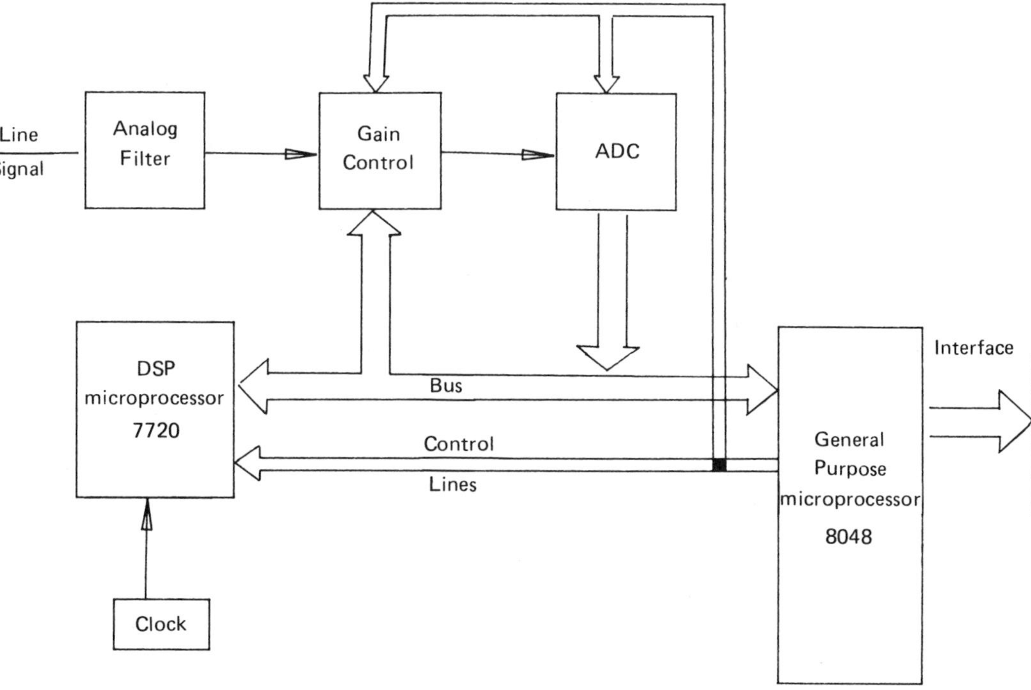


**Figure 22.2** Microprocessor-based architecture of a voiceband modem.

## 2.2 Automatic Gain Control

In a multiamplitude system, the received signal must remain normalized. Otherwise, proper detection of various levels is impossible. The principles of automatic gain control are quite simple; the received power is measured and compared with a reference. The difference is used for control of the received power level. Figure 22.3 represents an AGC for a digitally implemented modem. The received signal is passed through an antialiasing filer, which eliminates out-of-band noise, and then is sampled with an analog-to-digital (A/D) converter. The samples are squared and then low-pass-filtered to provide an estimate of the power. The power is compared with a reference and the difference is used to adjust the step size of the A/D converter. In this way, the range of the received signal is adjusted with average received power, and the digital representation of the received signal remains independent of power fluctuations.

## 2.3 Phase Recovery

To achieve lower error rates, modems usually adopt coherent modulation techniques which require a phase reference from the transmitted carrier. The easiest way to provide a reference phase is to transmit some part of the carrier at the modulator and detect it at the receiver with a narrowband filter. This method wastes parts of the transmission power. To avoid this waste, the received signal should be used

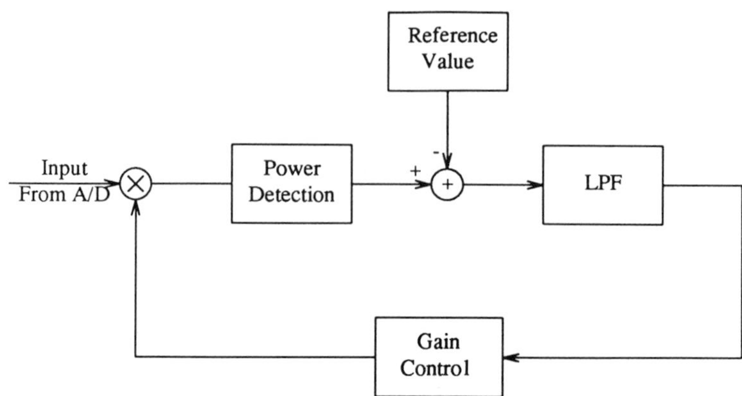


**Figure 22.3** Block diagram of the AGC circuit for a digitally implemented modem.

for recovering the phase reference. A phase-locked loop determines the carrier phase and corrects it either before the decision directly on the received point in the constellation or at the demodulator. In the first approach the loop is shorter, which results in faster tracking of the phase variations. At present, decision-aided phase-locked loops (DAPLLs) are used in most modems. The principle of operation of DAPLLs is explained in the following.

Let $\theta$ be the difference in phase between the received complex symbol after equalization and the detected complex symbol, and $\hat{\theta}$ be the input of the DAPLL used for phase adjustment of the next received symbol in the constellation. Then a second-order DAPLL adjusts the phase according to the following algorithm:

$$\varphi(n) = \varphi(n-1) + \gamma_2 e(n)$$
$$\hat{\theta}(n+1) = \hat{\theta}(n) + \gamma_1 e(n) + \varphi(n)$$

where

$$e(n) = \theta(n) - \hat{\theta}(n)$$

$\gamma_1, \gamma_2$ are the loop parameters, and $\hat{\theta}(n)$ and $\varphi(n)$ are the memories of the system. Figure 22.4 shows a block diagram of the DAPLL described above. In practice, if the phase correction is performed after equalization rather than at the demodulator, the error signal is given by

$$e(n) = \mathrm{Im}\left[\frac{s(nT)a_n}{|a_n|^2}\right]$$

where $s(nT)$ is sampled output of the equalizer and $a_n$ is the received complex symbol. More details of DAPLLs are available in [23].

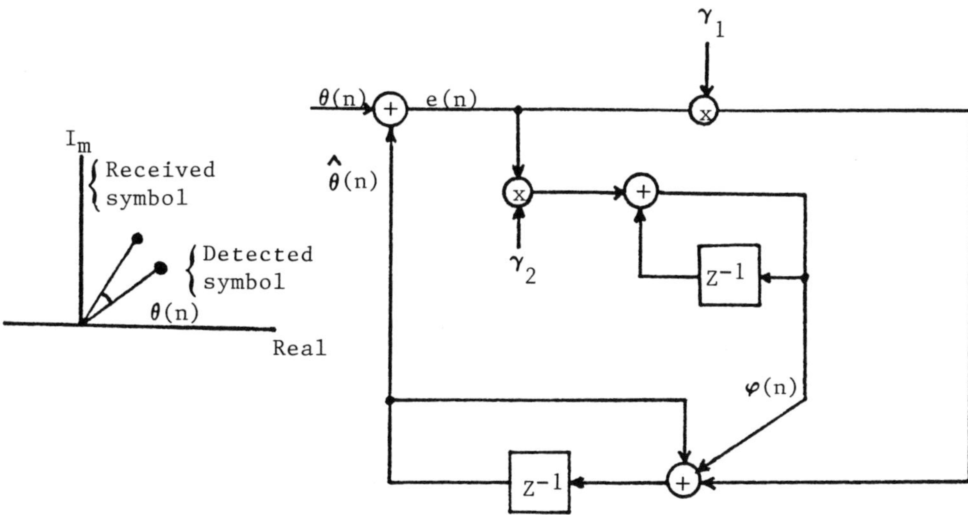


**Figure 22.4** Second-order decision-aided phase-locked loop (DAPLL) working on the phase difference between the received equalized sample and the detected point in the signal constellation.

## 2.4 Timing Recovery

In synchronous modems, proper recovery and tracking of the symbol timing are crucial for proper operation. Again, the simplest method is to transmit a pilot tone which is a harmonic of the symbol clock. The use of a pilot tone to transfer timing information is a waste of transmission power; for this reason, significant work has been done to enable derivation of timing information directly from the information-bearing signal. Several techniques are suggested and examined for this purpose.

One approach is to adopt the minimum mean square error at the equalizer output as the criterion for a decision-aided algorithm for finding the optimum sampling phase [24]. This method generally involves relatively complicated implementation and slow convergence and, consequently, has not had widespread acceptance.

A second approach is to use narrowband filters tuned to half the baud rate, followed by a square-law device and a passband filter [25]. The relationship of this method to the band-edge component maximization which yields the optimum sampler phase for infinite tap equalizers is analyzed in [26, 27]. Compared to the method suggested earlier, this technique yields more economical implementation while giving comparable performance. Extracting timing from band-edge components of the received signal is widely used in modems employing standard baud rate equalizers.

Figure 22.5a shows a block diagram of the timing recovery circuit described above. The received signal is passed through two filters centered at $f_c - 1/2T$ and $f_c + 1/2T$; the outputs of the filters are multiplied, and the frequency component at $1/T$ is filtered to extract timing information. Figure 22.5b shows a digital implementation of the circuit. The clock used for sampling is driven from the output

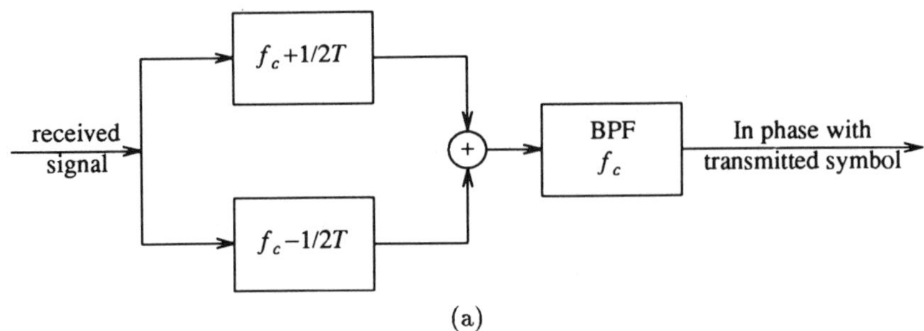


(a)

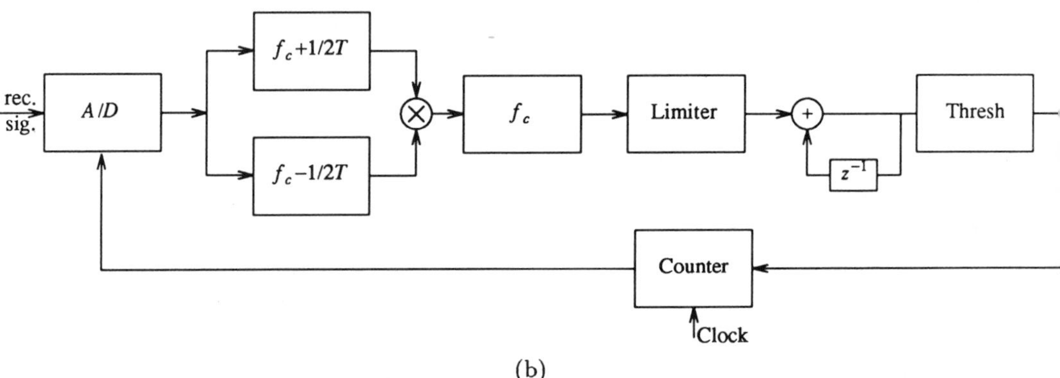


(b)

**Figure 22.5** Band-edge timing recovery circuit: (a) conceptual structure; (b) circuit for digital implementation.

of a long counter. The output of the filter at $f_c$ is quantized in 1 bit to cancel fluctuations of the amplitude due to the multilevel received symbols. The quantized output is integrated and compared with a reference value, and the difference is used to adjust the size of the counter. The size of the counter controls the sampling of the A/D converter.

A third approach is to use a fractionally spaced equalizer (FSE) rather than standard symbol-spaced equalizers [28–31]. A modem using FSE has minimal sensitivity to the timing error, and simple timing recovery circuits would result in satisfactory performance. A detailed analysis of the behavior of FSEs will be given later in this chapter. Other interesting work on timing recovery is available in [32–37].

## 2.5 Modulation and Coding

Signal processing parts of a modem are sensitive to the type of modulation and coding used for communications; as will be shown later, the designs of pulse-shaping filters and equalizers for several popular modulations are different. This section provides an introductory overview of the modulation and coding problems that are involved in the analysis of signals processing aspects of modems. The communication theory aspects of modulation techniques deal with calculation of the probability of error for various channel characteristics and are treated in the statistical communication theory literature [38].

The information digits are modulated into various amplitudes, frequencies, phases, or their combinations. Today, standard 300 and 1200 bps voiceband modems over four-wire telephone lines use frequency shift keying (FSK); 2400 bps modems uses noncoherent phase shift keying (PSK); 4800 bps and higher rate modems use quadrature amplitude modulation (QAM). For speeds above 14,400 bps, combined coding and QAM modulation is applied, and the highest data rate is 19,200 bps. As the data rate increases, the effects of channel impairments become more visible; to counteract these impairments, more sophisticated application of signal processing algorithms will be required. As a result, the most interesting signal processing algorithms are used in very high speed modems in which QAM is the standard modulation technique.

## QAM Modulation

Direct translation of the multiamplitude and multiphase signals from baseband to the carrier frequency results in two sidebands, which wastes the bandwidth. In double-sideband pulse amplitude modulation, the maximum normalized symbol rate is 1 (symbol/sec)/Hz. To increase the bandwidth efficiency, single-sideband and vestigial-sideband modulations may be used, which can have up to 2 (symbols/sec)/Hz. Another modulation technique to transmit 2 (symbols/sec)/Hz is quadrature amplitude modulation, in which two orthogonal channels are formed over the same bandwidth. Orthogonality of the channels is achieved by modulating over a sine and a cosine waveform. QAM is easier to implement and, consequently, is the most popular bandwidth-efficient modulation technique for voiceband and radio modems.

For QAM modulation the transmitted symbols can be regarded as complex numbers whose real and imaginary parts are transmitted over two orthogonal channels. Figure 22.6 represents the standard CCITT (Consultative Committee on International Telegraph and Telephone) signal constellations for 14,400 bps modems for four-wire telephone lines; each 6 data bits form a complex symbol. The symbol rate is 2400 symbols/sec, which results in a data rate of 14,400 bps.

Partial response signaling is also very popular in modem design, especially for radio communications. Transmission of 2 (symbol/sec)/Hz by QAM requires ideal pulse-shaping filters; in partial response signaling, the same bandwidth efficiency is achieved with a realizable filter. In partial response signaling, not only is the pulse shaping different from that in standard QAM but also, as will be shown later, the equalizer tap gains are different.

In the implementation of modems, the real and imaginary parts of the transmitted symbols are modulated over sine and cosine functions to preserve their orthogonality. When both of the orthogonalized channels are transmitted at the same time, the peak transmitting power is at the peak of the pulse-shaping filter. In many radio communication channels it is desirable to transmit at full power; at the same time, the power amplifiers have certain nonlinearities at their peak power. Therefore, in many cases the modulation is staggered so that the transmitted pulses for the real and imaginary parts of the symbols have a $T/2$ time delay. Staggering will reduce the peak power in transmission, allowing more room for transmission of power in the linear part of the amplifiers. As will be shown

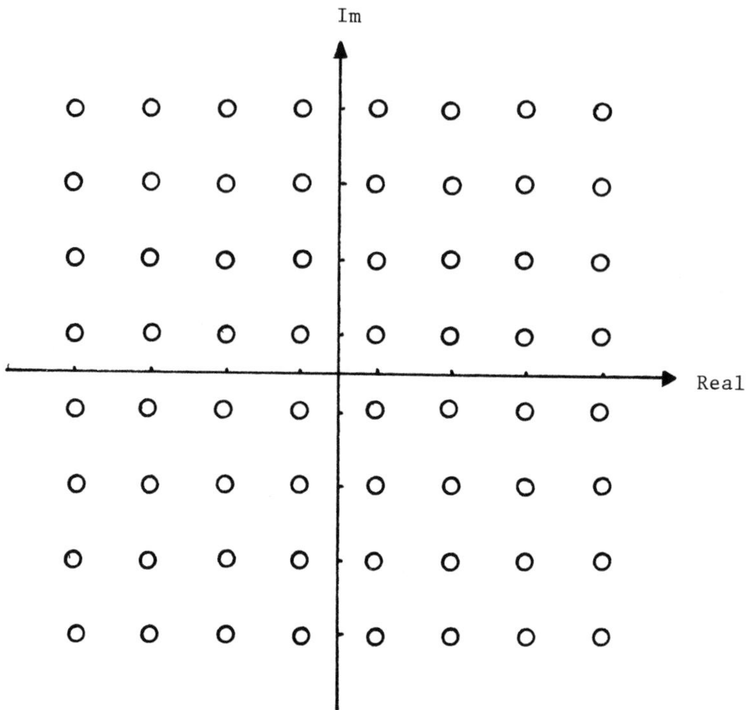


**Figure 22.6** The CCITT recommended signal constellation for 14400 bps modems over four-wire telephone lines.

later, the equalizers used for staggered modulations are different from standard equalizers.

## Trellis Code Modulation

In classical communication systems, error control is accomplished by coding the input data bits and then modulating the coded signal. To keep the data rate unchanged, one should compensate for the bits used for error correction by increasing the transmission rate. In band-limited channels, such as voiceband channels, an increase in transmission rate requires an increase in the number of points in the constellation, resulting in a higher transmission error rate. Until recently, it was believed that if the data rate remains the same, practical codes cannot compensate for the loss caused by an increase in the number of points. As a result, coding techniques were used mainly for radio communications over fading channels or satellite links, where the bandwidth is not the major constraint.

Recently, there has been renewed interest in coding of band-limited channels. The motivation was a new combined modulation and coding technique which is referred to as trellis code modulation (TCM) [39]. Various versions of TCM can improve the performance of a modem by 3–6 dB. The eight-state trellis code with a nominal gain of 4 dB is the most attractive, because more complex codes offer a little improvement with extensive additional implementation complexity. A version of TCM which can resolve 90° phase ambiguity [40] has been adopted by CCITT as an standard for QAM voiceband modems.

TCM is an extension of QAM in which the number of points in the constellation is increased to create a redundancy. These extra symbols enable the transmitter to create dependence between successive transmitted symbols. In this way, only a certain sequence of symbols is valid for communications. The received sequence of symbols is compared with all possible sequences, and the sequence with maximum likelihood is detected. The efficient search method under the maximum likelihood criterion is the Viterbi algorithm [41]. Implementation of the Viterbi algorithm for TCM is computationally complex, and the CCITT-recommended TCM requires a processing power almost equivalent to that of a TMS320 processor.

Comprehensive coverage of applied coded and uncoded constellations is available in [42]. Standard coded and uncoded QAM signal constellations can be modified to improve the performance of the modem in the presence of nonuniform noises, such as those resulting from phase jitter or nonlinear quantization [43, 44].

## 2.6 Pulse-Shaping Filters

Pulse shaping is used in modems to confine the transmission frequencies to a specific band, to control the ISI, and to minimize the effects of the noise. For the transmitted signal given by Eq. (22.1), if there is no distortion in the channel, the received baseband signal before equalization is

$$z(t) = \sum_k a_k h(t - kT)$$

where $h(t)$ is the overall pulse-shaping filter given by

$$h(t) = \int_{-\infty}^{\infty} h_1(t) h_2(t - \tau)\, d\tau$$

with $h_1(t)$ and $h_2(t)$ transmitter and receiver pulse-shaping filters.

The transmission medium is always band-limited, and a waveform cannot be band-limited and time-limited at the same time. Therefore, the overall impulse response of the pulse-shaping $h(t)$ designed for band-limited channels must extend beyond the time interval assigned for transmission of one symbol, which results in ISI. On the other hand, to optimize the signal-to-noise ratio after sampling at the receiver, it is desirable to have $h_1(t)$ and $h_2(t)$ as a pair of matched filters.

Raised cosine pulses are the traditional zero-ISI spectra, and approximation to raised cosine pulses is the conventional technique for designing these filters. The frequency domain characteristic of a raised cosine pulse is given by

$$H(f) = \begin{cases} T, & 0 \le |f| \le \dfrac{(1-\beta)}{2T} \\ \dfrac{T}{2}\left[1 - \sin \pi T \left(f - \dfrac{1}{2T}\right) / \beta\right], & \dfrac{1-\beta}{2T} \le |f| \le \dfrac{1+\beta}{2T} \end{cases} \tag{22.2a}$$

where $\beta$ is the so-called rolloff parameter, which controls the shape of the filter. The time domain representation of raised cosine pulses is

$$h(t) = \frac{\sin \pi t/T}{\pi t/T} \frac{\cos \beta \pi t/T}{1 - 4\beta^2 t^2/T^2} \tag{22.2b}$$

Figure 22.7 provides time domain and frequency domain representations of these pulses. As the rolloff factor $\beta$ increases, the sharpness of the filter decreases, resulting in an easier design. However, increasing the rolloff factor increases the required bandwidth while the data rate remains constant. A factor of two difference exists between the bandwidth efficiency of filters with minimum and maximum rolloff factors. Therefore, selection of the rolloff parameter requires a compromise between bandwidth efficiency and design complexity. In practice, pulse-shaping filters used for voiceband data communications have a rolloff factor of about 0.2. Because

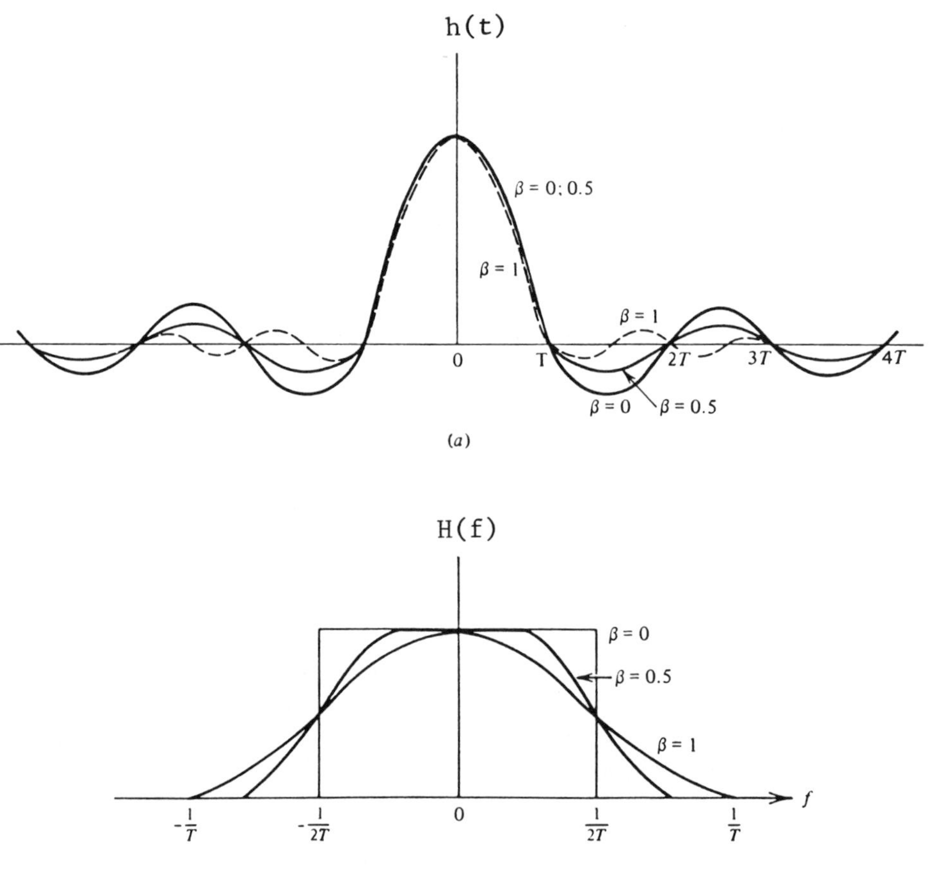


**Figure 22.7** Time and frequency domain representations of raised cosine pulses.

of design problems at higher frequencies, usually higher rolloff factors are used for radio modems.

The maximum 1 (symbol/sec)/Hz for QAM is associated with the case $\beta = 0$, which is not practically feasible. To preserve the same bandwidth efficiency with a realizable filter, one may use partial response signaling.

The impulse response of a partial response pulse shaping filter is one for two consecutive samples at $T$ and zero at all other samples. As a result, in the absence of channel distortions the received samples are $a_k + a_{k-1}$ rather than $a_k$. This forced ISI is known and does not create any problem for detection, as long as the decisions are based on the last two received sampled pulses. The frequency domain representation of the half-cosine pulse-shaping filters used in partial response systems is given by

$$H(f) = \begin{cases} 2T \cos \pi T f, & |f| \leq 1/2T \\ 0, & |f| \geq 1/2T \end{cases} \tag{22.3a}$$

The time domain representation for this waveform is

$$h(t) = \frac{4}{\pi} \cdot \frac{\cos \pi t/T}{1 - 4t^2/T^2} \tag{22.3b}$$

Figure 22.8 shows time and frequency domain representations of half-cosine pulses.

## Design Issues

Pulse shaping requires design of a pair of matched filters whose overall spectrum is given by either Eq. 22.2 or Eq. 22.3. The frequency response of each of the matched filters is the square root of the desired overall frequency response. The design of this filter requires maximization of the power in the bandwidth and minimization of the ISI. The design of standard filters does not involve the second constraint; therefore, standard filter design techniques must be modified to include minimization of the ISI. In [45] a method for designing a pulse-shaping filter that minimizes the stopband gain and ISI simultaneously is proposed. Using this technique, one can design a better filter than the direct approximation of raised cosine pulses.

Finite impulse response (FIR) filters are used for digital implementation of pulse-shaping filters to provide more flexibility in shaping the spectrum. A simple but relatively accurate method of designing zero-ISI digital filters is the standard windowing technique. Considering raised cosine pulses, the impulse response of the square root of the raised cosine spectrum is

$$h_2(-t) = h_1(t) = \frac{\sin[\pi(1-\beta)t] + 4\beta t \cos[\pi(1+\beta)t]}{\pi[1-(4\beta t)^2]t}$$

The impulse response of the FIR digital pulse-shaping filter consists of finite samples of the above waveform at the sampling rate used for digital processing. After

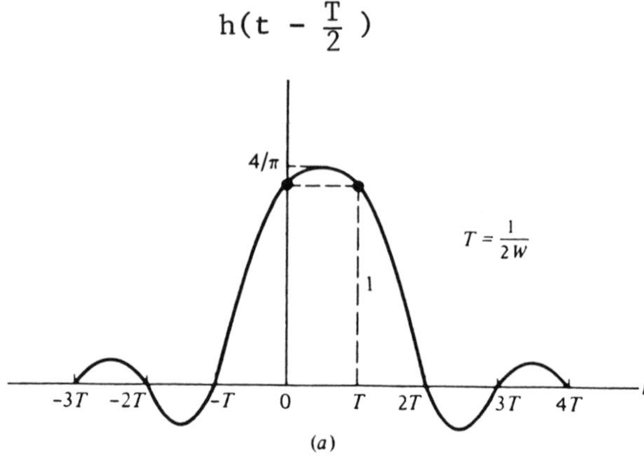


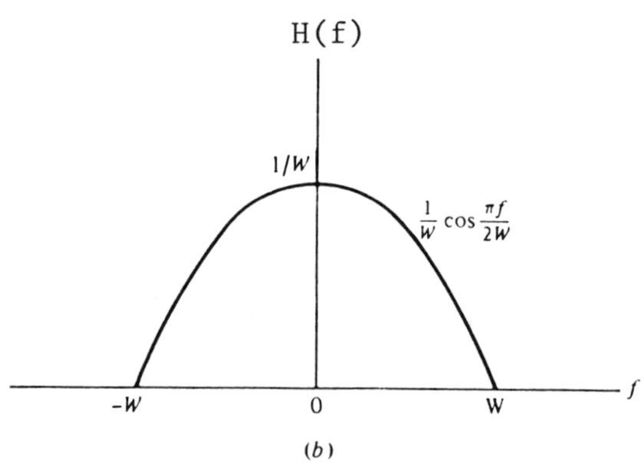


**Figure 22.8** Time and frequency domain representations of half-cosine pulses used in partial response signaling.

windowing, the rolloff factor will change from the value given in the above equation according to the length and type of the window.

The Remez exchange algorithm [46] can be used for the design of digital pulse-shaping filters in the frequency domain. The Remez exchange designs the optimal spectrum to match the defined passbands and stopbands. However, in the design of pulse-shaping filters the rolloff of the pulse is also important and cannot easily be entered into the algorithm; besides, this design does not minimize the ISI.

Form the above discussion, we may conclude that standard techniques such as the Remez exchange algorithm or the windowing technique by no means minimize the ISI and stopband attenuation at the same time. As a result, a designer should examine several designs to determine the right compromise between sideband attenuation and sharpness of the transient without any mathematical optimization.

Several approaches have been used for the optimal design of digital pulse-shaping filters. A digital linear phase FIR filter with special attention to zero ISI and minimum stopband attenuation is given in [47]. An iterative technique using the steepest descent algorithm to design a pair of zero-ISI matched filters with maximum spectral power in the passband is available in [48]. Other methods using linear programming [49, 50] and a modified Remez exchange algorithm [51] are also available.

## 2.7 Adaptive Processor

Three basic adaptive filters are used in modems. One is an adaptive equalizer in which the linear transversal filter attempts to minimize the ISI by adapting its sampled impulse response to compensate for the channel amplitude and phase distortion. The second is a channel estimator which adapts to the channel variation to provide an estimate of the amount of ISI due to the past and future symbols. This estimator is used in conjunction with maximum likelihood sequence estimation (MLSE). The third is an echo canceller which eliminates echoes in the telephone network.

### Adaptive Equalization

For a perfect pair of matched filters with zero ISI, the optimum transmission can be obtained if the channel does not suffer from amplitude and phase distortions. However, for channels with amplitude and phase distortions, a pair of matched filters is not the optimal solution. These channels suffer from ISI as well as additive noise. To improve the performance in additive noise channels the transmitted power can be increased; however, in ISI channels, increasing the power does not improve the performance because it amplifies the desired sample and the ISI at the same rate.

The traditional method of compensating for channel amplitude and phase distortion is to equalize these impairments by employing a filter at the receiver. In general, channel characteristics are subject to variations in time, resulting in a demand for adaptive equalizers. Tapped delay line equalizers are most commonly used, and their detailed analysis and theory of operation will be given in the next chapter. For more details, readers may refer to [38, 52], or [8].

### Adaptive MLSE

The sampled channel impulse response is measured with an adaptive tapped delay line filter similar to an equalizer. Given the samples of channel impulse response, a received sequence could be compared with all possible received sequences. The maximum likelihood procedure [53] is to determine the distance of the received sequence from all possible received sequences and detect the transmitted sequence of symbols with the minimum distance. The format of this search is similar to that of the detector used in TCM, and the Viterbi algorithm [41] can be used as a computationally optimal and efficient solution. In the literature, this method is referred to as MLSE.

MLSE is the optimal method for canceling ISI; however, the complexity of the receiver grows exponentially with the length of the channel impulse response, whereas the complexity of equalizers grows linearly with the length of the channel impulse response. For this reason, MLSE is interesting for channels with short-duration impulse responses; for longer channel impulse responses, equalizers are more practical. In the literature, MLSE is usually compared with decision feedback equalizers (DFEs), which have the best performance among all equalizers. This comparison for telephone line modems is given in [54, 55], for high-frequency (HF) radio in [56], and for troposcatter radio links in [57, 58]. The adaptive channel measurement techniques used with MLSE and applied to slow time-varying channels are given in [59, 60], and analysis of adaption techniques used for fading multipath radio channels is available in [61].

### Echo Cancellations

Echoes are the result of impedance mismatch in communication circuits. A substantial amount of echo energy comes from mismatching of impedance at the hybrid couplers which are located at two-wire/four-wire interfaces in telephone circuits. One coupler is at each end of a two-wire line; consequently, each user suffers from near and distant echoes created by a coupler at its side and a coupler at the other side of the telephone line. Most efforts in echo cancellation for telephone networks are devoted to improving the quality of voice transmission; for this application, shorter echoes are actually desirable and their existence keeps the telephone voice from sounding dead. Therefore, cancellers in the networks are normally located near the two-wire/four-wire interface, which does not eliminate the near echoes.

In full-duplex two-wire data communication modems, the near echoes are as damaging as the distant echoes. For this reason, and also because echo cancellers are not deployed throughout the telephone network, echo cancellers have been proposed [62–64] to improve the performance of full-duplex two-wire modems. Full-duplex operation of two-wire modems up to 2400 bps is feasible without echo cancellation; for higher data rates, 4800 and 9600 bps, satisfactory operation requires an echo canceller.

Echo cancellers used in full-duplex modems are adaptive transversal filters similar to those used for channel measurement, and their principle of operation and mathematical structure are similar to those of adaptive equalizers. Details of existing problems and an overview of echo cancellation techniques are given in [8–10].

## 3 ANALYSIS OF TAPPED-DELAY-LINE (TDL) EQUALIZERS

The principles of operation of TDL equalizers, channel estimators, and echo cancellers are similar and are considered as special cases of adaptive filtering problems. Equalizers have wider applications in telecommunications and were introduced and applied earlier than the other two. For these reasons, the detailed analysis in this chapter is devoted to equalizers. Issues in the application of TDL equalizers to modem design, such as sensitivity to timing recovery, variations according to modulation technique, and fast start-up equalizers, are discussed.

### 3.1 Equalizer Architectures

Equalizer architectures used in modem design technology are the following: linear transversal equalizers (LTEs), linear fractionally spaced equalizers (FSEs), decision feedback equalizers (DFEs), passband equalizers, and blind equalizers. This section introduces these equalizers with detailed emphasis on LTE, FSE, and DFE. Fast start-up equalizers are discussed separately in the next section.

## Linear Transversal Equalizer

The LTE, shown in Figure 22.9, is the earliest TDL equalizer. The received signal is passed through a tapped delay line with a tap spacing of $\Delta = T$. The tapped signals are weighted and added to form the equalized output signal. The optimum tap gains are determined under either the zero forcing criterion or the minimum mean square error (MMSE) criterion. In the zero forcing algorithm, the tap gains are determined so that the overall sampled impulse response after equalization is the same as the sampled impulse response when there are no channel amplitude and phase distortions [65, 66, 38]. For MMSE equalization, tap gains are determined to minimize the mean square of the error signal between equalized samples and actual transmitted symbols.

The equalizer is a sampled time filter intended to compensate for the amplitude and phase distortions of the channel. Intuitively, for an infinite tap equalizer, the sampled frequency response of the equalizer should be the inverse of the frequency response of the channel, which is a correct intuition for zero forcing equalization. In MMSE equalization, the tap gains are determined to minimize both channel distortions and additive noise, and the spectrum of the equalizer is also dependent on the variance of the noise.

Today, the MMSE is the dominant criterion for equalization, and various versions of the equalizer architectures are developed with this criterion. As a result, this chapter is mainly focused on MMSE equalization.

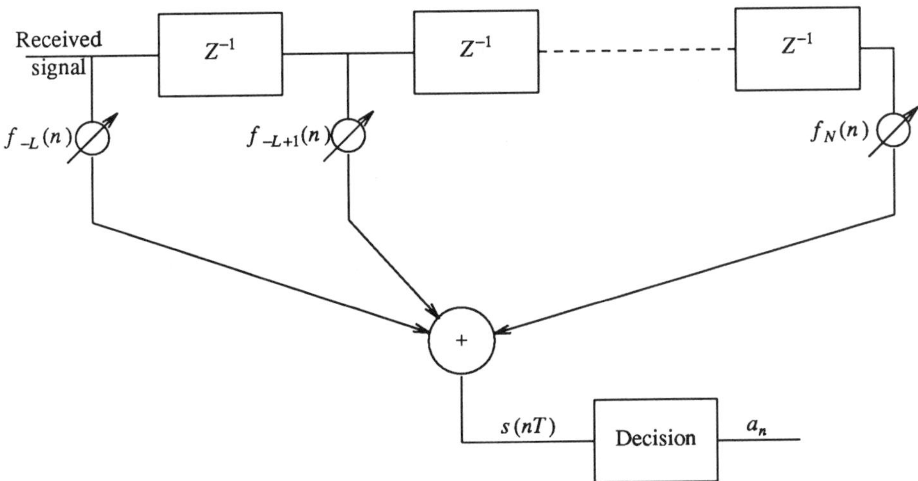


**Figure 22.9** Tapped-delay-line architecture for linear transversal equalizer (LTE).

## Fractionally Spaced Equalizer

The structure of FSE is the same as that of LTE, as shown in Figure 22.10, except that in FSE the tap spacing is $\Delta = (k/n)T$, with $k$ and $n$ integers and $k < n$. This minor difference results in three basic advantages: an FSE is insensitive to timing error, for FSE accurate pulse shaping is less important, and FSE can handle channel phase distortions in the corners of the spectrum better than standard LTE. For these reasons, FSEs are widely used for both telephone lines and radio modems. An intuitive explanation of the reason for the difference between LTEs and FSEs is given below.

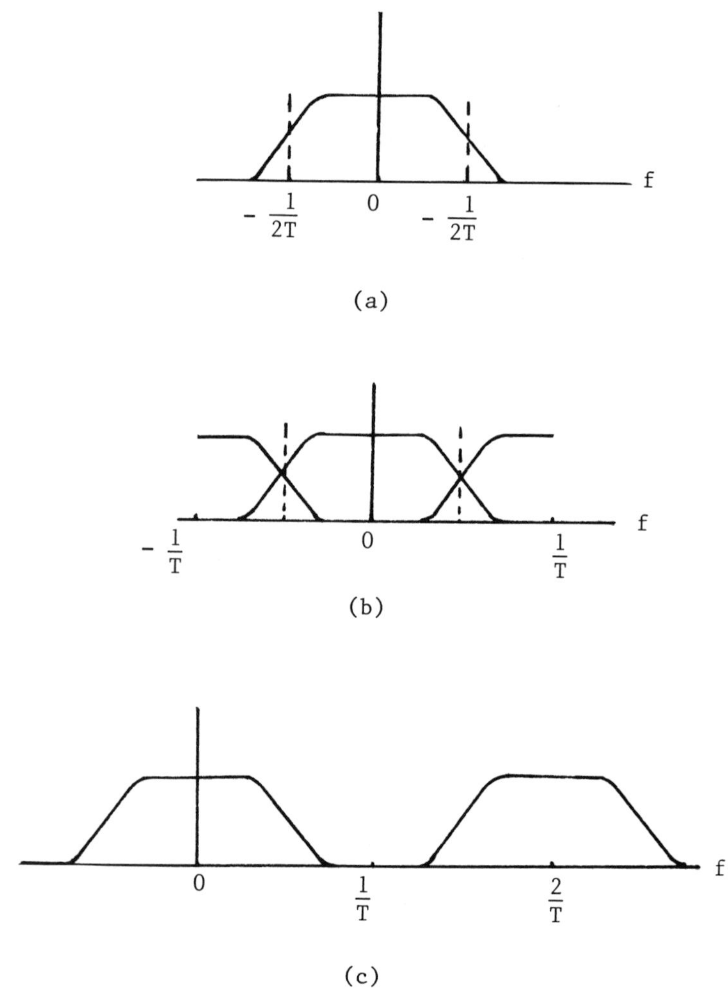


**Figure 22.10** Spectrum of received samples used for equalization: (a) spectrum of the received continuous time signal; (b) spectrum for FSE with tap spacing of $T/2$.

The spectrum of the received equivalent low-pass signal at the input of the equalizer is band-limited to $1/2T$, as shown in Figure 22.10a. The LTE operates at the baud rate and, as shown in Figure 22.10b, the spectrum to be equalized is the folded form of the input spectrum at $1/T$, which includes excessive aliasing. However, as shown in Figure 22.10c, for an FSE the sampled received spectrum for equalization does not suffer from aliasing. In an LTE, if there is a normalized sampling error $\tau$ between the transmitter and receiver timing, the received in-band sampled signal for equalization is given by

$$R\left(\omega - \frac{2\pi}{T}\right) e^{j(\omega - 2\pi/T)\tau T} + R(\omega)e^{j\omega\tau T} + R\left(\omega + \frac{2\pi}{T}\right) e^{j(\omega + 2\pi/T)\tau T}$$

where $|\omega| < 1/2T$, $R(\omega)$ is the power spectrum of the received signal with normalized sampling error $\tau$, and the first and third terms are caused by aliasing. Therefore, the LTE has no chance to work directly on the received signal with timing error, $R(\omega)e^{j\omega\tau}$, and the amplitude spectrum at each edge is the addition of the two terms which depend on $\tau$. For particular values of $\tau$ a 180° phase shift between the added aliased signals results in a deep null in the magnitude of the aliased signal used for equalization. Then the equalizer needs high gains in the amplitude spectrum to compensate for the null, which results in noise enhancement. The FSE has no aliasing, and the in-band sampled spectrum of the received signal is represented only by the middle term of the above equation. This term is the spectrum of the received signal plus an additional linear phase term caused by the timing error $\tau$, and the amplitude spectrum of the received signal is independent of $\tau$. As a result, the equalizer can compensate for timing error without suffering from noise enhancement.

Similarly, if the channel has severe delay distortion at the edge of the passband, an LTE has no chance to work on the distortion directly; as a result, compensation of these distortions after aliasing is more difficult. Since the FSE works directly on the received signal, it can also play the role of the matched filter. In these cases an accurate design for the matched filter at the receiver is unnecessary.

Continuing with the same sense of intuition in the frequency domain, one may conclude that in the FSE case an infinite number of solutions exists for the tap gains. The equalizer could be thought of as the inverse of the channel in the frequency domain; therefore, for areas with zeros in the spectrum of the nonaliased signal, an infinite number of inverse spectra may exist, which makes one think of the existence of an infinite set of optimal tap gains. For the MMSE criterion this statement is false, and it can be shown that the solution to Eq. (22.10) has a unique answer [30, 31].

The FSE performs well under computer simulation with floating-point arithmetic. However, in real-time implementation with fixed-point arithmetic and limited word length, the tap gains tend to diverge. Analysis and implementations to relate tap gain blowup to characteristics of the channel are given in [31]. They show that a digitally implemented FSE generally has many sets of tap weights which result in nearly equal values of the mean square error. Some of these solutions are large, and in a digital implementation, bias and round-off errors in the tap updating algorithm can cause these large solutions to occur, causing overflow of the registers and deterioration of performance. In general, the simple solution to tap gain blowup is the so-called tap leakage algorithm [28, 31]. This ad hoc solution introduces small corrections in tap gains to prevent their excessive growth.

Another aspect of the comparison between LTEs and FSEs is that with the same amount of hardware the LTE spans a longer time interval, giving an intuitive basis for superior performance of the LTE. Computer simulations reveal that, with the same number of taps, the FSE performance is at least equivalent to that of the LTE and becomes significantly better for channels with severe phase distortions in the corners of the spectrum [29].

## Decision Feedback Equalizer

A DFE [67, 68], shown in Figure 22.11, consists of two tapped-delay lines, referred to as forward and backward equalizers. The input to the forward equalizer is the received signal, and it operates similarly to the linear equalizers discussed earlier. In an ideal situation, the tap gains of this part cancel the ISI due to the future symbols, and their Fourier transform is the inverse of the channel frequency response. The input to the backward transversal filter is the detected symbols. The tap gains of this part are the estimates of the channel sampled impulse response including the forward equalizer, and they cancel the ISI due to past samples. For finite-tap equalizers the forward equalizer reduces the ISI due to future symbols and the backward equalizer of length $M$ eliminates the ISI due to $M$ past symbols.

Linear equalizers are unable to equalize the channels properly when they have a deep null in the passband. For these channels, the equalizer needs high gain in its frequency response to compensate for the null, which results in noise enhancement. The backward filter of a DFE does not suffer from the noise enhancement problem because it estimates the channel rather than its inverse; as a result, for channels with a deep null in the passband, DFEs are superior to linear equalizers.

In frequency-selective fading radio channels, occasionally a channel experiences deep nulls in the passband, resulting in unsatisfactory performance for linear equalizers. For these channels, DFEs are the rational choice, and they have been used for troposcatter [57], HF [56], and microwave line-of-sight [60, 70] channels. In a slowly time-varying telephone channel, significant amplitude and delay distortion correspond to the edges of the passband, while a DFE is more effective for the nulls in the middle of the passband. As a result, a DFE has little to offer over linear equalizers with a large number of taps; and linear equalizers are dominant in voiceband data communications over telephone channels.

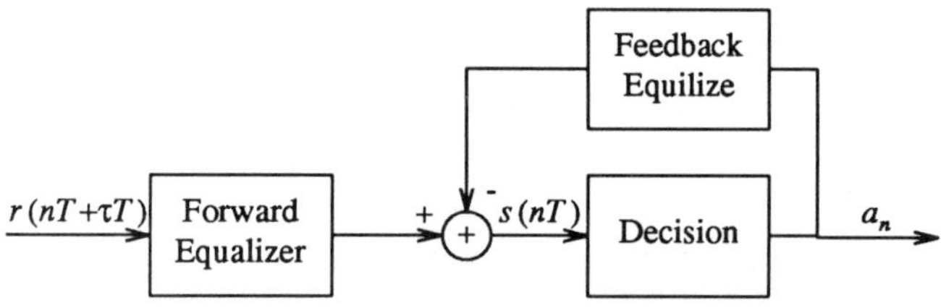


**Figure 22.11** Architecture of the decision feedback equalizer (DFE). Forward and backward equalizers are both tapped-delay-line filters.

### Blind Equalization

Standard equalizers are trained with a known sequence at the start of the data transmission. After training, with a high probability, the decisions at the output of the equalizer are the same as the transmitted symbols and are used as the reference for the tap gain adjustment. In many cases, both radio and telephone line modems need retraining without interruption of the normal data transmission. The retraining algorithm should be independent of any particular information about the sequence of the transmitted symbols. This equalization is usually referred to as blind equalization. A simple algorithm for blind equalization, which is based on the square of the error signal used for normal adjustments of the tap weights, is given in [55]; today, it is widely used to retrain the equalizers after interrupts. Since in these algorithms the reference to transmitted symbols is unavailable at the receiver, the convergence is about 10 times slower than with standard LMS algorithms. Additional analysis of the transient behavior of blind equalizers, applied to radio communications, is available in [71, 72].

### Bandpass Equalization

Bandpass equalization [73, 74] is performed before demodulation, as opposed to standard baseband equalization, which is done after demodulation. When the phase reference of the oscillator in the demodulator is obtained from the detected data, the delay between demodulator and phase recovery is smaller in passband equalizers, resulting in faster tracking of phase variations and, consequently, a more robust modem. However, in modern digital designs it is possible to demodulate without a phase reference and to adjust the phase by multiplying the demodulated symbol by a numerical phasor which shifts the point in the constellation into its proper place. For these implementations, no difference between baseband and bandpass equalization exists. In radio modems working on high-frequency carriers, sometimes surface acoustic wave (SAW) devices are a cost-efficient solution for implementation of the equalizer. SAW devices work at high frequencies; therefore, SAW device implementation of the equalizer at bandpass (either at intermediate frequency (IF) or radio frequency (RF)) is preferred to baseband equalization.

## 3.2 Analysis of Linear Equalizers

A block diagram of a linear equalizer is shown in Figure 22.9. This figure represents both LTEs and FSEs. The received baseband signal at the input of the equalizer is

$$r(t) = \sum_{p} a_p g(t - pT) + \eta(t) \tag{22.4}$$

where $g(t)$ is the overall impulse response, which is a convolution of the impulse responses of the transmitter filter, receiver filter, and channel, and $\eta(t)$ denotes the additive noise. In practice, the received noise at the input of the pulse-shaping filter is white, but the noise after the receiver pulse-shaping filter, $\eta(t)$, is no longer white. However, with an ideal zero-ISI receiver matched filter, the samples of $\eta(t)$ for sampling at the symbol interval are white, while those for sampling at fractions of the symbol interval are not. In this chapter, to simplify calculations the samples of $\eta(t)$, at any rate, are assumed to be white with a variance of $N_0$.

The received signal after pulse shaping, $r(t)$, is passed through a TDL with tap spacing $\Delta$. The output of each delay element is multiplied by an adjustable parameter called the tap gain, and the weighted output of the TDL is added to generate the output of the equalizer. This output is sampled at the transmitted symbol rate with a normalized delay difference of $\tau$ between the transmitted symbol timing and the receiver sampler. The sampled output of the equalizer is then given by

$$s(kT) = \sum_{j=-L}^{N} f_j r(kT + \tau T - j\Delta) \tag{22.5}$$

where $\{f_j\}$ is the set of the $L + N + 1$ equalizer tap gains. The error signal after equalization is defined as

$$e(kT) = a_k - s(kT) \tag{22.6}$$

where $a_k$ is the detected symbol and is assumed to be the same as the transmitted symbol. In normal operation of a modem, the probability of error is very low, and this assumption is accurate.

## Tap Gains Under the MMSE Criterion

Under the MMSE criterion, the tap gains of linear equalizers are determined to minimize the MSE function $\xi$ defined as

$$\xi = \overline{|e(kT)|^2} \tag{22.7}$$

where the overbar represents the time average.

The optimum tap gains are determined when the derivatives of the MSE function $\xi$ with respect to all taps gains are set to zero. Therefore, to obtain the optimal tap gains, one must have

$$\frac{\delta \xi}{\delta f_l} = 0, \qquad -L \leq l \leq N \tag{22.8}$$

The error function $\xi$ is a quadratic complex function; for quadratic functions, the above condition is satisfied when

$$\overline{e(kT)\frac{\delta e^*(kT)}{\delta f_l}} = -2\overline{\left[a_k - \sum_{i=-L}^{N} f_i r(kT + \tau T - i\Delta)\right] r^*(kT + \tau T - l\Delta)]} = 0 \tag{22.9a}$$

Substituting Eq. (22.4) into (22.8), with minor manipulations one has

$$\sum_{j=-L}^{N}\left[f_j\overline{|a_k|^2}\sum_p g(pT+\tau T-j\Delta)g^*(pT+\tau T-l\Delta)+N_0\delta_{jl}\right]$$
$$=\overline{|a_k|^2}g^*(\tau T-l\Delta), \qquad -L\le l\le N \quad (22.9b)$$

where $\delta_{jl}=1$ for $j=l$, and zero otherwise. Equation (22.9b) gives a set of $L+N+1$ linear equations with $L+N+1$ unknowns, and solutions are the set of optimal tap gains to minimize the MSE function, defined in Eq. (22.7).

Defining elements of the $(N+L+1)(N+L+1)$ matrix $A$ as

$$a_{jl}=\overline{|a_k|^2}\sum_p g(pT+\tau T-j\Delta)g^*(pT+\tau T-l\Delta)+N_0\delta_{jl}, \qquad -L\le j,l\le N \quad (22.10a)$$

elements of the $N+L+1$ vector $B$ as

$$b_l=\overline{|a_k|^2}g^*(\tau T-l\Delta), \qquad -L\le l\le N \quad (22.10b)$$

and $F_{\text{opt}}$ as the $L+N+1$ vector of optimal tap gains, one has the following simple vector representation:

$$AF_{\text{opt}}=B \quad (22.10c)$$

The above equations work for both LTEs and FSEs. The covariance matrix $A$ is always Hermitian symmetric, which means that elements with switched indices are complex conjugate. However, for LTEs it is also a Toeplitz matrix, which implies that all elements of the diagonals of the matrix are the same. As a result of this difference, all eigenvalues of the covariance matrix for an LTE are nonzero, while in an FSE a good fraction of the eigenvalues are zero; for example, for $\Delta = T/2$ one-half of the eigenvalues are zero.

## Calculation of the Minimum MSE

The minimum value of the MSE function, $\xi_{\text{min}}$, is attained when one substitutes the optimum tap gains in the calculation of the MSE function. The MMSE is the inverse of the signal-to-noise ratio after equalization and can be used as a performance criterion to compare various MMSE equalizer architectures. Substituting Eq. (22.5) in (22.6) and Eq. (22.6) in (22.7), the following general expression for the MSE function is obtained:

$$\xi = \overline{|a_k|^2} - \sum_{j=-L}^{N} f_j \overline{a_k^* r(kT + \tau T - j\Delta)} - \sum_{i=-L}^{N} f_i^* \overline{a_k r^*(kT + \tau T - i\Delta)}$$

$$+ \sum_{i=-L}^{N} f^{*_i} \sum_{j=-L}^{N} f_j \overline{r(kT + \tau T - j\Delta) r^*(kT + \tau T - i\Delta)} \quad (22.11)$$

Substituting the optimum value of the tap gains from Eq. (22.9a) in the above equation, the minimum MSE is determined. For optimum tap gains the last two terms of the above equation cancel, and simple manipulation of the first two terms results in

$$\xi_{\min} = \overline{|a_k^2|} \left[ 1 - \sum_{i=-L}^{N} f_i g^*(\tau T - i\Delta) \right] \quad (22.12a)$$

where the $f_i$'s are the optimal values of the tap gains found from Eq. (22.9b). In the vector notation defined in Eq. (22.10), the minimum MSE is

$$\xi_{\min} = \overline{|a_k^2|}[1 - B^* F_{\text{opt}}] = |a_k^2|[1 - B^* A^{-1} B] \quad (22.12b)$$

### 3.3 Analysis of the DFE

The DFE shown in Figure 22.11 consists of two TDLs, forward and feedback equalizers. The output of the DFE is given by

$$s(kT) = \sum_{j=-L}^{0} f_j r(kT + \tau T - j\Delta) - \sum_{j=1}^{M} b_j a_{k-j} \quad (22.13)$$

where the first term represents the output of the forward equalizer, which corrects for the future ISI, and the second term represents the output of the feedback equalizer, which operates on the decisions to eliminate the ISI due to the previous $M$ symbols. The tap gains of the forward and feedback equalizers are denoted by $f_i$ and $b_j$, respectively.

Definition of the MSE function for DFEs is the same as for linear equalizers, given in Eqs. (22.7) and (22.6) with $s(kT)$ defined in Eq. (22.13). The optimum tap gains are the values of $f_j$ and $b_j$ that minimize the error function. To determine the optimum tap gains one takes derivative of the MSE function with respect to both forward and feedback taps. For the forward tap gains one has

$$\overline{e(kT)\frac{\delta e^*(kT)}{\delta f_l}} = 0, \qquad -L \leq l \leq 0$$

and derivations similar to those for Eq. (22.9) of the last section result in

$$\sum_{j=-L}^{0} f_j \left[ \overline{|a_k|^2} \sum_p g(pT + \tau T - j\Delta) g^*(pT + \tau T - l\Delta) + N_0 \delta_{jl} \right]$$

$$- \overline{|a_k|^2} \sum_{j=1}^{M} b_j g^*(jT + \tau T - l\Delta) = \overline{|a_k|^2} g^*(\tau T - l\Delta), \qquad -N \le l \le 0 \quad (22.14a)$$

For the feedback taps one has

$$\overline{e(kT) \frac{\delta e^*(kT)}{\delta b_l}} = 0, \qquad 0 \le l \le M$$

which result in

$$f_l = \sum_{j=-N}^{0} f_j g(lT - j\Delta + \tau T), \qquad 1 \le l \le M \qquad (22.14b)$$

The right-hand side of Eq. (22.14b) is the convolution of the sampled channel impulse response from the transmitter to the input of the forward equalizer at $\Delta$, and the discrete time impulse response of the forward equalizer. The results are sampled at $T$, and only values on the right-hand side of the center are calculated; these samples are associated with ISI due to the past $M$ transmitted symbols. Therefore, the optimum tap gains completely eliminate the ISI due to the past $M$ samples.

Calculation of the MMSE is also very similar to that for linear equalizers, with the new error signal defined as

$$e(kT) = a_k - \sum_{j=-L}^{0} f_j r(kT + \tau T - j\Delta) + \sum_{j=1}^{M} b_j a_{k-j}$$

Using this equation as the error, the MSE function defined in Eq. (22.7) contains nine terms. Substituting optimum values of the tap gains from Eq. (22.15) cancels the last six terms, and the MMSE is then given by the first three terms:

$$\xi_{\min} = \overline{|a_k^2|} \left[ 1 - \sum_{j=-L}^{0} f_j g^*(\tau T - j\Delta) \right] + \sum_{j=1}^{M} f_j \overline{a_k^* a_{k-j}}$$

For the given values of $j$, the third term is also zero, and the final value of the minimum MSE function is

$$\xi_{\min} = \overline{|a_k^2|}\left[1 - \sum_{i=-L}^{0} f_i g^*(\tau T - i\Delta)\right] \tag{22.15}$$

At first glance, this equation suggests that the MMSE is independent of the feedback tap gains. This conclusion is incorrect; in Eq. (22.15), when the optimum set of tap gains are determined, one jointly minimizes for both sets of tap gains. As a result, optimum values of forward tap gains are affected by the optimum values of feedback tap gains, which indirectly change the MMSE. Numerical examples to support this statement will be given later.

## 3.4 Equalization for Other Modulations

The equalizers introduced so far are suitable for PAM (pulse amplitude modulation), PSK, and QAM modulation techniques. For other modulation techniques, the setup for optimum tap gains and MMSE could be different. Here, equalization of partial response signaling and staggered modulations which are widely used in radio modems are discussed.

### Equalizations for Partial Response Modulation

The equalizer architectures given earlier work for partial response signaling, when the reference in the error function is $a_k + a_{k-1}$, rather than $a_k$. To observe the difference in calculations of the optimum tap gains and MMSE, the linear equalizer for partial response modulation will be discussed here.

With the new reference signal for partial response modulation, the error after equalization is

$$e(kT) = a_k + a_{k-1} - s(kT)$$

where $s(kT)$ is the received sampled signal after equalization, given by Eq. (22.5). For the pulse-shaping filters used for partial response signaling, given in Eq. (22.3), the received signal is

$$r(t) = \sum_p a_p g\left(t - pT - \frac{T}{2}\right) + v(t)$$

where the $T/2$ delay adjusts the impulse response with transmitted symbols. Derivations similar to those in the preceding section for the new error function and received signal result in the following equations for the tap gains and the MMSE of a linear equalizer:

$$\sum_{j=-L}^{N}\left[f_j|a_k|^2\sum_p g\left(pT+\tau T-j\Delta-\frac{T}{2}\right)g^*\left(pT+\tau T-l\Delta-\frac{T}{2}\right)+N_0\delta_{jl}\right]$$

$$=\overline{|a_k|^2}\hat{g}^*(\tau T-l\Delta) \tag{22.16a}$$

$$\xi_{\min}=\overline{|a_k|^2}\left[2-\sum_{j=-L}^{N}f_j\hat{g}^*(\tau T-J\Delta)\right] \tag{22.16b}$$

where

$$\hat{g}(t)=g\left(t-\frac{T}{2}\right)+g\left(t+\frac{T}{2}\right) \tag{22.16c}$$

Similar derivation can be extended to DFEs; the general structure is the same as before, but the relation between impulse response and tap gains is different. Numerical examples of equalization of partial response modems will be given later.

## Equalization for Staggered Modulations

Equalizations of staggered signals needs more insight for implementation and derivation of the optimum taps and MMSE. The issue in implementation is how to handle the $T/2$ delay between transmitted in-phase and quadrature phase symbols. Since real and imaginary parts of the decision are detected with a $T/2$ delay, the feedback equalizer must use a $T/2$, rather than a $T$, delay line, resulting in doubling of the number of taps. However, these taps are alternately real and imaginary and in terms of hardware complexity each pair of them can be viewed as equivalent to one of the taps of equalizers for nonstaggered modulations.

The received signal after prefiltering at the receiver is

$$r(t)=\sum_p \hat{a}_p g\left(t-p\frac{T}{2}\right)$$

where $\hat{a}_p$ is the sequence of the real and imaginary parts of the transmitted digits, alternating each $T/2$ second between pure real and pure imaginary.

After linear equalization the sampled signal at $T/2$ is

$$s\left(k\frac{T}{2}\right)=\sum_{j=-L}^{N}f_j\, r\left(k\frac{T}{2}+\tau T-i\Delta\right)$$

and the real and imaginary parts of this signal are used alternately for detecting the real and imaginary parts of the transmitted symbols.

The new error function is defined as

$$\hat{e}(kT) = \text{Real}\left[e\left(k\frac{T}{2}\right)\right] + j\,\text{Im}\left[e\left((k+1)\frac{T}{2}\right)\right]$$

where

$$e\left(k\frac{T}{2}\right) = \hat{a}_k - s\left(k\frac{T}{2}\right)$$

The fundamental problem is that the MSE

$$\xi = \overline{|\hat{e}(kT)|^2}$$

is no longer an analytic function of the tap gains. Therefore, calculation of the tap gains must be done for their real and imaginary parts separately, which results in a more complex set of $2(L + N + 1)$ equations with $2(L + N + 1)$ unknowns for real and imaginary parts of the tap gains. Details of these calculations for staggered quadrature PSK and quadrature partial response modulations and a comparison of the performance of staggered and nonstaggered equalized systems are given in [69, 70].

### 3.5 Numerical Examples

In this subsection some numerical examples are provided to give an insight for comparing various equalization techniques. Figure 22.12 compares inverse of $\xi_{\min}$ for LTE and FSE as a function of normalized timing error $\tau$ over a telephone channel. The inverse of $\xi_{\min}$ is a measure of the signal-to-noise ratio after equalization, which can be viewed as a performance criterion for different equalizers. The FSE is insensitive to the timing errors while the MMSE of the LTE changes almost periodically with a period of $T$. The $T$ second change in the timing alignment of an LTE is equivalent to shifting the center tap of the equalizer to the next neighboring tap. Therefore, as long as the center tap is almost at the center of the equalizer and the equalizer is sufficiently long, the $\xi_{\min}$ repeats for every $T$ seconds error in timing.

The telephone channel does not suffer from deep nulls in its passband, while in frequency-selective fading radio channels it is possible to have deep nulls in the passband. As an example for radio communications, consider the frequency-selective microwave line-of-sight (LOS) channel whose frequency response is given by

$$H(j\omega) = A + B(j\omega)$$

where $A$ and $B$ are complex functions of time.

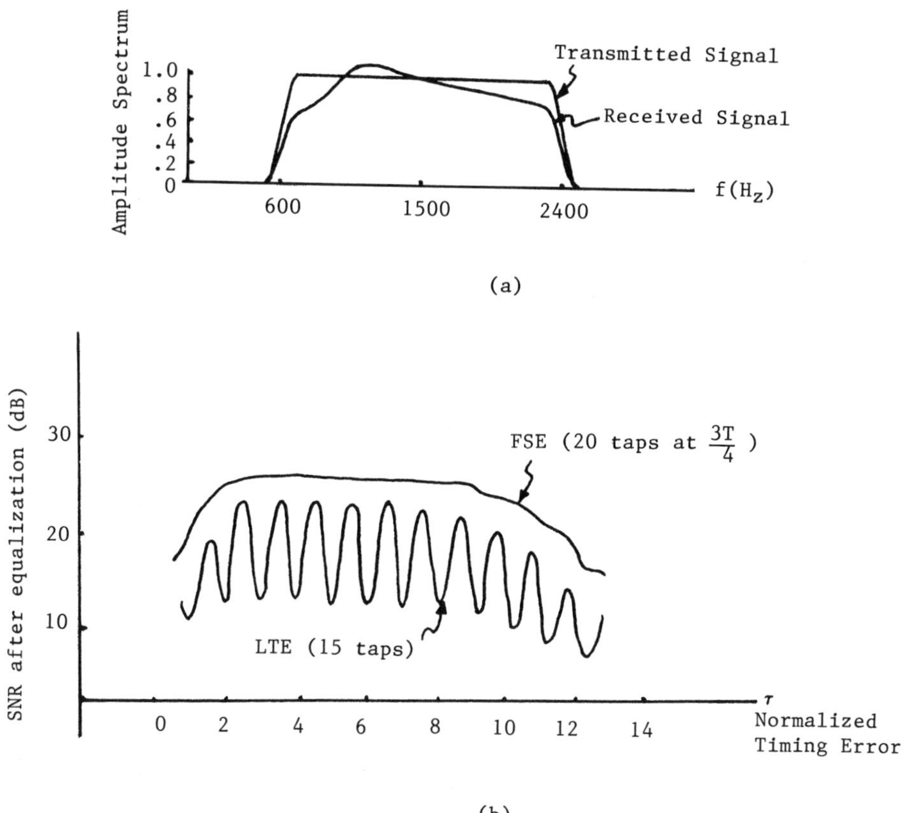


**Figure 22.12** Comparison of performance for an LTE and an FSE ($3T/4$) for SNR = 27 dB and 15 taps. The horizontal axis represents the timing error and the vertical axis the SNR in dB; (a) spectrum of the transmitted and received signals; (b) SNR after equalization—the periodic curve corresponds to the LTE.

The transfer function of this channel has one zero at $s = -A/B$ which results in a null at the frequency $\omega = \text{Im}(-A/B)$. The depth of the notch is determined by Real $-A/B$.

Figure 22.13 shows plots of the inverse of $\xi_{\text{min}}$ versus normalized timing error for $A = 1$, $B = 0.4$, and various numbers of taps for a DFE working with a staggered partial response modem. The SNR before equalization is 23 dB. The forward tap gains are associated with $\Delta = T/2$, and plots include various numbers of forward, $N$, and backward, $M$, tap gains. As the number of forward tap gains increases, sensitivity to timing error decreases, while an increase in the number of backward taps has an insignificant effect on the sensitivity to the timing.

Figure 22.14 shows the probability of error versus location of a deep null in the passband of the channel for a standard two-dimensional modulation, QPSK, over microwave LOS channels. Plots include a modem without equalizer, and FSE with five $T/2$ taps, and a DFE modem with three forward and three feed-

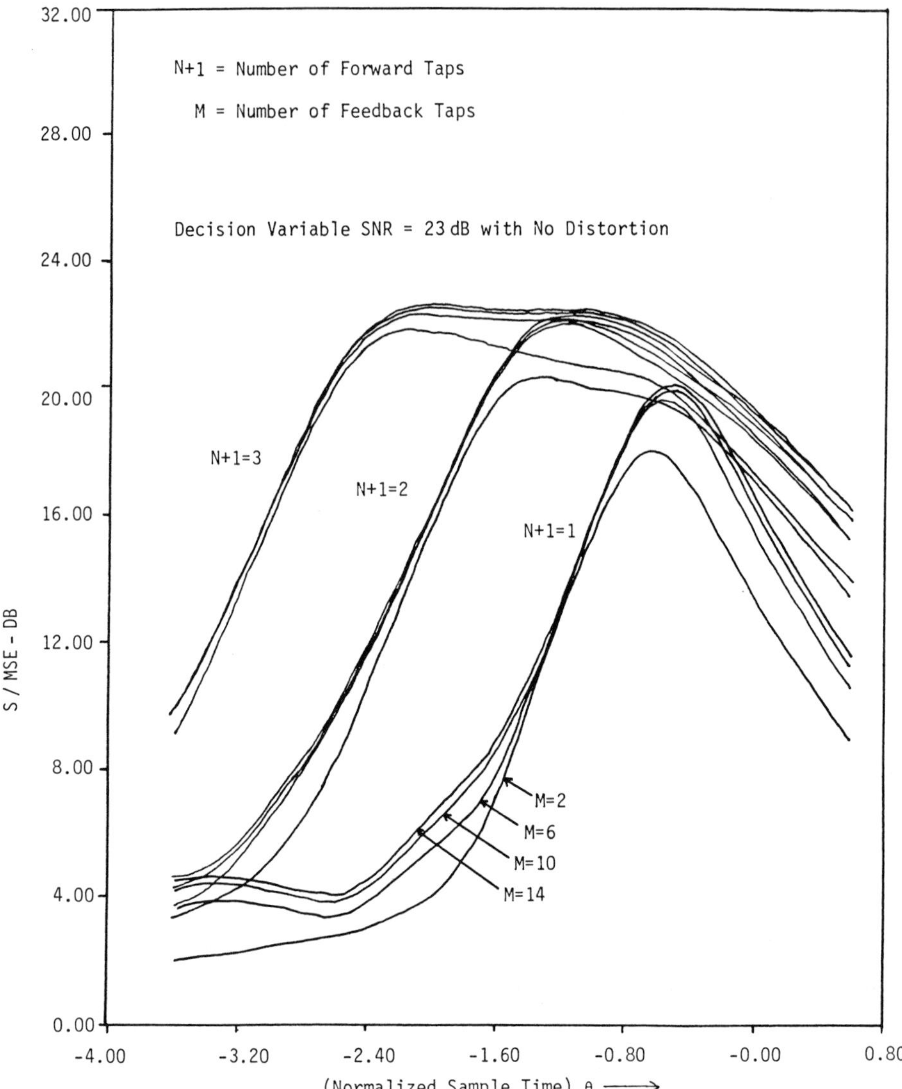


**Figure 22.13** SNR versus normalized timing error for a staggered partial response DFE over a frequency-selective microwave line-of-sight channel. The received SNR before equalization is 23 dB; the channel frequency response is $H(j\omega) = 1 + 0.4(j\omega)$; the forward taps, $N+1$, and backward taps, $M$, are both at $T/2$. Backward taps are pure real and pure imaginary alternately.

back taps. For the deep null around the center, only the DFE shows reasonable performance. As the null shifts to the corners of the passband, the other two modems improve in performance rapidly. As a result, one may conclude that for robust performance for all possible channel snapshots a DFE is required.

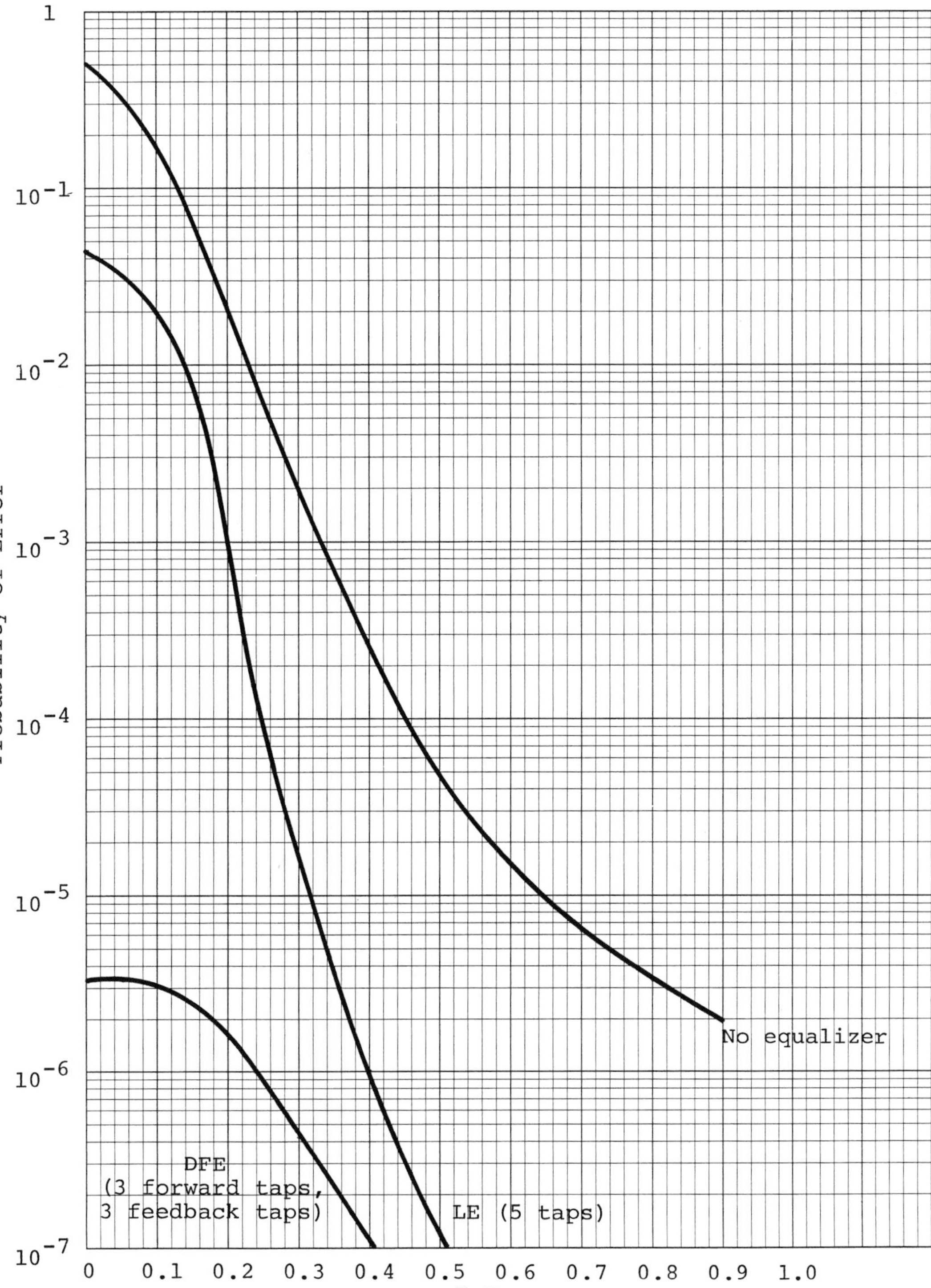


**Figure 22.14** Probability of error versus location of a null in the passband of a microwave LOS channel for FSE with five taps, DFE with three forward and three backward taps, and modem with no equalization. The frequency response of the channel is represented with a power series.

## 4 ADAPTIVE ALGORITHMS FOR EQUALIZERS

The last section analyzed the effectiveness of various equalizers under the $\xi_{\min}$ criterion and found that the optimal tap gain in all cases is determined from a set of linear equations. Direct calculation of these equations requires measurement of the instantaneous overall channel impulse response, measurement of the variance of the noise, formation of the covariance matrix, and solution of a large set of linear equations. This method requires a great many numerical calculations, which is by no means economically attractive for real-time applications, and demands a numerically efficient solution. Fortunately, both peak distortion and MSE are convex functions of the tap gains which allow application of the steepest descent algorithm (SDA). These types of algorithms are computationally attractive and can easily adapt to the channel variations. The least mean square algorithm (LMS) is the practical implementation of the steepest descent-based, and is the most successful algorithm applied to equalizers.

Application of the steepest descent algorithm to adaptive equalization was first employed for peak distortion criteria [65, 66] and then examined for adaptive MMSE equalization. Later, these algorithms turned out to be slow for some applications, and fast converging algorithms for fast start-up equalization and fast tracking of rapidly changing channels were investigated. In this section, convergence of the steepest descent algorithm used for $\xi_{\min}$ equalization and the relation between channel characteristics and convergence rate of the equalizer will be discussed. Various techniques applied to fast equalization will then be reviewed. Details of these discussions are mainly restricted to LTE, while other equalizers are referenced.

### 4.1 Steepest Descent Algorithm (SDA)

Since the MSE $\xi$, defined in Eq. (22.7), is a convex function of the tap gains, the solution to tap gains can be obtained by use of the steepest descent algorithm. In this method, tap gains are assumed to have an arbitrary initial setup and are moved in the direction of the optimum value when the MSE is minimized. The direction is determined by the gradient of the objective function $\xi$ with respect to the tap gains. For the $l$th tap of an LTE, formulation of the SDA is as follows:

$$f_l(k+1) = f_l(k) - \mu \frac{\delta \xi}{\delta f_l(k)} \tag{22.17}$$

where $\mu$ is the step size in the algorithm. Since

$$\frac{\delta \xi}{\delta f_l(k)} = -2\overline{e(kT)r(fT + \tau T - l\Delta)} \tag{22.18}$$

one has

$$f_l(k+1) = f_l(k) + 2\mu\overline{e(kT)r(kT + \tau T - l\Delta)} \tag{22.19}$$

This simple iterative algorithm converges to the optimum tap gains where the derivatives with respect to the tap gains are very small. A similar procedure could be applied to maximum distortion criteria for the zero forcing algorithm [66].

### Convergence of SDA

In vector form, Eq. (22.19) is

$$F(k+1) = F(k) + 2\mu[B - AF(k)]$$

where $[B - AF(k)]$ is equivalent to the right-hand side of Eq. (22.18) in vector from. Substituting $B$ with $A^{-1}F_{\text{opt}}$, with minor manipulations one can show that

$$F(k+1) - F_{\text{opt}} = [I - 2\mu A][F(k) - F_{\text{opt}}]$$

Defining $V'(k) = F(k) - F_{\text{opt}}$, the last equation reduces to

$$V'(k+1) = [I - 2\mu A]V'(k)$$

Matrix A is Hermitian symmetric; therefore, it can be decomposed to $A = U\Lambda U_t^*$, where $\Lambda$ is a diagonal matrix whose main elements are the eigenvalues of $A$, and $U$ is a matrix whose columns are the eigenvectors of $A$; the $t$ and $*$ stand for the transpose and complex conjugate, respectively. Defining a new tap gain vector $V(k) = U^* \cdot V'(k)$, the above equation reduces to

$$V(k+1) = [I - 2\mu\Lambda]V(k)$$

Only elements of the main diagonal of $\Lambda$ are nonzero; therefore, the elements of the above vector equation are independent of one another, and each can be solved separately. These are first-order difference equations, and the solution for each element is

$$\nu_l(k) = \nu_l(0)[1 - 2\mu\lambda_l]^k$$

The condition for convergence of this equation is that the absolute value of the term in the brackets be less than one, which implies that one must have

$$0 < \mu < \frac{1}{\lambda_{\max}}$$

If $\mu = 1/\lambda_{\max}$ one has the fastest convergence for the transformed tap, $v_l(k)$, associated with the maximum eigenvalue, and the convergence of this particular tap for all equalizers is the same. In this way, all taps converge and the last transformed tap to converge is the one associated with $\lambda_{\min}$. Therefore, convergence of the equalizer is proportional to $\lambda_{\max}/\lambda_{\min}$, which is a measure of the eigenvalue spread of the channel. The fastest convergence is achieved for orthogonalized channels for which all eigenvalues are the same, and as the eigenvalues start to spread the convergence becomes slower.

### Least Mean Square (LMS) Algorithm

Implementation of the steepest descent algorithm given in Eq. (22.19) requires evaluation of the samples of the cross correlation function of the error signal $e(kT)$ and the received signal $r(t)$. At the arrival of each data symbol, one needs a set of $N + L + 1$ samples of the new correlation function for tap gain adjustments. The estimate of the sampled correlation function can be obtained by time averaging of the sampled error signal and sampled received signal. The roughest estimate of this correlation function is the sample estimate in which

$$\overline{e(kT)r(kT + \tau T - l\Delta)} = e(kT)r(kT + \tau T - l\Delta)$$

This roughest estimate is the least computationally complex estimate of the correlation function, and if it is used in Eq. (22.19) the algorithm is referred to as the LMS algorithm [75]. In an equalizer using the LMS algorithm tap gains are adjusted by

$$f_l(k+1) = f_l(k) + \mu e(kT)r(kT + \tau T - l\Delta)$$

The convergence in mean of the LMS algorithm is the same as that of the steepest descent algorithm. However, convergence of the individual tap gains in each trial also depends on the length of the equalizer [52, 83]. Convergence of the LMS algorithm is also sensitive to the transmitted sequence, and certain cyclic codes result in faster convergence, in the order of two [76, 77].

## 4.2 Fast-Converging Equalizers

The main engineering motivations for studying fast-converging algorithms are fast start-up equalization and fast tracking of the channel variations. Fast start-up equalization is used for voiceband data modems in multipoint polling networks, where a central site modem should adapt to short bursts of data from several users over different channels. The time required by the equalizer to set up its tap coefficients represents the main part of the modem start-up time. Therefore, it is desirable to find methods for rapid setup of the equalizers. Fast tracking of the channel variations is used for communication over rapidly fading frequency-selective multipath channels, such as HF radio. For this application, the LMS algorithm cannot track the channel closely; as a result, the modem is unable to compensate properly for the effects of channel distortions.

Fast start-up equalization can be performed in the frequency domain by using fast Fourier transform (FFT) algorithms, by applying tap storage, and by using fast adaptive algorithms in the time domain. The first two are dominant in fast start-up equalization while the third is applicable to both fast start-up and fast tracking problems.

## FFT Base Equalization

The FFT algorithm can be used to determine the taps of the equalizer so that the FFT of the taps is one over the FFT of the channel frequency response. A cyclic training sequence, $a_n$, with a flat spectrum is transmitted. The receiver takes the FFT of the received signal:

$$R(k) = \sum_{j=-L}^{N} r(nT + \tau T) \exp\left(\frac{-2\pi}{N+L+1} nk\right), \qquad -L \le k \le N$$

and forms the FFT of the channel

$$G(k) = \frac{R(k)}{A(k)}, \qquad -L \le k \le N$$

where $A(k)$ is the FFT of the known transmitted training sequence. The FFT of the equalizer tap gains is

$$\hat{F}(k) = \frac{1}{G(k)} = \frac{A(k)}{R(k)} = \frac{A(k)R^*(k)}{|R(k)|^2}, \qquad -L \le k \le N$$

and the tap gains of the equalizer are the inverse FFT of $\hat{F}(k)$.

The minimum start-up time for this technique is the time required for collecting $L+N+1$ samples of the received signal at the baud rate. This method is straightforward, and the details of the implementation are given in [78–80]. Windowing has proved to result in better performance in the presence of frequency offset; in practice, a triangular window has been adopted [79]. This window operates on the received data with a length twice that of the equalizer. For a voiceband modem with a symbol rate of 2400 symbols per second and a typical linear equalizer with a length of about 30, the start-up time for the modem using FFT techniques falls to around 25 ms.

Calculation of the tap gains using this method requires 150–200 complex multiplications for a typical LTE used in voiceband modems. This calculation requires a real multiplication time of around 500–600 ns if one wants to update the tap gains at the symbol rate (typically 2400 symbols/sec), which is feasible with state-of-the-art VLSI DSP processors. For applications with higher symbol rates, the computational complexity of this method would be a major drawback.

As compared with the LMS algorithm, this algorithm requires about one order of magnitude more multiplications. In addition, in the LMS algorithm multiplications are not sensitive to the accuracy of the error function; and even with an error quantized to 1 bit, results are acceptable [38]. Multiplication with 1 bit reduces to addition or subtraction, which makes the LMS algorithm more attractive for applications with higher speeds.

The FFT base algorithm is unsuitable for steady-state operations because the transmitted data patterns are arbitrary, and not all combinations of the data patterns have a flat spectrum in short times. Therefore, as opposed to the LMS algorithm, here the estimate of the taps depend on the transmitted sequence. Application of this algorithm to FSEs is also doubtful because the FFT of the channel at $T/2$ has zero amplitude in the sides of the passband and cannot be invested to obtain the FFT of the tap gains.

### Tap Storage Algorithm

In tap storage, the tap gains related to each connection in a multipoint connection are stored in the receiver along with a unique identifying code. On reception of the code for a particular connection, the receiver recovers and updates the tap gains used for that connection.

In tap storage one needs to recover timing and phase and identify the modem code before updating the stored taps. In practice, a duration of about 10 symbols is adequate to prepare the modem, which is about one-third of the length of the linear equalizers normally used in telephone channels. For 2400 symbols/sec telephone line modems, this duration represents a start-up time of less than 5 ms. This approach is faster than the frequency domain approach; however, it requires identification numbers for various modems, while the frequency domain approach is independent of the particular modem. Obviously, this method is not suitable for fast tracking of channel variations.

### Fast Equalization in the Time Domain

The LMS algorithm could be modified to find faster algorithms in the time domain. As discussed earlier, convergence of the LMS algorithms depends on the eigenvalue spread of the channel covariance matrix and the length of the equalizer [81–83]. To increase the convergence rate, the tap updating algorithm could be modified to be independent of the eigenvalue spread of the channel. The new algorithm should modify the LMS algorithm so that the eigenvalues look the same. For this reason, these algorithms are sometimes referred to as orthogonalization algorithms. The introductory work in orthogonalization is discussed in [84, 85].

Two approaches to self-orthogonalization exist. The first is to change the adjustment parameter $\mu$ to a vector, so that the adjustment of each tap is performed independently of the others. The vector is determined to adjust all taps for their maximum convergence rate by applying Kalman filtering principles [86]. The second approach is to orthogonalize the received data and apply the LMS algorithm to the orthogonalized data by using lattice base algorithms [87, 88].

Neither of these techniques is currently used in a successful voiceband modem product. However, a variation of the Kalman filtering algorithm has been used experimentally for fast tracking of a radio modem working over fading multipath HF channels [89]. The work on fast converging equalization in the time-domain motivated extensive research on the general structure of adaptive filters [90]. A brief description of Kalman filtering and lattice algorithms, as applied to the LTE, follows.

## Equalization Based on Kalman Filters

In [86] the application of Kalman filtering to adaptive self-orthogonalized equalization is introduced, and it is shown that the convergence rate of this algorithm is independent of the eigenvalue spread of the channel. Convergence of this algorithm is proportional to the length of the equalizer; however, computational complexity is on the order of the square of the length of the equalizer. Another algorithm that has reduced computational complexity while retaining the same speed of convergence is introduced in [91], and a concept of fast Kalman filtering which takes advantage of the structure of the data vector and reduces the complexity of the computation to multiples of the equalizer length is discussed in [92]. In the basic Kalman algorithm [86] for fast start-up equalization, the adjustment parameter is replaced by a vector $K(n)$, usually referred to as the Kalman gain. If one defines the error signal as in Eq. (22.6), the tap gain adjustments for this algorithm are given by

$$F(n) = F(n-1) + K(n)e(nT) \tag{22.20a}$$

where $F(n)$ is the vector of $L + N + 1$ tap gains, and

$$K(n) = \frac{P(n-1)R(n)}{N_0 + R^T(n)P(n-1)R(n)} \tag{22.20b}$$

Here $P(n)$ is an auxiliary matrix representing the inverse of the channel covariance matrix $A$, $R(n)$ is the vector of the last $L+N+1$ samples of the received signal, and $N_0$ is a constant representing residual noise form the background and the algorithm. Results of computer simulations show that the algorithm is insensitive to $N_0$ [86].

The auxiliary matrix $P(n)$ is updated with the recursion

$$P(n) = P(n-1) - K(n)R^T(n)P(n-1) \tag{22.20c}$$

with initial conditions

$$F(0) = 0$$

and

$$P(0) = \delta^{-1}I$$

where $\delta$ is a very small positive number.

The deterministic counterpart of Kalman filtering is the so-called recursive least square (RLS) algorithm [38, 90], which is applied to all adaptive filters. Kalman filters minimize the mean square error in a recursive manner based on the ensemble average. In least square algorithms, the sum of all received errors is minimized at each iteration. For the RLS algorithm we intend to minimize the weighted received samples of error signal

$$\xi(nT) = \sum_{0}^{n} \lambda^{n-k} |e(kT)|^2$$

where $\lambda$ defines the memory of the system. For $\lambda = 1$ all the past observed samples have the same weight in contributing in the performance criterion $\xi(nT)$. For very small values of $\lambda$ only the last few samples contribute significantly in the calculations. The basic RLS algorithm which minimizes the above criterion is very similar to the Kalman filter algorithm, with $\lambda = N_0$. The only difference between the two algorithms is that in the RLS algorithm the left-hand side of Eq. (22.20c) is multiplied by $\lambda$ in each iteration. The numerical stability of this algorithm could be adjusted with parameter $\lambda$.

### Lattice-Based Equalization

Lattice structures have been studied for linear prediction coding for application in speech processing [93]. A lattice is a structure for implementing an adaptive linear predictor filter. The standard implementation of a linear predictor filter is a TDL FIR filter. The advantage of the lattice over TDL implementations is that the lattice consists of identical stages. In the linear prediction literature the two symmetric outputs of the lattice (see Fig. 22.15), $f_j(n)$ and $b_j(n)$, are termed the forward and backward prediction errors, respectively, and the coefficients $c_{jj}$ used to relate upper and lower branches are called reflection coefficients. The reflection coefficients are related to the forward and backward errors by the following equation [93]:

$$c_{jj} = \frac{\overline{f_j(n) b_j(n-1)}}{\overline{f_j^2(n)} + \overline{b_j^2(n-1)}}$$

The principle of orthogonality in lattices [93] indicates that the input and the backward prediction errors form a set of orthogonal processes. This property motivates one to use this set rather than the tapped delay version of the received signal for the purpose of equalization.

Figure 22.15 shows a lattice equalizer; the orthogonalized set of backward prediction errors are multiplied by the tap gains and added to form the output of the equalizer. This form has two advantages. First, the signal used for the equalization is orthogonal, resulting in faster convergence than for the LMS algorithm. Second, the modularity of the architecture simplifies manipulation of the hardware and facilitates the implementation. Since the error at each stage is available as a separate output, one can measure the optimum practical length for a given application.

For this structure the output signal is

$$\hat{d}(n) = \sum_{j=1}^{L} c_j(n) b_j(n)$$

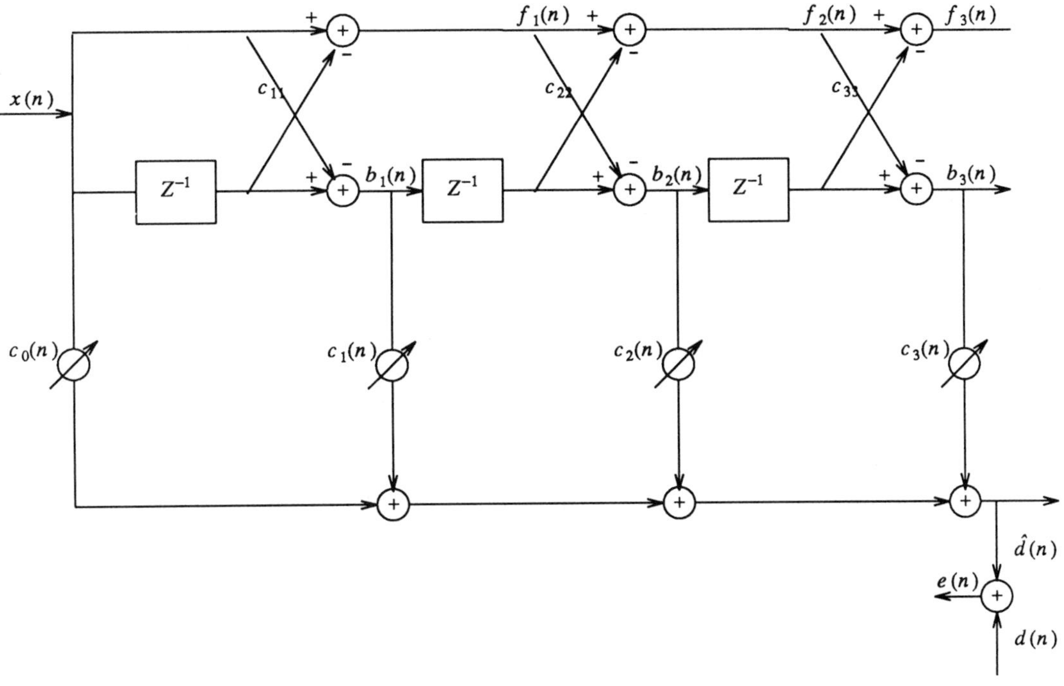


**Figure 22.15** A lattice implementation for LTE.

where $\hat{d}$ is the estimate of the transmitted symbol $d(n)$, and the $c_j(n)$'s are the tap gains. With the MSE function

$$\xi = \overline{|e(n)|^2} = \overline{|d(n) - \hat{d}(n)|^2}$$

the derivative used in the steepest descent algorithm is

$$\frac{\delta \xi}{\delta c_l(n)} = -2\overline{e(n)b_j(n)}$$

and estimation of the average with one sample, similar to the LMS algorithm, results in

$$c_j(n) = c_j(n-1) + 2\mu_j(n)e(n)b_j(n)$$

It can be shown that $\mu$ could be updated recursively with the following algorithm [87]:

$$\frac{1}{2\mu_j(n)} = \lambda \frac{1}{2\mu_j(n-1)} + b_j^2(n)$$

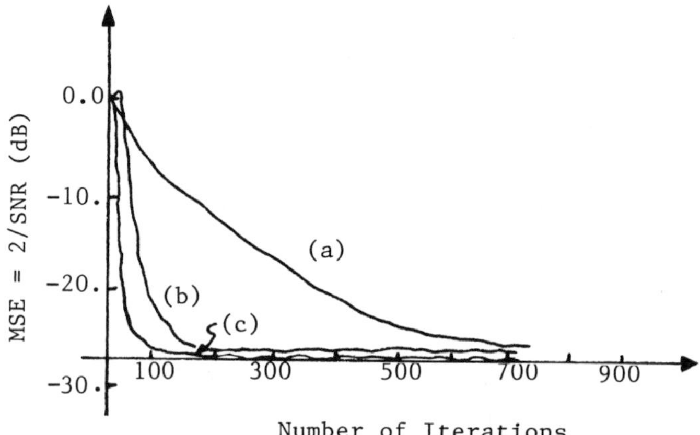


**Figure 22.16** The ensembled average of squared error (MSE) versus number of iterations for different implementations of a LTE: (a) the standard LMS algorithm, (b) lattice implementation, (c) Kalman base algorithm.

where $\lambda$ the parameter defining the memory of the system.

Figure 22.16 compares the convergence rates of the LMS equalizer, Kalman equalizer, and lattice equalizer for a channel with an eigenvalue spread factor of 11. More complex equalizers such as the FSE or DFE and more complex algorithms such as the RLS algorithm have been examined for equalization but are beyond the scope of this chapter. The interested reader is referred to [38, 90].

## ACKNOWLEDGMENTS

Throughout my professional career in signal processing and telecommunications, I have had the opportunity to work and learn from a number of people whom I would like to acknowledge. I appreciate Dr. J. L. Holsinger, who introduced me to the real issues in voiceband data communications; Dr. P. A. Bello, who improved my understandings in fading channel communications: and Prof. J. G. Proakis, from whom I received more insight into the analysis of the adaptive algorithms applied in telecommunication. I also appreciate Prof. J. W. Matthews, my Ph.D. thesis advisor, with whom I started my work in the general area of communications. Finally, I thank my graduate students, K. Zhang, Tom Sexton, and R. Ganesh, for helping me in preparing some of the graphs and diagrams for this chapter.

## REFERENCES

1. R. E. Cochiere and J. L. Flanagan, Current perspectives in digital speech, *IEEE Comm. Soc. Mag.*, 32–40 (January 1983).
2. A. Gresho and C. Cuperman, Vector quantization: A pattern-matching technique for speech coding, *IEEE Comm. Soc. Mag.*, pp. 15–21 (December 1983).

3. J. B. O'Neal, Jr., Waveform encoding of voice-band data signals, *Proc. IEEE*, 232–246 (February 1980).
4. N. S. Jayant and P. Noll, *Digital Coding of Waveforms*, Prentice-Hall, Englewood Cliffs, New Jersey (1984).
5. A. N. Natraveli and J. O. Lim, Picture coding: A review, *Proc. IEEE*, 366–406 (March 1980).
6. H. Kaneko and T. Ishiguro, Digital television transmission using bandwidth compression techniques, *IEEE Commun. Soc. Mag.*, 14–22 (July 1980).
7. G. Held, *Data Compression*, Wiley, New York (1984).
8. K. Feher (editor), *Advanced Digital Communications: Systems and Signal Processing Techniques*, Prentice-Hall (1987).
9. S. B. Weinstein, Echo cancellation in the telephone network, *IEEE Commun. Soc. Mag.*, 9–15 (January 1977).
10. D. G. Messerschmitt, Echo cancellation in speech and data transmission, *IEEE J. of Sci. Areas in Comm.*, 283–298 (March 1984).
11. K. Pahlavan and J. L. Holsinger, Voice-band data communication modems, a historical review: 1919–1988, *IEEE Commun. Soc. Mag.* (Jan. 1988).
12. H. Harris, T. Saliga, and D. Walsh, An all digital 9600 bps LSI modem, *National Telecomm. Conf.*, pp. 279–284 (December 1974).
13. H. L. Logan and G. D. Forney, A MOS/LSI multiple configuration 9600 bps data modem, *Int. Conf. on Comm.*, 48.7–48.12 (June 1976).
14. P. J. VanGerwen, N. A. M. Verhoeckx, H. A. VanEssen and F. A. M. Snijders, Microprocessor implementation of high-speed data modems, *IEEE Trans. Commun.*, 238–250 (February 1977).
15. D. Godard and D. Pilost, A 2400 bps microprocessor-based modem, *IBM J. Res. Dev.*, 17–24 (January 1981).
16. M. Kaya, K. Ishizuka, and N. Maeda, High speed data modem using digital signal processor, *Int. Conf. Comm.*, pp. 14.7.1–14.7.5 (June 1981).
17. K. Watanabe, K. Inoue, and Y. Sato, A 4800 bps microprocessor data modem *IEEE Trans. Commun.*, 493–498 (May 1978).
18. H. Haas, E. A. Fuchs, H. Sailer, and H. Schenk, Digital high speed modem using only a few standard components, *Int. Conf. on Comm.*, p. A5.8.1 (June 1983).
19. S. U. H. Qureshi and H. M. Ahmed, *VLSI Signal Processing*, IEEE Press, (1984), New York, N. Y.
20. Y. Mochida, S. Unagami, K. Murano, N. Fujimura, T. Kinoshita, and Y. Tanaka, VLSI high speed data modem, *IEEE Globecom*, 45.8.1–45.8.6 (1983).
21. W. Twaddell, Modem IC's, EDN, 160–172 (March 1985).
22. R. H. Cushman, Third-generation DSP's put advanced function on chip, EDN, 59–67 (July 1985).
23. M. Y. Levy and C. L. Poinas, "Adaptive Phase Correctors for Data Transmission Receivers," Proc. International Conference on Communications, pp. 45.5.1–45.5.5 (June 1979).
24. H. Koboyaski, Simultaneous adaptive estimation and detection algorithm for carrier modulated data communication systems, *IEEE Trans. Commun. Technol.* 268–280 (June 1971).
25. L. E. Franks and J. P. Bubrouski, Statistical properties of timing jitter in a PAM timing recovery scheme, *IEEE Trans. Commun.*, 913–920 (July 1974).

26. D. L. Lyon, Envelope-driven timing recovery in QAM and SQAM systems, *IEEE Trans. Commun.* 1327–1331 (November 1975).
27. J. E. Mazo, Optimum timing phase for an infinite equalizer, *Bell Syst. Tech. J.*, 189–201 (January 1975).
28. G. Ungerboeck, Fractional tap-spacing equalizer and consequences for clock recovery in data modems, *IEEE Trans. Commun.*, 856–864 (August 1976).
29. S. H. Qureshi and G. D. Forney, Performance and properties of $T/2$ equalizers, NTC, 11:1–1, 11:1–9 (December 1977).
30. R. D. Gitlin and S. B. Weinstein, Fractionally-spaced equalization: An improved digital transversal equalizer, *Bell Syst. Tech. J.*, 275–296 (February 1981).
31. R. D. Gitlin, H. C. Meadors, and S. B. Weinstein, The tap-leakage algorithm: An algorithm for the stable operation of a digitally implemented, fractionally-spaced adaptive equalizer, *Bell Syst. Tech. J.*, 1817–1839 (October 1982).
32. R. D. Gitlin and J. Salz, Timing recovery in PAM systems, *Bell Syst. Tech. J.*, 1645 (May 1971).
33. R. W. Chang, Joint equalization, carrier acquisition and timing recovery for data communications, *Int. Conf. of Comm.*, 34–43 (June 1970).
34. K. H. Mueller and M. Muller, Timing recovery in digital synchronous data receivers, *IEEE Trans. Commun.* 561–521 (May 1976).
35. D. N. Godard, Passband timing recovery in all-digital modem receiver, *IEEE Trans. Commun.*, 517–523 (May 1978).
36. S. H. Qureshi, Timing recovery for equalized partial-response systems, *IEEE Trans. Commun.* 1326–1330 (December 1976).
37. J. Steel and B. M. Smith, Carrier and clock recovery from transversal equalizer tap settings for a partial response system, *IEEE Trans. Commun.*, 976–979 (September 1975).
38. J. G. Proakis, *Digital Communications*, McGraw-Hill, New York (1983).
39. G. Ungerboeck, Trellis-coded modulation with redundant signal sets—Part I and II, *IEEE Commun. Soc. Mag.*, 5–21 (February 1987).
40. L. F. Wei, Rotationally invariant convolutional channel coding with expanded signal space—Part II: Nonlinear codes, *IEEE J. Selected Areas Commun.*, 672–686 (September 1984).
41. A. J. Viterbi, Error bounds for convolutional codes and an asymptotically optimum decoding algorithm, *IEEE Trans. Inf. Theory*, 260–269 (April 1967).
42. G. D. Forney, Jr., R. G. Gallager, G. R. Lang, F. M. Longstaff, and S. U. Qureshi, Efficient modulations for band-limited channels, *IEEE J. of Select. Areas in Comm.*, 632–647 (September 1984).
43. K. Pahlavan and J. L. Holsinger, "Signal Constellation for Voice-Band Data Communication over Channels with Nonlinear Quantizers," Proc. IEEE Int. Conf. on Commun., 50.2.1.–50.2.5 (June 1986).
44. K. Pahlavan and J. L. Holsinger, "Trellis Code Modulation with Expanded Constellation," Proc. IEEE Int. Conf. on Commun., 12.3.1–12.2.5 (June 1987).
45. D. A. Spaulding, Synthesis of pulse-shaping networks in the time domain, *Bell Syst. Tech. J.*, *48*: 2425–2444 (September 1969).
46. L. R. Rabiners and B. Gold, *Theory and Application of Digital Signal Processing*, Prentice-Hall, Englewood Cliffs, New Jersey (1975).

47. K. Muller, A new approach to optimum pulse shaping in sampled systems using time-domain filtering *Bell Syst. Tech. J.*, *52*: 723–729 (May–June 1973).
48. P. Chevillat and G. Ungerboeck, Optimum FIR transmitter and receiver filters for data transmission over band-limited channels, *IEEE Trans. Commun.*, *COM-30*: 1909–1915 (August 1982).
49. A. C. Salazar and V. B. Lawrence, "Design and Implementation of Transmitter and Receiver Filters with Periodic Coefficient Nulls for Digital Systems," Proc. IEEE International Conference on Acoustics, Speech, and Signal Processing, Paris, pp. 306–310 (May 1982).
50. J. K. Liang, R. J. P. DeFigueiredo, and F. C. Lu, Designing of optimum Nyquist partial response, $N$th band, and nonuniform tap spacing FIR digital filters using linear programming technique, *IEEE Trans. Circuits Syst.*, *CAS-32*: 386–392 (April 1985).
51. T. Saramaki and Y. Neuvo, A class of FIR filters with zero intersymbol interference, *IEEE Trans. Circuits Syst.*, (1986).
52. S. H. Qureshi, Adaptive equalization, *IEEE Proc.*, 1349–1387 (September 1985).
53. G. D. Forney, Jr., Maximum-likelihood sequence estimation of digital sequences in the presence of intersymbol interference, *IEEE Trans. Inf. Theory*, 363–378 (May 1972).
54. D. D. Falconer and F. R. Magee, Evaluation of decision feedback equalization and Viterbi algorithm detection for voiceband data transmission—Part I, *IEEE Trans. Commun.*, 1130–1139 (October 1976).
55. D. D. Falconer and F. R. Magee, Evaluation of decision feedback equalization and Viterbi algorithm detection for voiceband data transmission—Part II," *IEEE Trans. Commun.* 1238–1245 (November 1976).
56. D. D. Falconer, A. U. H. Sheikh, E. Eleftheriou, and M. Tobis, Comparison of DFE and MLSE receiver performance on HF channels, *IEEE Trans. Commun.*, 484–486 (May 1985).
57. P. Monsen, Theoretical and measured performance of a DFE modem on a fading multipath channel, *IEEE Trans. Commun.*, 1144–1153 (October 1977).
58. D. Chase and P. A. Bello, A combined coding and modulation approach for high speed data transmission over troposcatter channel, NTC, 28–32 (December 1975).
59. R. F. Magee and J. G. Proakis, Adaptive maximum likelihood sequence estimation for digital signaling in the presence of ISI, *IEEE Trans. Inf. Theory*, 120–124 (January 1973).
60. G. Ungerboeck, Adaptive maximum likelihood receiver for carrier modulated data transmission systems, IEEE Trans. Commun., 624–636 (May 1974).
61. K. Pahlavan, Channel measurements for wideband digital communications over fading channels, Ph.D. dissertation, Worcester Polytechnic Institute (June 1979).
62. V. G. Koll and S. B. Weinstein, Simultaneous two-way data transmission over a two wire line, *IEEE Trans. Commun.*, 143–147 (February 1973).
63. K. M. Mueller, A new digital echo canceller for two-wire full duplex data transmission, *IEEE Trans. Commun.*, 956–967 (September 1976).

64. S. B. Weinstein, A passband data-driven echo canceller for full duplex transmission on two-wire circuits, *IEEE Trans.*, 654–666 (July 1976). *EDN*, 59–67 (July 1985).
65. R. W. Lucky, Automatic equalization for digital communication, *Bell Syst. Tech. J.*, 547–588 (April 1965).
66. R. W. Lucky, Techniques for adaptive equalization of digital communication, *Bell Syst. Tech. J.*, 255–286 (February 1966).
67. C. A. Belfiore and J. H. Park, Jr., Decision feedback equalization, *Proc. IEEE*, 1143–1156 (August 1979).
68. J. Salz, Optimum mean-square decision feedback equalization, *Bell Syst. Tech. J.* 1341–1373 (October 1973).
69. P. A. Bello and K. Pahlavan, Adaptive equalization for SQPSK and SQPR over frequency selective microwave LOS channels, *IEEE Trans. Commun.*, 609–615 (May 1984).
70. K. Pahlavan, Comparison between the performance of QPSK, SQPSK, QPR, and SQPR systems over microwave LOS channels, *IEEE Trans. Commun.*, (March 1985).
71. D. Godard, Self-recovering equalization and carrier tracking in two-dimensional data communication systems, *IEEE Trans. Commun.*, *COM-28*: 1867–1875 (November 1980).
72. G. J. Foscihini, Equalizing without alternating or detecting data, *Bell Syst. Tech. J.* 1885–1911 (October 1985).
73. R. D. Gitlin, E. Y. Ho, and J. E. Mazo, Passband equalization for differentially phased-modulated data signals, *Bell Syst. Tech. J.* *52*(2): 219–238 (1973)
74. D. D. Falconer, Analysis of a gradient algorithm for simultaneous passband equalization and carrier phase recovery, *Bell Syst. Tech. J.*, 409–428 (April 1976).
75. B. Widrow and M. E. Hoff, "Adaptive Switching Circuits," *IRE WESCON Convention Record*, pp. 96–104 (August 1960).
76. S. U. H. Qureshi, "Fast start-up equalization with periodic training sequences, *IEEE Trans. Inf. Thy.* 553–563 (September 1977).
77. K. H. Mueller and D. A. Spaulding, Cyclic equalization—A new rapidly converging equalization technique for synchronous data communication, *Bell Syst. Tech. J.*, 369–406 (February 1975).
78. M. Choquet, Channel equalization apparatus and method using Fourier transform technique, U. S. Patent 4,152,649 (May, 1979).
79. D. N. Godard, A 9600 bps modem for multipoint communication systems, *National Telecomm. Conf.*, B3.3.1–B3.3.5 (1981).
80. A. Milewski, Periodic sequences with optimal properties for channel estimation and fast start-up equalization, *IBM J. Res. Dev.*, 426–431 (September 1983).
81. J. G. Proakis and J. H. Miller, An adaptive receiver for digital signaling through channels with intersymbol interference, *IEEE Trans. Inf. Theory*, 484–497 (July 1969).
82. A. Gresho, Adaptive equalization of highly dispersive channels for data transmission, *Bell Syst. Tech. J.*, 48–55 (1969).
83. G. Ungerboeck, Theory on the speed of convergence in adaptive equalizers for digital communications, *IBM J. Res. Dev.*, 546–555 (November 1972).

84. R. W. Chang, A new equalizer structure for fast start-up digital communication, *Bell Syst. Tech. J.*, 50 (1971).
85. H. Koboyashi, Application of Hestens-Stiefel algorithm to channel equalization, ICC, 21:25–21:30 (1971).
86. D. Godard, Channel equalization using a Kalman filter for fast data transmission, *IBM J. Res. Dev.*, 267–273 (May 1974).
87. E. H. Satorius and S. T. Alexander, Channel equalization using adaptive lattice algorithms, *IEEE Trans. Commun.*, 899–905 (June 1979).
88. E. H. Satorius and J. D. Pack, Application of least squares lattice algorithms to adaptive equalization, *IEEE Trans. Commun.*, 136–142 (February 1981).
89. F. M. Hsu, Square root Kalman filtering for high-speed data received over fading dispersive HF channels, *IEEE Trans. Inf. Theory*, 753–763 (1982).
90. S. Haykin, *Adaptive Filter Theory*, Prentice-Hall, Englewood-Cliffs, New Jersey (1986).
91. R. D. Gitlin and F. R. Magee, Jr., Self-orthogonalizing algorithms for accelerated convergence of adaptive equalizers, *IEEE Trans. Commun.*, 666–672 (July 1977).
92. D. D. Falconer and L. Ljung, Application of fast Kalman equalization to adaptive equalization, *IEEE Trans. Commun.*, 1439–1446 (October 1978).
93. J. Makhoul, A class of all-zero lattice digital filters: Properties and applications, *IEEE Trans. Acoust. Speech Signal Process.*, 304–314 (1978).

# 23

# Recent Advances in Signal Processing in Medical Tomography

M. IBRAHIM SEZAN* Photographic Products Group, Eastman Kodak Company, Rochester, New York

HENRY STARK, PEYMA OSKOUI-FARD, HUI PENG, and EITAN YUDILEVICH† Electrical, Computer and Systems Engineering Department, Rensselaer Polytechnic Institute, Troy, New York

## 1 INTRODUCTION

This chapter is devoted to some of the important signal processing problems that arise in recent medical imaging modalities, namely computed tomography (CT) and magnetic resonance (MR) imaging.

In a single chapter of reasonable length it is obviously impossible to review all of the signal processing activities in medical imaging. Even to yield a useful survey of all of the techniques in use in CT, MR imaging, nuclear medicine imaging, positron emission tomography, etc., would take more space than is available. Thus, we have decided to focus on a few recent activities, all still in the experimental stage, that can extend the range of application of medical imaging methods. Conventional signal processing methods for image reconstruction such as convolution back projection (CBP) and algebraic reconstruction techniques (ART) are well covered in the literature. The reader has available a broad range of basic and advanced textbooks that discuss nearly all aspects of conventional medical imaging.

Our emphasis will be on advances in four areas: 1) image reconstruction from limited view data, 2) reconstruction by Fourier methods, 3) geometry-free reconstruction from projections, and 4) reconstruction from sampling on a spiral scan in MR imaging.

Present affiliations:

*Photographic Research Laboratories, Eastman Kodak Company, Rochester, New York

†RAFAEL, Ministry of Defense, Haifa, Israel

The problem of image reconstruction from projections arises in diverse disciplines such as medicine, geophysics, astronomy, and electron microscopy [1]. Irrespective of the method of reconstruction, it is generally assumed that 360° view data are available for processing. However, the situation often arises where the number of views, their angular range, or the number of raysums (we use the term *raysum* rather than *projection* for reasons given below) within a view is restricted by various constraints. In medical CT there may be insufficient time for full-view data acquisition of time-varying anatomical situations. In other applications, e.g., on-site industrial inspection, incomplete view data can result when obstructions prevent full-view data collection or when imaging a moving part or a time-varying event. In geophysical exploration, view data may be restricted by economic constraints.

Initially, in this chapter, we consider the signal processing problems associated with tomographic image reconstruction from angularly view-limited raysum data. We combine the method of projections onto convex sets (POCS) [2–5] with the direct Fourier method (DFM) [6] of reconstructing an image from raysums into a single restoration/reconstruction algorithm. The resulting algorithm is then used to reconstruct good-quality imagery from angularly view-limited parallel-beam raysum data. In DFM, an image is reconstructed by taking the inverse fast Fourier transform (FFT) of the spectrum of the raysum data. The method of POCS is a general technique for determining a feasible solution consistent with a number of convex-type constraints. The solution is restricted to lie in convex sets associated with the a priori constraints. In the literature, there are numerous examples of POCS being widely used in signal/image restoration [7–11]. In this context, the main role of POCS is to recover the information missing in the view data.

At this point, we would like to clarify the double meaning of *projection*. In POCS, the operation of projection refers to a mathematical operation of finding a point, i.e., a function which lies in a convex set and yet is nearest to another point generally not lying in the set. In tomography, projection refers to a line integral or sequence of line integrals in which the integration paths are uniformly spaced parallel straight lines (as in parallel-beam projections) or lines that radiate out fan-style from a source (as in fan-beam projections). To avoid the ambiguity in the use of the term *projection*, we shall use the term *raysum* to denote a CT-type line integral and reserve the term *projection* for the set-restricted nearest neighbor of an arbitrary function.

We next develop a reconstruction algorithm entirely based on the method of POCS. The resulting technique is a *geometry-free* algorithm. By this we mean that the algorithm does not require a particular raysum geometry such as in parallel-beam or fan-beam CT. The direct use of POCS in actual reconstruction was first suggested in [12]. However, to the best of our knowledge this is the first time convex projections have actually been used to invert raysum data. We show that the algebraic reconstruction technique (ART) [13], known to be the only other geometry-free algorithm, is an elementary convex projections algorithm that fails to incorporate a priori information about the image. We demonstrate how a priori constraints can be used to stabilize the basic ART algorithm and present results demonstrating the feasibility of this method.

Modern X-ray CT systems use the fan-beam geometry owing to its fast scanning ability compared with the parallel-beam geometry. The standard algorithm for reconstructing from fan-beam data is filtered convolution back projection (CBP) [13, 14]. We investigate the extension of DFM to the fan-beam data case for the following reasons: 1) previous work involving parallel-beam data showed that good-quality reconstructions could be achieved using DFM, 2) DFM uses fewer computations, on the order of $N^2 \log N$ versus $N^3$ for CBP, in the case of $N$ views, and 3) because DFM involves operations in the space and the spatial frequency domain, it is well suited for problems involving incomplete data situations in which image recovery is achieved by enforcing a priori space and spatial frequency constraints. The first step of this extension is the rebinning process [13–15], where fan-beam raysums are uniquely associated with the corresponding raysums in a parallel-beam geometry. The resulting complication, however, is that the rebinned discrete raysum data are not on a uniform lattice in the space domain. Therefore, before DFM can be applied, an interpolation from unequally spaced samples to uniformly spaced samples must be performed. Moreover, given the extreme sensitivity of the quality of the reconstructed image to interpolation errors, the interpolation must be exact, or at least as exact as practical considerations allow. It turns out that the required interpolation of the rebinned fan-beam data into equally spaced parallel-beam data requires a two-dimensional (2-D) interpolation that can be efficiently done by two one-dimensional (1-D) interpolations. One interpolation makes use of the interpolation formula derived in [6] for computing values from uniformly spaced samples on a circle. The other interpolation is based on a recently published formula [16] that enables interpolation of 1-D signals from nonuniformly spaced samples. We derive an exact 2-D interpolation formula and determine the bounds on the parameter settings of fan-beam CT that enable direct Fourier reconstruction. The complete algorithm necessary for processing the fan-beam data is described.

Finally, we consider an important interpolation problem that arose in echo-planar MR imaging [17]. Originally proposed in [17], echo-planar imaging has been extended and modified to make it attractive as a high-speed imaging technique [18–21]. Such a technique opens the possibility of real-time human cardiac imaging [22]. An encouraging extension was suggested in [21] and implemented in [20], namely the use of a *linear spiral lattice* in the Fourier plane. We derive an exact interpolation formula for interpolating samples on a 2-D linear spiral lattice over a Cartesian lattice. The image is then reconstructed by computing the inverse FFT.

In the following, we summarize the principles of tomographic image reconstruction as applied to CT and MR imaging. We then briefly review the method of POCS and present the constraint sets of importance in tomography. These two sections set the stage for the rest of the chapter.

## 2 PRINCIPLES OF TOMOGRAPHIC IMAGE RECONSTRUCTION

In this section we summarize the principles of tomographic image reconstruction in CT and MR imaging.

### 2.1 X-Ray Computed Tomography

In X-ray CT, the 2-D attenuation coefficient distribution over a selected cross section of a 3-D object is reconstructed and imaged from the raysum data obtained by X-rays. Consider the configuration shown in Figure 23.1. The object is penetrated by a beam of width $\Delta$ and profile $w(\cdot)$. The raysum $p_\theta(s)$, at a view angle $\theta$, of the

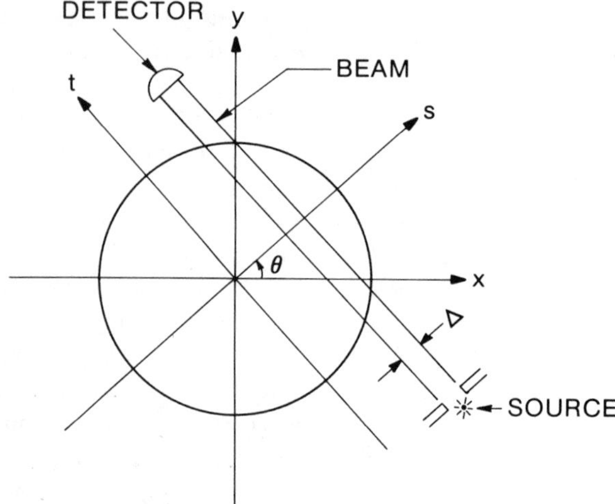


**Figure 23.1** Raysum configuration for a finite-width beam penetrating the object over a 2-D cross section.

cross-sectional distribution $f(x, y)$ in the rotated Cartesian system $(s, t)$ is given by

$$p_\theta(s) \triangleq \int_{s-\Delta/2}^{s+\Delta/2} ds' w(s - s') \int_{t_A(s',\theta)}^{t_B(s',\theta)} \hat{f}(s, t; \theta)\, dt \tag{23.1}$$

where

$$\begin{pmatrix} s \\ t \end{pmatrix} = \begin{pmatrix} \cos\theta & \sin\theta \\ -\sin\theta & \cos\theta \end{pmatrix} \begin{pmatrix} x \\ y \end{pmatrix} \tag{23.2a}$$

$$\begin{pmatrix} x \\ y \end{pmatrix} = \begin{pmatrix} \cos\theta & -\sin\theta \\ \sin\theta & \cos\theta \end{pmatrix} \begin{pmatrix} s \\ t \end{pmatrix}$$

$$\hat{f}(s, t; \theta) \triangleq f(s\cos\theta - t\sin\theta, s\sin\theta + t\cos\theta), \qquad -\infty < s < \infty,\ 0 < \theta \le \pi \tag{23.2b}$$

and $t_A(s', \theta)$ and $t_B(s', \theta)$ are the lower and upper limits of integration along the $t$ direction, respectively. The profile function $w(s)$ describes the X-ray beam intensity across its width and satisfies

$$\int_{-\infty}^{+\infty} w(s)\, ds = 1 \tag{23.3}$$

In the case where the beam is infinitely thin, $w(s) = \delta(s)$, the Dirac delta function, Eq. (23.1) reduces to

$$p_\theta(s) = \int_{t_A(s,\theta)}^{t_B(s,\theta)} \hat{f}(s,t;\theta)\,dt \tag{23.4}$$

Thus the integration is carried out along a segment of a single line $\mathcal{L}$ parametrized by $s$ and $\theta$. When $p_\theta(s) \triangleq p(s,\theta)$ is regarded as a function of two variables, it is sometimes called the *shadow* of $f(x,y)$. When $\theta$ is held fixed and $p_\theta(s)$ is regarded as a function of the single variable $s$, it is called the *projection function* of $f$ at view angle $\theta$. However, because we wish to avoid the ambiguity resulting from the two meanings of the word projection, we shall call $p_\theta(s)$ the shadow function of $f$ at angle $\theta$. When $s$ and $\theta$ are both fixed, $p(s,\theta)$ is the raysum of $f$ at displacement $s$ (from the origin) and angle $\theta$.

The fundamental problem of CT is to reconstruct $f$ from its shadow. In practice, we do not measure all of $p(s,\theta)$ but rather its samples over a set of discrete displacement values $\{s_n\}$, $n = 1, 2, \ldots, N_s$, and at a finite set of view angles $\{\theta_k\}$, $k = 1, 2, \ldots, N_\theta$. Hence, the practical problem of CT is to reconstruct $f$ at a discrete set of values $\{(x_l, y_m)\}$ from a finite set of raysums $\{p(s_i, \theta_i)\}$, $i = 1, 2, \ldots, N_{s\theta}$, $(N_{s\theta} \triangleq N_s N_\theta)$.

Three major reconstruction algorithms are 1) the algebraic reconstruction technique [13], 2) the convolution back-projection algorithm [13, 14], and 3) the direct Fourier method [6]. ART is a special case of the POCS-based algorithm that will be discussed in Section 5. The CBP algorithm is discussed at length in the literature [13, 14] and therefore it will not be discussed here. On the other hand, DFM is central to Sections 4, 5, and 6 and hence it will discussed in detail.

## The Direct Fourier Method

The DFM is based on the central slice projection theorem (CSPT). To derive the mathematical statement of the CSPT, we consider the 1-D Fourier transform (FT) of the shadow function of a view angle $\theta$, i.e.,

$$P_\theta(\omega) \triangleq \mathcal{F}[p_\theta(s)]\omega = \int_{-\infty}^{\infty} p_\theta(s) e^{-j\omega s}\,ds \tag{23.5}$$

where $\mathcal{F}[\cdot]$ denotes the Fourier transform operator. From Eq. (23.4) the Fourier integral can be written as

$$\int_{-\infty}^{\infty} p_\theta(s) e^{-j\omega s}\,ds = \int_{-\infty}^{\infty} \int_{t_A(s,\theta)}^{t_B(s,\theta)} \hat{f}(s,t;\theta) e^{-j\omega s}\,dt\,ds \tag{23.6}$$

Using Eq. (23.2), we have

$$\int_{-\infty}^{\infty} p_\theta(s)e^{-j\omega s}\, ds = \iint_{\Omega} f(x,y)e^{-j\omega(x\cos\theta + y\sin\theta)}\, dx\, dy \tag{23.7}$$

where $\Omega$ is the support of $f$. If we restrict $\omega \geq 0$ and $0 < \theta \leq 2\pi$ and let $\omega_x \triangleq \omega\cos\theta$, $\omega_y \triangleq \omega\sin\theta$, then $\omega = (\omega_x^2 + \omega_y^2)^{1/2}$ and $\theta = \tan^{-1}(\omega_y/\omega_x)$ denote a set of polar coordinates. Thus Eq. (23.7) can be rewritten as

$$P_\theta(\omega) = F(\omega,\theta) \qquad \omega \geq 0,\ 0 \leq \theta < 2\pi \tag{23.8}$$

where $F(\omega,\theta)$ is the 2-D FT of $f(x,y)$ in polar coordinates and $P_\theta(\omega)$ is the 1-D FT of $p_\theta(s)$. Equation (23.8) is the mathematical statement of the CSPT. Each function $p_\theta(s)$ enables us to compute the FT of $f(x,y)$ along a line at angle $\theta$ (Figure 23.2). In words, Eq. (23.8) says that the Fourier transform of the shadow function at angle $\theta$ yields a central cross section of the Fourier transform of the object function. In practice, $p_\theta(s)$ is the data actually obtained at the discrete detector locations $\{s_i\}$. In place of Eq. (23.5) the discrete Fourier transform is used to obtain $P_\theta(\omega)$ at the discrete set of spatial frequencies $\{\omega_n\}$. Since the view angles are also a discrete set, $F(\omega,\theta)$ is known at points on a polar lattice $\{\omega_n, \theta_k\}$. Here we immediately see that if we could compute the 2-D inverse FT of $F(\omega_n, \theta_k)$ we

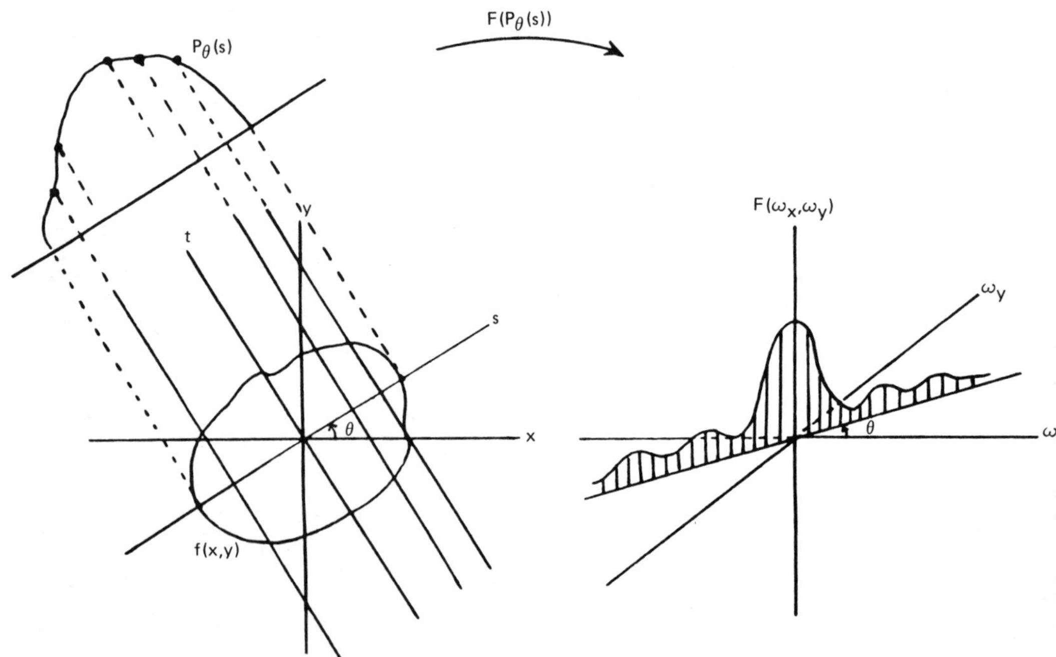


**Figure 23.2** Illustration of the CSPT for the parallel-beam geometry.

would obtain the desired image. In practice, however, the 2-D Fourier transform is realized via the fast Fourier transform, which requires knowledge of the data on a rectangular lattice of Cartesian points $\{u_l, v_m\}$. The data, however, are known on a polar lattice of points $\{\omega_n, \theta_k\}$ which, in general, do not coincide with the Cartesian points $\{u_l, v_m\}$ (see Figure 23.3). Hence, interpolation is required, and specifically interpolation from points on a polar lattice to points on a Cartesian lattice, to use the FFT for inverting $F(\omega_n, \theta_k)$. However, inexact interpolation is a major source of error in reconstruction by DFM even when the interpolation errors are not large (Lewitt [14] discussed this point in some detail). Attempts to implement DFM using nearest-neighbor and linear interpolation were made by Mersereau and Oppenheim [23]. However, we found out that these fail to give good results in CT. Recently, an exact interpolation procedure for angularly band-limited functions was derived [6, 24] which, when combined with direct Fourier imaging, produced images equal to or surpassing in quality those produced by the CBP algorithm. Moreover, these results were obtained under practical constraints, such as approximating exact interpolation with a finite sample formula and prefiltering non-band-limited signals to avoid aliasing. The interpolation formula is given in the following theorem.

**Theorem.** Let $f(x, y)$ be space-limited to $2A$. Let its FT in polar coordinates $F(\omega, \theta)$ be angularly band-limited to $K$. Then $F(\omega, \theta)$ can be reconstructed from its polar samples $F(\omega_n, \theta_k) \triangleq F(n/2A, 2\pi k/N)$ over a uniform lattice from

$$F(\omega,\theta) = \sum_{n=-\infty}^{\infty} \sum_{k=0}^{N-1} \tilde{F}\left(\frac{n}{2A}, \frac{2\pi k}{N}\right) \operatorname{sinc}\left[A\left(\frac{\omega}{\pi} - \frac{n}{A}\right)\right] \frac{\sin[\frac{1}{2}(N-1)(\theta - 2\pi k/N)]}{N \sin[(\frac{1}{2})(\theta - 2\pi k/N)]} \tag{23.9}$$

where $N$ is the number of views (assumed even) satisfying $N \geq 2K + 2$, $K$ being the *angular bandwidth of the object.* [This means that the Fourier series of $F(\omega, \theta)$ has no more than $2K + 1$ terms, that is, $F(\omega, \theta) = \sum_{k=-K}^{K} C_k e^{jk\theta}$.]

In Eq. (23.9) $A$ is the radius of the circle of support of the object function $f(x, y)$, and $\tilde{F}(n/2A, 2\pi k/N)$ is identical to the polar function $F(n/2A, 2\pi k/N)$ except that it allows for negative values of $n$ through the following artifice:

$$\tilde{F}\left(\frac{n}{2A}, \frac{2\pi k}{N}\right) = \begin{cases} F\left(\frac{n}{2A}, \frac{2\pi k}{N}\right), & n \geq 0 \\ F\left(-\frac{n}{2A}, \frac{2\pi k}{N} + \pi\right), & n < 0 \end{cases} \tag{23.10}$$

For $N$ odd a slightly different formula applies [24]. However, we shall assume, for specificity in what follows, that $N$ is even. To compute $F(u_l, v_m)$, Eq. (23.9) is used with

$$\omega_{lm} = \sqrt{u_l^2 + v_m^2} \tag{23.11a}$$

$$\theta_{lm} = \cos^{-1}\left(\frac{u_l}{\omega_{lm}}\right) \tag{23.11b}$$

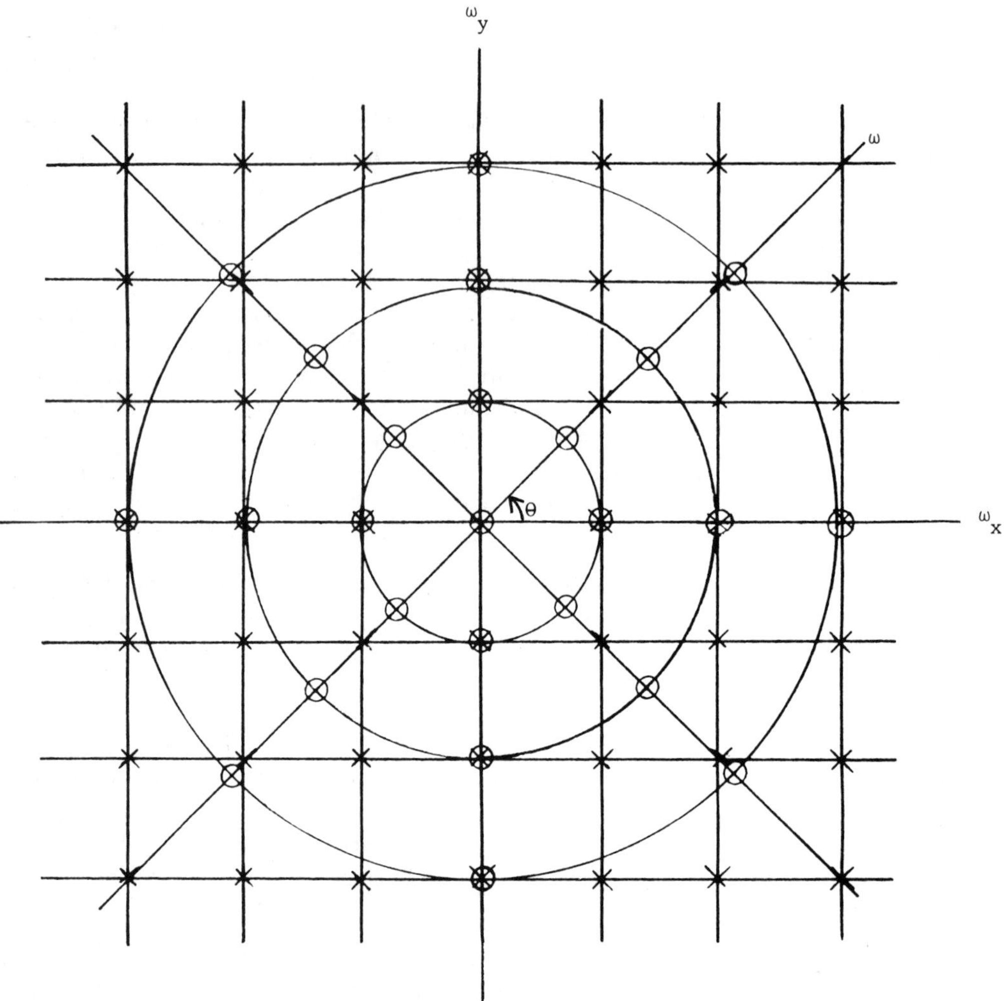


**Figure 23.3** The direct Fourier method requires interpolation from the polar lattice to a Cartesian lattice. (∘) Known points; (×) required points for FFT.

Thus the set $\{u_l, v_m\}$ can be computed directly from the set $\{\omega_n, \theta_k\}$. In practice, a truncated interpolation involving a finite number of terms is used. Thus, Eq. (23.9) is replaced by

$$\hat{F}(\omega, \theta) = \sum_{n=n_\omega - L_\omega}^{n_\omega + L_\omega} \sum_{k=k_\theta - L_\theta}^{k_\theta + L_\theta} \tilde{F}\left(\frac{n}{2A}, \frac{2\pi k}{N}\right) \operatorname{sinc}\left[A\left(\frac{\omega}{\pi} - \frac{n}{A}\right)\right] \times \frac{\sin[\frac{1}{2}(N-1)(\theta - 2\pi k/N)]}{N \sin[\frac{1}{2}(\theta - 2\pi k/N]} \tag{23.12}$$

where $[2A\omega] \triangleq n_\omega$ and $[N\theta/2\pi] \triangleq k_\theta$ are the nearest neighbors of the point about which the expansion is done. The notation $[z]$ here means the largest integer less than or equal to $z$. If $L_\omega = L_\theta = 0$, the above approximation reduces to the nearest-neighbor approximation. It is possible to adjust the relative proportions of radical and azimuthal interpolation by adjusting the parameters $L_\omega$ and $L_\theta$, respectively. In practice, image quality is improved when the truncation is tapered rather than abruptly truncated, the latter meaning that every point in the truncation window is given unit weight while those outside are given zero weight. The taper normally used is the Cartesian product of two identical 1-D triangular windows $\Delta(n)$ of the form

$$\Delta(n) = \max\left(1 - \frac{|n|}{\tau}, 0\right), \qquad n = 0, \pm 1, \pm 2, \ldots \tag{23.13}$$

where $\tau$ is a parameter. Then $\tau = \infty$ is the abrupt-truncation case. After interpolating the 2-D Fourier spectrum from a polar to a Cartesian lattice, a 2-D inverse FFT is computed to obtain the desired image. The direct Fourier method is illustrated in Figure 23.4.

Note that the interpolation theorem is valid provided that the data samples are available over a polar lattice in the frequency domain. In parallel-beam CT where the raysum data are available over a uniform lattice in the $(s, \theta)$ system, the direct 1-D FFT of the raysums $p_\theta(s)$ will yield frequency plane data over a polar lattice. However, in the case of fan-beam CT, the raysum data $p_\theta(s)$ are obtained over an irregular lattice in the $(s, \theta)$ system. Therefore, an interpolation preceding the 1-D FFT and the polar-to-Cartesian interpolation must be performed in order to use the DFM for reconstructing from fan-beam data. This is the topic of Section 6. The parallel-beam and fan-beam source-detector configurations are illustrated in Figures 23.5a and 23.26, respectively. By contrast, a geometry-free configuration (as used in ART, for example) is shown in Figure 23.5b.

## 2.2 Magnetic Resonance Tomography

Magnetic resonance imaging is a recent medical imaging modality. In contrast to CT imagery, MR images are based on more than one single property of the tissue, namely the spin density distribution, the spin-lattice relaxation time constant $T_1$, and the spin-spin relaxation time constant $T_2$. The MR system input consists of magnetic fields and radio frequency (RF) signals that can be tailored to produce strikingly diverse images. These images may be obtained by exciting single slices (two dimensions) or volumes (three dimensions) of the object being examined. The general principles of magnetic resonance and magnetic resonance imaging may be found in [25–31]. References [25–27] are excellent sources for the physical principles of magnetic resonance and relaxation phenomena. The basic MR concepts, the nature of the RF pulse sequences, various imaging techniques, and MR instrumentation are discussed in [27–31].

Our starting point here will be the MR imaging system output which is called the free induction decay (FID) signal. The FID for a 2-D experiment is given

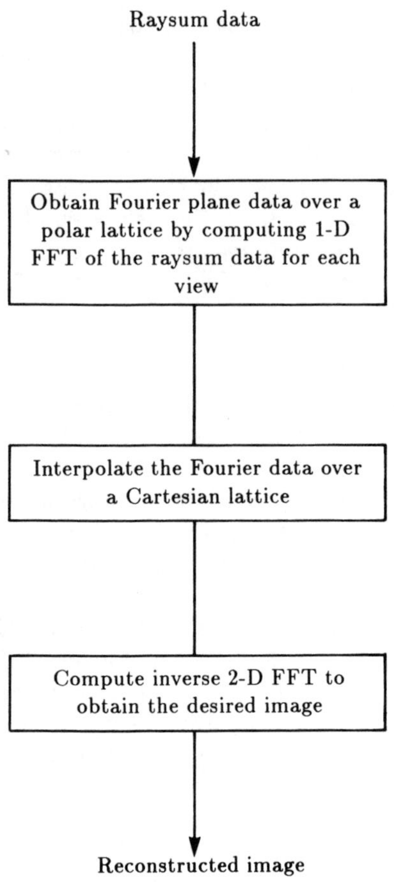


**Figure 23.4** The direct Fourier method (DFM).

by [20]

$$S(t) = \iint_{\Omega} f(x, y) \exp\{i(u(t)x + v(t)y)\} \exp\left\{\frac{-t}{T_2}\right\} dx\, dy \tag{23.14}$$

where

$$u(t) \triangleq \gamma \int_0^t G_x(t')\, dt' \tag{23.15a}$$

and

$$v(t) \triangleq \gamma \int_0^t G_y(t')\, dt' \tag{23.15b}$$

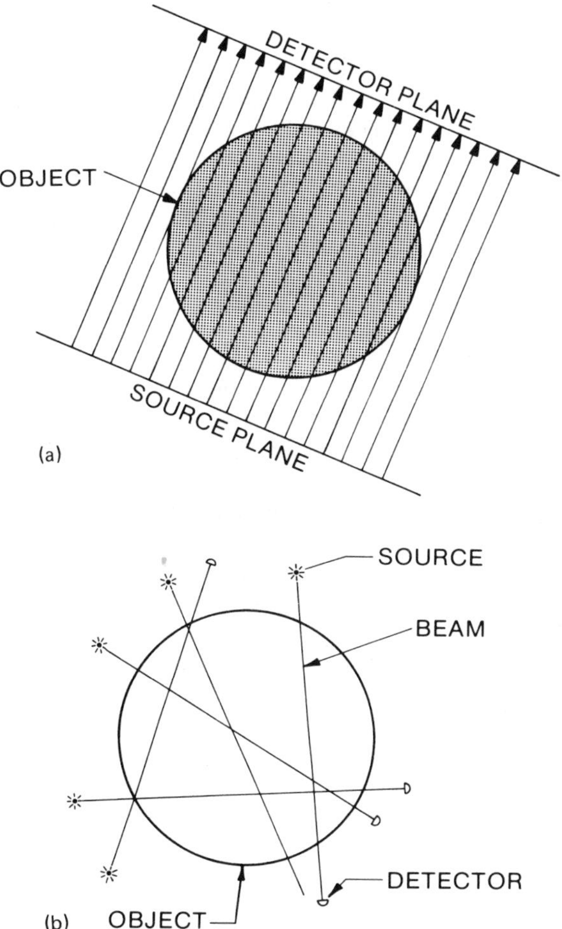


**Figure 23.5** Source-detector configurations for (a) parallel-beam geometry and (b) geometry-free data gathering.

In Eq. (23.14) $S(t)$ is the FID signal, $f(x, y)$ is the spin density distribution of the selected slice (by selective excitation [27], tomographic imaging of the spin density distribution over a particular slice is possible), and $G_x(t)$ and $G_y(t)$ are the time-varying gradient fields of the $x$ and $y$ coordinates, respectively. The quantity $\gamma$ is the gyromagnetic ratio. The exponential term $\exp(-t/T_2)$ is the so-called $T_2$-decay term. If the effect of this term is neglected [20], then the integral in Eq. (23.14) becomes the Fourier integral and $S$ and $f$ constitute a Fourier transform pair. Thus the data collected in an MR imaging experiment are given in frequency space. Different choices of gradients $G_x(t)$ and $G_y(t)$ yield different sampling strategies for the Fourier plane [21].

Our particular interest in this chapter is related to the *echo-planar* imaging technique [17]. We address and solve the interpolation problem associated with a recent extension of this technique [20, 21].

### Echo-Planar MR Imaging

Mansfield [17] utilized the echo-planar method to afford the possibility of high-speed imaging using the Fourier plane data collected in a time frame of tens of milliseconds (duration of the FID signal). This method has been modified to make it more feasible as a practical high-speed imaging technique [18, 21]. The main motivation for all this activity was to reduce the data collection time, originally limited by the spin-lattice relaxation time constant ($T_1$), which is of the order of 1 second. However, these methods require large and high-speed gradient fields, which are difficult to realize in practice and result in images with nonisotropic resolution.

Fortunately, because wide latitude exists in the manner in which the Fourier plane may be scanned during the FID period, it was possible to design schemes which retain the advantages of the echo-planar technique while avoiding some of its drawbacks. An encouraging scheme of this type is that suggested by Ljunggren [21] and implemented by Ahn et al. [20], namely the use of a *linear spiral scan.* With a linear spiral scan, image data are obtained on a linear spiral lattice in the Fourier plane. This type of scan has several advantages: 1) it is practically easier to realize, 2) it oversamples near the origin of the Fourier plane, where most of the signal energy is present, and 3) it results in isotropic coverage of the Fourier plane (Section 7.1).

A brief outline of the linear spiral scan is given in the following. Consider the 2-D Fourier imaging equation in Eq. (23.14). Using the gradients suggested by Ljunggren [21], that is,

$$G_x(t) = \frac{1}{\gamma}\frac{d}{dt}[u(t)] = \rho\cos\xi t - \rho\xi t\sin\xi t \tag{23.16a}$$

$$G_y(t) = \frac{1}{\gamma}\frac{d}{dt}[v(t)] = \rho\sin\xi t + \rho\xi t\cos\xi t \tag{23.16b}$$

we can sample the FT of $f(x, y)$ over a linear spiral lattice in the Fourier plane. In Eq. (23.16) $\rho$ and $\xi$ are constants [20]. If these samples are interpolated over a regular Cartesian lattice, then a discrete approximation to the spin density distribution can be computed via an inverse FFT. Derivation of a formula that enables exact interpolation from spiral to Cartesian samples is the subject of Section 7.

## 3 METHOD OF PROJECTIONS ONTO CONVEX SETS

### 3.1 Brief Review of POCS

As stated earlier, POCS has been applied in CT [9–12], electron microscopy [32], pattern recognition [33], and numerous other fields [4–8]. The basic theory of POCS was developed in the 1960s [2] but Youla and Webb [3] were the first to adapt it for general use in image processing. For the reader's convenience, we review here only the central idea of POCS and give the convex constraint sets that are important in tomographic image reconstruction. Much more on POCS can be obtained in the recent book on image recovery [5].

The image to be reconstructed, $f(x,y)$, is assumed to be a member of $\mathcal{H} \triangleq \mathcal{L}_{2\times 2}(\Omega)$, the Hilbert space of real-valued functions square integrable over $\Omega \subset \mathcal{R}^2$. In $\mathcal{H}$ the inner product and the norm are defined by, respectively,

$$\langle f_1, f_2 \rangle \triangleq \iint_{\Omega} f_1(x,y,)f_2(x,y)\,dx\,dy \tag{23.17a}$$

and

$$\|f_1\| \triangleq (\langle f_1, f_1 \rangle)^{1/2} \tag{23.17b}$$

where $f_1, f_2 \in \mathcal{H}$.

The basic idea of POCS is as follows: every known property of the unknown $f \in \mathcal{H}$ will restrict $f$ to lie in a closed convex set $\mathcal{C}_i$ in $\mathcal{H}$. Thus, for $m$ known properties there are $m$ closed convex sets $\mathcal{C}_i$, $i = 1, 2, \ldots, m$, and $f \in \mathcal{C}_0 \triangleq \cap_{i=1}^{m} \mathcal{C}_i$. Then, the problem is to find a point of $\mathcal{C}_0$ given the sets $\mathcal{C}_i$ and projection operators $P_i$ projecting onto $\mathcal{C}_i$, $i = 1, 2, \ldots, m$. The convergence properties of the sequence $\{f_k\}$ generated by the recursion relation

$$f_{k+1} = P_m P_{m-1} \cdots P_1 f_k, \qquad k = 0, 1, \ldots \tag{23.18}$$

or more generally by

$$f_{k+1} = T_m T_{m-1} \cdots T_1 f_k, \qquad k = 0, 1 \ldots \tag{23.19}$$

with $T_i \triangleq I + \lambda_i(P_i - I)$, $0 < \lambda_i < 2$, are based on fundamental theorems given by Gubin et al. [2]. The $\lambda_i$'s, $i = 1, \ldots, m$, are relaxation parameters and can be used to accelerate the rate of convergence of the algorithm. However, the $\lambda$'s that are effective in the absence of noise will often be ineffective when noise is present [34]. Thus, a single set of $\lambda$'s is not effective at all signal-to-noise ratios.

## 3.2 Sets of Importance in Tomography

The following convex constraint sets are important in tomographic image reconstruction. To prevent the overuse of the symbol $p$ we denote, in the following discussion, the $i$th raysum by the symbol $R_i$.

1. $\mathcal{C}_i \triangleq \{h : h \in \mathcal{H} \text{ and } \mathcal{P}_i h = R_i\}$, $i = 1, 2, \ldots, M$, where $\mathcal{P}_i$ is the raysum operator and $R_i$ is the raysum of $f$ at coordinates $(s_i, \theta_i)$. In words, $\mathcal{C}_i$ is the set of all functions $h(x,y)$ whose raysum at $(s_i, \theta_i)$ is $R_i$. The projection $g$ of an arbitrary $q \in \mathcal{H}$ is given, in the *continuous* case, by [12, 35]

$$g \triangleq P_i q = \begin{cases} q(x,y), & \hat{R}_i \triangleq \mathcal{P}_i q = R_i \\ q(x,y) + \dfrac{R_i - \hat{R}_i}{\|\eta_i\|^2}\eta_i, & \text{otherwise} \end{cases} \tag{23.20}$$

where $\eta_i \triangleq \eta_i(t, s_i, \theta_i)$ is the characteristic function of $\mathcal{L}_i$, the line along which the raysum is computed, and $||\eta_i||^2$ is easily shown to be the length $d_i$ of $\mathcal{L}_i$. In the continuous case, $R_i$ is given by Eq. (23.4) for $(s, \theta) \triangleq (s_i, \theta_i)$. In words, Eq. (23.20) states the following:

> If the raysum $\hat{R}_i$ of the computed image function $q(x, y)$ equals $R_i$, then $q(x, y)$ remains unchanged; otherwise all the points of $q(x, y)$ along the path of line $\mathcal{L}_i$ are modified to $q(x, y) + \Delta_i$, where $\Delta_i \triangleq (R_i - \hat{R}_i)/||\eta_i||^2$.

It is instructive to illustrate Eq. (23.20) using vectors. In Figure 23.6 let $f$ denote the desired image vector of which we have only the raysum $R_i$. From Eqs. (23.14) and (23.17) we can write $R_i = \langle f, \eta_i \rangle$ and $\hat{R}_i = \langle q, \eta_i \rangle$. Vectors ending on line $b$ represent the set $\mathcal{C}_i$. Now, given $q$, what is the nearest point $g \in \mathcal{C}_i$ to $q$? Clearly, $g = q + x$. To compute $x$, we write

$$g = P_i q = q + x \tag{23.21}$$

and

$$\langle g, \eta_i \rangle = \langle q + x, \eta_i \rangle = \langle q, \eta_i \rangle + \langle x, \eta_i \rangle \tag{23.22}$$

or

$$\langle x, \eta_i \rangle = R_i - \hat{R}_i \tag{23.23}$$

Since

$$x = ||x|| \frac{\eta_i}{||\eta||} \tag{23.24}$$

we obtain, from Eqs. (23.23) and (23.24),

$$||x|| = \frac{R_i - \hat{R}_i}{||\eta_i||}.$$

Thus

$$x = \frac{R_i - \hat{R}_i}{||\eta_i||^2} \eta_i$$

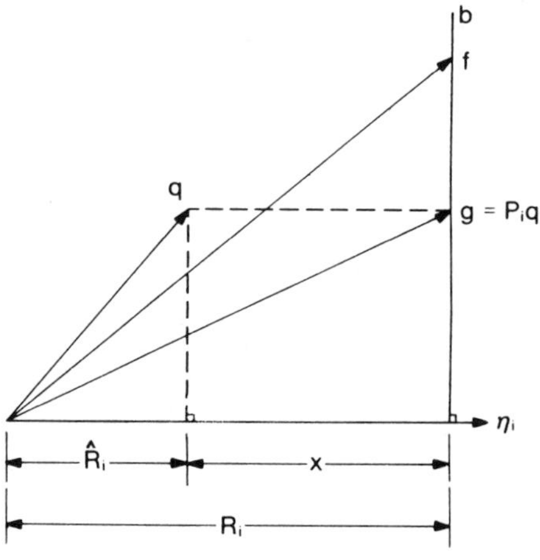


**Figure 23.6** Illustration of projection onto the raysum-constraint set. There is one such set for each raysum.

and
$$g = q + \frac{R_i - \hat{R}_i}{\|\eta_i\|^2}\eta_i \tag{23.25}$$

which is an equivalent form of Eq. (23.20).

Equation (23.20) is the correct projection for the continuous case, i.e., where pixels are points. However, in the discrete case, where pixels have finite dimensions, the correct projection is given by

$$g(x,y) = \begin{cases} q(x,y), \text{ for all } (x,y) \text{ if } \hat{R}_i = R_i \\ q(x,y) + \dfrac{R_i - \hat{R}_i}{\sum\sum_{\text{all } x,y} r_i^2(x,y)} r_i(x,y), \\ \qquad \text{for all } (x,y) \text{ if } \hat{R}_i \neq R_i. \end{cases} \tag{23.26}$$

In Eq. (23.26) $\hat{R}_i$ is the raysum for the prior image; $R_i$ is the correct (i.e., measured) raysum and is given by

$$R_i \triangleq \sum_{\text{all } x,y}\sum f(x,y) r_i(x,y), \tag{23.27}$$

and $r_i(x,y)$ is the length of the portion of line $\mathcal{L}(s_i, \theta_i)$ passing through the pixel at $(x,y)$.

2. $\mathcal{C}_{SL} \triangleq \{h: h \in \mathcal{H} \text{ and } h(x,y) = 0 \text{ for } (x,y) \notin \mathcal{A}\}$. In words: $\mathcal{C}_{SL}$ is the set of all functions in $\mathcal{H}$ that vanish outside a region $\mathcal{A}$, for example, functions of compact support. The projection $g$ onto $\mathcal{C}_{SL}$ is given by

$$g \triangleq P_{SL} q = \begin{cases} q(x,y), & (x,y) \in \mathcal{A} \\ 0, & (x,y) \notin \mathcal{A} \end{cases} \tag{23.28}$$

3. $\mathcal{C}_p \triangleq \{h: h \in \mathcal{H} \text{ and } h(x,y) \geq 0 \text{ for all } (x,y) \in \Omega\}$. In words: $\mathcal{C}_p$ is the set of all functions in $\mathcal{H}$ that are nonnegative. The projection $g$ onto $\mathcal{C}_p$ is

$$g \triangleq P_p q = \begin{cases} q(x,y), & \text{if } q \geq 0 \\ 0, & \text{if } q < 0 \end{cases} \tag{23.29}$$

4. $\mathcal{C}_E \triangleq \{h: h \in \mathcal{H} \text{ and } \|h\|^2 \leq E = \rho^2\}$. In words: $\mathcal{C}_E$ is the set of all functions in $\mathcal{H}$ that are upper-bounded in energy by $E$. The projection $g$ onto $\mathcal{C}_E$ is

$$g \triangleq P_E q = \begin{cases} q, & \|q\| \leq \rho \\ \sqrt{\left(\dfrac{E}{E_q}\right)} q, & \|q\| > \rho \end{cases} \tag{23.30}$$

where $E_q \triangleq \|q\|^2$. The intersection of $\mathcal{C}_E$ and $\mathcal{C}_p$ is also a closed convex set defined by

$$\mathcal{C}_{EP} \triangleq \{h: h \in \mathcal{H}, h \geq 0 \text{ and } \|h\|^2 \leq E = \rho^2\}$$

The projection $g$ onto $\mathcal{C}_{EP}$ is realized by

$$g \triangleq P_{EP}q = \begin{cases} P_p q, & E^+ \leq E \\ \sqrt{\dfrac{E}{E^+}} P_p q, & E^+ > E \end{cases} \tag{23.31}$$

where $E^+ \triangleq \|P_p q\|^2$.

5. $\mathcal{C}_R = \{h: h \in \mathcal{H} \text{ and } \|h - f_R\| \leq \epsilon_R\}$. In words: $\mathcal{C}_R$ is the set of all functions in $\mathcal{H}$ that are within a distance $\epsilon_R$ of a reference function $f_R$. The projection $g$ onto $\mathcal{C}_R$ (as well as onto $\mathcal{C}_E$) can easily be visualized using Figure 23.7.

In Figure 23.7 the functions $q(x,y)$, $g(x,y)$, and $f_R(x,y)$ are shown as vectors $q$, $g$, and $f_R$, respectively. It is clear that $g$, the nearest point to $q$ in $\mathcal{C}_R$, must lie on the line connecting $q$ to $f_R$. A unit vector on this line is

$$\frac{q - f_R}{\|q - f_R\|} \tag{23.32}$$

Thus, $g = P_R q$ is given by

$$g = f_R + \epsilon_R \frac{q - f_R}{\|q - f_R\|} \tag{23.33}$$

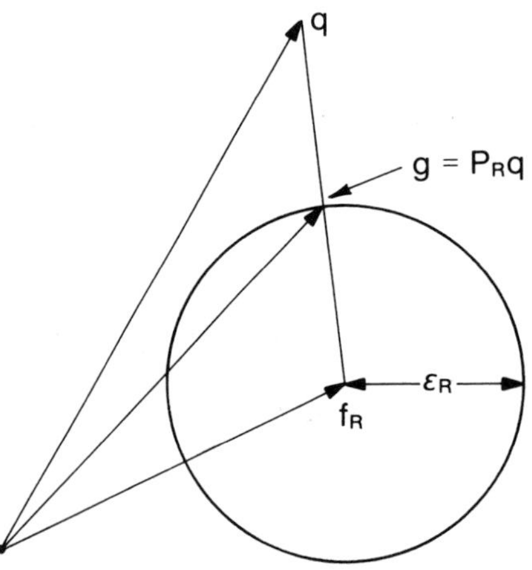


**Figure 23.7** Illustration of projection onto the spherical set $\mathcal{C}_R$.

which implicitly defines the projector $P_R$.

The reference image $f_R$ can be a CT image of the object taken at some time in the past or an average over CT images of a population of similar objects. The constraint set $\mathcal{C}_R$ not only reduces the size of the feasible solution set but also, as shown later, furnishes regularization in an entirely natural fashion.

6. $\mathcal{C}_F \triangleq \{h: h \in \mathcal{H}$ and $H(u,v) = G(u,v)$ for $(u,v) \in \Lambda\}$. In words: $\mathcal{C}_F$ is the set of all functions whose Fourier transform agrees with a prescribed value $G$ over a predetermined region $\Lambda$ in the Fourier plane. The projection onto $\mathcal{C}_F$ is defined by

$$g \triangleq P_F q \longleftrightarrow \begin{cases} G(u,v), & (u,v) \in \Lambda \\ Q(u,v), & (u,v) \notin \Lambda \end{cases} \tag{23.34}$$

The set $\mathcal{C}_F$ is of particular importance if the parallel-beam view data can be obtained only over a limited angular range. From the CSPT (Section 2), the angularly limited view data are equivalent to knowing the Fourier transform over a 2-D cone (actually a wedge-shaped angular region in the range $\theta_1 < \theta < \theta_2$), the so-called data cone (see Figure 23.11). The angularly limited view data problem will be discussed in the next section.

7. $\mathcal{C}_{AL} = \{h: h \in \mathcal{H}$ and $a \le h \le b\}$. In words: $\mathcal{C}_{AL}$ is the set of all amplitude-limited functions in $\mathcal{H}$ whose values must lie in the range $[a,b]$. The projection onto $\mathcal{C}_{AL}$ is realized by the following rule:

$$g \triangleq P_{AL} q = \begin{cases} a, & q(x,y) < a \\ q(x,y), & a \le q(x,y) \le b \\ b, & q(x,y) > b \end{cases} \tag{23.35}$$

## 4 APPLICATIONS OF THE METHOD OF POCS TO TOMOGRAPHIC IMAGE RECONSTRUCTION FROM INCOMPLETE VIEW DATA

In medical CT, the usual situation is that 360° view data are available for processing. However, we mentioned earlier that in other situations full-view data may not be available. What can one do in the way of reconstructing good-quality imagery when less than 360° view data are available? This is the subject of this section. Assuming reconstruction based on DFM, we seek to recover the information in the missing view data by POCS. The technique is also valid for MR imaging, ultrasound imaging, and imaging by synthetic aperture radar [5, Chapter 11].

In the following we combine the method of POCS with the DFM into a single recovery/reconstruction algorithm which we call POCS-DFM. In the case of CT, this algorithm is limited to parallel-beam geometry. DFM as applied to fan-beam geometry will be discussed in Section 6.

## 4.1 The POCS-DFM Algorithm

If parallel-beam view data are available over a limited angular range, then, from CSPT, the image Fourier transform can be obtained over a polar lattice within the data cone. The Fourier data within the data cone can be interpolated over a Cartesian lattice by the interpolation formula discussed in Section 2.1. The frequency values remaining outside the data cone are set to zero as a first approximation. The missing frequency information is then recovered by POCS and the desired image is obtained by inverse transforming. The complete algorithm—the POCS-DFM algorithm—is illustrated in Figure 23.8.

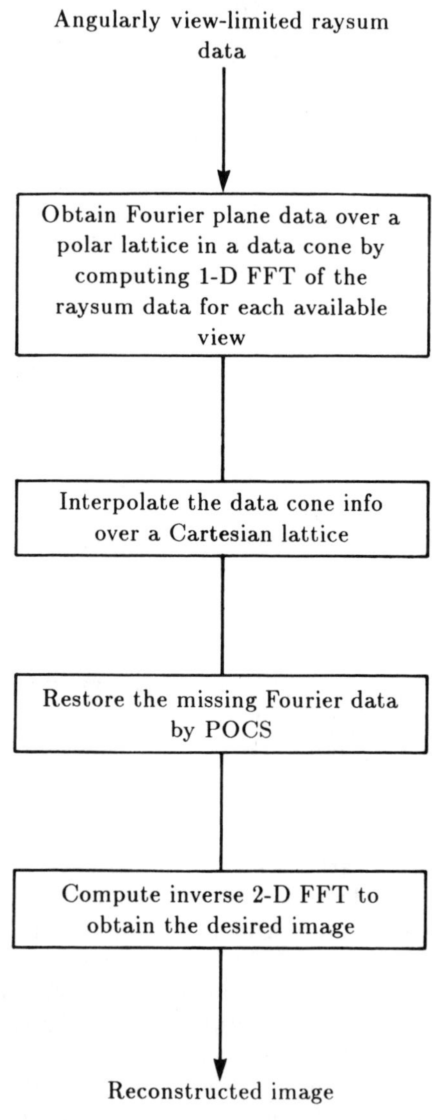


**Figure 23.8** The POCS-DFM algorithm.

## 4.2 Case Study: Reconstruction of a Thorax Phantom Cross Section from Angularly Limited X-Ray View Data

### Description of the Data

The $128 \times 128$ pixel human thorax is shown in Figure 23.9. The simulated parallel-beam projection data assume an X-ray energy of 70 keV, 128 detectors spaced apart by 0.3 cm, and 360 views over 360°. The interpolated Fourier spectrum with $L_\omega = 3$, $L_\theta = 1$, that is, 21 interpolation points using tapering corresponding to $\tau = 5$, and the resulting reconstruction from complete view data using the direct Fourier method are shown in Figure 23.10.

Reconstruction by POCS-DFM from angularly limited projection data is attempted for the three cases in which projection data are limited to the view range of $[-80°, 80°]$, $[-67°, 67°]$, or $[-45°, 45°]$. The formula of Eq. (23.12) with $L_\omega = 3$, $L_\theta = 1$ is used to interpolate $F(\omega, \theta)$ from polar to Cartesian points. The Fourier spectra for full-view data and the above three cases are shown after interpolation in Figure 23.11.

### Results

The basis for comparing results was the percent error $e_k$ defined by

$$e_k \triangleq 100 \frac{\|f - f_k\|}{\|f\|} \tag{23.36}$$

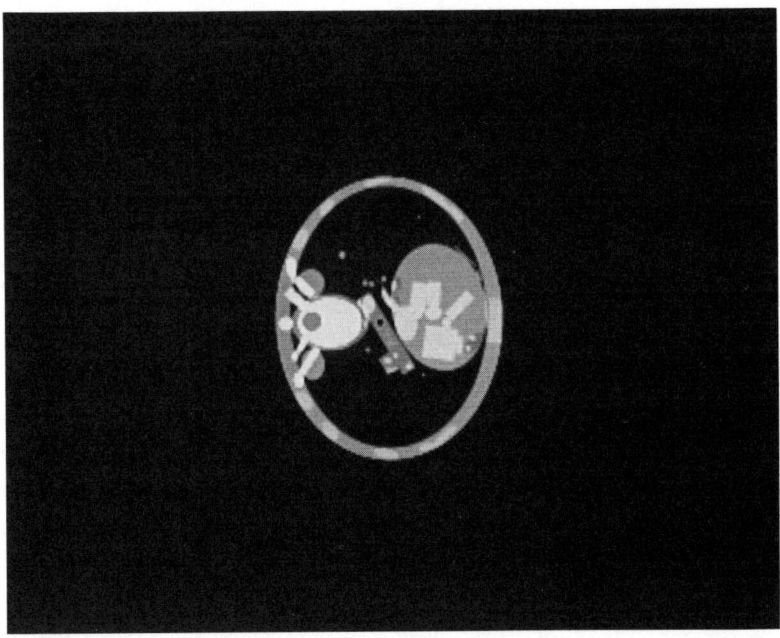

**Figure 23.9** Thorax phantom (through eighth vertebra). (From Sezan and Stark [11].)

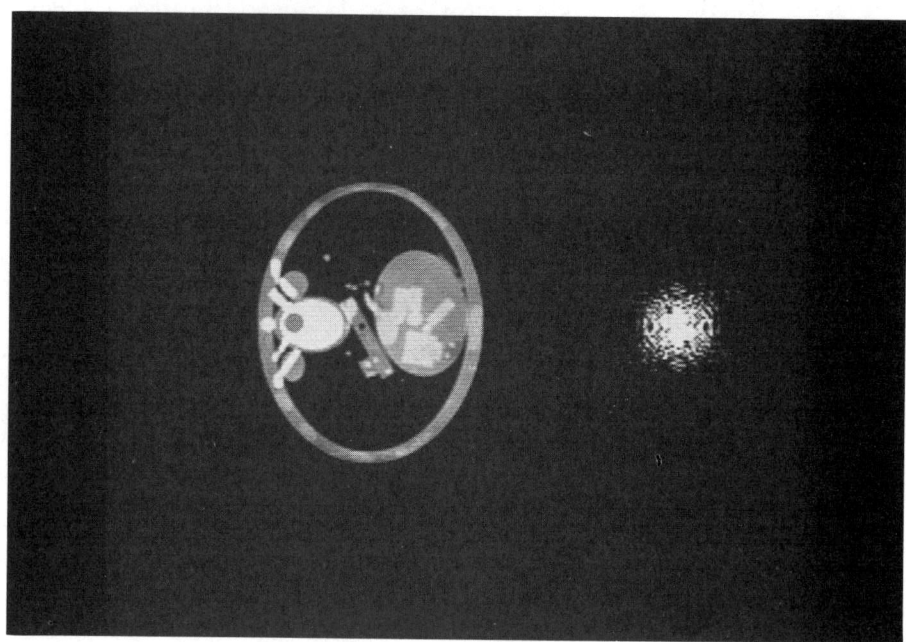

**Figure 23.10** DFM reconstruction (left) and interpolated spectrum. (From Sezan and Stark [11].)

where $f$ denotes the image reconstructed from *complete* view data.

A summary of the a priori known facts and assumptions made about the image is given below:

| A priori constraints | Actual |
|---|---|
| 1) Image support confined to rectangular region of length 124 pixels and width 103 pixels. | 1) Image support is elliptical (Figure 23.9). |
| 2) Gray levels $f$ satisfy $0 \leq f \leq 0.4$. | 2) Gray levels $f$ satisfy $0 \leq f \leq 0.38$. |
| 3) Energy over 128 x 128 pixel$^2$ field cannot exceed $\rho^2 = 284.0$. | 3) Energy over 128 x 128 pixel$^2$ field is 282.7. |

For this particular problem the sets $\mathcal{C}_{SL}$, $\mathcal{C}_F$, $\mathcal{C}_{PE}$, and $\mathcal{C}_{AL}$ are as follows:

1. $\mathcal{C}_{SL}$ is the set of all functions that vanish outside a rectangular region of length 124 pixels and width 103 pixels.
2. $\mathcal{C}_F$ is the set of all functions whose Fourier transforms agree with the values obtained by measurement in the spectral data cone. There are three cases: $[-45°, 45°]$, $[-67°, 67°]$, and $[-80°, 80°]$. The data cone of $[-45°, 45°]$ is associated with view data limited to 180°, the data cone of $[-67°, 67°]$ with view

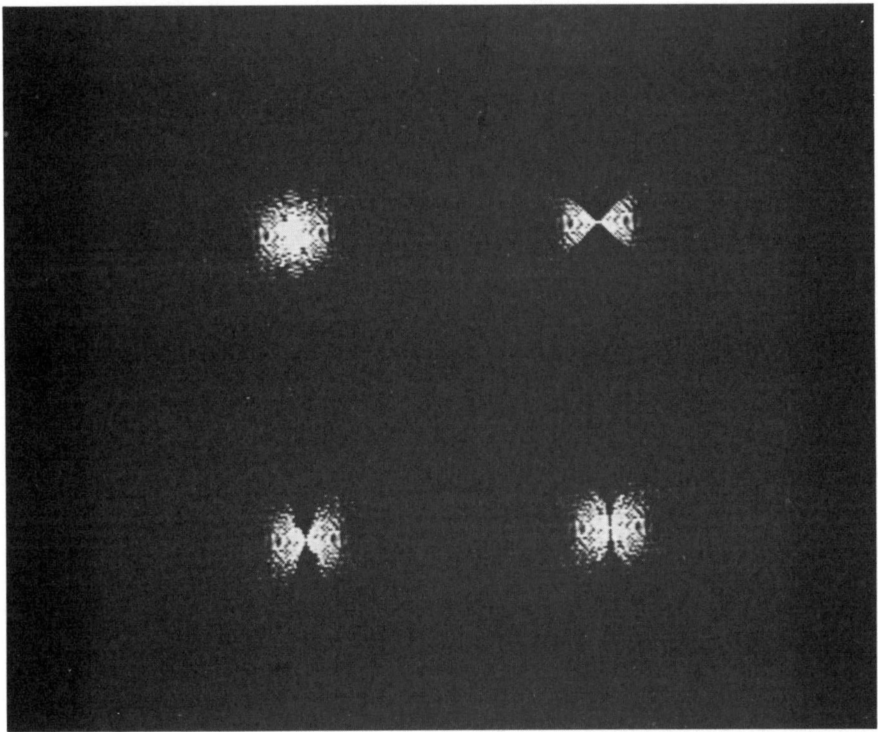

**Figure 23.11** Interpolated Fourier spectrum. Clockwise from upper left: full-view projection data, view limited to [−45°, 45°], view limited to [−80°, 80°], view limited to [−67°, 67°]. (From Sezan and Stark [11].)

data limited to 268°, and the data cone of [−80°, 80°] with view data limited to 320°.

3. $\mathcal{C}_{PE}$ is the set of all nonnegative functions whose norm (energy) is less than or equal to 284 pixels squared.
4. $\mathcal{C}_{AL}$ is the set of all functions whose gray level $f$ is confined to the range $0 \leq f \leq 0.4$.

The associated projection operators are $P_{SL}$, $P_F$, $P_{PE}$, and $P_{AL}$. Table 23.1 furnishes a summary of the experimental results.

The first (leftmost) column of Table 23.1 gives the angular extent of the spectral data cone over which data are available; the second column is the name of the restoration algorithm; in the third and fourth columns are the recursion relations and the associated values of the relaxation parameters, respectively; the fifth column is the error between the full scan image and the image immediately reconstructed from the limited view data (the so-called zero-order solution); and the last column is the error at the 30th iteration for the three cases. Note that the algorithm with $P_F P_{SL} f_k = f_{k+1}$ is equivalent to the well-known Gerchberg–Papoulis (G-P) algorithm [36, 37].

**Table 23.1** Summary of Results

| Available view range | Restoration algorithm: Name | Recursion | Relaxation parameters | Percent error: $e_0$ | $e_{30}$ |
|---|---|---|---|---|---|
| | G-P | $P_F P_{SL} f_k = f_{k+1}$ | $\lambda_{SL} = \lambda_F = 1.0$ | | 47.511 |
| $[-45°, 45°]$ | UNIRELAXL | $P_{AL} P_F P_{EP} P_{SL} f_k = f_{k+1}$ | $\lambda_{SL} = \lambda_F = 1.0$ <br> $\lambda_{EP} = \lambda_{AL} = 1.0$ | 52.419 | 52.057 |
| | G-P | $P_F P_{SL} f_k = f_{k+1}$ | $\lambda_{SL} = \lambda_F = 1.0$ | | 22.203 |
| $[-67°, 67°]$ | UNIRELAX | $P_F P_{EP} P_{SL} f_k = f_{k+1}$ | $\lambda_{SL} = \lambda_F = \lambda_{EP} = 1.0$ | 45.000 | 17.837 |
| | RELAX | $T_F T_{EP} T_{SL} f_k = f_{k+1}$ | $\lambda_{SL} = \lambda_{EP} = 1.9995$ <br> $\lambda_F = 1.0$ | | 16.184 |
| | G-P | $P_F P_{SL} f_k = f_{k+1}$ | $\lambda_{SL} = \lambda_F = 1.0$ | | 15.485 |
| $[-80°, 80°]$ | UNIRELAX | $P_F P_{EP} P_{SL} f_k = f_{k+1}$ | $\lambda_{SL} = \lambda_F = \lambda_{EP} = 1.0$ | 41.361 | 12.100 |
| | RELAX | $T_F T_{EP} T_{SL} f_k = f_{k+1}$ | $\lambda_{SL} = \lambda_{EP} = 1.9995$ <br> $\lambda_F = 1.0$ | | 9.352 |

**Results for the [−45°, 45°] Data Cone.** In this case, the reconstructed image at the end of 30 iterations is of low visual quality for all restoration algorithms. The reconstruction with UNIRELAXL restoration (i.e., where level constraints are also applied) outperforms the reconstruction with the Gerchberg–Papoulis restoration. The error performances of the reconstructions are compared in Figure 23.12.

**Results for the [−67°, 67°] Data Cone.** In this case, the three different restoration algorithms G-P, UNIRELAX, and RELAX are each, in turn, combined with the direct Fourier method of reconstruction to realize a better image than the zero-order solution. The zero-order solution is the image generated by inverse Fourier-transforming the available data without extrapolation. It results in an image whose error, as computed by Eq. (23.36) is 45%. In Figure 23.13 the zero-order solution is compared with full-view data reconstruction as well as reconstructions using G-P, UNIRELAX, and RELAX restorations. With the RELAX restoration, an error less than 17% is reached at the end of 30 iterations and the resulting reconstruction has fairly good visual quality.

**Results for the [−80°, 80°] Data Cone.** The zero-order image (for which the error is $e_0 = 42\%$), the full-view image, and the reconstructed images at the end of 30 iterations using RELAX, UNIRELAX, and G-P restorations are compared in Figure 23.14. As in the previous case, reconstructions using a priori restorations are significantly better than the zero-order reconstruction.

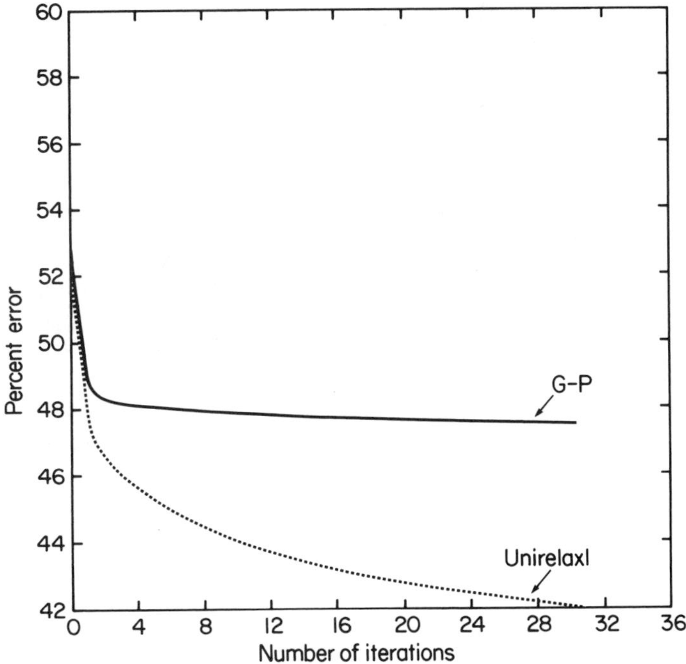


**Figure 23.12** Error performance of G-P and UNIRELAXL in reconstructing from projection data available over [−45°, 45°]. (From Sezan and Stark [11].)

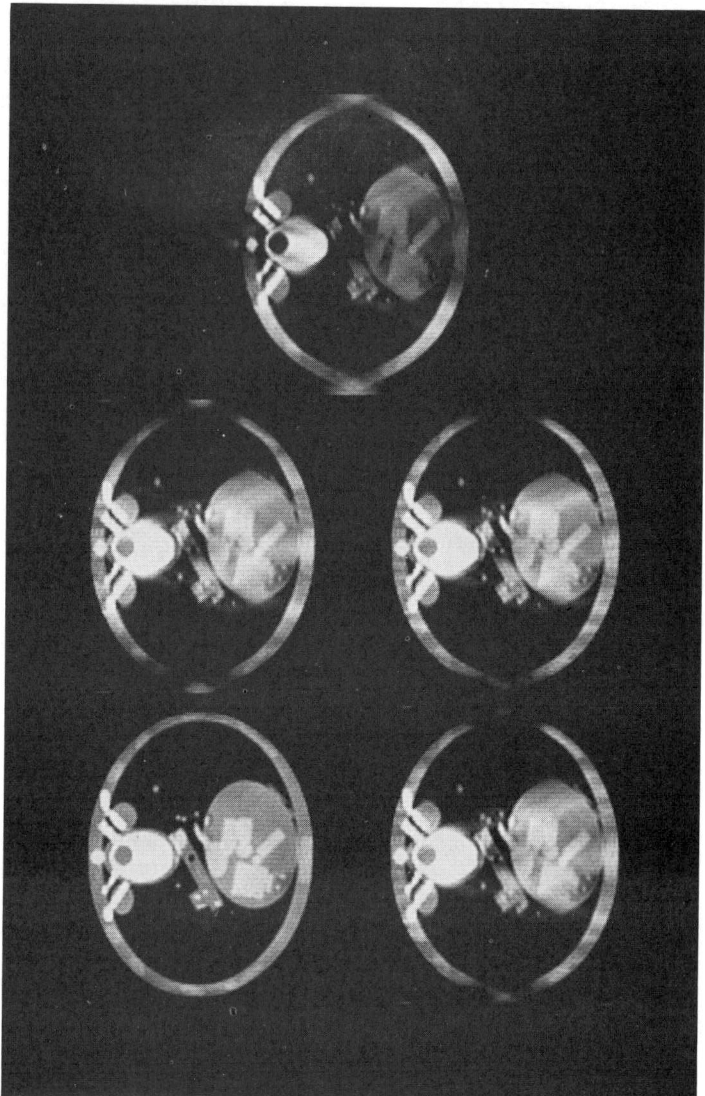

**Figure 23.13** Reconstruction when projection data are limited to [−67°, 67°]. (Top) Zero-order reconstruction; (below) clockwise from upper left: G-P, RELAX, UNIRELAX, and full-view data reconstruction.

### 4.3 Conclusions

The main conclusion drawn from the results of our case study is the following: when the projection data are angularly view-limited, the images obtained by direct Fourier reconstruction incorporating a priori POCS restoration are superior to the zero-order image obtained by direct inversion of the incomplete Fourier spectrum.

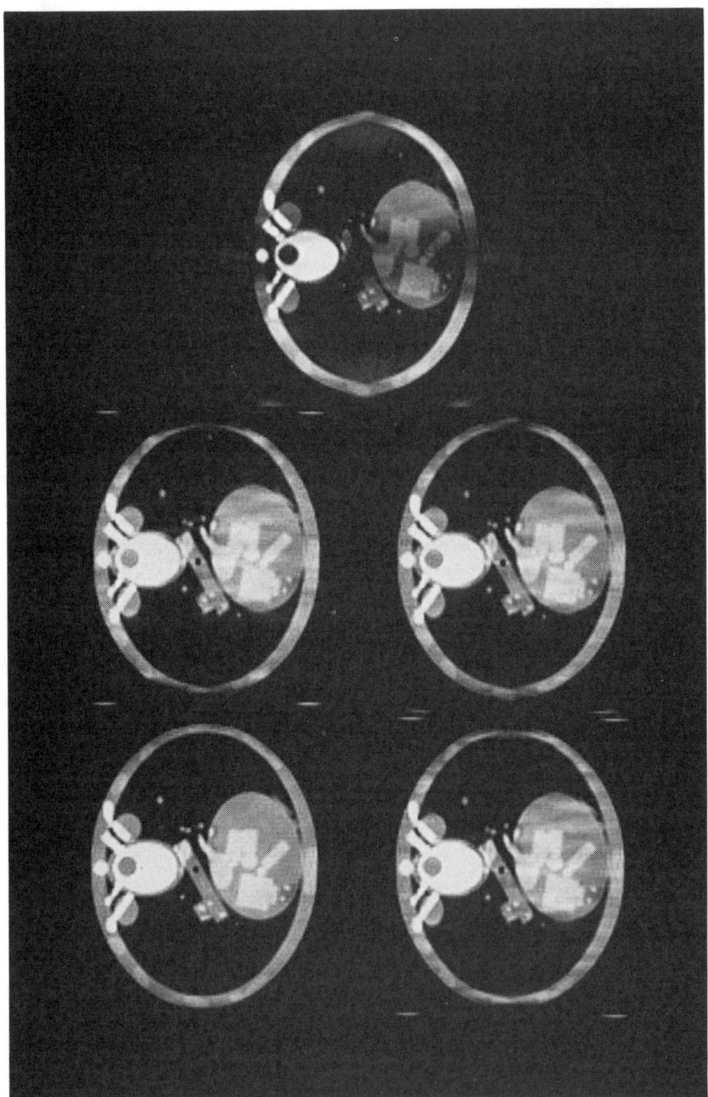

**Figure 23.14** Reconstruction when projection data are limited to [−80°, 80°]. (Top) Zero-order reconstruction; (below) clockwise from upper left: G-P, RELAX, UNIRELAX, and full-view data reconstruction.

Because of the additional energy constraint, the more general algorithms, UNIRELAX and RELAX, outperformed the G-P algorithm. In the [−45°, 45°] (data cone) case, however, none of the algorithms was able to produce an image with good visual quality. The practical significance of this result is that there is a level of data insufficiency that prevents recovery of the missing data regardless of how sophisticated the recovery algorithm is. However, when this is not the case, then further improvement is best obtained by using relaxed projections, in agreement with the results obtained in [4].

## 5 GEOMETRY-FREE TOMOGRAPHIC IMAGE RECONSTRUCTION USING THE METHOD OF POCS

In this section we derive a geometry-free reconstruction algorithm based on the theory of convex projections. By *geometry-free* we mean that the algorithm does not require a particular source-detector geometry such as in parallel-beam or fan-beam CT (Figure 23.5). Indeed our point of view is that any particular raysum restricts the image function to lie in a convex constraint set $\mathcal{C}_i$ defined in Section 3. The only other geometry-free reconstruction algorithms known to the authors are those variously described as algebraic reconstruction technique [13]. In the following, we demonstrate that ART is in fact an elementary convex projection algorithm and that therefore it converges to an image consistent with its raysum constraints.

### 5.1 Relation Between Convex Projections and ART

The basic conventional ART algorithm is an iterative process that does the following: At the end of the $k$th cycle raysums are computed using the estimated image. Let $\hat{R}_i^{(k)}$ denote the $i$th raysum, $i = 1, \ldots, N$, computed from the $k$th (i.e., the current) estimated image. The updating of the $k$th image to form the $(k+1)$st image is done by subtracting $\hat{R}_i^{(k)}$ from $R_i$ and distributing the difference in a weighted fashion among the pixels intercepted by the $i$th ray. Specifically:

$$f_{k+1}(x,y) = f_k(x,y) + \frac{R_i - \hat{R}_i^k}{\sum\sum_{\text{all } x,y} r_i^2(x,y)} r_i(x,y) \qquad \text{for all } (x,y) \in \mathcal{L}_i,\ i = 1, \ldots, N \tag{23.37}$$

**Fundamental Remarks.**

1. Comparing Eq. (23.37) with Eq. (23.26), we find that they are identical although they are written in slightly different forms. Thus the interpretation of the basic ART algorithm is that each full cycle of corrections of line integrals is a series of $N$ alternating projections onto $N$ distinct convex sets. In terms of the convex projection algorithm, we can describe the ART algorithm by

$$f_{k+1} = P f_k \tag{23.38}$$

where $P \triangleq P_N \cdots P_2 P_1$, that is, a composition of $N$ projection operators that project onto the closed convex sets $\mathcal{C}_i$, $i = 1, \ldots, N$. The projection operator $P_i$ that operates on, say, $f_k$ to produce the next estimate is implicitly defined by Eq. (23.37). Thus, basic ART and convex projections, in which all the other a priori constraints are ignored, are identical.

Having identified ART as a convex projection algorithm, we can now use the theory of convex projections to make the following two categorical statements regarding ART:

2. Ignoring noise and questions of stability, the algorithm given by Eq. (23.38) with the operator $P_i$ implicitly defined by Eq. (23.26) converges at least weakly [3] to a feasible solution (i.e., one that is consistent with all the constraints $\mathcal{C}_i$, $i = 1, \ldots, N$).
3. In a finite-dimensional setting, ART converges strongly (i.e., in the m.s. sense) to a feasible solution.

The numerical evaluation of ART-like algorithms may present problems. In the continuous (i.e., theoretical) case unregularized ART involves the solution of a Fredholm integral equation of the first kind which is known to be ill-posed; i.e., small perturbations in the raysum data can give rise to unacceptably large changes in the solution for $\hat{f}(t, s; \theta)$ [38]. A stabilization algorithm based on quadratic optimization was discussed by Herman [13].

By approaching the reconstruction problem from the point of view of POCS, we can use a robust stabilization technique that doesn't alter the convex-projection structure of the algorithm. The natural choice is to introduce a constraint that is readily available and can also form a closed convex set. Such a constraint was introduced in Section 3 as the normed distance from a reference image function $f_R$. To restate it: if $\mathcal{C}_R$ is the set of functions in $\mathcal{H}$ within a distance $\epsilon_R$ from a reference image $f_R$, then $\mathcal{C}_R$ is a closed convex set with the projector described by Eq. (23.33). Using $\mathcal{C}_R$, the desired image is constrained to have a distance shorter than or equal to $\epsilon_R$ from $f_R$.

We define an *iteration* as a full cycle of all available projection operators being applied to the current image, as in Eqs. (23.18) and (23.19). At each iteration the projector $P_R$ must be applied after the series of raysum-constraint operators in order to stabilize the solution. Let $P \triangleq P_n \cdots P_2 P_1$ represent the composition of projection operators due to $N$ raysums. The involvement of $P_R$ can then be written explicitly as

$$f_k = P_R P f_{k-1} \tag{23.39}$$

where $f_{k-1}$ and $f_k$ are reconstructions after $k-1$ and $k$ iterations, respectively. To prove that projecting onto $\mathcal{C}_R$ prevents divergence of the iterates, we write

$$||f_k - f|| = ||f_k - f_R + f_R - f|| \leq ||f_k - f_R|| + ||f_R - f|| \leq 2\epsilon_R \tag{23.40}$$

since $f_k \in \mathcal{C}_R$, and by assumption $f \in \mathcal{C}_R$.

What about the role of the other constraints? As stated earlier, $\mathcal{C}_R$ is identical to $\mathcal{C}_E$ when we set $f_R = 0$ and $\epsilon_R = \sqrt{E}$. Thus $\mathcal{C}_R$ is to be viewed as a generalization of $\mathcal{C}_E$ and, while projecting onto $\mathcal{C}_E$ can also be used to stabilize the reconstruction, $\mathcal{C}_E$ is generally a much "larger" set than $\mathcal{C}_R$ and therefore does much less to constrain the set of feasible solutions. In a practical sense, the other constraint sets $\mathcal{C}_{SL}$, $\mathcal{C}_P$, $\mathcal{C}_F$, and $\mathcal{C}_{AL}$ can significantly reduce the diameter of the set of feasible solutions and thereby force the reconstructed image to be closer to $f$. Recent work on stabilization (regularization) of ill-posed recovery problems using the theory of convex projections is discussed in [8].

A final remark regarding the set $\mathcal{C}_R$ is in order. Clearly, the effectiveness of projecting onto $\mathcal{C}_R$ is directly related to the size of $\epsilon_R$. The smaller $\epsilon_R$, the tighter is the constraint imposed on the set of a feasible solutions. If we make $\epsilon_R$ too small, there is the danger that the intersection of $\mathcal{C}_R$ with the other sets is empty. Conversely, if $\epsilon_R$ is too large, $\mathcal{C}_R$ represents a feeble constraint that may achieve little. We stated earlier that in medical CT, $f_R$ might represent the most recent image of a patient available to the radiologist, or it might represent an average over some population. In inspection, $f_R$ might represent the correct reference against which samples are compared. In any case, the use of a reference image $f_R$ for stabilization of the algorithm represents, in our view, a feasible alternative to other stabilization techniques [13].

## 5.2 Experimental Results

### Description of Model and Data

The experiment uses a simulated image of two concentric squares with uniform gray levels of 0.8 (inner square) and 0.4 (outer square). The inner square contains $8 \times 8$ pixels squared and the outer square has an area of $16 \times 16$ pixels squared. The squares are centrally located in a $32 \times 32$ pixels squared field with a black background. The image is shown in Figure 23.15.

The projection data are presented as a series of raysums (discrete line integrals) over the image. In this treatment the rays are infinitely thin and a parallel-beam geometry is used as a means of conveniently indexing the raysums, although, in general, the projection data and hence the constraints are geometry-free. The parallel-beam geometry is not exploited in the reconstruction. Each raysum $R_i \triangleq p(s_i, \theta_i)$ is specified by $s_i$, the distance of line $\mathcal{L}_i$ from the origin of the image coordinate system, and by $\theta_i$, the angle of rotation of $\mathcal{L}_i$ with respect to the vertical. Because of the size of the image and spacing between raysums there are at most 23 nonzero raysums in each view; the distance between any 2 adjacent raysums in a view is preset to the length of a unit-length pixel. We have used weighted raysums in the sense that the length of the path within a pixel is considered (see [13]).

### Constraints

In addition to the raysum constraints obtained from the projection data, five other a priori constraints are defined in Section 3 and consist of the sets $\mathcal{C}_{SL}$, $\mathcal{C}_P$, $\mathcal{C}_R$, $\mathcal{C}_F$, and $\mathcal{C}_{AL}$. Here we summarize the parameters of these sets.

1. For the space-limited constraint $\mathcal{C}_{SL}$, the region $\mathcal{A}$ is set to be a $16 \times 16$ square.
2. No parameter need be specified for the positivity constraint $\mathcal{C}_P$ defined by Eq. (23.29).

**Figure 23.15** The test image.

3. The reference image $f_R$ used in Eq. (23.32) is set to be a $32 \times 32$ square image with uniform gray level of 0.6 in the inner $16 \times 16$ region and 0.0 outside. In this case $\epsilon_R$ is set to be $\|f - f_R\|$.
4. For the Fourier constraint defined in Eq. (23.34), the region $\Lambda$ in the Fourier plane is taken to be a low-pass region of dimension $9 \times 9$ pixels squared centered at the origin of the frequency plane. This specific Fourier constraint is not practical in CT. Indeed, in CT, a more logical Fourier constraint would be knowledge of the Fourier transform of the image in a wedge-shaped region (data cone) of the frequency plane. This was the case in Section 4, where parallel-beam geometry was assumed. However, this practical refinement would have required additional computations that we felt were unnecessary in a study of feasibility.
5. The range $[a, b]$ used for the amplitude constraint of Eq. (23.35) is set to $[0, 1]$ as in most reconstruction cases.

While $\mathcal{C}_{AL} \subset \mathcal{C}_P$ and therefore applying $\mathcal{C}_P$ is redundant if applied back to back with $\mathcal{C}_{AL}$, we projected onto $\mathcal{C}_P$ as the $(N+5)$th operation within a cycle while projecting onto $\mathcal{C}_{AL}$ as the $(N+1)$st operation. On the other hand, since $\mathcal{C}_R$ is a generalization of $\mathcal{C}_E$ and both involve parameters somewhat more difficult to specify, we projected only onto $\mathcal{C}_R$ and omitted $\mathcal{C}_E$.

### Algorithm

Equation (23.18) defines the general pure-projection algorithm used in POCS. The specific POCS algorithm used in this experiment can be written as

$$f_{k+1} = P_P P_{SL} P_F P_R P_{AL} P_N \cdots P_1 f_k \tag{23.41}$$

where $P_N \cdots P_1$ is the series of $N$ projections onto $N$ closed convex sets, $\mathcal{C}_i (i = 1, \ldots, N)$, defined by $N$ raysums as in the set $\mathcal{C}_i$. The other projectors in Eq. (23.41) have been defined in Section 3.

To determine the effect of various a priori constraints, we applied variations of Eq. (23.41) with one or more of the a priori constraint projectors absent. For example, if we omitted the constraint imposed by membership in $\mathcal{C}_R$, then Eq. (23.41) reduced to

$$f_{k+1} = P_P P_{SL} P_F P_{AL} P_N \cdots P_1 f_k \tag{23.42}$$

To compare POCS with the basic ART algorithm, we defined the basic ART algorithm as

$$f_{k+1} = P_{AL} P_N P_{N-1} \cdots P_1 f_k \tag{23.43}$$

We did not, in this study, consider the use and optimization of relaxed projectors $T_i \triangleq I + \lambda_i (P_i - I)$, $0 < \lambda_i < 2$.

### Results

In our simulation two objectives were pursued: 1) to show the effects and advantages of using a priori constraints other than raysum constraints on the reconstruction and 2) to establish the individual effects of each constraint on the overall performance of POCS.

Figure 23.16 shows the reconstruction from 32 raysums after nine iterations. Since there are 1024 unknowns, the system is grossly underdetermined. Nevertheless, POCS gives a fair reconstruction, while basic ART yields a poor image. This experiment demonstrates the importance of the support constraint.

In all cases the starting image $f_0$ is the zero image; i.e., it has zero intensity everywhere. The percent error is defined as in Eq. (23.36) of Section 4.2.

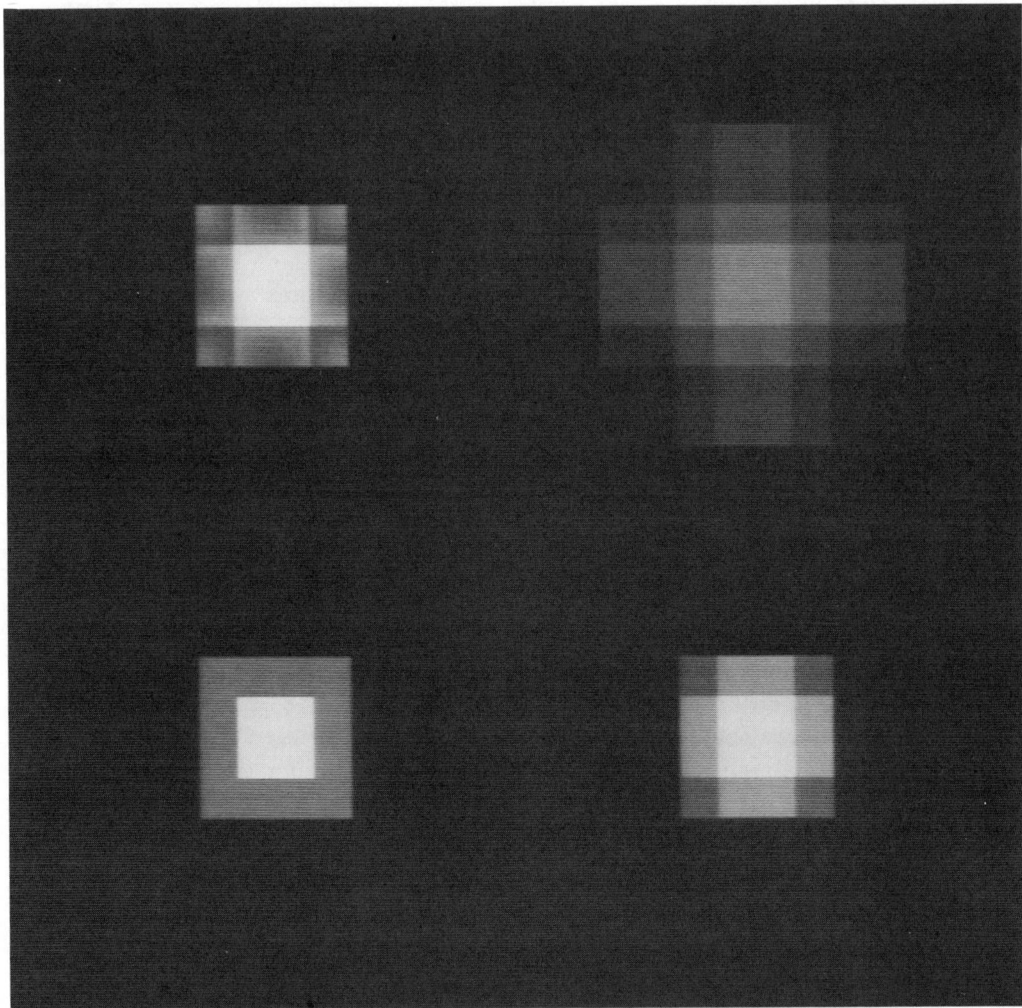

**Figure 23.16** Reconstruction after nine iterations from 32 raysums. Clockwise from upper left: POCS reconstruction using all constraints, ART reconstruction from raysums only and the 0–1 amplitude constraint, ART with support constraint added as a convex set, and true image.

The results in Figure 23.17 are similar to those in Figure 23.16 except that a different set of constraints are involved. Once again we see the relative importance of the support constraint, but the main result is that the Fourier constraint dominates the others. Indeed, use of the Fourier constraint and basic ART reconstructs an image of the same quality as POCS. The ART algorithm using the reference image as a convex constraint yields a poorer quality image.

A second series of experiments deal with overdetermined data involving 1352 raysums (60 views, one every 6°). Figure 23.18 shows reconstructions after 30 iterations. Starting in the upper left and going clockwise, we see the reconstruction from constraints only, the POCS reconstruction using 1352 raysums and all constraints (it is virtually undistinguishable from the true image), and the ART reconstruction in the lower right. From these results we see that for a full set of data both POCS and ART yield comparable, excellent results. In Fig. 23.19 we compare the POCS and ART reconstructions when the number of measured raysums is small. Here we expect that the role of prior information is more important. Indeed we see that the POCS reconstruction is clearly superior to the ART reconstruction.

The last experiment tested the importance of using the correct operator for projection. In the upper left panel of Fig. 23.20 is the ART reconstruction using the correct projector in Eq. (23.26). In the upper right panel is the reconstruction using the projector of Eq. (23.20) which is *incorrect for finite-size pixels*. The reconstruction is catastrophically bad. In the lower right is the POCS reconstruction using the incorrect raysum projector. We see that the systematic use of prior information ameliorates the effects of using the incorrect raysum projector. In this case, the reference image constraint played a major role.

Figure 23.21 compares the error versus iteration number for the reconstruction shown in Fig. 23.19. Note that changing the order of projections hardly affects the quality of the reconstructions after 50 iterations.

## 5.3 Conclusions

In this section we used the method of convex projections to achieve a geometry-free reconstruction of an image $f$ from a set of raysums and a second set of relatively typical a priori known constraints. Each raysum restricted the image to the convex set of all functions that yield that raysum for the prescribed path of integration. We found that the projection onto the raysum-constraint set is identical to the ART algorithm's correction procedure. It thus turns out that basic ART is a convex projection algorithm that seeks a solution at the intersection of $N$ raysum-constraint sets. Thus ART is a convergent algorithm in theory but may diverge in practice because of the ill-posed nature of the problem.

To stabilize the algorithm we introduced regularization in a very natural manner (at least in our view) by introducing a reference image $f_R$, and it was not unreasonable to assume knowledge of an $\epsilon_R$ such that $\|f - f_R\| < \epsilon_R$. By projecting on this "sphere" we ensured that our iterates $f_k$, at some point in every cycle of the algorithm, satisfied $\|f - f_k\| < 2\epsilon_R$.

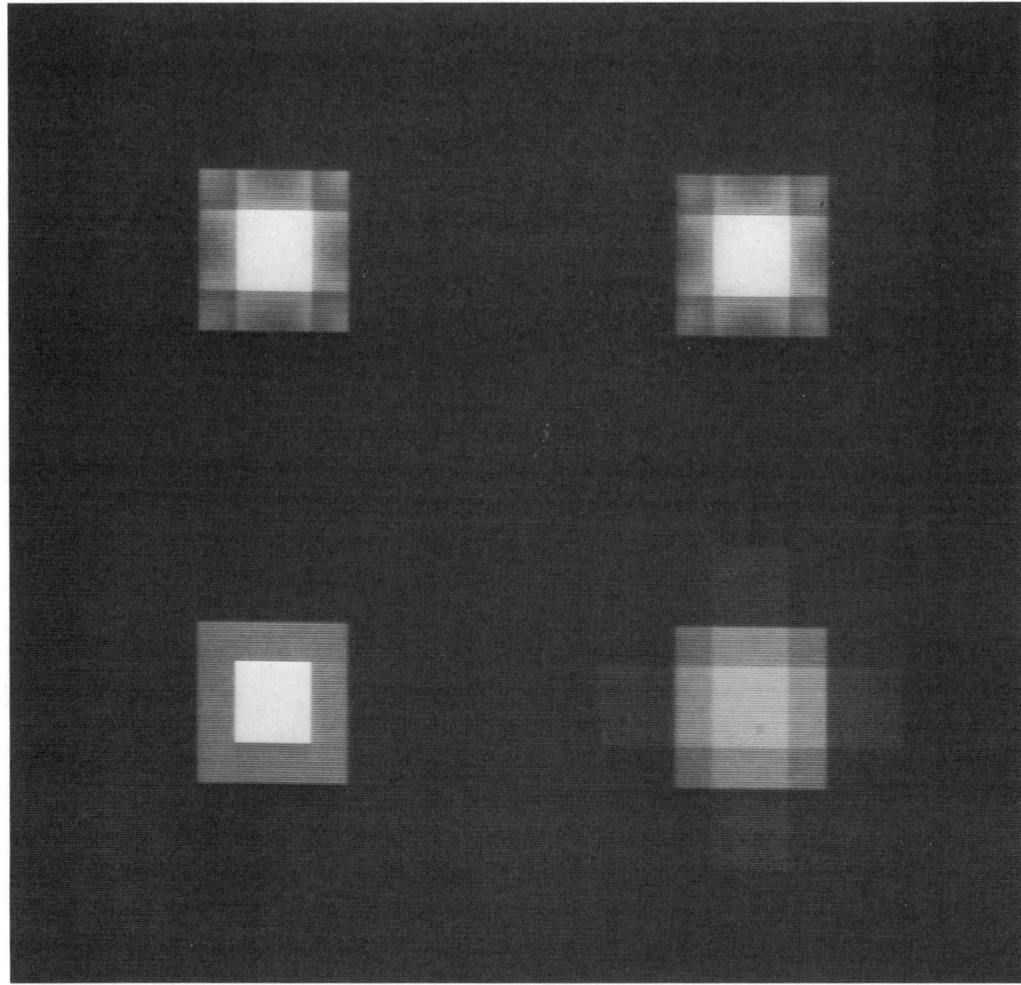

**Figure 23.17** Reconstruction after nine iterations from 32 raysums. Clockwise from upper left: POCS, ART with Fourier and support constraints, ART with reference image constraint only, and true image.

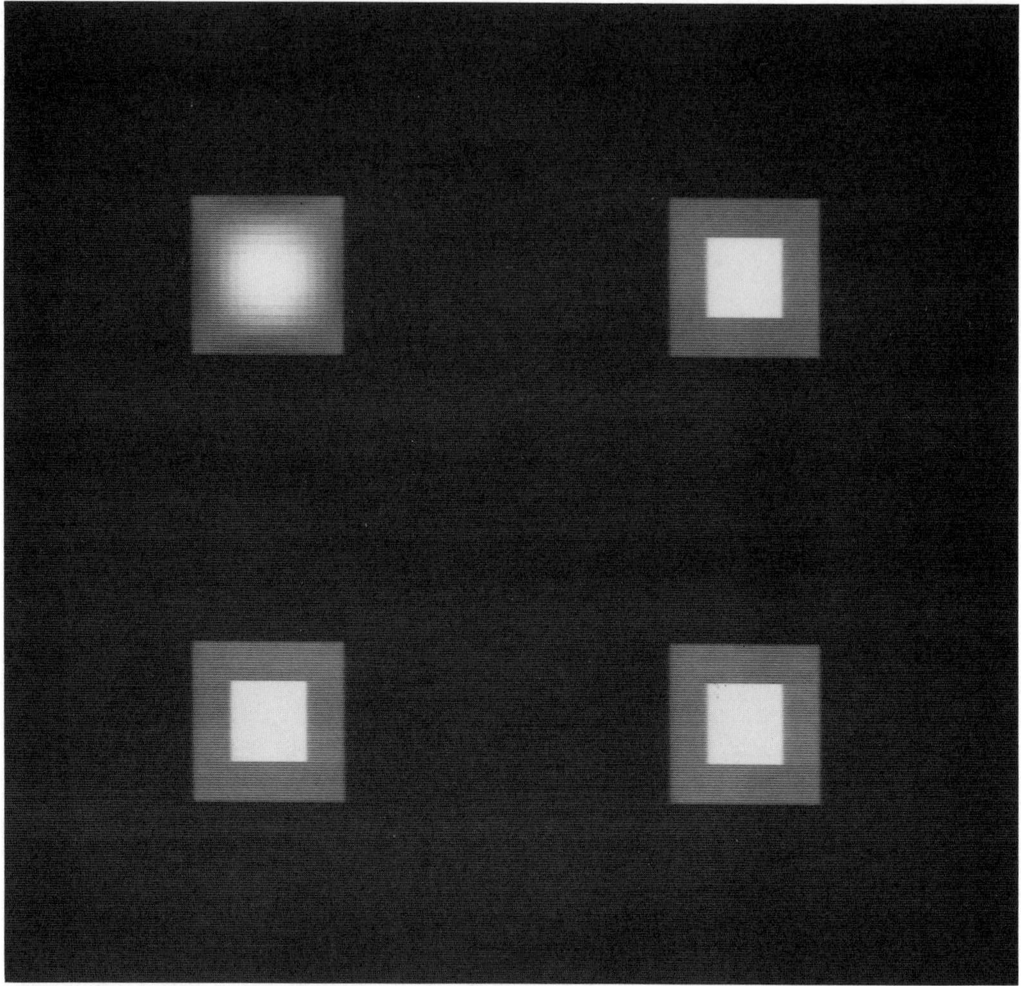

**Figure 23.18** Reconstruction after 30 iterations and 1352 raysums. Clockwise from upper left: reconstruction from constraints only, POCS, ART, and true image.

## 6 DIRECT FOURIER TOMOGRAPHIC IMAGE RECONSTRUCTION IN THE CASE OF FAN-BEAM GEOMETRY

Modern X-ray CT uses the fan-beam geometry because of its fast scanning ability compared with the parallel-beam geometry. The standard approach to reconstructing from fan-beam data is filtered convolution back projection [13, 14]. However, previous work involving parallel-beam projection data showed that good-quality reconstruction could be achieved using the direct Fourier method [6, 24] (Section 3). This method uses fewer computations, of the order of $N^2 \log N$ versus $N^3$ for convolution back projection. Moreover, because the direct Fourier method involves operations in the space and spatial frequency domain, it is well-suited for problems

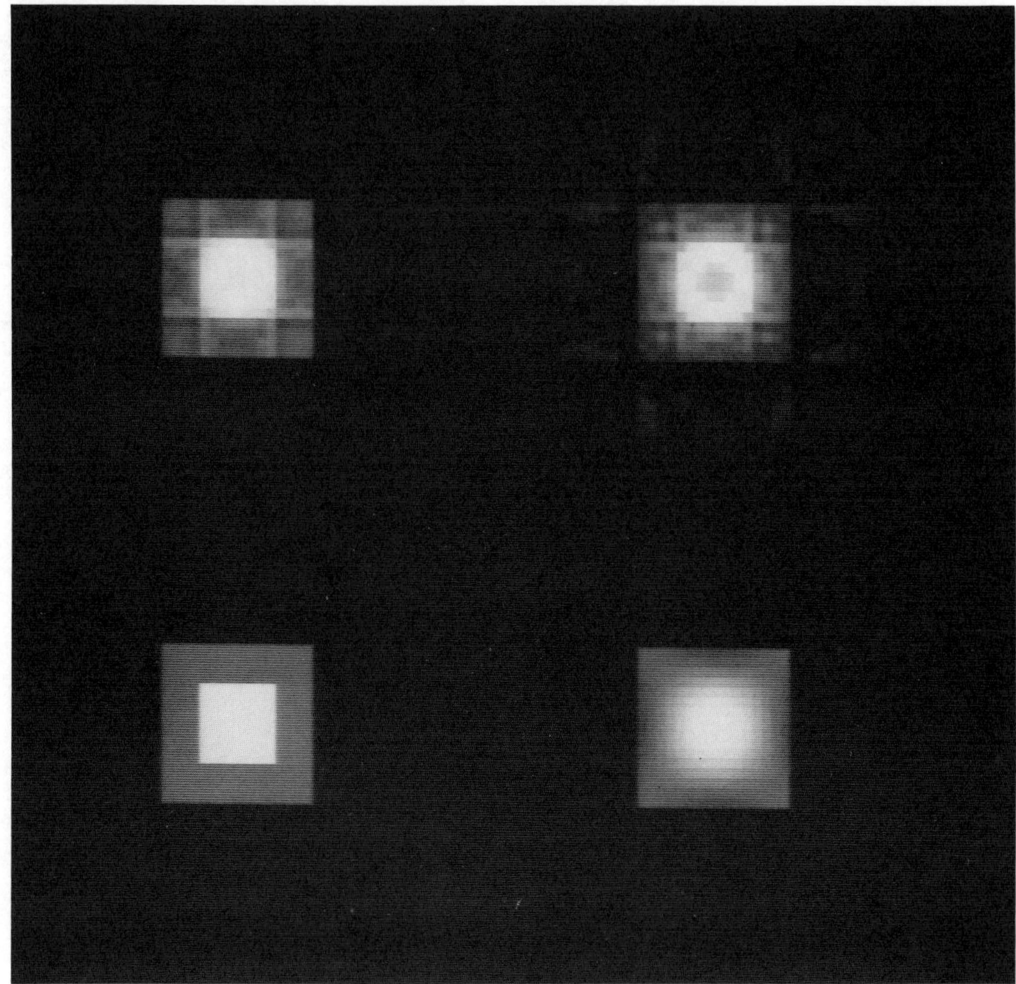

**Figure 23.19** Reconstruction after 50 iterations from 248 raysums (every 30°). Clockwise from upper left: POCS, ART, reconstruction from constraints only, and true image.

involving incomplete data in which image recovery is achieved by enforcing a priori space and spatial frequency constraints (Section 4).

Because of the stated reasons, we have investigated the extension of DFM to the fan-beam geometry. We will show that, provided that certain parameter constraints are satisfied, the direct Fourier method can be successfully applied to fan-beam CT and still maintain an $N^2 \log N$ order of computations.

The extension of direct Fourier reconstruction to the fan-beam geometry is based on the fact that each fan-beam raysum has a unique (one-to-one) mapping to a

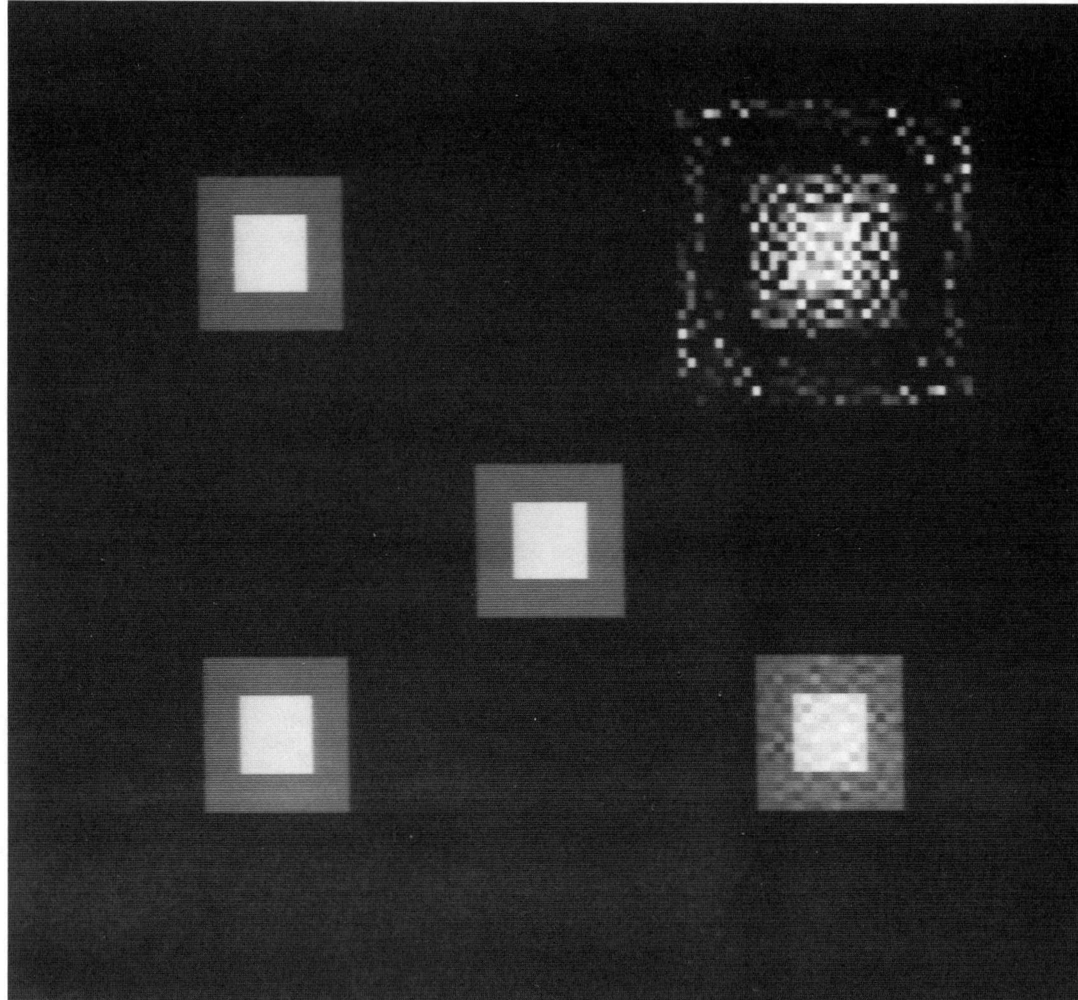

**Figure 23.20** Reconstruction after 50 iterations from 1352 raysums using two different projection operators. Upper left: weighted ART; upper right: unweighted ART; lower right: POCS using unweighted raysum projections; lower left: POCS using weighted raysum projectors; and middle: true image.

corresponding projection in a parallel-beam geometry. This was observed by Dreike and Boyd [15], Herman [13], and Lewitt [14]. The process of associating fan-beam projections with parallel-beam projections is called rebinning. When fan-beam data are rebinned into parallel-beam data, the resulting shadow functions $p_\theta(s)$ are not uniformly spaced in $s$, nor is the $\theta$ index uniformly offset for each $s$. The result is that the fan-beam data are distributed over an irregular lattice in the $(s, \theta)$ system. Thus, even before DFM can be applied, an interpolation from unequally spaced samples to uniformly spaced samples must be performed.

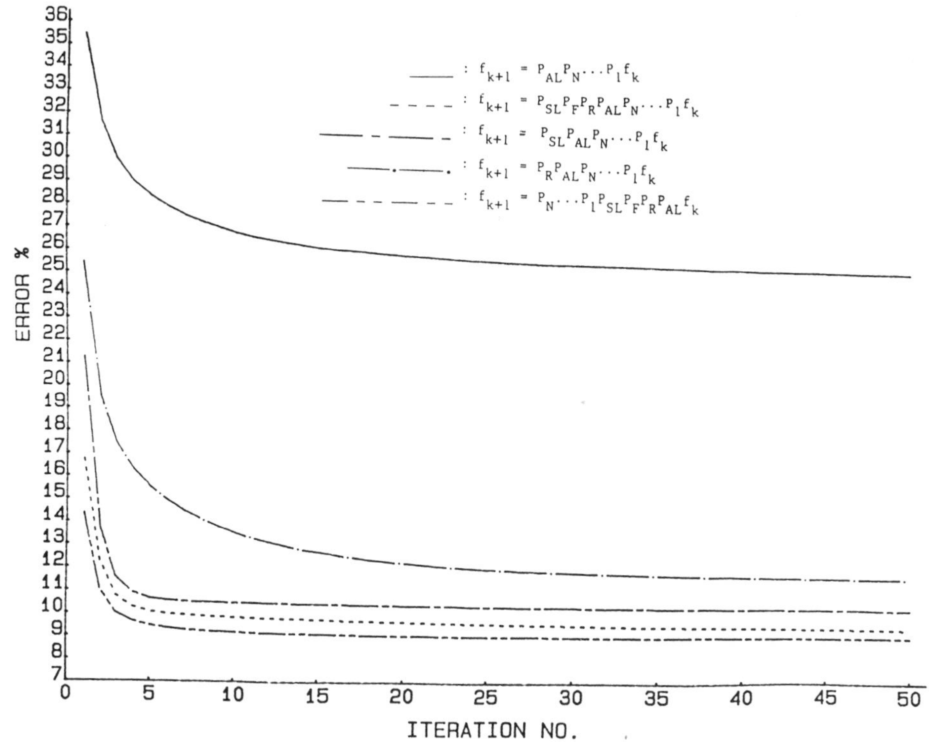


**Figure 23.21** Error histories from 248 raysums (every 30°).

The added complication of this nontrivial additional interpolation brings into question the value of doing direct Fourier reconstruction of fan-beam data. Why not simply use convolution back projection and avoid these difficulties? The answer is that even with the additional interpolation, we can still achieve a large computational saving over CBP when $N$ is large. Figure 23.22 shows a block diagram of DFM as applied to fan-beam data. The major computations are the two interpolations and the 2-D inverse FFT required for reconstruction. As we shall see, the first interpolation can be done in $k_1 N^2$ operations, where $k_1$ is a constant that depends on the number of neighboring samples needed to furnish a good result. Both FFTs require $O(N^2 \log N)$ multiplications and additions. The polar-to-Cartesian interpolation requires $k_2 N^2$ operations, where $k_2$ is like $k_1$. Thus the total number of operations involved in direct Fourier reconstruction of fan-beam data is $(k_1 + k_2)N^2 + O(N^2 \log N)$. This number could be small compared to the computational loading of filtered-convolution back projection requiring $O(N^3)$. Of course, the exact savings depend on $k_1$ and $k_2$ and the resolution of the image.

## 6.1 Interpolating from Nonuniformly Spaced Samples

Interpolation of nonuniformly spaced parallel-beam view data has been attempted earlier. Two-step linear interpolation was used in conjunction with standard par-

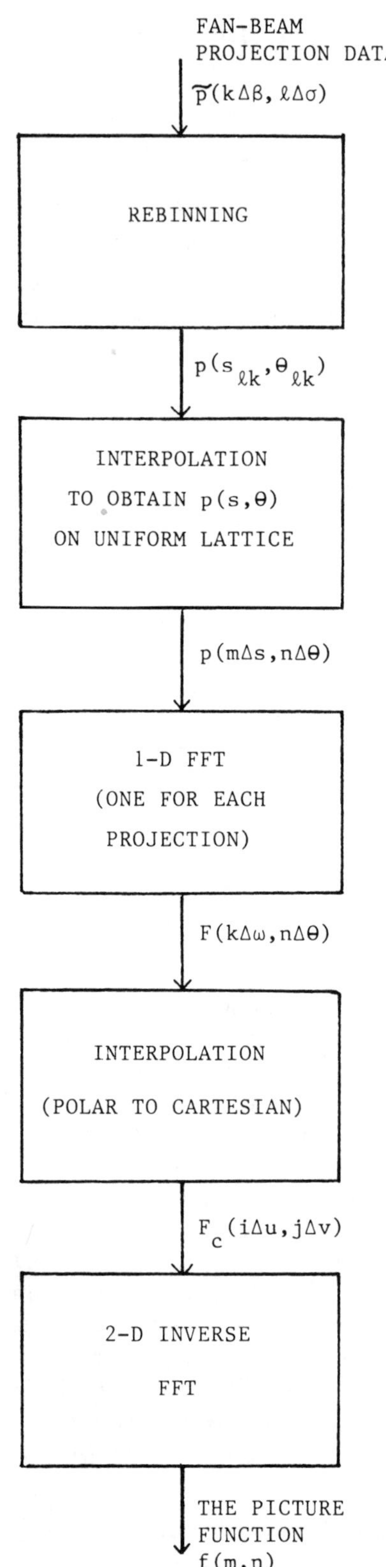


**Figure 23.22** Block diagram of DFM as applied to fan-beam data.

allel-beam convolution back projection in [13]. More recently, an interpolation method based on numerical approximation of low-pass filtering was given by Soumekh [39]. These methods are not exact and in this respect they are different from the procedure recently published by Clark, Palmer, and Lawrence [16]. Their approach (henceforth the CPL theorem) is based on finding a stretching/compression coordinate scaling transformation that maps nonuniformly sampled points onto a uniform sampling lattice. The stretching/compression function, if one can be found, is to be one-to-one and continuous and will give *exact interpolation* provided that the original function to be interpolated images as a band-limited function under the transformation. In general, a function, even a band-limited one, whose argument is subject to a nonlinear transformation does not remain a band-limited function of the new argument. In fan-beam tomography, however, an equivalent bandwidth can be found that, for all practical purposes, is sufficient.

The CPL theorem can be stated as follows for the one-dimensional case: Let $f(t)$ be a single variable function, sampled at points $t_n$, $n = \ldots, -2, -1, 0, 1, \ldots$, where $\{t_n\}$ is not necessarily a sequence of uniformly spaced numbers. If a one-to-one continuous mapping $\tau \triangleq \gamma(t)$ exists such that $nT = \gamma(t_n)$ and if $h(\tau) \triangleq f[\gamma^{-1}(\tau)]$ is band-limited to $\omega_0 = \pi/T$, then $f(t)$ can be reconstructed from its nonuniformly spaced samples $\{t_n\}$ according to

$$f(t) = \sum_{n=-\infty}^{\infty} f(t_n)\psi_n(t) \tag{23.44a}$$

where

$$\psi_n(t) \triangleq \frac{\sin[\omega_0(\gamma(t) - nT)]}{\omega_0(\gamma(t) - nT)} \tag{23.44b}$$

In fan-beam CT the band-limited requirement puts an upper bound on the detector spacing along the detector arc. This important result is discussed in quantitative terms in the next section and in [40]. Another result is that under the stretching/compression transformation that is relevant in fan-beam CT, one set of functions that are band-limited after the transformation but are not band-limited before the transformation is the set of finite-degree polynomials $P_N(s) \triangleq \{\sum_{i=0}^{N} a_i s^i \mid a_i \in \mathcal{R}\}$. The proof of this statement and the relation of the bandwidth to the degree are given in [40]. This result has important theoretical significance in CT because, according to the Weierstrass approximation theorem [41], any continuous function of one variable on a closed-bounded interval $[a, b]$ can be approximated there, within $\epsilon$, by a polynomial. The projection function $p_\theta(s)$, over the interval $[-A/2, A/2]$, can be approximated by functions (i.e., polynomials) that are band-limited after the nonlinear transformation that maps parallel-beam coordinates into fan-beam coordinates. Thus the practical implementation of the CPL formula in CT is ultimately based on solid theoretical foundations. We caution the reader that this remark, while true for the particular transformation involved in the interpolation of the rebinned CT data, should not be taken as true in general.

A final remark regarding the use of Eq. (23.44) is in order. In general, the reconstruction of $f(t)$ will only be approximate since practical considerations require that only a finite number of terms will be involved in the summation in Eq. (23.44). As we shall see in the next section, this results in a ringing artifact that ultimately requires correction. However, even when Eq. (23.44) is used in such a "truncated" mode, it will yield an exact reconstruction at a sampled point. To see this, let $t = t_j$ be a point at which $f(t_j)$ is sampled. Then, since $\gamma(t_j) = jT$, we obtain under a truncation of $2N$ terms:

$$\sum_{n=-N}^{N-1} f(t_n)\frac{\sin\{\omega_0[\gamma(t_j) - nT]\}}{\omega_0[\gamma(t_j) - nT]} = \sum_{n=-N}^{N-1} f(t_n)\frac{\sin[\omega_0(j-n)T]}{\omega_0(j-n)T} = f(t_j)$$

This observation will help explain the oscillatory ringing we observe in uncompensated data.

## 6.2 Application of Nonuniform Sampling Formula to Fan-Beam Tomography

The indexing of fan-beam projection data is done with a pair of angular coordinates $(\beta, \sigma)$, where $\beta$ is the angular position of the source and $\sigma$ is the angular position, *relative to the source*, of the line along which the raysum is obtained. Equivalently, it is the angular position of that detector and therefore could be called the detector index. We assume that both $\beta$ and $\sigma$ are uniformly incremented. Figure 23.23 shows the angular coordinates $\beta$ and $\sigma$. Comparing these with the parallel-beam

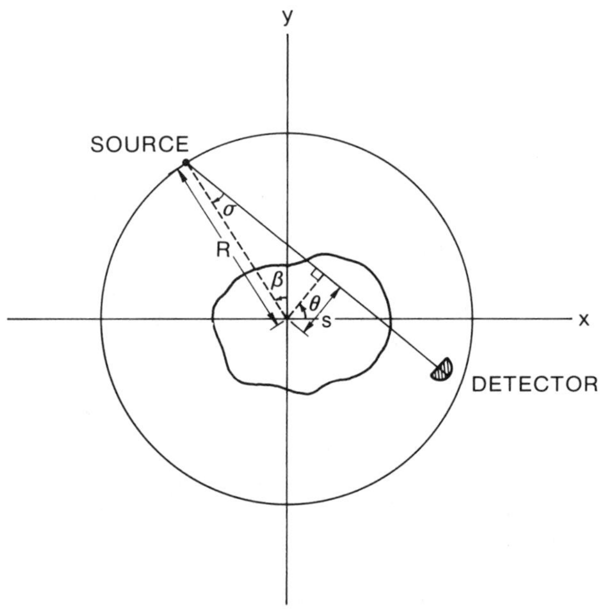


**Figure 23.23** Angular coordinates in a fan-beam geometry.

coordinates of Figure 23.1, we obtain the well-known relation between fan-beam coordinates $(\beta, \sigma)$ and parallel-beam coordinates $(s, \theta)$, i.e.,

$$s = R \sin \sigma \tag{23.45a}$$

and

$$\theta = \beta + \sigma \tag{23.45b}$$

Equation (23.45) says that a raysum can be described unambiguously in fan-beam or parallel-beam coordinates; in the former its coordinates are $(\beta, \sigma)$, while in the latter they are $(s, \theta)$. One way to reconstruct an image from fan-beam data is to first convert the raysums to parallel-beam coordinates and then use convolution back projection. This process requires rebinning, which, as stated earlier, is the association of fan-beam projections with an equivalent set of parallel-beam projections. A typical lattice on which one obtains the rebinned data is shown in Figure 23.24a. We note that for any given $s$, the data are equally spaced in $\theta$ but shifted by a different amount for each $s$. Along the $s$ direction there is a tendency for the data to appear closer and closer together as $|s|$ increases.

Figure 23.24b is a close-up view of the spacings between the rebinned fan-beam data. We observe that since $\theta = l\Delta\beta + \sin^{-1}(s/R)$, for any fixed value of $s$, the data in the $\theta$ direction are equally spaced by $\Delta\beta$ but vertically offset by an amount $\Delta\sigma$ between adjacent vertical parallel lines. The spacing between adjacent vertical lines that contain data is given by

$$s_{k+1} - s_k = R[\sin(k+1)\Delta\sigma - \sin k\Delta\sigma] \tag{23.46}$$

so that in the horizontal direction the data become more closely packed as one moves away from the origin in either direction.

From the discussion in the previous section, we can see that use of the CPL interpolation formula for interpolation of the rebinned data on a uniform lattice requires that the stretching/compression function $\gamma(\bullet)$ takes on the value

$$\gamma(s) = \sin^{-1}\left(\frac{s}{R}\right) \tag{23.47}$$

since $\sigma = \sin^{-1}(s/R)$ and the sampled $\sigma$ points are uniformly spaced. Thus

$$k\Delta\sigma = \sin^{-1}\left(\frac{s_k}{R}\right) \tag{23.48a}$$

and

$$\omega_0 = \frac{\pi}{\Delta\sigma} \tag{23.48b}$$

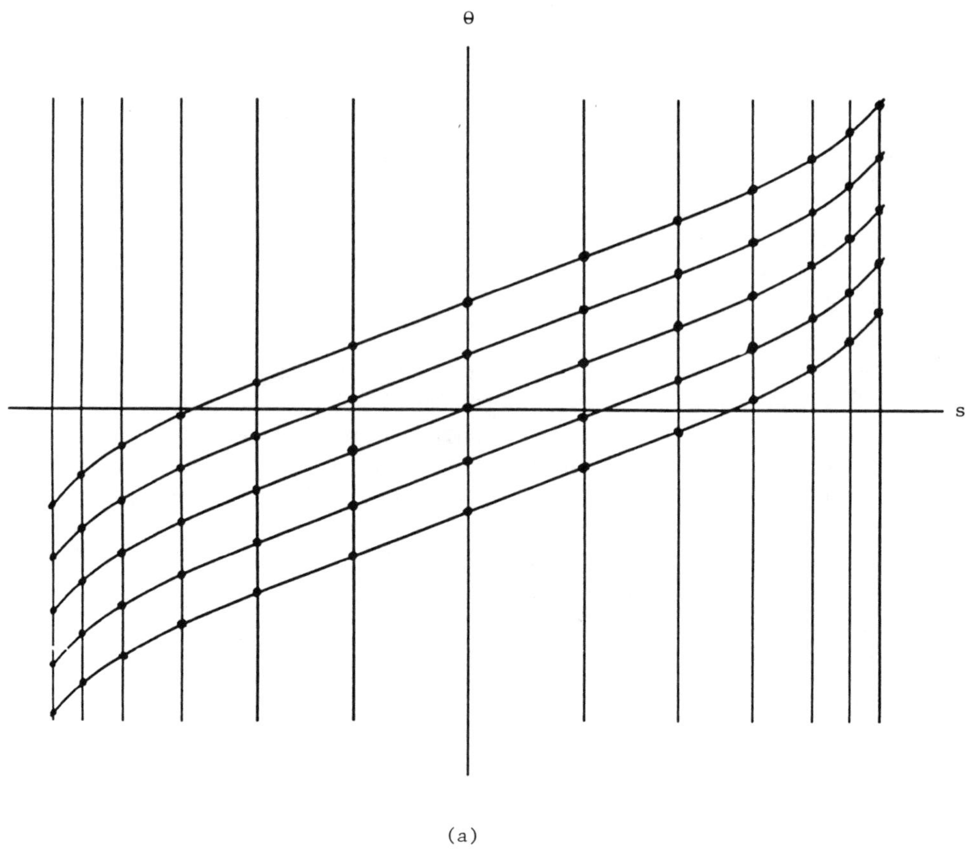


(a)

**Figure 23.24** (a) Rebinned fan-beam data on the $(s, \theta)$ lattice used. (b) Close-up showing displacements.

and the $s$-direction interpolation function is given by

$$\psi(s; \sigma_k) = \frac{\sin\{\pi/\Delta\sigma[\sin^{-1}(s/R) - k\Delta\sigma]\}}{(\pi/\Delta\sigma)[\sin^{-1}(s/R) - k\Delta\sigma]} \tag{23.49}$$

where $\sigma_k \triangleq k\Delta\sigma$.

To interpolate in the $\theta$ direction we need only use the angular interpolation functions in Eq. (23.9), i.e.,

$$\Phi(\theta; \beta_l, \sigma_k) = \frac{\sin\{\frac{1}{2}(N-1)[\theta - (l\Delta\beta + k\Delta\sigma)]\}}{N\sin\{\frac{1}{2}[\theta - (l\Delta\beta + k\Delta\sigma)]\}} \tag{23.50}$$

where $\Delta\beta = 2\pi/N$, the term $k\Delta\sigma$ accounts for the offset in the $\theta$ samples on the line $s = s_k$, and $N$ is the number of views.

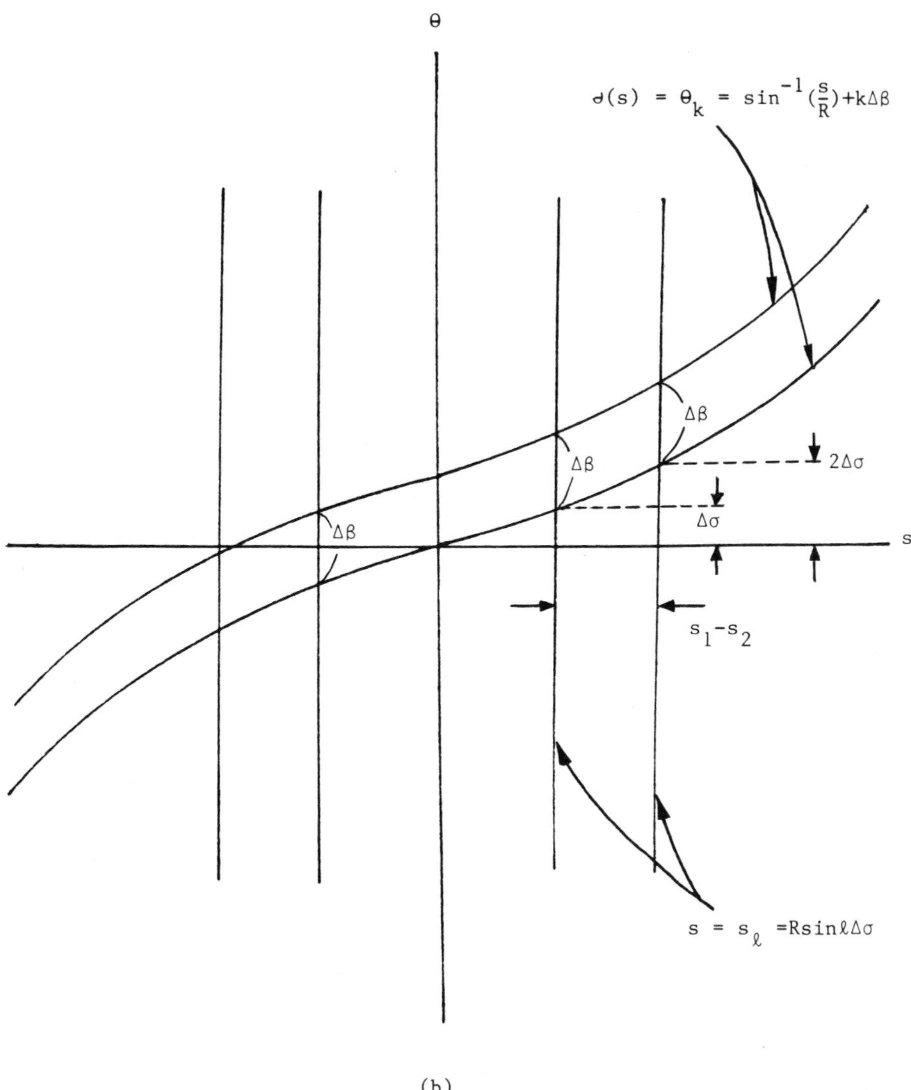


(b)

To make explicit the fact that we must now consider $p_\theta(s)$ as a function of two variables, let $p(s,\theta) \triangleq p_\theta(s)$, where $p_\theta(s)$ is given by Eq. (23.4). Likewise, we denote a projection in the $(\beta,\sigma)$ system by $\tilde{p}(\beta,\sigma)$. Clearly, $\tilde{p}(\beta,\sigma) \triangleq p(\beta+\sigma, R\sin\sigma)$. Then the whole interpolation process can be written as

$$p(s,\theta) = \sum_k \sum_l \tilde{p}(l\Delta\beta, k\Delta\sigma)\Phi(\theta; l\Delta\beta, k\Delta\sigma)\psi(s; k\Delta\sigma) \tag{23.51}$$

Ordinarily, a double sum as in Eq. (23.51) that involves $N^2$ points for each interpolated $(s,\theta)$ point would involve $O(N^4)$ operations in a (unrealistic) worst-case analysis, i.e., where all $N^2$ points on the irregular lattice are involved in interpolating a single point on a uniform lattice. However, because the interpolation in $s$ involves only the points $k\Delta\sigma$, $k = 1, 2, \ldots, N$, the worst-case analysis yields $O(N^3)$ operations, an order of operations fewer. That this is so is easy to demonstrate. Suppose we hold $k$ fixed and interpolate in the $\theta$ direction. Then there are $N^2$ operations for every $k$ and, since there are $N$ values of $k$, the conversion from an irregular $\theta$ lattice to a uniform $\theta$ lattice involves $N^3$ operations. Finally, for each fixed value of $\theta$, $N^2$ operations are required to convert from an irregular $s$ lattice to a uniform $s$ lattice, and since there are $N$ values of $\theta$, $N^3$ operations are required altogether to convert from an irregular to the uniform $s$ lattice. The total number of operations is then of the order of $2N^3$ rather than $N^4$. Of course, the direct Fourier method is based on the fact that far fewer than $N^2$ points are required to interpolate a single point from its neighbors. The purpose of this discussion was to show that Eq. (23.51) has a structure that admits an efficient interpolation process.

A second remark concerns the validity of Eq. (23.51) as an exact interpolation formula. Equation (23.51) furnishes an exact reconstruction if $\tilde{p}(\beta, \sigma)$ is band-limited. Assume that $p(s, \theta)$ is effectively radially band-limited to $B_s$ and angularly band-limited to $B_\theta$. What then are the angular bandwidths of $\tilde{p}(\beta, \sigma)$, that is, $B_\beta$ and $B_\sigma$? Since both $\theta$ and $\beta$ refer to angular displacement and differ only in the offset $\sigma$, we expect $B_\theta$ and $B_\beta$ to be the same. Thus if $\Delta\theta$ is the maximum allowable sampling interval in angle, then $\Delta\beta = \Delta\theta$ [40]. It is less obvious that the relation between $B_\sigma$ and $B_s$ is given by

$$B_\sigma \simeq B_s R + 1 \tag{23.52}$$

We proved this result in [40]. An upper bound for $B_\theta$ in terms of $B_s$ was furnished by Stark et al. [6].

### 6.3 Experimental Results

In this section we present some experimental results obtained from fan-beam data using the direct Fourier method. The procedure follows the steps shown in Figure 23.22. The rebinning is done as in Eq. (23.45) and the interpolation to a uniform $(s, \theta)$ lattice is done as in Eq. (23.51). The latter step in effect yields uniformly spaced parallel-beam data. The next step is to take a sequence of one-dimensional FFTs, one for each view, which leads to two-dimensional Fourier data on a polar lattice. The penultimate step is a polar-to-Cartesian interpolation as in Eq. (23.9). Finally a two-dimensional inverse FFT yields the image.

Truncation is unavoidable in practical interpolation. In earlier work on parallel-beam data [6], we found that the polar-to-Cartesian interpolation required only five radial and three azimuthal neighbors (15 points altogether) to yield excellent 256 × 256 pixels squared imagery. However, in the initial DFM reconstruction of fan-beam data, severe ringing artifacts were observed. These were due to the use of a finite number of terms in Eq. (23.51) in interpolating from nonuniformly spaced samples in the $s$ direction. The cause and amelioration of these artifacts are given below.

### Description of the Data and Simulation Parameters

Two image functions and their associated fan-beam projected data were synthesized analytically: 1) a uniform disk of radius 15 (a consistent set of units is assumed throughout) of unity intensity (DISK); and 2) two disks contained within a set of boundary ellipses (PHANTOM). These are shown in Figures 23.25a and 23.25b, respectively. The physical size of the pixel is 0.3 on a side and the picture size is 128 x 128 pixels. The data-collecting geometry is shown in Figure 23.26. Adjacent source positions were incremented by $\Delta\beta = 1.25°$.

The detector spacing $\Delta d$ was taken to be 0.2 except in one case involving Figure 23.25b, where a detector spacing $\Delta d = 0.4$ was attempted. These spacings gave rise to angular sampling intervals of

$$\Delta\sigma = 0.2/110.735 = 1.8 \times 10^{-3} \text{ radians} \tag{23.53a}$$

and

$$\Delta\sigma = 0.4/110.735 = 3.6 \times 10^{-3} \text{ radians} \tag{23.53b}$$

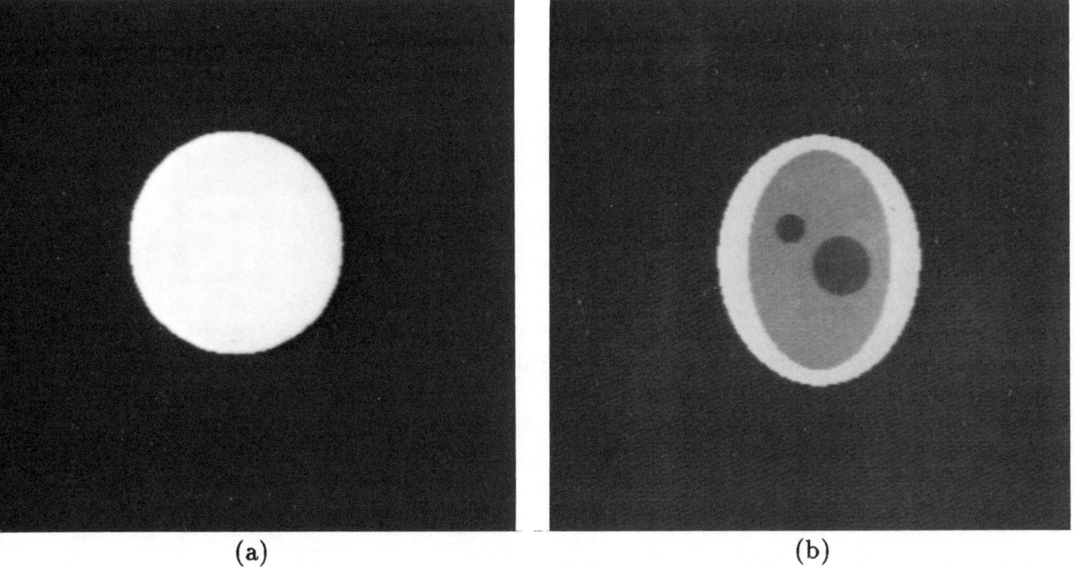

(a) (b)

**Figure 23.25** Objects to be reconstructed: (a) a uniform disk of radius 15 and unit intensity; (b) a phantom consisting of two disks at $(3, -1)$ and $(-4, 4)$ with radii 4 and 2, respectively, both of intensity 0.2. The framing ellipses are described by $(x/14)^2 + (y/17)^2 = 1$ (outer) and $(x/10)^2 + (y/15)^2 = 1$ (inner). The intensity between the ellipses is unity and in the internal background is 0.5. The external background has intensity zero.

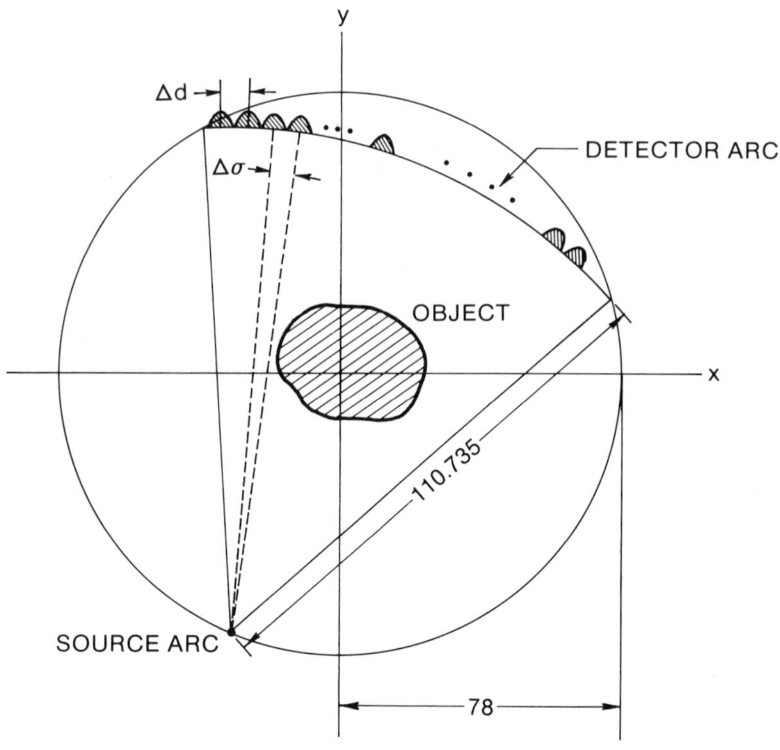


**Figure 23.26** Assumed fan-beam CT geometry.

We prove in [40] that the angular Nyquist interval $\Delta\sigma_{\max}$ is related to the Cartesian Nyquist interval $\Delta s_{\max}$ by

$$\Delta\sigma_{\max} \cong \frac{\Delta s_{\max}}{R} \tag{23.54}$$

Therefore, if the Cartesian Nyquist interval $\Delta s_{\max}$ (the pixel size) is 0.3, then from Eq. (23.54)

$$\Delta\sigma_{\max} = 0.3/78 = 3.8 \times 10^{-3} \text{ radians} \tag{23.55}$$

Hence, both of the sampling intervals given in Eq. (22.53) satisfy the Nyquist rate. The analytical relation between the required Cartesian sampling and the minimum angular sampling enabled us, in fact, to reconstruct excellent images of PHANTOM using half the number of detectors used by other researchers in their simulation [42] configurations.

In the following, we first consider reconstructing by truncating the interpolation formula given by Eq. (23.51). If the reconstruction is performed without compensating for the truncation effect, the result will be called an *uncompensated* reconstruction; otherwise it is a *compensated* reconstruction.

## Uncompensated Reconstructions

The initial reconstructions of DISK and PHANTOM, using the method diagrammed in Figure 23.22, are shown in Figure 23.27. The interpolation of the rebinned projection data used 12 points in azimuth $\theta$ (6 on each side) and 16 points in the linear direction $s$ (8 on each side). We denote this by $(12, 16)$. Severe ringing artifacts were observed in the reconstructions. When the numbers of azimuthal and linear neighbors were increased to $(16, 32)$, the ringing artifacts were reduced but were still noticeable (see Figure 23.28). In all cases, the interpolation in Fourier spaces in relocating data on a Cartesian lattice from data on a polar lattice involved 21 points (3 azimuthal and 7 radial).

Thus, allowing that the ringing artifacts are rooted in the truncation of Eq. (23.51), two questions arise. First, why does truncation of Eq. (23.51) produce an oscillatory artifact structure, and second, how can we compensate for it?

To answer the first question, we found that the ringing artifacts were the results of taking a finite number of terms in the $s$ direction (henceforth "$s$-truncation"). It will be recalled that practical interpolation in the $s$ direction involved the CPL formula [Eq. (23.44)] in an $s$-truncated mode. Since only a finite number of terms, say $2N$, are used, we expect that the interpolated point will be in error. The size of the error depends on the position of the interpolated point relative to the nearest

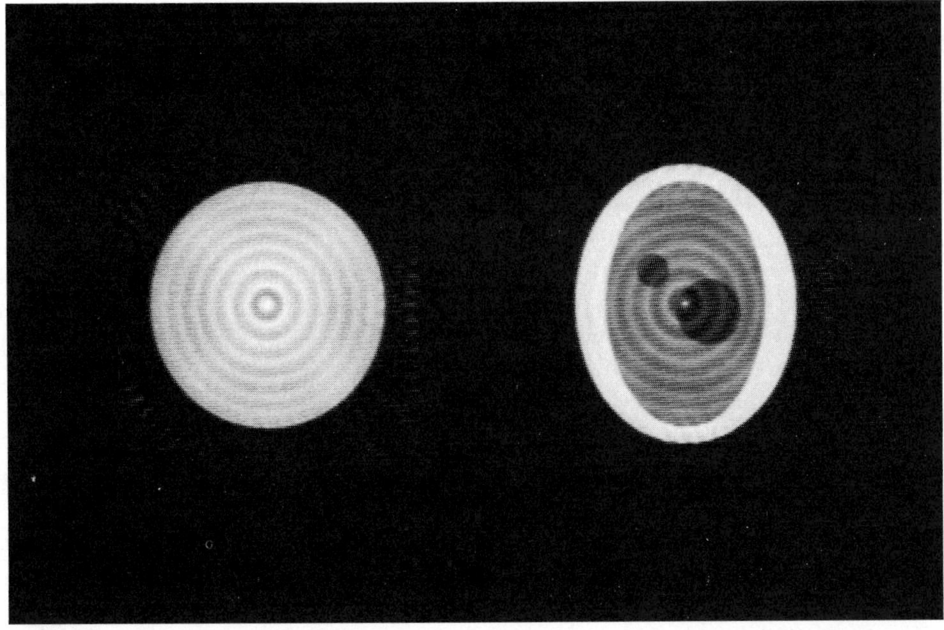

**Figure 23.27** Reconstructed DISK and PHANTOM using $(12, 16)$ interpolation (12 points in azimuth, 16 in linear direction $s$). Severe ringing artifacts are observed.

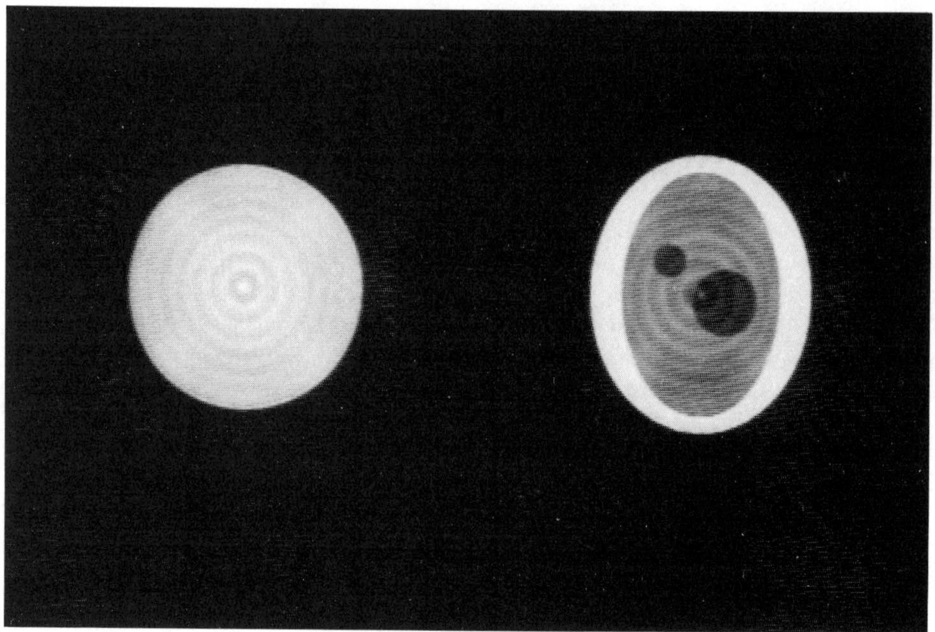

**Figure 23.28** Reconstructed DISK and PHANTOM using (16, 32) interpolation. Although there is an improvement over the results of Figure 23.27, significant ringing is still observed.

available data point. As discussed in Section 6.1, the $s$-truncated CPL formula gives an exact reconstruction when the interpolated point coincides with a data point; the interpolation error increases as the interpolated point moves away from a data point and then again decreases and becomes zero as the interpolated point approaches and coincides with the next data point. This pattern is repeated and the same for each view and leads, at least qualitatively, to an explanation of the ringing behavior when $\sigma = \sin^{-1}(s/R)$.

A quantitative analysis of this phenomenon will enable us to compensate for the ringing artifact. Since the latter stems from the $s$-truncation of the CPL formula, we need only consider this component of the interpolation rather than all of Eq. (23.51). Thus, with the variable $\theta$ submerged in arguments and/or subscripts, we write

$$p(s) = \sum_{k=M-N}^{M+N-1} p(s_k)\psi(s; k\Delta\sigma) \tag{23.56}$$

where

$$\psi(s; k\Delta\sigma) = \frac{\sin\{(\pi/\Delta\sigma)[\sin^{-1}(s/R) - k\Delta\sigma]\}}{(\pi/\Delta\sigma)[\sin^{-1}(s/R) - k\Delta\sigma]} \tag{23.57}$$

In Eq. (23.56) $M$ is the smallest integer greater than or equal to $[\sin^{-1}(s/R)]/\Delta\sigma$, that is, $M\Delta\sigma$ is the nearest right data point to the interpolated point $\sigma = \sin^{-1}(s/R)$, and $2N$ is the number of terms used in the interpolation. Note that there will be $N$ data points to the left of the interpolated point and $N$ to the right except when the interpolated point actually coincides with a data point; then there will be $N$ points to the left and $N-1$ points to the right of the interpolated point.

Since our interest is in interpolating at the points $s = m\Delta s$, $m = \ldots, -1, 0, 1, \ldots$, we write

$$p(m\Delta s) = \sum_{k=M-N}^{M+N-1} p(s_k)\psi(m\Delta s; k\Delta\sigma) \tag{23.58}$$

and assume, for convenience, that $p(s)$ is a wide-sense stationary (WSS) random process. Then with $E[\bullet]$ denoting the expectation operator, we get

$$\begin{aligned} E[p(m\Delta s)] &= \sum_{k=M-N}^{M+N-1} E[p(s_k)]\psi(m\Delta s; k\Delta\sigma) \\ &= E[p(s_k)]B_N(m\Delta s) \end{aligned} \tag{23.59}$$

where $E[p(s_k)]$ does not depend on $k$ because the process is WSS and

$$B_N(m\Delta s) \triangleq \sum_{k=M-N}^{M+N-1} \psi(m\Delta s; k\Delta\sigma) \tag{23.60}$$

Since $B_N(m\Delta s)$ is generally not equal to unity, we see that the average value of the interpolated points is in error. The function $B_N(m\Delta s)$ has introduced a *bias*. Thus to remove the bias we need to multiply the interpolated projection data $p(m\Delta s)$ by $B_N^{-1}(m\Delta s)$.

In practice, the bias function $B_N(m\Delta s)$ is responsible for the ringing observed in the reconstructed imagery. This is easily demonstrated by plotting $B_N(m\Delta s)$ versus $m$. Figure 23.29a shows $B_N(m\Delta s)$ versus $m$ for $\Delta s = 0.3$, $\Delta\sigma = 1.8 \times 10^{-3}$ radians, and $N = 4$. Figure 23.29b does the same for $N = 16$. We see that for $N = 16$ the bias is significantly less but is still there and oscillatory. Tests show that if a true set of projections $p_\theta(s)$ is multiplied by $B(m\Delta s)$ for each view, the distorted projection data will appear with vertical stripes and the reconstructed image will contain ringing artifacts [40].

## Compensated Reconstructions

In an initial series of experiments we compared compensated with uncompensated reconstructions for $(16, 32)$ interpolation and $\Delta\sigma = 1.8 \times 10^{-3}$ radians. The results

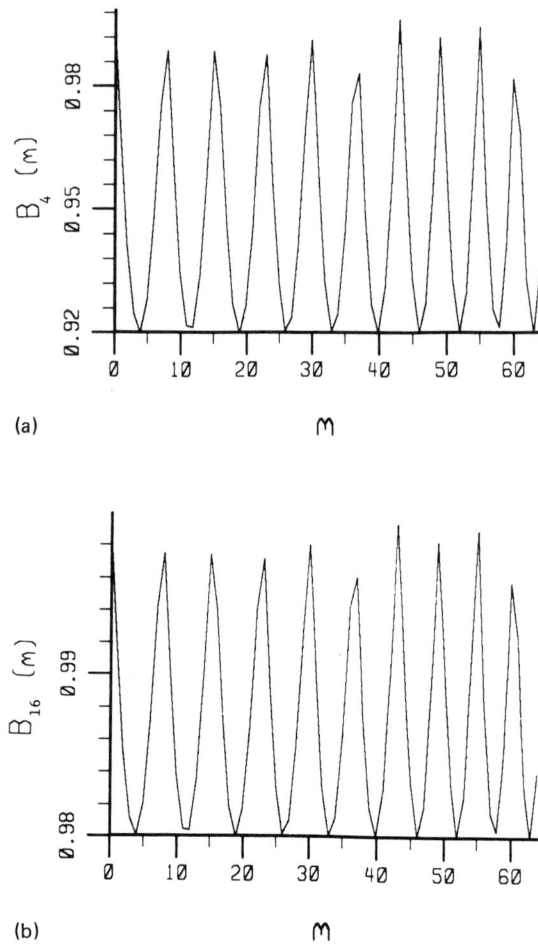


**Figure 23.29** Plot of the bias function $B_N(m\Delta s)$ versus $m$. (a) $\Delta s = 0.3$, $\Delta\sigma = 1.8 \times 10^{-3}$ radians, $N = 4$; (b) same as (a) except $N = 16$.

of the bias-uncompensated reconstruction of DISK with the compensated reconstruction, i.e., projections at each view are multiplied by $B_N^{-1}(m\Delta s)$ (prior to computing the FFT), is given in Figure 23.30. Figure 23.31 shows the uncompensated versus compensated reconstruction of PHANTOM. In both cases, the ringing artifacts have been largely eliminated.

In a final series of experiments we reconstructed PHANTOM by the direct Fourier method, starting with fan-beam data and compensating for the effect of $B_N(m\Delta\sigma)$. The most important parameters—number of terms used in interpolating the rebinned data and detector spacing—were both varied. The results are shown in Figure 23.32 for (8, 8) and (16, 32) interpolation and $\Delta\sigma = 1.8 \times 10^{-3}$ radians (255 detectors) and $\Delta\sigma = 3.6 \times 10^{-3}$ radians (127 detectors). Several important observations follow these simulations:

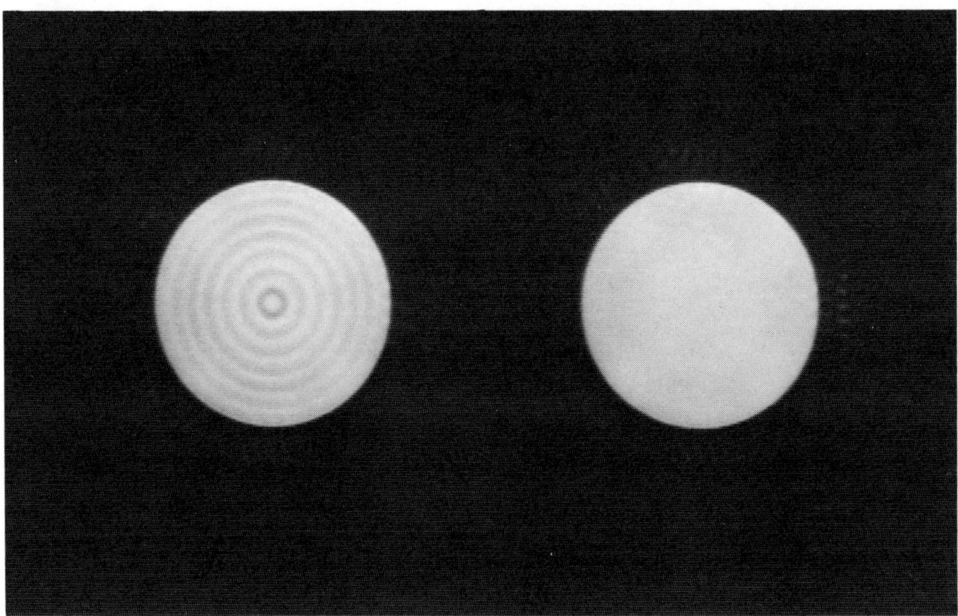

**Figure 23.30** Uncompensated versus compensated reconstruction of DISK.

1. When the reconstructions in Figure 23.32 are compared with those of Figure 23.27, we see that the ringing artifacts have largely vanished as a result of compensation for the bias $B_N(m\Delta s)$. All the images in Figure 23.32 are bias-compensated and all yield satisfactory reconstructions. Figure 23.32a shows an $(8, 8)$ interpolation reconstruction which is free from ringing but suffers some streak artifacts near the edges. When the interpolation is increased to $(16, 32)$, the resulting reconstruction is given in Figure 23.32c. There we see that the streak artifacts are significantly reduced. The improvement in mean-square error in going from Figure 23.32a to 23.32c is about 2%.

2. The number of terms required to interpolate the rebinned projection data is one-eighth of the total number of detectors per view. When $N$ increases, this fraction is reduced even further. Thus the interpolation does not negate the computational advantage of the direct Fourier method.

3. Reducing the number of detectors by half (but still keeping within the Nyquist rate in angular sampling) has little or no deleterious effect on the quality of the reconstructed image. This is seen in Figures 23.32b and 23.32d. Indeed, while the number of detectors has been halved and the angular sampling rate has been reduced by a factor of two, the larger distance between detectors now results in an interpolation in which neighbors physically farther away contribute to the interpolated point. We conjecture that this ameliorates the negative effects of reduced sampling density (the overall number of terms involved in the interpolation in either case—255 or 127 detectors—is the same).

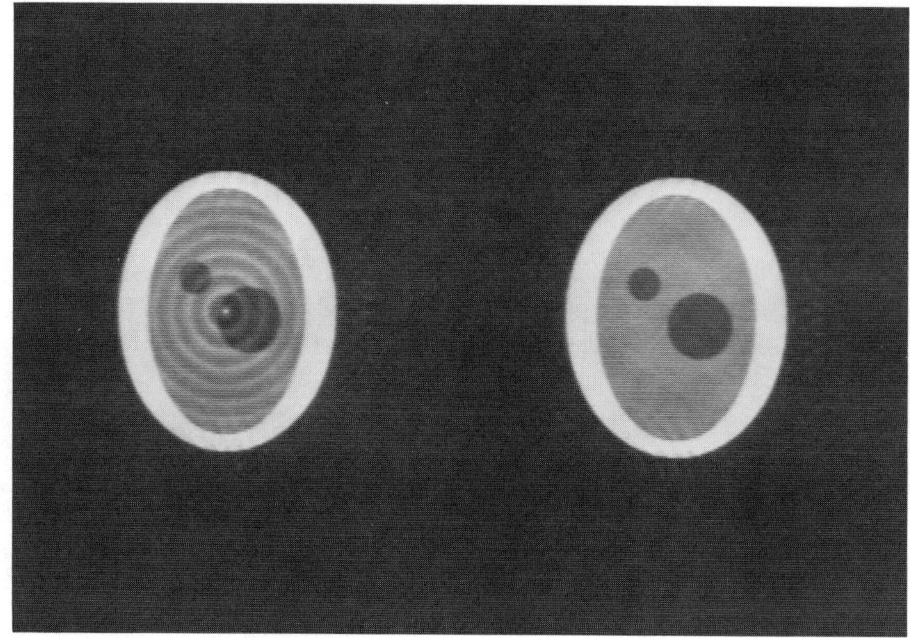

**Figure 23.31** Uncompensated versus compensated reconstruction of PHANTOM.

### 6.4 Conclusions

We have discussed an algorithm for reconstructing CT imagery from fan-beam projection (i.e., raysum) data using the direct Fourier method. We demonstrated that good-quality imagery can be obtained after compensation for truncation in the interpolation of rebinned projection data. Our method, which requires $O(N^2 \log N + aN^2)$ operations, is indeed more efficient than the CBP algorithm.

Finally, in the analytical relationship between Nyquist intervals in both $s$ and $\sigma$ domains enabled us to reconstruct excellent images using half the number of detectors used in the literature.

## 7 IMAGE RECONSTRUCTION FROM LINEAR SPIRAL SCAN IN MR TOMOGRAPHY

For the reasons given in Section 2.2, the linear spiral scan in frequency space is a useful trajectory for gathering data in MR tomography. Here we discuss a method that enables us to interpolate the data from a spiral lattice to a uniform Cartesian lattice. Once the data are on a Cartesian lattice, the inverse FFT can be used to reconstruct an image. In the following, we first present a theorem that enables exact interpolation from spiral samples to a Cartesian lattice. We then investigate two practical implementations of the theorem in which a finite number of interpolating

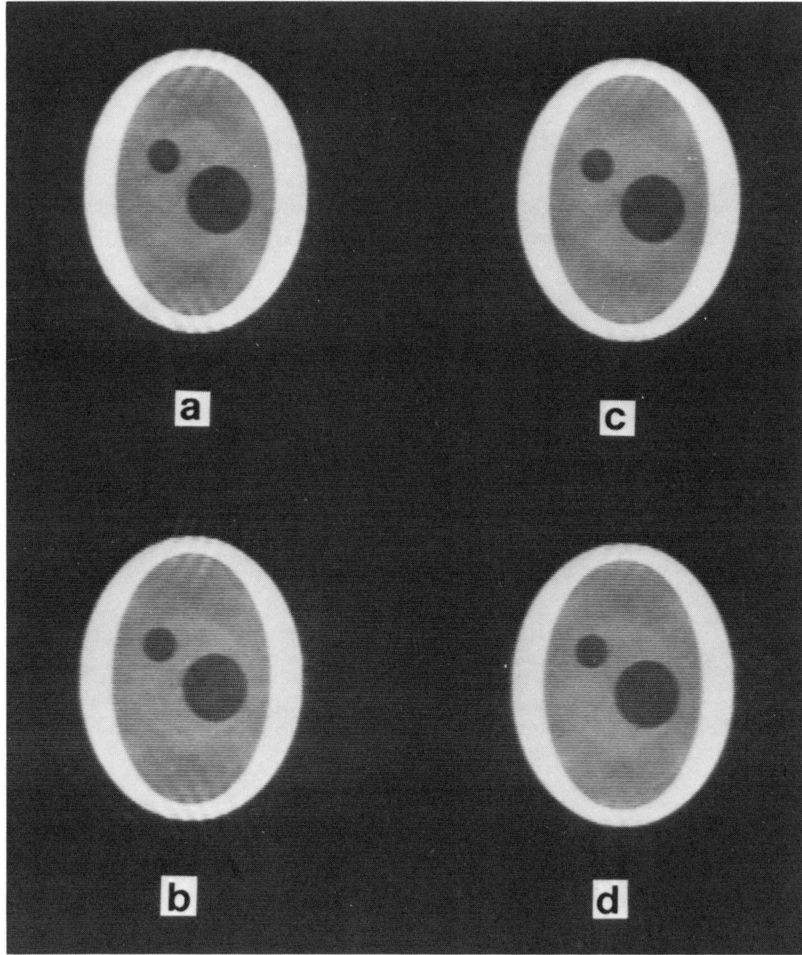


**Figure 23.32** Reconstruction of PHANTOM: (a) (8, 8) interpolation, 255 detectors; (b) (8, 8) interpolation, 127 detectors; (c) (16, 32) interpolation, 255 detectors; (d) (16, 32) interpolation, 127 detectors.

points are used to calculate the value at a new point. Finally, we consider some reconstruction examples using the technique.

## 7.1 Linear Spiral Sampling and Interpolation

A linear spiral sampling lattice can be defined by the following equations:

$$u_{jk} = \omega_{jk} \cos \theta_k \tag{23.61a}$$

$$v_{jk} = \omega_{jk} \sin \theta_k \tag{23.61b}$$

where $\omega_{jk}$, the radial displacement at $\theta_k$; $k = 0, 1, \ldots, N$, is a linear function of the *revolution number* $j$ and is given by

$$\omega_{jk} = \alpha(\theta_k + 2\pi j), \qquad 0 \leq \theta_k < 2\pi, \ j = 0, 1, 2, \ldots \tag{23.61c}$$

An example of a linear spiral scan trajectory is shown in Figure 23.33. The corresponding lattice for $N = 32$ is shown in Figure 23.34. This type of sampling can be obtained by applying gradients of the form of Eq. (23.16) (Section 2.2).

In Eq. (23.61c), $\alpha$ is a parameter related to the sampling requirements dictated by the size of the object. We will assume that $f(x, y)$ has finite support and space limited to $2A$. Then for a fixed $k$ the radial sampling interval $\Delta\omega$, given from Eq. (23.61c) by $\Delta\omega = 2\pi\alpha$, must satisfy the constraint $\Delta\omega \leq 1/2A$, so that

$$\alpha \leq \frac{1}{4\pi A} \tag{23.62}$$

$N$ is the number of samples in one revolution. In MR, as in CT, $N$ is most often even, in which case it should be at least $(2K + 2)$, where $K$ is the angular bandwidth of $F(\omega, \theta)$ (Section 2.1). If $F(\omega, \theta)$ is not angularly band-limited, then angular prefiltering before sampling must be done [6]. Even if $F(\omega, \theta)$ is angularly band-limited, the required sampling rate per revolution may be too large for the existing hardware. Here, too, prefiltering must be done.

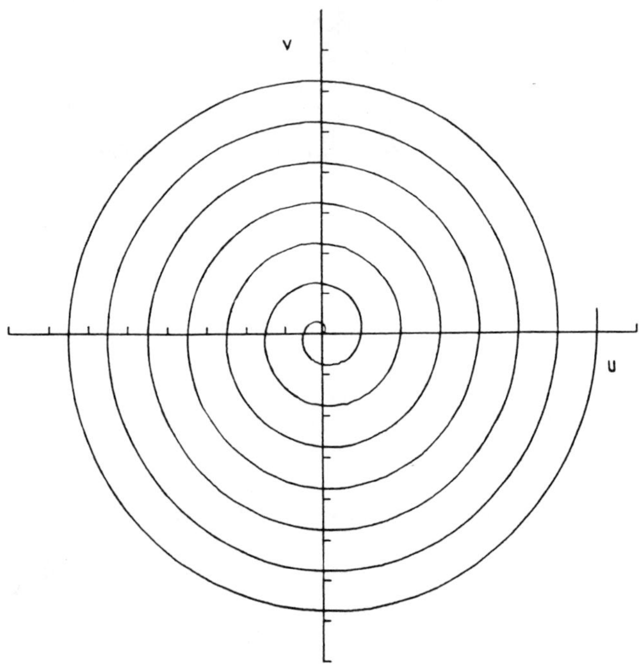


**Figure 23.33** Example of a linear spiral scan in the $(u, v)$ frequency plane.

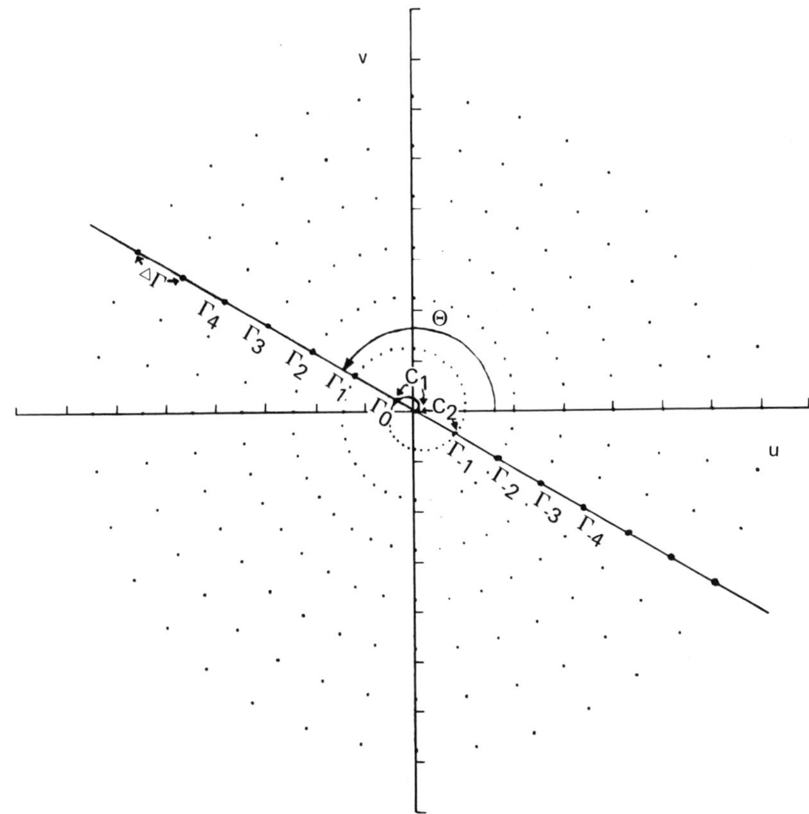


**Figure 23.34** Spiral sampling grid for $N = 32$. The notation used to describe the samples along a line at angle $\theta$ from the real axis is also shown.

Spiral sampling does not yield uniformly spaced samples. This can be understood with the help of Figure 23.34. Consider a line at some fixed angle $\varphi$ $(0 \leq \varphi < \pi)$, passing through the origin. Hereafter, we will use an alternative labeling for the points along a line that is advantageous in this discussion and for computational purposes. Let $(\omega, \theta)$ be the polar coordinates of a point on the line. We define $r$ as follows:

$$r \triangleq \begin{cases} \omega, & \theta = \varphi \\ -\omega, & \theta = \varphi + \pi \end{cases} \tag{23.63}$$

Also, the samples along a line will be denoted by $r_{lk}$, defined as

$$r_{lk} \triangleq \begin{cases} \omega_{lk}, & l \geq 0 \\ -\omega_{(-l-1)(k+K+1)}, & l < 0, \ (0 \leq k < K+1) \end{cases} \tag{23.64a}$$

and

$$r_{lk} = r_{l(k-K-1)}, \qquad K+1 \le k < 2K+2 \tag{23.64b}$$

Thus, the intercepts of the line with the spiral are labeled $\ldots r_{-2k}, r_{-1k}, r_{0k}, r_{1k}, \ldots$. The samples $r_{lk}$ $(l \ge 0)$ are located in the upper half-plane and the samples $r_{lk}$ $(l < 0)$ are located in the lower half-plane (see Figure 23.34); where we omit the index $k$ as we do in what follows, since $k$ is held constant). Now, the spacings between $r_{l+1}$ and $r_l$ $(l \ge 0)$ are uniform, as are the spacings between $r_l$ and $r_{l-1}$ for $l \le -1$. However, the spacing between $r_0$ and $r_{-1}$ is different from the other spacings. It is convenient to refer to the spacing between $r_0$ and $r_{-1}$ as the sum of two positive numbers $c_1 + c_2$ $(c_1 \ge 0,\ c_2 > 0)$, where $c_1$ and $c_2$ are the actual distances of samples $r_0$ and $r_{-1}$ from the origin $u = v = 0$. Referred to the origin, the actual position of $r_{-1}$ is at $-c_2$ and the position of $r_0$ is at $c_1$. Note that when $c_1 = 0$ and $c_2 = \Delta r = \Delta\omega$, the points are equally spaced and therefore the sampling is uniform.

For a fixed $\theta_k = \varphi$, $0 \le \varphi < \pi$, it is not hard to see using Eq. (23.61c) that

$$\begin{aligned} c_{1_k} &= \alpha\theta_k = \frac{\Delta r}{2\pi}\theta_k \\ c_{2_k} &= c_{1_k} + \frac{\Delta r}{2} \end{aligned} \tag{23.65}$$

since $\Delta r = \Delta\omega = 2\pi\alpha$. Also, the following relation defines the set of nonuniform samples $\{r_{lk}\}$ along a line:

$$r_{lk} = \begin{cases} c_{1_k} + l\Delta r, & l \ge 0 \\ -c_{2_k} + (l+1)\Delta r, & l < 0 \end{cases} \tag{23.66a}$$

for $\theta_k = \varphi$ $(0 \le \varphi < \pi)$, and

$$r_{lk} = r_{l(k-K-1)} \tag{23.66b}$$

for $\theta_k = \varphi + \pi$. Thus, the reconstruction of $F(\omega, \theta)$ from its linear spiral samples is equivalent to the following interpolation problem: given the set of nonuniform sample points $\{r_{lk}\}$, $l = 0, \pm 1, \pm 2, \ldots;\ k = 0, 1, \ldots, N-1$, specified in Eq. (23.64), find $F(\omega, \theta)$. Next we present the interpolation theorem to solve this problem.

## 7.2 Linear Spiral Sampling Theorem (LSST)

Let $f(x, y)$ be space-limited such that $f(x, y) = 0$ outside a disk of radius $A - \epsilon$, $0 < \epsilon < A$, centered at the origin. Let its FT in polar coordinates $F(\omega, \theta)$ be angularly band-limited to $K$. Then $F(\omega,\theta)$ can be reconstructed from its spiral samples $F(\omega_{jk},\theta_k)$, where

$$\omega_{jk} = \frac{\theta_k}{4\pi A} + j\frac{1}{2A}, \qquad j = 0,1,2,\ldots \tag{23.67a}$$

$$\theta_k = \frac{\pi k}{K+1}, \qquad k = 0,1,2,\ldots,2K+1 \tag{23.67b}$$

via

$$F(\omega,\theta) = \sum_{l=-\infty}^{\infty} \sum_{k=0}^{2K+1} F_1(r_{lk},\theta_k)\psi(r,r_{lk})\Phi(\theta-\theta_k) \tag{23.68}$$

and where the samples $F_1(r_{lk},\theta_k)$ are obtained from $F(\omega_{jk},\theta_k)$ as implied by Eq. (23.64). Likewise, the argument $r$ of the interpolation function $\psi(\bullet,\bullet)$ is obtained from Eq. (23.62) with $\theta = \theta_k$. The interpolation functions $\psi(\bullet,\bullet)$ are given by

$$\psi(r,r_{lk}) = B_{lk}(r)\,\mathrm{sinc}[2A(r-r_{lk})] \tag{23.69}$$

where

$$B_{lk}(r) = \begin{cases} \dfrac{\Gamma(2Ac_k+l)}{\Gamma(1+l)}\dfrac{\Gamma[1+2A(r-c_{1_k})]}{\Gamma[2A(r+c_{2_k})]}, & l \geq 0 \\[2ex] \dfrac{\Gamma(2Ac'_k-l)}{\Gamma(-l)}\dfrac{\Gamma[1-2A(r+c_{2_k})]}{\Gamma[-2A(r+c_{1_k})]}, & l < 0 \end{cases} \tag{23.70}$$

and $\Gamma(\bullet)$ is the standard gamma function. Also

$$r_{lk} = \begin{cases} \dfrac{l}{2A} + c_{1_k}, & l \geq 0 \\[2ex] \dfrac{l}{2A} - c'_{2_k}, & l < 0 \end{cases} \tag{23.71}$$

and

$$c_{1_k} = \begin{cases} \dfrac{\theta_k}{4\pi A}, & 0 \leq \theta_k < \pi \\[2ex] \dfrac{(\theta_k-\pi)}{4\pi A}, & \pi \leq \theta_k < 2\pi \end{cases} \tag{23.72a}$$

$$c_{2_k} = c_{1_k} + 1/4A \tag{23.72b}$$

$$c'_{2_k} \triangleq c_{2_k} - 1/2A \tag{23.72c}$$

$$c_k \triangleq c_{1_k} + c_{2_k} \tag{23.72d}$$

$$c'_k \triangleq c_k - 1/2A \tag{23.72e}$$

Finally, $\Phi(\bullet)$ is given by

$$\Phi(\theta) = \frac{\sin[\frac{1}{2}(2K+1)\theta]}{(2K+2)\sin(\frac{1}{2}\theta)} \tag{23.73}$$

The proof of this theorem is given in [43]. In the next section we will deal with some practical aspects and furnish a practical version of Eq. (23.68). In the last section we will show the performance of the practical version by means of some experiments.

We remark in passing that because $B_{lk}(\bullet)$ in Eq. (23.69) depends on $r$, the interpolating functions $\psi(r, r_{lk})$ can behave in a manner drastically different from the sinc($\bullet$) function. The behavior of $\psi(r, r_{lk})$ depends of the values of $c_{1_k}$ and $c'_{2_k}$ [see Eq. 23.72]. When $c_{1_k}$ and $c'_{2_k}$ have small values and $|r|$ becomes large, $\psi(r, r_{lk})$ behaves essentially like the sinc($\bullet$) function.

## Practical considerations

In practice, we consider a truncated version of Eq. (23.68), i.e.

$$\hat{F}(\omega, \theta) = \sum_{l=l_r-L_r}^{l_r+L_r} \sum_{k=k_\theta-L_\theta}^{k_\theta+L_\theta} F_1(r_{lk}, \theta_k)\psi(r, r_{lk})\Phi(\theta - \theta_k) \tag{23.74a}$$

where

$$k_\theta = \left[(K+1)\frac{\theta}{\pi}\right] \tag{23.74b}$$

and

$$l_r = \begin{cases} [2A(r - c_{1_k})], & r \geq 0 \\ [2A(r + c'_{2_k})], & r < 0 \end{cases} \tag{23.74c}$$

As before, $[\bullet]$ indicates rounding to the nearest integer. The reconstruction of $f(x, y)$ from the linear spiral samples of its FT calls for the following procedure. We let

$$\omega_{mn} \triangleq \sqrt{u_m^2 + v_n^2} \tag{23.75a}$$

$$\theta_{mn} \triangleq \cos^{-1} \frac{u_m}{\omega_{mn}} \tag{23.75b}$$

and apply Eq. 23.74 to $F(\omega_{mn}, \theta_{mn}) \triangleq F_c(u_m, v_n)$ ($F_c$ denotes the FT of $f$ in Cartesian coordinates). Then we use inverse FFT to reconstruct $f(x_m, y_n)$. This procedure follows what is done in practical applications of the polar-to-Cartesian interpolation formula (Section 2.1).

As a first example of the application of the LSST, we used Eq. (23.74) with relatively large values of $L_r$ and $L_\theta$ to reconstruct a function from the linear spiral samples of its FT. The 2-D function used in this experiment (as well as in others presented later) is

$$f(x,y) = \cos(\pi x)\cos(\pi y)\mathrm{rect}(x)\mathrm{rect}(y), \quad \text{in the square } -1 \le x, y \le 1 \tag{23.76}$$

The FT of this function was sampled in a spiral lattice with $\Delta\omega = 0.5$, $N = 72$. As our reference for comparisons we use a reconstruction of $f(x, y)$ from Cartesian samples of $F_c(u, v)$ using a 64 × 64 FFT (see Figure 23.35). This reconstruction is used to calculate the reconstruction error [defined as in Eq. (23.36)] in our experiments; i.e., Figure 23.35 is considered the original.

The result of this experiment is shown in Figure 23.36. The reconstruction error is MSE = 1.1%. This residual error is due mainly to the truncation of the radial summation. This result suggests that by taking a large number of samples we can reconstruct a function from its linear spiral samples to an arbitrarily high degree of accuracy using Eq. (23.74).

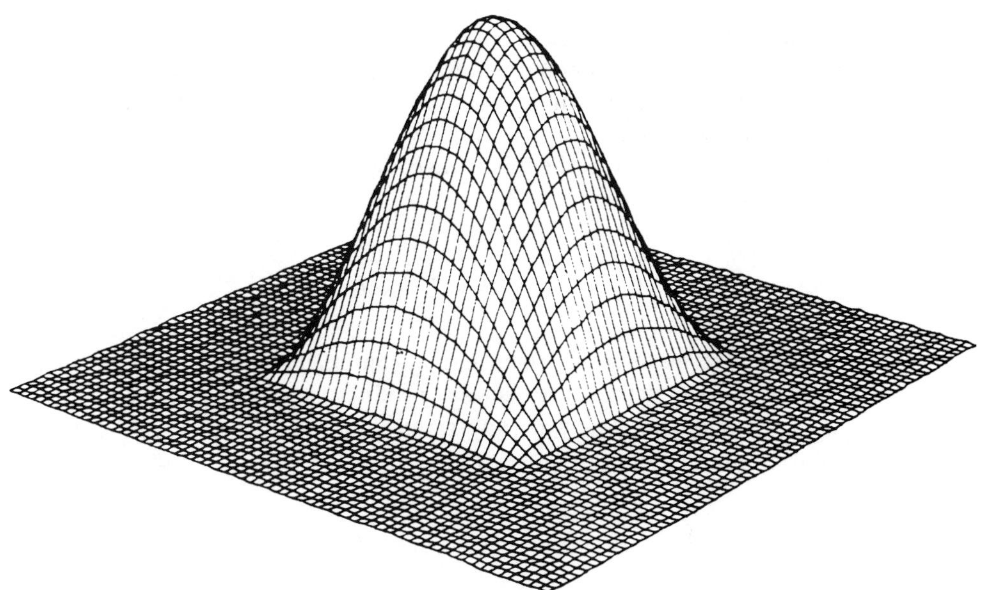

**Figure 23.35** Original image distribution: $f(x, y) = \cos(\pi x) \times \cos(\pi y) \times \mathrm{rect}(x) \times \mathrm{rect}(y)$.

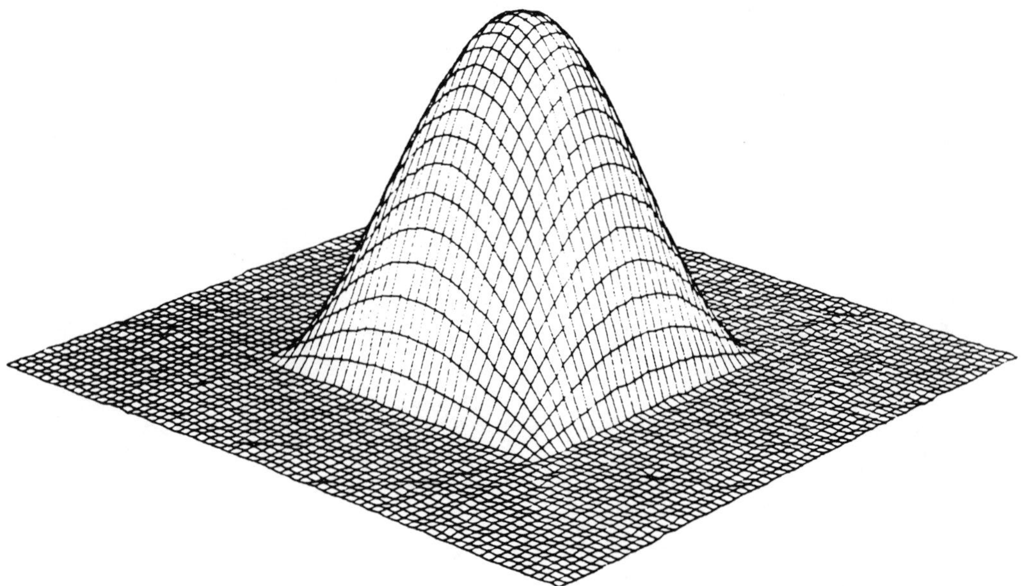

**Figure 23.36** Reconstruction of $f(x,y)$ from linear spiral samples, $\Delta\omega = 0.5$, $N = 72$ using the LSST (with a large number of terms in the summation). The reconstruction error is $e = 1.1\%$.

When Eq. (23.74) is significantly truncated, i.e., we do not use many points $r_{lk}$, then the reconstruction may exhibit an undesirable ringing artifact. A means of reducing the ringing is to use a tapered window over the sample points involved in the reconstruction. One possible window is the Cartesian product of two identical 1-D triangular windows, $\Delta(\bullet)$, defined by Eq. (23.13).

Equation (23.74) may be put in the following compact form, which may be useful in an implementation:

$$\hat{F}(\omega,\theta) = F_{d_1 \times d_2} * Y_{d_1 \times d_2} \tag{23.77}$$

where $F$ and $Y$ are matrices with $d_1$ rows and $d_2$ columns, and $*$ indicates a scalar matrix product. (We define the scalar product as $A_{d_1 \times d_2} * B_{d_1 \times d_2} \triangleq \sum_{i=1}^{d_1} \sum_{j=1}^{d_2} a_{ij} b_{ij}$.) According to Eq. (23.74), $d_1 = 2L_r + 1$, $d_2 = 2L_\theta + 1$. $Y$ is the interpolation matrix whose element $(l_1, k_1)$ is given by

$$Y_{l_1,k_1} = \psi(r, r_{lk})\Phi(\theta - \theta_k)\Delta(l_1, k_1) \tag{23.78}$$

where $l = l_r + l_1$, $k = k_\theta + k_1$, and $-L_r \le l_1 \le L_r$, $-L_\theta \le k_1 \le L_\theta$.

Note that the interpolation matrices required for the procedure do not depend on the specific image being reconstructed (except for the object size, which in most applications can be fixed to some value). As stated earlier, if $L_r$ and $L_\theta$ are chosen small ($L_r \approx 3$, $L_\theta \approx 2$), one can precalculate and store the values of $Y$ at the Cartesian points. The required storage would, of course, be a prerequisite for this approach.

Earlier we stated that the $\psi(\bullet,\bullet)$ interpolating functions in Eq. (23.68) do not behave like sinc($\bullet$) functions. This results from the fact that we are dealing with nonuniform samples. Indeed, when the spacing between $r_0$ and $r_{-1}$ is very different from $\Delta r$, the $\psi(\bullet,\bullet)$ functions can exhibit high-amplitude oscillatory behavior away from the point being interpolated. This means that Eq. (23.74) could lead to significant errors in the reconstruction. In such instances it might be preferable to replace Eq. (23.74) with an approximate interpolation formula where radial interpolation functions behave like sinc($\bullet$) functions. We furnish such an approximation below; the approach is motivated by the results of Clark et al. [16] (see the CPL theorem discussed in Section 6.1).

The key step is to replace the functions $\psi(\bullet,\bullet)$ of Eq. (23.69) with another function given by

$$\psi(r, r_{lk}) = \operatorname{sinc}\left\{2A\left[\gamma_k(r) - \frac{l}{2A}\right]\right\} \tag{23.79}$$

where

$$\gamma_k(r) = \begin{cases} r - c_{1_k}, & r \geq c_{1_k} \\ \dfrac{1}{2Ac_k}(r - c_{1_k}), & -c_{2_k} < r < c_{1_k} \\ r + c'_{2_k}, & r \leq -c_{2_k} \end{cases} \tag{23.80}$$

and $c_{1_k}$, $c_{2_k}$, $c'_{2_k}$, and $c_k$ are given in Eq. (23.72). The effect of applying Eq. (23.80) to the nonuniformly spaced samples along $r$ is to produce a sequence of uniformly spaced samples along $\gamma_k(r)$. Henceforth, the use of Eq. (23.68) with Eqs. (23.79) and (23.80) replacing the exact interpolation functions given in Eqs. (23.69) and (23.70) will be referred to as the *transformation method.*

The interpolation functions in Eq. (23.79) will not give an exact solution, since an exact reconstruction would require that the functions $F(\gamma_k^{-1}(r'), \theta_k)$ be space-limited. However, in practice the errors induced by truncating the exact interpolation formula [Eq. (23.68)] may exceed the aliasing errors introduced by the non-space-limitedness of $F(\gamma_k^{-1}(r'), \theta_k)$. Hence it is worth considering the transformation method as a practical alternative to the truncated exact version of Eq. (23.68), i.e., Eq. (23.74). Moreover, application of the well-known Whittaker–Shannon–Kotelnikov uniform sampling theorem [44] to the computation of the interpolating functions for the transformation method [Eq. (23.79)] is considerably simpler than that of the exact interpolation [Eq. (23.70)].

We close this section by summarizing the approximations considered for the practical implementation of the theorem stated in the previous section, namely:

1. Use a truncated form of the exact interpolation theorem and apply a tapered window over the neighborhood of the interpolated point if the ringing induced by the truncation is excessive. This is the approach taken in the limited-view CT problem in Section 4.
2. Remap the unequally spaced sample points on transformed coordinate axes, where the points appear equally spaced. Then apply the well-known Whittaker–Shannon–Kotelnikov uniform sampling theorem [44] to interpolate points in the radial direction. This approach is suggested by the results of Clark et al. [16].

## 7.3 Experimental Results

In this section, our goal is to study the performance of practical realizations of the exact LSST. Practical realizations always involve some degree of truncation, so our starting point is Eq. (23.74). From then on we proceed to use either the exact interpolation functions of the LSST or the approximation introduced by Eq. (23.79).

Our experiments involve a synthetic image shown in Figure 23.37a. This image was produced by sampling a (continuous) FT in a $64 \times 64$ point Cartesian grid and then applying the inverse FFT to reconstruct a discrete approximation of the original 2-D function. This discrete version is taken to be our reference for comparison. Figure 23.37b is the result using the exact interpolation functions ($L_r = 3$, $L_\theta = 1$) and Figure 23.37c is the result of using the transformation method, both without a tapered window. Both images contain visible differences when compared with the original (Figure 23.37a). These differences are more pronounced in Figure 23.37b, which has an error of 14%, while the error of Figure 23.37c is only 7.6%. In Figures 23.37d and 23.37e we show results using a tapered window with parameter $\tau = 5$ [Eq. (23.13)]. In this case, the error is significantly lowered by the introduction of the tapered window, and the differences in performance between using the exact interpolation functions (Figure 23.37d) and the transformation method (Figure 23.37e) practically disappear. In summary, this set of experiments shows that the transformation method performs better than the truncated exact interpolation formula when using only a few neighbors. The introduction of the tapered window significantly improves the performance in all cases and reduced the differences between the two methods to almost nil.

We close this section with the following remark. In [20] the reconstruction procedure takes advantage of the Hermitian symmetry of the Fourier plane data by reflecting samples from opposite quadrants. In this way, a higher sampling rate is obtained, twice the rate calculated by using Eq. (23.61c). Also, by exploiting this symmetry one needs to interpolate only half of the points to obtain the complete set of Cartesian points that are required for the inverse FFT. In order to use the Hermitian symmetry it is necessary to introduce only minor modifications in the theory developed above. These modifications are outlined in [43].

## 7.4 Conclusions

In this section we introduced an exact interpolation theorem to reconstruct an angularly band-limited 2-D function from its spiral samples in the Fourier plane. We also investigated the practical implementations of the theorem. The experimen-

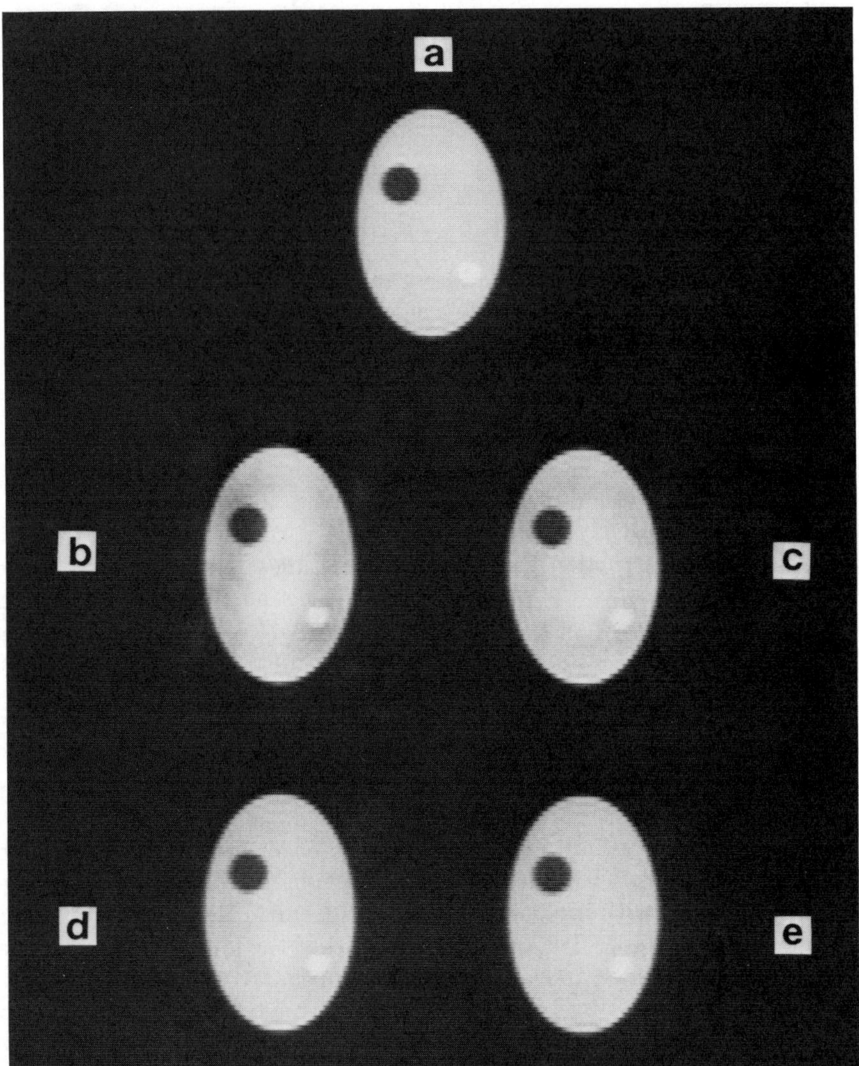


**Figure 23.37** Reconstruction of a synthetic image. (a) Original image. (b) Reconstruction from spiral samples $\Delta\omega = 1.0$, $N = 72$, $L_r = 3$, $L_\theta = 1$, without a tapered window using the exact interpolation functions, $e = 14\%$. (c) Same as (b), but using the transformation method, $e = 7.6\%$. (d) Reconstruction from spiral samples with same parameters as in (b), including a tapered window ($\tau = 5$), using the exact interpolation functions, $e = 6.8\%$. (e) Same as (d), but using the transformation method, $e = 5.8\%$.

tal results confirm the theorem's validity and also demonstrate that its practical implementations yield very good constructions. Therefore, our main conclusion is that the direct Fourier reconstruction can be used in the case of spiral scan MR imaging.

## 8 SUMMARY

Our aim in this chapter has been to address and solve some of the important signal processing problems associated with image reconstruction in medical tomography. First, we considered the problem of tomographic image reconstruction from angularly view-limited data as applied to parallel-beam CT. The POCS-DFM algorithm, synthesized from the method of POCS and the direct Fourier method, was used to reconstruct a simulated X-ray tomogram of a human thorax. Reconstructions from full-view and limited-view data were performed.

Next, we used the method of POCS to reconstruct an image from a set of raysums and a set of relatively typical a priori known constraints. The result was a geometry-free reconstruction algorithm. Interestingly, we found that the well-known ART algorithm was a primitive convex projections algorithm. Our formulation based on the theory of convex projections facilitated the incorporation of regularization constraints into the basic ART algorithm. We furnished simulations aimed at demonstrating the feasibility of this method.

We developed an algorithm to implement the reconstruction of CT imagery from fan-beam raysum data using the direct Fourier method. One of our key results was the interpolation formula we derived for interpolating nonuniformly spaced rebinned raysum data over a uniform lattice. We also derived the fundamental bandwidth and sampling constraints that must be satisfied in order to form an equivalent set of parallel-beam data from fan-beam raysums. We illustrated that good-quality imagery can be obtained with the proposed technique, which is computationally more efficient than the standard CBP algorithm. Note that the interpolation formula for mapping nonuniform $p_\theta(s)$ lattice data onto a uniform lattice can be incorporated into the POCS-DFM algorithm to extend the algorithm to the fan-beam data case. Thus, fan-beam data can also be reconstructed using the direct Fourier method.

Finally, we addressed the interpolation problem associated with the linear spiral scan echo-planar MR imaging. We introduced an exact interpolation theorem to reconstruct an angularly band-limited 2-D function from its Fourier samples over a spiral lattice in the Fourier plane. Using the interpolation formula, spiral-scan MR tomography allows for a direct Fourier reconstruction.

## REFERENCES

1. S. R. Deans, *The Radon Transform and Some of Its Applications*, Wiley, New York (1984).
2. L. G. Gubin, B. T. Polyak, and E. V. Raik, The method of projections for finding the common point of convex sets, *USSR Comput. Math. Math. Phys.*, *7*: 1–24 (1967).

3. D. C. Youla and H. Webb, Image restoration by the method of convex projections onto convex sets—Part 1, *IEEE Trans. Med. Imag.*, *MI-1*(2): 81–94 (1982).
4. M. I. Sezan and H. Stark, Image restoration by the method of convex projections: Part 2—Applications and numerical results, *IEEE Trans. Med. Imag.*, *MI-1*: 95–101 (1982).
5. H. Stark, ed., *Image Recovery, Theory and Applications*, Academic Press, New York (1987).
6. H. Stark, J. W. Woods, I. Paul, and R. Hingorani, Direct Fourier reconstruction in computer tomography, *IEEE Trans. Biomed. Eng.*, *BME-28*: 496–505 (1981).
7. H. J. Trussel and M. R. Civanlar, Feasible solution in signal restoration, *IEEE Trans. Acoust. Speech Signal Process.*, *32*: 201–212 (1984).
8. M. I. Sezan, A. M. Tekalp, and C. T. Chen, "Regularized Signal Restoration Using the Theory of Convex Projections," Proc. International Conference on Acoustics, Speech, and Signal Processing, Dallas, Texas (1987).
9. H. K. Tuy, "An Algorithm for Incomplete Range of Views Reconstruction," Technical Digest of Topical Meeting on Signal Recovery and Synthesis with Incomplete Information and Partial Constraints, pp. FAI-1–FAI-4 (1983).
10. B. P. Medoff, Image reconstruction from limited data, Ph.D. dissertation, Department of Geophysics, Stanford University, Stanford, California (1983).
11. M. I. Sezan and H. Stark, Tomographic image reconstruction from incomplete view data by convex projections and direct Fourier inversion, *IEEE Trans. Med. Imag.*, *MI-3*: 91–98 (1984).
12. M. I. Sezan and H. Stark, Incorporation of a priori moment information into signal recovery and synthesis problems, *J. Math. Anal. Appl.*, *121*: 172–186 (February 1987).
13. G. T. Herman, *Image Processing from Projections*, Academic Press, New York (1980).
14. R. M. Lewitt, Reconstruction algorithms: Transform methods, *Proc. IEEE*, *71*(3): 390–408 (1983).
15. P. Dreike and D. P. Boyd, Convolution reconstruction of fan-beam projections, *Comput. Graphics Image Process.*, *5*: 459–469 (1976).
16. J. J. Clark, M. R. Palmer, and P. D. Lawrence, A transformation method for the reconstruction of functions from nonuniformly spaced samples, *IEEE Trans. Acoust. Speech Signal Process.*, *ASSP-33*(4): 1151–1165 (1985).
17. P. Mansfield, Multiplanar image formation using NMR spin echoes, *J. Phys. C*, *10*: 55–58 (1977).
18. M. M. Tropper, Image reconstruction for the NMR echo-planar technique, and for a proposed adaptation to allow continuous data acquisition, *J. Magn. Reson.*, *42*: 193–202 (1981).
19. G. Johnson and J. H. S. Hutchinson, The limitations of NMR recalled-echo imaging techniques, *J. Magn. Reson.*, *63*: 14–30 (1985).
20. C. B. Ahn, J. H. Kim, and Z. H. Cho, High-speed spiral echo planar NMR imaging—I, *IEEE Trans. Med. Imag.*, *MI-5*: 2–7 (1986).
21. S. Ljunggren, A simple graphical representation of Fourier-based imaging method, *J. Magn. Reson.*, *54*: 338–343 (1983).

22. P. G. Morris, *Nuclear Magnetic Resonance Imaging in Medicine and Biology*, Clarendon Press, Oxford (1986).
23. R. M. Mersereau and A. Oppenheim, Digital reconstruction of multidimensional signals from their projections, *Proc. IEEE*, *62*: 1319–1338 (1974).
24. H. Stark and M. Wengrowitz, Comments and corrections on the use of polar sampling theorems, *IEEE Trans. Acoust. Speech Signal Process.*, *ASSP-31*: 1329–1331 (1983).
25. F. Bloch, Nuclear Induction, *Phys. Rev.*, *70*: 460–474 (1946).
26. E. R. Andrew, *Nuclear Magnetic Resonance*, Cambridge University Press, Cambridge, England (1955).
27. P. Mansfield and P. G. Morris, *NMR Imaging in Biomedicine*, Academic Press, New York (1982).
28. W. G. Bradley, T. H. Newton, and L. E. Crooks, Physical principles of nuclear magnetic resonance, *Advanced Imaging Techniques: Modern Neuroradiology*, vol. 2, Clavadel Press, San Anselmo, California (1983).
29. W. S. Hinshaw and A. H. Lent, An introduction to NMR imaging: From the Bloch equation to the imaging equation, *Proc. IEEE*, *71*: 338–350 (1983).
30. Z. H. Cho, H. S. Kim, H. B. Song, and J. Cumming, Fourier transform nuclear magnetic resonance tomographic imaging, *Proc. IEEE*, *70*: 1152–1173 (1982).
31. Z. H. Cho, O. Nalcioglu, J. C. Jeong, and H. B. Song, Direct Fourier reconstruction techniques in NMR tomography, *Selected Topics, Image Science, Lecture Notes in Medical Informatics*, vol. 23, Springer-Verlag, Berlin (1984).
32. J. M. Carazo and J. L. Carrascosa, Information recovery in missing angular data cases: An approach by the convex projection method in three dimensions, *J. Microsc.*, *145*: 23–43 (January 1987).
33. R. Kumaradjaja, Application of projections onto convex sets in pattern recognition, M.S. thesis, Department of Electrical, Computer, and Systems Engineering, Rensselaer Polytechnic Institute, Troy, New York (1986).
34. M. I. Sezan, and H. Stark, Image restoration by convex projections in the presence of noise, *Appl. Opt.*, *22*: 2781–2789 (1983).
35. P. Oskoui-Fard and H. Stark, Tomographic image reconstruction using the theory of convex projections, *IEEE Trans. Med. Imag.*, *MI-7*(1): 45–58 (1988).
36. A. Papoulis, A new algorithm in spectral analysis and band-limited signal extrapolation, *IEEE Trans. Circuits Syst.*, *CAS-22*: 735–742 (1975).
37. R. W. Gerchberg, Super resolution through error energy reduction, *Opt. Acta*, *21*: 790–720 (1974).
38. C. K. Rushforth, Signal restoration, functional analysis, and the Fredholm integral equations of the first kind, *Signal Recovery: Theory and Applications* (H. Stark, ed.), Academic Press, Orlando, Florida, chapter 1 (1987).
39. M. Soumekh, Image reconstruction techniques in tomographic image systems, *IEEE Trans. Acoust. Speech Signal Process.*, *ASSP-34*(4): 952–962 (1986).
40. H. Peng and H. Stark, Direct Fourier reconstruction in fan-beam tomography, *IEEE Trans. Med. Imag.*, *MI-6*(3): 209–219 (1987).
41. R. Buck, *Advanced Calculus*, McGraw-Hill, New York (1956).
42. R. Lewitt, and H. Tuy, private communication.

43. E. Yudilevich and H. Stark, Interpolation from samples on a linear spiral scan, *IEEE Trans. Med. Imag.*, *MI-6*(3): 193–200 (1987).
44. A. J. Jerri, The Shannon sampling theorem—its various extensions and applications: A tutorial review, *Proc. IEEE*, *65*: 1565–1596 (1977).

# Appendix A

## Survey of Signal Processing Software

A number of dedicated signal processing software packages are now available commercially for use in minicomputers and microcomputers. They offer a wide range of digital filtering. spectral analysis, and other signal processing functions. Availability of such software packages simplifies many signal processing tasks. No effort is made here to provide a complete survey. The few packages that are briefly discussed below, however, can give good insight into software capabilities.

1. The ILS (Interactive Laboratory System) package of Signal Technology Inc., Goleta, California, is probably the most popular signal processing software. It is now in version ILS V6.0 and is supported on a wide range of computers, workstations, and graphics terminals. Analog data can be input with an analog-to-digital (A/D) converter. The most popular way to store and input the data is as digital data on magnetic tape. Simplified menu operations offer easy access to ILS functions including digital filtering, numerical analysis and data manipulation, frequency analysis, speech processing, and pattern analysis. As the package was originally designed for speech processing, it still works best for speech signals. The graphics are excellent. It is not easy, however, to add new signal processing functions because of the special data format required and the unavailability of source codes. The pattern classification capability, not available in other packages, is quite limited.

2. The I*S*P (Interactive Signal Processor) of Bedford Research, Bedford, Massachusetts, is very largely equivalent to, but less extensive then, the ILS in signal processing functions. Its graphics are good. Available for the VAX and IBM PC

and other systems, I*S*P can easily be used to generate additional special-purpose programs and its versatility is further enhanced by its ability to call FORTRAN or MACRO programs.

3. The SPD (Signal Processing and Display) software package of Tektronix, Beaverton, Oregon, supports 196 signal processing, display, and analysis functions for IBM PCs. It has real-time data acquisition and analysis capability. Its graphics are very good but it has a much smaller selection of signal processing functions than ILS or I*S*P. Notable among Tektronix signal analysis software products is the ASYST software for scientific application. ASYST supports several different digitizers.

4. The TMS 320-based signal processing package of Texas Instruments, Dallas, Texas, represents a unique class of real-time signal processing systems, making use of a combination of digital circuits and software support. It provides low-cost signal processing for many applications. However, an all-software signal processing system can perform a greater variety of complex signal processing tasks. A number of TMS 320-based signal processors are now available.

5. The Whitman Engineering signal processing package has a very good fast Fourier transform capability, including FFT for almost any number of date points. Digital filtering is also part of the package.

6. The Los Alamos National Laboratory signal processing package is free of charge, with all the essential spectral analysis and digital filtering features as signal processing functions.

7. Information Research Laboratory, North Dartmouth, Massachusetts, provides a high-resolution maximum entropy spectral analysis software package, MESA/IRL, for one-channel, multichannel, and two-dimensional data.

The above software packages are all interactive and suitable for microcomputers as well as minicomputers. The real-time signal processing capability available in some packages is very important, but the trade-off for this capability is the limitation to work with reasonably simple signal processing algorithms. Such capability indeed has broadened the applications of signal processing to many parts of our daily life. The comprehensive signal processing software packages allow new users to use "off-the-shelf" software rather than to reinvent the wheels for standard yet sophisticated signal processing functions. The flexibility of such packages to add new and complicated signal processing functions is an important advantage.

The following are the addresses of companies surveyed above.

Bedford Research
4 De Angelo Drive
Bedford, MA 01730

Information Research Laboratory, Inc.
415 Bradford Place
North Dartmouth, MA 02747

P. S. Wahler or D. L. Payne, L-156
Lawrence Livermore National Laboratory
P.O. Box 5504
Livermore, CA 94550

Signal Technology Inc.
5951 Encina Road
Goleta, CA 93117

Tektronix, Inc.
Attn: SPD Programs Software Package
P.O. Box 1700
Beaverton, OR 97075

Texas Instruments
P.O. Box 225474
MS 8214
Dallas, TX 75265

Whitman Engineering Inc.
P.O. Box 1929
Maitland, FL 32751

# Appendix B

## Program for 2-D Power Spectrum Estimate—Hybrid Algorithm*

```
C*******************************************************************
C         THIS PROGRAM COMPUTES 2-D POWER SPECTRUM ESTIMATE USING
C         HYBRID ALGORITHM.
C
C         INPUT 2-D SIGNAL DATA MUST BE STORED IN "SIG.DAT".
C         OUTPUT POWER SPECTRUM DATA WILL BE STORED INTO 'PSE.DAT'.
C*******************************************************************
          COMPLEX*16  X(64,64),PSE(64,64)
          OPEN(UNIT=22,FILE='SIG.DAT',STATUS='OLD')
          WRITE(*,110)
110       FORMAT(6X,' ENTER # OF COLUMN SAMPLES N1 & # OF ROW SAMPLES N2')
          READ(*,111)N1,N2
111       FORMAT(2I3)
          WRITE(112,*)
112       FORMAT(6X,' ENTER COLUMN # OF PSD NPSD1 & NPSD2')
          READ(*,111)NPSD1,NPSD2
          WRITE(*,114)
114       FORMAT(6X,' ENTER COLUMN FFT (N1=2**NEXP1) NEXP1 & ROW NEXP2')
          READ(*,111)NEXP1,NEXP2
          WRITE(*,115)
115       FORMAT(6X," ENTER BURG # OF PEF COEFFICIENTS LA')
          READ(116,*)LA
116       FORMAT(I2)
          DO 210 K=1,N2
          READ(22,*)(X(J,K),J=1,N1)
210       CONTINUE
          CLOSE(UNIT=22)
          CALL HYBRID(N1,N2,NPSD1,NPSD2,64,LA,X,PSD)
          OPEN(UNIT=33,FILE='PSE.DAT',STATUS='NEW')
          DO 211 N=1,NPSD2
          WRITE(33,*)(PSD(K,N),K=1,NPSD1)
211       CONTINUE
          STOP
          END
```

*by C. H. Chen

```
C******************************************************************
C         THIS PROGRAM COMPUTES THE TWO-DIMENSIONAL SPECTRUM
C
C         INPUT  PARAMETERS:
C
C          N1    - NUMBER OF COLUMN SAMPLES
C          N2    - NUMBER OF ROW SAMPLES
C          NPSD1    - NUMBER OF PSD VALUES IN COLUMN DIRECTION
C          NPSD2    - NUMBER OF PSD VALUES IN ROW DIRECTION
C          N2MAX    - MAXIMUM EXPECTED SIZE FOR N2
C          LA       - NUMBER OF PEF COEFFICIENTS
C          X        - COMPLEX TWO-DIMENSIONAL ARRY OF N1 x N2 DATA SAMPLES
C
C         OUTPUT PARAMETERS:
C
C          PSD      - REAL TWO-DIMENSIONAL ARRAY OF NPSD1 x NPSD2 POWER
C          SPECTRAL DENSITY VALUES
C******************************************************************
          SUBROUTINE HYBRID (N1,N2,NPSD1,NPSD2,N2MAX,LA,X,PSD)
          COMPLEX X(64,64),Z1(64),Z2(64),TMP(64,64),A(64)
          COMPLEX W1(64),W2(64)
          REAL PSD(64,64)
C
          CALL PREFFT (NPSD1,0,NEXP1,W1)
          CALL PREFFT (NPSD2,0,NEXP2,W2)
C  COMPUTE THE COLUMN TRANSFORMS
          DO 100 K = 1,N2
            DO 110 J =1,N1
110           Z1(J)=X(J,K)
            DO 120 J=N+1,NPSD1
120           Z1(J)=(0.,0.)
          CALL FFT (NPSD1,0,1.,NEXP1,W1,Z1)
          DO 130 J=1,NPSD1
130         TMP(J,K)=Z1(J)
100       CONTINUE
C  COMPUTE THE ROW TRANFORMS AND SQUARED MAGNITUDE TO YIELD PSD
          DO 200 K=1,NPSD1
            DO 210 J=1,N2
210            Z2(J)=TMP(K,J)
          CALL CBURG(N2,LA,Z2,A,P)
          DO 215 J=1,LA
215         Z2(J)=A(J)
          DO 220 J=LA+1,NPSD2
220         Z2(J)=(0.,0.)
          CALL FFT (NPSD2,0,1.,NEXP2,W2,Z2)
          DO 230 J=1,NPSD2
            PSD(K,J)=(REAL(Z2(J))**2+AIMAG(Z2(J))**2)
230         PSD(K,J)=P/PSD(K,J)
200       CONTINUE
          RETURN
          END
```

```
C*********************************************************************
C        THESE TWO PROGRAMS SET UP THE COMPLEX EXPONENTIAL TABLE (PREFFT)
C        AND COMPUTE THE DISCRETE-TIME FOURIER SERIES OF AN ARRAY OF COMPLEX
C        DATA SAMPLES USING A DECIMATION-IN-FREQUENCY FAST FOURIER TRANSFORM
C        (FFT) ALGORITHM.
C
         SUBROUTINE PREFFT (N,MODE,NEXP,W)
C
C        INPUT PARAMETERS:
C         MODE  - SET TO 0 FOR DISCRETE-TIME FOURIER SERIES OR 1 FOR INVERSE
C
C        OUTPUT PARAMETERS:
C
C         NEXP  - INDICATES POWER-OF-2 EXPONENT SUCH THAT N = 2**NEXP
C                 WILL BE SET TO -1 TO INDICATE ERROR CONDITION IF N IS NOT
C                  A POWER OF 2 (THIS INTEGER USED BY SUB. FFT)
C         W     - COMPLEX EXPONENTIAL ARRAY
C
C        NOTES:
C
C         EXTERNAL ARRAY W MUST BE DIMENSIONE .GE. N BY CALLING PROGRAM
C*********************************************************************

         COMPLEX W(1),C1,C2
         NEXP=1
5        NT=2**NEXP
         IF (NT .GE. N) GO TO 10
         NEXP=NEXP+1
         GOTO5
10       IF (NT .EQ. N) GO TO 15
         NEXP=-1
         RETURN
15       S=8.*ATAN(1.)/FLOAT(NT)
         C1=CMPLX(COS(S),-SIN(S))
         IF (MODE .NE. 0) C1=CONJG(C1)
         C2=(1.,0.)
         DO 20 K=1,NT
           W(K)=C2
20         C2=C2*C1
         RETURN
         END
```

```
      SUBROUTINE FFT (N,MODE,T,NEXP,W,X)
C*******************************************************************
C     INPUT PARAMETERS:
C      N,MODE,NEXP,W - SEE PARAMETER LIST FOR SUBROUTINE PREFFT
C      T              - SAMPLE INTERVAL IN SECONDS
C      X              - ARRAY OF N COMPLEX DATA SAMPLES, X(1) TO X(N)
C
C     OUTPUT PARAMETERS:
C
C      X - N COMPLEX TRANSFORM VALUES REPLACE ORIGINAL DATA SAMPLES
C          INDICATED FROM K=1 TO K=N, REPRESENTING THE FREQUENCIES
C           (K-1)/NT HERTZ
C*******************************************************************
      COMPLEX X(1),W(1),C1,C2
      MM=1
      LL=N
      DO 70 K=1,NEXP
         NN=LL/2
         JJ=MM+1
         DO 40 I=1,N,LL
            KK=I+NN
             C1=X(I)+X(KK)
             X(KK)=X(I)-X(KK)
 40          X(I)=C1
       IF (NN .EQ. 1) GO TO 70
       DO 60 J=2,NN
         C2=W(JJ)
         DO 50 I=J,N,LL
            KK=I+NN
             C1=X(I)+X(KK)
             X(KK)=(X(I)-X(KK))*C2
 50          X(I)=C1
 60      JJ=JJ+MM
       LL=NN
       MM=MM*2
 70    CONTINUE
       NV2=N/2
       NM1=N-1
       J=1
       DO 90 I=1,NM1
         IF (I .GE. J) GO TO 80
         C1=X(J)
         X(J)=X(I)
         X(I)=C1
 80    K=NV2
 85    IF (K .GE. J) GO TO 90
       J=J-K
       K=K/2
       GO TO 85
 90    J=J+K
       IF (MODE .EQ. 0) S=T
       IF (MODE .NE. 0) S=1./(T*FLOAT(N))
       DO 100 I=1,N
 100     X(I)=X(I)*S
       RETURN
       END
```

```
C*******************************************************************
C        THIS SUBROUTINE COMPUTES THE PEF COEFFICIENTS FOR A TIME SERIERS
C        BY MEANS OF THE SO CALLED BURG'S TECHNIQUE OF MAXIMUM ENTROPY
C        SPECTRUM ANALYSIS.
C
C        PARAMETERS:
C
C        LX - NUMBER OF SAMPLES (INPUT)
C        LA - NUMBER OF PEF COEFFICIENTS (INPUT)
C        X  - COMPLEX ARRAY CONTAINING THE LX SAMPLES OF THE INPUT TIME SERIES
C        A  - COMPLEX ARRAY CONTAINING THE LA COEFFICIENTS OF PEF (OUTPUT)
C        G  - COMPLEX ARRAY CONTAINING THE LA REFLECTION COEFFICIENTS (OUTPUT)
C        P  - MEASURE OF THE FINAL PREDICTION ERROR ENERGY WHEN LA COEFF. ARE USED
C        PEFOR - WORK AREA FOR THE FORWARD LINEAR PREDICTION ERROR
C        PEBAC - WORK AREA FOR THE BACKWARD LINEAR PREDICTION ERROR
C        TEMP  - TEMPORARY WORK AREA
C*******************************************************************
         SUBROUTINE CBURG(LX,LA,X,A,P)
         COMPLEX*16  X(64),A(64),G(64),PEFOR(64),PEBAC(64),TEMP(64)
         REAL*8 P
         COMPLEX*16 Q,BOTTOM,TOP
         REAL*8 TMP

         P=0.0D0
         DO 10 I=1,LX
           P=P+X(I)*DCONJG(X(I))
10       CONTINUE
         P=P/LX
         DO 15 I=1,LA
           TEMP(I)=(0.0D0,0.0D0)
15       CONTINUE
         A(I)=(1.0D0,0.0D0)
         DO 20 I=1.LX
           PEFOR(I)=X(I)
           PEBAC(I)=X(I)
20       CONTINUE
         DO 100 M=2,LA
           BOTTOM=(0.0D0,0.0D0)
           TOP=(0.0D0,0.0D0)
           DO 30 I=M,LX
             TOP=TOP+PEFOR(I)*DCONJG(PEBAK(I-M+1))
             BOTTOM=BOTTOM+ABS(PEFOR(I))**2+ABS(PEBAC(I-M+1))**2
30       CONTINUE
         G(M)=-2.0D0*TOP/BOTTOM
         TMP=ABS(G(M))
         IF (TMP .GE. 1.0D0) TMP=0.99999D0
         P=(1.0D0-TMP**2)*P
         DO 40 I=M,LX
           Q=PEFOR(I)
           PEFOR(I)=PEFOR(I)+G(M)*PEBAC(I-M+1)
           PEBAC(I-M+1)=PEBAC(I-M+1)-Q*DCONJG(-G(M))
40       CONTINUE
         A(M)=0.0D0
         DO 50 I=1,M
           TEMP(I)=A(I)+G(M)*DCONJG(A(M-I+1))
50       CONTINUE
         DO 60 I=1,M
           A(I)=TEMP(I)
60       CONTINUE
100      CONTINUE
         RETURN
         END
```

# Appendix C

## Program for Time Domain Deconvolution Using Spiked Filter

```
  TYPE SPIKE.FOR
C***********************************************************************C
C                                                                       C
C         TIME DOMAIN DECONVOLUTION BY WEI-LIEN HSU                       C
C           USING SPIKING FILTER                                          C
C             Y(F)*H(F)=X(F)                                              C
C             Y:A SPIKED SIGNAL VECTOR , DIMENSION = N                    C
C             H:INVERSE FILTER MATRIX OF Y , DIMESION = N,M               C
C             X:SPIKING VECTOR OF Y ,DIMENSION = M                        C
C           PROGRAM PROCEDURE                                             C
C             1.) INPUT SPIKED SIGNAL AND MAX VALUE OF SIGNAL             C
C             2.) FIND INVERSE FILTER  MATRIX AND PERFORMANCE P           C
C             3.) FIND THE OPTIMA INVERSE FILTER VECTOR THAT HAS OPTIMA   C
C                 PERFORMANCE ,BY SELECTING MAX DIAGONAL VALUE OF P       C
C             4.) CONVOLUTE Y AND OPTIMA INVERSE FITER VECTOR             C
C             5.) OUTPUT RESULT X                                         C
C           NOTE                                                          C
C             ESP COEFFICIENT USED TO DE NOISE                            C
C                                                                       C
C***********************************************************************C
          CHARACTER*15 NAM
        DIMENSION EE(200),Y(200),R(200)
          DIMENSION P(400,400),H(200,400),YY(200,70),PL(200,200)
C
C         INPUT DATA
C
          WRITE(*,150)
150       FORMAT(' INPUT FILE NAME :',$)
          READ(*,160)NAM
160       FORMAT(A15)
        OPEN(UNIT=91,FILE=NAM,STATUS='OLD')
```

```
         WRITE(*,10)
10       FORMAT('INPUT N,M,EPS,max,po :',$)
           READ(*,*)N,M,EPS,xm,ipo
C
C          EPS IS A COEEICIENT USED TO DENOISE
C
         DO 30 I=1,N
             READ(91,*)RDR
             write(*,*)rdr
             y(i)=rdr/xm
30         CONTINUE
           CALL CLOSE(91)
           write(*,101)
101        format(' pas 1')
C
C
           DO 1 I=0,N+M
             DO 1 J=0,M
1              YY(I,J)=0.
           DO 2 J=0,M
             DO 2 I=J,N+J
2              YY(I,J)=Y(I-J)
           DO 3 K=0,M
             R(K)=0.
             DO 3 I=K,N+M
3              R(K)=R(K)+YY(I,0)*YY(I,K)
           R(0)=(1+EPS)*R(0)
c
c          Levise ALGORITHM USED TO FIND INVERSE MATRIX
c
           DO 51 I=0,M
           DO 51 J=I+1,M
51             PL(I,J)=0.
           PL(0,0)=1.
           PL(1,1)=1.
           PL(1,0)=-R(1)/R(0)
           WRITE(*,*)R(0)
           ER=R(0)
           PPP=PL(1,0)*PL(1,0)
           EE(1)=R(0)*(1.-PPP)
           DO 52 I=2,M
             GAP=0.
              DO 53 K=0,I-1
53              GAP=GAP+R(K+1)*PL(I-1,K)
             GAMMA=GAP/EE(I-1)
             PL(I,0)=-GAMMA
             DO 54 K=1,I-1
54             PL(I,K)=PL(I-1,K-1)-GAMMA*PL(I-1,I-1-K)
               PL(I,I)=1.
52           EE(I)=EE(I-1)*(1-GAMMA*GAMMA)
           DO 4 I=0,M
             DO 4 J=0,M+N
               H(I,J)=0.
                 DO 4 K1=I,M
                   DO 4 K2=0,K1
                     IF(K1 .NE. 0)GO TO 41
                     H(I,J)=H(I,J)+PL(K1,I)*PL(K1,K2)*YY(J,K2)/ER
                     GO TO 4
41                   H(I,J)=H(I,J)+PL(K1,I)*PL(K1,K2)*YY(J,K2)/EE(K1)
4          CONTINUE
           DO 5 I=0,M+N
             DO 5 J=0,M+N
               P(I,J)=0.
                 DO 5 K=0,M
5          P(I,J)=P(I,J)+YY(I,K)*H(K,J)
```

```
C
C         DISPLAY DIAGONAL VALUE OF P AND MAX OF THESE VALUES
C       STORE PERFORMANCE P
C
          write(*,150)
          read(*,160)NAM
          open(unit=92,file=nam,status='new')
          IMM=0
          XM=P(0,0 )
        DO 152 I=0,M+N
          XXM=P(I,I)
            write(92,*)XXM
            WRITE(*,*)i,XXM
            IF(XXM .Le. XM)GO TO 152
            XM=XXM
            IMM=i
152       CONTINUE
c
c         CONVOLUTION Y,H
c
          inn=0
          knn=0
          xnn=0.
          xn=0.
          do 1001 k=0,m
            do 989 i=0,m+n
              r(i)=0.
              do 987 j=0,i
                r(i)=r(i)+y(j)*h(n-j,k)
987           continue
989         continue
            do 999 i=0,n
              xnn=r(i)
              if(xnn .le. xn)go to 999
              inn=i
              knn=k
              xn=xnn
999         continue
1001      continue
          CALL CLOSE(92)
        write(*,*)imm,inn,knn,xn,y(30)
          WRITE(*,150)
          READ(*,160)NAM
C
C         OUTPUT
C
        OPEN(UNIT=93,FILE=NAM,STATUS='NEW')
          write(*,998)
998       format(' input col',$)
          read(*,*)imm
        do 110 i=0,m
            WRITE(93,*)H(I,imm)
110       CONTINUE
          CALL CLOSE(93)
        STOP
          END
$
```

# Appendix D

## Time Domain Deconvolution Algorithm

$i = 0$

I $\quad i = i + 1$

$D^i = 0$

DO III $\quad k = 1,\ N - M + 1$

$$\left.\begin{aligned} G_k^i &= \sum_{l=1}^{N-K+1} I_l T_{k+l-1}^i \\ D_k^i &= G_k^i - E_k \\ D^i &= D^i + D_k^i * D_k^i \end{aligned}\right\} \text{Compare } I * T \text{ with the real data } E$$

DO II $\quad m = 1,\ N - k + 1$

$$\left.\begin{aligned} C_k^i &= D_k^i/(A * (G_k^i + E_k) + 1.) \\ T_{k+m-1}^i &= T_{k+m-1}^i * (1. - C_k^i * I_m) \end{aligned}\right\} \text{Modify the deconvolution data } T_i$$

II $\quad$ Continue

III $\quad$ Continue

If $D^i <$ Error limit, then END

GO TO I

Notes: $N$, number of data points in the output; $M$, peak location of reference data; $E_k$, measured data; $T$, deconvolved data (result); $I_i$, reference data.

# Index